D1474515

FIRST COURSE IN ALGEBRA AND NUMBER THEORY

FIRST COURSE IN ALGEBRA AND NUMBER THEORY

Edwin Weiss

Boston University

ACADEMIC PRESS NEW YORK AND LONDON

ACADEMIC PRESS, INC.
111 Fifth Avenue, New York, New York 10003

United Kingdom Edition published by
ACADEMIC PRESS, INC. (LONDON) LTD.
Berkeley Square House, London W1X 6BA

LIBRARY OF CONGRESS CATALOG CARD NUMBER: 73-158813

PRINTED IN THE UNITED STATES OF AMERICA

Cover design by Jorge Hernandez

FOR JACK AND FLORENCE

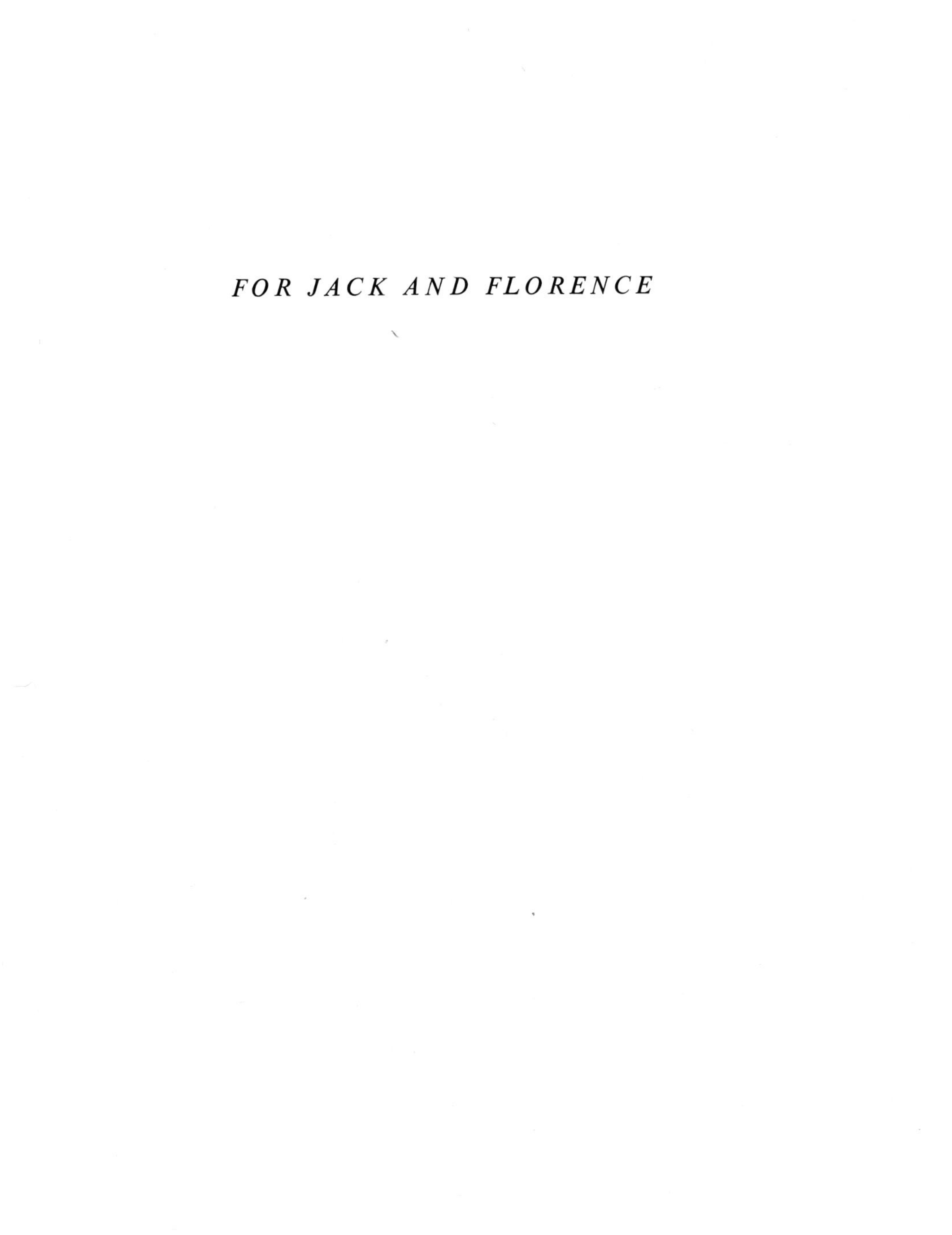

CONTENTS

Preface ix

I. **Elementary Number Theory**

1-1. Divisibility 2
1-2. The Division Algorithm 6
1-3. The Greatest Common Divisor 13
1-4. Unique Factorization 21
1-5. A Convenient Notation 28
1-6. Linear Diophantine Equations 42
1-7. Congruence 50
1-8. Radix Representation 64
Miscellaneous Problems 83

II. **Rings and Domains**

2-1. Rings: Elementary Properties 91
2-2. Examples 104
2-3. Ordered and Well-Ordered Domains 138
2-4. Computation Rules 160
2-5. Characterization of the Integers 174
Miscellaneous Problems 199

III. Congruences and Polynomials

3-1. Linear Congruences 211
3-2. Units and Fields 230
3-3. Polynomials and Polynomial Functions 253
3-4. Factorization in $F[x]$ 274
3-5. Roots of Polynomials 294
3-6. Solving Polynomials in $\mathbf{Z}_m[x]$ 322
3-7. Quadratic Reciprocity 340
Miscellaneous Problems 365

IV. Groups

4-1. Basic Facts and Examples 381
4-2. Subgroups and Cosets 401
4-3. Cyclic Groups 421
4-4. Normal Subgroups; Factor Groups; Homomorphisms 439
4-5. Permutation Groups 460
4-6. The Group $\mathbf{Z}_m^*$ 490
Miscellaneous Problems 508

Selected Answers and Comments 521

Subject Index 543

PREFACE

There is a well-known bit of advice for an author which says that preferably before beginning, but certainly before finishing a book, he should ask himself why he is writing it. This fundamental question and the author's answer to it are of interest to the potential user of a mathematical text within the context of a web of related questions. Is this book really necessary? Does it have any novel features that distinguish it from other books? For whom is it designed? What are the prerequisites? Where does it take the reader? What is the subject matter? How is the material organized and structured?

In recent years it has become fairly standard practice for colleges to offer a one-semester course in linear algebra at the sophomore level, the justification and purpose being the connection with certain topics normally covered in second year calculus. After teaching such a course on a number of occasions, with singular lack of success, I emerged with the strongly held view that not only was the justification for the course insufficient, but even more that the nature of the material was inappropriate for students at the given level of mathematical experience. Although the available texts were more than adequate, only the most gifted students were in a position to develop successfully, in the time-span of a single semester, the required combination of geometric insight, capacity for abstract reasoning, and technique of formal proof. The conclusion to be drawn, almost inevitably, was that the study of linear algebra should be deferred, reserving it for students with some mathematical maturity.

Once the preceding line of argument is accepted, the problem to be addressed is how to introduce the interested student to the spirit, style, and content of modern (some prefer to call it "abstract") algebra. There are many fine texts of this genre available, but they tend to be aimed at the junior level and are consequently not quite suitable for our purpose. The true beginner usually finds that they go too far too fast and are too abstract; he is unable to assimilate the many new definitions and concepts, and does not possess the mathematical sophistication to manipulate them in a meaningful way.

Out of such considerations there evolved, over a period of several years, a book with the following characteristics:

It is designed for a full year course at the freshman or the sophomore level. Since there are no prerequisites other than familiarity with the standard elementary topics of high school mathematics (especially algebra), this course may be taken before, after, simultaneously with, or even in place of first-year calculus. Therefore, this text can also be used for well-prepared and motivated high school seniors.

The style is slow-moving, detailed, quite wordy, and occasionally repetitive, as dictated by a struggle to provide the student (insofar as space and other constraints permit) with a book he can read with a minimum of outside help. In this connection it may be usefully noted that preliminary versions of this text have been used, and found readable, by such diverse groups as: students of liberal arts intending to major in mathematics, students in a school of education preparing to become mathematics specialists, and high school mathematics teachers (in a NSF institute) anxious to enlarge their mathematical horizons by learning material that impinges upon and enriches the subject matter in their own curriculum.

No attempt has been made to produce an encyclopedic book, or to throw masses of material at the student, or to probe deeply. Rather, I have tried to treat a small number of topics, selected, in the main, for their contribution toward the larger objective of studying algebraic questions arising out of number theory, with care, precision of thought and expression, and a certain amount of thoroughness.

Experience provides ample verification of the assumption that in teaching mathematics it is pedagogically essential to motivate concepts carefully, and to proceed from the familiar to the unfamiliar, from the specific to the general, from the concrete to the abstract. Adherence to these guidelines is a central feature of this book. Thus, we begin by studying the most familiar mathematical objects, the integers, and then use them to motivate, introduce, and illustrate some of the basic concepts of modern algebra. The underlying thread or connective tissue for the book is the interweaving of certain questions of elementary number theory with the tools, techniques and concepts of modern, abstract algebra. Topics from linear algebra are consciously excluded since they would constitute a diversion from the main thrust of the book.

Each section (with the exception of the first few, which are relatively short) centers on a single topic and develops it rather fully—thus the sections tend to be considerably longer than is customary. Roughly speaking, the material in each section is arranged according to increasing complexity and with decreasing detail. Many examples are included, mostly of a numerical or computational nature. They serve to illustrate and reinforce the discussion in the text and to provide previews of coming attractions. Each section closes with a large number of problems, ranging from the purely mechanical to the theoretical and abstract. While knowledge of the problems is not needed for subsequent developments in the text, it is essential that the student work on lots of them, even if his success is limited. This is the best way really to understand and master the subject matter. Selected answers to the problems appear at the back of the book. At the end of each chapter, there is a collection of "miscellaneous problems." These are usually of a more advanced type; some of them, if expanded fully, could make up an entire section in this, or any other, book. Only the very best students can be expected to make serious inroads in the miscellaneous problems.

A few additional comments are in order. I have preferred to plunge right in and deal with mathematical questions instead of starting with a preliminary section on sets and functions. These are discussed at the appropriate time—namely, when they are needed. Some may consider the selection of topics perverse on the grounds of being too pedestrian or too difficult. I would merely observe that the amount of space devoted to a topic in the text is not necessarily a measure of its mathematical importance. Moreover, the precise contents, and the emphasis, of a given course are determined according to the personal tastes of the instructor. This book attempts to give him some leeway in his choices.

It remains for me to express my deep appreciation to: Sandra Spinacci, who typed the manuscript with her usual skill and efficiency; Bill Adams, who read the manuscript carefully and found a number of errors; my family and S. Shufro, who provided encouragement and assistance; my students, who participated (unknowingly) in all kinds of pedagogical experiments; the people at Academic Press, individually and collectively, for their competence, courtesy, and unfailing cooperation.

1

ELEMENTARY NUMBER THEORY

In this chapter, we shall be concerned exclusively with the set of all integers —positive, negative, and zero. This set is denoted by

$$\mathbf{Z} = \{0, \pm 1, \pm 2, \pm 3, \ldots\}$$

and whenever we write any of the symbols $a, b, c, d, \ldots, x, y, z, \ldots$ they shall represent elements of $\mathbf{Z}$.

It is taken for granted that we are familiar with the integers; however, in order not to complicate matters unnecessarily at the start, we choose to be imprecise as to exactly what we know about them. Of course, it is assumed that we know how to operate, or compute, with integers. As the discussion progresses, additional rather obvious properties of the integers will be used, and we shall try to point these up explicitly at the appropriate time. Thus, at the current stage, our attitude toward the integers will be relatively informal as compared to the formal axiomatic approach to algebraic objects that will be adopted in later chapters.

1-1. Divisibility

It is assumed that the facts about addition, subtraction, and multiplication of integers are known; however, "officially" we do not recognize the existence of "division." We consider only objects that belong to $\mathbf{Z}$, so symbols like "5/2" or "2/5" are outside our realm of discourse. From this point of view, "fractions," that is, symbols "m/n," where $m \in \mathbf{Z}$ (we read $m \in \mathbf{Z}$ as: m **is an element of** $\mathbf{Z}$, m **is in** $\mathbf{Z}$, or m **belongs to** $\mathbf{Z}$) and $n \in \mathbf{Z}$ have no meaning. Our purpose here is to begin the investigation of questions of "divisibility" in $\mathbf{Z}$, without going outside of $\mathbf{Z}$ to do so.

1-1-1. Definition. Given $a, b \in \mathbf{Z}$ (this notation asserts that both a and b are elements of $\mathbf{Z}$) with $a \neq 0$; we say that a **divides** b (and write this symbolically as $a \mid b$) when there exists $c \in \mathbf{Z}$ such that $ac = b$. There are alternate ways to express this—namely, a is a **divisor** of b, a is a **factor** of b, b is a **multiple** of a, b is **divisible** by a.

If no such c exists, we say that a **does not divide** b (or: a is not a divisor of b, a is not a factor of b, b is not a multiple of a, b is not divisible by a) and write $a \nmid b$.

It should be noted that the statement $a \mid b$ includes the fact that $a \neq 0$; thus, according to the definition, 0 does not divide any integer.

The definition of $a \mid b$ is concerned only with the theoretical question of the existence of the desired element c—it says nothing about how to decide, in practice, if such an integer c exists or how to find it. For example, suppose we wish to decide if $a = 7$ divides $b = 91$. The reader would settle this quickly by cheating—that is, by using long division. For us, unfortunately, the restriction on division remains in force (in particular, we do not know about long division), so we must somehow search through $\mathbf{Z}$ seeking an element c for which $7c = 91$. By testing $c = 1, 2, 3, \ldots$, in turn, we see eventually that $7 \cdot 13 = 91$, so that $7 \mid 91$.

In the same spirit, let us decide if $a = 7$ divides $b = 99$. Having observed that $7 \cdot 13 = 91$, we note further that

$$7 \cdot 14 = 98 < 99 < 7 \cdot 15 = 105.$$

Therefore, if $c \leq 14$, then $7c < 99$, and if $c \geq 15$, then $7c > 99$; it follows that $7 \nmid 99$.

The discussion above is designed to emphasize our avoidance of long division, and surely it would not be very exciting to test other pairs of integers a and b for divisibility. Instead, we turn to some of the immediate theoretical consequences of the definition of divisibility.

1-1-2. Facts. For elements of **Z**, the following properties hold:

(*i*) $a \mid 0$ for every $a \neq 0$.

Proof: Here $b = 0$, so by taking $c = 0$ we have $ac = b$. ▮

(*ii*) If $a \mid b$ and $b \mid c$, then $a \mid c$.

Proof: By hypothesis, there exist $d, e \in \mathbf{Z}$ such that $b = ad$ and $c = be$. Therefore, $c = a(de)$, so that de is an integer whose existence guarantees that $a \mid c$. ▮

(*iii*) $(\pm 1) \mid a$ for all a, and $(\pm a) \mid a$ for all $a \neq 0$.

Proof: Both statements follow immediately from the equation $a = (a)(1) = (-a)(-1)$. ▮

(*iv*) The following four assertions are equivalent:

(1) $a \mid b$,
(2) $a \mid (-b)$,
(3) $(-a) \mid b$,
(4) $(-a) \mid (-b)$.

Proof: By equivalence of the four assertions, we mean that any two of them are equivalent. Of course, in general, two assertions are said to be equivalent when each one implies the other. Thus, equivalence of our four assertions amounts to saying that each assertion implies the other three—so there are a total of 12 implications to be proved. However, by proceeding "cyclically," it clearly suffices to prove only the following four implications

$$(1) \Rightarrow (2), \qquad (2) \Rightarrow (3), \qquad (3) \Rightarrow (4), \qquad (4) \Rightarrow (1).$$

The proofs of all four of these implications proceed the same way, so we content ourselves with proving just one of them—namely, (2) ⇒ (3); the reader can easily provide proofs for the three remaining implications.

When (2) is given we have $a \mid (-b)$, so there exists $c \in \mathbf{Z}$ such that $ac = -b$. By the properties of "minus," we have, therefore, $b = (-a)(c)$. This says that $(-a) \mid b$ and completes the proof that (2) ⇒ (3). ▮

A common notation used to express all the equivalences just proved is

$$a \mid b \Leftrightarrow a \mid (-b) \Leftrightarrow (-a) \mid b \Leftrightarrow (-a) \mid (-b)$$

or simply,

$$(1) \Leftrightarrow (2) \Leftrightarrow (3) \Leftrightarrow (4).$$

One immediate, but important, consequence of these equivalences is that in considering questions of divisibility there is nothing lost in assuming that $a > 0$—for after all, $a \mid b \Leftrightarrow (-a) \mid b$.

(*v*) If $a \mid b$ and $a \mid c$, then $a \mid (bx + cy)$ for all choices of $x, y \in \mathbf{Z}$. In particular, if $a \mid b$ and $a \mid c$, then $a \mid (b + c)$ and $a \mid (b - c)$.

Proof: We must show that $bx + cy$ can be written in the form a times an integer. The proof is straightforward. By hypothesis, we may write $b = ad$, $c = ae$, where $d, e \in \mathbf{Z}$. Therefore.

$$bx + cy = adx + aey = a(dx + ey)$$

which says that $a \mid (bx + cy)$. In particular, by making a specific choice of x and y—namely, $x = 1, y = 1$—we see that $a \mid (b + c)$. Similarly, by putting $x = 1, y = -1$, we see that $a \mid (b - c)$. ▮

It is convenient to call any element of form $bx + cy$ (or $xb + yc$), where $x, y \in \mathbf{Z}$, a **linear combination** of b and c. Note that the following numbers are (or more precisely, can be expressed as) linear combinations of b and c—b (take $x = 1, y = 0$), $-c$ (take $x = 0, y = -1$), 0 (take $x = 0, y = 0$), $2b + 3c$, $3b - 5c = 3b + (-5)c$, $-3b + 5c, \ldots$. With this terminology, the result proved above may be restated in the following form: If a divides both b and c, then it divides any linear combination of b and c; in particular, if a divides b and c, then it divides their sum and their difference.

(*vi*) $|a| = |b| \Leftrightarrow a = \pm b$.

Proof: Many of us may well be prepared to consider this assertion as obvious, but a somewhat more convincing argument is required. For this, it is necessary to recall the definition of "absolute value."

First of all, one defines

$$|0| = 0.$$

Then, for $a \neq 0$, one notes that the integers a and $-a$ are distinct, with one of them positive and the other negative. (Naturally, we are assuming that the basic facts about positive and negative integers are known. It may also be noted in passing that 0 is, by common usage, neither positive nor negative.) One then defines, for $a \neq 0$,

$$|a| = \text{the positive member of the set } \{a, -a\}.$$

Thus, $|a|$ is either a or $-a$, and we "abbreviate" this by writing $|a| = \pm a$. Of course, this definition of absolute value is the same as the one commonly used for real numbers; some of the properties of absolute value are given in the problems—see 1-1-3, Problem 7.

Returning to the proof, let us suppose that $a = \pm b$—that is, a equals plus or minus b. The trivial case occurs when $a = b = 0$, as it is then immediate that $|a| = |b| = 0$. Upon discarding the trivial case, we have both $a \neq 0$ and $b \neq 0$. In this case, $a = \pm b$ and also $-a = \pm b$; so it follows that $\{a, -a\} = \{b, -b\}$—meaning that the two sets $\{a, -a\}$ and $\{b, -b\}$, each of which consists of two distinct elements, are identical. The definition of absolute value now says that $|a| = |b|$—thus proving the implication $\Leftarrow$.

Conversely, to prove the implication $\Rightarrow$, let us suppose that $|a| = |b|$. Excluding the trivial case where $|a| = |b| = 0$ (as then $a = b = 0$), we may assume that $|a| = |b| \neq 0$, so $a \neq 0$ and $b \neq 0$. Since $|b| = \pm b$ we have $|a| = \pm b$, which says that $|a|$ belongs to the two element set $\{b, -b\}$. Hence, the positive member of $\{a, -a\}$ is equal to the positive member of $\{b, -b\}$. There are four possibilities for the two positive members—namely, a and b, a and $-b$, $-a$ and b, $-a$ and $-b$—and in each case, $a = \pm b$. This completes the proof. ▮

(*vii*) If $a \mid b$ and $b \neq 0$, then $|a| \leq |b|$.

Proof: By hypothesis there exists $c \in \mathbf{Z}$ such that $b = ac$, and $c \neq 0$ because $b \neq 0$. Taking absolute values yields $|b| = |ac| = |a| \cdot |c|$, and because c is a nonzero integer we know that $1 \leq |c|$. Consequently,

$$|a| = |a| \cdot (1) \leq |a| \cdot |c| = |b|. \quad ▮$$

It should be noted that this proof makes use of the following "facts" about integers:The absolute value of a product equals the product of the absolute values; there is no integer between 0 and 1; multiplying an inequality by a positive number preserves the inequality. It may also be noted that the proof makes heavy use of the assumption that $b \neq 0$; in fact, if $b = 0$ the conclusion of (*vii*) is false.

(*viii*) If $a \mid (\pm 1)$, then $a = \pm 1$.

Proof: One way to verify this is to observe that, according to (*vii*), $a \mid (\pm 1)$ implies $|a| \leq |\pm 1| = 1$. Because $a \neq 0$, it follows that $|a| = 1 = |1|$, and hence, in virtue of (*vi*), $a = \pm 1$. ▮

(*ix*) If $a \mid b$ and $b \mid a$, then $a = \pm b$.

Proof: By the definition of divisibility, we know that $a \neq 0$ and $b \neq 0$. Applying (*vii*) yields $|a| \leq |b|$ and $|b| \leq |a|$. Consequently, $|a| = |b|$, and by (*vi*) $a = \pm b$. ▮

1-1-3 / *PROBLEMS*

1. How many integers between 20 and 200 are divisible by 7? How many are divisible by 11?

2. Prove that if $ac \mid bc$, then $a \mid b$. If $a \mid b$ and $c \mid d$, then $ac \mid bd$.

3. Show that if $a \mid b$ and $0 \leq b < a$, then $b = 0$.

4. For $a = 10$, how many integers d are there such that $d \mid a$ and $1 \leq d \leq a$? What is the sum of all such divisors d? Answer the same questions for $a = 17, 24, 72, 77, 79, 97, 210, 420$.

5. Can you find an integer greater than 1 which divides both $a = 5311$ and $b = 7571$? Can you find a positive integer less than ab which is a multiple of both a and b?

6. Do Problem 5 for $a = 4181$, $b = 7663$.

7. For $a, b, r_1, r_2 \in \mathbf{Z}$ prove that:
(*i*) $|ab| = |a|\,|b|$,
(*ii*) $|a + b| \leq |a| + |b|$,
(*iii*) $|a - b| \leq |a| + |b|$,
(*iv*) If $0 \leq r_1 < a$ and $0 \leq r_2 < a$, then $|r_1 - r_2| < a$.
Of course, the proofs will apply also when a, b, r_1, r_2 are real numbers.

8. (*i*) List some integers that can be expressed as linear combinations of $b = 3$ and $c = 6$, and others which cannot be so expressed. In which class does the integer 1 fall?
(*ii*) Answer the same questions for $b = 3$, $c = 5$.
(*iii*) Answer the same questions for $b = 4$, $c = 6$.

9. Write out, in words, the definition of each of the following sets:
(*i*) $\{2n \mid n \in \mathbf{Z}\}$,
(*ii*) $\{2m + 1 \mid m \in \mathbf{Z}\}$,
(*iii*) $\{2x + 3y \mid x, y \in \mathbf{Z}\}$,
(*iv*) $\{ax + by \mid x, y \in \mathbf{Z}\}$,
(*v*) $\{b - xa \mid x \in \mathbf{Z}\}$,
(*vi*) $\{d \in \mathbf{Z} \mid d \mid a\}$,
(*vii*) $\{d > 0 \mid d \mid a\}$,
(*viii*) $\left\{n > 0 \;\middle|\; \sum_{i=1}^{n} i = \frac{n(n+1)}{2}\right\}$.

1-2. The Division Algorithm

Now that the elementary properties of divisibility are understood, we turn to a crucial result which is normally taken for granted. Given $a, b \in \mathbf{Z}$ with $a > 0$, then a may divide b or it may not; however, there always exists an expression for b in terms of a, and this expression also settles the question of divisibility—namely,

1-2-1. Division Algorithm. Given $a, b \in \mathbf{Z}$ with $a > 0$, then there exist unique integers q and r such that

$$b = qa + r, \qquad 0 \leq r < a.$$

Proof: This result is, of course, quite familiar. To illustrate—if $a = 7$ is fixed, then taking in turn $b = 10, 5, 28, -5, -91, 94, 1, 0, -1$ we have

$$\begin{aligned}
10 &= 1 \cdot 7 + 3, \\
5 &= 0 \cdot 7 + 5, \\
28 &= 4 \cdot 7 + 0, \\
-5 &= (-1) \cdot 7 + 2, \\
-91 &= (-13) \cdot 7 + 0, \\
94 &= 13 \cdot 7 + 3, \\
1 &= 0 \cdot 7 + 1, \\
0 &= 0 \cdot 7 + 0, \\
-1 &= (-1) \cdot 7 + 6.
\end{aligned}$$

As for the division algorithm itself, there are two assertions to be proved—firstly, the existence of q and r, and secondly, that there is only one pair of integers q and r which satisfy the conditions.

Existence: The proof of existence is somewhat formal, so it is useful to sketch the geometric idea on which the formal proof rests. Suppose for illustrative purposes that $b > a$. Let us take the point 0, that is the origin, on the real line (also known as the "number line," and which is just another, more geometric, way of referring to the set $\mathbf{R}$ of all real numbers) and then mark off the point a. We can then mark off the point $2a$ by "adding" the segment of size a to itself. This process may be repeated as many times as desired, giving $a, 2a, 3a, 4a, \ldots$. Clearly, if a is added to itself enough times, we arrive at a number bigger than b. (This is, of course, an example of the "Archimedean" property of the real numbers.) In particular, there is a last, or biggest, multiple of a, call it qa, such that $qa \leq b$; and the next multiple of a, $(q + 1)a$, is greater than b. In other words, we have

$$qa \leq b < (q + 1)a$$

and the picture on the real line is

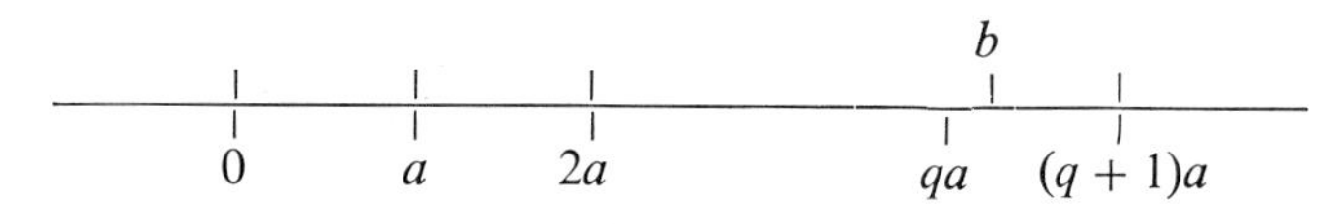

We may now put $r = b - qa$—so that $b = qa + r$ and $0 \leq r < a$.

Suppose that for any integer n, we let $[na, (n + 1)a)$ denote the set of all real numbers x such that $na \leq x < (n + 1)a$—so

$$[na, (n + 1)a) = \{x \in \mathbf{R} \mid na \leq x < (n + 1)a\}.$$

Geometrically (that is, on the real line), $[na, (n + 1)a)$ is an "interval" which includes its endpoint on the left but not its endpoint on the right. The geometric idea underlying the discussion above is that as n runs over $\mathbf{Z}$, b falls in an interval (in fact, exactly one) of form $[na, (n + 1)a)$.

The preceding sketch of a proof is defective in that it seems to make use of real numbers and of a geometric "picture" We turn, therefore, to a more "satisfactory" proof—namely, a formal algebraic proof in which objects outside of $\mathbf{Z}$ never occur.

Given $a, b \in \mathbf{Z}$ with $a > 0$, let

$$S = \{b - xa \geq 0 \mid x \in \mathbf{Z}\}.$$

In words, S is defined to be the set of all those integers of form $b - xa$ which are greater than or equal to 0, when x is allowed to run over $\mathbf{Z}$. (Note that the vertical stroke "$\mid$" which appears in the definition of S has nothing to do with divisibility.) To put it another way, as x runs over $\mathbf{Z}$, then presumably some of the integers $b - xa$ are greater than or equal to 0 and some are less than 0—we consider only those $b - xa$ which are greater than or equal to 0. Of course, S may equally well be written as

$$S = \{b - xa \mid x \in \mathbf{Z}, \quad b - xa \geq 0\}.$$

We assert first of all that $S \neq \varnothing$ (the proper way to read this notation is: S is nonempty). To show this, it suffices to exhibit an x for which $b - xa \geq 0$. Suppose we try $x = -|b|$. Because $a \geq 1$ we have $|b| \cdot a \geq |b| \cdot 1 = |b|$, and then adding b to both sides gives

$$b + |b| \cdot a \geq b + |b|.$$

But $b + |b| \geq 0$, so that $b + |b| \cdot a \geq 0$, and hence $x = -|b|$ works. (By retracing the steps of our argument, it is not hard to see why we chose to try $x = -|b|$—it was not pulled out of a hat.)

Since S is a nonempty set of nonnegative integers, it contains a smallest element (the existence of such a smallest element appears obvious, but it is really an assumption about $\mathbf{Z}$)—call it r. Denote a value of x for which r arises, by q. (In other words, q is a value of x for which $b - qa = r$.) Thus, $r \geq 0$, $r = b - qa$, and r is less than any other element of S.

It remains to show that $r < a$. Suppose this is false—that is, suppose $r \geq a$. Now consider the integer

$$r - a = (b - qa) - a = b - (q + 1)a.$$

Since $r - a \geq a - a = 0$, we see that $(r - a) \in S$. Furthermore, $r - a < r$ because $a > 0$. This says that $r - a$ is an element of S which is smaller than r—contradicting the choice of r. Our assumption that $r \geq a$ having led to a contradiction, we conclude that $r < a$. The existence part of the proof is now complete.

Uniqueness: We show that there cannot exist two distinct expressions of the desired form for b. More precisely, suppose

$$b = q_1 a + r_1, \qquad 0 \le r_1 < a,$$
$$b = q_2 a + r_2, \qquad 0 \le r_2 < a$$

are two expressions for b; we must show that $r_1 = r_2$, $q_1 = q_2$.

If $r_1 = r_2$, then $q_1 a = q_2 a$, and since $a \neq 0$ we obtain, by cancellation, $q_1 = q_2$—so the two expressions for b are the same. Suppose, therefore, that $r_1 \neq r_2$—say $r_1 < r_2$. From the two expressions we obtain

$$(q_1 - q_2)a = r_2 - r_1.$$

Since $0 \le r_1 < r_2 < a$, it follows immediately that $0 < r_2 - r_1 < a$—so

$$0 < (q_1 - q_2)a < a.$$

Because $a > 0$, $0 < (q_1 - q_2)a$ implies $(q_1 - q_2) > 0$—so $(q_1 - q_2) \ge 1$ and $(q_1 - q_2)a \ge a$, which contradicts $(q_1 - q_2)a < a$. The assumption that $r_1 \neq r_2$ is therefore untenable, and the proof of the division algorithm is finally complete. ▮

It is worth noting that our uniqueness proof shows that if there are two expressions for b, then they are identical; no use is really made of existence. Thus, it would have been perfectly feasible to prove uniqueness first.

There is still another way to visualize the existence part of the division algorithm. Suppose both a and b are positive, and consider any set of b objects. Start to rearrange these b objects in rows of a elements each. In the end, we have a certain number of complete rows, and there may be some objects left over. If we let q denote the number of complete rows and r denote the number of leftovers, then clearly $b = qa + r$ and $0 \le r < a$.

1-2-2. Remark. In the statement of the division algorithm we assumed that $a > 0$. What happens if $a < 0$? Well, in this situation $-a$ is positive, so we can apply the division algorithm to b and $-a$. Consequently, there exist q and r such that

$$b = q(-a) + r, \qquad 0 \le r < |a|.$$

But this may be rewritten as $b = (-q)a + r$, and, of course, $-q$ is an integer. The conclusion is that the division algorithm can also be stated in more general form, as follows:

Given $a, b \in \mathbf{Z}$ with $a \neq 0$, then there exist unique integers q and r such that

$$b = qa + r, \qquad 0 \le r < |a|.$$

In the future, when referring to the division algorithm, it is to be understood that we have this general version in mind.

To illustrate the division algorithm when $a < 0$, let us fix $a = -37$ and take in turn $b = 1, 0, -1, 111, 63, -63$. The procedure mentioned above leads to the expressions

$$\begin{aligned}
1 &= 0 \cdot (-37) + 1, \\
0 &= 0 \cdot (-37) + 0, \\
-1 &= 1 \cdot (-37) + 36, \\
111 &= (-3) \cdot (-37) + 0, \\
63 &= (-1) \cdot (-37) + 26, \\
-63 &= (2) \cdot (-37) + 11.
\end{aligned}$$

1-2-3. Remark. In connection with the division algorithm, it is customary to call the integers q and r, which are determined uniquely by b and a the **quotient** and **remainder**, respectively. The remainder r is intimately related to the question of divisibility; in fact, it is not hard to see that

$$a \mid b \Leftrightarrow r = 0$$

and also, the logically equivalent statement,

$$a \nmid b \Leftrightarrow r \neq 0.$$

These assertions are so obvious that it is worthwhile to prove them in detail. If $r = 0$, then $b = qa + r = qa + 0 = qa$, so that $a \mid b$. Conversely, if $a \mid b$, then there exists $c \in \mathbf{Z}$ such that $b = ac$; thus, $b = c \cdot a + 0$, and by uniqueness of the division algorithm, $r = 0$. (Incidentally, this also shows that if $a \mid b$, then the integer $c \in \mathbf{Z}$ such that $b = ac$ is unique!)

The division algorithm says nothing about actually finding q and r in concrete cases. In this sense, it is not really an algorithm—a term usually reserved for a straightforward, mechanical, often repetitive procedure for carrying out a computation. However, we already "know" how to compute q and r; this is what the process of long division, which we all learned long ago, is all about. Strictly speaking, it needs to be proved that long division really results in q and r—but this would take us too far afield, so we take in on faith and leave the reader some food for thought.

1-2-4. Example. Suppose we fix $a = 2$. Then according to the division algorithm, each $b \in \mathbf{Z}$ can be expressed uniquely in exactly one of the forms $2q$ or $2q + 1$. If $b = 2q$ (that is, if $2 \mid b$) we say b is **even**; if $b = 2q + 1$ (that is, if $2 \nmid b$) we say that b is **odd**. It is now easy to verify the simple rules

$$\begin{aligned}
\text{even} + \text{even} &= \text{even}, & \text{even} \cdot \text{even} &= \text{even}, \\
\text{even} + \text{odd} &= \text{odd}, & \text{even} \cdot \text{odd} &= \text{even}, \\
\text{odd} + \text{odd} &= \text{even}, & \text{odd} \cdot \text{odd} &= \text{odd}.
\end{aligned}$$

Suppose that $a, b \in \mathbf{Z}$ with $a \neq 0$ are given. Applying the division algorithm to b and a determines q and r such that $b = qa + r$. Now, assuming that $r \neq 0$, the division algorithm may be applied to a and r. Of course, this process may then be repeated. Because several equations arise in this way, it is convenient to organize the notation and index the q's and r's. Thus, instead of $b = qa + r$ we shall write $b = q_1 a + r_1$; the next equation will be written as $a = q_2 r_1 + r_2$, and so on. All this leads to the following result.

1-2-5. Euclidean Algorithm. Given $a, b \in \mathbf{Z}$ with $a \neq 0$, we have equations of form

$$\begin{array}{lll} & b = q_1 a + r_1, & 0 < r_1 < |a|, \\ & a = q_2 r_1 + r_2, & 0 < r_2 < r_1, \\ & r_1 = q_3 r_2 + r_3, & 0 < r_3 < r_2, \\ & \vdots & \vdots \\ (i\text{th equation}) & r_{i-2} = q_i r_{i-1} + r_i, & 0 < r_i < r_{i-1}, \\ & \vdots & \vdots \\ & r_{n-2} = q_n r_{n-1} + r_n, & 0 < r_n < r_{n-1}, \\ & r_{n-1} = q_{n+1} r_n + 0. & \end{array}$$

Proof: There really is nothing to prove. The Euclidean algorithm asserts that our process of division stops after a finite number of steps—in other words, eventually we get a remainder of 0. But this is clear since $r_1 > r_2 > r_3 > \cdots$ with all terms greater than or equal to 0, and any strictly decreasing sequence of nonnegative integers must reach 0. ▌

The notation is chosen so that n is the index for which r_n is the last nonzero remainder—and then $r_n \mid r_{n-1}$. Of course, if $a \mid b$, then $r_1 = 0$, and there is not much point in discussing the Euclidean algorithm.

1-2-6. Example. Consider $a = 5530$, $b = 145{,}299$. For these integers the Euclidean algorithm is

$$\begin{aligned} 145{,}299 &= 26 \cdot 5530 + 1519, \\ 5530 &= 3 \cdot 1519 + 973, \\ 1519 &= 1 \cdot 973 + 546, \\ 973 &= 1 \cdot 546 + 427, \\ 546 &= 1 \cdot 427 + 119, \\ 427 &= 3 \cdot 119 + 70, \\ 119 &= 1 \cdot 70 + 49, \\ 70 &= 1 \cdot 49 + 21, \\ 49 &= 2 \cdot 21 + 7, \\ 21 &= 3 \cdot 7. \end{aligned}$$

The reader who is interested in the bookkeeping aspects will note that $q_1 = 26, r_1 = 1519, q_2 = 3, r_2 = 973, q_3 = 1, r_3 = 546, q_4 = 1, r_4 = 427, q_5 = 1,$ $r_5 = 119,\ q_6 = 3,\ r_6 = 70,\ q_7 = 1,\ r_7 = 49,\ q_8 = 1,\ r_8 = 21,\ q_9 = 2\ r_9 = 7,$ $q_{10} = 3.$

It is of some interest to observe that 7, the last nonzero remainder, divides both $a = 5530$ and $b = 145{,}299$. Can you find any additional properties of $r_9 = 7$?

1-2-7 | *PROBLEMS*

1. Describe completely the set $S = \{b - xa \geq 0 \mid x \in \mathbf{Z}\}$ when:
(*i*) $a = 7, b = 12$, (*ii*) $a = -7, b = 12$,
(*iii*) $a = 7, b = -12$, (*iv*) $a = 9, b = 5$,
(*v*) $a = -9, b = -5$,

2. Carry out the division algorithm and the Euclidean algorithm in each of the following cases:
(*i*) $a = 63{,}020$, $b = 76{,}084$, (*ii*) $a = 76{,}084$, $b = 63{,}020$,
(*iii*) $a = -63{,}020$, $b = 76{,}084$, (*iv*) $a = -76{,}084$, $b = 63{,}020$,
(*v*) $a = -63{,}020$, $b = -76{,}084$, (*vi*) $a = 113$, $b = -10{,}961$,
(*vii*) $a = 5311$, $b = 7571$.

3. In 1-2-2, the general form of the division algorithm was stated as follows: Given $a, b \in \mathbf{Z}$ with $a \neq 0$, then there exist unique integers q and r such that $b = qa + r$ and $0 \leq r < |a|$. At that time, we showed that q and r exist. Prove that q and r are indeed unique.

4. Show that the sum of any positive integer and its square is even. What if the integer in question is not positive?

5. Prove that the product of two integers of form $4n + 1$ is also of form $4n + 1$; the product of two integers of form $4n + 3$ is of form $4n + 1$; the product of an integer of form $4n + 1$ and one of form $4n + 3$ is of form $4n + 3$.

6. Why must the last digit in the square of an integer be one of 0, 1, 4, 5, 6, 9?

7. Prove: If both a and b are odd then $a^2 + b^2$ is even but not divisible by 4. Moreover, $a^2 + b^2$ is not a perfect square—that is, there is no $c \in \mathbf{Z}$ for which $a^2 + b^2 = c^2$.

8. Show that there is no integer n for which 4 divides $n^2 + 2$.

9. Prove that if $5 \nmid n$, then n^4 is of form $5m + 1$.

10. Show that an integer of form $6n + 5$ is also of form $3n - 1$, but the converse is false.

11. Can 100 be expressed as the sum of two integers one of which is divisible by 7 and the other by 11? How about 99? 101?

12. In the formal proof of the Division Algorithm 1-2-1, why does one begin by considering the set

$$S = \{b - xa \geq 0 \mid x \in \mathbf{Z}\}?$$

1-3. The Greatest Common Divisor

We turn next to a topic whose results can be arranged in such a way that they may be viewed as consequences of the Euclidean algorithm.

1-3-1. Definition. Suppose $a \neq 0$ and $b \neq 0$ are given. Any integer d which satisfies the following condition

(*i*) $d \mid a$ and $d \mid b$

is said to be a **common divisor** of a and b. If, in addition, d satisfies the following two conditions

(*ii*) If $c \mid a$ and $c \mid b$, then $c \mid d$,
(*iii*) $d > 0$,

then d is said to be a **greatest common divisor** (we abbreviate this to gcd) of a and b.

Expressed in words: A greatest common divisor of a and b is a common divisor which is positive and is also divisible by every common divisor.

As an example, we note that $a = 210$ and $b = 280$ have among their common divisors ± 1, ± 2, ± 7, ± 10, ± 14, ± 35 (are there any others?). Of course, 1 and -1 are common divisors for any choice of a and b. The situation regarding gcd is more complicated at this stage. It is not hard for the reader to convince himself that 6 is a greatest common divisor of $a = 18$ and $b = 24$ (after all, the only common divisors are ± 1, ± 2, ± 3, ± 6), nor is it much more difficult to see that 70 is a greatest common divisor of $a = 210$ and $b = 280$. On the other hand, for $a = 5530$, $b = 145{,}299$ we know from 1-2-6 that 7 is a common divisor, but the problem of locating a gcd looks messy—even more, are we really certain that these integers have a gcd?

As a matter of fact, for general a and b, a positive number is defined to be a greatest common divisor solely by virtue of its divisibility properties; its "size" does not enter into consideration, so from this point of view the use of the word "greatest" is probably deceptive and surely awkward. Thus, the definition of gcd raises several questions immediately—namely:

(1) Does a gcd exist?
(2) If a gcd exists, how many are there?
(3) If a gcd exists, can we find one?

These questions are answered by the following result.

1-3-2. Theorem. Given $a \neq 0$, $b \neq 0$, then their greatest common divisor exists and is unique. Moreover, we can find it explicitly via the Euclidean algorithm.

Proof: Let us treat uniqueness first. Suppose both d and d' are greatest common divisors of a and b, Since d' is a common divisor, condition (2) guarantees that it divides the gcd d—thus, $d' \mid d$. By reversing the roles of d and d', we have $d \mid d'$. Therefore, according to 1-1-2, part (*ix*), $d = \pm d'$. But they are both positive, so $d = d'$. We have shown that a and b can have *at most one* greatest common divisor.

Next, let us show that a greatest common divisor does exist. Consider the Euclidean Algorithm for the integers a and b—with the notation as in 1-2-5. Thus, r_n is the last nonzero remainder. We assert that r_n is a gcd! To prove this, it is necessary to verify the three conditions in the definition of gcd.

To show that r_n is a common divisor, note first that according to the last equation (that is, the one for r_{n-1}) $r_n \mid r_{n-1}$. From the equation immediately preceding this one (that is, the equation for r_{n-2}), it follows that $r_n \mid r_{n-2}$. In this way, by moving upward through the equations of the Euclidean algorithm, one by one, it follows that r_n divides each of the preceding r_i, and a and b, too. (The reader may find it instructive here to carry out this discussion for the concrete example—in which $r_n = 7$—treated in 1-2-6.) Thus, r_n is indeed a common divisor of a and b.

Since $r_n > 0$, in order to complete the proof that r_n is a gcd, it remains to show that if c is a common divisor of a and b, then $c \mid r_n$. For this, consider the first equation of the Euclidean Algorithm. Since $c \mid a$ and $c \mid b$, we see that $c \mid r_1$. Turning to the second equation, we know that $c \mid a$ and $c \mid r_1$—so it follows that $c \mid r_2$. In this way, by moving downwards through the Euclidean algorithm, one equation at a time, we conclude at the end that $c \mid r_n$. (The reader may wish to carry out this argument in the concrete example 1-2-6 and verify that if c divides both $a = 5530$ and $b = 145{,}299$, then $c \mid 7$.) Consequently, r_n is a gcd; in fact, it is the only one. The entire proof is now complete. ▮

In virtue of this theorem, we now have the right to speak of *the* greatest common divisor of a and b; we will denote it by (a, b). Since the definition of gcd makes no distinction between a and b—that is, the roles of a and b are interchangeable—it is clear that $(a, b) = (b, a)$. Furthermore, the sign of a or of b does not matter, for it is clear that

$$(a, b) = (-a, b) = (a, -b) = (-a, -b).$$

Consequently, in dealing with questions about the gcd, there is no loss of generality in assuming that $a > 0$ and $b > 0$.

1-3-3. Example. Suppose $a = 258$, $b = 354$, and let us compute $(258, 354)$. The Euclidean algorithm says

$$\begin{aligned} 354 &= 1 \cdot 258 + 96, \\ 258 &= 2 \cdot 96 \ + 66, \\ 96 &= 1 \cdot 66 \ + 30, \\ 66 &= 2 \cdot 30 \ + \ 6, \\ 30 &= 5 \cdot 6. \end{aligned}$$

Therefore, $(258, 354) = 6$. Furthermore, as observed above, we now know that $(-258, 354) = (258, -354) = (-258, -354) = 6$. Of course, these assertions could also be proved directly; for example, applying the Euclidean algorithm for $a = -258$, $b = -354$ gives

$$\begin{aligned} -354 &= (2)(-258) + 162, \\ -258 &= (-2)(162) + \ 66, \\ 162 &= 2 \cdot 66 \qquad + \ 30, \\ 66 &= 2 \cdot 30 \qquad + \ \ 6, \\ 30 &= 5 \cdot 6, \end{aligned}$$

so, once again, $(-258, -354) = 6$.

1-3-4. Remark. Suppose we compute the Euclidean algorithm for given a and b; then, as has already been proved, the last nonzero remainder r_n is precisely the gcd (a, b). If the first equation is now discarded the remaining equations obviously constitute the Euclidean algorithm for r_1 and a. Hence, $(r_1, a) = r_n$. Thus, by discarding the equations (from the top) one at a time, it follows that

$$r_n = (a, b) = (r_1, a) = (r_2, r_1) = \cdots = (r_n, r_{n-1}).$$

1-3-5. Remark. For given a and b, their gcd (a, b) (which is often denoted by d) is the biggest positive integer which divides them both. To see this, one observes that if c is any positive common divisor of a and b, then $c \mid d$, and since $d > 0$ it follows [by 1-1-2, part (*vii*)] that $c \leq d$. This property of the gcd—namely, biggest positive common divisor—may be used as the definition of gcd; in fact, it is probably the more natural definition. However, in view of generalizations and analogies to be discussed later, the definition we have given is really the "proper" one.

1-3-6. Theorem. Suppose $a \neq 0$, $b \neq 0$ are given and let $d = (a, b)$, then:

(*i*) Every linear combination $xa + yb$ of a and b is divisible by d.
(*ii*) The gcd d can be expressed as a linear combination of a and b.
(*iii*) The smallest positive integer of form $xa + yb$ is the gcd d.
(*iv*) The set of all linear combinations of a and b is identical with the set of all multiples of d—in symbols.

$$\{xa + yb \mid x, y \in \mathbf{Z}\} = \{nd \mid n \in \mathbf{Z}\}.$$

Proof: We proved (*i*) long ago, in 1-2-1, part (*v*). To prove (*ii*) we rewrite the equations of the Euclidean algorithm, as given in 1-2-5, by solving for the remainder in each case. We have then

$$\begin{aligned}
r_1 &= b - q_1 a, \\
r_2 &= a - q_2 r_1, \\
r_3 &= r_1 - q_3 r_2, \\
&\vdots \\
r_i &= r_{i-2} - q_i r_{i-1}, \\
&\vdots \\
r_{n-1} &= r_{n-3} - q_{n-1} r_{n-2}, \\
r_n &= r_{n-2} - q_n r_{n-1}.
\end{aligned}$$

Of course, the first equation implies that r_1 is a linear combination of a and b. Substituting the expression for r_1 in the second equation gives

$$r_2 = (1 + q_2 q_1)a + (-q_2)b$$

so r_2 is a linear combination of a and b. Now, when the expressions for r_1 and r_2 are substituted in the third equation, it follows that r_3 is a linear combination of a and b. In this way, by working downward through our system of equations, we arrive finally at an explicit expression for $d = r_n$ as a linear combination of a and b.

One can also prove (*ii*) and find values of x and y for which $d = xa + yb$ by working upward through our system of equations arising from the Euclidean algorithm. More precisely, starting from the last equation, we substitute in it the expression for r_{n-1} given by the next to last equation; thus,

$$\begin{aligned} r_n &= r_{n-2} - q_n r_{n-1} \\ &= r_{n-2} - q_n(r_{n-3} - q_{n-1} r_{n-2}) \\ &= (1 + q_n q_{n-1}) r_{n-2} - q_n r_{n-3}. \end{aligned}$$

Now, substitute $r_{n-2} = r_{n-4} - q_{n-2} r_{n-3}$ in this expression. The process may be repeated until, at the end, $d = r_n$ is expressed explicitly as a linear combination of a and b.

To prove (*iii*), let x_0 and y_0 be integers for which $x_0 a + y_0 b$ is the smallest positive integer of form $xa + yb$. Let us write $d_0 = x_0 a + y_0 b$; we must show that $d_0 = d$. Since the positive integer d is, according to (*ii*), a linear combination of a and b, it follows from the definition of d_0 that $d_0 \leq d$. On the other hand, according to (*i*), $d \mid d_0$. We conclude that $d_0 = d$.

To prove (*iv*), note first that, according to (*i*), every linear combination of a and b is a multiple of $d = (a, b)$. Symbolically, we may write

$$\{xa + yb \mid x, y \in \mathbf{Z}\} \subset \{nd \mid n \in \mathbf{Z}\}.$$

On the other hand, since $d = x_0 a + y_0 b$ (with x_0, y_0 as above), any multiple of d is of form

$$nd = n(x_0 a + y_0 b) = (nx_0)a + (ny_0)b$$

—so it is a linear combination of a and b. This shows that

$$\{nd \mid n \in \mathbf{Z}\} \subset \{xa + yb \mid x, y \in \mathbf{Z}\}.$$

The proof is now complete. ▌

1-3-7. Example. As in 1-3-3, consider $a = 258$, $b = 354$. Let us express $(258, 354) = 6$ as a linear combination of 258 and 354. The procedure is, of course, the one used in the proof of 1-3-6, part (*ii*). We rewrite the Euclidean algorithm, which was computed in 1-3-3, in the form

$$\begin{aligned} 96 &= 354 - 1 \cdot 258, \\ 66 &= 258 - 2 \cdot 96, \\ 30 &= 96 - 1 \cdot 66, \\ 6 &= 66 - 2 \cdot 30. \end{aligned}$$

Working downward from the top, we have

$$\begin{aligned} 96 &= b - a, \\ 66 &= a - 2(b - a) = 3a - 2b, \\ 30 &= (b - a) - (3a - 2b) = 3b - 4a, \\ 6 &= (3a - 2b) - 2(3b - 4a) = 11a - 8b. \end{aligned}$$

Consequently,

$$6 = (11)(258) + (-8)(354).$$

On the other hand, we can also obtain an expression for 6 as a linear combination of 258 and 354 by working upward from the bottom, as follows.

$$\begin{aligned} 6 &= 66 - (2)(30) \\ &= 66 - 2(96 - 66) \\ &= 3(66) - 2(96) \\ &= 3(258 - 2(96)) - 2(96) \\ &= 3(258) - 8(96) \\ &= 3(258) - 8(354 - 258) \\ &= 11(258) - 8(354). \end{aligned}$$

It may also be noted that to express the gcd of $a = 258$, $b = -354$ as a linear combination of a and b, we simply take note of the work above, and write

$$(258, -354) = 6 = (11)(258) + (8)(-354).$$

In similar fashion, we have

$$\begin{aligned} (-258, 354) &= (-11)(-258) + (-8)(354), \\ (-258, -354) &= (-11)(-258) + (8)(-354). \end{aligned}$$

1-3-8. Proposition. The greatest common divisor satisfies the following properties:

(*i*) If $m > 0$, then $(ma, mb) = m(a, b)$.
(*ii*) If $(a, b) = d$ and we write $a = da'$, $b = db'$, then $(a', b') = 1$.

Proof: (*i*) In virtue of 1-3-6, part (*iii*) and the fact that m is positive, it is clear that (ma, mb) is equal to the smallest positive integer of form $x(ma) + y(mb)$, which is equal to m times the smallest positive integer of form $xa + yb$, which in turn is equal to $m(a, b)$.

Furthermore, it is clear that if the condition $m > 0$ is dropped, this result takes the form

$$(ma, mb) = |m|\,(a, b).$$

(*ii*) This follows from (*i*), since

$$d = (a, b) = (da', db') = d(a', b')$$

implies $(a', b') = 1$. ∎

As illustrations of these facts, we note that $(258, 354) = (2 \cdot 129, 2 \cdot 177) = 2(129, 177)$, $(258, 354) = (3 \cdot 86, 3 \cdot 118) = 3(86, 118)$, and $(258, 354) = (6 \cdot 43, 6 \cdot 59) = 6(43, 59)$; moreover, because it is already known that $(258, 354) = 6$, it follows that $(43, 59) = 1$.

1-3-9. Definition. The integers $a \neq 0$, $b \neq 0$ are said to be **relatively prime** when $(a, b) = 1$.

We conclude this section with several elementary properties of relative primeness, the second of which will play a crucial role in the next section.

1-3-10. Proposition. For integers a, b, c we have:

(*i*) a and b are relatively prime $\Leftrightarrow$ there exist integers x_0, y_0 such that $x_0 a + y_0 b = 1$.
(*ii*) If $a \mid bc$ and $(a, b) = 1$, then $a \mid c$.
(*iii*) If $a \mid c$, $b \mid c$, and $(a, b) = 1$, then $ab \mid c$.
(*iv*) If $(a, b) = 1$ and $(a, c) = 1$, then $(a, bc) = 1$.

Proof: (*i*) This is a trivial consequence of 1-3-6. In fact, if $(a, b) = 1$, then surely there exist x_0, y_0 such that $x_0 a + y_0 b = 1$. Conversely, if $x_0 a + y_0 b = 1$, then $(a, b) = d$ must divide 1—so $(a, b) = d = 1$.

(*ii*) According to (*i*), there exist x_0, y_0 such that $x_0 a + y_0 b = 1$; and then $x_0 ac + y_0 bc = c$, which implies that $a \mid c$.

(*iii*) Since $a \mid c$, we may write $c = aa'$ for some $a' \in \mathbf{Z}$. But then $b \mid aa'$ and $(b, a) = 1$, so in virtue of (*ii*) we conclude that $b \mid a'$. Consequently, $ab \mid aa'$.

(*iv*) The hypotheses imply that there exist x_0, y_0, x_1, y_1 such that $x_0 a + y_0 b = 1$ and $x_1 a + y_1 c = 1$. Therefore,

$$\begin{aligned} 1 &= x_0 a + (y_0 b) \cdot 1 \\ &= x_0 a + (y_0 b)(x_1 a + y_1 c) \\ &= (x_0 + y_0 b x_1)a + (y_0 y_1)bc \end{aligned}$$

which guarantees that $(a, bc) = 1$. ▌

1-3-11. Exercise. There is a flaw in our discussion of gcd. It was defined for $a \neq 0$, $b \neq 0$ and its existence was proved by making use of the Euclidean algorithm. What happens if $a \mid b$? In this situation, $b = qa + 0$, so there is no nonzero remainder and the existence proof for gcd, which was given in 1-3-2, breaks down. The way to avoid this difficulty is to include in the proof of 1-3-2 a direct proof that if $a \mid b$, then $|a|$ satisfies the requirements for a gcd of a and b. (The proof is left to the reader.) Thus, if $a \mid b$ then $(a, b) = |a|$; and in this case too we could say that the Euclidean algorithm determines the gcd. More precisely, in all cases, the gcd is the absolute value of the divisor in the last equation (meaning the one with remainder 0) of the Euclidean algorithm.

Furthermore, if we look carefully at the various results about gcd, especially 1-3-6 and 1-3-8, it becomes apparent that their proofs do not take care of the case where $a \mid b$. We leave it to the reader to provide the missing proofs.

The upshot is that all the results about gcd are valid, as stated, for arbitrary $a \neq 0$, $b \neq 0$.

1-3-12. Exercise. What happens if instead of defining gcd only when both integers are not equal to 0, we permit one of them to be 0—say $a \neq 0$, $b = 0$? In this situation (which is related to the special case $a \mid b$ as treated in 1-3-11) it is clear that $(a, 0) = |a|$ It is easy to show that the results about gcd remain valid in this case too.

1-3-13 / *PROBLEMS*

1. For $a \neq 0$, $b \neq 0$, prove carefully that $(a, b) = (-a, b) = (a, -b) = (-a, -b)$.
2. In 1-2-6 we saw that $7 = (5530, 145{,}299)$ Use the Euclidean algorithm, as given there, to derive the following expression for 7 as a linear combination of $a = 5530$ and $b = 145{,}299$
$$7 = (-6122)a + 233b$$
3. In each of the following cases find (a, b) and express it as a linear combination of a and b:
   (*i*) $a = 91, b = 143$,
   (*ii*) $a = 143, b = 91$,
   (*iii*) $a = -143, b = -91$,
   (*iv*) $a = 5311, b = 7571$,
   (*v*) $a = 5311, b = -7571$
4. Prove in detail that $(a, b) = 1 \Leftrightarrow$ there exist integers x_1, y_1 such that $x_1 a + y_1 b = -1$ (Of course, this result should be compared with 1-3-10, part (*i*).)
5. Prove that the following properties of gcd hold:
   (*i*) If $(a, b) = 1$ and $c \mid a$, then $(c, b) = 1$.
   (*ii*) If $(a, bc) = 1$, then $(a, b) = (a, c) = 1$.
   (*iii*) If $(a, b) = 1$, then $(ca, b) = (c, b)$.
6. Suppose $(a, b) = d$, and let x_0, y_0 be integers for which $x_0 a + y_0 b = d$. Show that
   (*i*) $(x_0, y_0) = 1$,
   (*ii*) x_0 and y_0 are not unique.

7. If $c > 0$, $c \mid a$, $c \mid b$ show that $(a/c, b/c) = (a, b)/c$. Note that since we deal only with integers, the reader must first clarify the meaning of a/c, b/c, and $(a, b)/c$.

8. Show that for every integer $a \neq 0$, $(a, a + 2) = 1$ or 2. Moreover, for every n and every $a \neq 0$, $(a, a + n)$ divides n.

9. Prove: If $(a, 4) = 2 = (b, 4)$, then $(a + b, 4) = 4$.

10. Suppose $(a, b) = 1$—then
(*i*) $(a + b, b) = 1$; in fact, $(a + mb, b) = 1$ for every $m \in \mathbf{Z}$.
(*ii*) $(a, b^2) = 1$; in fact, $(a^s, b^t) = 1$ for all positive integers s and t.

11. Suppose a, b, c, d, m, n, u, and v are nonzero integers with $ad - bc = \pm 1$, $u = am + bn$, $v = cm + dn$; show that $(m, n) = (u, v)$.

12. If $(a, b) = 1$, show that
(*i*) $(a + b, a - b) = 1$ or 2,
(*ii*) $(a^2 + b^2, a + b) = 1$ or 2,
(*iii*) $(a^3 + b^3, a^2 + b^2)$ divides $a - b$,
(*iv*) $(a + b, a^2 - ab + b^2) = 1$ or 3.

13. Consider the set $S = \{1, 2, \ldots, n\}$, and let 2^r be the highest power of 2 which belongs to S. Show that 2^r does not divide any element of S other than itself.

14. Show that $(a, b + ma) = (a, b)$ for all $m \in \mathbf{Z}$.

15. In the proof of part (*iii*) of 1-3-6, we started by taking integers x_0, y_0 for which $d_0 = x_0 a + y_0 b$ is the smallest positive integer of form $xa + yb$. How do we know that such integers x_0, y_0, d_0 exist?

16. State the various parts of 1-3-10 "in words."

1-4. Unique Factorization

Given a nonzero integer, we may try to "break it up" with respect to multiplication—that is, to factor it—into "simpler" integers. From this point of view, the simplest integers are those which cannot be factored further; this is what underlies the following definition.

1-4-1. Definition. An integer $p > 1$ is said to be **prime** when 1 and p are its only positive divisors; if $p > 1$ is not prime, we say it is **composite.**

Thus, to say that $n > 1$ is composite means that we can write $n = ab$ with $1 < a < n$ and $1 < b < n$, while saying that $n > 1$ is prime means that n cannot be written in this form.

For example, 91, 5311, 7571 are composite because $91 = 7 \cdot 13$, $5311 = 47 \cdot 113$, $7571 = 67 \cdot 113$. On the other hand, it is easy to see that 2, 3, 7, 13, 19,

37, 67, 79, 97, 113, 119, 223, 229 are primes. Of course, according to the definition, 1 is not a prime—nor is it composite. The reason for leaving 1 in limbo is that it clearly plays no role in questions of factorization (even though 1 is a divisor or factor of any integer); 1 can be included or removed from any factorization, so it should appropriately be ignored.

What about the factorization of negative numbers? If $n < -1$, then $n = (-1)(-n)$ with $-n > 1$, so the question of factoring the negative number n reduces to factoring the positive integer $-n$. In addition, $+1$ and -1 make no meaningful contribution to a factorization, and they may be safely ignored. Thus, we shall restrict ourselves, without loss of generality, to factorization questions for integers greater than 1.

In this connection, it should be noted that if p is prime, then $-p$ satisfies the same divisibility properties as p—but it contributes no new information. It would be awkward and confusing, at best, to keep track of both p and $-p$, or to view them as distinct primes. It is because of this that our definition requires that a prime be positive.

1-4-2. Proposition. If p is prime, then

(*i*) If $p \nmid a$, then $(p, a) = 1$.
(*ii*) If $p \mid ab$ and $p \nmid a$, then $p \mid b$.
(*iii*) If $p \mid a_1 a_2 a_3 \cdots a_s$, then p divides one or more of the a_i.

Proof: (*i*) The only positive divisors of p are 1 and p, so $p \nmid a$ implies that 1 is the greatest common divisor of p and a. Thus, for example, since $7 \nmid 99$ and 7 is prime, we know that 7 and 99 are relatively prime.

(*ii*) In virtue of (*i*) we have $p \mid ab$ and $(p, a) = 1$, and then 1-3-10, part (*ii*) implies $p \mid b$. This shows that if a prime divides the product of two integers and does not divide the first factor, then it divides the second.

(*iii*) Proceed inductively. If $p \mid a_1$ we are finished; if not, then according to (*ii*), $p \mid a_2 a_3 \cdots a_s$. Now if, $p \mid a_2$ we are finished; if not, then $p \mid a_3 \cdots a_s$. By repeating the process, we arrive eventually at an a_i which is divisible by p—thus showing that if a prime divides a product, then it must divide at least one of the factors. ▌

1-4-3. Theorem. Every integer greater than 1 can be expressed as a finite product of primes.

Proof: This result may appear obvious, but it does require proof. We give a proof by contradiction.

Suppose the theorem is false; so there exists an integer, greater than 1, that cannot be expressed as a finite product of primes (call it m). Let S denote the set of all integers greater than 1 that cannot be written as a finite product of primes. Then $S \neq \varnothing$ (as $m \in S$). Since S is a nonempty collection of positive integers, it contains a smallest element—call it n; so n is less than any other element of S. Now, $n > 1$ (because $n \in S$ and every element of S is greater than 1) and n is composite (for if n is prime, then it is already written as a finite product of primes—with only one factor—and, therefore, n could not belong to S). Therefore, we may write

$$n = ab, \qquad 1 < a < n, \qquad 1 < b < n.$$

But then, because n is the smallest element of S, we have $a \notin S$, $b \notin S$; so both a and b can be expressed as a finite product of primes. Consequently, their product $n = ab$ can also be written as a finite product of primes—which contradicts $n \in S$. Our original supposition that the theorem is false having led to a contradiction, we conclude that the theorem is true. ▮

We digress from the main theme of this section to deal with two related results whose proofs make use of 1-4-3.

1-4-4. Theorem. (Euclid: circa 300 B.C.). The number of primes is infinite.

Proof: Suppose the number of primes is finite, and denote all of them by

$$p_1 < p_2 < p_3 < \cdots < p_n.$$

Thus, $p_1 = 2, p_2 = 3, p_3 = 5, p_4 = 7$, and so on. Now consider the integer

$$N = p_1 p_2 p_3 \cdots p_n + 1. \tag{$*$}$$

In words, N is obtained by adding 1 to the product of all the primes. Since N is not a prime (because, according to our supposition, p_n is the biggest prime, and N is bigger than p_n), it can therefore be expressed as a product of primes. But there are no primes other than $p_1, p_2, \ldots, p_n$, so at least one p_i appears in the prime factorization of N. In particular, this p_i divides N. Because p_i divides the product $p_1 p_2 \cdots p_n$, it follows from $(*)$ that $p_i \mid 1$—a contradiction. The initial assumption that the number of primes is finite having led to a contradiction, we conclude that the number of primes is infinite. ▮

Suppose we apply the division algorithm for an arbitrary integer b and $a = 4$; then b is clearly of exactly one of the following forms:

$$\begin{aligned} \text{I}&: \ 4n,\\ \text{II}&: \ 4n+1,\\ \text{III}&: \ 4n+2,\\ \text{IV}&: \ 4n+3. \end{aligned}$$

Obviously, any integer of type I or III is even, while any integer of type II or IV is odd. Conversely, any even integer must be of type I or III, while any odd integer must be of type II or IV.

There are no primes of type I, and there is exactly one prime of type III—namely, the prime 2. The remaining primes, which by 1-4-4 are infinite in number, are odd—so each of them is of type II or IV. One may ask, how many primes are there of form $4n + 3$ (that is, of type IV)? In other words, how many primes appear in the arithmetic progression

$$3, 7, 11, 15, 19, 23, 27, 31, \ldots$$

with initial term 3 and difference 4? As is expected perhaps, the answer is as follows:

1-4-5. Theorem. The number of primes of form $4n + 3$ is infinite.

Proof: Our proof is by contradiction and involves a slight refinement of the argument used to prove 1-4-4.

Suppose there are only a finite number of primes of form $4n + 3$. Denote them by

$$p_1 < p_2 < p_3 < \cdots < p_m.$$

In particular, $p_1 = 3, p_2 = 7, p_3 = 11, p_4 = 19, p_5 = 23, \ldots$ and we are assuming that p_m is the biggest prime of form $4n + 3$. Consider the integer

$$M = 4p_2 p_3 \cdots p_m + 3,$$

which is surely of form $4n + 3$. We observe that M is composite (because p_m is the largest prime of form $4n + 3$), $p_1 \nmid M$ (because $p_1 = 3$ does not divide $4p_2 p_3 \cdots p_m$), and $p_i \nmid M$ for $i = 2, \ldots, m$ (because any such p_i does not divide 3). Therefore, the factorization of M cannot contain any primes of form $4n + 3$. Furthermore, M is odd, so its factorization cannot include the prime 2. It follows that all the prime factors of M are of form $4n + 1$. But the product of two, or more, integers of form $4n + 1$ is again of form $4n + 1$, which implies that M must be of form $4n + 1$—contradiction. Hence, the number of primes of form $4n + 3$ is infinite. ▌

It is also known that the number of primes of form $4n + 1$ is infinite. The reader may wish to undertake the proof, but he will probably fail because our present techniques are insufficient to win through to the end. However, a proof of this fact will be given in Chapter III.

There is a deep and famous theorem of Dirichlet (1805–1859) that settles all questions of this sort. It says, roughly, that any arithmetic progression contains an infinite number of primes—more precisely, if $c \neq 0$, $d \neq 0$, and $(c, d) = 1$, then the sequence

$$c, c + d, c + 2d, c + 3d, \ldots,$$

contains an infinite number of primes. In our terminology this says that for $c \neq 0$, $d \neq 0$, $(c, d) = 1$ there are an infinite number of primes of form $c + md$.

Let us return to the main theme. We have seen that any integer greater than 1 can be written as a product of primes. It is easy, for example, to factor 7200—namely, $7200 = 2 \cdot 2 \cdot 2 \cdot 3 \cdot 3 \cdot 2 \cdot 2 \cdot 5 \cdot 5$; on the other hand, trying to factor 24,523 is considerably more difficult—it turns out that $24{,}523 = 137 \cdot 179$. In fact, the theorem on existence of a prime factorization says nothing about how to find a factorization, and there really is not much we can do about it. However, another interesting question arises. In how many ways can an arbitrary integer greater than 1 be factored into primes? The fact that we are "certain" that there is essentially only one factorization (after all, this is what experience tells us) does not necessarily make it so. Of course, in virtue of the commutative law for multiplication, the order in which the prime factors appear does not matter—for example, we should clearly not consider

$$2 \cdot 2 \cdot 2 \cdot 3 \cdot 3 \cdot 2 \cdot 2 \cdot 5 \cdot 5 \quad \text{and} \quad 2 \cdot 3 \cdot 2 \cdot 3 \cdot 2 \cdot 5 \cdot 2 \cdot 5 \cdot 2$$

as distinct factorizations. Except for this unavoidable limitation, we have the best of all possible worlds—namely:

1-4-6. Theorem. The factorization of any integer greater than 1 into primes is unique up to order. In other words, the order in which the terms of the prime factorization are listed does not matter.

Proof: Suppose the theorem is false; so the set of S of all integers greater than 1 which have more than one prime factorization is nonempty. Then S has a smallest element—call it n. Thus, n has at least two prime factorizations, and any integer greater than 1 and less than n has a unique prime factorization. Consider any two distinct factorizations

$$n = p_1 p_2 \cdots p_r = q_1 q_2 \cdots q_s,$$

where every p_i and q_j is prime. (We do not assert that all the p's are distinct or that all the q's are distinct; any of the primes may well appear several times.)

We must have $r \geq 2$ and $s \geq 2$; in other words, each side of the equation must have at least two factors. For if one side has only a single factor, we have, for example,

$$n = p_1 = q_1 q_2 \cdots q_s,$$

which provides a factorization of the prime p_1. This being impossible, we must have $s = 1$ and $p_1 = q_1$, which says that the two factorizations are identical and contradicts the basic property of n. So indeed $r \geq 2$ and $s \geq 2$.

Now consider the prime p_1. It divides the right side so, according to 1-4-2, it divides at least one of the q's. Reindexing the q's if necessary (here is where the question of order of the factors enters) we may assume that p_1 divides q_1. But q_1 is prime, so it is immediate that $p_1 = q_1$. Consequently,

$$n = p_1 p_2 \cdots p_r = p_1 q_2 \cdots q_s,$$

and if we write $n = p_1 n'$, then

$$n' = p_2 \cdots p_r = q_2 \cdots q_s.$$

Surely, $n' > 1$ (because $r \geq 2$, which implies $n \neq p_1$) and $n' < n$. Therefore, it follows from the choice of n that the factorization of n' into primes is unique; so with reordering, if necessary, we have

$$p_2 = q_2, \qquad p_3 = q_3, \ldots, p_r = q_r \quad \text{and} \quad r = s.$$

Because $p_1 = q_1$, we see that our two distinct factorizations of n are the same—contradiction. The proof of the theorem is now complete. ▮

1-4-7. Remark. Although existence and uniqueness of factorization were proved for integers $n > 1$, these results clearly apply also for negative n—one simply takes out a factor -1 at the start. Our general result, therefore, reads as follows:

Any integer $n \neq 0$ can be expressed uniquely in the form

$$n = (\pm 1) p_1 p_2 \cdots p_r,$$

where the p_i are primes and exactly one of $+1$, -1 occurs.

It should be noted that the proofs of existence of a prime factorization 1-4-3 and of uniqueness of prime factorization are independent of each other; either theorem could be proved first. The existence proof makes use of none of the information we have collected about "number theory." On the other hand, the uniqueness proof uses 1-4-2, which in turn depends on facts about the gcd.

1-4-8 | *PROBLEMS*

1. Find all primes less than 300. [Hint: One way to do this mechanically is by the so-called "sieve of Eratosthenes." First write all the integers from 2 to 299. Then starting from 2, cross out every second number, while

keeping 2. This eliminates all numbers divisible by 2. Then, 3 being the smallest number (in fact, prime) remaining, keep it, and cross out every third number starting from 3 (note that the numbers that have already been crossed out are counted.) Next, go on to 5 and cross out every fifth integer. This process is to be continued as long as necessary.]

2. If the positive integer n is composite, show that it has a prime factor p satisfying $p \leq \sqrt{n}$. Thus, to decide if n is prime it suffices to check that it is not divisible by any prime less than $\sqrt{n}$. This also tells us how far to go in the sieve of Eratosthenes.

3. Express every prime less than 225 which is of form $4n + 1$ as a sum of two squares.

4. Prove that any prime of form $3n + 1$ is of form $6n + 1$.

5. Show that any positive integer of form $3n + 2$ has a prime factor of the same form.

6. In the proof of 1-4-4—that the number of primes is infinite—use is made of the integer $N = p_1 p_2 p_3 \cdots p_n + 1$. What happens if we take

$$N = p_n! + 1?$$

7. We proved 1-4-4 and 1-4-5 (the infinitude of primes and the infinitude of primes of form $4n + 3$) immediately after proving the existence of prime factorization 1-4-3, but before proving uniqueness 1-4-6. Show that uniqueness is not needed for the proof of 1-4-4 or 1-4-5.

8. Prove carefully the statement in 1-4-7; namely, any integer $n \neq 0$ can be expressed uniquely in the form $n = (\pm 1)p_1 p_2 \cdots p_r$ where the p_i are primes and exactly one of $+1$, -1 occurs, What happens when $n = \pm 1$?

9. Suppose that $(a, p^2) = p$, $(b, p^4) = p^2$ and p is prime. Evaluate:
(i) (ab, p^5), *(ii)* $(a + b, p^4)$,
(iii) $(a - b, p^5)$, *(iv)* $(pa - b, p^5)$.

10. Suppose that $(a, b) = p$, p prime. Evaluate
(i) (a^2, b), *(ii)* (a^2, b^2),
(iii) (a^3, b), *(iv)* (a^2, b^3).
In each case, give numerical examples to illustrate your answer.

11. For any positive integer n, show that
(i) $2 \mid (n^2 - n)$, *(ii)* $6 \mid (n^3 - n)$,
(iii) $30 \mid (n^5 - n)$.

12. Show that the product of three consecutive integers is divisible by 6. Moreover, if the first one is even, the product is divisible by 24.

13. Show that the product of four consecutive integers is divisible by 24.

14. Prove: If the sum of two fractions a/b and c/d in lowest terms (meaning that $(a, b) = 1$, $(c, d) = 1$) is an integer, then $b = \pm d$.

15. Prove that the number of primes of form $6n + 5$ is infinite.

16. If $d = (a, b)$ and $b > 0$, show that the finite sequence

$$a, 2a, 3a, \ldots, ba$$

has exactly d of its terms divisible by b.

17. For p prime, decide if the following statements are true or false; if true, prove the statement; if false, give a counterexample.
(*i*) $p \mid (a^2 + b^2), p \mid (b^2 + c^2) \Rightarrow p \mid (a^2 - c^2)$.
(*ii*) $p \mid a^7 \Rightarrow p \mid a$.
(*iii*) $p \mid a, p \mid (a^2 + b^2) \Rightarrow p \mid b$.
(*iv*) $p \mid (a^2 + b^2), p \mid (b^2 + c^2) \Rightarrow p \mid (a^2 + c^2)$.
(*v*) $a^3 \mid b^3 \Rightarrow a \mid b$.
(*vi*) $a^3 \mid b^2 \Rightarrow a \mid b$.
(*vii*) $a^2 \mid b^3 \Rightarrow a \mid b$.

18. Consider the set

$$2\mathbf{Z} = \{2n \mid n \in \mathbf{Z}\},$$

which is, of course, the set of all even integers. We can add, substract, or multiply two elements of $2\mathbf{Z}$, and the result is an element of $2\mathbf{Z}$. In analogy with $\mathbf{Z}$, we can discuss questions of divisibility and factorization in $2\mathbf{Z}$. (Note, for example, that in $2\mathbf{Z}$, 2 does not divide $6 = 2 \cdot 3$, but 2 divides $4 = 2 \cdot 2$). We can also introduce the notion of a prime in $2\mathbf{Z}$—namely, a positive element of $2\mathbf{Z}$ whose only positive divisor in $2\mathbf{Z}$ is itself, (Note that $1 \notin 2\mathbf{Z}$, so we ignore it as a possible divisor of a prime.)
(*i*) List some primes of $2\mathbf{Z}$. Can you describe all the primes of $2\mathbf{Z}$?
(*ii*) Prove that every positive element of $2\mathbf{Z}$ can be expressed as a product of primes of $2\mathbf{Z}$.
(*iii*) Show that this factorization into primes need not be unique.
(*vi*) What about negative primes and the factorization of negative elements of $2\mathbf{Z}$?

19. Is 999,991 prime?

20. Show that $n^4 + 4$ is composite for every $n > 1$.

I-5. A Convenient Notation

Consider an integer $a > 1$ and its unique factorization into primes. A given prime p may appear more than once in this factorization and, as is perfectly natural, we combine such repeated terms. In this way, the factorization of a is expressed in terms of distinct primes raised to appropriate powers. For example, if $a = 7200 = 2 \cdot 3 \cdot 2 \cdot 3 \cdot 2 \cdot 5 \cdot 2 \cdot 5 \cdot 2$ we would write

$$a = 7200 = 2^5 \cdot 3^2 \cdot 5^2.$$

Similarly, for $b = 9996 = 2 \cdot 3 \cdot 7 \cdot 2 \cdot 17 \cdot 7$ we would write

$$b = 9996 = 2^2 \cdot 3 \cdot 7^2 \cdot 17.$$

Of course, these factorizations of $a = 7200$ and $b = 9996$ in terms of prime powers are unique up to order. Now, we can compute the product ab, but of much greater interest here is the fact that we can immediately produce the factorization of ab in terms of prime powers—namely,

$$ab = (7200)(9996) = 2^7 \cdot 3^3 \cdot 5^2 \cdot 7^2 \cdot 17.$$

How is this done? For a prime which appears in both factorizations we add its two exponents; for a prime which appears in just one factorization, we view it as appearing to the power 0 in the other factorization and again add exponents. Naturally, primes which appear in neither factorization are ignored. In other words, we "really" write

$$a = 7200 = 2^5 \cdot 3^2 \cdot 5^2 \cdot 7^0 \cdot 17^0, \qquad b = 9996 = 2^2 \cdot 3^1 \cdot 5^0 \cdot 7^2 \cdot 17^1,$$

and add the exponents for each of the primes.

Let us return to the general situation. Any $a > 1$ can be expressed uniquely (up to order, of course) as a product of prime powers—that is,

$$a = p_1^{r_1} p_2^{r_2} \cdots p_m^{r_m},$$

where $p_1, p_2, \ldots, p_m$ are distinct primes and all the exponents $r_1, r_2, \ldots, r_m$ are greater than 0. Naturally, another integer $b > 1$ has an expression of form

$$b = q_1^{s_1} q_2^{s_2} \cdots q_n^{s_n},$$

where $q_1, q_2, \ldots, q_n$ are distinct primes and the exponents $s_1, s_2, \ldots, s_n$ are greater than 0. If we try to write out the prime power factorization of ab, things turn out awkwardly because the notation gets in the way. In general, some primes appear among both the p's and the q's, other primes occur only among the p's, and still others only among the q's. It is obviously a mess to carry out, in a theoretical context, the procedure employed above in the numerical example $a = 7200$, $b = 9996$—namely, when a prime occurs in just one of the factorizations, insert it into the other factorization with exponent 0. After juggling and renaming the p's and q's the result is, in the general case, expressions for a and b such that exactly the same primes appear in the two expressions, except that 0 exponents can occur. Then to multiply a and b we simply add exponents for corresponding primes. Note that we do not even try to set up the required notation.

Suppose further that we consider the same element a and replace b by an arbitrary integer $c > 1$. As above, by inserting appropriate primes with exponent 0, we arrive at expressions for a and c in which exactly the same primes occur. However, since c and b need not have the same primes appearing in their original factorizations (that is, when all exponents are greater than 1) the expression we obtain here for a is not, under normal circumstances, identical with our previous expression for a. For example, we have seen that for $a =$ 7200, $b = 9996$ we obtain expressions

$$a = 2^5 \cdot 3^2 \cdot 5^2 \cdot 7^0 \cdot 17^0, \qquad b = 2^2 \cdot 3^1 \cdot 5^0 \cdot 7^2 \cdot 17^1.$$

If we then take $c = 80{,}465 = 5 \cdot 7 \cdot 11^2 \cdot 19$, we obtain for a and c the expressions

$$a = 2^5 \cdot 3^2 \cdot 5^2 \cdot 7^0 \cdot 11^0 \cdot 19^0, \qquad c = 2^0 \cdot 3^0 \cdot 5^1 \cdot 7^1 \cdot 11^2 \cdot 19^1.$$

Clearly, the notation (especially because it is not unique) leaves something to be desired, and it would be nice to have a uniform notation which is a help rather than a hindrance. Our approach will be to look at all primes and simultaneously to keep track of the exponents.

1-5-1. Definition. For any prime p and any integer $a > 1$, we let $v_p(a)$ (read this as "nu sub p of a") denote the exponent to which p appears in the factorization of a.

Of course, it is understood that when p is not a factor of a, so that it does not appear in the factorization of a, we write $v_p(a) = 0$. This amounts to taking the formal view that p^0 is the power of p which appears in the factorization of a.

Some concrete examples should help to clarify the definition. For $a =$ $7200 = 2^5 \cdot 3^2 \cdot 5^2$, we have

$$v_2(7200) = 5, \qquad v_3(7200) = 2, \qquad v_5(7200) = 2,$$

Furthermore, if p is any prime other than 2, 3, or 5, then it does not divide 7200 and does not appear in the factorization; so

$$v_p(7200) = 0 \qquad \text{for all } p \neq 2, 3, 5.$$

In similar fashion, if $b = 9996 = 2^2 \cdot 3^1 \cdot 7^2 \cdot 17^1$, then

$$v_2(9996) = 2, \qquad v_3(9996) = 1, \qquad v_7(9996) = 2, \qquad v_{17}(9996) = 1$$

and, as for the remaining primes,

$$v_p(9996) = 0 \qquad \text{for all } p \neq 2, 3, 7, 17.$$

Finally for $c = 80{,}465 = 5 \cdot 7 \cdot 11^2 \cdot 19$ we have

$$v_5(80{,}465) = 1, \quad v_7(80{,}465) = 1, \quad v_{11}(80{,}465) = 2, \quad v_{19}(80{,}465) = 1,$$
$$v_p(80{,}465) = 0, \quad p \neq 5, 7, 11, 19.$$

It should be noted that the definition of $v_p(a)$ depends on two "variables": the integer $a > 1$ and the prime p. Our discussion will emphasize fixing a, and studying the collection of all $v_p(a)$ as p runs over the set of all primes.

Suppose we fix $a > 1$ and look at the prime factorization of a. For the primes p that appear in this factorization we have clearly $v_p(a) > 0$; these are, of course, the primes which divide a, and there are only a finite number of such primes. On the other hand, for the primes p that do not appear in the factorization of a we have $v_p(a) = 0$; these are the primes that do not divide a, and there are an infinite number of such primes. This information may be summarized and restated as follows:

1-5-2. Proposition. Given $a > 1$ we have

(*i*) $v_p(a) \geq 0$ for all p,
(*ii*) $v_p(a) = 0$ for almost all p,
(*iii*) $v_p(a) > 0 \Leftrightarrow p$ divides a,
(*iv*) $v_p(a) = 0 \Leftrightarrow p$ does not divide a.

Proof: The proof has already been given. It is only necessary to note that the phrase "for almost all p" is common mathematical parlance—its meaning is "for all but a finite number of p." ▮

Suppose $a > 1$ is fixed, what profit is there in knowing $v_p(a)$ for every prime p? A concrete example will help us understand the answer. Consider $a = 142{,}025 = 5^2 \cdot 13 \cdot 19 \cdot 23$, so that

$$v_5(142{,}025) = 2, \quad v_{13}(142{,}025) = 1, \quad v_{19}(142{,}025) = 1, \quad v_{23}(142{,}025) = 1,$$
$$v_p(142{,}025) = 0 \quad \text{for all } p \neq 5, 13, 19, 23.$$

Now, given all the exponents $v_p(142{,}025)$, let us consider the expression

$$2^0 \cdot 3^0 \cdot 5^2 \cdot 7^0 \cdot 11^0 \cdot 13^1 \cdot 17^0 \cdot 19^1 \cdot 23^1 \cdot 29^0 \cdot 31^0 \cdots.$$

It is formed by writing each prime p with the exponent $v_p(a)$; of course, the three dots at the end indicate that we run through the infinite collection of all primes and that the missing primes occur with exponent 0.

This expression may be viewed as a formal expression in which we have a term $p^{v_p(a)}$ for each prime p; however, we prefer to view it as a product. The fact that it seems to be an infinite product is no problem—in fact, because $v_p(a) = 0$ for almost all p, we see that almost all terms are $p^{v_p(a)} = p^0 = 1$, so only a finite number of terms in the product are not equal to 1, and we are really dealing with a product of a finite number of terms. As the reader has probably observed already, this product is precisely $a = 142{,}025$ itself; after all, what we have effectively done is take the factorization $5^2 \cdot 13 \cdot 19 \cdot 23$ of a and insert every other prime with exponent 0. Thus, knowing the set of values $\{v_p(142{,}025) \mid p \text{ prime}\}$ enables us to recapture $a = 142{,}025$; we simply take the product of all terms of form $p^{v_p(142{,}025)}$.

Next, let us turn to the case of an arbitrary integer $a > 1$. Consider the infinite product

$$\prod_p p^{v_p(a)}$$

where p runs over all primes. This is a compact notation for the infinite product, and is obviously more economical than writing it out as was done above. Of course, the product is really finite; to see this one notes that because $v_p(a) = 0$ for almost all p, there are only a finite number of factors $p^{v_p(a)}$ which are not equal to 1. Moreover, it is clear that these factors $p^{v_p(a)} \neq 1$ are precisely the terms of the prime power factorization of a—so

$$a = \prod_p p^{v_p(a)}. \tag{$*$}$$

Furthermore, we note that the set of all $v_p(a)$, $\{v_p(a) \mid p \text{ prime}\}$, determines a, in the sense that if a is lost, then it can be recaptured from the $v_p(a)$'s by writing $\prod p^{v_p(a)}$. In particular, $p^{v_p(a)}$ is the highest power of p which divides a.

1-5-3. Proposition. Any $a > 1$ has an expression of form

$$a = \prod_p p^{v_p(a)}.$$

Moreover if, in addition, $b > 1$, then the following are equivalent:

(*i*) $a = b$,

(*ii*) $v_p(a) = v_p(b)$ for all p,

(*iii*) $\prod_p p^{v_p(a)} = \prod_p p^{v_p(b)}$.

Proof: The validity of the expression for a was proved above. Of course, b has an expression of form

$$b = \prod_p p^{v_p(b)}$$

and we must show that

$$(i) \Leftrightarrow (ii) \Leftrightarrow (iii).$$

From the definition of v_p and the foregoing discussion we have

$$\begin{aligned} a = b \quad &\Rightarrow \quad v_p(a) = v_p(b) \text{ for all } p \\ &\Rightarrow \quad \prod p^{v_p(a)} = \prod p^{v_p(b)} \\ &\Rightarrow \quad a = b\,. \end{aligned}$$

This shows that

$$(i) \Rightarrow (ii) \Rightarrow (iii) \Rightarrow (i)$$

and completes the proof. ▌

The importance of this result for us is that it provides a criterion for deciding when two integers, both greater than 1, are equal—namely, when the corresponding v_p's are equal for every p.

1-5-4. Remark. We may extend the definition of v_p to other integers. Since 1 may be written in the form

$$1 = 2^0 \cdot 3^0 \cdot 5^0 \cdot 7^0 \cdot 11^0 \cdots,$$

it is clear that for any prime p we should define

$$v_p(1) = 0$$

and hence,

$$1 = \prod_p p^{v_p(1)}$$

—which tells us that the expression (*) is valid also for $a = 1$.

What about v_p of negative numbers? Any negative number is of form $-a$ where $a \geq 1$. But then

$$-a = (-1)a = (-1) \prod_p p^{v_p(a)}$$

is essentially a factorization of $-a$. Consequently, we define $v_p(-a)$ as the exponent to which p appears in the factorization of $-a$ (which is identical with the way in which $v_p(a)$ is defined for positive a), and it follows immediately that

$$v_p(-a) = v_p(a) \qquad \text{for all } p. \tag{**}$$

It is now clear that the assertions of 1-5-2 hold for all $a \neq 0$, and 1-5-3 holds for $a > 0$, $b > 0$.

As for $v_p(0)$, at this stage, it is best to brush it aside saying that it is undefined.

1-5-5. Proposition. If $a \neq 0$, $b \neq 0$, then

$$v_p(ab) = v_p(a) + v_p(b) \qquad \text{for all } p.$$

Proof: In virtue of (**) there is no loss of generality in taking $a > 0$, $b > 0$. We have the expressions

$$a = \prod p^{\nu_p(a)}, \quad b = \prod p^{\nu_p(b)}, \quad ab = \prod p^{\nu_p(ab)}.$$

On the other hand,

$$ab = \left(\prod p^{\nu_p(a)}\right)\left(\prod p^{\nu_p(b)}\right) = \prod p^{\nu_p(a) + \nu_p(b)}$$

because multiplication involves simply adding exponents for each p. (Here is one place where we benefit from our notation which treats all primes equally.) We have, therefore,

$$\prod p^{\nu_p(ab)} = \prod p^{\nu_p(a) + \nu_p(b)}.$$

Now, both $\nu_p(a)$ and $\nu_p(b)$ are equal to 0 for almost all p, so $\nu_p(a) + \nu_p(b)$ equals 0 for almost all p. Thus, the right side is indeed a finite product, as is the left side. We have, therefore, two prime power factorizations of the same integer ab. By uniqueness, the factorizations are identical. Putting everything together, we conclude that $\nu_p(ab) = \nu_p(a) + \nu_p(b)$ for all p. ▮

The above proof involves a good deal of fussing, and there remain a number of details that were glossed over; however, the statement could easily be considered as obvious from the start—after all it simply says that for each prime p, its exponent in the product equals the sum of its exponents in the two terms.

Let us turn to questions of divisibility. Among the divisors of $7200 = 2^5 \cdot 3^2 \cdot 5^2$ we certainly have 2, $2 \cdot 3$, $3 \cdot 5^2$, $2^5 \cdot 3$, $2^4 \cdot 3^2 \cdot 5$, and so on. Even more, in virtue of unique factorization it is clear that an integer of form

$$2^{r_1} 3^{r_2} 5^{r_3}, \quad r_1 \geq 0, \quad r_2 \geq 0, \quad r_3 \geq 0$$

is a divisor of 7200 if and only if $r_1 \leq 5$, $r_2 \leq 2$, $r_3 \leq 2$, and also that every divisor of 7200 is of this form. Although we have not mentioned the primes $p \neq 2, 3, 5$ explicitly, there is no doubt that the set of all exponents tells the story. The general result is as follows:

1-5-6. Proposition. If $a \neq 0$, $b \neq 0$, then

$$a \mid b \Leftrightarrow \nu_p(a) \leq \nu_p(b) \text{ for all } p.$$

Proof: This is obvious from the way multiplication works, but we choose to give a formal proof. As before, there is no loss of generality in assuming $a > 0$, $b > 0$. Now

$$\begin{aligned} a \mid b &\Rightarrow \text{there exists } c > 0 \text{ such that } ac = b \\ &\Rightarrow \nu_p(a) + \nu_p(c) = \nu_p(b) \text{ for all } p \\ &\Rightarrow \nu_p(a) \leq \nu_p(b) \text{ for all } p. \end{aligned}$$

As for the converse, if $v_p(a) \leq v_p(b)$ for all p, then, first of all, $v_p(a) = v_p(b) = 0$ for almost all p. Therefore, $v_p(b) - v_p(a) \geq 0$ for all p, and $v_p(b) - v_p(a) = 0$ for almost all p. This permits us to define

$$\prod p^{v_p(b) - v_p(a)}$$

(which is really a finite product) and to call this positive integer c. Consequently,

$$ac = (\prod p^{v_p(a)}) (\prod p^{v_p(b) - v_p(a)}) = \prod p^{v_p(b)} = b$$

and the proof is complete. ▌

We have just seen that the set of all exponents provides a tool for dealing with questions of divisibility. It is not surprising, therefore, that the set of all exponents enables us to describe and analyze the gcd. For example, suppose $a = 7200 = 2^5 \cdot 3^2 \cdot 5^2$, $b = 9996 = 2^2 \cdot 3^1 \cdot 7^2 \cdot 17^1$, $c = 80{,}465 = 5^1 \cdot 7^1 \cdot 11^2 \cdot 19^1$. (To keep the notation simple we leave out the terms with 0 exponent.) Among the common divisors of a and b we clearly find 2, 2^2, 3; it is also obvious that

$$(a, b) = 2^2 \cdot 3^1.$$

In similar fashion, because the story is in the exponents, we see that

$$(a, c) = 5^1, \qquad (b, c) = 7^1.$$

The general result is as follows:

1-5-7. Proposition. If $a \neq 0$, $b \neq 0$, then

$$(a, b) = \prod_p p^{\min\{v_p(a), v_p(b)\}}.$$

Proof: After having done the concrete examples, the proof is obvious—but we give the details.

By $\min\{v_p(a), v_p(b)\}$ we mean, as is customary, the minimum of the two numbers $v_p(a)$, $v_p(b)$. First of all, we need to know that

$$\prod p^{\min\{v_p(a), v_p(b)\}}$$

has meaning—in other words, that it represents an integer. For this, we must verify the two conditions

(*i*) $\min\{v_p(a), v_p(b)\} \geq 0$ for all p,
(*ii*) $\min\{v_p(a), v_p(b)\} = 0$ for almost all p.

Now, (*i*) is immediate because both $v_p(a) \geq 0$ and $v_p(b) \geq 0$ for all p, and (*ii*) follows from the fact that both $v_p(a) = 0$ and $v_p(b) = 0$ for almost all p. Consequently, we may consider the integer

$$d = \prod p^{\min\{v_p(a), v_p(b)\}}$$

and check that it satisfies the requirements for the gcd (as given in the definition 1-3-1).

Since $v_p(d) = \min\{v_p(a), v_p(b)\} \le v_p(a)$ and $v_p(d) = \min\{v_p(a), v_p(b)\} \le v_p(b)$ for all p, it follows from 1-5-6 that $d \mid a$ and $d \mid b$. Furthermore, if $c \mid a$ and $c \mid b$, then, according to 1-5-6, $v_p(c) \le v_p(a)$ and $v_p(c) \le v_p(b)$ for all p. Therefore,

$$v_p(c) \le \min\{v_p(a), v_p(b)\} = v_p(d) \quad \text{for all } p.$$

and, by another application of 1-5-6, $c \mid d$. Of course, $d > 0$—which completes the proof that d is the gcd of a and b. ▮

We turn next to a notion that is analogous ("dual," is a better word) to the greatest common divisor.

1-5-8. Definition. Suppose $a \neq 0$ and $b \neq 0$ are given. Any integer m which satisfies the following condition

(*i*) $a \mid m$ and $b \mid m$

is said to be a **common multiple** of a and b. If, in addition, m satisfies the following two conditions

(*ii*) if $a \mid c$ and $b \mid c$, then $m \mid c$,
(*iii*) $m > 0$,

then m is said to be a **least common multiple** (we abbreviate this to lcm) of a and b.

Expressed in words: A least common multiple of a and b is a common multiple which is positive and is also a divisor of every common multiple.

For any $a \neq 0$, $b \neq 0$, the element ab is a common multiple, and in fact, so is 0. It is easy to see that 2 and 3 have least common multiple 6, and that 4 and 6 have least common multiple 12. The full story about lcm is given by our next result.

1-5-9. Theorem. Given $a \neq 0$, $b \neq 0$, then their least common multiple exists and is unique; we denote it by $[a, b]$. Furthermore, $[a, b]$ is the smallest positive common multiple of a and b, and it can be expressed in the form

$$[a, b] = \prod_p p^{\max\{v_p(a), v_p(b)\}}.$$

Proof: Uniqueness is easy, as in the case of gcd. In fact, if both m and m' are least common multiples of a and b, then $m \mid m'$ and $m' \mid m$. Hence $m = m'$, because both are positive.

Existence is also easy. Because the proof uses the same techniques as in 1-5-7, we merely sketch it. It is clear that $\max\{v_p(a), v_p(b)\}$ (by which we mean the maximum of the two numbers $v_p(a)$, $v_p(b)$) is greater than or equal to 0 for all p, and equal to 0 for almost all p. Therefore,

$$\prod p^{\max\{v_p(a),\, v_p(b)\}}$$

represents an integer—call it m. Of course, $m > 0$. Since

$$v_p(a) \leq \max\{v_p(a), v_p(b)\} = v_p(m) \quad \text{for all } p,$$

it follows that $a \mid m$; similarly $b \mid m$; so m is a common multiple of a and b. Finally,

$$\begin{aligned} a \mid c,\ b \mid c &\Rightarrow v_p(a) \leq v_p(c) \text{ and } v_p(b) \leq v_p(c) \text{ for all } p \\ &\Rightarrow v_p(m) = \max\{v_p(a), v_p(b)\} \leq v_p(c) \text{ for all } p \\ &\Rightarrow m \mid c \end{aligned}$$

which shows that m is a least common multiple—and hence m is *the* lcm.

Obviously, m is the smallest positive common multiple of a and b. Note also that $[a, b] = [b, a]$ because everything about lcm is symmetric in a and b. ▮

1-5-10. Example. To illustrate 1-5-7, 1-5-9, and how one operates with exponents, suppose

$$\begin{aligned} a &= 2 \cdot 3^4 \cdot 19^3 \cdot 37^2 \cdot 97^5, \\ b &= 3^3 \cdot 5 \cdot 11^5 \cdot 19^2 \cdot 79^6 \cdot 97, \\ c &= 2 \cdot 3^2 \cdot 5 \cdot 11 \cdot 13^3 \cdot 37^2 \cdot 97 \cdot 113. \end{aligned}$$

By working accurately, we see that

$$\begin{aligned} ab &= 2 \cdot 3^7 \cdot 5 \cdot 11^5 \cdot 19^5 \cdot 37^2 \cdot 79^6 \cdot 97^6, \\ ac &= 2^2 \cdot 3^6 \cdot 5 \cdot 11 \cdot 13^3 \cdot 19^3 \cdot 37^4 \cdot 97^6 \cdot 113, \\ bc &= 2 \cdot 3^5 \cdot 5^2 \cdot 11^6 \cdot 13^3 \cdot 19^2 \cdot 37^2 \cdot 79^6 \cdot 97^2 \cdot 113, \\ (a, b) &= 3^3 \cdot 19^2 \cdot 97, \\ (a, c) &= 2 \cdot 3^2 \cdot 37^2 \cdot 97, \\ (b, c) &= 3^2 \cdot 5 \cdot 11 \cdot 97, \\ [a, b] &= 2 \cdot 3^4 \cdot 5 \cdot 11^5 \cdot 19^3 \cdot 37^2 \cdot 79^6 \cdot 97^5, \\ [a, c] &= 2 \cdot 3^4 \cdot 5 \cdot 11 \cdot 13^3 \cdot 19^3 \cdot 37^2 \cdot 97^5 \cdot 113, \\ [b, c] &= 2 \cdot 3^3 \cdot 5 \cdot 11^5 \cdot 13^3 \cdot 19^2 \cdot 37^2 \cdot 79^6 \cdot 97 \cdot 113. \end{aligned}$$

1-5-11. Proposition. If $a > 0$, $b > 0$, then

$$ab = (a, b)[a, b].$$

Proof: By virtue of 1-5-3, it suffices to show that for every prime p,

$$v_p(ab) = v_p\{(a, b)[a, b]\}.$$

Since $v_p\{(a, b)\} = \min\{v_p(a), v_p(b)\}$ and $v_p\{[a, b]\} = \max\{v_p(a), v_p(b)\}$, all that has to be shown is

$$v_p(a) + v_p(b) = \min\{v_p(a), v_p(b)\} + \max\{v_p(a), v_p(b)\}.$$

But this is trivial; because the roles of a and b are interchangeable, we may assume that $v_p(a) \leq v_p(b)$, and then $\min\{v_p(a), v_p(b)\} = v_p(a)$, $\max\{v_p(a), v_p(b)\} = v_p(b)$. This completes the proof. ▮

Essentially the same result holds under the more general hypotheses $a \neq 0$, $b \neq 0$; the left side of the equation must then be replaced by $|ab|$.

1-5-12. Exercise. Consider $a_1 \neq 0$, $a_2 \neq 0, \ldots, a_n \neq 0$, where $n \geq 2$. In keeping with what has gone before, define a **gcd** of $a_1, a_2, \ldots, a_n$ to be a number d satisfying the conditions: (*i*) $d \mid a_i$ for $i = 1, \ldots, n$ (that is, d is a common divisor of the a_i's); (*ii*) if $c \mid a_i$ for $i = 1, \ldots, n$, then $c \mid d$ (that is, if c is a common divisor of the a_i's, then $c \mid d$); (*iii*) $d > 0$. In similar fashion, define a **lcm** of $a_1, \ldots, a_m$ to be a number m satisfying the conditions: (*i*) $a_i \mid m$ for $i = 1, \ldots, n$ (*ii*) if $a_i \mid c$ for $i = 1, \ldots, n$, then $m \mid c$; (*iii*) $m > 0$.

(1) The gcd of $a_1, a_2, \ldots, a_n$ exists and is unique; in fact, if we denote it by $(a_1, a_2, \ldots, a_n)$, then

$$(a_1, a_2, \ldots, a_n) = \prod_p p^{\min_{i=1,\ldots,n} \{v_p(a_i)\}}.$$

[If $(a_1, a_2, \ldots, a_n) = 1$, we say that $a_1, a_2, \ldots, a_n$ are **relatively prime.**]

(2) The lcm of $a_1, a_2, \ldots, a_n$ exists and is unique; in fact, if we denote it by $[a_1, a_2, \ldots, a_n]$, then

$$[a_1, a_2, \ldots, a_n] = \prod_p p^{\max_{i=1,\ldots,n} \{v_p(a_i)\}}.$$

(3) The greatest common divisor $(a_1, a_2, \ldots, a_n)$ is the biggest positive divisor of $a_1, a_2, \ldots, a_n$. The least common multiple $[a_1, a_2, \ldots, a_n]$ is the smallest positive multiple of $a_1, a_2, \ldots, a_n$.

(4) We have

$$(a_1, a_2, a_3) = ((a_1, a_2), a_3),$$
$$(a_1, a_2, a_3, a_4) = ((a_1, a_2, a_3), a_4),$$
$$\vdots$$
$$(a_1, a_2, \ldots, a_{n-1}, a_n) = ((a_1, a_2, \ldots, a_{n-1}), a_n),$$

and

$$[a_1, a_2, a_3] = [[a_1, a_2], a_3],$$
$$(a_1, a_2, a_3, a_4] = [[a_1, a_2, a_3,], a_4],$$
$$\vdots$$
$$[a_1, a_2, \ldots, a_{n-1}, a_n] = [[a_1, a_2, \ldots, a_{n-1}], a_n].$$

It is worth elaborating on the significance of these relations. If the factorizations of $a_1, \ldots, a_n$ are all known, then, by making use of parts (1) and (2), we can easily find $(a_1, \ldots, a_n)$ and $[a_1, \ldots, a_n]$. However, if the factorizations of $a_1, \ldots, a_n$ are not known, these relations may be used to determine $(a_1, \ldots, a_n)$ and $[a_1, \ldots, a_n]$. In more detail: First, we compute $(a_1, a_2) = d$ by the methods of Section 1-3, and then use the relation $|a_1 a_2| = (a_1, a_2)[a_1, a_2]$ to find $[a_1, a_2] = m$. Once the procedure for computing the gcd and lcm of two integers is known, our relations call for repeating the process as many times as necessary. Thus, (a_1, a_2, a_3) is found by computing $((a_1, a_2), a_3) = (d, a_3)$, and $[a_1, a_2, a_3]$ is found by computing $[[a_1, a_2], a_3] = [m, a_3]$, and so on. Of course, it is a tedious process, but it works.

Note that in either case, it does not matter in what order the numbers $a_1, a_2, \ldots, a_n$ are listed.

(5) The gcd $(a_1, \ldots, a_n)$ is the smallest positive integer which can be expressed as a linear combination of $a_1, \ldots, a_n$; notationally,

$$(a_1, \ldots, a_n) = \min\{x_1 a_1 + x_2 a_2 + \cdots + x_n a_n > 0 \mid x_1, x_2, \ldots, x_n \in \mathbf{Z}\}.$$

1-5-13 / *PROBLEMS*

1. Factor the numbers $a = 28{,}050$ and $b = 9555$, then use the factorizations to find (a, b) and $[a, b]$.

2. Find the least common multiple of all integers from 1 through 15.

3. Suppose $a = 7200$, $b = 9996$, $c = 80{,}465$. Find
(*i*) $(a, b), (a, c), (b, c)$,
(*ii*) $[a, b], [a, c], [b, c]$,
(*iii*) $(a, b, c), [a, b, c]$,
(*iv*) $(a, [b, c]), [a, (b, c)], [(a, b), (a, c)], ([a, b], [a, c])$.

4. Find the least common multiple of $a = 5311$, $b = 7571$. Can you express it as a linear combination of a and b?

5. Suppose $a = 24{,}523$, $b = 31{,}373$, $c = 44{,}929$. Without factoring (which would take too long), find
(*i*) (a, b) and (a, b, c),
(*ii*) $[a, b]$ and $[a, b, c]$,
(*iii*) $[(a, b), c]$ and $([a, b], c)$.

6. Discuss the least common multiple of a and b when one of them is 0.

7. Let x, y, z be any real numbers. Show that:
(*i*) $\min\{\max\{x, y\}, z\} = \max\{\min\{x, z\}, \min\{y, z\}\}$,
(*ii*) $\max\{\min\{x, y\}, z\} = \min\{\max\{x, z\}, \max\{y, z\}\}$,
(*iii*) $\min\{\max\{x, y\}, \max\{x, z\}, \max\{y, z\}\}$
$= \max\{\min\{x, y\}, \min\{x, z\}, \min\{y, z\}\}$,
(*iv*) $\max\{x, y, z\} + \min\{x + y, x + z, y + z\} = x + y + z$.

8. If a is a positive integer, what are $(a, a + 1)$ and $[a, a + 1]$?

9. If a and b are positive integers with $a \mid b$, what are (a, b) and $[a, b]$?

10. For positive integers a and b, show that
(*i*) $(a, b) = [a, b] \Leftrightarrow a = b$,
(*ii*) $[a, b] = ab \Leftrightarrow (a, b) = 1$.

11. Show that if $a_1, a_2, \ldots, a_n$ are relatively prime in pairs (meaning that any two are relatively prime), then they are relatively prime.

12. By making use of the v_p's prove that if $m > 0$ and a, b are arbitrary, then

$$(ma, mb) = m(a, b).$$

What if m is an arbitrary integer not equal to 0?

13. Prove in two different ways that if $m > 0$, then

$$[ma, mb] = m[a, b]$$

for all a, b. What happens if m is an arbitrary integer not equal to 0?

14. Suppose a and b are positive with $(a, b) = 1$, then ab is a perfect square $\Leftrightarrow$ both a and b are perfect squares. Even more, for any $n \geq 2$, ab is an nth power $\Leftrightarrow$ both a and b are nth powers.

15. Suppose we define $v_p(0) = \infty$ for all primes p, and consider ∞ as bigger than any integer. Show that for all $a, b \in \mathbf{Z}$

$$v_p(a + b) \geq \min\{v_p(a), v_p(b)\}.$$

Moreover, if $v_p(a) \neq v_p(b)$, then

$$v_p(a+b) = \min\{v_p(a), v_p(b)\}.$$

16. This problem gives a general formulation for a type of argument that occurs more than once in the text.

(*i*) Suppose that for each p we are given an integer $m_p \geq 0$ and that $m_p = 0$ for almost all p, then the infinite product

$$\prod p^{m_p}$$

represents a positive integer.

(*ii*) There exists a positive integer a such that

$$v_p(a) = m_p \qquad \text{for all } p.$$

(*iii*) Suppose, in addition, that for each p we have an integer $n_p \geq 0$ such that almost all $n_p = 0$. Show that if

$$\prod p^{m_p} = \prod p^{n_p},$$

then $m_p = n_p$ for all p.

17. Decide if the following assertions are true or false: if true, give a proof; if false, give a counterexample.

(*i*) $(a, b) = (a, c) \Rightarrow [a, b] = [a, c]$.
(*ii*) $(a, b) = d \Rightarrow (a^2, b^2) = d^2$.
(*iii*) $(a, b) = (a, c) \Rightarrow (a^2, b^2) = (a^2, c^2)$.
(*iv*) $(a, b) = (a, c) \Rightarrow (a, b) = (a, b, c)$.
(*v*) $(a, b) = 1 \Rightarrow (a^2, ab, b^2) = 1$.
(*vi*) $[a^2, ab, b^2] = [a^2, b^2]$.
(*vii*) $b \mid (a^2 - 1) \Rightarrow b \mid (a^4 - 1)$.
(*viii*) $b \mid (a^2 + 1) \Rightarrow b \mid (a^4 + 1)$.
(*ix*) $a^3 \mid b^3 \Rightarrow a \mid b$.
(*x*) $a^3 \mid b^2 \Rightarrow a \mid b$.
(*xi*) $a^2 \mid b^3 \Rightarrow a \mid b$.
(*xii*) $(a, b, c) = ((a, b), (a, c))$.

18. (*i*) Show that

$$([a, b], c) = [(a, c), (b, c)],$$
$$[(a, b), c] = ([a, c], [b, c]).$$

(*ii*) Even more, show inductively that

$$([a_1, a_2, \ldots, a_{n-1}], a_n) = [(a_1, a_n), \ldots, (a_{n-1}, a_n)],$$
$$[(a_1, a_2, \ldots, a_{n-1}), a_n] = ([a_1, a_n], \ldots, [a_{n-1}, a_n]).$$

19. In terms of the v_p's, how can one decide if $(a, b) = 1$?

20. Show that the positive integers $a_1, a_2, \ldots, a_n$ are relatively prime in pairs $\Leftrightarrow$ $a_1 a_2 \cdots a_n = [a_1, a_2, \ldots, a_n]$.

21. Use the v_p's to do 1-4-8, Problem 10.

1-6. Linear Diophantine Equations

Given integers $a \neq 0$, $b \neq 0$, and c, we want to solve the equation

$$ax + by = c.$$

We must first settle the question of what is meant by a solution. Graphing this equation in the plane gives a straight line, and the pair of coordinates x, y of any point on this line satisfy the equation. From this point of view, there are an infinite number of solutions; for every choice of x we can solve for the corresponding y. However, this is not what we have in mind here. Our concern is with solutions where both x and y are integers. Geometrically, this means that we want to know if the straight line whose equation is $ax + by = c$ passes through any points both of whose coordinates are integers (such a point is often referred to as an **integral point**).

Thus, by a **solution** of $ax + by = c$ we shall mean a pair of integers $\{x_0, y_0\}$ for which $ax_0 + by_0 = c$. The equation is known as a **linear diophantine equation**—"linear" signifies that the unknowns x and y appear only to the first power, "diophantine" signifies that we are concerned solely with integral solutions (the name derives from that of Diophantus of Alexandria who considered the problem of finding integral solutions of equations during the third or fourth century A.D.).

In connection with the linear diophantine equation $ax + by = c$ there are several questions about its solutions that we can ask. Does a solution exist? Is there a simple criterion for deciding if a solution exists? If there is a solution, how many solutions are there? Can we find all the solutions explicitly? A strictly geometric approach to these questions is not satisfactory. It lacks precision and the ingredients necessary for giving proofs. For example, by drawing the line $ax + by = c$ we can not always decide if it passes through an integral point. However, the algebraic approach (with a geometric "picture" kept at the back of one's mind) provides complete and definitive answers to all our questions.

1-6-1. Theorem. Given integers $a \neq 0$, $b \neq 0$, and c, let $d = (a, b)$. Then the linear diophantine equation $ax + by = c$ has a solution $\Leftrightarrow$ $d \mid c$. Moreover, in the case where a solution exists,

(*i*) We can find an explicit solution $\{x_0, y_0\}$ by use of the Euclidean algorithm.
(*ii*) There are an infinite number of solutions.
(*iii*) If $\{x_0, y_0\}$ is any known solution, then all solutions are given by the pairs $\{\bar{x}, \bar{y}\}$ where

$$\bar{x} = x_0 + t\left(\frac{b}{d}\right)$$
$$\bar{y} = y_0 - t\left(\frac{a}{d}\right). \qquad t = 0, \pm 1, \pm 2, \pm 3, \ldots$$

Proof: According to 1-3-6, we know that the set

$$S = \{ax + by \mid x, y \in \mathbf{Z}\} = \{xa + yb \mid x, y \in \mathbf{Z}\}$$

of all linear combinations of a and b is equal to the set

$$T = \{nd \mid n \in \mathbf{Z}\}$$

of all multiples of $d = (a, b)$. Therefore,

$ax + by = c$ has a solution $\Leftrightarrow$ c is a linear combination of a and b,

$$\Leftrightarrow c \in S,$$

$$\Leftrightarrow c \in T,$$

$$\Leftrightarrow d \mid c,$$

which proves the first statement.

In this situation, that is, when $(a, b) = d$ divides c, let us write

$$a = da', \qquad b = db', \qquad c = dc'.$$

We can find an explicit solution $\{x_0, y_0\}$ of $ax + by = c$ as follows. Since $d = (a, b)$, d can be expressed as a linear combination of a and b; in fact, as seen in 1-3-2, the Euclidean algorithm may be used to find such a linear combination. In this way, we locate integers x', y' such that

$$ax' + by' = d.$$

(Note that in the case $a \mid b$—where, for all practical purposes, the Euclidean algorithm breaks down—we can still locate x' and y'; namely, because here $d = |a|$, we may take $y' = 0$, $x' = \pm 1$, with the sign of x' depending on whether a is positive or negative.) Then multiplying through by c' gives

$$ac'x' + bc'y' = dc' = c,$$

which says that for $x_0 = c'x'$, $y_0 = c'y'$ the pair $\{x_0, y_0\}$ is a solution of $ax + by = c$. This proves (*i*).

To prove (*ii*) suppose $\{x_0, y_0\}$ is any solution; it does not have to be the solution we found above. For any integer t, let us put

$$\bar{x} = x_0 + tb', \qquad \bar{y} = y_0 - ta'.$$

It is easy to check, by substituting in the diophantine equation, that $\{\bar{x}, \bar{y}\}$ is a solution—in detail,

$$\begin{aligned} a\bar{x} + b\bar{y} &= a(x_0 + tb') + b(y_0 - ta') \\ &= ax_0 + by_0 + t(ab' - ba') \\ &= c + t(da'b' - db'a') \\ &= c. \end{aligned}$$

[Note that the choice of $\bar{x}$ and $\bar{y}$ is quite natural; it is based on the observation that $ab' + b(-a') = 0$.] Consequently, we have an infinite number of solutions, one for each choice of $t \in \mathbf{Z}$.

To prove (*iii*), it remains to show that if a solution $\{x_0, y_0\}$ is known and $\{x_1, y_1\}$ is an arbitrary solution, then there exists an integer t such that

$$x_1 = x_0 + tb', \qquad y_1 = y_0 - ta'.$$

(Note that in the statement of the theorem we found it convenient to write a/d for a' and b/d for b', rather than introduce the additional elements a' and b'. Of course, the properties of fractions are never used in the proof, nor do the fractions a/d, b/d appear.) Because $\{x_0, y_0\}$ and $\{x_1, y_1\}$ are solutions, we have $ax_0 + by_0 = c = ax_1 + by_1$ which implies

$$a(x_1 - x_0) = b(y_0 - y_1).$$

Recalling that $a = da'$, $b = db'$, we obtain

$$a'(x_1 - x_0) = b'(y_0 - y_1).$$

Now, according to 1-3-8, $(a', b') = 1$, so it follows from 1-3-10, part (*ii*) that $a' \mid (y_0 - y_1)$. Thus, there exists $t \in \mathbf{Z}$ with $ta' = y_0 - y_1$. Substituting this back gives

$$a'(x_1 - x_0) = b'ta'.$$

We conclude that $x_1 = x_0 + tb'$ and $y_1 = y_0 - ta'$, which completes the proof. ▌

The restrictions $a \neq 0$, $b \neq 0$ were included in the statement of the theorem, because if one of them is 0 our linear diophantine equation becomes $ax = c$ or $by = c$—and we have no interest in these equations because they are old hat, involving nothing more than questions of divisibility. The reader will note that if one of a or b is 0 the assertions of the theorem still hold (except for the reference to the Euclidean algorithm).

Of course, it is quite possible that $c = 0$. In this situation, everything goes quickly. One solution of $ax + by = 0$ is $x_0 = b$, $y_0 = -a$, and all the solutions are given by

$$\begin{aligned} \bar{x} &= b + t\left(\frac{b}{d}\right), \\ \bar{y} &= -a - t\left(\frac{a}{d}\right), \end{aligned} \qquad t = 0, \pm 1, \pm 2, \ldots .$$

1-6-2. Example. Consider the linear diophantine equation

$$258x + 354y = 18. \tag{$*$}$$

In order to solve this equation, we pattern our discussion on the proof of 1-6-1 with $a = 258$, $b = 354$, $c = 18$.

To decide if a solution exists, we must check if $d = (258, 354)$ divides $c = 18$. It is easy to see that $(258, 354) = 6$ (in fact, this was done in 1-3-3), so a solution exists.

To find an explicit solution $\{x_0, y_0\}$, we begin by expressing $d = 6$ as a linear combination of $a = 258$ and $b = 354$ via the Euclidean algorithm. This has already been done in 1-3-7—with the result

$$(11)(258) + (-8)(354) = 6.$$

(In the notation of the proof of 1-6-1, this means that $x' = 11$, $y' = -8$.) Multiplying this equation by 3 (of course, $3 = d/c = c'$) we have

$$(33)(258) + (-24)(354) = 18,$$

so $x_0 = 33$, $y_0 = -24$ is a solution of $(*)$.

It remains to list all the solutions of $(*)$. Using the fact that $a' = a/d = 258/6 = 43$ and $b' = b/d = 343/6 = 59$, it follows that all solutions of $(*)$ are given by

$$\begin{aligned} \bar{x} &= 33 + 59t, \\ \bar{y} &= -24 - 43t, \end{aligned} \qquad t = 0, \pm 1, \pm 2, \ldots .$$

Once this example is done, there are a number of related linear diophantine equations which can be solved with a minimum of effort. Let us list a few.

(i) Consider the equation

$$258x - 354y = 18.$$

Here, $a = 258$, $b = -354$, $c = 18$. From previous results we know that $d = 6$, and

$$(33)(258) + (24)(-354) = 18.$$

Therefore, $x_0 = 33$, $y_0 = 24$ is a solution, and all solutions are given by

$$\begin{aligned} \bar{x} &= 33 - 59t, \\ \bar{y} &= 24 - 43t, \end{aligned} \qquad t = 0, \pm 1, \pm 2, \ldots.$$

(*ii*) Consider the equation

$$354x - 258y = 18.$$

In this case, according to the way our notation is set up, $a = 354$, $b = -258$, $c = 18$, $d = (354, -258) = 6$, $a/d = 59$, $b/d = -43$. The previous expression for 18 can be rewritten in the form

$$(-24)(354) + (-33)(-258) = 18.$$

Hence, $x_0 = -24$, $y_0 = -33$ is a solution, and all solutions are given by

$$\begin{aligned} \bar{x} &= -24 - 43t, \\ \bar{y} &= -33 - 59t, \end{aligned} \qquad t = 0, \pm 1, \pm 2, \pm 3, \ldots.$$

Note that here, as elsewhere, we can change signs, because t runs over $\mathbf{Z}$, and hence write all solutions in the form

$$\begin{aligned} \bar{x} &= -24 + 43t, \\ \bar{y} &= -33 + 59t, \end{aligned} \qquad t = 0, \pm 1, \pm 2, \pm 3, \ldots.$$

(*iii*) The linear diophantine equation

$$354x + 258y = 35$$

has no solution, because $d = (a, b) = (354, 258) = 6$ does not divide $c = 35$.

(*iv*) Consider the linear diophantine equation

$$354x + 258y = 36.$$

Here $a = 354$, $b = 258$, $c = 36$, $d = 6$. Since $6 \mid 36$, a solution exists. Starting from the relation

$$(-8)(354) + (11)(258) = 6,$$

which was proved earlier, it follows that

$$(-48)(354) + (66)(258) = 36.$$

Thus, $x_0 = -48$, $y_0 = 66$ is a solution, and all solutions are of form

$$\begin{aligned} \bar{x} &= -48 + 43t, \\ \bar{y} &= 66 - 59t, \end{aligned} \qquad t \in \mathbf{Z}.$$

1-6-3. Euler's Method. There is another, completely elementary, method for finding a solution of an explicit linear diophantine equation. We illustrate the method, which is due to Euler (1707–1783), by finding a solution to the same equation

$$258x + 354y = 18 \tag{$*$}$$

that was discussed in 1-6-2.

The objective is to find integers x and y for which $258x + 354y = 18$. Rewriting this equation, we have (using our knowledge of fractions)

$$x = \frac{18 - 354y}{258} = \frac{18 - 96y}{258} - y.$$

If we put

$$z = \frac{18 - 96y}{258},$$

then we are looking for an integer y for which z is an integer; as then x will be an integer and $\{x, y\}$ will be a solution of ($*$). Now consider the expression for z and rewrite it as $258z = 18 - 96y$, which leads to

$$y = \frac{18 - 258z}{96} = \frac{18 + 30z}{96} - 3z.$$

(Note that one may also write

$$y = \frac{18 - 66z}{96} - 2z,$$

and proceed as we shall do in a moment; we have chosen to use the multiple of 96 closest to -258.) Thus, we seek an integer z for which

$$w = \frac{18 + 30z}{96}$$

is an integer. Rewriting this as

$$z = \frac{96w - 18}{30} = 3w - 1 + \frac{6w + 12}{30}$$

we seek an integer w for which

$$t = \frac{6w + 12}{30} = \frac{w + 2}{5}$$

is an integer. But now the numbers are small, and it is trivial to find a w of the desired type—for example, $w = -2, 3, 8, -7, \ldots$. Once a w is chosen for which t is an integer, we proceed through the relations

$$z = 3w - 1 + t,$$
$$y = w - 3z,$$
$$x = z - y,$$

in order to find x and y.

In particular, starting with $w = -2$, we obtain $t = 0$, $z = -7$, $y = 19$, $x = -26$; so $\{-26, 19\}$ is a solution of $(*)$, and according to 1-6-1 all solutions are of form

$$\begin{aligned} \bar{x} &= -26 + 59t, \\ \bar{y} &= 19 - 43t, \end{aligned} \qquad t \in \mathbf{Z}.$$

Of course, this is the same set of solutions that was found in 1-6-2. The difference in appearance is explained by the fact that we have two distinct "parametrizations" of the same set.

Similarly, if we start with $w = 3$, then $t = 1$, $z = 9$, $y = -24$, $x = 33$—so, in this case, we obtain the solution $\{33, -24\}$, and the set of all solutions becomes identical in appearance (that is, in notation) with the set of solutions found in 1-6-2.

In all this, we used Euler's method to find a particular solution of the linear diophantine equation and, in virtue of 1-6-1 and the fact that the gcd $d = 6$ was known, were then able to list all the solutions. However, only slight modifications in the organization of Euler's method are needed to find all solutions directly—that is, without finding d or using 1-6-1. In detail, we proceed as before until we arrive at the relation

$$t = \frac{w + 2}{5}.$$

Rewriting this in the form $w = 5t - 2$, we have, therefore, the system of relations

$$w = 5t - 2,$$
$$z = 3w - 1 + t;$$
$$y = w - 3z,$$
$$x = z - y,$$

in which all the unknowns t, w, z, y, and x are to be integers. Now one may obviously view $t \in \mathbf{Z}$ as a "parameter" and express the other variables—especially x and y— in terms of t. In particular, if $t = 0$, then $w = -2$, $z = -7$, $y = 19$, $x = -26$ so $\{-26, 19\}$ is a solution, and if $t = 1$, then $w = 3$, $z = 9$,

$y = -24$, $x = 33$ so $\{33, -24\}$ is a solution. In general, we have

$$\begin{aligned} w &= 5t - 2, \\ z &= 3(5t - 2) - 1 + t = 16t - 7, \\ y &= (5t - 2) - 3(16t - 7) = -43t + 19, \\ x &= (16t - 7) - (-43t + 19) = 59t - 26, \end{aligned}$$

so it follows that

$$\begin{aligned} x &= -26 + 59t, \\ y &= 19 - 43t, \end{aligned} \qquad t \in \mathbf{Z}$$

is the set of all solutions!

1-6-4 / *PROBLEMS*

1. Find all solutions of the following linear diophantine equations:
(*i*) $91x + 33y = 147$, (*ii*) $24x + 30y = 14$,
(*iii*) $30x - 43y = 97$, (*iv*) $93x - 81y = 15$,
(*v*) $17x + 646y = 51$, (*vi*) $91x + 56y = 0$,
(*vii*) $874x - 19y = 1052$, (*viii*) $84x - 91y = 11$,
(*ix*) $4147x + 10{,}672y = 58$.

2. Solve the equations of Problem (1) by use of Euler's method.

3. Find all positive solutions (meaning solutions $\{\bar{x}, \bar{y}\}$ with $\bar{x} > 0$, $\bar{y} > 0$) of
(*i*) $18x + 7y = 302$, (*ii*) $18x - 7y = 302$,
(*iii*) $54x - 38y = 82$, (*iv*) $11x + 13y = 47$,
(*v*) $10x + 28y = 1240$.
A geometric picture is helpful here.

4. Show that there exist no integers a and b such that $(a, b) = 7$ and $a + b = 100$. On the other hand, there exist an infinite number of pairs of integers a and b such that $(a, b) = 5$ and $a + b = 100$.

5. Which of the following sets are the same?
(*i*) $\{17 + 157t \mid t \in \mathbf{Z}\}$, (*ii*) $\{1744 + 157t \mid t \in \mathbf{Z}\}$,
(*iii*) $\{-768 + 157t \mid t \in \mathbf{Z}\}$, (*iv*) $\{100 - 157t \mid t \in \mathbf{Z}\}$,
(*v*) $\{51 - 157t \mid t \in \mathbf{Z}\}$, (*vi*) $\{-57 + 157t \mid t \in \mathbf{Z}\}$.

6. Find two fractions having 5 and 7 for denominators whose sum is equal to 26/35.

7. Find a number that leaves the remainder 16 when divided by 39 and the remainder 27 when divided by 56.

8. Show that $a = 14t + 3$ and $b = 21t + 1$ are relatively prime for each choice of $t \in \mathbf{Z}$.

9. If the diophantine equation $ax + by = c$ has no solutions (that is, when $(a, b) \nmid c$), then Euler's method must break down somewhere. How does this occur?

10. Use Euler's method to solve
(*i*) $6x - 10y + 15z = 2$, (*ii*) $28x + 24y - 92z = 202$.

11. Solve the simultaneous linear diophantine equations

$$8x + 5y = -7,$$
$$15x - 12y = 393.$$

12. Four men and a monkey spent the day on a tropical island collecting coconuts. At night, while the others slept, one of the men arose and divided the nuts into four equal piles. There was one coconut left over, so he gave it to the monkey. He then hid his share, put the rest of the nuts into a single pile, and went back to sleep. In turn, each of the other men went through the same procedure, and each time there was one nut for the monkey. In the morning, the four men divided the remaining coconuts equally, and again there was one left over for the monkey. What is the smallest number of nuts which could have been in the original pile?

13. Do the "coconut problem" when there are
(*i*) 3 men, (*ii*) 5 men.

14. Suppose in the original coconut problem that the fourth man, after taking his share and giving one to the monkey, leaves the island. In the morning, the three remaining men give one nut to the monkey (as usual) and take equal shares. What happens?

15. If a and b are relatively prime positive integers, show that the diophantine equation

$$ax - by = c$$

has an infinite number of positive solutions.

I-7. Congruence

The notion of congruence was introduced by Gauss (1777–1855; a leading candidate for the title of greatest mathematician of all time), and it has turned out to be very fruitful. In this section, we merely make the definition, and point out a few of its immediate consequences. A detailed study of congruences will be undertaken in Chapter III.

1-7-1. Definition. Fix an integer $m > 0$. For any $a, b \in \mathbf{Z}$ we write

$$a \equiv b \pmod{m}$$

and read this as "a is **congruent** to b modulo m," when m divides $a - b$.

For example, $3 \equiv -4 \pmod{7}$, $76 \equiv 21 \pmod{11}$, $2^{11} \equiv 1 \pmod{23}$.

1-7-2. Properties. Congruence modulo m behaves like equality in many ways. The key properties of congruence which parallel properties of equality are the following.

(*i*) $a \equiv a \pmod{m}$ for all a.

This is known as the **reflexive** property; it is trivial since m divides $(a - a) = 0$.

(*ii*) If $a \equiv b \pmod{m}$, then $b \equiv a \pmod{m}$.

This is known as the **symmetric** property of congruence. It requires proof because the definition distinguishes between the left side and the right side of the "$\equiv$" sign. Of course, the proof is trivial since $m \mid (a - b)$ implies $m \mid (b - a)$.

(*iii*) If $a \equiv b \pmod{m}$ and $b \equiv c \pmod{m}$, then $a \equiv c \pmod{m}$.

This is known as the **transitive** property of congruence. It holds because $m \mid (a - b)$ and $m \mid (b - c)$ together imply that m divides $(a - b) + (b - c) = (a - c)$.

1-7-3. Proposition. For $a, b \in \mathbf{Z}$, $a \equiv b \pmod{m} \Leftrightarrow$ both a and b have the same remainder upon division by m.

Proof: The division algorithm gives unique expressions

$$a = q_1 m + r_1, \qquad 0 \le r_1 < m,$$
$$b = q_2 m + r_2, \qquad 0 \le r_2 < m,$$

and upon subtraction

$$a - b = (q_1 - q_2)m + (r_1 - r_2). \tag{$*$}$$

Since r_1 and r_2 are the remainders upon division by m, we must show that $a \equiv b \pmod{m} \Leftrightarrow r_1 = r_2$.

If $a \equiv b \pmod{m}$, then $m \mid (a - b)$, and from $(*)$ we see that $m \mid (r_1 - r_2)$. Because of the bounds on r_1 and r_2, it follows that $r_1 = r_2$. Conversely, if $r_1 = r_2$, then $(*)$ implies $m \mid (a - b)$, which says that $a \equiv b \pmod{m}$. ▌

This criterion for when two integers are congruent could have been taken as the definition for congruence at the start. In other words, the definition would be that two numbers are congruent (mod m) when their remainders (upon division by m) are equal. Of course, it does not matter which definition is given initially; the important thing is that

$$a \equiv b \pmod{m} \Leftrightarrow m \mid (a-b) \Leftrightarrow a \text{ and } b \text{ have equal remainders.}$$

The remainder criterion for congruence is often convenient for use in a proof; for example, it makes it crystal clear that congruence mod m satisfies the reflexive, symmetric, and transitive properties.

Although the definition of congruence was given for any modulus $m > 0$, we really have no interest in the case $m = 1$—for any two integers are congruent (mod 1), and there is no information about integers that could possibly be derived from this congruence.

Pursuing the analogy between equality and congruence that was pointed up in 1-7-2, we recall that for integers, adding equals to equals gives equals and multiplying equals by equals gives equals. It is a central fact about congruences that the corresponding statements hold for them also. More precisely, we have:

1-7-4. Proposition. If $a \equiv b \pmod{m}$ and $a' \equiv b' \pmod{m}$, then

$$a + a' \equiv b + b' \pmod{m} \quad \text{and} \quad aa' \equiv bb' \pmod{m}.$$

Proof: The hypotheses say that $m \mid (a-b)$ and $m \mid (a'-b')$. It follows that m divides the sum $(a-b) + (a'-b') = (a+a') - (b+b')$, which means that $a + a' \equiv b + b' \pmod{m}$.

As for the second part, we wish to show that

$$m \mid (aa' - bb').$$

The way to accomplish this is to make use of a fairly common "trick"—which involves adding and subtracting a well-chosen term. Namely,

$$aa' - bb' = aa' - ba' + ba' - bb' = (a-b)a' + b(a'-b'),$$

and since m divides both $(a-b)$ and $(a'-b')$, it follows that m divides $aa' - bb'$. This completes the proof. ▮

This result says, for example, that $7 \equiv 53 \pmod{23}$ and $18 \equiv -5 \pmod{23}$ together imply $7 + 18 \equiv 53 - 5 \pmod{23}$ and $(7)(18) \equiv (53)(-5) \pmod{23}$. Naturally, one may also check directly that $25 \equiv 48 \pmod{23}$ and $126 \equiv -265 \pmod{23}$.

Since we now have the "right" to add and multiply two congruences mod m, the process may be repeated for more than two congruences, and the following statements are clearly valid.

1-7-5. Corollary. If we have several congruences

$$a_i \equiv b_i \pmod{m}, \qquad i = 1, \ldots, r,$$

then

$$\sum_{i=1}^{r} a_i \equiv \sum_{i=1}^{r} b_i \pmod{m}$$

and

$$\prod_{i=1}^{r} a_i \equiv \prod_{i=1}^{r} b_i \pmod{m}$$

In particular, if $a = a_1 = a_2 = \cdots = a_r$ and $b = b_1 = b_2 = \cdots = b_r$, these assertions take the following form:

If $a \equiv b \pmod{m}$, then $ra \equiv rb \pmod{m}$ and $a^r \equiv b^r \pmod{m}$ for every positive integer r.

1-7-6. Example. Congruences can sometimes be used as labor-saving devices in certain types of computations. As an example, consider the following question: What is the remainder when 2^{50} is divided by 7? According to theory, we can write

$$2^{50} = q \cdot 7 + r, \qquad 0 \leq r < 7,$$

but it would be foolish to compute 2^{50} and then divide by 7. Instead, we observe from this equation that the remainder r satisfies

$$2^{50} \equiv r \pmod{7}$$

so our problem is really to find the unique integer r with $0 \leq r < 7$ which is congruent to 2^{50} modulo 7. For this, we may start with

$$2^1 \equiv 2 \pmod{7}, \qquad 2^2 \equiv 4 \pmod{7}, \qquad 2^3 \equiv 1 \pmod{7}.$$

Raising the last congruence to the 16th power gives $2^{48} \equiv 1 \pmod{7}$; then multiplying this by 2^2, we arrive at

$$2^{50} \equiv 4 \pmod{7},$$

so the remainder is 4.

Let us consider another example in the same vein: Is $2^{30} \cdot 14^{50} - 1$ divisible by 11? We treat the powers of 2 and 14 separately. Starting with $2^2 \equiv 4 \pmod{11}$ and $2^3 \equiv 8 \equiv -3 \pmod{11}$, we see that

$$2^5 = 2^2 \cdot 2^3 \equiv (4)(-3) = -12 \equiv -1 \pmod{11}.$$

Consequently, raising this congruence to the 6th power,

$$2^{30} = (2^5)^6 \equiv (-1)^6 = 1 \pmod{11}.$$

Furthermore, $14 \equiv 3 \pmod{11}$ so $14^{50} \equiv 3^{50} \pmod{11}$. Now $3^2 \equiv -2 \pmod{11}$ so $3^4 \equiv 4 \pmod{11}$, and by multiplying these two $3^6 \equiv -8 \equiv 3 \pmod{11}$. Therefore,

$$3^{10} = 3^6 \cdot 3^4 \equiv 3 \cdot 4 \equiv 1 \pmod{11}$$

and then

$$14^{50} \equiv 3^{50} \equiv 1 \pmod{11}.$$

We conclude that $2^{30} \cdot 14^{50} - 1 \equiv (1)(1) - 1 = 0 \pmod{11}$—which says that the given number is divisible by 11.

Needless to say, other arrangements of the foregoing computations are possible; they are all equally valid, so long as the amount of work is kept under control.

1-7-7. Application. A much more interesting application of congruences than the preceding one involves find a criterion for divisibility by 11. For this, it is necessary to recall the underlying meaning of the standard notation for integers. If we write, for example, the number 728, this is simply a compact notation for

$$7(10)^2 + 2(10) + 8 = 8 + 2(10) + 7(10)^2$$

and by the same token 31,059 is "shorthand" for

$$9 + 5(10) + 0(10)^2 + 1(10)^3 + 3(10)^4 = 3(10)^4 + 1(10)^3 + 0(10)^2 + 5(10) + 9$$

In general, we write

$$a = a_n a_{n-1} \cdots a_2 a_1 a_0,$$

where $0 \leq a_i \leq 9$ for $i = 0, 1, \ldots, n$ and $a_n \neq 0$, to signify that

$$a = a_n(10)^n + a_{n-1}(10)^{n-1} + \cdots + a_2(10)^2 + a_1(10) + a_0.$$

[It is often convenient to write this in ascending powers of 10; for example, if $a = 306{,}394{,}230{,}450$ it is easier to start with $a = 0 + 5(10) + \cdots$ and work one's way up, than to start at the front with 3 and locate the power of 10 that must be attached to 3.] This is simply another version of the "well-known fact" (more about this in Section 1-8) that we write integers using only the digits 0, 1, 2, 3, 4, 5, 6, 7, 8, 9 and with the front digit not equal to 0.

Turning to the powers of 10, we note that $10 \equiv -1 \pmod{11}$; squaring this congruence gives $(10)^2 \equiv 1 \pmod{11}$, while cubing it gives $(10)^3 \equiv -1 \pmod{11}$. The general rule is obviously

$$(10)^{\text{odd}} \equiv -1 \pmod{11}, \qquad (10)^{\text{even}} \equiv 1 \pmod{11}.$$

According to the rule for multiplication of congruences,

$$a_1(10) \equiv -a_1 \pmod{11}, \quad a_2(10)^2 \equiv a_2 \pmod{11}, \quad a_3(10)^3 \equiv -a_3 \pmod{11},$$

and so on. Now by the rule for adding several congruences, 1-7-5. we see that

$$a = a_0 + a_1(10) + a_2(10)^2 + \cdots + a_n(10)^n$$
$$\equiv a_0 - a_1 + a_2 + \cdots + (-1)^n a_n \pmod{11}.$$

This says that a and the alternating sum of the digits,

$$a_0 - a_1 + a_2 - a_3 + \ldots + (-1)^n a_n = \sum_{i=0}^{n} (-1)^i a_i,$$

are congruent to each other mod 11; so, in particular, by 1-7-3, they have the same remainder upon division by 11. Consequently, we have shown

$$11 \mid a \Leftrightarrow 11 \mid \sum_{i=0}^{n} (-1)^i a_i$$

or in words, an integer a is divisible by 11 $\Leftrightarrow$ the alternating sum of its digits is divisible by 11.

As an illustration, consider $a = 31{,}059$; since the alternating sum of the digits is $9 - 5 + 0 - 1 + 3 = 6$, we conclude that 31,059 is not divisible by 11. In similar fashion, the reader may verify that 11 divides 306,394,230,450.

With regard to the alternating sum, it should be noted that

$$11 \mid (a_0 - a_1 + \cdots + (-1)^n a_n) \Leftrightarrow 11 \mid -(a_0 - a_1 + \cdots + (-1)^n a_n).$$

Since exactly one of $(a_0 - a_1 + \cdots + (-1)^n a_n)$ and $-(a_0 - a_1 + \cdots + (-1)^n a_n)$ contains the term $+a_n$, it follows that in taking the alternating sum there is no harm (as far as the question of divisibility by 11 is concerned) in starting from $+a_n$ and then alternating signs as one moves through the digits of a from left to right. For example, if $a = 306{,}394{,}230{,}450$ we take the alternating sum $3 - 0 + 6 - 3 + 9 - 4 + 2 - 3 + 0 - 4 + 5 - 0$; since this equals 11, a is divisible by 11.

1-7-8. Definition. Given any modulus m, let us introduce the following notation; for any $a \in \mathbf{Z}$ we write

$$\lfloor a \rfloor_m = \{b \in \mathbf{Z} \mid b \equiv a \pmod{m}\}.$$

In words, $\lfloor a \rfloor_m$ denotes the set of all integers which are congruent to $a \pmod{m}$; so it is appropriate to call $\lfloor a \rfloor_m$ the **congruence class** of $a \pmod{m}$.

Since two integers are congruent modulo m $\Leftrightarrow$ they have the same remainder upon division by m, we may view $\lfloor a \rfloor_m$ in another way—namely, $\lfloor a \rfloor_m$ is the set of all integers which have the same remainder as a upon division by m. This point of view underlies the fact that $\lfloor a \rfloor_m$ is also referred to as the **residue class** of $a \pmod{m}$.

What does a residue class (mod m) look like in a concrete situation? Suppose, for example, $m = 7$, and consider $\lfloor 3 \rfloor_7$. We want to decide which integers belong to the set $\lfloor 3 \rfloor_7$. According to the definition, $b \in \lfloor 3 \rfloor_7 \Leftrightarrow b \equiv 3 \pmod 7$, so the way to decide if $b \in \lfloor 3 \rfloor_7$ is to check if $b \equiv 3 \pmod 7$. Thus, $3 \in \lfloor 3 \rfloor_7$ because $3 \equiv 3 \pmod 7$, $10 \in \lfloor 3 \rfloor_7$ because $10 \equiv 3 \pmod 7$, $14 \notin \lfloor 3 \rfloor_7$ because $14 \not\equiv 3 \pmod 7$ $-4 \in \lfloor 3 \rfloor_7$ because $-4 \equiv 3 \pmod 7$, $-11 \in \lfloor 3 \rfloor_7$ because $-11 \equiv 3 \pmod 7$, and so on. By now, it is more or less clear that

$$\lfloor 3 \rfloor_7 = \{\ldots, -18, -11, -4, 3, 10, 17, 24, \ldots\}.$$

In the same way, the reader may convince himself that

$$\lfloor 12 \rfloor_7 = \{\ldots, -16, -9, -2, 5, 12, 19, 26, 33, \ldots\}.$$

These examples seem to say that to describe $\lfloor a \rfloor_7$ one starts with a and keeps adding or subtracting 7. Indeed, we have the following general result which settles the question.

1-7-9. Proposition. For any $a \in \mathbf{Z}$, the residue class $\lfloor a \rfloor_m$ consists of all integers that arise by adding a multiple of m to a—in symbols,

$$\lfloor a \rfloor_m = \{a + tm \mid t \in \mathbf{Z}\}.$$

Proof: Making use of the definitions of residue class, congruence and divisibility, we have

$$\begin{aligned} b \in \lfloor a \rfloor_m &\Leftrightarrow b \equiv a \pmod m \\ &\Leftrightarrow m \mid (b - a) \\ &\Leftrightarrow b - a = tm, \text{ for some } t \in \mathbf{Z} \\ &\Leftrightarrow b = a + tm, \text{ for some } t \in \mathbf{Z} \\ &\Leftrightarrow b \in \{a + tm \mid t \in \mathbf{Z}\}. \end{aligned}$$

This does it. ▌

In virtue of this result we can write down residue classes (mod 7) at will; for example,

$$\begin{aligned} \lfloor 5 \rfloor_7 &= \{\ldots, -23, -16, -9, -2, 5, 12, 19, 26, 33, \ldots\}, \\ \lfloor -1 \rfloor_7 &= \{\ldots, -22, -15, -8, -1, 6, 13, 20, 27, 34, \ldots\}, \\ \lfloor 6 \rfloor_7 &= \{\ldots, -22, -15, -8, -1, 6, 13, 20, 27, 34, \ldots\}, \\ \lfloor 0 \rfloor_7 &= \{\ldots, -28, -21, -14, -7, 0, 7, 14, 21, 28, \ldots\}, \\ \lfloor 1 \rfloor_7 &= \{\ldots, -27, -20, -13, -6, 1, 8, 15, 22, 29, \ldots\}, \end{aligned}$$

$$\lfloor 7 \rfloor_7 = \{\ldots, -28, -21, -14, -7, 0, 7, 14, 21, 28, \ldots\},$$
$$\lfloor 15 \rfloor_7 = \{\ldots, -20, -13, -6, 1, 8, 15, 22, 29, 36, 43, \ldots\}.$$

These numerical examples suggest various properties of congruence classes; some of the significant ones are given in the next result.

1-7-10. Proposition. The residue classes mod m satisfy the following properties:

(*i*) $a \in \lfloor a \rfloor_m$ for all $a \in \mathbf{Z}$,

(*ii*) $b \in \lfloor a \rfloor_m \Leftrightarrow a \in \lfloor b \rfloor_m$,

(*iii*) $b \in \lfloor a \rfloor_m \Rightarrow \lfloor b \rfloor_m = \lfloor a \rfloor_m$,

(*iv*) $\lfloor a \rfloor_m \cap \lfloor b \rfloor_m \neq \varnothing \Rightarrow \lfloor a \rfloor_m = \lfloor b \rfloor_m$.

Proof: Our proof will be based on the interpretation of $\lfloor a \rfloor_m$ as the set of all integers which have the same remainder as a upon division by m; the operative form of this is

$$b \in \lfloor a \rfloor_m \Leftrightarrow b \text{ has the same remainder as } a.$$

Now, (*i*) is trivial because a has the same remainder as a. In words, (*i*) may be stated as: Any integer belongs to the residue class it determines.

To prove (*ii*), one simply observes that b has the same remainder as $a \Leftrightarrow a$ has the same remainder as b. In words, (*ii*) says that b belongs to the residue class of $a \Leftrightarrow a$ belongs to the residue class of b.

As for (*iii*), if b has the same remainder as a, then any number that has the same remainder as b has the same remainder as a (which implies $\lfloor b \rfloor_m \subset \lfloor a \rfloor_m$) and conversely (which implies $\lfloor a \rfloor_m \subset \lfloor b \rfloor_m$); hence, $\lfloor a \rfloor_m = \lfloor b \rfloor_m$. Thus (*iii*) tells us that any element (that is, b) of a residue class (that is, $\lfloor a \rfloor_m$) determines the residue class.

Finally, (*iv*) says that if two residue classes have an element in common (other ways to express this are: have nonempty intersection, meet, intersect, are not disjoint) they are identical. From this we conclude that two residue classes are disjoint or else they are identical.

To prove (*iv*), suppose $\lfloor a \rfloor_m \cap \lfloor b \rfloor_m \neq \varnothing$, so there exists an integer c in the intersection; then

$$\begin{aligned} c \in \lfloor a \rfloor_m \cap \lfloor b \rfloor_m &\Rightarrow c \in \lfloor a \rfloor_m \quad \text{and} \quad c \in \lfloor b \rfloor_m \\ &\Rightarrow \lfloor c \rfloor_m = \lfloor a \rfloor_m \quad \text{and} \quad \lfloor c \rfloor_m = \lfloor b \rfloor_m, \qquad \text{(by } iii\text{)} \\ &\Rightarrow \lfloor a \rfloor_m = \lfloor b \rfloor_m. \end{aligned}$$

This completes the proof. ▮

1-7-11. Corollary. For $a, b \in \mathbf{Z}$ the following are equivalent:

(1) $a \equiv b \pmod{m}$,

(2) a and b have the same remainder upon division by m,

(3) $a \in \lfloor b \rfloor_m$,

(4) $b \in \lfloor a \rfloor_m$,

(5) $\lfloor a \rfloor_m = \lfloor b \rfloor_m$.

Proof: We already know that (1) $\Leftrightarrow$ (2), (2) $\Leftrightarrow$ (3), (3) $\Leftrightarrow$ (4), (4) $\Leftrightarrow$ (5), so there is nothing to prove. ▌

It is sometimes convenient to make use of the contrapositive form of the equivalence of (5), (1), and (2); namely, for $a, b \in \mathbf{Z}$, $\lfloor a \rfloor_m \neq \lfloor b \rfloor_m \Leftrightarrow a \not\equiv b \pmod{m} \Leftrightarrow a$ and b have different remainders upon division by m. Of course, these equivalences do not require proof.

Let us illustrate some of the consequences of these results. Suppose we take $m = 7$ and consider the residue classes

$$\lfloor 0 \rfloor_7, \quad \lfloor 1 \rfloor_7, \quad \lfloor 2 \rfloor_7, \quad \lfloor 3 \rfloor_7, \quad \lfloor 4 \rfloor_7, \quad \lfloor 5 \rfloor_7, \quad \lfloor 6 \rfloor_7.$$

No two of these residue classes are the same because no two of the integers 0, 1, 2, 3, 4, 5, 6 have the same remainder upon division by 7; so these seven residue classes are distinct. Now, consider any integer a. By the division algorithm, $a = q \cdot 7 + r$ where $0 \leq r < 7$, and therefore $\lfloor a \rfloor_7 = \lfloor r \rfloor_7$ which is one of the 7 residue classes already listed. We conclude that there are exactly 7 residue classes modulo 7, and any integer belongs to one (in fact, because residue classes are distinct, to *exactly* one) residue class. Note that because a residue class has many "names" (for example, $\lfloor 0 \rfloor_7 = \lfloor 7 \rfloor_7 = \lfloor 14 \rfloor_7 = \lfloor 21 \rfloor_7$) the 7 residue classes can be denoted in various ways; for example,

$$\lfloor 21 \rfloor_7, \quad \lfloor 92 \rfloor_7, \quad \lfloor -5 \rfloor_7, \quad \lfloor 3 \rfloor_7, \quad \lfloor -31 \rfloor_7, \quad \lfloor 40 \rfloor_7, \quad \lfloor 20 \rfloor_7$$

is a perfectly valid way to list the 7 residue classes.

It is clear that the same arguments and results apply for an arbitrary modulus m, and we may summarize as follows:

1-7-12. Theorem. For any integer $m > 0$, $\mathbf{Z}$ decomposes (that is, breaks up) into exactly m disjoint residue classes (mod m)—namely,

$$\lfloor 0 \rfloor_m, \quad \lfloor 1 \rfloor_m, \quad \lfloor 2 \rfloor_m, \ldots, \lfloor m-1 \rfloor_m.$$

Every integer belongs to exactly one residue class, and every residue class contains an infinite number of integers.

Proof: By now this is obvious—so the details may be left to the reader. ∎

1-7-13. Remark. The key ingredient in arriving at the decomposition of **Z** into disjoint residue classes is 1-7-10, in which four basic properties of residue classes (mod m) are listed. These properties were proved by viewing residue classes in terms of remainders upon division by m. It is highly instructive to arrive at the decomposition of **Z** into disjoint congruence classes by proving 1-7-10 in another way—specifically, by using only the basic properties (namely the reflexive, symmetric, and transitive properties) of congruence (mod m). The importance of this approach lies in the fact that it is a special case of a very general phenomenon that will be discussed in Chapter III.

We start from the definition of $\lfloor a \rfloor_m$ in the form

$$b \in \lfloor a \rfloor_m \iff b \equiv a \pmod{m}.$$

The reflexive property says that $a \equiv a \pmod{m}$, so $a \in \lfloor a \rfloor_m$ and part (*i*) of 1-7-10 holds. Using the symmetric property we have: $b \in \lfloor a \rfloor_m \iff b \equiv a \pmod{m} \iff a \equiv b \pmod{m} \iff a \in \lfloor b \rfloor_m$, which proves (*ii*). As for (*iii*), $b \in \lfloor a \rfloor_m$ implies $b \equiv a \pmod{m}$, and using the transitive property, we see that for $c \in \mathbf{Z}$

$$c \in \lfloor a \rfloor_m \iff c \equiv a \pmod{m} \iff c \equiv b \pmod{m} \iff c \in \lfloor b \rfloor_m$$

which shows that $\lfloor a \rfloor_m = \lfloor b \rfloor_m$. The proof of (*iv*) given earlier depends only on (*iii*), so there is no need to modify it.

1-7-14. Discussion. For each $m > 0$, we let $\mathbf{Z}_m$ denote the set whose elements are the congruence classes (mod m). Thus, $\mathbf{Z}_m$ is a set with m elements, and we use the symbols $A, B, C, \ldots$ to represent elements of $\mathbf{Z}_m$. It would be nice to be able to combine elements of $\mathbf{Z}_m$—more precisely, for any two congruence classes A and B to be able to "add" and "multiply" them. Naturally, the results of addition and multiplication would be denoted by $A + B$ and $A \cdot B$, respectively.

How might one go about this? One possible method is as follows. Since A and B are congruence classes we may write $A = \lfloor a \rfloor_m$, $B = \lfloor b \rfloor_m$, where $a, b \in \mathbf{Z}$. We know how to add or multiply a and b, so it seems natural to take $A + B$ and $A \cdot B$ as the congruence classes $\lfloor a + b \rfloor_m$ and $\lfloor ab \rfloor_m$, respectively. In other words, we make the definitions

$$\lfloor a \rfloor_m + \lfloor b \rfloor_m = \lfloor a + b \rfloor_m,$$

$$\lfloor a \rfloor_m \cdot \lfloor b \rfloor_m = \lfloor ab \rfloor_m.$$

For example, if $A = \lfloor 5 \rfloor_7$ and $B = \lfloor 6 \rfloor_7$, then, according to the definition, $A + B = \lfloor 11 \rfloor_7 = \lfloor 4 \rfloor_7$ and $A \cdot B = \lfloor 30 \rfloor_7 = \lfloor 2 \rfloor_7$.

There is, however, a possible flaw in these definitions; they depend on the choice of the integers $a \in A$ and $b \in B$. What if, instead of writing $A = \lfloor a \rfloor_m$, $B = \lfloor b \rfloor_m$, we write $A = \lfloor a' \rfloor_m$, $B = \lfloor b' \rfloor_m$ where $a', b' \in \mathbf{Z}$? Do we obtain the same answers for $A + B$ and $A \cdot B$ when a' and b' (instead of a and b) are used as the "representatives" of A and B, respectively? In other words, when a' and b' are used, the definitions say that

$$\lfloor a' \rfloor_m + \lfloor b' \rfloor_m = \lfloor a' + b' \rfloor_m,$$
$$\lfloor a' \rfloor_m \cdot \lfloor b' \rfloor_m = \lfloor a'b' \rfloor_m.$$

Do these definitions give the same answers as before? The critical question is, therefore, to decide if

$$\lfloor a' + b' \rfloor_m = \lfloor a + b \rfloor_m \quad \text{and} \quad \lfloor a'b' \rfloor_m = \lfloor ab \rfloor_m.$$

Let us illustrate what is troubling us with a numerical example. Suppose $A = \lfloor 5 \rfloor_7$, $B = \lfloor 6 \rfloor_7$, so $A + B = \lfloor 5+6 \rfloor_7$ and $A \cdot B = \lfloor 5 . 6 \rfloor_7$. Now, A and B are sets

$$A = \{\ldots, -16, -9, -2, 5, 12, 19, 26, \ldots\},$$

$$B = \{\ldots, -15, -8, -1, 6, 13, 20, 27, 34, \ldots\},$$

and can be expressed in the $\lfloor \ \rfloor_7$ notation in many ways. In fact, for any $x \in A$, $y \in B$ we have $A = \lfloor x \rfloor_7$, $B = \lfloor y \rfloor_7$. For concreteness, suppose we write $A = \lfloor -9 \rfloor_7$, $B = \lfloor 27 \rfloor_7$; then $A + B = \lfloor -9+27 \rfloor_7$ and $A \cdot B = \lfloor (-9)(27) \rfloor_7$. Are these the same congruence classes $A + B$ and AB as before? So we need to know if

$$\lfloor 5 + 6 \rfloor_7 = \lfloor -9 + 27 \rfloor_7 \quad \text{and} \quad \lfloor 5 \cdot 6 \rfloor_7 = \lfloor (-9)(27) \rfloor_7.$$

Of course, these equalities do hold here—the sums both equal $\lfloor 4 \rfloor_7$, and the products both equal $\lfloor 2 \rfloor_7$. But all this involves specific choices. What if the choices are arbitrary? The reader should try to convince himself that for any $x \in A$, $y \in B$ we still obtain $\lfloor x + y \rfloor_7 = \lfloor 4 \rfloor_7$ and $\lfloor xy \rfloor_7 = \lfloor 2 \rfloor_7$.

Returning to the general situation, we note that the mathematical language commonly used to express the question facing us is to ask: are the definitions of addition and multiplication of congruence classes **well defined**? The term "well defined" refers to the choices that enter into the definitions—namely, $a \in A$ and $b \in B$; we need to guarantee that for all possible choices of a and b the definitions give the same result. In our case, this boils down to showing: if $\lfloor a' \rfloor_m = \lfloor a \rfloor_m$ and $\lfloor b' \rfloor_m = \lfloor b \rfloor_m$, then $\lfloor a' + b' \rfloor_m = \lfloor a + b \rfloor_m$ and $\lfloor a'b' \rfloor_m = \lfloor ab \rfloor_m$.

Once the question has been formulated, it is easy to provide the answer. If $\lfloor a' \rfloor_m = \lfloor a \rfloor_m$ and $\lfloor b' \rfloor_m = \lfloor b \rfloor_m$, then, by 1-7-11, $a' \equiv a \pmod{m}$ and $b' \equiv b \pmod{m}$. According to 1-7-4, congruences may be added or multiplied; so $a' + b' \equiv a + b \pmod{m}$ and $a'b' \equiv ab \pmod{m}$—which says that $\lfloor a' + b' \rfloor_m = \lfloor a + b \rfloor_m$ and $\lfloor a'b' \rfloor_m = \lfloor ab \rfloor_m$.

We have shown that the rather natural definitions of addition and multiplication of congruence classes (mod m) are indeed well defined.

1-7-15. Example. Consider the set

$$\mathbf{Z}_7 = \{\lfloor 0 \rfloor_7, \lfloor 1 \rfloor_7, \lfloor 2 \rfloor_7, \lfloor 3 \rfloor_7, \lfloor 4 \rfloor_7, \lfloor 5 \rfloor_7, \lfloor 6 \rfloor_7\}.$$

It is surely preferable, especially psychologically, to write the elements of $\mathbf{Z}_7$ in this way rather than in an outlandish form like

$$\mathbf{Z}_7 = \{\lfloor 91 \rfloor_7, \lfloor -6 \rfloor_7, \lfloor 37 \rfloor_7, \lfloor -25 \rfloor_7, \lfloor -31 \rfloor_7, \lfloor 54 \rfloor_7, \lfloor 6 \rfloor_7\}.$$

We know how to add and multiply elements of $\mathbf{Z}_7$; for example,

$$\lfloor 3 \rfloor_7 + \lfloor 5 \rfloor_7 = \lfloor 1 \rfloor_7, \qquad \lfloor 2 \rfloor_7 + \lfloor 5 \rfloor_7 = \lfloor 0 \rfloor_7, \qquad \lfloor 2 \rfloor_7 + \lfloor 3 \rfloor_7 = \lfloor 5 \rfloor_7,$$
$$\lfloor 2 \rfloor_7 \cdot \lfloor 3 \rfloor_7 = \lfloor 6 \rfloor_7, \qquad \lfloor 3 \rfloor_7 \cdot \lfloor 4 \rfloor_7 = \lfloor 5 \rfloor_7, \qquad \lfloor 2 \rfloor_7 \cdot \lfloor 6 \rfloor_7 = \lfloor 5 \rfloor_7.$$

More generally, we can make tables for addition and multiplication in $\mathbf{Z}_7$. For this it is convenient to drop the cumbersome $\lfloor\ \rfloor_7$ notation and write

$$\mathbf{Z}_7 = \{0, 1, 2, 3, 4, 5, 6\},$$

with the understanding that each of these integers represents its congruence class. Thus, the examples above take the form

$$3 + 5 = 1, \qquad 2 + 5 = 0, \qquad 2 + 3 = 5,$$
$$2 \cdot 3 = 6, \qquad 3 \cdot 4 = 5, \qquad 2 \cdot 6 = 5$$

and the desired tables are as follows:

addition in $\mathbf{Z}_7$

+	0	1	2	3	4	5	6
0	0	1	2	3	4	5	6
1	1	2	3	4	5	6	0
2	2	3	4	5	6	0	1
3	3	4	5	6	0	1	2
4	4	5	6	0	1	2	3
5	5	6	0	1	2	3	4
6	6	0	1	2	3	4	5

multiplication in $\mathbf{Z}_7$

·	0	1	2	3	4	5	6
0	0	0	0	0	0	0	0
1	0	1	2	3	4	5	6
2	0	2	4	6	1	3	5
3	0	3	6	2	5	1	4
4	0	4	1	5	2	6	3
5	0	5	3	1	6	4	2
6	0	6	5	4	3	2	1

Another way to represent $\mathbf{Z}_7$ is:

$$\mathbf{Z}_7 = \{\lfloor -3 \rfloor_7, \lfloor -2 \rfloor_7, \lfloor -1 \rfloor_7, \lfloor 0 \rfloor_7, \lfloor 1 \rfloor_7, \lfloor 2 \rfloor_7, \lfloor 3 \rfloor_7\}$$
$$= \{-3, -2, -1, 0, 1, 2, 3\}$$

and with this notation the addition and multiplication tables for $\mathbf{Z}_7$ take the form

addition in $\mathbf{Z}_7$

+	0	1	2	3	−3	−2	−1
0	0	1	2	3	−3	−2	−1
1	1	2	3	−3	−2	−1	0
2	2	3	−3	−2	−1	0	1
3	3	−3	−2	−1	0	1	2
−3	−3	−2	−1	0	1	2	3
−2	−2	−1	0	1	2	3	−3
−1	−1	0	1	2	3	−3	−2

multiplication in $\mathbf{Z}_7$

·	0	1	2	3	−3	−2	−1
0	0	0	0	0	0	0	0
1	0	1	2	3	−3	−2	−1
2	0	2	−3	−1	1	3	−2
3	0	3	−1	2	−2	1	−3
−3	0	−3	1	−2	2	−1	3
−2	0	−2	3	1	−1	−3	2
−1	0	−1	−2	−3	3	2	1

1-7-16 / *PROBLEMS*

1. Consider the integers 3, 19, 87, −15, −71, 96, 240, −113, 69, 378, −91, −14, 500, −312, 153; which of them are congruent to each other (mod m) when
(*i*) $m = 7$, (*ii*) $m = 11$, (*iii*) $m = 13$.

2. Show that
(*i*) If $a \equiv b \pmod{m}$ and $n \mid m$, $n > 0$, then $a \equiv b \pmod{n}$.
(*ii*) Suppose that $m_1, m_2, \ldots, m_r$ are all positive, then $a \equiv b \pmod{m_i}$ for $i = 1, 2, \ldots, r \Leftrightarrow a \equiv b \pmod{[m_1, m_2, \ldots, m_r]}$.

3. How would you express $a \equiv 0 \pmod{m}$ in another way?

4. What is the remainder when $b = 2^{50} - 1$ is divided by $a = 31 = 2^5 - 1$? What if $b = 2^{40} + 17$?

5. Find and prove a criterion for deciding if an integer is divisible by 9. Do the same thing for 3. Is 748,052,301,472 divisible by 9? by 3? by 11?

6. How can one decide easily if an integer is divisible by 2? by 4? by 8? by 10? by 5? by 25?

7. Show that for every $n \geq 0$
(*i*) $10^n + 3 \cdot 4^{n+2}$ leaves a remainder of 4 upon division by 9.
(*ii*) 24 divides $2 \cdot 7^n + 3 \cdot 5^n - 5$,
(*iii*) $3^{4n+2} + 5^{2n+1}$ is divisible by 14.

8. (*i*) Find the remainder when a^7 is divided by 7 for each a satisfying $0 \le a < 7$.
(*ii*) For each integer a satisfying $0 \le a < 11$ find the integer less than 11 which is congruent to a^{11} (mod 11).
(*iii*) Do the same thing with 11 replaced by 12.

9. Prove: If p is prime and $(a, p) = 1$, then
(*i*) $a^2 \equiv 1 \pmod{p} \Rightarrow a \equiv \pm 1 \pmod{p}$,
(*ii*) $ab \equiv ac \pmod{p} \Rightarrow b \equiv c \pmod{p}$

10. Show that if $a \equiv b \pmod{m}$, then $(a, m) = (b, m)$.

11. Choose three explicit elements a_1, a_2, a_3 in $\lfloor 3 \rfloor_7$ and explicit three elements b_1, b_2, b_3 in $\lfloor 9 \rfloor_7$. Consider the nine numbers $a_i + b_j$ $i, j = 1, 2, 3$, and verify that they all belong to the same congruence class—namely, $\lfloor 5 \rfloor_7$. Furthermore, show that the nine products $a_i b_j$, $i, j = 1, 2, 3$ all belong $\lfloor 6 \rfloor_7$.

12. Which of the following congruence classes are equal?—

$\lfloor 76 \rfloor_9$, $\lfloor 84 \rfloor_5$, $\lfloor 9 \rfloor_6$, $\lfloor 43 \rfloor_7$, $\lfloor 19 \rfloor_8$, $\lfloor 36 \rfloor_5$, $\lfloor 147 \rfloor_7$, $\lfloor -32 \rfloor_8$, $\lfloor -14 \rfloor_9$, $\lfloor -36 \rfloor_6$, $\lfloor 184 \rfloor_6$, $\lfloor 11 \rfloor_7$, $\lfloor -97 \rfloor_9$, $\lfloor -57 \rfloor_5$, $\lfloor 171 \rfloor_8$, $\lfloor 171 \rfloor_7$, $\lfloor 53 \rfloor_5$, $\lfloor 66 \rfloor_8$, $\lfloor 47 \rfloor_9$, $\lfloor 47 \rfloor_5$, $\lfloor -72 \rfloor_7$, $\lfloor 12 \rfloor_8$, $\lfloor 87 \rfloor_9$, $\lfloor -16 \rfloor_5$, $\lfloor -18 \rfloor_8$, $\lfloor 50 \rfloor_7$, $\lfloor 38 \rfloor_6$.

13. Make addition and multiplication tables for:
(*i*) $\mathbf{Z}_2 = \{0, 1\}$, (*ii*) $\mathbf{Z}_6 = \{0, 1, 2, 3, 4, 5\}$
(*iii*) $\mathbf{Z}_5 = \{-2, -1, 0, 1, 2\}$.

14. Suppose $f(x) = c_0 + c_1 x + c_2 x^2 + \cdots + c_n x^n$ where $c_0, c_1, \ldots, c_n$ are integers, prove that if $a \equiv b \pmod{m}$, then $f(a) \equiv f(b) \pmod{m}$.

15. How are the residue classes in $\mathbf{Z}_6$ related to the residue classes in $\mathbf{Z}_{12}$?

16. Show that if $ab \equiv ac \pmod{m}$, then $b \equiv c \pmod{m/(a, m)}$.

17. Discuss the solutions of each of the following:
(*i*) $4x \equiv 2 \pmod{6}$, (*ii*) $4x \equiv 1 \pmod{6}$,
(*iii*) $4x \equiv 4 \pmod{6}$, (*iv*) $x^2 \equiv 1 \pmod{15}$,
(*v*) $3x \equiv 800 \pmod{11}$.

18. Does there exist a positive integer n for which
(*i*) $2^n \equiv 1 \pmod{13}$, (*ii*) $3^n \equiv 1 \pmod{13}$,
(*iii*) $2^n \equiv 1 \pmod{17}$, (*iv*) $3^n \equiv 1 \pmod{17}$?
In each case, find the smallest n that will do.

19. Show by example that $a^n \equiv b^n \pmod{m}$ need not imply $a \equiv b \pmod{m}$.

20. A person makes a 48¢ purchase in a store. The purchaser has a one dollar bill and three pennies. The storekeeper has six dimes and seven nickels. How should the transaction be arranged? In how many ways can it be done? What if the purchase is for 49¢?

I-8. Radix Representation

The major results of this chapter—for example, the properties of greatest common divisor, uniqueness of factorization, solving linear diophantine equations—all depend, in part, on the division algorithm. In this section, we discuss one more extremely important consequence of the division algorithm.

Suppose we fix an integer $a > 1$. Consider any $b > 0$. According to the division algorithm we can write, uniquely, $b = qa + r$ where $0 \leq r < a$. In the Euclidean algorithm, we then apply the division algorithm to a and r and repeat the process until we arrive at a remainder of 0. Here, instead, we apply the division algorithm to q and a and repeat the process until we arrive at a quotient q which is 0. More precisely, rewriting the equation we already have as $b = q_0 a + r_0$, and then indexing in the obvious fashion, we have

$$\begin{aligned}
b &= q_0 a + r_0, & 0 &\leq r_0 < a,\\
q_0 &= q_1 a + r_1, & 0 &\leq r_1 < a,\\
q_1 &= q_2 a + r_2, & 0 &\leq r_2 < a,\\
&\vdots & &\vdots\\
q_{n-2} &= q_{n-1} a + r_{n-1}, & 0 &\leq r_{n-1} < a,\\
q_{n-1} &= 0 \cdot a + r_n, & 0 &\leq r_n < a.
\end{aligned}$$

The index n is taken here as the smallest integer greater than or equal to 0 for which $q_n = 0$. Of course, it is necessary to show that our process always gets to a quotient q_n which equals 0. This is quite easy. Because $a > 1$ and $b > 0$, the first equation implies that $b > q_0 \geq 0$. If $q_0 = 0$, we are finished; so suppose $q_0 > 0$. The second equation then exists, and it implies $q_0 > q_1 \geq 0$. Thus, as our process unfolds, we obtain

$$b > q_0 > q_1 > q_2 > \cdots \geq 0$$

which is a strictly decreasing sequence of nonnegative integers. After a finite number of steps such a sequence must reach 0; that is, eventually we obtain $q_n = 0$.

Note that once we locate the n for which $q_n = 0$, our choice of notation guarantees that $q_{n-1} > 0$ and $r_n = q_{n-1} \neq 0$.

What good is all this? Let us substitute for q_0 in the first equation—this gives

$$\begin{aligned} b &= q_0 a + r_0 \\ &= (q_1 a + r_1)a + r_0 \\ &= q_1 a^2 + r_1 a + r_0 . \end{aligned}$$

Substituting for q_1 in this expression, we obtain

$$\begin{aligned} b &= (q_2 a + r_2)a^2 + r_1 a + r_0 \\ &= q_2 a^3 + r_2 a^2 + r_1 a + r_0 . \end{aligned}$$

This procedure is repeated until, at the end, we have

$$\begin{aligned} b &= q_{n-2} a^{n-1} + r_{n-2} a^{n-2} + \cdots + r_2 a^2 + r_1 a + r_0 \\ &= (q_{n-1} a + r_{n-1})a^{n-1} + r_{n-2} a^{n-2} + \cdots + r_1 a + r_0 \\ &= q_{n-1} a^n + r_{n-1} a^{n-1} + r_{n-2} a^{n-2} + \cdots + r_1 a + r_0 \\ &= r_n a^n + r_{n-1} a^{n-1} + \cdots + r_2 a^2 + r_1 a + r_0 , \end{aligned}$$

where $0 \leq r_0, r_1, \ldots, r_n < a$ (meaning that $r_0, r_1, \ldots, r_n$ are all greater than or equal to 0 and less than a) and $r_n \neq 0$.

This shows, in rough terms, that every positive integer b can be expanded in powers of a; and even more, we have a recipe for finding such an expansion for b. We illustrate with a numerical example. Immediately, thereafter, we shall give the precise formulation of our general result.

1-8-1. Example. Consider $a = 2$, $b = 489$. The division algorithm procedure gives

$$\begin{aligned} 489 &= 244 \cdot 2 + 1, \\ 244 &= 122 \cdot 2 + 0, \\ 122 &= 61 \cdot 2 + 0, \\ 61 &= 30 \cdot 2 + 1, \\ 30 &= 15 \cdot 2 + 0, \\ 15 &= 7 \cdot 2 + 1, \\ 7 &= 3 \cdot 2 + 1, \\ 3 &= 1 \cdot 2 + 1, \\ 1 &= 0 \cdot 2 + 1. \end{aligned}$$

Although it is not essential, we list the values of the q's and r's. Starting from $b = 489$, $a = 2$ and working from the top down we have: $q_0 = 244$, $r_0 = 1$, $q_1 = 122$, $r_1 = 0$, $q_2 = 61$, $r_2 = 0$, $q_3 = 30$, $r_3 = 1$, $q_4 = 15$, $r_4 = 0$, $q_5 = 7$, $r_5 = 1, q_6 = 3, r_6 = 1, q_7 = 1, r_7 = 1, q_8 = 0, r_8 = 1$, Moreover, the r's provide the coefficients in the expansion of 489 in terms of powers of 2; in fact, we have immediately

$$489 = 1 \cdot 2^8 + 1 \cdot 2^7 + 1 \cdot 2^6 + 1 \cdot 2^5 + 0 \cdot 2^4 + 1 \cdot 2^3 + 0 \cdot 2^2 + 0 \cdot 2^1 + 1.$$

Note that running through the list of remainders in our process from top to bottom (that is, from r_0 to r_n, where $n = 8$) determines the coefficients of the powers of 2, going from right to left.

Similarly, for $a = 7$, $b = 489$ we have

$$\begin{aligned} 489 &= 69 \cdot 7 + 6, \\ 69 &= 9 \cdot 7 + 6, \\ 9 &= 1 \cdot 7 + 2, \\ 1 &= 0 \cdot 7 + 1, \end{aligned}$$

so that

$$\begin{aligned} 489 &= 6 + 6 \cdot 7^1 + 2 \cdot 7^2 + 1 \cdot 7^3 \\ &= 1 \cdot 7^3 + 2 \cdot 7^2 + 6 \cdot 7^1 + 6 \end{aligned}$$

and for $a = 3$, $b = 543$ we obtain

$$\begin{aligned} 543 &= 181 \cdot 3 + 0, \\ 181 &= 60 \cdot 3 + 1, \\ 60 &= 20 \cdot 3 + 0, \\ 20 &= 6 \cdot 3 + 2, \\ 6 &= 2 \cdot 3 + 0, \\ 2 &= 0 \cdot 3 + 2, \end{aligned}$$

so that

$$\begin{aligned} 543 &= 0 + 1 \cdot 3^1 + 0 \cdot 3^2 + 2 \cdot 3^3 + 0 \cdot 3^4 + 2 \cdot 3^5 \\ &= 2 \cdot 3^5 + 0 \cdot 3^4 + 2 \cdot 3^3 + 0 \cdot 3^2 + 1 \cdot 3^1 + 0. \end{aligned}$$

1-8-2. Theorem. If $a > 1$ is fixed, then any $b > 0$ can be expressed uniquely in the form

$$b = r_n a^n + r_{n-1} a^{n-1} + \cdots + r_2 a^2 + r_1 a + r_0,$$

where

$$0 \leq r_0, r_1, \ldots, r_n < a \qquad \text{and} \qquad r_n \neq 0.$$

Proof: It is understood that n is not fixed; it depends on the choice of b. Of course, $n \geq 0$. The case $n = 0$ occurs when $b < a$, as then $b = 0 \cdot a + r_0$.

To keep the notation uniform one might prefer to write r_0 as $r_0 a^0$ in the statement of the theorem, as then there is no need to point up the case $n = 0$.

The existence of an expression for b of the desired form was dealt with earlier. The reader can easily supply the missing details, so we shall consider this part proved.

It remains to prove the uniqueness of the expansion for every $b > 0$. For this, suppose b is an integer with two expressions

$$b = r_n a^n + \cdots + r_1 a + r_0 = s_m a^m + \cdots + s_1 a + s_0,$$

where $0 \leq r_0, r_1, \ldots, r_n < a$, $r_n \neq 0$, $0 \leq s_0, s_1, \ldots, s_m < a$, $s_m \neq 0$. We must show that $r_0 = s_0, r_1 = s_1, \ldots, r_n = s_n$ and $m = n$. As a first step, let us rewrite the two expressions for b as

$$b = (r_n a^{n-1} + \cdots + r_1)a + r_0 = (s_m a^{m-1} + \cdots + s_1)a + s_0.$$

Now, both of these express the division algorithm for b and a; so, by uniqueness of the division algorithm, we have $r_0 = s_0$ and

$$r_n a^{n-1} + \cdots r_2 a + r_1 = s_m a^{m-1} + \cdots + s_2 a + s_1.$$

Applying the same procedure to these two expressions gives $r_1 = s_1$ and

$$r_n a^{n-2} + \cdots + r_2 = s_m a^{m-2} + \cdots + s_2.$$

This procedure may be repeated as many times as necessary. In particular, if $m = n$, then we obtain, sequentially, $r_0 = s_0, r_1 = s_1, \ldots, r_m = s_m$, and the proof of uniqueness is complete in this case. On the other hand, if $m \neq n$, we may suppose without loss of generality, that $m < n$. Then after $m + 1$ steps of our process, we have $r_0 = s_0, r_1 = s_1, \ldots, r_m = s_m$, and

$$r_n a^{n-m} + \cdots + r_{m+2} a^2 + r_{m+1} a = 0.$$

Each term on the left side is greater than or equal to 0. Furthermore, $r_n > 0$ and $n - m > 0$ (because $n > m$), so the leading term $r_n a^{n-m}$ is greater than 0. Thus, the left side is greater than 0, a contradiction. Because the assumption that $m \neq n$ leads to a contradiction, we conclude that $m = n$, and the proof is complete. ▌

This theorem, which is concerned with what is known as the **representation** of b in **base** a (or with **radix** a), is of immense importance. Among other things, it underlies our standard notation for integers, and this notation determines the rules according to which we compute with integers. This is not a trivial matter. The Romans had a rather cumbersome notation for integers (nowadays, we refer to its a "Roman numerals") but it was completely unsatisfactory for treating addition or multiplication; for example, in this system, how does one compute XVIII + LIX or XVIII · LIX? By contrast, our "modern" notation is so incisive that we can teach elementary school children how to add, subtract, multiply, and divide integers, even though they (and, perhaps, their teachers) may not understand why the techniques are valid.

Historically, man learned how to count and deal with numbers long ago. Consider, for example, the following translation of a passage from Genesis:

> ... And Methuselah lived seven and eighty years and one hundred years, and begot Lamech. And Methuselah lived after he begot Lamech two and eighty years and seven hundred years, and begot sons and daughters. And all the days of Methuselah were nine and sixty years and nine hundred years, and he died. And Lamech lived two and eighty years and one hundred years, and begot a son. And he called his name Noah, saying; this one shall comfort us in our work and in the toil of our hands, which comes from the earth which the Lord has cursed. And Lamech lived after he begot Noah five and ninety years and five hundred years, and begot sons and daughters. And all the days of Lamech were seven and seventy years and seven hundred years, and he died ....

Thus, we see that even in "ancient" times it was known how to express numbers (that is, positive integers) in words, and also how to add. Unfortunately, there was no notation available that facilitated computation with integers. Presumably, computations had to be done by brute force. For example, to add one hundred eighty two and five hundred ninety five one takes a pile of one hundred eighty two stones and another pile of five hundred ninety five stones, combines them into a single pile and counts—ending up with seven hundred seventy seven. Actually, one does not have to do the counting in so primitive a fashion. As the biblical terminology indicates, numbers involve certain groupings, and these simplify the addition. Thus, the first collection of stones consists of: one group of a hundred, one group of eighty, and one group of two, while the second collection of stones consists of: five groups of a hundred, one group of ninety, and one group of five. Combining these gives: six groups of a hundred, one group of eighty plus ninety, and one group of seven. Now the group of eighty plus ninety can be organized into one group of a hundred and one group of seventy. So in the end there are seven groups of a hundred, one group of seventy, and one group of seven!

What does multiplication involve when there is no satisfactory notation for numbers and they are given solely in words? A single illustration should suffice. Suppose we want to multiply twenty three by thirty seven; this means we must count the number of objects in a collection consisting of thirty seven groups with twenty three objects in each group—or equivalently, we need to count the number of elements in a rectangular array with twenty three rows and thirty seven columns. (Of course, the roles of rows and columns may be interchanged.) This is an onerous task, especially for large numbers, but it can be simplified by "cutting-up" the rectangle judiciously.

1-8-3. Notation. Having fussed a bit with numbers when a satisfactory notation is lacking, we turn to a quick sketch of how and why our standard notation for numbers works.

We start from **numbers given only by words**. Let us fix $a =$ ten. The theorem on radix representation says that any positive integer b can be written uniquely in the form

$$b = r_n a^n + r_{n-1} a^{n-1} + \cdots + r_1 a + r_0 \qquad (*)$$

where $0 \leq r_0, r_1, \ldots, r_n < a$ and $r_n \neq 0$. Of course, all the symbols here represent words and integers. The sequence of r's clearly determines b completely, and each of the r's is a number less than ten. We introduce, therefore, the symbols "0, 1, 2, 3, 4, 5, 6, 7, 8, 9" for the numbers from zero through nine, respectively. (Actually, zero, which indicates an "absence of number," requires special treatment, but we are careless and lump it in with the others.) Thus, when we associate with b the sequence

$$r_n r_{n-1} \cdots r_1 r_0$$

as shorthand for $(*)$, we are writing b as a sequence of symbols of type 0, 1, 2, 3, 4, 5, 6, 7, 8, 9 and such that the first one (that is, the one on the left) is not equal to 0. This sequence of symbols is normally referred to as the number itself, whereas it is really a representation (expression, or expansion) of the number b in the base a. Every number has a unique expression of this form. The expansion for one, two, ..., nine are clear—namely, 1, 2, ..., 9, respectively. What is the expansion for $a =$ ten? Since

$$a = 1 \cdot a + 0,$$

we see that

$$a = \text{ten} = 10.$$

What is the notation for one hundred? Since one hundred is the number of objects in a ten by ten array, we have

$$\text{one hundred} = \text{ten} \cdot \text{ten} = 10 \cdot 10 = a \cdot a = a^2 = 1 \cdot a^2 + 0 \cdot a + 0$$

so that

$$\text{one hundred} = 10 \cdot 10 = 100.$$

Similarly, we may see that the notation for one thousand is 1000, and so on.

In general, $r_n r_{n-1} \cdots r_1 r_0$ represents

$$r_n(10)^n + r_{n-1}(10)^{n-1} + \cdots + r_1(10) + r_0$$

so it follows that $(10)^n$ whose expansion is surely

$$(10)^n = 1(10)^n + 0(10)^{n-1} + \cdots + 0(10) + 0$$

is denoted by

$$(10)^n = 100\cdots 0, \qquad n \geq 1$$

where n zeros appear on the right.

We may note in passing that we are now "entitled" to read numbers (or better, symbols representing numbers) in the usual way, For example, 4352 (which is identified as four, three, five, two, ... the way one lists a telephone number) represents

$$4(10)^3 + 3(10)^2 + 5(10) + 2 = 4(1000) + 3(100) + 5(10) + 2$$

so it is four thousand three hundred fifty two.

Let us turn to the question of how to compute. First of all, we must know how to add and multiply numbers from 0 to 9. This is accomplished by primitive means—that is, by counting. For example, we obtain $6 + 8 = 14$, $9 + 7 = 16$, $6 \cdot 8 = 48$, $9 \cdot 7 = 63$. All these facts may be listed in two tables, one for addition and the other for multiplication; these are the tables that elementary school children used to spend so much time memorizing in the not too distant past.

In one leap forward we can now add or multiply any two integers—this involves use of the tables, and keeping the meaning of our notation clearly in mind. For example.

$$\begin{aligned}
35{,}874 + 6196 &= [3(10)^4 + 5(10)^3 + 8(10)^2 + 7(10) + 4] \\
&\qquad + [6(10)^3 + 1(10)^2 + 9(10) + 6] \\
&= 3(10)^4 + (5 + 6)(10)^3 + (8 + 1)(10)^2 + (7 + 9)(10) + (4 + 6) \\
&= 3(10)^4 + \big(1(10) + 1\big)(10)^3 + 9(10)^2 + \big(1(10) + 6\big)(10) + 1(10) \\
&= 3(10)^4 + 1(10)^4 + 1(10)^3 + 9(10)^2 + 1(10)^2 + 6(10) + 1(10) \\
&= (3 + 1)(10)^4 + 1(10)^3 + (9 + 1)(10)^2 + (6 + 1)(10) \\
&= 4(10)^4 + 1(10)^3 + (10)(10)^2 + 7(10) \\
&= 4(10)^4 + 1(10)^3 + 1(10)^3 + 7(10) \\
&= 4(10)^4 + 2(10)^3 + 0(10)^2 + 7(10) + 0 \\
&= 42{,}070.
\end{aligned}$$

The standard way to write this addition is

$$\begin{array}{r}
35874 \\
+\ \ 6196 \\
\hline
42070
\end{array}$$

and the reader should convince himself that the way he performs this addition is just a compact version of the gory details given above. The same procedure obviously works for any addition problem.

As for multiplication, when we write, for example.

$$\begin{array}{r} 726 \\ \times\ \ 354 \\ \hline 2904 \\ 3630 \\ 2178 \\ \hline 257004 \end{array}$$

this is just a telescoped version (as the reader may verify) of

$$\begin{aligned}
(726) \cdot (354) &= [7(10)^2 + 2(10) + 6] \cdot [3(10)^2 + 5(10) + 4] \\
&= (7 \cdot 3)(10)^4 + (2 \cdot 3)(10)^3 + (6 \cdot 3)(10)^2 \\
&\quad + (7 \cdot 5)(10)^3 + (2 \cdot 5)(10)^2 + (6 \cdot 5)(10) \\
&\quad + (7 \cdot 4)(10)^2 + (2 \cdot 4)(10) + (6 \cdot 4) \\
&= \big(2(10) + 1\big)(10)^4 + 6(10)^3 + \big(1(10) + 8\big)(10)^2 \\
&\quad + \big(3(10) + 5\big)(10)^3 + \big(1(10)\big)(10)^2 + \big(3(10)\big)(10) \\
&\quad + \big(2(10) + 8\big)(10)^2 + 8(10) + 2(10) + 4 \\
&= 2(10)^5 + 1(10)^4 + 7(10)^3 + 8(10)^2 \\
&\quad + 3(10)^4 + 6(10)^3 + 3(10)^2 \\
&\quad + 2(10)^3 + 9(10)^2 + 4 \\
&= 2(10)^5 + (1 + 3)(10)^4 + (7 + 6 + 2)(10)^3 \\
&\quad + (8 + 3 + 9)(10)^2 + 4 \\
&= 2(10)^5 + 4(10)^4 + \big(1(10) + 5\big)(10)^3 + \big(2(10)\big)(10)^2 + 4 \\
&= 2(10)^5 + 4(10)^4 + 1(10)^4 + 5(10)^3 + 2(10)^3 + 4 \\
&= 2(10)^5 + 5(10)^4 + 7(10)^3 + 0(10)^2 + 0(10) + 4 \\
&= 257{,}004.
\end{aligned}$$

The same procedure obviously works for any multiplication problem.

Subtraction presents no serious difficulties; it works very much like addition. The reader can easily choose two integers, use the expanded notation (meaning that the powers of 10 are kept) to subtract the smaller from the larger one, and observe how this leads to the standard notation and procedure for performing the subtraction. Of course, the key thing is to understand "borrowing."

Division is more difficult; it is not quite straightforward, and involves some trial and error. The reader may wish, as a challenge, to explain in detail how the division

$$\begin{array}{r|l} & 124 \\ \hline 37 & 4589 \\ & 37 \\ \hline & 88 \\ & 74 \\ \hline & 149 \\ & 148 \\ \hline & 1 \end{array}$$

comes from something like

$$\begin{aligned} 4(10)^3 + 5(10)^2 + 8(10) + 9 &= (3(10) + 7)(1(10)^2) + 8(10)^2 + 8(10) + 9, \\ 8(10)^2 + 8(10) + 9 &= (3(10) + 7)(2(10)) + 1(10)^2 + 4(10) + 9, \\ 1(10)^2 + 4(10) + 9 &= (3(10) + 7)(4) + 1, \end{aligned}$$

so

$$4(10)^3 + 5(10)^2 + 8(10) + 9 = (3(10) + 7)\big(1(10)^2 + 2(10) + 4\big) + 1.$$

1-8-4. Remarks. The notation discussed above is known as **positional notation** or as a **place value system** because each "digit" (depending on its position or place) is associated with an appropriate power of ten. The final step in the historical development of this notation was the use of a symbol "0" for missing terms. Of course, the choice of $a =$ ten as the base is fairly natural; through the ages, man usually did his counting in terms of tens because he had ten fingers. However, any other choice of $a > 1$ as a base is valid and feasible—it leads to the same kind of notation and computational procedures.

Let us illustrate by taking $a =$ seven. There is no harm in writing this as $a = 7$; in general, instead of naming numbers by words, we find it convenient to write them in the common, base ten, notation. According to the theorem on radix representation 1-8-2, every positive number b has a unique expression

$$b = r_n \cdot 7^n + r_{n-1} \cdot 7^{n-1} + \cdots + r_1 \cdot 7 + r_0$$

where $0 \le r_0, r_1, \ldots, r_n < 7$, $r_n \ne 0$. We abbreviate this to

$$b = (r_n r_{n-1} \cdots r_1 r_0)_{\text{seven}}$$

or

$$b = (r_n r_{n-1} \cdots r_1 r_0)_7$$

and call it the **representation of** (or **expansion of**, or **expression for**) b **in base 7.** Note that in the base 7 expansion of a number only the symbols (or digits) 0, 1, 2, 3, 4, 5, 6 can appear; an expression like $(207{,}185)_7$ is meaningless.

The subscript seven, or 7, in the notation is essential. It signifies in which base we are working. As a general rule, when an expression is given without subscript, the base ten will be understood. Thus 2156 means $2(10)^3 + 1(10)^2 + 5(10) + 6$, while $(2156)_7$ means $2(7)^3 + 1(7)^2 + 5(7) + 6$—clearly, these are different numbers. As a matter of fact $(2156)_7 = 2(343) + 1(49) + 5(7) + 6 = 776$ (by which is meant $7(10)^2 + 7(10) + 6$.

We know that every number has a unique representation in base 7; can we actually find it? Fortunately, to settle this question it is only necessary to recall some things that have already been done. The existence part of the theorem on radix representation 1-8-2 was proved in a "constructive" fashion—namely, we produced (that is, constructed) the r's of the expansion by listing the remainders that arise by repeated application of the division algorithm. This gives a detailed recipe for finding the base 7 (or any other base) representation of a number. Even more, this recipe has already been illustrated in 1-8-1, where in one of the examples we arrived at

$$489 = 1 \cdot 7^3 + 2 \cdot 7^2 + 6 \cdot 7 + 6$$

—which says that

$$489 = (1266)_7 .$$

In other words, the number four hundred eighty nine (which is written as 489 in the base ten) has the representation $(1266)_7$ in base 7.

As a trivial example of the recipe, we apply it to 6 and obtain $6 = 0 \cdot 7 + 6$; hence,

$$6 = (6)_7 .$$

Of course, this is also obvious directly—after all, the unique expression for 6 in terms of powers of 7 is clearly

$$6 = \cdots + 0 \cdot 7^2 + 0 \cdot 7 + 6.$$

In exactly the same way, we see that

$$1 = (1)_7, \qquad 2 = (2)_7, \qquad 3 = (3)_7, \qquad 4 = (4)_7, \qquad 5 = (5)_7 .$$

Let us go one step further and find the expressions for 7 and 8 in base 7. There is no need to use the recipe because the expressions are obviously

$$7 = 1 \cdot 7 + 0 \quad \text{and} \quad 8 = 1 \cdot 7 + 1$$

so

$$7 = (10)_7 \quad \text{and} \quad 8 = (11)_7 .$$

In order to compute in base 7 we must first know how to operate with the digits 0, 1, 2, 3, 4, 5, 6, which are the only ones used in base 7. But this is easy; for example,

$$(5)_7 + (6)_7 = 5 + 6 = 11 = 1 \cdot 7 + 4 = (14)_7$$
$$(5)_7 \cdot (6)_7 = 5 \cdot 6 = 30 = 4 \cdot 7 + 2 = (42)_7$$

In fact, it is easy to make tables for addition and multiplication which contain all the required data.

addition in base 7

+	0	1	2	3	4	5	6
0	0	1	2	3	4	5	6
1	1	2	3	4	5	6	10
2	2	3	4	5	6	10	11
3	3	4	5	6	10	11	12
4	4	5	6	10	11	12	13
5	5	6	10	11	12	13	14
6	6	10	11	12	13	14	15

multiplication in base 7

·	0	1	2	3	4	5	6
0	0	0	0	0	0	0	0
1	0	1	2	3	4	5	6
2	0	2	4	6	11	13	15
3	0	3	6	12	15	21	24
4	0	4	11	15	22	26	33
5	0	5	13	21	26	34	42
6	0	6	15	24	33	42	51

It is understood that all the entries in the tables should really include the subscript 7. As for the digits 0, 1, 2, 3, 4, 5, 6, these can be viewed either in base ten or base seven without affecting anything, because if a is one of these digits, then, as seen above, $a = (a)_7$. In particular, both $5 \cdot 5 = (34)_7$, and $(5)_7 \cdot (5)_7 = (34)_7$ may be read off from the table.

1-8-5. Examples. We give several examples concerning numbers in various bases.

(1) What numbers are represented by $(23014)_5$ and $(10110101)_2$?

Answer:

$$
\begin{aligned}
(23014)_5 &= 4 + 1 \cdot 5 + 0 \cdot 5^2 + 3 \cdot 5^3 + 2 \cdot 5^4 = 1634, \\
(10110101)_2 &= 1 + 0 \cdot 2 + 1 \cdot 2^2 + 0 \cdot 2^3 \\
&\qquad + 1 \cdot 2^4 + 1 \cdot 2^5 + 0 \cdot 2^6 + 1 \cdot 2^7 = 181.
\end{aligned}
$$

(2) Find the expression for 40152 in base 7, and also in base 6.

Answer: The division algorithm recipe takes the form

$$
\begin{aligned}
40152 &= 5736 \cdot 7 + 0, \\
5736 &= 819 \cdot 7 + 3, \\
819 &= 117 \cdot 7 + 0, \\
117 &= 16 \cdot 7 + 5, \\
16 &= 2 \cdot 7 + 2, \\
2 &= 0 \cdot 7 + 2,
\end{aligned}
$$

so that

$$40152 = (225030)_7 .$$

It is convenient to arrange the repeated divisions as follows:

$$
\begin{array}{r|r:l}
7 & 40152 & \\
\hline
 & 5736 & 0 \\
 & 819 & 3 \\
 & 117 & 0 \\
 & 16 & 5 \\
 & 2 & 2 \\
 & 0 & 2
\end{array}
$$

where the remainders are placed to the right of the dotted line. Thus, to express 40152 in base 6, we compute

$$
\begin{array}{r|r:l}
6 & 40152 & \\
\hline
 & 6692 & 0 \\
 & 1115 & 2 \\
 & 185 & 5 \\
 & 30 & 5 \\
 & 5 & 0 \\
 & 0 & 5
\end{array}
$$

and conclude that

$$40152 = (505520)_7 .$$

(3) Perform the following additions:
(*i*) $(10011101)_2 + (1110110)_2$, (*ii*) $(12212)_3 + (21021)_3$,
(*iii*) $(2163)_7 + (4565)_7$.

Solution: We have in the first instance

$$\begin{array}{r} (10011101)_2 \\ +\ (1110110)_2 \\ \hline (100010011)_2 \end{array}$$

which is a reflection of what happens when everything is written out in terms of powers of 2. One may then check this addition by transferring to the more familiar base 10—and see that indeed $157 + 118 = 275$.

In case (*ii*), we have

$$\begin{array}{r} (12212)_3 \\ +\ (21021)_3 \\ \hline (111010)_3 \end{array}$$

which, in base 10, says that $158 + 196 = 354$.

Turning to case (*iii*),

$$\begin{array}{r} (2163)_7 \\ +\ (4565)_7 \\ \hline (10061)_7 \end{array}$$

which is another version of $780 + 1664 = 2444$.

(4) Perform the following subtractions:
(*i*) $(23014)_5 - (4123)_5$, (*ii*) $(505520)_6 - (41432)_6$,
(*iii*) $(4562)_7 - (2165)_7$.

Solution: We have

$$\begin{array}{r} (23014)_5 \\ -\ (4123)_5 \\ \hline (13341)_5 \end{array} \qquad \begin{array}{r} (505520)_6 \\ -\ (41432)_6 \\ \hline (424044)_6 \end{array} \qquad \begin{array}{r} (4562)_7 \\ -\ (2165)_7 \\ \hline (2364)_7 \end{array}$$

and in base 10 these become $1634 - 538 = 1096$, $40152 - 5564 = 34588$, $1661 - 782 = 879$, respectively.

(5) Perform the following multiplications:
(*i*) $(11001)_2 \cdot (1101)_2$, (*ii*) $(1342)_5 \cdot (314)_5$,
(*iii*) $(564)_7 \cdot (403)_7$.

Solution: In case (*i*), we write (in base 2)

$$\begin{array}{r}
11001 \\
\times \quad 1101 \\
\hline
11001 \\
110010 \\
11001 \\
\hline
101000101
\end{array}$$

while the corresponding multiplication in base 10 is $25 \cdot 13 = 325$.

The work for (*ii*) looks as follows (with base 5 understood)

$$\begin{array}{r}
1342 \\
\times \quad 314 \\
\hline
12023 \\
1342 \\
10131 \\
\hline
1044043
\end{array}$$

and in base 10 this translates to $222 \cdot 84 = 18648$.

As for (*iii*), working in base 7, we have

$$\begin{array}{r}
564 \\
\times \quad 403 \\
\hline
2355 \\
32520 \\
\hline
330555
\end{array}$$

and in base 10 this translates to $(291)(199) = 57909$.

(6) Carry out the division algorithm (that is, divide) for the pairs of numbers

(*i*) $(11101)_2$ and $(111011000111)_2$
(*ii*) $(23)_7$ and $(225030)_7$.

Solution: The division in base 2 looks like

$$\begin{array}{r|l}
 & 10000010 \\
\cline{2-2}
11101 & 111011000111 \\
 & 11101 \\
\cline{2-2}
 & 00100011 \\
 & 11101 \\
\cline{2-2}
 & 1101
\end{array}$$

Thus we have

$$(111011000111)_2 = (10000010)_2 \cdot (11101)_2 + (1101)_2$$

which is another way of saying

$$3783 = (130)(29) + 13.$$

As for part (*ii*), the division in base 7 takes the form

```
         6612
   23 | 225030
        204:::
         210::
         204::
          33:
          23:
          100
           46
           21
```

We have, therefore,

$$(225030)_7 = (6612)_7 \cdot (23)_7 + (21)_7$$

which indeed corresponds to the base 10 equation

$$40152 = (2361) \cdot (17) + 15$$

1-8-6. Remark. Instead of looking only at bases smaller than 10, as has been done so far, let us consider a base greater than 10—namely, the base 12. In order to represent numbers in base 12 we need symbols for the integers which are greater than or equal to 0 and less than twelve. Let us use the standard digits 0, 1, . . . , 9 for the integers from zero through nine, respectively, and then introduce the symbols

$$\tau \quad \text{for ten,} \qquad \varepsilon \quad \text{for eleven.}$$

Maintaining our convention that numbers without subscripts are in base 10, we may note that

$$\begin{array}{llll} 10 = (\tau)_{12}, & 11 = (\varepsilon)_{12}, & 12 = (10)_{12}, & 13 = (11)_{12}, \ldots \\ 21 = (19)_{12}, & 22 = (1\tau)_{12}, & 23 = (1\varepsilon)_{12}, & 24 = (20)_{12}, \end{array}$$

and so on. Of course, the unadorned symbols τ and ε have no meaning in base ten; they exist only in the context of base twelve.

To compute in base 12, one must know how to operate with the digits 0, 1, 2, . . . , 9, τ, ε. This information may be gathered into tables for addition and multiplication, among whose entries the following facts will be found.

$$(\tau)_{12} + (\tau)_{12} = (18)_{12}, \qquad (\tau)_{12} + (\varepsilon)_{12} = (19)_{12}, \qquad (\varepsilon)_{12} + (\varepsilon)_{12} = (1\tau)_{12},$$
$$(\tau)_{12} \cdot (\tau)_{12} = (84)_{12}, \qquad (\tau)_{12} \cdot (\varepsilon)_{12} = (92)_{12}, \qquad (\varepsilon)_{12} \cdot (\varepsilon)_{12} = (\tau 1)_{12}.$$

These are easy to derive; for example,

$$\begin{aligned}(\tau)_{12} + (\varepsilon)_{12} &= 10 + 11 = 21 = 1 \cdot 12 + 9 = (19)_{12},\\ (\tau)_{12} \cdot (\tau)_{12} &= 10 \cdot 10 = 100 = 8 \cdot 12 + 4 = (84)_{12},\\ (\varepsilon)_{12} \cdot (\varepsilon)_{12} &= 11 \cdot 11 = 121 = 10 \cdot 12 + 1 = (\tau 1)_{12}.\end{aligned}$$

The connections between base 10 and base 12 work just like they did earlier when bases other than 12 were used. For example, to find the number $(\tau 5 \varepsilon 7)_{12}$ we write,

$$\begin{aligned}(\tau 5 \varepsilon 7)_{12} &= \tau(12)^3 + 5(12)^2 + \varepsilon(12) + 7\\ &= 10(1728) + 5(144) + 11(12) + 7\\ &= 18139.\end{aligned}$$

Thus, $(\tau 5 \varepsilon 7)_{12} = 18139$, and we may check this by finding the expression for 18139 in base 12, namely,

$$\begin{array}{r|r|l} 12 & 18139 & \\ & 1511 & 7 \\ & 125 & 11 \\ & 10 & 5 \\ & 0 & 10 \end{array}$$

so because the remainders are 7, 11, 5, 10 the base 12 notation for 18139 is indeed $(\tau 5 \varepsilon 7)_{12}$.

As illustrations of how one adds and subtracts in base 12, we have

$$\begin{array}{r} (\tau 5 \varepsilon 7)_{12} \\ +\ (469\tau)_{12}, \\ \hline (13095)_{12} \end{array} \qquad \begin{array}{r} (\tau 5 \varepsilon 7)_{12} \\ -\ (469\tau)_{12}. \\ \hline (5\varepsilon 19)_{12} \end{array}$$

Incidentally, the reader may check that in base 10 these become 18139 + 7894 = 26033 and 18139 − 7894 = 10245.

Somewhat more effort is required to do a multiplication in base 12; for example (with subscripts dropped momentarily for convenience)

$$\begin{array}{r} 27\varepsilon\tau \\ \times \quad \tau 3 \varepsilon \\ \hline 253\tau 2 \\ 7\varepsilon\varepsilon 6 \\ 227\tau 4 \\ \hline 2363742 \end{array}$$

so that $(27\varepsilon\tau)_{12} \cdot (\tau 3\varepsilon)_{12} = (2363742)_{12}$. In base 10 this says: $4606 \cdot 1487 = 6849122$.

Finally, let us divide $(\tau 5\varepsilon 7)_{12}$ by $(2\varepsilon)_{12}$, which corresponds to dividing 18139 by 35. The work looks like

$$\begin{array}{r|l} & 372 \\ \hline 2\varepsilon & \tau 5\varepsilon 7 \\ & 89 \\ \hline & 18\varepsilon \\ & 185 \\ \hline & 67 \\ & 5\tau \\ \hline & 9 \end{array}$$

so that

$$(\tau 5\varepsilon 7)_{12} = (372)_{12} \cdot (2\varepsilon)_{12} + (9)_{12}.$$

In base 10 this becomes

$$18139 = 518 \cdot 35 + 9.$$

1-8-7. Nim. Consider the following game (known as Nim) for two players. At the start, the players are presented with three separate piles of chips, stones, or what have you. The number of objects in each pile is arbitrary, but greater than 0. Each player, in turn, then removes as many objects as he wishes (possibly even all of them) from *exactly one* of the piles, and throws them away. Of course, at his turn, a player cannot stand pat—he must remove one or more objects. The player who removes the last object wins.

This game is far from trivial, as the reader will surely discover by playing it. However, it is, surprisingly perhaps, subject to precise mathematical analysis.

At his turn, a player is confronted with a triplet of nonnegative integers $\{n_1, n_2, n_3\}$, where n_i represents the number of objects in the ith pile for $i = 1, 2, 3$. His move amounts to changing one of the n_i to a strictly smaller nonnegative integer and keeping the other two fixed. Now, let us write out the base 2 expansions for n_1, n_2, n_3, one under the other, with the columns lined up, and then count the number of 1's in each column. For example, if $\{n_1, n_2, n_3\} = \{46, 27, 35\}$ we write

$$\begin{array}{r} 46 = 101110 \\ 27 = 11011 \\ 35 = 100011 \\ \hline \text{count of 1's: } 212132 \end{array}$$

(the count of 1's is simply a list which, for every column, tells how many 1's it contains) and if $\{n_1, n_2, n_3\} = \{50, 20, 38\}$, we write

$$\begin{aligned} 50 &= 110010 \\ 20 &= 10100 \\ 38 &= 100110 \\ \hline \text{count of 1's: } & 220220. \end{aligned}$$

We say that $\{n_1, n_2, n_3\}$ is an **even position** when the count of 1's contains only even integers (0 is considered to be even). By an **odd position** we mean one that is not even; thus, an odd position is one for which the count of 1's contains at least one odd number. For example, {46, 27, 35} is an odd position, {50, 20, 38} is an even position, and {0, 0, 0} is an even position. Clearly, a position is either odd or even, but not both.

Once we have a classification for positions, it is useful to classify moves. We say that a player has made a **good move** if he leaves his opponent with an even position, and a **bad move** if he leaves his opponent with an odd position. For example, a move that leaves the opponent in the position {18, 3, 17} is a good move, while a move that leaves him with the position {18, 2, 17} is a bad move. Clearly, a move is either good or bad, but not both.

Now that the terminology is in place, we turn to two crucial facts, whose proofs are left to the reader.

(1) If a player is confronted with an even position, then any move he makes is a bad move. (Before undertaking the proof, the reader should find it helpful to try this out on the even position {50, 20, 38}.)

(2) If a player is confronted with an odd position, then he can make a good move—in other words, by selecting his move judiciously he can leave his opponent with an even position. (Before undertaking the proof, the reader should find it helpful to try this out on the odd position {46, 27, 35}. A move to the position {46, 13, 35} is a good move; in fact, it is the only good move a player can make when confronted with {46, 27, 35}.)

Suppose a game is in progress, and at some point one of the players (call him A) is confronted with an odd position. According to (2), he can make a good move and leave B with an even position. Then, according to (1), B is forced to make a bad move; in other words, he cannot avoid leaving A in an odd position. Therefore, if A knows what he is doing, he can "guarantee" that every move he makes is a good move and that every move B makes is a bad move. The last move in the game, namely the winning move, is one that leaves the opponent with the even position {0, 0, 0}—so it is a good move. Since B is forced to make only bad moves, he never gets the opportunity to remove the last object (because this is a good move). This means that eventually A wins.

The conclusion is that a player who understands the game is almost certain to win against a player who does not. We have "solved" the game of Nim, an old Chinese game.

1-8-8 / *PROBLEMS*

1. Make addition and multiplication tables for each of the following bases: (*i*) 2, (*ii*) 3, (*iii*) 5, (*iv*) 6, (*v*) 12.

2. Find the expression for 234,567 in each of the following bases: (*i*) 2, (*ii*) 3, (*iii*) 5, (*iv*) 6, (*v*) 7, (*vi*) 12.

3. Find the following numbers:
(*i*) $(1101001101)_2$, (*ii*) $(211020121)_3$,
(*iii*) $(32414)_5$, (*iv*) $(150423)_6$,
(*v*) $(32414)_7$, (*vi*) $(1\tau\varepsilon93)_{12}$.

4. What are the advantages and disadvantages of the binary system (meaning: base 2) as compared with the duodecimal system (meaning: base 12)?

5. Find the expression for
(*i*) $(1101001101)_2$ in base 3, (*ii*) $(32414)_5$ in base 6,
(*iii*) $(32414)_7$ in base 12, (*iv*) $(\tau 3\varepsilon)_{12}$ in base 5,
(*v*) $(41503)_6$ in base 3, (*vi*) $(202021)_3$ in base 7.

6. Perform the following additions, and check the results by transferring to base 10:
(*i*) $(11010101)_2 + (101011111)_2$, (*ii*) $(41343)_5 + (24312)_5$,
(*iii*) $(301246)_7 + (123456)_7$, (*iv*) $(2\tau 7\varepsilon)_{12} + (805\tau)_{12}$.

7. Perform the following subtractions, and check the results by transferring to base 10:
(*i*) $(202021)_3 - (12202)_3$, (*ii*) $(150423)_6 - (41503)_6$,
(*iii*) $(301246)_7 - (123456)_7$, (*iv*) $(805\tau)_{12} - (2\tau 7\varepsilon)_{12}$.

8. Perform the following multiplications, and check the results by transferring to base 10:
(*i*) $(5423)_6 \cdot (453)_6$, (*ii*) $(3246)_7 \cdot (564)_7$,
(*iii*) $(85\tau)_{12} \cdot (\tau 7\varepsilon)_{12}$.

9. Divide (and check your results)
(*i*) $(111)_2$ into $(101101101)_2$, (*ii*) $(43)_5$ into $(41343)_5$
(*iii*) $(38)_{12}$ into $(2\tau 7\varepsilon)_{12}$

10. Find the greatest common divisor of
(*i*) $(212)_3$ and $(20102)_3$, (*ii*) $(416)_7$ and $(3025)_7$.

11. How is the game of Nim affected if there are more than three piles at the start?

12. Can you find a criterion which tells if $(r_n r_{n-1} \cdots r_1 r_0)_7$ is divisible by $(11)_7$? What about divisibility by $(6)_7$?

13. I once knew a farmer who knew how to add and subtract, but had never been taught the multiplication table. However, he did know how to multiply by 2 and divide by 2. To multiply two integers he followed a procedure which we illustrate in the case $97 \cdot 113$.

Divide 97 by 2 to get a quotient of 48, and throw the remainder away. Repeat this for 48, and keep going until a quotient 1 is reached. Starting from 97 write all the quotients, in order, in a column. To the right of this column, write another column with exactly the same number of entries as the first one. This column starts with 113, and each of the subsequent entries is obtained from the one preceding it by multiplying by 2. For each of the even entries in the left-hand column, cross out the corresponding entry of the right-hand column. Adding the remaining entries of the right-hand column gives the product $97 \cdot 113$. In full detail, we have:

97	113
48	~~226~~
24	~~452~~
12	~~904~~
6	~~1808~~
3	3616
1	7232
	10961

so $97 \cdot 113 = 10961$. Explain why this method (which used to be standard operating procedure for Russian peasants) works.

14. Given an integer in base 10, we know how to apply the division algorithm repeatedly in order to find its expression in any other base. Explain how the same idea may be used to translate a number given in an arbitrary base ($\neq 10$) into base 10. For example, given a number in base 7, one divides it, in base 7, by $(13)_7$ (which is the base 7 expression for 10) and keeps using this divisor $(13)_7$ in all the "divisions."

Do Problem 3 by this method.

Miscellaneous Problems

1. Show that for any $n \geq 1$,

(*i*) $3 \mid (2^{2n} - 1)$ (*ii*) $3 \mid (2^{2n-1} + 1)$

(*iii*) $(a - b) \mid (a^n - b^n)$ (*iv*) $(a + b) \mid (a^{2n} - b^{2n})$

(*v*) $(a + b) \mid (a^{2n-1} + b^{2n-1})$ (*vi*) $(a^2 - b^2) \mid (a^{2n} - b^{2n})$

2. If $n \geq 1$ and $(a, b) = 1$ show that $(a^{n+1} + b^{n+1}, a^n + b^n)$ divides $a - b$.

3. Let m and n be positive integers; show that
(*i*) $2^n + 1$ is prime $\Rightarrow n$ is a power of 2;
(*ii*) $2^n - 1$ is prime $\Rightarrow n$ is prime;
(*iii*) $(2^m - 1) \mid (2^n - 1) \Leftrightarrow m \mid n$.

4. Prove that the square of an integer not divisible by 2 or 3 is of the form $12n + 1$.

5. If both a and b are prime to 3, show that $a^2 + b^2$ cannot be a perfect square.

6. If n is odd, prove that there are no prime triplets n, $n + 2$, $n + 4$ (meaning that all three are prime) except 3, 5, 7.

7. Express the cube of any positive integer as the difference of the squares of two integers.

8. Do there exist distinct positive integers x, y, z for which $x + y + z = xyz$?

9. Prove that $n^2 - 6n + 10$ is positive for every integer n.

10. Find the square root of each of the following, in the appropriate base:
(*i*) $(11010010001)_2$, (*ii*) $(2022021)_3$,
(*iii*) $(11441)_6$, (*iv*) $(4621)_7$.

11. Consider the set $T = \{1, 3, 5, \ldots, 2n - 1\}$ and suppose 3^r is the highest power of 3 that belongs to T. Show that 3^r does not divide any other element of T.

12. If a_1, a_2, b_1, b_2 are integers with $a_1 b_2 - a_2 b_1 = \pm 1$ show that $a_1 + a_2$, and $b_1 + b_2$ are relatively prime [in other words, the fraction $(a_1 + a_2)/(b_1 + b_2)$ is in lowest terms].

13. Suppose a and b are positive integers with $a \mid b^2$, $b^2 \mid a^3$, $a^3 \mid b^4, \ldots$, $b^{2n} \mid a^{2n+1}$, $a^{2n+1} \mid b^{2n+2}, \ldots$ (ad infinitum). Show that $a = b$.

14. Show that $3n^5 + 5n^3 + 7n$ is divisible by 15 for every $n \in \mathbf{Z}$.

15. State and prove a criterion that tells when an integer is divisible by 7.

16. Can you find an integer n such that $7n$ has
(*i*) all its digits equal to 1? (*ii*) all its digits equal to 3?

17. Consider the number $(800)! = 800 \cdot 799 \cdot 798 \cdots 2 \cdot 1$. When this number is multiplied out, how many zeros are there at the end?

18. Find 25 consecutive integers all of which are composite. More generally, show that there exist consecutive primes (meaning that all integers between them are composite) whose difference is arbitrarily large.

19. Use the v_p's to prove:
(*i*) If $a \mid c$, $b \mid c$, and $(a, b) = 1$ then $ab \mid c$.
(*ii*) If $(a, b) = 1$ and $(a, c) = 1$ then $(a, bc) = 1$.
(*iii*) If $(a, b) = 1$ and $c \mid a$ then $(c, b) = 1$.
(*iv*) If $(a, bc) = 1$ then $(a, b) = (a, c) = 1$.
(*v*) If $(a, b) = 1$ then $(ca, b) = (c, b)$.
(*vi*) If $(a, b) = 1$ then $(ab, c) = (a, c)(b, c)$.

20. (*i*) Show that there are no nonzero integers m and n for which $m^2 = 2n^2$. Why does this imply that $\sqrt{2}$ is "irrational"?
(*ii*) More generally, if $a > 1$ is not the square of an integer, then $m^2 = an^2$ cannot be solved in nonzero integers m and n; hence, $\sqrt{a}$ is irrational.

21. (*i*) Find all pairs of positive integers $\{a, b\}$ such that $(a, b) = 10$ and $[a, b] = 100$.
(*ii*) Find all triples of positive integers $\{a, b, c\}$ such that $(a, b, c) = 10$ and $[a, b, c] = 100$.

22. Let two positive integers d and m be given. Show that there exists a pair of positive integers $\{a, b\}$ for which $(a, b) = d$ and $[a, b] = m$ if and only if $d \mid m$. Moreover, in this situation, the number of such pairs is 2^r, where r is the number of distinct prime factors of m/d.

23. Suppose two positive integers c and d are given. Show that there exists a pair of positive integers $\{a, b\}$ for which:
(*i*) $a + b = c$ and $(a, b) = d \Leftrightarrow d \mid c$, (*ii*) $ab = c$ and $(a, b) = d \Leftrightarrow d^2 \mid c$.
In each case, can you count the number of such pairs?

24. For any integers a, b, c show that
(*i*) $([a, b], [a, c], [b, c]) = [(a, b), (a, c), (b, c)]$
(*ii*) $[a, b, c](ab, bc, ac) = |abc|$
(*iii*) $[a, b, c](a, b, c) \leq |abc|$, and equality holds if and only if a, b, c are relatively prime in pairs.
(*iv*) $(ab, cd) = (a, c)\,(b, d)\left(\dfrac{a}{(a, c)}, \dfrac{d}{(b, d)}\right)\left(\dfrac{c}{(a, c)}, \dfrac{b}{(b, d)}\right)$

25. Exhibit three integers a, b, c such that $(a, b, c) = 1$ and
(*i*) exactly one (*ii*) exactly two (*iii*) all three
of the pairs of integers $\{a, b\}$, $\{a, c\}$, $\{b, c\}$ are relatively prime.

26. (*i*) For $s = |a_1 a_2 \cdots a_n| \neq 0$ show that

$$s = [a_1, a_2, \ldots, a_n]\left(\frac{s}{a_1}, \frac{s}{a_2}, \ldots, \frac{s}{a_n}\right).$$

(*ii*) If $m > 0$ is a common multiple of $a_1, a_2, \ldots, a_n$ (all nonzero) show that:

$$m = [a_1, a_2, \ldots, a_n] \Leftrightarrow \left(\frac{m}{a_1}, \frac{m}{a_2}, \ldots, \frac{m}{a_n}\right) = 1.$$

27. (*i*) Consider the polynomial $f(x) = x^2 + x + 17$. Verify that $f(n)$ is prime for $n = 1, 2, \ldots, 16$ but $f(17)$ is not prime,

(*ii*) Consider the polynomial $f(x) = x^2 - x + 41$. Verify that $f(n)$ is prime for $n = 1, 2, \ldots, 40$ but $f(41)$ is not prime.

(*iii*) Find the smallest positive integer n for which $f(n) = n^2 - 79n + 1601$ is composite.

28. Prove that there exists no polynomial $f(x)$ with integer coefficients such that $f(n)$ is prime for every positive integer n.

29. Discuss the problem of solving the two simultaneous linear diophantine equations

$$a_1 x + b_1 y = c_1,$$
$$a_2 x + b_2 y = c_2.$$

30. Prove that the linear diophantine equation $a_1 x_1 + a_2 x_2 + \cdots + a_n x_n = c$ has a solution $\Leftrightarrow$ $(a, \ldots, a_n)$ divides c. Sketch a method for finding all solutions.

31. State and prove a necessary and sufficient condition that the linear diophantine equation $ax + by = c$ have an infinite number of positive solutions.

32. Suppose a and b are relatively prime positive integers, and consider the linear diophantine equation $ax + by = c$.

(*i*) Give two proofs, one geometric, one algebraic, that there cannot be an infinite number of positive solutions.

(*ii*) If $c > ab$ show that a positive solution exists.

(*iii*) For each $n \geq 0$ show how to construct a nontrivial example of this type that has exactly n positive solutions.

33. Find the smallest and the biggest integer c for which the linear diophantine equation $5x + 7y = c$ has exactly nine positive solutions.

34. For $n \geq 1$, let $\tau(n)$ denote the number of positive divisors of n and let $\sigma(n)$ denote the sum of all the positive divisors of n.

(*i*) Find $\tau(n)$ and $\sigma(n)$ for $n = 11, 11^5, (11)(13), (11)^5(13)^7, 6840$.

(*ii*) For $m = 220$, $n = 284$ show that $\sigma(m) = \sigma(n) = m + n$. Such integers m and n are said to be **amicable numbers**.

(*iii*) If m and n are any relatively prime positive integers then

$$\tau(mn) = \tau(m) \cdot \tau(n), \qquad \sigma(mn) = \sigma(m) \cdot \sigma(n).$$

(*iv*) Prove the following formulas for any $n \geq 1$:

$$\tau(n) = \prod_p (1 + v_p(n)), \quad \sigma(n) = \prod_p \left(\frac{p^{1+v_p(n)} - 1}{p - 1}\right).$$

35. For $n \geq 1$ show that: $\tau(n)$ is even $\Leftrightarrow$ n is not a perfect square. Moreover, in this situation

$$\sigma(n) = n^{\tau(n)/2}.$$

What happens in the case where $\tau(n)$ is odd?

36. (*i*) For every integer $r > 1$ there are an infinite number of positive integers n with $\tau(n) = r$.
(*ii*) What is the smallest positive n satisfying $\tau(n) = 15$?

37. (*i*) Find all $n \geq 1$ which satisfy $\sigma(n) = s$ when $s = 10$.
(*ii*) Do the same thing for $s = 24$ and $s = 72$.
(*iii*) Does there exist an $s > 2$ for which $\sigma(n) = s$ has no solution?

38. (*i*) Can you express 1547 as a difference of two squares? In how many ways can you do so?
(*ii*) Do the same for 1768.
(*iii*) Describe a general procedure for settling this question for an arbitrary integer n.
(*iv*) Show that $n > 0$ can be expressed as a difference of two squares if and only if it is of form $2k + 1$ or $4k$. Moreover, if n is an odd prime its expression as a difference of squares is unique.

39. Find $n \geq 1$ for which $n(n + 180)$ is a perfect square. How many such values of n can you find?

40. Show that the positive integer n can be expressed as a sum of consecutive integers [that is, $n = m + (m + 1) + \cdots + (m + k)$] if an only if n is not a power of 2.

41. If the positive integer n is odd, show that the sum of the first n integers divides their product. What happens to this result when n is even?

42. Do there exist integers $a > b > 0$, $n > 1$ for which $a^n - b^n$ divides $a^n + b^n$?

43. Prove that $1 + \frac{1}{2} + \cdots + \frac{1}{n}$ is not an integer for any $n > 1$.

44. Suppose $a > 1$ and $n > 1$. If $a^n - 1$ is prime, show that $a = 2$ and n is prime. Such primes $a^n - 1$ are known as **Mersenne primes.** Find a few of them. [Note: A number of form $2^p - 1$ with p prime need not be prime; for example, $2^{11} - 1$ is composite.]

45. (*i*) Suppose $a > 1$ and $n \geq 1$. If $a^n + 1$ is prime, show that a is even and n is a power of 2.

(*ii*) A prime of form $2^{2^m} + 1$ is said to be a **Fermat prime**. Find a few of them. [Note: a number of form $2^{2^m} + 1$ need not be prime; for example, $2^{2^5} + 1$ is composite—show, in fact, that it is divisible by 641.]

46. (*i*) Suppose $a > 1$ and $m > n > 1$. Show that

$$(*) \quad a^{2^n} + 1 \text{ divides } a^{2^m} - 1,$$

$$(**) \quad (a^{2^m} + 1, a^{2^n} + 1) = \begin{cases} 1 & \text{if } a \text{ is even}, \\ 2 & \text{if } a \text{ is odd}. \end{cases}$$

(*ii*) Use the infinite sequence

$$u_1 = 2^{2^1} + 1, u_2 = 2^{2^2} + 1, u_3 = 2^{2^3} + 1, \ldots, u_n = 2^{2^n} + 1, \ldots$$

to prove that the number of primes is infinite.

47. Given two integers a and b, it is customary to compute their greatest common divisor via the Euclidean algorithm. This involves repeated use of the division algorithm. Show that the number of steps (meaning: uses of the division algorithm) needed to compute (a, b) is at most five times the number of digits in (the base 10 expression for) the smaller number.

48. (*i*) Suppose we are given a balance scale along with the five known weights 1, 2, 4, 8, 16. Show that we can determine the weight of any object whose weight is an integer less than or equal to 31 by placing an appropriate choice of the given weights in one pan.

(*ii*) If we are given a sixth known weight, 32, show that any integral weight less than or equal to 63 can be determined.

(*iii*) Generalize these facts to n weights. How is this related to 1-8-2?

49. (*i*) Suppose we are given a balance scale and the four known weights 1, 3, 9, 27. Show that by placing some of these judiciously in one pan, or in both, we can determine the weight of any object whose weight is an integer less than or equal to 40.

(*ii*) If we are given a fifth known weight, 81, show that, by using both pans, any integral weight less than or equal to 121 can be determined.

(*iii*) Generalize these facts to n weights.

50. If $a > 1$ is fixed show, that any $b \in \mathbf{Z}$ can be expressed uniquely in the form $b = qa + r$ where

$$-\frac{a-1}{2} \leq r \leq \frac{a-1}{2} \quad \text{when } a \text{ is odd},$$

$$-\frac{a}{2} < r \leq \frac{a}{2} \quad \text{when } a \text{ is even}.$$

This may be referred to as the "least absolute value remainder" version of the division algorithm.

51. (*i*) By repeated use of the least absolute value version of the division algorithm, starting with integers a and b, one is led to a new version of the Euclidean algorithm. [Note: If a remainder r_i is negative, one uses $|r_i|$ as the divisor in the next step.] Prove that this process also determines (a, b).

(*ii*) Use this "new" Euclidean algorithm to compute (a, b) when $a = 63020$, $b = 76084$.

52. If $a > 1$ show that any $b > 0$ can be expressed uniquely in the form

$$b = r_n a^n + r_{n-1} a^{n-1} + \cdots + r_1 a + r_0,$$

with $r_n > 0$ and

$$-\frac{a-1}{2} \leq r_0, r_1, \ldots, r_n \leq \frac{a-1}{2} \qquad \text{when } a \text{ is odd,}$$

$$-\frac{a}{2} < r_0, r_1, \ldots, r_n \leq \frac{a}{2} \qquad \text{when } a \text{ is even.}$$

In particular, if $a = 3$ then any positive integer b can be expressed uniquely as a sum of distinct powers of 3 with coefficients 1, 0, or -1, and with leading coefficient 1; more precisely,

$$b = 3^n + r_{n-1} \cdot 3^{n-1} + r_{n-2} \cdot 3^{n-2} + \cdots + r_1 \cdot 3 + r_0,$$

where $-1 \leq r_0, r_1, \ldots, r_{n-1} \leq 1$. How is this related to Problem 49 concerning the use of integral weights on both sides of a balance scale?

II

RINGS AND DOMAINS

In the preceding chapter, we assumed that the integers were known and proceeded to investigate additional properties of **Z** such as: division algorithm, Euclidean algorithm, unique factorization, greatest common divisor, least common multiple, congruence, and radix representation. At appropriate places in a number of proofs, we made use of certain commonly accepted properties of **Z**. Thus, our approach was somewhat informal in the sense that the axioms or assumptions about the integers were never formalized or made explicit.

In this chapter, we adopt another point of view and examine the facts about **Z** that, heretofore, were assumed to be known. As a first step we introduce several algebraic axioms about addition and multiplication in an arbitrary set—axioms which are satisfied in **Z**, in $\mathbf{Z}_m$, and also in many other familiar systems. Step by step, we introduce additional axioms and discuss some of their consequences. In the end, we arrive at a set of axioms that describes **Z** completely!

2-1. Rings: Elementary Properties

In this section, we begin the formal axiomatic approach to algebraic systems; this is the point of view of modern algebra. As a first step, we introduce a very general algebraic object whose definition is motivated, in part, by the properties of $\mathbf{Z}$ and $\mathbf{Z}_m$.

2-1-1. Definition. A **ring** is a nonempty set R, whose elements we denote by $a, b, c, \ldots, x, y, z, \ldots$, together with two operations, **addition** (usually denoted by $+$) and **multiplication** (usually denoted by $\cdot$), and such that the following properties hold:

A1: closure under addition. To any pair of elements a and b of R (in the given order) there is associated an element $a + b$ in R. This element $a + b$ is called the **sum** of a and b; in words, $a + b$ is read as a **plus** b.

A2: associative law for addition. For any elements a, b, c of R we have

$$(a + b) + c = a + (b + c).$$

A3: identity for addition. There exists an element 0 in R such that

$$a + 0 = a \qquad \text{for all } a \in R.$$

A4: inverse for addition. Given any $a \in R$ there exists an element $\bar{a} \in R$ such that

$$a + \bar{a} = 0.$$

A5: commutative law for addition. For any elements a and b of R we have

$$a + b = b + a.$$

M1: closure under multiplication. To any pair of elements a and b of R (in the given order) there is associated an element $a \cdot b$ in R. This element $a \cdot b$ (which we shall usually write simply as ab) is called the **product** of a and b; in words, $a \cdot b$ is read as a **times** b.

M2: associative law for multiplication. For any elements a, b, c of R we have

$$(ab)c = a(bc).$$

D1: distributive laws. For any a, b, c in R we have

$$a(b + c) = ab + ac \qquad \text{and} \qquad (b + c)a = ba + ca.$$

2-1-2. Discussion. The axioms for a ring may seem numerous and even mysterious, but with constant use they will eventually seem fairly natural. At this stage, it is necessary to make a number of remarks designed to elaborate upon and clarify the definition.

(1) Operations, as we have called them in the very first sentence of the definition, are often referred to as " binary operations." To explain the meaning of this term, let $R \times R$ denote the set of all ordered pairs (a, b) of elements of R; in symbols,

$$R \times R = \{(a, b) \mid a \in R, b \in R\}.$$

Of course, (a, b) has nothing to do with greatest common divisor. Rather, the notation is entirely analogous with the way the familiar x–y plane is formed from two copies of the number line, and each point in the plane is given by the ordered pair of its coordinates (x, y). By a **binary operation** on R we mean any mapping of

$$R \times R \to R$$

(the arrow is read as " into ")—that is, a rule which assigns to each ordered pair $(a, b) \in R \times R$ an element of R. We have chosen to denote the two binary operations on a ring R by "$+$" and "$\cdot$" respectively, in order to maintain the analogy with $\mathbf{Z}$, and also because this is the standard notation. On rare occasions other notation will be used.

(2) Strictly speaking, it is not correct to refer to a ring R. After all, to talk about a ring, the two operations must also be specified. Thus, a more accurate way to refer to a ring would be as $\{R, +, \cdot\}$, where, by convention, the addition operation is listed first. However, we shall almost always use the abbreviated notation, R, because there will be no doubt about the notation for and meaning of addition and multiplication in the ring.

(3) It is really part of the definition of a binary operation that we have closure under the operation. However, we have listed the closure axioms A1 and M1, for addition and multiplication, explicitly, in order to provide emphasis; without these axioms closure tends to be ignored or forgotten.

(4) The sum $a + b$ and the product ab are defined for all choices of a and b in R; in particular, a and b may be the same element of R.

(5) We have $a + b$ defined for all $a, b \in R$, but the expression $a + b + c$ has no meaning at this stage. The binary operation $+$ tells us about adding two elements of R, but it says nothing about adding three elements. One might try adding a and b, and then adding this result and c; the end product would be expressed as $(a + b) + c$ in keeping with the customary usage for parentheses. On the other hand, there is one more way possible for adding a, b, c (in this order)—namely $a + (b + c)$. The associative law for addition, A2,

asserts that $(a + b) + c = a + (b + c)$, so the two computations give the same result. Thus, the associative law allows us to use the notation $a + b + c$, without any ambiguity, for both $(a + b) + c$ and $a + (b + c)$.

Clearly, similar remarks apply for the symbol abc in virtue of the associative law for multiplication, M2.

(6) The symbol "=" (which we read as "equals") appears throughout the definition. What does it mean? More precisely, what is the meaning of $a = b$? For us, the meaning is simply that they are the "same." The symbols may look different from each other, but they refer to the same element of R—they are different names for the same element. Because of this, it is a rule of logic that if a appears in an expression, then it may be replaced by b without affecting the "value" of the expression. This procedure is often carelessly called "substitution"; "replacement" is a much more appropriate term. For example, if $a = b$, then, by replacement, $a + c$ is the same element as $b + c$—that is, $a + c = b + c$.

It may be noted that the reflexive, symmetric, and transitive laws are valid for the notion of equality—that is: $a = a$ for all a; if $a = b$, then $b = a$; if $a = b$ and $b = c$, then $a = c$.

One final illustration: Suppose $a = b$ and $c = d$, then, by replacement, $a + c = b + c$ and $b + c = b + d$, so, by transitivity, $a + c = b + d$. Of course, this is the familiar statement that adding equals to equals gives equals. Once this kind of thing has been stated in a single instance, we will not fuss about such "commonly accepted rules of logic" in the future.

(7) Axiom A3 postulates the existence of an element, denoted by 0 and called **zero** or **zero element**, which may be added (on the right) to any element of R with no effect. It is this property which is the basis for calling 0 an **identity for addition**. Such an element 0 behaves like the element zero of $\mathbf{Z}$, and this accounts for the notation "0" and the name "zero."

Incidentally, we do not yet know how many such zero elements there are in R. The phrase "there exists" in Axiom A3 asserts only that there is at least one zero element in R; quite possibly there are several zero elements. The true state of affairs will be clarified soon.

(8) Once an element 0 has been fixed there exists (according to A4) for each $a \in R$ an element $\bar{a} \in R$, called an **inverse** of a, such that $a + \bar{a} = 0$. The question of how many inverses an element a has for addition will be answered shortly.

(9) In virtue of the commutative law for addition, A5, we know that $0 + a = a$ for all $a \in R$ and that $\bar{a} + a = 0$. These properties were not stated as parts of A3 and A4, respectively, precisely because they are immediate consequences of the commutative law.

Note that the axioms say nothing about the commutative law for multiplication; it may hold or it may not.

(10) The five axioms A1–A5 are concerned solely with addition, while the two axioms M1 and M2 are concerned solely with multiplication. There is some parallelism between the axioms for multiplication and addition in the sense that M1 corresponds to A1 and M2 corresponds to A2. This parallelism could be extended by introducing additional Axioms M3–M5 for multiplication. At a later stage, we shall see what happens when one or more of the axioms M3–M5 is added to the axioms for a ring.

The only connection between addition and multiplication in a ring is provided by the distributive laws, D1. Note that because multiplication in a ring need not be commutative, we require two distributive laws.

(11) As stated at the beginning of this section, the axioms for a ring are motivated by properties which hold in $\mathbf{Z}$ and $\mathbf{Z}_m$. Why this choice of axioms? The proof of the pudding is presumably in the eating. As the discussion progresses, it will be seen that our choice of axioms for a ring is not a bad one. There are all kinds of rings, and computations in a ring behave according to many of the standard rules. If the axioms did not give "suitable results" (whatever this means) we would change them.

2-1-3. Remark. We have made a big point about the axioms for a ring being chosen as properties that hold in $\mathbf{Z}$ and $\mathbf{Z}_m$. While it is clear that $\mathbf{Z}$ is a ring, it is not quite so obvious that $\mathbf{Z}_m$ is a ring—the axioms need to be verified. Of course, it is meaningless to attempt to verify that $\mathbf{Z}_m$ is a ring until the operations of addition and multiplication have been specified. Naturally, the operations are taken to be those defined in 1-7-14; namely,

$$\lfloor a \rfloor_m + \lfloor b \rfloor_m = \lfloor a + b \rfloor_m, \qquad \lfloor a \rfloor_m \cdot \lfloor b \rfloor_m = \lfloor ab \rfloor_m.$$

These definitions say, in particular, that we have closure for both addition and multiplication.

The verification of the remaining ring axioms is straightforward. The associative law for addition says that

$$(\lfloor a \rfloor_m + \lfloor b \rfloor_m) + \lfloor c \rfloor_m = \lfloor a \rfloor_m + (\lfloor b \rfloor_m + \lfloor c \rfloor_m)$$

for all $\lfloor a \rfloor_m, \lfloor b \rfloor_m, \lfloor c \rfloor_m$ in $\mathbf{Z}_m$; it is valid because

$$(\lfloor a \rfloor_m + \lfloor b \rfloor_m) + \lfloor c \rfloor_m = \lfloor a + b \rfloor_m + \lfloor c \rfloor_m = \lfloor (a + b) + c \rfloor_m,$$

and

$$\lfloor a \rfloor_m + (\lfloor b \rfloor_m + \lfloor c \rfloor_m) = \lfloor a \rfloor_m + \lfloor b + c \rfloor_m = \lfloor a + (b + c) \rfloor_m.$$

Similarly, the associative law for multiplication says that

$$(\lfloor a \rfloor_m \cdot \lfloor b \rfloor_m) \cdot \lfloor c \rfloor_m = \lfloor a \rfloor_m \cdot (\lfloor b \rfloor_m \cdot \lfloor c \rfloor_m)$$

for all $\lfloor a \rfloor_m$, $\lfloor b \rfloor_m$, $\lfloor c \rfloor_m$ in $\mathbf{Z}_m$; it is valid because

$$(\lfloor a \rfloor_m \cdot \lfloor b \rfloor_m) \cdot \lfloor c \rfloor_m = \lfloor ab \rfloor_m \cdot \lfloor c \rfloor_m = \lfloor (ab)c \rfloor_m$$

and

$$\lfloor a \rfloor_m \cdot (\lfloor b \rfloor_m \cdot \lfloor c \rfloor_m) = \lfloor a \rfloor_m \cdot \lfloor bc \rfloor_m = \lfloor a(bc) \rfloor_m.$$

(Note that the proofs of the two associative laws in $\mathbf{Z}_m$ rest on the fact that the corresponding associative laws hold in $\mathbf{Z}$. As a matter of fact, the verification of each of the axioms in $\mathbf{Z}_m$ boils down to the use of the corresponding axiom in $\mathbf{Z}$.)

The element $\lfloor 0 \rfloor_m$ of $\mathbf{Z}_m$ is an identity for addition, since $\lfloor a \rfloor_m + \lfloor 0 \rfloor_m = \lfloor a \rfloor_m$ for all $\lfloor a \rfloor_m \in \mathbf{Z}_m$. Every element of $\mathbf{Z}_m$ has an inverse for addition—in fact, an inverse of an arbitrary element $\lfloor a \rfloor_m$ is the element $\lfloor -a \rfloor_m$, since $\lfloor a \rfloor_m + \lfloor -a \rfloor_m = \lfloor 0 \rfloor_m$. Of course, the commutative law for addition holds. Only the distributive laws remain. To show that

$$\lfloor a \rfloor_m \cdot (\lfloor b \rfloor_m + \lfloor c \rfloor_m) = \lfloor a \rfloor_m \cdot \lfloor b \rfloor_m + \lfloor a \rfloor_m \cdot \lfloor c \rfloor_m$$

for all $\lfloor a \rfloor_m$, $\lfloor b \rfloor_m$, $\lfloor c \rfloor_m$ in $\mathbf{Z}_m$, we observe that

$$\begin{aligned} \lfloor a \rfloor_m \cdot (\lfloor b \rfloor_m + \lfloor c \rfloor_m) &= \lfloor a \rfloor_m \cdot \lfloor b + c \rfloor_m \\ &= \lfloor a(b + c) \rfloor_m \\ &= \lfloor ab + ac \rfloor_m \end{aligned}$$

and

$$\begin{aligned} \lfloor a \rfloor_m \cdot \lfloor b \rfloor_m + \lfloor a \rfloor_m \cdot \lfloor c \rfloor_m &= \lfloor ab \rfloor_m + \lfloor ac \rfloor_m \\ &= \lfloor ab + ac \rfloor_m. \end{aligned}$$

The other distributive law goes exactly the same way. This concludes the verification that $\mathbf{Z}_m$ is a ring.

It is rather cumbersome to denote an element of $\mathbf{Z}_m$ by $\lfloor a \rfloor_m$ and to use such symbols for computation. The customary way to simplify the notation is to represent each congruence class by its smallest nonnegative element. Thus for $\lfloor 0 \rfloor_m, \lfloor 1 \rfloor_m, \ldots, \lfloor m-1 \rfloor_m$ we write $0, 1, 2, \ldots, m - 1$, respectively. This is in keeping with what was done for $\mathbf{Z}_7$ in 1-7-15. Addition and multiplication in $\mathbf{Z}_m$ may then be given completely by tables. For example, in $\mathbf{Z}_6$, addition and multiplication are given by:

addition in $\mathbf{Z}_6$

+	0	1	2	3	4	5
0	0	1	2	3	4	5
1	1	2	3	4	5	0
2	2	3	4	5	0	1
3	3	4	5	0	1	2
4	4	5	0	1	2	3
5	5	0	1	2	3	4

multiplication in $\mathbf{Z}_6$

·	0	1	2	3	4	5
0	0	0	0	0	0	0
1	0	1	2	3	4	5
2	0	2	4	0	2	4
3	0	3	0	3	0	3
4	0	4	2	0	4	2
5	0	5	4	3	2	1

Similarly, the addition and multiplication tables for $\mathbf{Z}_{11}$ are:

addition in $\mathbf{Z}_{11}$

+	0	1	2	3	4	5	6	7	8	9	10
0	0	1	2	3	4	5	6	7	8	9	10
1	1	2	3	4	5	6	7	8	9	10	0
2	2	3	4	5	6	7	8	9	10	0	1
3	3	4	5	6	7	8	9	10	0	1	2
4	4	5	6	7	8	9	10	0	1	2	3
5	5	6	7	8	9	10	0	1	2	3	4
6	6	7	8	9	10	0	1	2	3	4	5
7	7	8	9	10	0	1	2	3	4	5	6
8	8	9	10	0	1	2	3	4	5	6	7
9	9	10	0	1	2	3	4	5	6	7	8
10	10	0	1	2	3	4	5	6	7	8	9

multiplication in $\mathbf{Z}_{11}$

·	0	1	2	3	4	5	6	7	8	9	10
0	0	0	0	0	0	0	0	0	0	0	0
1	0	1	2	3	4	5	6	7	8	9	10
2	0	2	4	6	8	10	1	3	5	7	9
3	0	3	6	9	1	4	7	10	2	5	8
4	0	4	8	1	5	9	2	6	10	3	7
5	0	5	10	4	9	3	8	2	7	1	6
6	0	6	1	7	2	8	3	9	4	10	5
7	0	7	3	10	6	2	9	5	1	8	4
8	0	8	5	2	10	7	4	1	9	6	3
9	0	9	7	5	3	1	10	8	6	4	2
10	0	10	9	8	7	6	5	4	3	2	1

2-1-4. Definition. A ring R is said to be a **ring with unity** (or **with identity** or **with "one"**) when there exists an element $e \in R$ such that

$$ea = ae = a \qquad \text{for all } a \in R.$$

A ring R is said to be **commutative** when

$$ab = ba \quad \text{for all } a, b \in R$$

—that is, when the commutative law for multiplication holds.

Under our scheme for numbering axioms, the commutative law for multiplication would be M5. The existence of an identity for multiplication would be Axiom M3. Because one often deals with a ring with unity in which the commutative law for multiplication is not valid (such a ring would be called a noncommutative ring with unity) we require that the element e should serve as an identity both on the left and on the right.

It is clear that $\mathbf{Z}$ and $\mathbf{Z}_m$ are commutative rings with unity. Of course, $\lfloor 1 \rfloor_m$ is a unity element of $\mathbf{Z}_m$.

It is easy to exhibit many rings, but we shall defer the discussion of examples to the next section. Here we shall undertake to derive a few of the elementary consequences of the ring axioms. In particular, we shall see that a number of standard properties of $\mathbf{Z}$ are valid in any ring. Thus let us consider an arbitrary ring $\{R, +, \cdot\}$. It is understood that henceforth, in this section, all elements come from R and all statements are about R.

2-1-5. Proposition. The zero element, 0, is unique.

Proof: We know that $a + 0 = a$ for all a. Suppose there is another zero element—call it $0'$; so $a + 0' = a$ for all a. We must show $0 = 0'$. Since both 0 and $0'$ are zero elements and since the commutative law for addition holds, we have:

$$0 = 0 + 0' = 0' + 0 = 0'. \quad \blacksquare$$

2-1-6. Proposition. If $a + b = a + c$, then $b = c$.

Proof: This is known as the **cancellation law** for addition. Because addition is commutative it is not necessary to state the cancellation law on the other side; more precisely, once the cancellation law stated above is proved then it is immediate that $b + a = c + a$ implies $b = c$.

As for the proof, making use of replacement, we have

$$\begin{aligned} a + b = a + c &\Rightarrow \bar{a} + (a + b) = \bar{a} + (a + c) \\ &\Rightarrow (\bar{a} + a) + b = (\bar{a} + a) + c \\ &\Rightarrow 0 + b = 0 + c \\ &\Rightarrow b = c. \ \blacksquare \end{aligned}$$

2-1-7. Proposition. If $a + b = a$, then $b = 0$.

Proof: The axiom about the zero element, A3, says that $a + 0 = a$ for all a. The assertion here is a sort of weak converse; it says that if b is an element which serves as an additive identity for just one element a, then $b = 0$. The proof itself is trivial. We have $a + b = a = a + 0$, so by the cancellation law, $b = 0$. ▮

2-1-8. Proposition. For each a, its additive inverse $\bar{a}$ is unique.

Proof: We have $a + \bar{a} = 0$ and must show that if $a + a' = 0$ (meaning that a' is an additive inverse of a), then $a' = \bar{a}$. But this is trivial, since $a + a' = a + \bar{a}$ implies $a' = \bar{a}$, by the cancellation law. ▮

2-1-9. Proposition. The zero element is its own inverse under addition; in symbols,

$$\bar{0} = 0.$$

Proof: By definition of additive inverse, $0 + \bar{0} = 0$. On the other hand, $0 + 0 = 0$. Hence, by cancellation, $\bar{0} = 0$. ▮

2-1-10. Proposition. For each a, we have

$$\overline{(\bar{a})} = a.$$

Proof: In other words, with respect to addition, the inverse of the inverse of an element is the element itself. As for the proof, by definition of inverse, $\bar{a} + \overline{(\bar{a})} = 0$. Of course, $\bar{a} + a = a + \bar{a} = 0$, so by cancellation, $\overline{(\bar{a})} = a$. ▮

2-1-11. Proposition. For any $a, b \in R$ there exists a unique element $x \in R$ such that $a + x = b$; in fact, $x = \bar{a} + b$.

Proof: A much more colloquial way to express this is to say: the equation $a + x = b$ has a unique solution in R, namely $x = \bar{a} + b$.

It is clear that $\bar{a} + b$ is a solution, since $a + (\bar{a} + b) = (a + \bar{a}) + b = b$. As for uniqueness, if $y \in R$ is also a solution, then $a + y = b = a + x$, so by cancellation $x = y$. ▮

2-1-12. Proposition. For any a,

$$a \cdot 0 = 0 = 0 \cdot a.$$

Proof: This tells us that 0 behaves "correctly" (meaning, as expected) with respect to multiplication. Because the basic characteristic of 0 is additive, and 0 is multiplied here by an arbitrary element a, one might expect the proof to depend on the connections between addition and multiplication—that is, on the distributive laws. In fact, we have

$$a0 = a(0 + 0) = a0 + a0$$

and then 2-1-7 permits the conclusion $a0 = 0$. The proof of $0a = 0$ proceeds in similar fashion. ▮

2-1-13. Proposition. For any a and b, we have

$$\bar{a}b = a\bar{b} = \overline{ab} \quad \text{and} \quad \bar{a}\bar{b} = ab.$$

Proof: In virtue of the preceding result,

$$0 = a0 = a(b + \bar{b}) = ab + a\bar{b}.$$

This says that $a\bar{b}$ is the additive inverse of ab; in symbols, $a\bar{b} = \overline{ab}$. In the same way,

$$0 = 0\,b = (a + \bar{a})b = ab + \bar{a}b$$

so that $\bar{a}b = \overline{ab}$. This proves the first part.

For any a and b we now know that $\bar{a}b = a\bar{b}$. Applying this to the elements a and $\bar{b}$, we have

$$\bar{a}\bar{b} = a\bar{\bar{b}} = ab$$

which proves the second part. ▮

2-1-14. Proposition. For any a and b, we have

$$\overline{a + b} = \bar{a} + \bar{b}.$$

Proof: In words, this says that the inverse of a sum is the sum of the inverses. Let us give the proof in full detail. Consider

$$\begin{aligned}(a+b)+(\bar{a}+\bar{b}) &= ((a+b)+\bar{a})+\bar{b}\\ &= (\bar{a}+(a+b))+\bar{b}\\ &= ((\bar{a}+a)+b)+\bar{b}\\ &= (0+b)+\bar{b}\\ &= b+\bar{b}\\ &= 0,\end{aligned}$$

which says that $\overline{a+b}=\bar{a}+\bar{b}$.

The reason for the fussing at the beginning of the proof is that the symbol $a+b+\bar{a}+\bar{b}$ has no meaning because we have no associative law for adding four elements. Instead, we must keep the parentheses and apply the associative law to three elements at a time. ▌

2-1-15. Proposition. If R has an identity for multiplication, then it is unique.

Proof: By hypothesis, there exists an element $e \in R$ such that $ea = a = ae$ for all $a \in R$. Suppose $e' \in R$ is also an identity for multiplication, so $e'a = a = ae'$ for all $a \in R$. Now consider ee'. Since e is an identity on the left, this equals e', and since e' is an identity on the right, this equals e. Thus,

$$e' = ee' = e.$$

Note that the proof here goes just like the proof for uniqueness of the zero element. ▌

2-1-16. Proposition. If R has a unity e, then $(\bar{e})(\bar{e}) = e$ and $\bar{e}a = \bar{a}$ for all $a \in R$.

Proof: According to 2-1-13, $\bar{e}\cdot\bar{e} = e\cdot e = e$, and for any a, $\bar{e}a = e\bar{a} = \bar{a}$. ▌

2-1-17. Remark. Now, let us revert to the more common notation and write $-a$ (called **minus** a) instead of $\bar{a}$, and $b - a$ (which is read as b **minus** a) instead of $b + (-a) = b + \bar{a}$. Thus, what is normally called "subtracting a from b" is really adding the additive inverse of a to b.

Rewriting the preceding properties (namely, 2-1-9, 2-1-10, 2-1-13, 2-1-14, 2-1-16) in terms of the minus notation, we have:

(*i*) $-0 = 0$,
(*ii*) $-(-a) = a$,
(*iii*) $(-a)b = a(-b) = -(ab)$,
(*iv*) $(-a)(-b) = ab$,
(*v*) $-(a + b) = -a - b$,
(*vi*) $(-e)(-e) = e$ and $(-e)a = -a$ when e is an identity for multiplication.

We chose not to introduce the minus notation at the start solely for pedagogical reasons—so the notation should not lead the reader to assume or expect that certain properties of minus hold before they have been proved carefully. In general, the standard and familiar rules for computation with the minus sign also hold in a ring. For example, the meaning of $a - b + c$ in a ring is clear, and it is equal to $a + c - b$. Additional examples will appear in the problems.

Thus far we have seen that many properties of **Z** also hold in any ring. On the other hand, there are a number of extremely important properties of **Z** that need not hold in an arbitrary ring. For example, if $ab = 0$, where a and b are elements of the ring R, we cannot conclude that at least one of these elements is 0. This seemingly pathological situation, in which the product of two nonzero elements is zero, occurs in $\mathbf{Z}_6$—for here, $\lfloor 2 \rfloor_6 \neq \lfloor 0 \rfloor_6$, $\lfloor 3 \rfloor_6 \neq \lfloor 0 \rfloor_6$ and $\lfloor 2 \rfloor_6 \cdot \lfloor 3 \rfloor_6 = \lfloor 0 \rfloor_6$. In order to focus on this kind of situation, we make a definition.

2-1-18. Definition. A nonzero element a in the commutative ring R is said to be a **zero-divisor** when there exists $b \neq 0$ in R such that $ab = 0$. A commutative ring with unity element $e \neq 0$ and no zero-divisors is known as an **integral domain**; it is usually denoted by D.

A few comments about the definition are in order: (1) According to the definition, only a nonzero element can be a zero-divisor; the element 0 is not a zero-divisor. In particular, $\lfloor 2 \rfloor_6$ is a zero-divisor in $\mathbf{Z}_6$ and $\lfloor 0 \rfloor_6$ is not. (2) Note further that a zero-divisor was defined only in a commutative ring. If the ring were not commutative, it would be necessary to deal with $ab = 0$ and $ba = 0$ separately, and to distinguish between right and left—a distinction which we choose to avoid. (3) The reason for the requirement $e \neq 0$ in an integral domain D is not mysterious at all. If $e = 0$, then for every $a \in D$

$$0 = 0a = ea = a,$$

which means that D consists of just the one element 0—and this ring is so trivial that it may as well be excluded from the discussion. (4) The basic condition for an integral domain may be stated as:

$$\text{if } ab = 0 \text{ and } a \neq 0, \text{ then } b = 0.$$

In fact, if $a \neq 0$, then a is not a zero-divisor precisely when $ab = 0$ implies $b = 0$.

Obviously, $\mathbf{Z}$ is an integral domain, and it is easy to verify (by trial and error) that $\mathbf{Z}_5$ and $\mathbf{Z}_7$ are integral domains. On the other hand, $\mathbf{Z}_6$ is not an integral domain. What about the commutative ring with unity $\mathbf{Z}_m$? The full story is given by the next result.

2-1-19. Theorem. The ring $\mathbf{Z}_m$ is an integral domain $\Leftrightarrow$ m is prime.

Proof: In order to be completely accurate and avoid possible confusion we use the $\lfloor \ \rfloor_m$ notation for elements of $\mathbf{Z}_m$.

Suppose m is composite. Then there exist integers a and b such that $m = ab$, $1 < a < m$, $1 < b < m$. We have, therefore, $\lfloor a \rfloor_m \lfloor b \rfloor_m = \lfloor m \rfloor_m = \lfloor 0 \rfloor_m$ with $\lfloor a \rfloor_m \neq \lfloor 0 \rfloor_m$ and $\lfloor b \rfloor_m \neq \lfloor 0 \rfloor_m$—so $\mathbf{Z}_m$ is not an integral domain. This proves: m composite $\Rightarrow$ $\mathbf{Z}_m$ is not an integral domain. Of course, this implication is logically equivalent to: $\mathbf{Z}_m$ is an integral domain $\Rightarrow m$ is prime.

To prove the other half of the theorem, suppose m is a prime p; we must show that $\mathbf{Z}_p$ is an integral domain. Suppose $\lfloor a \rfloor_p \neq \lfloor 0 \rfloor_p$ and $\lfloor a \rfloor_p \lfloor b \rfloor_p = \lfloor 0 \rfloor_p$; then

$$\begin{aligned} \lfloor a \rfloor_p \lfloor b \rfloor_p = \lfloor 0 \rfloor_p &\Rightarrow \lfloor ab \rfloor_p = \lfloor 0 \rfloor_p \\ &\Rightarrow ab \equiv 0 \pmod{p} \\ &\Rightarrow p \mid ab. \end{aligned}$$

Since $\lfloor a \rfloor_p \neq \lfloor 0 \rfloor_p$ we know that $p \nmid a$. Because p is prime, we conclude that $p \mid b$—or what is the same thing, $\lfloor b \rfloor_p = \lfloor 0 \rfloor_p$. Consequently, the arbitrary nonzero element $\lfloor a \rfloor_p$ is not a zero-divisor (because there is no nonzero $\lfloor b \rfloor_p$ for which $\lfloor a \rfloor_p \lfloor b \rfloor_p = \lfloor 0 \rfloor_p$) and $\mathbf{Z}_p$ is an integral domain. This completes the proof. ▌

2-1-20. Remark. Suppose R is a commutative ring with unity $e \neq 0$ in which the following property, known as the **cancellation law** for multiplication, holds:

(1) If $ab = ac$ and $a \neq 0$, then $b = c$.

It is easy to see that the cancellation law is equivalent to the condition

(2) If $ab = 0$ and $a \neq 0$, then $b = 0$.

In fact, if (1) holds, then $ab = 0 = a0$, and by cancellation, $b = 0$—so (2) holds. Conversely, if (2) holds then $ab = ac$ implies $a(b - c) = 0$, and by (2) we have $b - c = 0$—so $b = c$, and (1) holds.

Of course, condition (2) says that an element $a \neq 0$ cannot be a zero-divisor—so condition (2) holds if and only if R is an integral domain. Consequently, we have proved: if R is a commutative ring with unity $e \neq 0$, then R is an integral domain $\Leftrightarrow$ the cancellation law for multiplication holds.

2-1-21 / *PROBLEMS*

1. In any ring, prove the following rules:
(*i*) $a + (b + c) = c + (a + b)$,
(*ii*) $(a(b + c))d = (ab)d + a(cd)$,
(*iii*) $a + ((b + c) + d) = ((a + b) + c) + d = (a + b) + (c + d)$,
(*iv*) $(a + b)(c + d) = (ac + bc) + (ad + bd) = (ac + ad) + (bc + bd)$.

2. In a commutative ring, show that
(*i*) $a(bc) = c(ba)$,
(*ii*) $a((b + c)d) = (ad)b + (ad)c$,
(*iii*) $(a + b)(a + b) = (a^2 + b^2) + (ab + ab)$.

3. Construct addition and multiplication tables for $\mathbf{Z}_2$, $\mathbf{Z}_3$, $\mathbf{Z}_8$, and $\mathbf{Z}_{10}$.

4. In $\mathbf{Z}_{11}$, compute $5 \cdot 8$, $5 \cdot 9$, $5 \cdot (8 + 9)$, $(5 \cdot 8)9$, $5(8 \cdot 9)$, $9(8 \cdot 5)$, $8(5 \cdot 9)$, $8(9 + 5)$, $9(8 + 5)$.

5. In $\mathbf{Z}_{11}$, check the rules given in Problems 1 and 2 when $a = 3$, $b = 7$, $c = 2$, $d = 9$.

6. Construct addition and multiplication tables for $\mathbf{Z}_{12}$. Find all pairs $a, b \in \mathbf{Z}_{12}$ for which $ab = 0$, and all pairs $a, b \in \mathbf{Z}_{12}$ for which $ab = 1$.

7. Consider the set $\mathbf{Z} = \{0, \pm 1, \pm 2, \ldots\}$. Define addition of two integers as usual, but define the product of any two integers to be 0. Is $\mathbf{Z}$ a ring with respect to these operations?

8. Exhibit a ring that has exactly two elements. If n is a positive integer, does there exist a ring with exactly n elements?

9. In an arbitrary ring, does $\bar{a} = \bar{b}$ imply $a = b$?

10. In any ring, verify that
(*i*) $(a - b) + (c - d) = (a + c) - (b + d)$,
(*ii*) $-(a - b) = b - a$,
(*iii*) $(a - b) - (c - d) = (a + d) - (b + c)$,
(*iv*) $a(b - c) = ab - ac$.

11. In any ring, verify that
  (*i*) $-((-a)b) = ab = -(a(-b))$,
  (*ii*) $(a+b)(c-d) = (ac-bd)+(bc-ad) = (ac+bc)-(ad+bd)$,
  (*iii*) $(a-b)(c+d) = (ac-bd)+(ad-bc) = (ac+ad)-(bc+ad)$,
  (*iv*) $(a-b)(c-d) = (ac+bd)-(bc+ad)$.

12. In $\mathbf{Z}_{13}$ verify the assertions of Problems 10 and 11 when $a=5$, $b=10$, $c=7$, $d=11$.

13. In any ring, evaluate:
  (*i*) $(a+b)^2$, (*ii*) $(a-b)^2$, (*iii*) $(a-b+c)(a+b-c)$,
  (*iv*) $(a+b)^3$, (*v*) $(a-b)^3$, (*vi*) $(a-b+c)^3$.

14. If the ring R is commutative, evaluate all the expressions listed in Problem 13.

15. Find a necessary and sufficient condition on the ring R that
$$(a+b)(a-b) = a^2 - b^2 \qquad \text{for all } a, b \in R.$$
What is meant by a necessary and sufficient condition?

16. For $n \geq 4$, discuss the possible meanings for $a_1 + a_2 + \cdots + a_n$ and $a_1 \cdot a_2 \cdots a_n$ in a ring.

17. In an integral domain, if a is an **idempotent** (meaning that $a^2 = a$), then a is 0 or e. Can you give an example of a commutative ring with unity in which there is an idempotent not equal to 0 or e.

2-2. Examples

For us, the rings of primary interest are $\mathbf{Z}$ and $\mathbf{Z}_m$. Nevertheless, it is useful to give many examples of rings—some of which may never be referred to again. These should deepen our understanding of the concept of a ring and provide us with a certain amount of intuition about rings in general.

2-2-1. Example. Let us examine several number systems that are especially familiar.

(*i*) Consider the set $\mathbf{R}$ of all real numbers. It would take us too far afield to discuss exactly what is meant by a real number. (As a matter of fact, to do it properly would probably require a full semester.) Instead, we shall assume here that the real numbers are "known"—just as we assumed in the preceding chapter that the integers $\mathbf{Z}$ are known. One may wish to think of real numbers as the points on the so-called **number line**. Under the customary operations of addition and multiplication, it is immediate that $\mathbf{R}$ is a ring because the required axioms are simply "known" properties of real numbers. Even more,

R is surely a commutative ring with identity 1, and since the product of two nonzero real numbers cannot equal 0, it follows that **R** is an integral domain.

(*ii*) Consider the set **Q** of all rational numbers. By this is meant the set of all numbers that can be expressed as the ratio of two integers—in symbols,

$$\mathbf{Q} = \left\{ \frac{m}{n} \,\middle|\, m, n \in \mathbf{Z},\ n \neq 0 \right\}.$$

Of course, we are assuming that the rational numbers are "known"; because of this words like "fraction" or "ratio," and symbols like m/n, have meaning. Naturally, we know how to add and multiply rational numbers (that is, "fractions"). The usual rules are given by

$$\frac{m_1}{n_1} + \frac{m_2}{n_2} = \frac{m_1 n_2 + m_2 n_1}{n_1 n_2},$$

$$\frac{m_1}{n_1} \cdot \frac{m_2}{n_2} = \frac{m_1 m_2}{n_1 n_2}.$$

These rules may be taken as the definitions of addition and multiplication in **Q**. Under these operations, **Q** is a ring because all the axioms for a ring are properties that have always been taken for granted in the set of rational numbers. However, one can be much more formal and actually verify the ring axioms. Let us sketch this quickly.

To verify closure for addition we must show that if m_1/n_1 and m_2/n_2 are elements of **Q**, then so is $(m_1/n_1) + (m_2/n_2)$. Because m_1, m_2, n_1, n_2 are integers, so are $m_1 n_2 + m_2 n_1$ and $n_1 n_2$. In addition, $n_1 n_2 \neq 0$ because $n_1 \neq 0$ and $n_2 \neq 0$. Therefore,

$$\frac{m_1 n_2 + m_2 n_1}{n_1 n_2}$$

is indeed an element of **Q**, and Axiom A1 holds. In similar fashion, $m_1 m_2$ is an integer and $n_1 n_2$ is a nonzero integer, so $m_1 m_2 / n_1 n_2 \in \mathbf{Q}$, and it follows that Axiom M1, closure for multiplication, holds. Note that the verification of these axioms (and the remaining ones also) depends on properties of **Z**.

The associative law for addition, A2, is valid because

$$\begin{aligned}
\left(\frac{m_1}{n_1} + \frac{m_2}{n_2}\right) + \frac{m_3}{n_3} &= \left(\frac{m_1 n_2 + m_2 n_1}{n_1 n_2}\right) + \frac{m_3}{n_3} \\
&= \frac{(m_1 n_2 + m_2 n_1) n_3 + m_3 (n_1 n_2)}{(n_1 n_2) n_3} \\
&= \frac{m_1 n_2 n_3 + m_2 n_1 n_3 + m_3 n_1 n_2}{n_1 n_2 n_3}
\end{aligned}$$

is equal to

$$\frac{m_1}{n_1} + \left(\frac{m_2}{n_2} + \frac{m_3}{n_3}\right) = \frac{m_1}{n_1} + \left(\frac{m_2 n_3 + m_3 n_2}{n_2 n_3}\right)$$

$$= \frac{m_1(n_2 n_3) + n_1(m_2 n_3 + m_3 n_2)}{n_1(n_2 n_3)}$$

$$= \frac{m_1 n_2 n_3 + m_2 n_1 n_3 + m_3 n_1 n_2}{n_1 n_2 n_3}.$$

The associative law for multiplication, M2, is clear because

$$\left(\frac{m_1}{n_1} \cdot \frac{m_2}{n_2}\right) \cdot \frac{m_3}{n_3} = \frac{m_1 m_2 m_3}{n_1 n_2 n_3} = \frac{m_1}{n_1} \cdot \left(\frac{m_2}{n_2} \cdot \frac{m_3}{n_3}\right).$$

The remaining axioms for addition, A3–A5, are easy: $0/1$ is clearly a zero element, an inverse of m/n is surely $(-m)/n$, and the commutative law holds since

$$\frac{m_1}{n_1} + \frac{m_2}{n_2} = \frac{m_1 n_2 + m_2 n_1}{n_1 n_2} = \frac{m_2}{n_2} + \frac{m_1}{n_1}.$$

Finally, it is straightforward to verify the distributive laws, D1—and this is left to the reader.

Thus, **Q** is a ring. In addition, $1/1$ is an identity for multiplication and the commutative law for multiplication holds, so **Q** is a commutative ring with unity. Even more, **Q** is an integral domain because the product of two nonzero rational numbers is not zero.

All this has gone very smoothly, but there are some difficulties that have been glossed over. For example, $0/1$ is a zero element, but so is $0/2$, or $0/3$, or $0/n$ for any $n \neq 0$. Since **Q** is a ring, this apparently contradicts the uniqueness of the zero element (see 2-1-5). The difficulty is caused by our carelessness in defining the elements of **Q**. A rational number is not just a ratio m/n—we know for example that

$$\frac{1}{2} = \frac{2}{4} = \frac{3}{6} = \cdots \quad \text{and} \quad \frac{0}{1} = \frac{0}{2} = \cdots = \frac{0}{n} = \cdots.$$

In other words, there are really many distinct names or expressions for the same rational number. Consequently, it is necessary to go back to the beginning, define **Q** carefully, and verify all the axioms. At a later stage, we shall develop **Q** with great care and precision. For the time being, we shall take it for granted that **Q** is an integral domain and that the difficulties mentioned above can be overcome.

(*iii*) Consider the set $\mathbf{C}$ of all complex numbers. By a complex number we mean a symbol or expression $a + bi$ where a and b are real numbers and $i = \sqrt{-1}$. Thus

$$\mathbf{C} = \{a + bi \mid a \in \mathbf{R}, b \in \mathbf{R}\}.$$

Of course, every complex number has a unique expression; in other words,

$$a + bi = a' + b'i \Leftrightarrow a = a' \text{ and } b = b'.$$

The standard definitions for addition and multiplication in $\mathbf{C}$ are

$$(a + bi) + (c + di) = (a + c) + (b + d)i,$$

$$(a + bi)(c + di) = (ac - bd) + (bc + ad)i.$$

Naturally, we choose these definitions of the operations in $\mathbf{C}$ precisely because they express the way we have always added or multiplied complex numbers. Clearly, $\mathbf{C}$ satisfies closure for both of these operations. Furthermore, it is straightforward to verify that addition and multiplication are commutative, that both associative laws hold, and that the distributive laws hold; all these verifications rest ultimately on the properties of $\mathbf{R}$. For example, the associative law for multiplication follows from

$$\begin{aligned}((a + bi)(c + di))(e + fi) &= ((ac - bd) + (ad + bc)i)(e + fi)\\ &= ((ac - bd)e - (ad + bc)f)\\ &\quad + ((ac - bd)f + (ad + bc)e)i\\ &= (ace - bde - adf - bcf)\\ &\quad + (acf - bdf + ade + bce)i\end{aligned}$$

and

$$\begin{aligned}(a + bi)((c + di)(e + fi)) &= (a + bi)((ce - df) + (cf + de)i)\\ &= (ace - adf - bcf - bde)\\ &\quad + (bce - bdf + acf + ade)i.\end{aligned}$$

The element $0 = 0 + 0i$ is an identity for addition. An additive inverse of $a + bi$ is $(-a) + (-b)i$. [Strictly speaking, the symbols $-(a + bi)$ or $-a - bi$ have no meaning. After $\mathbf{C}$ has been shown to be a ring, the minus can then be introduced, and both of these symbols represent the inverse of $a + bi$.] Moreover, the element $1 = 1 + 0i$ is an identity for multiplication. We see, therefore, that $\mathbf{C}$ is a commutative ring with unity.

Now, let us prove that $\mathbf{C}$ is an integral domain. (Surely everyone "knows" that if the product of two complex numbers is 0, then at least one of them must be 0. The point is that we can prove this "fact" using only properties of **R**.) For this consider

$$(a + bi)(c + di) = 0 \qquad \text{with } a + bi \neq 0. \tag{$*$}$$

We must show $c + di = 0$—in other words, $c = d = 0$. Multiplying out, we have

$$(ac - bd) + (bc + ad)i = 0 = 0 + 0i$$

which says

$$\begin{aligned} ac - bd &= 0, \\ bc + ad &= 0. \end{aligned} \tag{$**$}$$

Consequently, $a(ac - bd) = a0 = 0$ and $b(bc + ad) = 0$. Because multiplication in **R** is commutative, adding these expressions yields

$$(a^2 + b^2)c = 0. \tag{$\#$}$$

Similarly, from ($**$) we obtain $b(ac - bd) = 0$ and $a(bc + ad) = 0$—so subtraction yields

$$(a^2 + b^2)d = 0. \tag{$\#\#$}$$

Now, $a + bi \neq 0$, by hypothesis; so not both a and b are 0. Therefore, by well-known properties of real numbers, it follows that $a^2 + b^2 \neq 0$. Since **R** is an integral domain, (#) and (# #) imply that $c = d = 0$. Thus, **C** is an integral domain.

An alternative proof that **C** has no zero-divisors goes as follows. From ($*$) we have

$$(a - bi)\big((a + bi)(c + di)\big) = (a - bi)0 = 0$$

and hence

$$(a^2 + b^2)(c + di) = 0.$$

Since $a^2 + b^2 \neq 0$, there is an element $1/(a^2 + b^2)$ (by which is meant an inverse of $a^2 + b^2$ for multiplication) in **R**. Of course, $\mathbf{R} \subset \mathbf{C}$, so, in particular, $1/(a^2 + b^2) \in \mathbf{C}$. Consequently, multiplication by $1/(a^2 + b^2)$ gives $c + di = 0$. The key to this version of the proof is some insight as to what goes on in **C**.

2-2-2. Example. Consider the set of all even integers. It may be denoted by

$$2\mathbf{Z} = \{2n \mid n \in \mathbf{Z}\}.$$

With respect to the usual operations of addition and multiplication of integers, $2\mathbf{Z}$ is clearly a commutative ring with no zero-divisors. Note that in verifying the required axioms, several of them—for example, the two associative laws, the two commutative laws, and the distributive laws—are automatic, because they hold for any integers and hence surely for even integers. However, $2\mathbf{Z}$ does not have an identity for multiplication, so according to the definition 2-1-18, $2\mathbf{Z}$ is not an integral domain.

In similar fashion, for any integer $m > 1$, consider the set of all integers which are divisible by m. This set is denoted by

$$m\mathbf{Z} = \{mn \mid n \in \mathbf{Z}\},$$

and it consists of all multiples of m in $\mathbf{Z}$. Again, with respect to the usual operations with integers, $m\mathbf{Z}$ is a commutative ring with no unity element and with no zero-divisors.

2-2-3. Example. Suppose we have a set consisting of two elements, $R = \{0, 1\}$. Let us define addition and multiplication in R by the rules

$$0 + 1 = 1 + 0 = 1, \qquad 0 + 0 = 1 + 1 = 0,$$

$$0 \cdot 1 = 1 \cdot 0 = 0 \cdot 0 = 0, \qquad 1 \cdot 1 = 1.$$

It is easy to verify, by explicit enumeration of the various cases for each axiom, that R is an integral domain. This is not surprising because R is "really" $\mathbf{Z}_2$ which, according to 2-1-19, is an integral domain.

In somewhat analogous fashion, let R be the three-element set $R = \{a, b, c\}$ and define addition and multiplication in R according to the following tables.

addition in R

+	a	b	c
a	c	a	b
b	a	b	c
c	b	c	a

multiplication in R

·	a	b	c
a	c	b	a
b	b	b	b
c	a	b	c

It is tedious to verify that $\{R, +, \cdot\}$ is a ring—for example, each associative law has 27 different cases, each of which must be checked. Furthermore, multiplication is commutative, c is an identity for multiplication, and b is the zero element. In fact, R is an integral domain.

The incisive way to look at this example is to set up the correspondence

$$a \leftrightarrow 2, \qquad b \leftrightarrow 0, \qquad c \leftrightarrow 1.$$

Then R becomes the set $\{0, 1, 2\}$, and the operations as given by the tables become those of $\mathbf{Z}_3$. In other words, this example is simply a disguised form of the integral domain $\mathbf{Z}_3$.

2-2-4. Example. Consider the complex number $i = \sqrt{-1}$, and the set of all complex numbers $a + bi$ for which both a and b are integers. This subset of $\mathbf{C}$ may be written as

$$\{a + bi \mid a, b \in \mathbf{Z}\}.$$

Under the standard operations of addition and multiplication of complex numbers, it is straightforward to verify that we have an integral domain. It is known as the domain of **Gaussian integers**. We shall denote it by $\mathbf{Z}[i]$, or $\mathbf{Z}[\sqrt{-1}]$, or $\mathbf{Z} + \mathbf{Z}\sqrt{-1}$.

In the same way, the reader may check that

$$\mathbf{Q}[i] = \{a + bi \mid a, b \in \mathbf{Q}\},$$

which we also denote by $\mathbf{Q}[\sqrt{-1}]$ or $\mathbf{Q} + \mathbf{Q}\sqrt{-1}$, is an integral domain. Of course,

$$\mathbf{Z}[i] \subset \mathbf{Q}[i] \subset \mathbf{C}.$$

2-2-5. Example. Consider the real number $\sqrt{2}$ and the subset $\mathbf{Z}[\sqrt{2}] = \mathbf{Z} + \mathbf{Z}\sqrt{2}$ of $\mathbf{R}$ defined by

$$\mathbf{Z}[\sqrt{2}] = \{a + b\sqrt{2} \mid a, b \in \mathbf{Z}\}.$$

The usual way to add and multiply these elements is

$$(a + b\sqrt{2}) + (c + d\sqrt{2}) = (a + c) + (b + d)\sqrt{2},$$

$$(a + b\sqrt{2})(c + d\sqrt{2}) = (ac + 2bd) + (bc + ad)\sqrt{2},$$

and, of course

$$a + b\sqrt{2} = c + d\sqrt{2} \Leftrightarrow a = c \text{ and } b = d.$$

One checks easily that with these standard operations $\mathbf{Z}[\sqrt{2}]$ is an integral domain.

In the same way, we see that

$$\mathbf{Q}[\sqrt{2}] = \{a + b\sqrt{2} \mid a, b \in \mathbf{Q}\}$$

(which contains $\mathbf{Z}[\sqrt{2}]$ and is contained in $\mathbf{R}$) is an integral domain.

We did not carry out the verification of the ring axioms for the preceding examples in detail, nor did we even indicate how to show that $\mathbf{Z}[i]$ and $\mathbf{Z}[\sqrt{2}]$ have no zero-divisors. In part, the intention was to induce the reader to supply the details himself. However, as we shall see momentarily, most of the mechanical work involved in verifying the axioms was really superfluous.

2-2-6. Definition. Suppose $\{R, +, \cdot\}$ is a ring and let S be a nonempty subset of R. We say that S is a **subring** of R if, with respect to the operations $+$ and $\cdot$ given in R, S becomes a ring in its own right—in other words, S is a subring of R when $\{S, +, \cdot\}$ is a ring.

To illustrate, R itself is always a subring of R, and so is (0), the subset consisting of the zero element alone. Furthermore, let us consider the inclusions $\mathbf{Z}[i] \subset \mathbf{Q}[i] \subset \mathbf{C}$, $2\mathbf{Z} \subset \mathbf{Z} \subset \mathbf{Q} \subset \mathbf{R} \subset \mathbf{C}$, $\mathbf{Z} \subset \mathbf{Z}\sqrt{2} \subset \mathbf{Q}\sqrt{2} \subset \mathbf{R}$. In virtue of the earlier examples, each term or symbol which appears is a ring. Because in each and every case we use the "standard" addition and multiplication, it follows that each of the rings listed is a subring of all the listed rings which contain it.

Consider further the ring $\mathbf{Z}$, and let S be the subset of all odd integers,

$$S = \{2n + 1 \mid n \in \mathbf{Z}\}.$$

Then S is not a subring of $\mathbf{Z}$; in fact, S is not closed under addition, nor does it have a zero element.

Finally, suppose S is the set of all nonnegative real numbers

$$S = \{a \mid a \in \mathbf{R},\ a \geq 0\}.$$

Is S a subring of $\mathbf{R}$? One checks easily that S satisfies all the axioms for a ring except the existence of inverses for addition, so S is not a subring of $\mathbf{R}$.

To decide, in general, if a subset S in a ring R is a subring it is not necessary to verify all the ring axioms in S. Our next result provides criteria for deciding whether or not S is a subring.

2-2-7. Proposition. Suppose S is an nonempty subset of the ring $\{R, +, \cdot\}$ then the following are equivalent:

(*i*) S is a subring of R,
(*ii*) $a, b \in S \Rightarrow a + b \in S,\ -a \in S,\ ab \in S$,
(*iii*) $a, b \in S \Rightarrow a - b \in S,\ ab \in S$.

Proof: According to condition (*ii*), in order to show that S is a subring it suffices to verify that S is closed under the operations, in R, of addition, taking of additive inverse, and multiplication. According to condition (*iii*), it suffices to verify that S is closed under subtraction and multiplication in R. Now for the proof.

(*i*) $\Rightarrow$ (*ii*). By hypothesis, S is a ring when addition and multiplication of elements are carried over from R. If $a, b \in S$, then $a + b$ and ab (the sum and product of elements of R) are in S, because S is a ring. It remains to show that $-a$, which is the additive inverse of a in R, belongs to S.

The difficulty here is this. The zero element of R is $0 \in R$. Since S is a ring, it has a zero element, but we do not know that it is the element 0. Thus let $0' \in S$ denote the zero element of S, and in similar fashion let $a' \in S$ denote the additive inverse of a in S. We must show that $-a = a' \in S$.

For $a \in S$ we have

$$a + 0 = a = a + 0'$$

so, by cancellation in R, $0 = 0'$. In other words, the zero element of R belongs to S, and R and S have the same zero element. Consequently,

$$a + a' = 0' = 0 = a + (-a)$$

and we conclude (again, by cancellation in R) that $a' = -a$—thus completing this part of the proof.

(ii) ⇒ *(iii)*. Suppose we are given $a, b \in S$. Because *(ii)* holds, we have $ab \in S$ and also $-b \in S$. In order for *(iii)* to hold, it remains to show that $a - b \in S$. But now, $a, b \in S$ implies $a, -b \in S$, and then by *(ii)* their sum $a + (-b)$ is in S—that is, $a - b \in S$, and this part of the proof is complete.

(iii) ⇒ *(i)*. We must verify all the ring axioms for S, under the assumption that *(iii)* holds. The associative laws for addition and multiplication hold for elements of S because they already hold for arbitrary elements of R. Similarly, the commutative law for addition and the distributive laws hold for elements of S. In addition, according to *(iii)*, S is closed under multiplication. It remains to show: A1, S is closed under addition; A3, S has a zero element; A4, every element of S has an additive inverse in S. To do all this, consider any $a, b \in S$. Applying *(iii)* for the special case $b = a$ we have $0 = a - a = a - b \in S$—so A3 holds. Now, applying *(iii)* for the elements $0, a \in S$ gives $-a = 0 - a \in S$—so A4 holds. Finally, returning to arbitrary $a, b \in S$ we have $-b \in S$ and by *(iii)*, $a + b = a - (-b) \in S$— so A1 holds. This completes the entire proof. ▮

2-2-8. Remark. The utility of the preceding result may be illustrated by using it to show that $\mathbf{Z}[\sqrt{2}]$ is a ring. Taking it for granted that $\mathbf{R}$ is a ring with respect to the usual operations, we note first that $\mathbf{Z}[\sqrt{2}] = \{a + b\sqrt{2} \mid a, b \in \mathbf{Z}\}$ is a nonempty subset of $\mathbf{R}$. Moreover, the operations of $+$ and $\cdot$ defined for $\mathbf{Z}[\sqrt{2}]$ (see 2-2-5) are precisely those for real numbers. Now, for elements $\alpha = a + b\sqrt{2}$ and $\beta = c + d\sqrt{2}$ of $\mathbf{Z}[\sqrt{2}]$ it is immediate that $\alpha - \beta \in \mathbf{Z}[\sqrt{2}]$ and $\alpha\beta \in \mathbf{Z}[\sqrt{2}]$—so, according to condition *(iii)* of 2-2-7, $\mathbf{Z}[\sqrt{2}]$ is a subring of $\mathbf{R}$, and in particular $\mathbf{Z}[\sqrt{2}]$ is a ring in its own right.

Furthermore, $1 = 1 + 0\sqrt{2}$ is clearly an identity for multiplication, and multiplication in $\mathbf{Z}[\sqrt{2}]$ is commutative because the commutative law for multiplication holds in $\mathbf{R}$. Thus, $\mathbf{Z}[\sqrt{2}]$ is a commutative ring with unity; but how we know that it is an integral domain? To settle this question it is not necessary to imitate the procedure used in 2-2-1 to verify that $\mathbf{C}$ has no zero-divisors. Instead, we note that because the product of two nonzero real numbers cannot be zero, the same property holds for elements of the subset $\mathbf{Z}[\sqrt{2}]$ of $\mathbf{R}$. Consequently, $\mathbf{Z}[\sqrt{2}]$ has no zero-divisors and it is an integral domain.

In general, this argument shows that any subring of an integral domain has no zero-divisors. However, the subring need not be an integral domain because (just as for $2\mathbf{Z} \subset \mathbf{Z}$) the subring may not have an identity for multiplication—and according to the definition of integral domain, 2-1-18, it must have a unity element.

2-2-9. Examples. Consider $\mathbf{Z} \times \mathbf{Z}$, the set of all ordered pairs of integers; in other words,

$$\mathbf{Z} \times \mathbf{Z} = \{(a, b) \mid a, b \in \mathbf{Z}\}.$$

This is a special case of a general notation according to which for any sets X and Y we write

$$X \times Y = \{(x, y) \mid x \in X, y \in Y\}.$$

There are many ways to define addition and multiplication in $\mathbf{Z} \times \mathbf{Z}$. For example, let us put

$$\begin{aligned} (a, b) + (c, d) &= (a + c, b + d), \\ (a, b) \cdot (c, d) &= (ac - bd, bc + ad). \end{aligned} \tag{$*$}$$

The burden of performing the arithmetic work required for verifying that $\mathbf{Z} \times \mathbf{Z}$ is a ring with respect to these operations is left to the reader. For example, the associative law for multiplication says that for any $\alpha = (a, b)$, $\beta = (c, d)$, $\gamma = (e, f)$ in $\mathbf{Z} \times \mathbf{Z}$ we have

$$(\alpha\beta)\gamma = \alpha(\beta\gamma)$$

and this relation holds because both sides turn out to equal

$$(ace - bde - bcf - adf, bce + ade + acf - bdf).$$

Even more, the reader may wish to verify that this is an integral domain.

In the same way, there is no difficulty in checking that $\mathbf{Z} \times \mathbf{Z}$ is a ring with respect to the operations

$$\begin{aligned} (a, b) + (c, d) &= (a + c, b + d), \\ (a, b) \cdot (c, d) &= (ac + 2bd, bc + ad). \end{aligned} \tag{$**$}$$

For example, one of the distributive laws says that if $\alpha = (a, b)$, $\beta = (c, d)$, $\gamma = (e, f)$ in $\mathbf{Z} \times \mathbf{Z}$ we have

$$\alpha(\beta + \gamma) = \alpha\beta + \alpha\gamma$$

and this relation holds because both sides equal

$$(ac + ae + 2bd + 2bf, bc + be + ad + af).$$

Here, as in the preceding case, $(0, 0)$ is the zero element and the additive inverse of (a, b) is $(-a, -b)$. Again, the reader may try to verify that $\mathbf{Z} \times \mathbf{Z}$ with the operations $(**)$ is an integral domain. Clearly, $(1, 0)$ is an identity for multiplication and multiplication is commutative. What about the property of "no zero-divisors"? For this, given $(a, b) \neq (0, 0) = 0$ (meaning that not both a and b equal 0), it must be shown that if

$$(a, b) \cdot (c, d) = (0, 0),$$

then $(c, d) = (0, 0)$. The reader is challenged to do this.

Still another way to make $\mathbf{Z} \times \mathbf{Z}$ into a ring is to define

$$\begin{aligned} (a, b) + (c, d) &= (a + c, b + d), \\ (a, b) \cdot (c, d) &= (ac - bd, ad + bc - bd). \end{aligned} \tag{$***$}$$

In this case, verification of the ring axioms is still straightforward, but it is somewhat more difficult to show that $\mathbf{Z} \times \mathbf{Z}$ is an integral domain.

It may well be asked what these examples, in which $\mathbf{Z} \times \mathbf{Z}$ is made into a ring, are about? Where do they come from? The explanation is surprisingly simple. For example, consider the integral domain $\mathbf{Z}[i]$. If we choose to associate with an element $a + bi$ the ordered pair (a, b), this determines a 1–1 correspondence

$$a + bi \leftrightarrow (a, b)$$

between the sets $\mathbf{Z}[i]$ and $\mathbf{Z} \times \mathbf{Z}$. Upon rewriting the rules for addition and multiplication in $\mathbf{Z}[i]$ in terms of this new ordered-pair notation, we observe that they turn out to be precisely the rules given by $(*)$. Consequently, $\mathbf{Z} \times \mathbf{Z}$, with operations defined by $(*)$, is just $\mathbf{Z}[i]$ in disguise—so it is an integral domain, and all the work the reader may have done in showing that $\mathbf{Z} \times \mathbf{Z}$ with operations $(*)$ is an integral domain was superfluous!

Similarly, $\mathbf{Z} \times \mathbf{Z}$ with operations given by (**) is simply another version of $\mathbf{Z}[\sqrt{2}]$—that is, with (a, b) corresponding to $a + b\sqrt{2}$—so it is an integral domain.

The story underlying the situation (***) is a bit more complicated. Consider the complex number

$$\omega = \frac{-1 + \sqrt{-3}}{2}.$$

Then

$$\omega^2 = \frac{-1 - \sqrt{-3}}{2} \quad \text{and} \quad \omega^3 = 1.$$

Thus ω is a cube root of unity (the other cube roots of unity being ω^2 and 1), and because $0 = \omega^3 - 1 = (\omega - 1)(\omega^2 + \omega + 1)$ we know that

$$\omega^2 + \omega + 1 = 0$$

(a fact which may also be verified by a direct computation). Let us write

$$\mathbf{Z}[\omega] = \mathbf{Z} + \mathbf{Z}\omega = \{a + b\omega \mid a, b \in \mathbf{Z}\}.$$

This is surely a nonempty subset of the integral domain $\mathbf{C}$, and we already know how to add and multiply elements of $\mathbf{Z}[\omega]$; namely,

$$(a + b\omega) + (c + d\omega) = (a + c) + (b + d)\omega$$

and using the fact that $\omega^2 = -1 - \omega$,

$$(a + b\omega)(c + d\omega) = (ac - bd) + (ad + bc - bd)\omega.$$

Thus, $\mathbf{Z}[\omega]$ is closed under the addition and multiplication from $\mathbf{C}$. In addition, $-(a + b\omega) = (-a) + (-b)\omega$ is an element of $\mathbf{Z}[\omega]$, so it follows from 2-2-7 that $\mathbf{Z}[\omega]$ is a subring of $\mathbf{C}$. Of course, $1 = 1 + 0\omega$ is an identity for multiplication, and because $\mathbf{Z}[\omega]$ is contained in $\mathbf{C}$, multiplication is commutative and there are no zero-divisors. Therefore, $\mathbf{Z}[\omega]$ is an integral domain. Now, instead of $a + b\omega$ let us write (a, b), thereby setting up a 1–1 correspondence between $\mathbf{Z}[\omega]$ and $\mathbf{Z} \times \mathbf{Z}$. When addition and multiplication is $\mathbf{Z}[\omega]$ are rewritten in the notation of $\mathbf{Z} \times \mathbf{Z}$, the result is precisely the rules (***). This proves that (and indicates why) $\mathbf{Z} \times \mathbf{Z}$ is an integral domain in the situation (***)!

2-2-10. Example. Let $\mathscr{M}(\mathbf{Z}, 2)$ denote the set of all 2×2 matrices with entries from $\mathbf{Z}$. Of course, by an element $A \in \mathscr{M}(\mathbf{Z}, 2)$ we mean an array

$$A = \begin{pmatrix} a_{11} & a_{12} \\ a_{21} & a_{22} \end{pmatrix}, \qquad a_{11}, a_{12}, a_{21}, a_{22} \in \mathbf{Z}$$

with two rows and two columns. The notation is arranged so that a_{ij} is the entry in the ith row and the jth column of A (of course, here the only choices for i and j are 1 and 2). If B is also an element of $\mathcal{M}(\mathbf{Z}, 2)$, then we write

$$B = \begin{pmatrix} b_{11} & b_{12} \\ b_{21} & b_{22} \end{pmatrix}$$

and define the sum $A + B$ and product AB as

$$\begin{pmatrix} a_{11} & a_{12} \\ a_{21} & a_{22} \end{pmatrix} + \begin{pmatrix} b_{11} & b_{12} \\ b_{21} & b_{22} \end{pmatrix} = \begin{pmatrix} a_{11} + b_{11} & a_{12} + b_{12} \\ a_{21} + b_{21} & a_{22} + b_{22} \end{pmatrix},$$

$$\begin{pmatrix} a_{11} & a_{12} \\ a_{21} & a_{22} \end{pmatrix} \cdot \begin{pmatrix} b_{11} & b_{12} \\ b_{21} & b_{22} \end{pmatrix} = \begin{pmatrix} a_{11}b_{11} + a_{12}b_{21} & a_{11}b_{12} + a_{12}b_{22} \\ a_{21}b_{11} + a_{22}b_{21} & a_{21}b_{12} + a_{22}b_{22} \end{pmatrix}.$$

Thus, addition involves simply adding corresponding entries; for example,

$$\begin{pmatrix} 2 & 3 \\ 5 & 7 \end{pmatrix} + \begin{pmatrix} -1 & 1 \\ -1 & 1 \end{pmatrix} = \begin{pmatrix} 1 & 4 \\ 4 & 8 \end{pmatrix}.$$

Multiplication is more complicated; the scheme for carrying it out is often called row-by-column multiplication. To obtain the i, j entry of AB—that is, the element in the ith row and jth column of the product matrix AB—one takes the ith row of A and the jth column of B, multiplies term by term and takes the sum of these products. An example may further clarify the meaning of these words.

$$\begin{pmatrix} 2 & 3 \\ 5 & 7 \end{pmatrix} \cdot \begin{pmatrix} -1 & 1 \\ -1 & 1 \end{pmatrix} = \begin{pmatrix} -2-3 & 2+3 \\ -5-7 & 5+7 \end{pmatrix} = \begin{pmatrix} -5 & 5 \\ -12 & 12 \end{pmatrix}.$$

It is understood, of course, that two matrices are equal when they are identical term-by-term; in other words,

$$A = B \Leftrightarrow a_{11} = b_{11}, a_{12} = b_{12}, a_{21} = b_{21}, a_{22} = b_{22}.$$

It is clear from the definitions that $\mathcal{M}(\mathbf{Z}, 2)$ is closed under both addition and multiplication. The associative and commutative laws for addition hold in $\mathcal{M}(\mathbf{Z}, 2)$ because they hold in $\mathbf{Z}$. The matrix

$$0 = \begin{pmatrix} 0 & 0 \\ 0 & 0 \end{pmatrix}$$

is clearly a zero-element—since $A + 0 = A$ for all $A \in \mathcal{M}(\mathbf{Z}, 2)$. The matrix A has an additive inverse—namely,

$$-A = \begin{pmatrix} -a_{11} & -a_{12} \\ -a_{21} & -a_{22} \end{pmatrix}$$

—for surely $A + (-A) = 0$.

The associative law for multiplication says

$$(AB)C = A(BC)$$

for all $A, B, C \in \mathcal{M}(\mathbf{Z}, 2)$. It is not obvious at all, and the only way to verify it is by brute force. Let us carry out the details.

$$\left[\begin{pmatrix} a_{11} & a_{12} \\ a_{21} & a_{22} \end{pmatrix}\begin{pmatrix} b_{11} & b_{12} \\ b_{21} & b_{22} \end{pmatrix}\right]\begin{pmatrix} c_{11} & c_{12} \\ c_{21} & c_{22} \end{pmatrix}$$

$$= \begin{pmatrix} a_{11}b_{11} + a_{12}b_{21} & a_{11}b_{12} + a_{12}b_{22} \\ a_{21}b_{11} + a_{22}b_{21} & a_{21}b_{12} + a_{22}b_{22} \end{pmatrix}\begin{pmatrix} c_{11} & c_{12} \\ c_{21} & c_{22} \end{pmatrix}$$

$$= \begin{pmatrix} (a_{11}b_{11} + a_{12}b_{21})c_{11} + (a_{11}b_{12} + a_{12}b_{22})c_{21} & (a_{11}b_{11} + a_{12}b_{21})c_{12} + (a_{11}b_{12} + a_{12}b_{22})c_{22} \\ (a_{21}b_{11} + a_{22}b_{21})c_{11} + (a_{21}b_{12} + a_{22}b_{22})c_{21} & (a_{21}b_{11} + a_{22}b_{21})c_{12} + (a_{21}b_{12} + a_{22}b_{22})c_{22} \end{pmatrix}$$

while

$$\begin{pmatrix} a_{11} & a_{12} \\ a_{21} & a_{22} \end{pmatrix}\left[\begin{pmatrix} b_{11} & b_{12} \\ b_{21} & b_{22} \end{pmatrix}\begin{pmatrix} c_{11} & c_{12} \\ c_{21} & c_{22} \end{pmatrix}\right]$$

$$= \begin{pmatrix} a_{11} & a_{12} \\ a_{21} & a_{22} \end{pmatrix}\begin{pmatrix} b_{11}c_{11} + b_{12}c_{21} & b_{11}c_{12} + b_{12}c_{22} \\ b_{21}c_{11} + b_{22}c_{21} & b_{21}c_{12} + b_{22}c_{22} \end{pmatrix}$$

$$= \begin{pmatrix} a_{11}(b_{11}c_{11} + b_{12}c_{21}) + a_{12}(b_{21}c_{11} + b_{22}c_{21}) & a_{11}(b_{11}c_{12} + b_{12}c_{22}) + a_{12}(b_{21}c_{12} + b_{22}c_{22}) \\ a_{21}(b_{11}c_{11} + b_{12}c_{21}) + a_{22}(b_{21}c_{11} + b_{22}c_{21}) & a_{21}(b_{11}c_{12} + b_{12}c_{22}) + a_{22}(b_{21}c_{12} + b_{22}c_{22}) \end{pmatrix}$$

and then one checks easily that these two end-result matrices are equal.

The distributive laws, both of which must be verified, say that

$$A(B + C) = AB + AC \qquad \text{and} \qquad (B + C)A = BA + CA$$

for all $A, B, C \in \mathcal{M}(\mathbf{Z}, 2)$. The details may safely be left to the reader.

Furthermore, if we write

$$I = \begin{pmatrix} 1 & 0 \\ 0 & 1 \end{pmatrix},$$

then a straightforward computation gives $IA = AI = A$ for all $A \in \mathcal{M}(\mathbf{Z}, 2)$—so I is an identity for multiplication.

From all this we conclude that $\mathcal{M}(\mathbf{Z}, 2)$ is a ring with unity. Is this ring commutative? Roughly speaking, if the reader chooses any two elements $A, B \in \mathcal{M}(\mathbf{Z}, 2)$ at random, he will find that $AB \neq BA$. For example, taking

$$A = \begin{pmatrix} 1 & 0 \\ 0 & 0 \end{pmatrix}, \qquad B = \begin{pmatrix} 0 & 0 \\ 1 & 0 \end{pmatrix}$$

we have

$$AB = 0, \qquad BA = B.$$

Except for showing that multiplication in $\mathcal{M}(\mathbf{Z}, 2)$ does not satisfy the commutative law, this example touches on other questions—such as zero-divisors or cancellation—but we shall not pursue such matters here.

The ring $\mathcal{M}(\mathbf{Z}, 2)$ has many subrings. Consider, for example, the non-empty subset

$$S = \{A \in \mathcal{M}(\mathbf{Z}, 2) \mid a_{12} = a_{21} = a_{22} = 0\}$$

of $\mathcal{M}(\mathbf{Z}, 2)$. In other words, S consists of all elements of form

$$A = \begin{pmatrix} a & 0 \\ 0 & 0 \end{pmatrix}, \qquad a \in \mathbf{Z}.$$

If we also take an element

$$B = \begin{pmatrix} b & 0 \\ 0 & 0 \end{pmatrix}, \qquad b \in \mathbf{Z}$$

from S, then clearly

$$A - B = \begin{pmatrix} a - b & 0 \\ 0 & 0 \end{pmatrix}, \qquad AB = \begin{pmatrix} ab & 0 \\ 0 & 0 \end{pmatrix}.$$

Thus, $A - B$ and AB have the "proper form" and are therefore elements of S. This proves that S is a subring. In fact, S behaves very much like the ring $\mathbf{Z}$.

Consider next the set T of all elements of $\mathcal{M}(\mathbf{Z}, 2)$ which are of form

$$\begin{pmatrix} a & -b \\ b & a \end{pmatrix}, \qquad a, b \in \mathbf{Z}.$$

Thus,

$$\begin{pmatrix} 3 & 5 \\ -5 & 3 \end{pmatrix}$$

is an element of T, and

$$\begin{pmatrix} 3 & -5 \\ 5 & -3 \end{pmatrix}$$

is not. Taking the difference of two elements of T, we have

$$\begin{pmatrix} a & -b \\ b & a \end{pmatrix} - \begin{pmatrix} c & -d \\ d & c \end{pmatrix} = \begin{pmatrix} a - c & -(b - d) \\ b - d & a - c \end{pmatrix}$$

—so the result is an element of T (because it is of the proper form), and T is, therefore, closed under subtraction. As for the product of two elements of T, we have

$$\begin{pmatrix} a & -b \\ b & a \end{pmatrix} \cdot \begin{pmatrix} c & -d \\ d & c \end{pmatrix} = \begin{pmatrix} ac - bd & -ad - bc \\ bc + ad & bd + ac \end{pmatrix}.$$

The resulting matrix is an element of T—so T is closed under multiplication. It follows that T is a subring of $\mathcal{M}(\mathbf{Z}, 2)$. Even more, the reader can easily check that T is a commutative ring with unity.

Now, $\mathscr{M}(\mathbf{Z}, 2)$ can be generalized in several ways. One way is to take the entries of our 2×2 matrices from some arbitrary ring R, instead of from $\mathbf{Z}$. This set may be denoted by $\mathscr{M}(R, 2)$. It is a ring—the verification is exactly as for $\mathscr{M}(\mathbf{Z}, 2)$.

In addition, we may consider $\mathscr{M}(R, n)$—by which is meant the set of all $n \times n$ matrices with entries from R—for any n greater than or equal to 2. Thus, an element of $\mathscr{M}(R, n)$ is of form

$$A = \begin{pmatrix} a_{11} & a_{12} & \cdots & a_{1j} & \cdots & a_{1n} \\ a_{21} & a_{22} & \cdots & a_{2j} & \cdots & a_{2n} \\ \vdots & \vdots & & \vdots & & \vdots \\ a_{i1} & a_{i2} & \cdots & a_{ij} & \cdots & a_{in} \\ \vdots & \vdots & & \vdots & & \vdots \\ a_{n1} & a_{n2} & \cdots & a_{nj} & \cdots & a_{nn} \end{pmatrix}, \qquad a_{ij} \in R, \quad i = 1, \ldots, n, \quad j = 1, \ldots, n.$$

Upon defining addition of $n \times n$ matrices to be componentwise and multiplication to be "row-by-column" it can be verified (a lot of bookkeeping is involved in keeping the subscripts straight, especially in the associative law for multiplication) that $\mathscr{M}(R, n)$ is a ring.

2-2-11. Example. Suppose a ring R is given and consider the product set $R \times R = \{(a, b) \mid a, b \in R\}$. If we define addition and multiplication "componentwise" in $R \times R$—namely,

$$(a, b) + (c, d) = (a + c, b + d),$$
$$(a, b) \cdot (c, d) = (ac, bd),$$

then it is trivial to check that $R \times R$ becomes a ring. Everything goes smoothly precisely because R is a ring. We denote this new ring by $R^{(2)}$ or $R \oplus R$, and refer to it as the **direct sum** (or as the **direct product**) of R with itself. The zero element of $R \oplus R$ is obviously $(0, 0)$ and the additive inverse of (a, b) is $(-a, -b)$.

Does $R \oplus R$ have an identity for multiplication? This depends on R. If R has an identity e, then obviously (e, e) is an identity for multiplication in $R \oplus R$. On the other hand, if R has no identity, then neither does $R \oplus R$. Turning to the commutative law for multiplication, it is immediate that $R \oplus R$ is a commutative ring if and only if R is a commutative ring. Note, however, that under no circumstances can $R \oplus R$ be an integral domain—in fact, if $a \neq 0$ is an element of R, then $(a, 0)$ is a zero-divisor in $R \oplus R$, since $(a, 0)(0, a) = (0, a)(a, 0) = (0, 0)$.

Now, let us look at the set

$$R' = \{(a, b) \in R \oplus R \mid b = 0\}.$$

In other words, R' is the subset of $R \oplus R$ consisting of all elements of form $(a, 0)$. Because R' is obviously closed under subtraction and multiplication, we see that R' is a subring of $R \oplus R$. In a sense, R' is just a copy of R; for the second coordinate of elements of R' is always 0, and one might choose to ignore the second coordinate.

Similarly,

$$R'' = \{(a, b) \in R \oplus R \mid a = 0\},$$

the subset of $R \oplus R$ consisting of all elements whose first coordinate is 0, is a subring of $R \oplus R$. It too is a "version" of R under the correspondence $(0, b) \leftrightarrow b$.

Still another subring of $R \oplus R$ which behaves like R is the **diagonal** Δ; by this we have in mind

$$\Delta = \{(a, b) \in R \oplus R \mid a = b\} = \{(a, a) \mid a \in R\}.$$

What about generalizing the direct sum $R \oplus R = R^{(2)}$? For any integer $n \geq 2$, the set of all n-tuples of elements of R becomes a ring, which we denote by $R^{(n)}$, when the operations are defined componentwise. In other words,

$$R^{(n)} = \{(a_1, a_2, \ldots, a_n) \mid a_i \in R, i = 1, \ldots, n\}$$

and the operations are

$$(a_1, a_2, \ldots, a_n) + (b_1, b_2, \ldots, b_n) = (a_1 + b_1, a_2 + b_2, \ldots, a_n + b_n),$$
$$(a_1, a_2, \ldots, a_n) \cdot (b_1, b_2, \ldots, b_n) = (a_1 b_1, a_2 b_2, \ldots, a_n b_n).$$

Then $R^{(n)}$ is called the **direct sum** (or the **direct product**) of n copies of R. One often denotes it by $R \oplus \cdots \oplus R$, but then it is necessary to specify exactly how many components (that is, copies of R) are involved.

Incidentally, in talking about $R^{(n)}$ the requirement that $n \geq 2$ is not essential. One may surely speak also of $R^{(1)}$, the ring of 1-tuples, which is clearly R itself.

One may go a step beyond n-tuples and make the set of all infinite sequences of elements from R into a ring by defining addition and multiplication componentwise. This ring is denoted by

$$R^{(\infty)} = \{(a_1, a_2, \ldots, a_n, \ldots) \mid a_i \in R, \quad i = 1, 2, 3, \ldots\}$$

with operations

$$(a_1, a_2, \ldots, a_n, \ldots) + (b_1, b_2, \ldots, b_n, \ldots) = (a_1 + b_1, a_2 + b_2, \ldots, a_n + b_n, \ldots)$$

$$(a_1, a_2, \ldots, a_n, \ldots) \cdot (b_1, b_2, \ldots, b_n, \ldots) = (a_1 b_1, a_2 b_2, \ldots, a_n b_n, \ldots).$$

Note that for any n, $R^{(n)}$ may be viewed as a subring of $R^{(\infty)}$. More precisely, for fixed n, we consider the set

$$\{(a_1, a_2, \ldots, a_n, a_{n+1}, \ldots) \in R^{(\infty)} \mid a_{n+1} = a_{n+2} = \cdots = 0\}$$

—in other words, we take those elements of $R^{(\infty)}$ for which all coordinates after the nth are 0. This is clearly a subring of $R^{(\infty)}$, and surely it is a copy of $R^{(n)}$.

There is really no need to restrict the discussion of direct sum to the case of a single R. Thus, if we have n rings $R_1, R_2, \ldots, R_n$, which may or may not be distinct, then by the **direct sum** (or **direct product**) $R_1 \oplus R_2 \oplus \cdots \oplus R_n$ we mean the ring whose underlying set is the product set

$$R_1 \times R_2 \times \cdots \times R_n = \{(a_1, a_2, \ldots, a_n) \mid a_i \in R_i, \quad i = 1, \ldots, n\}$$

(in other words, we take all n-tuples in which the ith coordinate comes from R_i, for each $i = 1, \ldots, n$) and in which the operations are componentwise. Of course, we also write, simply,

$$R_1 \oplus \cdots \oplus R_n = \{(a_1, a_2, \ldots, a_n) \mid a_i \in R_i, \quad i = 1, \ldots, n\}.$$

Naturally, the same thing can be done for an infinite number of rings R_i, $i = 1, 2, 3, \ldots$.

2-2-12. Example. In the preceding example we have seen one method for constructing new rings from old ones. Here we shall describe another procedure for constructing new rings. The method, which is extremely important, turns out to have certain connections with the method used in 2-2-11.

Suppose there are given an arbitrary ring $R = \{a, b, c, \ldots\}$ and an arbitrary set $X = \{x, y, z, w, \ldots\}$. Consider the set of all mappings from X into R. We denote this set by Map(X, R). An element of Map(X, R) is a mapping or function from X into R and is denoted by $f\colon X \to R$. Of course, a mapping or function here is a "rule," f say, according to which there is assigned to each $x \in X$ an element $f(x) \in R$; symbolically,

$$f\colon x \to f(x), \qquad x \in X.$$

This "rule" f may be given by some kind of a "formula" or it may simply be a set of words that permit us to determine $f(x)$ when x is given.

A quick illustration: Suppose the set X consists of three elements

$$X = \{x_1, x_2, x_3\}$$

and the ring R is $\mathbf{Z}_2 = \{0, 1\}$. What is Map(X, $\mathbf{Z}_2$)? First of all, an element $f \in$ Map(X, $\mathbf{Z}_2$) is determined by specifying elements $f(x_1)$, $f(x_2)$ and $f(x_3)$ of $\mathbf{Z}_2$. Since each of $f(x_1), f(x_2), f(x_3)$ must be either 0 or 1, it follows that there are eight mappings from X into $\mathbf{Z}_2$. These may be labeled by $f_1, f_2, f_3, f_4, f_5, f_6, f_7, f_8$ (or in any other way the reader prefers) with the following definitions.

$$
\begin{array}{llll}
f_1: & f_1(x_1) = 0, & f_1(x_2) = 0, & f_1(x_3) = 0. \\
f_2: & f_2(x_1) = 0, & f_2(x_2) = 0, & f_2(x_3) = 1. \\
f_3: & f_3(x_1) = 0, & f_3(x_2) = 1, & f_3(x_3) = 0. \\
f_4: & f_4(x_1) = 1, & f_4(x_2) = 0, & f_4(x_3) = 0. \\
f_5: & f_5(x_1) = 0, & f_5(x_2) = 1, & f_5(x_3) = 1. \\
f_6: & f_6(x_1) = 1, & f_6(x_2) = 0, & f_6(x_3) = 1. \\
f_7: & f_7(x_1) = 1, & f_7(x_2) = 1, & f_7(x_3) = 0. \\
f_8: & f_8(x_1) = 1, & f_8(x_2) = 1, & f_8(x_3) = 1.
\end{array}
$$

It needs to be emphasized that functions are not strangers to us. We have been dealing with them in all courses in mathematics. The functions under consideration usually came from Map(**R**, **R**)—for example, the sine function $x \to \sin x$, or any polynomial function such as $x \to 3x^5 - 2x^4 + x^2 - x + 4$, or the exponential function $x \to e^x$.

Returning to the arbitrary set X and ring R, we observe that for $f, g \in \text{Map}(X, R)$ the definition of equality for functions says that $f = g$ if and only if $f(x) = g(x)$ for all $x \in X$—in other words, two functions are considered equal (or the same) $\Leftrightarrow$ they take the same values for every choice of $x \in X$.

Our objective is to make Map(X, R) into a ring in a natural way, so for any $f, g \in \text{Map}(X, R)$ we must define $f + g$ and $f \cdot g$. To do this, we need to specify what $f + g$ and $f \cdot g$ do to every $x \in X$—that is, we must give the "value" of $f + g$ and $f \cdot g$ at every $x \in X$. Naturally, we do this precisely in the expected manner; first, one "evaluates" both f and g at x and then adds or multiplies the results. In symbols

$$
\left.\begin{aligned}
(f + g)(x) &= f(x) + g(x), \\
(f \cdot g)(x) &= f(x) \cdot g(x).
\end{aligned}\right. \qquad \text{for all } x \in X.
$$

Note that both $f(x)$ and $g(x)$ are in R, so they can indeed be added and multiplied. On the other hand, there is no need for any kind of structure on the set X.

Now, let us verify that Map(X, R) is a ring with respect to these operations. In view of the definitions, Map(X, R) is clearly closed under both addition and multiplication. The two associative laws, $(f + g) + h = f + (g + h)$ and $(f \cdot g) \cdot h = f \cdot (g \cdot h)$ for all $f, g, h \in \text{Map}(X, R)$, are trivial. For example, to prove the latter, we observe that for $x \in X$,

$$
\begin{aligned}
((f \cdot g) \cdot h)(x) &= ((f \cdot g)(x)) \cdot h(x) = (f(x) \cdot g(x)) \cdot h(x), \\
(f \cdot (g \cdot h))(x) &= f(x) \cdot ((g \cdot h)(x)) = f(x) \cdot (g(x) \cdot h(x)),
\end{aligned}
$$

and these are equal because multiplication in R is associative. [Note that we are using $+$ and $\cdot$ for the operations in both R and Map(X, R). There is little danger of confusion, especially because everything goes smoothly.]

If we define the map $0: X \to R$ by

$$0(x) = 0 \qquad \text{for all } x \in X$$

—meaning that the mapping 0 takes the value 0 in R for every $x \in X$—then for any $f \in \text{Map}(X, R)$

$$(f + 0)(x) = f(x) + 0(x) = f(x) + 0 = f(x)$$

for all $x \in X$, so $f + 0 = f$ and 0 is a zero element.

For $f \in \text{Map}(X, R)$, we seek an additive inverse $-f$ (or $\bar{f}$). If we define $-f$ by

$$(-f)(x) = -(f(x)) \qquad \text{for all } x \in X$$

—meaning that we find $f(x)$ and then take its additive inverse $-(f(x))$ in R—then $-f \in \text{Map}(X, R)$, and $f + (-f) = 0$ because

$$\big(f + (-f)\big)(x) = f(x) + (-f)(x) = f(x) - f(x) = 0 = 0(x)$$

for all $x \in X$. (The reader should justify each equal sign carefully.)

The commutative law for addition, $f + g = g + f$, is trivial. Finally, the two distributive laws, $f(g + h) = fg + fh$ and $(g + h)f = gf + hf$, hold; for example, the first one is valid because

$$\begin{aligned}
\big(f(g + h)\big)(x) &= f(x)\big(g(x) + h(x)\big) \\
&= f(x)g(x) + f(x)h(x) \\
&= (fg)(x) + (fh)(x) \\
&= (fg + fh)(x)
\end{aligned}$$

for all $x \in X$. This completes the verification of the ring axioms in $\text{Map}(X, R)$; the reader should note that the verification of each axiom depends ultimately on the fact that R is a ring.

To illustrate, let us look at the ring $\text{Map}(X, R)$ in the special case mentioned earlier where $X = \{x_1, x_2, x_3\}$, $R = \mathbf{Z}_2 = \{0, 1\}$. It is easy to perform the operations in this ring. For example, to compute $f_3 + f_4$ we observe that

$$\begin{aligned}
(f_3 + f_4)(x_1) &= f_3(x_1) + f_4(x_1) = 0 + 1 = 1, \\
(f_3 + f_4)(x_2) &= f_3(x_2) + f_4(x_2) = 1 + 0 = 1, \\
(f_3 + f_4)(x_3) &= f_3(x_3) + f_4(x_3) = 0 + 0 = 0,
\end{aligned}$$

and consequently, $f_3 + f_4 = f_7$. Similarly, to find $f_3 \cdot f_4$ we observe that

$$\begin{aligned}
(f_3 f_4)(x_1) &= f_3(x_1) f_4(x_1) = 0 \cdot 1 = 0, \\
(f_3 f_4)(x_2) &= f_3(x_2) f_4(x_2) = 1 \cdot 0 = 0, \\
(f_3 f_4)(x_3) &= f_3(x_3) f_4(x_3) = 0 \cdot 0 = 0,
\end{aligned}$$

and consequently, $f_3 \cdot f_4 = f_1$. More generally, we may safely leave it to the reader to check the accuracy of the following tables for addition and multiplication in the eight-element ring under consideration.

addition in Map $(X, \mathbf{Z}_2)$

+	f_1	f_2	f_3	f_4	f_5	f_6	f_7	f_8
f_1	f_1	f_2	f_3	f_4	f_5	f_6	f_7	f_8
f_2	f_2	f_1	f_5	f_6	f_3	f_4	f_8	f_7
f_3	f_3	f_5	f_1	f_7	f_2	f_8	f_4	f_6
f_4	f_4	f_6	f_7	f_1	f_8	f_2	f_3	f_5
f_5	f_5	f_3	f_2	f_8	f_1	f_7	f_6	f_4
f_6	f_6	f_4	f_8	f_2	f_7	f_1	f_5	f_3
f_7	f_7	f_8	f_4	f_3	f_6	f_5	f_1	f_2
f_8	f_8	f_7	f_6	f_5	f_4	f_3	f_2	f_1

multiplication in Map $(X, \mathbf{Z}_2)$

·	f_1	f_2	f_3	f_4	f_5	f_6	f_7	f_8
f_1	f_1	f_1	f_1	f_1	f_1	f_1	f_1	f_1
f_2	f_1	f_2	f_1	f_1	f_2	f_2	f_1	f_2
f_3	f_1	f_1	f_3	f_1	f_3	f_1	f_3	f_3
f_4	f_1	f_1	f_1	f_4	f_1	f_4	f_4	f_4
f_5	f_1	f_2	f_3	f_1	f_5	f_2	f_3	f_5
f_6	f_1	f_2	f_1	f_4	f_2	f_6	f_4	f_6
f_7	f_1	f_1	f_3	f_4	f_3	f_4	f_7	f_7
f_8	f_1	f_2	f_3	f_4	f_5	f_6	f_7	f_8

We close out our discussion with a matter of real interest—namely, how $\text{Map}(X, R)$ can be interpreted as generalizing the direct sum rings $R^{(n)}$ or $R^{(\infty)}$. More precisely, suppose the set X has n elements—so there is no harm in writing $X = \{1, 2, 3, \ldots, n\}$. Consider any $f \in \text{Map}(X, R)$, where the ring R is arbitrary. For each $i = 1, 2, \ldots, n$, $f(i)$ is an element of R, and we may associate with f the n-tuple $(f(1), f(2), \ldots, f(n))$ which is, of course, an element of $R^{(n)}$. Conversely, given any element $(a_1, a_2, \ldots, a_n) \in R^{(n)}$ we can use it to define the element $f \in \text{Map}(X, R)$ for which $f(i) = a_i$, $i = 1, \ldots, n$; and then the element of $R^{(n)}$ associated with f is precisely $(a_1, a_2, \ldots, a_n)$. Therefore, there is a "natural" 1–1 correspondence between sets

$$\text{Map}(X, R) \leftrightarrow R^{(n)}$$

according to which

$$f \leftrightarrow (a_{,1}\, a_2, \ldots, a_n),$$

when $f(i) = a_i$ for $i = 1, 2, \ldots, n$. Suppose further that $g \in \text{Map}(X, R)$ corresponds to the n-tuple $(b_1, b_2, \ldots, b_n)$ in $R^{(n)}$—thus

$$g \leftrightarrow (b_1, b_2, \ldots, b_n),$$

and $g(i) = b_i$ for $i = 1, \ldots, n$. Which elements of $R^{(n)}$ correspond to $f + g$ and fg, respectively? Since for each $i = 1, \ldots, n$

$$(f + g)(i) = f(i) + g(i) = a_i + b_i \qquad \text{and} \qquad (fg)(i) = f(i)g(i) = a_i b_i$$

it follows immediately that

$$f + g \leftrightarrow (a_1 + b_1, a_2 + b_2, \ldots, a_n + b_n),$$

and

$$fg \leftrightarrow (a_1 b_1, a_2 b_2, \ldots, a_n b_n).$$

The significance of this is that under the 1–1 correspondence between the rings Map(X, R) and $R^{(n)}$, addition and multiplication also correspond—in other words, the ring operations in the two rings correspond to each other. This leads to the following conclusion.

When $X = \{1, 2, \ldots, n\}$ (or any set with n elements), the rings Map(X, R) and $R^{(n)}$ are "essentially" the same. $(*)$

In particular, for the concrete example discussed earlier where $X = \{x_1, x_2, x_3\}$ and $R = \mathbf{Z}_2$, we see that the rings Map$(X, \mathbf{Z}_2)$ and $\mathbf{Z}_2 \oplus \mathbf{Z}_2 \oplus \mathbf{Z}_2$ are essentially the same. The correspondence between them is given by

$$f_1 \leftrightarrow (0, 0, 0), \quad f_2 \leftrightarrow (0, 0, 1), \quad f_3 \leftrightarrow (0, 1, 0), \quad f_4 \leftrightarrow (1, 0, 0),$$
$$f_5 \leftrightarrow (0, 1, 1), \quad f_6 \leftrightarrow (1, 0, 1), \quad f_7 \leftrightarrow (1, 1, 0), \quad f_8 \leftrightarrow (1, 1, 1).$$

The result $(*)$ can be extended. Namely, if X is the set of all positive integers, $X = \{1, 2, 3, \ldots\}$ and the ring R is arbitrary, then clearly the rings Map(X, R) and $R^{(\infty)}$ are essentially the same. The correspondence is

$$f \leftrightarrow (f(1), f(2), f(3), \ldots).$$

2-2-13. Example. Consider an arbitrary set $X = \{a, b, c, \ldots, x, y, \ldots\}$, which may be finite or infinite. Let $\mathscr{S}(X)$ denote the set of all subsets of X. Elements of $\mathscr{S}(X)$, which are subsets of X, are denoted by $A, B, C, \ldots$. In particular, the empty set $\varnothing$ and the set X itself are elements of $\mathscr{S}(X)$.

Can we make $\mathscr{S}(X)$ into a ring? Perhaps the most obvious approach is to define for $A, B \in \mathscr{S}(X)$.

$$A + B = A \cup B,$$
$$A \cdot B = A \cap B.$$

We recall that the **union** $A \cup B$ consists of those elements x of X which belong to A, or to B, or to both, and the **intersection** $A \cap B$ consists of all those elements x of X which belong to both A and B.

Now, $\mathscr{S}(X)$ is surely closed under addition and multiplication. Both associative laws clearly hold; in fact, $A \cup (B \cup C) = (A \cup B) \cup C$ because each side consists of those elements x which belong to at least one of the sets A, B, C, and $A \cap (B \cap C) = (A \cap B) \cap C$ because each side consists of those elements x which belong to all three of the sets A, B, C. Since $A + \varnothing = A$ for all $A \in \mathscr{S}(X)$, we see that $\varnothing$ is a zero element. However, given $A \neq \varnothing$ there is no B for which $A + B = A \cup B = \varnothing$—so additive inverses do not exist, and out attempt to make $\mathscr{S}(X)$ into a ring fails.

Let us try another approach. As is customary, we let $A - B$ (which is read as "A minus B") denote the set of all elements which belong to A but not to B; in symbols,

$$A - B = \{x \in X \mid x \in A, x \notin B\}.$$

Then define operations in $\mathscr{S}(X)$ by

$$A + B = (A - B) \cup (B - A),$$
$$A \cdot B = A \cap B.$$

The meaning of $A + B$, which is known as the **symmetric difference** of A and B, may be understood from the accompanying Venn diagram,

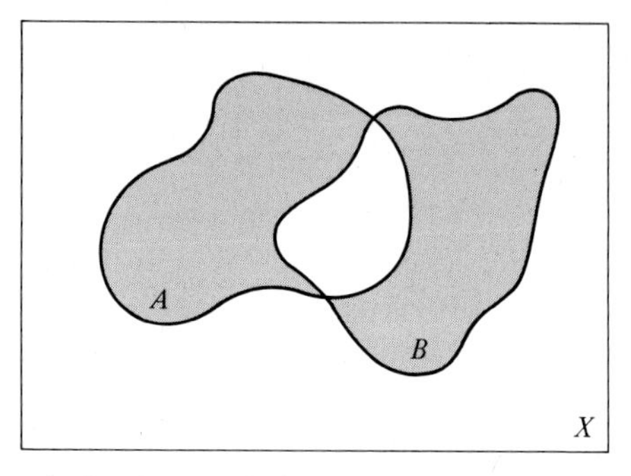

where $A + B$ is precisely the total shaded area, which consists of the union of $A - B$ and $B - A$. Another way to characterize the shaded area $A + B$ is as the set of all elements x which are in A or in B but not in both—so clearly

$$A + B = (A - B) \cup (B - A) = (A \cup B) - (A \cap B).$$

Now, let us examine the ring axioms for $\mathscr{S}(X)$. Obviously we have closure for both addition and multiplication. The associative law for multiplication $A \cdot (B \cdot C) = (A \cdot B) \cdot C$ is clear, because it asserts that $A \cap (B \cap C) = (A \cap B) \cap C$. The associative law for addition, $A + (B + C) = (A + B) + C$, is rather difficult to prove in a formal way; however, the reader may convince himself of its validity by use of the accompanying Venn diagram

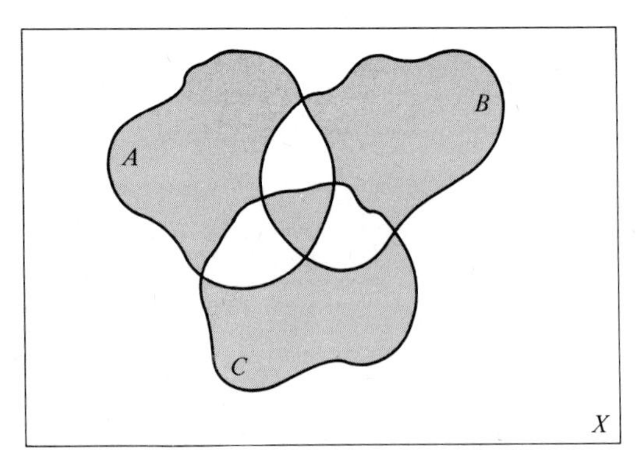

in which both $A + (B + C)$ and $(A + B) + C$ turn out to be the shaded area. Since $A + \varnothing = A$ for all $A \in \mathscr{S}(X)$, $\varnothing$ is a zero element. Then given

$A \in \mathscr{S}(X)$ we seek B such that $A + B = \varnothing$; trying $B = A$, we have

$$A + A = (A - A) \cup (A - A) = \varnothing \cup \varnothing = \varnothing$$

so A is always its own inverse under addition. [Note: One may choose to denote the additive inverse of A by $\bar{A}$, so $\bar{A} = A$, but it would be a mistake to denote it by $-A$ because all kinds of things can go wrong. For example, once $\mathscr{S}(X)$ is shown to be a ring, our notational conventions call for writing $B + (-A)$ (which equals $B + A$) as $B - A$, which is not the same as the difference $B - A$ defined above. Thus, for us the unadorned symbol $-A$ will have no meaning; it will be permissible only when it appears in a difference such as $B - A$.] Of course, addition and multiplication are commutative, and X is an identity for multiplication since

$$AX = XA = X \cap A = A \qquad \text{for all } A \in \mathscr{S}(X).$$

Finally, because multiplication is commutative, we need only check one distributive law—say $A(B + C) = AB + AC$. This amounts to showing that

$$A \cap \big((B - C) \cup (C - B)\big) = (A \cap B - A \cap C) \cup (A \cap C - A \cap B)$$

—a nontrivial task at this stage. However, the reader may convince himself that the distributive law holds because both $A(B + C)$ and $AB + AC$ represent the shaded area in the following Venn diagram.

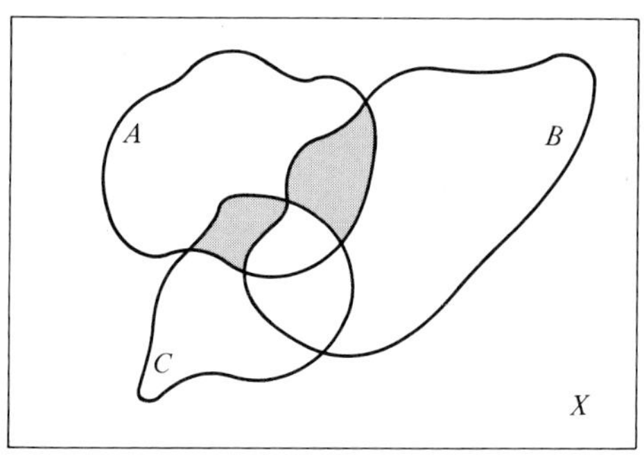

IA-2-sf3

All this shows that $\mathscr{S}(X)$ is a commutative ring with unity.

In connection with the foregoing discussion two questions may be raised. Firstly, where do the definitions of the operations, especially addition, in $\mathscr{S}(X)$ come from? Why do these operations work, in the sense that $\mathscr{S}(X)$ becomes a ring? What is really going on in $\mathscr{S}(X)$? Secondly, even though a picture may be worth a thousand words, the use of Venn diagrams to prove the associative law for addition and the distributive law should not be construed as providing a formal algebraic proof. Can we give more "honest" proofs? These questions will be answered momentarily by showing how $\mathscr{S}(X)$ can be viewed in another, more incisive, way.

Consider our fixed set X and the two-element ring $\mathbf{Z}_2 = \{0, 1\}$. For each given $A \in \mathscr{S}(X)$, let us associate with it the element χ_A (to be read as chi sub A) of the ring Map(X, $\mathbf{Z}_2$) which is defined by

$$\chi_A(x) = \begin{cases} 1 & \text{if } x \in A, \\ 0 & \text{if } x \notin A. \end{cases}$$

In particular, $\chi \in A \Leftrightarrow \chi_A(x) = 1$. We call χ_A the **characteristic function** of the set A; it takes the value 1 for $x \in A$ and the value 0 for $x \in X - A$. On the other hand, given $f \in$ Map(X, $\mathbf{Z}_2$), let us associate with it the set

$$A = \{x \in X \mid f(x) = 1\}$$

—in other words, A (which is more appropriately denoted by A_f) is the set of all elements x for which f takes the value 1. What is the characteristic function χ_A of this set A? The answer is obvious from the definitions—namely, $\chi_A = f$. We see, therefore, that there is a 1–1 correspondence

$$A \leftrightarrow \chi_A$$

between the sets

$$\mathscr{S}(X) \leftrightarrow \text{Map}(X, \mathbf{Z}_2).$$

Consequently, because Map(X, $\mathbf{Z}_2$) is a ring (in fact, it is a commutative ring with unity) a natural way to make $\mathscr{S}(X)$ into a ring is to "transfer" the operations bodily from Map(X, $\mathbf{Z}_2$). To do this, we must [because every element of Map(X, $\mathbf{Z}_2$) is of form χ_A for some $A \in \mathscr{S}(X)$] look first at $\chi_A + \chi_B$ and $\chi_A \chi_B$; that is, we must examine addition and multiplication in Map(X, $\mathbf{Z}_2$) in terms of this new notation for elements of Map(X, $\mathbf{Z}_2$). According to the definition of multiplication in Map(X, $\mathbf{Z}_2$) we have, for $x \in X$,

$$(\chi_A \chi_B)(x) = \chi_A(x)\chi_B(x).$$

So, it follows from the rules for multiplication in $\mathbf{Z}_2$ that the right-hand side is 1 only when both $\chi_A(x)$ and $\chi_B(x)$ are 1. Thus,

$$(\chi_A \chi_B)(x) = \begin{cases} 1 & \text{if } x \in A \cap B, \\ 0 & \text{if } x \notin A \cap B, \end{cases}$$

and, therefore,

$$\chi_A \chi_B = \chi_{A \cap B}. \tag{$*$}$$

In other words, because $\chi_A \chi_B$ is the function which takes the value 1 for $x \in A \cap B$ and 0 elsewhere, it is none other than $\chi_{A \cap B}$, the characteristic function of $A \cap B$.

Furthermore, for $x \in X$, we have by the definition of addition in $\mathrm{Map}(X, \mathbf{Z}_2)$,

$$(\chi_A + \chi_B)(x) = \chi_A(x) + \chi_B(x)$$

and the right-hand side equals 1 if and only if one of the terms is 0 and the other is 1. It follows that

$$(\chi_A + \chi_B)(x) = \begin{cases} 1 & \text{if } x \in (A - B) \cup (B - A), \\ 0 & \text{if } x \notin (A - B) \cup (B - A), \end{cases}$$

so that $\chi_A + \chi_B$ is precisely the characteristic function of the set $(A - B) \cup (B - A)$; in symbols,

$$\chi_A + \chi_B = \chi_{(A-B)\cup(B-A)}. \qquad (**)$$

Now, we are ready to transfer the operations from the ring $\mathrm{Map}(X, \mathbf{Z}_2)$ to the set $\mathscr{S}(X)$. To transfer addition, $A + B$ must be taken as the subset of X, which corresponds to the element $\chi_A + \chi_B$ of the ring $\mathrm{Map}(X, \mathbf{Z}_2)$. In virtue of (**), the "correct" definition is

$$A + B = (A - B) \cup (B - A).$$

Similarly, to transfer multiplication, AB must be taken as the subset of X which corresponds to the element $\chi_A \chi_B$ of the ring $\mathrm{Map}(X, \mathbf{Z}_2)$. But in virtue of (*), the subset corresponding to $\chi_A \chi_B = \chi_{A \cap B}$ is precisely $A \cap B$—so the "correct" definition is

$$A \cdot B = A \cap B.$$

Because of the way the operations of addition and multiplication were transferred from $\mathrm{Map}(X, \mathbf{Z}_2)$ to $\mathscr{S}(X)$, it is clear that $\mathscr{S}(X)$ becomes a ring which is essentially the same as $\mathrm{Map}(X, \mathbf{Z}_2)$.

2-2-14. Exercise. This exercise is concerned with the so-called "calculus of sets." All our sets, $A, B, C, \ldots$, are assumed to be contained in a universal set X. In addition to using the symbols $\cup$, $\cap$, $-$ dealt with in 2-2-13, we write A^c for $X - A$ and call it the **complement** of A.

To prove relations such as inclusion or equality between sets, Venn diagrams are suggestive, but they do not constitute a proof in themselves. One must give a formal, logical argument using words and symbols. For example, the elemental way to show that $A \subset B$ involves verifying that every element of A belongs to B (that is, $x \in A \Rightarrow x \in B$)—a picture will not do. Also, a standard way to prove $A = B$ is to show that $A \subset B$ and $B \subset A$. (Our notation $A \subset B$, it should be noted, does not exclude the possibility $A = B$). Of course, after the elementary properties have been proved the more complicated ones can often be proved directly (that is, by "computational" techniques) rather than having to deal with individual elements. Now prove:

For any sets A, B, C we have

(*i*) If $A \subset B$ and $B \subset C$, then $A \subset C$,
(*ii*) $A \cap B \subset A$ and $A \cap B \subset B$,
(*iii*) $A \subset A \cup B$ and $B \subset A \cup B$,
(*iv*) $A \subset B \cap C \Leftrightarrow A \subset B$ and $A \subset C$,
(*v*) $A \cup B \subset C \Leftrightarrow A \subset C$ and $B \subset C$,
(*vi*) $A \cap B = A \Leftrightarrow A \subset B$,
(*vii*) $A \cup B = A \Leftrightarrow B \subset A$,
(*viii*) $A \subset B \Leftrightarrow B^c \subset A^c$,
(*ix*) $A \cup A^c = X$ and $A \cap A^c = \varnothing$,
(*x*) $(A^c)^c = A$,
(*xi*) $A \subset B^c \Leftrightarrow A \cap B = \varnothing$,
(*xii*) $A \supset B^c \Leftrightarrow A \cup B = X$,
(*xiii*) $A - B = A - (A \cap B) = A \cap B^c$,
(*xiv*) $A \cap (B \cup C) = (A \cap B) \cup (A \cap C)$,
(*xv*) $A \cup (B \cap C) = (A \cup B) \cap (A \cup C)$,
(*xvi*) de Morgan's laws: $(A \cap B)^c = A^c \cup B^c$, $(A \cup B)^c = A^c \cap B^c$,
(*xvii*) $(A - B) \cap (A - C) = A - (B \cup C)$,
(*xviii*) $(A - B) \cup (A - C) = A - (B \cap C)$,
(*xix*) $(A - B) \cup (B - A) = (A \cup B) - (A \cap B) = (A \cap B)^c \cap (A \cup B)$.

2-2-15 / *PROBLEMS*

1. By verifying the axioms, decide which of the following are rings with respect to the standard operations:

(*i*) $\{a + b\sqrt[3]{3} \mid a, b \in \mathbf{Z}\}$,

(*ii*) $\{a + b\sqrt[3]{3} + c\sqrt[3]{9} \mid a, b, c \in \mathbf{Z}\}$,

(*iii*) $\{a + b\sqrt{2} \mid a, b \in 2\mathbf{Z}\}$,

(*iv*) $\left\{a + b\left(\frac{1 + \sqrt{5}}{2}\right) \middle| a, b \in \mathbf{Z}\right\}$,

(*v*) $\left\{a + b\left(\frac{1 + \sqrt{7}}{2}\right) \middle| a, b \in \mathbf{Z}\right\}$.

2. By verifying the axioms, decide which of the following are rings with respect to the standard operations:

(*i*) $\left\{\dfrac{m}{n}\,\middle|\, m, n \in \mathbf{Z}, (m, n) = 1, n > 0, n \text{ odd}\right\}$,

(*ii*) $\left\{\dfrac{m}{p^r}\,\middle|\, m \in \mathbf{Z}, p \text{ a fixed prime}, r \geq 1\right\}$,

(*iii*) $\left\{\dfrac{m}{n}\,\middle|\, m, n \in \mathbf{Z}, n \neq 0, (m, n) = 1, (n, p) = 1\right\}$, p is a fixed prime.

What is the significance of the fact that rational numbers m/n can be written in more than one way?

3. Suppose $\{R, +, \cdot\}$ is a ring. If we define a new "multiplication" operation $\circ$ in R by: $a \circ b = 0$ for all $a, b \in R$, is $\{R, +, \circ\}$ a ring?

4. In $\mathbf{Z} \times \mathbf{Z}$ define addition and multiplication by

$$(a, b) + (c, d) = (a + c, b + d),$$
$$(a, b) \cdot (c, d) = (ac + bd, ad + bc + bd).$$

Verify that this is a commutative ring with unity. Is it an integral domain?

5. Consider $\mathbf{Z}$ with the usual operations $+$ and $\cdot$. Now, let us define new operations $\perp$ and $\circ$ (to serve as addition and multiplication, respectively) by

$$a \perp b = a + b + 1, \qquad a \circ b = a + b + ab.$$

Show that $\{\mathbf{Z}, \perp, \circ\}$ is a commutative ring with unity—the zero element is -1, and the identity for multiplication is 0. Is it an integral domain?

6. Suppose $\{R, +, \cdot\}$ is a ring.

(*i*) If we define $a \perp b = a \cdot b$ and $a \circ b = a + b$, is $\{R, \perp, \circ\}$ a ring?

(*ii*) If we define $a \circ b = a + b + a \cdot b$, is $\{R, +, \circ\}$ a ring? Does it have a unity?

(*iii*) If $\{R, +, \cdot\}$ has a unity e, and we define

$$a \perp b = a + b - e, \qquad a \circ b = a + b - ab$$

is $\{R, \perp, \circ\}$ a ring? Is it commutative? Does it have a unity?

7. Let S denote the set of all positive integers. Is S a subring of $\mathbf{Z}$?

8. Is $\mathbf{Z}_2$ a subring of $\mathbf{Z}$? Why? How about $\mathbf{Z}_m$?

9. All the sets considered in Problems 1 and 2 above may be viewed as subsets of $\mathbf{R}$. Which ones are subrings of $\mathbf{R}$? This provides a simple way to settle the question raised in Problems 1 and 2—namely, which ones are rings with respect to the standard operations.

10. In each of the following cases, find as many subrings of the given ring as you can:

$$(i)\ \mathbf{Z}_7, \qquad (ii)\ \mathbf{Z}_6, \qquad (iii)\ \mathbf{Z}_{12}.$$

Are there any other subrings? In each case, can you prove that you have found all subrings?

11. If m is an integer greater than or equal to 0 show that $m\mathbf{Z} = \{mn \mid n \in \mathbf{Z}\}$ is a subring of $\mathbf{Z}$. Show further that every subring of $\mathbf{Z}$ is of this form—or to put it another way, there are no other subrings of $\mathbf{Z}$.

12. (*i*) Let m be a nonzero integer. Is $m\mathbf{Q} = \{ma \mid a \in \mathbf{Q}\}$ a subring of $\mathbf{Q}$ under the usual operations?

(*ii*) Except for the subrings (0) and $\mathbf{Q}$ of the ring $\mathbf{Q}$, can you find any other subrings?

13. (*i*) Let $\omega = (1 + \sqrt{5})/2$ and consider the subset

$$\mathbf{Z}[\omega] = \{a + b\omega \mid a, b \in \mathbf{Z}\}$$

of $\mathbf{R}$. Show that $\mathbf{Z}[\omega]$ is a subring of $\mathbf{R}$, and even more that $\mathbf{Z}[\omega]$ is an integral domain.

(*ii*) Observe that $\mathbf{Z}[\omega]$ is essentially the same as the ring of Problem 4 above, and conclude thereby that the ring of Problem 4 is an integral domain.

14. (*i*) Let R_1 and R_2 be subrings of the ring R, then $R_1 \cap R_2$ is a subring of R.

(*ii*) Furthermore, the intersection

$$\bigcap_{i=1}^{n} R_i = R_1 \cap R_2 \cap \cdots \cap R_n$$

of a finite collection of subrings $R_1, R_2, \ldots, R_n$ is a subring.

(*iii*) Does this result carry over to the intersection of an infinite collection of subrings?

15. If R_1 and R_2 are subrings of the ring R show by example that $R_1 \cup R_2$ need not be a ring. In fact, $R_1 \cup R_2$ is a subring $\Leftrightarrow R_1 \subset R_2$ or $R_2 \subset R_1$.

16. Fix an element $x \neq 0$ of the ring R.

(*i*) Is the set $xR = \{xa \mid a \in R\}$ a subring of R? How about $Rx = \{ax \mid a \in R\}$?

(*ii*) Is $A_x = \{a \in R \mid ax = 0\}$ a subring of R?

(*iii*) Consider the set $A = \{a \in R \mid aR = (0)\}$—so $a \in A$ if and only if $ab = 0$ for all $b \in R$. Is A a subring of R? How is A related to the A_x's?

17. The **center** $\mathfrak{Z}$ of the ring R is defined by

$$\mathfrak{Z} = \{a \in R \mid ab = ba \quad \text{for all } b \in R\}.$$

Show that $\mathfrak{Z}$ is a subring of R.

18. (*i*) Suppose R_1 and R_2 are subrings of the ring R and we write

$$R_1 + R_2 = \{a_1 + a_2 \mid a_1 \in R_1,\ a_2 \in R_2\}.$$

Thus, $R_1 + R_2$, which is called the **sum** of R_1 and R_2, consists of all elements of R which arise by adding an element of R_1 and an element of R_2. (Of course, $R_1 + R_2$ has nothing to do with the direct sum or with the sum, $+$, of subsets of R.) Is $R_1 + R_2$ a subring of R?

(*ii*) In the case where $R = \mathbf{Z}$, consider two subrings $m\,\mathbf{Z}$ and $n\,\mathbf{Z}$. Describe the set $m\,\mathbf{Z} + n\,\mathbf{Z}$. Is it a subring of $\mathbf{Z}$?

19. Use the fact that $\mathbf{Z}[i]$ is a subset of $\mathbf{C}$ to show it is an integral domain.

20. Suppose D is an integral domain with unity e, D' is an integral domain with unity e', and $D' \subset D$—or more precisely, D' is a **subdomain** of D. Show that $e' = e$. In other words, an integral domain and any subdomain have the same unity.

21. Suppose S is a nonempty subset of an integral domain D. What properties must one check in order to guarantee that S is an integral domain (that is, a subdomain of D)?

22. Given the following matrices in $\mathscr{M}(\mathbf{Z}, 2)$,

$$A = \begin{pmatrix} 1 & 2 \\ -2 & 3 \end{pmatrix}, \qquad B = \begin{pmatrix} -1 & 6 \\ -5 & 4 \end{pmatrix},$$

$$C = \begin{pmatrix} 3 & -3 \\ 1 & -1 \end{pmatrix}, \qquad D = \begin{pmatrix} -1 & 0 \\ -4 & 1 \end{pmatrix},$$

compute:

(*i*) $A + B$, $A + C$, $A + D$, $B + C$,

(*ii*) $A + B + C$, $B + C + D$,

(*iii*) $A - B$, $B - C$, $C - D$,

(*iv*) AB, BA, CD, DC,

(*v*) ABC, BCD,

(*vi*) $A(B + C)$, $AB + AC$, $B(C - D)$, $BC - BD$.

23. Verify the two distributive laws in $\mathscr{M}(\mathbf{Z}, 2)$.

24. If R is a ring, verify that $\mathscr{M}(R, 2)$ is a ring with respect to the "natural" operations.

25. Prove, in detail, that $\mathscr{M}(\mathbf{Z}, 3)$ is a ring.

26. Consider the following elements of $\mathcal{M}(\mathbf{Z}, 4)$

$$A = \begin{pmatrix} 1 & -1 & 2 & -2 \\ 0 & 3 & 1 & -1 \\ 2 & -3 & 0 & 4 \\ -1 & -1 & -4 & 1 \end{pmatrix}, \quad B = \begin{pmatrix} 2 & 3 & 4 & 1 \\ -3 & -1 & 4 & 1 \\ 1 & 0 & 0 & -1 \\ 3 & -5 & 5 & 0 \end{pmatrix},$$

$$C = \begin{pmatrix} 3 & 0 & 0 & -3 \\ 1 & 2 & 3 & -2 \\ 0 & -2 & 2 & 1 \\ 4 & -4 & 0 & 1 \end{pmatrix},$$

and compute:

(*i*) $A + B$, $A + C$, $B + C$, $A + (B + C)$, $(A + B) + C$,
(*ii*) $A - B$, $A - C$, $B - C$,
(*iii*) AB, AC, BC, BA,
(*iv*) $(AB)C$, $A(BC)$,
(*v*) $B(A + C)$, $BA + BC$,
(*vi*) $(C - A)B$, $CB - AB$.

27. In 2-2-10, we showed that

$$T = \left\{ \begin{pmatrix} a & -b \\ b & a \end{pmatrix} \middle| \; a, b \in \mathbf{Z} \right\}$$

is a subring of $\mathcal{M}(\mathbf{Z}, 2)$. Show that T is a commutative ring with unity. Is it an integral domain?

28. Which of the following are subrings of $\mathcal{M}(\mathbf{Z}, 2)$:

(*i*) $\left\{ \begin{pmatrix} a & b \\ 0 & 0 \end{pmatrix} \middle| \; a, b \in \mathbf{Z} \right\}$,

(*ii*) $\left\{ \begin{pmatrix} a & 0 \\ b & 0 \end{pmatrix} \middle| \; a, b \in \mathbf{Z} \right\}$,

(*iii*) $\left\{ \begin{pmatrix} 0 & a \\ 0 & 0 \end{pmatrix} \middle| \; a \in \mathbf{Z} \right\}$,

(*iv*) $\left\{ \begin{pmatrix} a & b \\ 0 & c \end{pmatrix} \middle| \; a, b, c \in \mathbf{Z} \right\}$,

(*v*) $\left\{ \begin{pmatrix} a & 0 \\ 0 & b \end{pmatrix} \middle| \; a, b \in \mathbf{Z} \right\}$,

(vi) $\left\{\begin{pmatrix} a & 0 \\ 0 & a \end{pmatrix} \,\middle|\, a \in 3\,\mathbf{Z}\right\}$,

(vii) $\left\{\begin{pmatrix} 0 & b \\ a & 0 \end{pmatrix} \,\middle|\, a, b \in \mathbf{Z}\right\}$,

(viii) $\left\{\begin{pmatrix} 0 & 0 \\ a & b \end{pmatrix} \,\middle|\, a, b \in \mathbf{Z}\right\}$,

(ix) $\left\{\begin{pmatrix} a & 1 \\ 0 & b \end{pmatrix} \,\middle|\, a, b \in \mathbf{Z}\right\}$.

Can you find additional subrings of $\mathscr{M}(\mathbf{Z}, 2)$?

29. For any complex number $\alpha = a + bi$ let us write $\bar{\alpha} = a - bi$. Show that

$$Q = \left\{\begin{pmatrix} \alpha & \beta \\ -\bar{\beta} & \bar{\alpha} \end{pmatrix} \,\middle|\, \alpha, \beta \in \mathbf{C}\right\}$$

is a subring of $\mathscr{M}(\mathbf{C}, 2)$. It is known as the ring of **quaternions**.
Consider the four elements

$$e = \begin{pmatrix} 1 & 0 \\ 0 & 1 \end{pmatrix}, \qquad \lambda = \begin{pmatrix} i & 0 \\ 0 & -i \end{pmatrix}, \qquad \mu = \begin{pmatrix} 0 & 1 \\ -1 & 0 \end{pmatrix}, \qquad \nu = \begin{pmatrix} 0 & i \\ i & 0 \end{pmatrix}$$

of Q, and verify that

$$\lambda^2 = \mu^2 = \nu^2 = -e,$$
$$\lambda\mu = \nu, \qquad \mu\nu = \lambda, \qquad \nu\lambda = \mu,$$
$$\mu\lambda = -\nu, \qquad \nu\mu = -\lambda, \qquad \lambda\nu = -\mu.$$

30. For any matrix $A = \begin{pmatrix} a & b \\ c & d \end{pmatrix}$ in $\mathscr{M}(\mathbf{Z}, 2)$ define $A^* \in \mathscr{M}(\mathbf{Z}, 2)$ by

$$A^* = \begin{pmatrix} a & c \\ b & d \end{pmatrix}$$

and call it the **transpose** of A. Show that for all $A, B \in \mathscr{M}(\mathbf{Z}, 2)$,

$$(A + B)^* = A^* + B^*,$$
$$(AB)^* = B^*A^*.$$

31. Consider the set $X = \{0, 1, 2, 3, 4, 5, 6, 7, 8, 9\}$ and the subsets

$$A = \{0, 1, 2, 3\}, \qquad B = \{2, 3, 5, 7, 9\}, \qquad C = \{4, 6\},$$
$$D = \{3, 4, 6, 8, 9\}, \qquad E = \{1, 3, 5, 8\}, \qquad F = \varnothing,$$
$$G = \{0\}, \qquad H = \{4, 6\}, \qquad I = \{6\}.$$

(For any subset Y of X we shall write Y^c for $X - Y$, and refer to it as the **complement** of Y.) Find the following sets:

(1) $A \cup B, B \cup C, E \cup F, A \cup G, H \cup C, F \cup G, X \cup E$.
(2) $A \cap D, A \cap C, B \cap F, A \cap G, H \cap I, E \cap G, X \cap E$.
(3) $A \times H, A \times I, C \times G, C \times C, H \times C, G \times I$.
(4) $A - B, B - C, E - F, F - G, C - G, H - C, I - H$.
(5) $A^c, C^c, F^c, G^c, I^c, (B^c)^c, (E^c)^c$.
(6) $A \cup A^c, D \cup D^c, F \cup F^c, G \cup G^c, I \cup I^c$.
(7) $B \cap B^c, C \cap C^c, F \cap F^c, G \cap G^c, H \cap H^c$,
(8) $B - (B \cap E), A - (B - C), D - (C \cup H), C \cup (A - G)$.
(9) $(A \cap B) \cup C, A \cap (B \cup C), (A \cup C) \cap (B \cup C)$.
(10) $A \cap (B \cup C), (A \cap B) \cup (A \cap C)$.
(11) $(H \cap I)^c, H^c \cup I^c$.
(12) $(E \cup G)^c, E^c \cap G^c$,
(13) $A - (B \cup C), (A - B) \cap (A - C)$.
(14) $A - (B \cap D), (A - B) \cup (A - D)$.

32. Suppose A, B, C are sets for which $A \cup B = A \cup C$ and $A \cap B = A \cap C = \varnothing$; show that $B = C$.

33. If X is a set with n elements, show that Map(X, $\mathbf{Z}_2$) has exactly 2^n elements. How many elements are there in $\mathscr{S}(X)$, the ring of all subsets of X? How many elements are there in Map(X, $\mathbf{Z}_3$)?

34. Let X be a set with three elements. Give names to the eight subsets of X, and construct addition and multiplication tables for the ring $\mathscr{S}(X)$.

35. Suppose a set X and a ring R are given, each with more than one element —$\#(X) > 1$, $\#(R) > 1$. Then

(*i*) R has a unity element $\Leftrightarrow$ Map(X, R) has a unity element.
(*ii*) R is a commutative $\Leftrightarrow$ Map(X, R) is commutative; and in this situation, Map(X, R) always has divisors of zero.

36. (*i*) Consider the ring Map(X, R) and fix an element $x_0 \in X$. Let

$$M_{x_0} = \{ f \in \text{Map}(X, R) \mid f(x_0) = 0 \}.$$

Show that M_{x_0} is a subring of Map(X, R). Can you find other subrings?

(*ii*) When $R = \mathbf{Z}_2$, Map(X, $\mathbf{Z}_2$) may be interpreted as $\mathscr{S}(X)$. Consequently, M_{x_0} may be interpreted as a subring of $\mathscr{S}(X)$; which one?

37. Describe some subrings of $\mathscr{S}(X)$.

38. If R_1 and R_2 are rings, verify carefully that $R_1 \oplus R_2$ is a ring. Moreover, $R_1 \oplus R_2$ has an identity $\Leftrightarrow$ both R_1 and R_2 have identities, and $R_1 \oplus R_2$ is commutative $\Leftrightarrow$ both R_1 and R_2 are commutative.

39. Suppose X is a set with two elements. Discuss the "sameness" of the rings Map(X, $\mathbf{Z}_2$), $\mathscr{S}(X)$, and $\mathbf{Z}_2 \oplus \mathbf{Z}_2$.

40. (*i*) Show that $S = \{0, 2, 4, 6\}$ is a subring of $\mathbf{Z}_8$. Construct addition and multiplication tables for S.
(*ii*) Show that $T = \{0, 4, 8, 12\}$ is a subring of $\mathbf{Z}_{16}$. Construct addition and multiplication tables for T.
(*iii*) Construct addition and multiplication tables for the four-element ring $\mathbf{Z}_2 \oplus \mathbf{Z}_2$.
(*iv*) In $\mathbf{Z}_2 \oplus \mathbf{Z}_2$, keep the same addition, and take multiplication to be trivial—that is, the product of any two elements is $0 = (0, 0)$. This is a ring (see Problem 3), which we denote by $(\mathbf{Z}_2 \oplus \mathbf{Z}_2)^0$. Construct addition and multiplication tables for this ring.
(*v*) The ring $\mathscr{M}(\mathbf{Z}_2, 2)$ of all 2×2 matrices with entries from $\mathbf{Z}_2$ has 16 elements. Find a four-element subring which is not commutative. Make addition and multiplication tables for it.
(*vi*) Of course, $\mathbf{Z}_4$ is also a ring with four elements. Make tables for it.
(*vii*) Can you find any additional rings with four elements?

41. Construct several distinct rings with 72 elements.

42. For every positive integer m, there exists at least one ring with m elements. Produce as many such rings as you can.

43. Give an example of a ring with 37 elements which does not have a unity.

44. Suppose R is a commutative ring in which there exist elements a and b such that $ab \neq 0$; then $\mathscr{M}(R, 2)$ is not commutative.

45. Suppose S_1 is a subring of R_1 and S_2 is a subring of R_2; prove that $S_1 \oplus S_2$ is a subring of $R_1 \oplus R_2$.

46. Interpret "geometrically" the ring $\mathbf{R} \oplus \mathbf{R}$ and the subring $\mathbf{Z} \oplus \mathbf{Z}$.

47. In 2-2-11, we discussed the ring $R \oplus R$ and the three subrings

$$R' = \{(a, b) \in R \oplus R \mid b = 0\},$$
$$R'' = \{(a, b) \in R \oplus R \mid a = 0\},$$
$$\Delta = \{(a, b) \in R \oplus R \mid a = b\},$$

each of which is essentially the same as R. What are $R' \cap R''$, $R' \cap \Delta$, $R'' \cap \Delta$? Using the notation $+$ introduced in Problem 18, what are $R' + R''$, $R' + \Delta$, $R'' + \Delta$?

48. (*i*) We observed in 2-2-11 one way in which $R^{(n)}$ could be viewed as a subring of $R^{(\infty)}$. Find other ways of viewing $R^{(n)}$ as a subring of $R^{(\infty)}$.
(*ii*) Find other subrings of $R^{(\infty)}$.

49. (*i*) Suppose for each $i = 1, 2, 3, \ldots$ we are given a ring R_i. The product set of all these rings, $R_1 \times R_2 \times \cdots \times R_n \times \cdots$, consists of all infinite sequences $(a_1, a_2, a_3, \ldots, a_n, \ldots)$ where $a_i \in R_i$ for each i. When addition and multiplication are defined componentwise this becomes a ring—called the **direct product** of the R_i's and denoted by $\Pi_{i=1}^{\infty} R_i$.

Consider the set of all elements $(a_1, a_2, a_3, \ldots)$ of $\Pi_1^{\infty} R_i$ which have only a finite number of nonzero components a_i (in other words, $a_i = 0$ for almost all i). This set is a subring of $\Pi_1^{\infty} R_i$, which is called the **direct sum** of the R_i's and denoted by $\Sigma_{i=1}^{\infty} \oplus R_i$.

Note that if this procedure is applied to a **finite** number of rings $R_1, R_2, \ldots, R_n$, then the direct product $\Pi_{i=1}^{n} R_i$ and the direct sum $\Sigma_{i=1}^{n} \oplus R_i$ are the same.

(*ii*) Suppose each of the rings R_i, $i = 1, 2, 3, \ldots$ is the same ring R, and we let $X = \{1, 2, 3, \ldots\}$. Then the direct product ring $\Pi_1^{\infty} R_i$ is essentially the same as Map(X, R). How would you interpret the direct sum $\Sigma_1^{\infty} \oplus R_i$ as a subring of Map(X, R)?

2-3. Ordered and Well-Ordered Domains

The definitions of ring and integral domain involved isolating certain properties of **Z**—namely, properties concerning addition and multiplication—and using them as "axioms" for a general system. However, there is much more going on in **Z**. For example, in **Z** we have such notions as: the size of an integer, comparing two integers as to size (that is, which is bigger), or ordering the integers according to their size. We shall carry such considerations over to a domain D by axiomatizing the notion of order in a rather indirect fashion—that is, by focusing on the set of "positive elements."

2-3-1. Definition. An integral domain D is said to be an **ordered domain** when there exists a nonempty subset P of D, called the **set of positive elements,** which satisfies the following conditions:

(*i*) $a, b \in P \Rightarrow a + b \in P$; in words: P is closed under addition.
(*ii*) $a, b \in P \Rightarrow ab \in P$; in words: P is closed under multiplication.
(*iii*) If we write $-P = \{-a \mid a \in P\}$ (the set of additive inverses of the elements of P), and call it the **set of negative elements,** then D is a disjoint union

$$D = P \cup \{0\} \cup -P.$$

According to the definition, an element of P is said to be **positive**, and an element of $-P$ is said to be **negative**. Because D is a disjoint union of the three set P, $\{0\}$, $-P$ we note that the element 0 is neither positive nor negative.

Condition (*iii*), which may be referred to as the **trichotomy law**, says that for each $a \in D$ *exactly* one of the following choices holds

$$a \in P, \qquad a = 0, \qquad a \in -P$$

—in other words, a is positive, or a is 0, or a is negative.

We observe that for $b \in D$,

$$b \in -P \Leftrightarrow -b \in P$$

or in words,

$$b \text{ is negative} \Leftrightarrow -b \text{ is positive}$$

since $b \in -P \Leftrightarrow b = -a$ for some $a \in P \Leftrightarrow -b = a$ for some $a \in P \Leftrightarrow -b \in P$. In addition, replacing b by $-b$ we conclude that

$$-b \in -P \Leftrightarrow b \in P$$

or in words,

$$b \text{ is positive} \Leftrightarrow -b \text{ is negative.}$$

2-3-2. Examples. Consider the integral domains $\mathbf{Z}$, $\mathbf{Q}$, or $\mathbf{R}$. In each case, let P be the set of all elements which are greater than 0—thus, P is what we are accustomed to calling the "set of all positive elements." Clearly, the requirements of the definition 2-3-1 are properties which we "know" for integers, rational numbers, or real numbers. Consequently, $\mathbf{Z}$, $\mathbf{Q}$, and $\mathbf{R}$ are ordered domains.

On the other hand, the domain $\mathbf{Z}_2 = \{0, 1\}$ cannot be made into an ordered domain. Obviously, there is no way to choose P so that the trichotomy law holds.

Moreover, the integral domain $\mathbf{C}$ cannot be made into an ordered domain. For suppose $\mathbf{C}$ is an ordered domain with set of positive elements P. Since $i = \sqrt{-1} \neq 0$, the trichotomy law says that $i \in P$ or $i \in -P$. If $i \in P$, then, using condition (*ii*), $-1 = i^2 \in P$. If $i \in -P$, then $-i \in P$ and, by condition (*ii*), $-1 = (-i)^2 \in P$. Thus, in either case, $-1 \in P$, which implies $1 \in -P$. On the other hand, $-1 \in P$ implies $1 = (-1)^2 \in P$. We have, therefore, $1 \in P \cap -P$, which contradicts the fact that the sets P and $-P$ are disjoint.

2-3-3. Definition. Suppose D is an ordered domain with P the set of positive elements. We shall express this situation in more concise fashion by saying that "$\{D, P\}$ is an ordered domain." For $a, b \in D$ we say that a is **less than** b, and write this symbolically as $a < b$ or as $b > a$, when $b - a \in P$. In such circumstances, following the usual terminology, we also say that a is

smaller than b, b is **greater than** a, or b is **bigger than** a. Thus, the symbol " $<$ " is to be read as "less than" and the symbol " $>$ " is to be read as "greater than."

According to the definition

$$a < b \Leftrightarrow b - a \in P \Leftrightarrow b - a \text{ is positive.}$$

So by taking $a = 0$ and changing b to a, or by taking $b = 0$, we have

$$\begin{aligned} &a > 0 \Leftrightarrow a \in P \Leftrightarrow a \text{ is positive,} \\ &a < 0 \Leftrightarrow -a \in P \Leftrightarrow a \in -P \Leftrightarrow a \text{ is negative.} \end{aligned} \tag{$*$}$$

Consequently, the notation $<$ (or $>$) and terminology regarding positive and negative elements conform to our expectations based on our experience with real numbers. In other words, things turn out exactly the way they should!

2-3-4. Remark. The requirements for an ordered domain as given in the definition 2-3-1, may be translated to the language and notation of "less than, $<$," or "greater than, $>$." They become:

(*i*) $a > 0, b > 0 \Rightarrow a + b > 0$.
(*ii*) $a > 0, b > 0 \Rightarrow ab > 0$.
(*iii*) For any $a \in D$ exactly one of the following holds:

$$a > 0, \qquad a = 0, \qquad a < 0.$$

Our next result provides additional properties of the relation $<$ in an ordered domain. Even more important, we show how these properties of $<$ can be used as an alternative set of axioms for making an integral domain into an ordered domain.

2-3-5. Theorem. Suppose D is an integral domain. Then D is an ordered domain $\Leftrightarrow$ there exists a relation $<$ on D with the properties:

(1) $a < b, b < c \Rightarrow a < c$; we say that the relation $<$ is **transitive**.
(2) $a < b \Rightarrow a + c < b + c$ for all $c \in D$; we say that the relation " $<$ " is **additive**.
(3) $a < b, c > 0 \Rightarrow ac < bc$; we say that the relation $<$ is **multiplicative**.
(4) For any $a, b \in D$ exactly one of the following holds:

$$a < b, \qquad a = b, \qquad b < a$$

—we say that the relation $<$ satisfies the **trichotomy law**.

(Obviously, this result may be restated in terms of a relation $>$.)

Proof: $\Rightarrow$. To prove this part, we suppose $\{D, P\}$ is an ordered domain and define $<$ (and $>$) as before, in 2-3-3; we must then prove properties (1)–(4).

The proof of (1) consists of

$$\begin{aligned} a < b, b < c &\Rightarrow b - a \in P, c - b \in P \\ &\Rightarrow c - a = (c - b) + (b - a) \in P \\ &\Rightarrow a < c. \end{aligned}$$

In similar fashion, for any $c \in D$ we have

$$\begin{aligned} a < b &\Rightarrow b - a \in P \\ &\Rightarrow (b + c) - (a + c) = b - a \in P \\ &\Rightarrow a + c < b + c \end{aligned}$$

which proves (2).

The proof of (3) consists of

$$\begin{aligned} a < b, \quad c > 0 &\Rightarrow b - a \in P, c \in P \\ &\Rightarrow bc - ac = (b - a)c \in P \\ &\Rightarrow ac < bc. \end{aligned}$$

Finally, for $a, b \in D$ exactly one of the following holds

$$b - a \in P, \qquad b - a = 0, \qquad b - a \in -P.$$

In virtue of the definition of $<$, and because $b - a \in -P \Leftrightarrow a - b \in P$, these three possibilities translate to

$$a < b, \qquad a = b, \qquad b < a$$

—so (4) holds, and this part of the proof is complete.

$\Leftarrow$. To prove this part, we suppose D is an integral domain with relation $<$ satisfying properties (1)–(4); we must show that there is a subset P of D for which $\{D, P\}$ is an ordered domain. To do this, let us put

$$P = \{a \in D \mid 0 < a\}.$$

It remains to verify the three conditions of the definition 2-3-1 of ordered domain for this P.

(*i*) If $a \in P$ and $b \in P$, then $0 < a$ and $0 < b$. Since $0 < a$ we have by property (2), $0 + b < a + b$. Combining $b < a + b$ with $0 < b$ we have by transitivity [property (1)] $0 < a + b$—so $a + b \in P$. This proves (*i*).

(*ii*) If $a \in P$ and $b \in P$, then $0 < a$ and $0 < b$. Applying property (3), we multiply through in $0 < a$ by the element $b > 0$, to obtain $0 = 0b < ab$—so $ab \in P$. This proves (*ii*).

(*iii*) Given $a \in D$, we apply property (4) for this a and $b = 0$. The result is that exactly one of the following holds

$$a < 0, \qquad a = 0, \qquad 0 < a.$$

Now, adding $-a$ to both sides of the case $a < 0$, we have according to (2), $0 = a + (-a) < 0 + (-a) = -a$. Thus, the three distinct possibilities are

$$0 < -a, \qquad a = 0, \qquad 0 < a,$$

and these become

$$-a \in P, \qquad a = 0, \qquad a \in P.$$

Defining $-P$ in the customary manner as $\{-b \mid b \in P\}$ we have exactly one of the following

$$a \in -P, \qquad a = 0, \qquad a \in P.$$

This proves (*iii*) and completes the proof. ▮

It is worthwhile for the reader to examine both parts of the preceding proof and see, in each case, where and how all the hypotheses are used.

The fact that there are two distinct but equivalent ways to axiomatize an ordered domain is not of crucial importance for us. We wish to derive additional properties of ordered domains, over and above the properties of the set P of positive elements and the relation $<$ which are already in hand.

2-3-6. Proposition. In an ordered domain $\{D, P\}$ we have:

(1) If $a \neq 0$, then $a^2 > 0$; in words, the square of any nonzero element is positive.

(2) $e > 0$ and $-e < 0$; in words, the identity e is positive, and its additive inverse $-e$ is negative.

(3) If $c < 0$, then $a < b \Leftrightarrow cb < ca$.

(4) $a < b \Leftrightarrow -b < -a$.

(5) If $a < 0$ and $b > 0$, then $ab < 0$; in words, the product of a positive element and a negative element is negative.

Proof: (1) If $a \neq 0$, then $a \in P$ or $a \in -P$, but not both. If $a \in P$, then $a^2 = a \cdot a \in P$, so $a^2 > 0$. If $a \in -P$, then $-a \in P$ and $a^2 = (-a) \cdot (-a) \in P$, so $a^2 > 0$. This takes care of the two possibilities and shows that the square of a nonzero element is positive.

(2) According to the definition of integral domain $e \neq 0$, so by the preceding $e = e^2$ is greater than 0. Thus $e \in P$, which implies $-e \in -P$; this means (see (*), for example) that $-e < 0$.

(3) $\Rightarrow$. This part is straightforward—namely

$$\begin{aligned} c < 0, \quad a < b &\Rightarrow -c \in P,\ b - a \in P \\ &\Rightarrow ca - cb = (-c)(b - a) \in P \\ &\Rightarrow cb < ca. \end{aligned}$$

$\Leftarrow$. The proof of this part is less direct. We are given $c < 0$ and $cb < ca$. Concerning a and b there are three distinct possibilities: $a < b$, $a = b$, $b < a$. If $a = b$, then $ca = cb$, which excludes the possibility that $a = b$. If $b < a$, then, using the first part of this proof, we have $ca < cb$. This contradicts the hypothesis $cb < ca$ and excludes the possibility that $b < a$. The only remaining possibility is $a < b$.

(4) By part (2), $-e < 0$, and we simply apply (3) with $c = -e$.

(5) Applying part (3) to $a < 0$ and $0 < b$ gives $ab < a0 = 0$. ▮

Of course, many other "natural" properties of $<$ hold. Some additional ones are listed in the problems.

2-3-7. Remark. Consider an element $a \neq 0$ in the ordered domain $\{D, P\}$. If $a \in P$, then $-a \in -P$, and we have $-a < 0 < a$. On the other hand, if $a \in -P$, then $-a \in P$, and we have $a < 0 < -a$. In either case, looking at a and $-a$, we see that one of these elements is positive and the other is negative. In particular, a and $-a$ are distinct—that is, $a \neq -a$—so the set $\{a, -a\}$ has two elements. We may now define $|a|$, the **absolute value** of a, for any $a \in D$, as follows

$$|a| = \begin{cases} 0, & \text{if } a = 0, \\ \text{the positive member of the set } \{a, -a\}, & \text{if } a \neq 0. \end{cases} \qquad (\#)$$

If $a \neq 0$, then, according to the definition, $|a|$ is a positive element of D; it follows, therefore, that for $a \in D$,

$$|a| = 0 \Leftrightarrow a = 0.$$

The definition of absolute value may be rephrased as

$$|a| = \text{the biggest element of } \{a, -a\} \qquad \text{for all } a \in D.$$

In fact, if $a \neq 0$, then $|a|$ is the positive element of $\{a, -a\}$, and the positive element is surely the bigger one. Even more, this version of the definition is applicable for $a = 0$ also—since then $-a = a = 0$, and 0 may still be considered as the biggest element of $\{a, -a\}$.

Instead of the word "biggest," it is, in general, common to use the word **maximum**—and consequently we write

$$|a| = \max\{a, -a\} \qquad \text{for all } a \in D. \qquad (\#\#)$$

For any $a \in D$ (including the possibility $a = 0$) the set $\{a, -a\}$ and the set $\{-a, -(-a)\}$ are identical—meaning that they have the same elements—so in virtue of (# #), we have

$$|-a| = |a| \qquad \text{for all } a \in D.$$

There is still another way to state the definition of absolute value. We write, $a \le b$ or $b \ge a$ (and read these in the standard way) to mean $a < b$ or $a = b$, and observe that the basic properties of $<$ carry over to $\le$. Then clearly

$$|a| = \begin{cases} a, & \text{if } a \ge 0, \\ -a, & \text{if } a \le 0. \end{cases} \qquad (\#\#\#)$$

The appearance of the situation $a = 0$ twice in this formulation causes no difficulty. We then have $a = -a = 0$ and $|a| = 0$ in both cases, so everything is consistent.

It is often convenient to write

$$|a| = \pm a$$

(the right side to be read as "plus or minus a"). The meaning is simply that $|a|$ is a or $-a$; we do not know which one, and often do not really care. Of course, this notation includes $|0| = \pm 0$.

We turn to several of the basic properties of absolute value.

2-3-8. Proposition. For any a, b in the ordered domain $\{D, P\}$, we have:

(1) $-|a| \le a \le |a|$.

(2) If $b \ge 0$ and $-b \le a \le b$, then $|a| \le b$.

(3) $|ab| = |a|\,|b|$; in words, the absolute value of a product is the product of the absolute values.

(4) $|a + b| \le |a| + |b|$; in words, the absolute value of a sum is less than or equal the sum of the absolute values. This is known as the **triangle inequality**.

Proof: (1) For any $a \in D$ we have clearly $a \le \max\{a, -a\} = |a|$. Applying this to the element $-a$ gives $-a \le |-a| = |a|$. The latter relation may be multiplied by $-e$ [that is, we apply 2-3-6, part (4), which carries over to the relation $\le$] to obtain $-|a| \le a$. Consequently, $-|a| \le a \le |a|$.

(2) Given $-b \le a \le b$ we multiply through by $-e$, as above, to obtain $-b \le -a \le -(-b) = b$. Thus, both a and $-a$ are less than or equal to b; hence $|a| = \max\{a, -a\} \le b$.

(3) We know that $|a| = \pm a$ and $|b| = \pm b$, so according to the rules for multiplication with $+$ and $-$ (see 2-1-17), it follows that $|a|\,|b| = \pm ab$. We do not know if $+ab$ or $-ab$ is the proper choice, but we do know that the

proper choice is $\max\{ab, -ab\}$. But this is precisely $|ab|$. Hence,

$$|a| \cdot |b| = \pm ab = \max\{ab, -ab\} = |ab|.$$

(4) We start from $-|a| \leq a \leq |a|$ and $-|b| \leq b \leq |b|$. It is easy to see that adding yields

$$-(|a| + |b|) \leq a + b \leq (|a| + |b|)$$

[see, for example, Problem 4, part (*xi*) at the end of this section]. Since $|a| + |b| \geq 0$ we may apply part (2) above, and conclude that $|a + b| \leq |a| + |b|$. ∎

This completes our discussion of what happens when, in analogy with **Z**, the notion of order is introduced in an integral domain. However, there is one more property of the integers which we shall transfer to integral domains —namely, the commonly accepted fact (which was used several times in Chapter I) that any nonempty collection of positive integers has a smallest element.

2-3-9. Definition. In the ordered domain $\{D, P\}$, the set of positive elements P is said to be **well ordered** when every nonempty subset of P has a smallest element, and in this case (to keep the terminology brief) $\{D, P\}$ is said to be **well ordered**.

Of course, by a smallest element of the nonempty subset S of P we mean an element $a \in S$ such that $a \leq b$ for all $b \in S$. Note that if a smallest element of S exists, then it is unique. Indeed, if $a' \in S$ is also a smallest element of S, then we have $a \leq a'$ and $a' \leq a$. This implies $a = a'$—for if $a \neq a'$, then $a < a'$ and $a' < a$, so by transitivity $a < a$, contradiction.

2-3-10. Examples. The most obvious example of a well-ordered domain is $\{\mathbf{Z}, P\}$ where P is the customary set of all positive integers. This is no surprise; after all, the requirements for a well-ordered domain were chosen precisely because they are satisfied in **Z**. However, as a preview of coming attractions it may be remarked that it is extremely hard to find additional examples of well-ordered domains. Thus, the rational numbers **Q** are an ordered domain when P is taken as the set of all positive rational numbers. However, $\{\mathbf{Q}, P\}$ is not well ordered—since if we take for S the set P itself, then S has no smallest element because, as is well known, there is no smallest positive rational number.

We turn now to an investigation of integer-like properties which hold in any well-ordered domain.

2-3-11. Theorem. If $\{D, P\}$ is a well-ordered domain with unity element e, then there are no elements between 0 and e; in other words, e is the smallest positive element.

Proof: According to 2-3-6, we have $0 < e$. Then, as expected, by the phrase "a is an element between 0 and e" we mean that $0 < a < e$. (Note that the meaning of "between" excludes the possibilities $a = 0$ or $a = e$.) Of course, such an a is automatically an element of P.

Turning to the proof, let us suppose the theorem is false. So there exists an element between 0 and e. Consequently, if we put

$$S = \{a \in D \mid 0 < a < e\}$$

(in words, S is the set of all elements between 0 and e), then S is a nonempty subset of P. By well ordering, S has a smallest element—call it c. We have then $0 < c < e$ and $c > 0$, so because $<$ is multiplicative (see 2-3-5) it follows that

$$c \cdot 0 < c \cdot c < c \cdot e.$$

Simplifying (and using $c < e$) we have

$$0 < c^2 < c < e.$$

This tells us that c^2 is an element of S which is smaller than c—contradicting the choice of c as the smallest element of S. Thus, our initial assumption that the theorem is false leads to a contradiction. Hence, the theorem is true.

Clearly, e is the smallest element of P—for if $a \in P$ is smaller than e, then $0 < a < e$, and according to the foregoing this is impossible. ▌

2-3-12. Corollary. If $\{D, P\}$ is a well-ordered domain, then, for any $a \in D$, there are no elements between a and $a + e$.

Proof: Since $e > 0$ we know that $a < a + e$. Now, suppose b is an element of D which lies between a and $a + e$—

$$a < b < a + e.$$

Adding $-a$ to each of these terms we obtain, according to 2-3-5,

$$a - a < b - a < (a + e) - a$$

so that

$$0 < b - a < e.$$

Thus, $b - a$ is an element between 0 and e, which contradicts 2-3-11—so the hypothesized element b which lies between a and $a + e$ cannot exist. ▌

In virtue of this result, if a is any element of a well-ordered domain, it is appropriate to refer to $a + e$ as the "next" element, and also to call a and $a + e$ "consecutive" elements. What happens if one keeps passing from an element to the next element over and over? Do things turn out as in **Z**? The answer will appear shortly.

2-3-13. Definition. Suppose $\{D, P\}$ is a well-ordered domain. A subset S of D is said to be **inductive** when it satisfies the condition

$$a \in S \Rightarrow a + e \in S.$$

The definition of an inductive set requires that for any $a \in S$ the next element $a + e$ must also belong to S. In other words, we might say (carelessly, perhaps) that S is closed under the operation of adding the unity element e, or that S is closed under "passage to the next element."

As examples of inductive sets in the well-ordered domain $\{\mathbf{Z}, P\}$, we may list $S = \{n \in \mathbf{Z} \mid n \geq 0\}$, or $S = \{n \in \mathbf{Z} \mid n > 5\}$, or $S = \{n \in \mathbf{Z} \mid n \geq -3\}$. In fact, the reader may easily convince himself that a subset S of **Z** is inductive $\Leftrightarrow$ $S = \mathbf{Z}$ or S is of form $\{n \in \mathbf{Z} \mid n \geq m\}$ for some fixed $m \in \mathbf{Z}$. Unfortunately, because we have given no examples of well-ordered domains other than $\{\mathbf{Z}, P\}$ we are in no positition to give additional examples of inductive sets.

The use of the name "inductive" for the property under consideration is suggestive of something quite familiar for the integers—namely, mathematical induction. The justification for our choice of this name appears in the next result and its consequences.

2-3-14. Theorem (*Mathematical Induction*). Suppose $\{D, P\}$ is a well-ordered domain. If S is a subset of P which satisfies the conditions

$$(i)\ e \in S, \qquad (ii)\ S \text{ is inductive,}$$

then $S = P$. In other words, an inductive set of positive elements which contains the identity e must be P itself. In particular, this theorem is applicable for the well-ordered domain **Z**.

Proof: Suppose the theorem is false—so $S \neq P$. Let us put

$$T = P - S$$

(by which is meant, as usual, that T consists of those elements of P which do not belong to S) so, because $S \neq P$, T is a nonempty subset of P. Clearly, S and T are disjoint sets whose union is P.

Because we are in a well-ordered domain, the nonempty subset T of P has a smallest element, call it t. Now, $t \notin S$ (since $t \in T$ and the sets S and T are disjoint) and hence $t \neq e$ (because $e \in S$). Furthermore, since $t \in P$, e is the smallest element of P, and $t \neq e$, it follows that $t > e$; and then $t - e > 0$. We observe that $t - e \notin T$ (because t is the smallest element of T, and $t - e < t$) and consequently $t - e \in S$. But S is inductive—so $t = (t - e) + e \in S$, contradiction. This completes the proof. ▮

The preceding formulation of mathematical induction may seem somewhat strange, but we can derive the more familiar versions from it in short order.

2-3-15. Corollary (*Mathematical Induction*). Suppose $\{D, P\}$ is a well-ordered domain and for every $a \in P$ we have a statement (that is, assertion or proposition) $\pi(a)$ which is either true or false. If the following two conditions are satisfied:

(i) $\pi(e)$ is true;
(ii) for any $a \in P$, $\pi(a)$ is true implies that $\pi(a + e)$ is true,

—then $\pi(a)$ is true for all $a \in P$. In particular, this result applies for the well-ordered domain $\mathbf{Z}$.

Proof: Let S be the set of all positive elements a for which $\pi(a)$ is true:

$$S = \{a \in P \mid \pi(a) \text{ is true}\}.$$

Condition (i) guarantees that $e \in S$. Using condition (ii) we have

$$a \in S \Rightarrow \pi(a) \text{ is true} \Rightarrow \pi(a + e) \text{ is true} \Rightarrow a + e \in S,$$

so that S is inductive. Hence, according to 2-3-14, $S = P$, and $\pi(a)$ is true for all $a \in P$. ▮

2-3-16. Corollary (*Mathematical Induction*). Suppose $\{D, P\}$ is a well-ordered domain and for every $a \in P$ we have a statement $\pi(a)$ which is either true or false. If the following condition is satisfied:

(i) For any $a \in P$, the truth of $\pi(b)$ for all $b \in P$ satisfying $b < a$ implies that $\pi(a)$ is true.

—then $\pi(a)$ is true for all $a \in P$. In particular, this result applies for the well-ordered domain $\mathbf{Z}$.

Proof: Suppose the corollary is false; so there exists an $a \in P$ for which $\pi(a)$ is false. Therefore, the set

$$T = \{a \in P \mid \pi(a) \text{ is false}\}$$

is a nonempty subset of P, and by well ordering T has a smallest element—call if c. The choice of c implies that $\pi(b)$ is true for all $b \in P$ with $b < c$. But now condition (i) tells us that $\pi(c)$ is true—contradiction. This completes the proof. ∎

2-3-17. Remark. The three preceding results are of interest to us primarily in the case of the well-ordered domain **Z**. For this reason we restate these versions of mathematical induction explicitly for integers in the more familiar forms.

(1) Suppose S is a set of positive integers such that
(i) $1 \in S$, (ii) $n \in S \Rightarrow n + 1 \in S$,
—then S is the set of all positive integers.

(2) Suppose that for each positive integer n we have a statement $\pi(n)$ which is either true or false. If
(i) $\pi(1)$ is true,
(ii) for any $n > 0$, $\pi(n)$ is true $\Rightarrow \pi(n + 1)$ is true
—then $\pi(n)$ is true for all $n > 0$.

(3) Suppose that for each positive integer n we have a statement $\pi(n)$ that is either true or false. Suppose
(i) for any $n > 0$, $\pi(m)$ is true for all m satisfying $0 < m < n$ implies that $\pi(n)$ is true
—then $\pi(n)$ is true for all $n > 0$.

A word of caution is necessary in connection with the use of (3) to prove something by mathematical induction. On the face of it, the use of (3) appears simpler than the use of (2), because (3) involves the verification of only one condition whereas (2) involves two conditions. However, the appearance is deceptive. More precisely, consider the verification of condition (i) of (3) for the case $n = 1$. The set of all m satisfying $0 < m < 1$ is empty, so the hypothesis part of condition (i), which says that $\pi(m)$ is true for all m satisfying $0 < m < 1$, is vacuously true. Hence, in order to verify condition (i) in the case of $n = 1$, we must show that this vacuously true hypothesis implies the truth of $\pi(1)$. Thus, a direct proof (without any inductive assumptions) of the truth of $\pi(1)$ must be given. This means that things turn out as in (2) where a direct proof of the truth of $\pi(1)$ is required.

2-3-18. Example. To illustrate a proof by induction, let us prove the formula, in $\mathbf{Z}$,

$$\sum_{i=1}^{n} i^2 = \frac{n(n+1)(2n+1)}{6} \qquad \text{for all } n \geq 1. \qquad (*)$$

This is, of course, an infinite number of formulas, one for each positive integer n. As is well known, the left-hand side of $(*)$ is just a shorthand notation for the sum of the squares of the first n integers, so we are concerned with proving

$$1^2 + 2^2 + 3^2 + \cdots + n^2 = \frac{n(n+1)(2n+1)}{6} \qquad \text{for all } n \geq 1.$$

We are not concerned here with how the expression on the right-hand side, $n(n+1)(2n+1)/6$, was derived; our sole interest is in verifying that the formula "works" for all $n \geq 1$.

Naturally, it is physically impossible to verify the formula for each n individually; instead we use the version of mathematical induction given in (1) of 2-3-17. Let

$$S = \{n > 0 \mid (*) \text{ is true for } n\}.$$

In other words, S is the set of all positive integers n for which the formula $(*)$ holds.

To prove that $1 \in S$, we need to check the validity of the formula $(*)$ when $n = 1$. But this is trivial, since for $n = 1$

$$1^2 + 2^2 + \cdots + n^2 \text{ equals } 1 \quad \text{and} \quad \frac{n(n+1)(2n+1)}{6} \text{ equals } \frac{1 \cdot 2 \cdot 3}{6} = 1.$$

It remains to prove that S is inductive—or what is the same, $n \in S \Rightarrow n + 1 \in S$. Thus, we assume (inductively) that $(*)$ is true for a fixed n—so for this n we know

$$\sum_{i=1}^{n} i^2 = \frac{n(n+1)(2n+1)}{6},$$

—and we seek to prove the validity of $(*)$ for $n + 1$. Of course, the right side of $(*)$ for $n + 1$ is obtained by replacing n by $n + 1$ in the right side of the expression $(*)$, so we must prove

$$\sum_{i=1}^{n+1} i^2 = \frac{(n+1)(n+2)(2n+3)}{6}.$$

But this is straightforward; in fact, overdoing the details, we have

$$\begin{aligned}\sum_{i=1}^{n+1} i^2 &= 1^2 + 2^2 + 3^2 + \cdots + (n+1)^2 \\ &= (1^2 + 2^2 + \cdots + n^2) + (n+1)^2 \\ &= \left(\sum_{i=1}^{n} i^2\right) + (n+1)^2 \\ &= \frac{n(n+1)(2n+1)}{6} + (n+1)^2 \qquad \text{(since } n \in S\text{)} \\ &= (n+1)\left(\frac{n(2n+1) + 6(n+1)}{6}\right) \\ &= \frac{(n+1)(n+2)(2n+3)}{6}.\end{aligned}$$

Now, according to version (1) of mathematical induction, S is the set of all positive integers—so $(*)$ is true for all $n > 0$.

It is also possible to prove the preceding result by using version (2) of mathematical induction. The details of the proof are very much like the foregoing, so we content ourselves with a sketch. For each $n > 0$, let $\pi(n)$ be the statement:

the formula $(*)$ holds for the n under consideration.

As before, $\pi(1)$ is clearly true. Next, suppose inductively that $\pi(n)$ is true, so

$$\sum_{i=1}^{n} i^2 = \frac{n(n+1)(2n+1)}{6}$$

for the n in question. Using this one evaluates $\Sigma_{i=1}^{n+1} i^2$, and just as before this becomes $(n+1)(n+2)(2n+3)/6$; so it follows that $\pi(n+1)$ is true. Consequently, according to version (2) of mathematical induction, $\pi(n)$ is true for all $n > 0$. This completes the proof. ▮

2-3-19. Examples. Consider the infinite sequence of integers $u_1, u_2, u_3, \ldots, u_n, \ldots$ defined as follows

$$u_1 = 1, \quad u_2 = 1, \quad u_3 = 2, \quad u_4 = 3, \quad u_5 = 5, \quad u_6 = 8, \ldots$$

and such that for $n \geq 2$, u_{n+1} is defined recursively by the rule

$$u_{n+1} = u_n + u_{n-1}.$$

In other words, each term, starting with the third, is the sum of the two preceding terms; so the sequence looks like

1, 1, 2, 3, 5, 8, 13, 21, 34, 55, 89, 144, 233, 377, 610, ...

It is sometimes convenient to put $u_0 = 0$. This does not harm the recursion, since clearly $u_2 = u_1 + u_0$.

This sequence is commonly known as the **Fibonacci sequence.** It has a number of interesting properties which may be proved by induction. We first list several such properties, and then prove (1), (2), (4), (9), and (10).

(1) $(u_n, u_{n+1}) = 1,$ for all $n \geq 1,$

(2) $u_{n+1} < \left(\frac{7}{4}\right)^n,$ for all $n \geq 1,$

(3) $\sum_{i=1}^{n} u_i = u_{n+2} - 1,$ for all $n \geq 1,$

(4) If we put $a = \frac{1+\sqrt{5}}{2}$ and $b = \frac{1-\sqrt{5}}{2}$, then

$$u_n = \frac{a^n - b^n}{\sqrt{5}}$$ for all $n \geq 1,$

(5) $\sum_{i=1}^{n} u_{2i-1} = u_{2n},$ for all $n \geq 1,$

(6) $\sum_{i=1}^{n} u_{2i} = u_{2n+1} - 1,$ for all $n \geq 1,$

(7) $\sum_{i=1}^{n} (-1)^{i+1} u_i = 1 + (-1)^n u_{n-1},$ for all $n \geq 1,$

(8) $\sum_{i=1}^{n} u_i^2 = u_n u_{n+1},$ for all $n \geq 1,$

(9) $\sum_{i=1}^{2n-1} u_i u_{i+1} = u_{2n}^2,$ for all $n \geq 1,$

(10) $u_{m+n+1} = u_m u_n + u_{m+1} u_{n+1},$ for all $m \geq 0, n \geq 0,$

(11) $u_n \mid u_{2n},$ for all $n \geq 1,$

(12) $u_n \mid u_{kn},$ for all $k \geq 1, n \geq 1,$

(13) $a^n = u_{n-1} + a u_n,$ for all $n \geq 1$ [a as in (4)]

Proof of (1): For each $n \geq 1$, let $\pi(n)$ be the statement

$$(u_n, u_{n+1}) \text{ is equal to } 1.$$

In other words, $\pi(n)$ is the assertion that u_n and u_{n+1} are relatively prime. In the first place, $\pi(1)$ is true because $(u_1, u_2) = (1, 1) = 1$. Now, suppose inductively that $\pi(n)$ is true for some given n—so $(u_n, u_{n+1}) = 1$. We must prove the truth of $\pi(n+1)$—so we must show that $(u_{n+1}, u_{n+2}) = 1$. To do this, suppose d is any positive divisor of (u_{n+1}, u_{n+2}); so d divides both u_{n+1} and $u_{n+2} = u_{n+1} + u_n$. It follows that $d \mid u_n$, and consequently d divides $(u_n, u_{n+1}) = 1$. Therefore, $d = 1$, and then $(u_{n+1}, u_{n+2}) = 1$—so $\pi(n+1)$ is indeed true. Hence, by mathematical induction, $\pi(n)$ is true for all n, which completes the proof that any two consecutive terms of the Fibonacci sequence are relatively prime. ▌

Proof of (2): For each $n \geq 1$, let $\pi(n)$ be the statement

$$u_{n+1} \text{ is less than } (7/4)^n.$$

For $n = 1$ we have $u_2 = 1$ and $(7/4)^1 = 7/4$, so $\pi(1)$ is true. For $n = 2$ we have $u_3 = 2$ and $(7/4)^2 = 49/16$, so $\pi(2)$ is true. Now, let us apply version (3) of mathematical induction (see 2-3-17). We must show

$$\text{for any } n > 0, \quad \pi(m) \text{ is true for all } m \text{ satisfying } 0 < m < n \text{ implies } \pi(n) \text{ is true.} \tag{\#}$$

This implication is surely valid when $n = 1$; for in this case, we have already shown directly that $\pi(1)$ is true—so it is of no consequence that there are no integers m for which $0 < m < n = 1$. Furthermore, the implication (#) is valid when $n = 2$. In this case, we have already shown directly that $\pi(2)$ is true, so it does not really matter if $\pi(m)$ is true for all m satisfying $0 < m < n = 2$—in other words, it does not really matter if $\pi(1)$ is true—the implication (#) would be valid even if $\pi(1)$ were false. (The argument used here is a fact of logic, and the reader should make every effort to understand it thoroughly.) Finally, let us show the validity of the implication (#) when $n \geq 3$. In this situation, we have by the hypothesis of (#), $u_{m+1} < (7/4)^m$ for $m = 1, 2, \ldots, n-1$, and then

$$\begin{aligned} u_{n+1} = u_n + u_{n-1} &< \left(\frac{7}{4}\right)^{n-1} + \left(\frac{7}{4}\right)^{n-2} = \left(\frac{7}{4}\right)^{n-2}\left(\frac{7}{4} + 1\right) \\ &= \left(\frac{7}{4}\right)^{n-2}\left(\frac{11}{4}\right) < \left(\frac{7}{4}\right)^{n-2}\left(\frac{7}{4}\right)^2 = \left(\frac{7}{4}\right)^n \end{aligned} \tag{$*$}$$

—so $\pi(n)$ is true. Thus, the implication (#) is valid for all $n > 0$, so $\pi(n)$ is true for all $n > 0$ and the proof is complete. ∎

In connection with this proof, it should be noted that a proof based on version (2) of mathematical induction (see 2-3-17) cannot be given. Because a term of the Fibonacci sequence depends on the two terms immediately preceding it, an induction which tries to pass from information about a single term of the sequence to the same kind of information about the next term cannot work. In particular, it is for this reason that we proved the truth of $\pi(n)$ for both $n = 1$ and 2 at the start—then an induction based on version (3) of mathematical induction could proceed

Proof of (4): Here again we use version (3) of mathematical induction. A brief sketch will suffice. One verifies in a straightforward manner that the relation

$$u_n = \frac{a^n - b^n}{\sqrt{5}}$$

is true for $n = 1$ and $n = 2$. Then under the inductive assumption that

$$u_m = \frac{a^m - b^m}{\sqrt{5}} \quad \text{for} \quad m = 1, 2, 3, \cdots, n-1 \quad (\text{where } n \geq 3)$$

one obtains (using $a^2 = a + 1$ and $b^2 = b + 1$)

$$\begin{aligned} u_n = u_{n-1} + u_{n-2} &= \frac{a^{n-1} - b^{n-1}}{\sqrt{5}} + \frac{a^{n-2} - b^{n-2}}{\sqrt{5}} \\ &= \frac{a^{n-2}(a+1) - b^{n-2}(b+1)}{\sqrt{5}} \\ &= \frac{(a^{n-2})(a^2) - (b^{n-2})(b^2)}{\sqrt{5}} \\ &= \frac{a^n - b^n}{\sqrt{5}}. \end{aligned}$$

Consequently, the desired relation is true for all $n \geq 1$. ▌

Proof of (9): We give a compact, but complete, proof. Let

$$S = \left\{ n \geq 1 \;\middle|\; \sum_{i=1}^{2n-1} u_i u_{i+1} = u_{2n}^2 \right\}.$$

First of all, $1 \in S$, because in this case $\Sigma_{i=1}^{2n-1} u_i u_{i+1}$ is $u_1 u_2$ while u_{2n}^2 is u_2^2 and $u_1 u_2 = 1 = u_2^2$. Then, suppose inductively that $n \in S$—so $\Sigma_{i=1}^{2n-1} u_i u_{i+1} = u_{2n}^2$; we seek to show that $n + 1 \in S$—since $2(n+1) - 1 = 2n + 1$ this means we want

$$\sum_{i=1}^{2n+1} u_i u_{i+1} = u_{2(n+1)}^2.$$

To do this, we compute

$$\begin{aligned} \sum_{i=1}^{2n+1} u_i u_{i+1} &= \sum_{i=1}^{2n-1} u_i u_{i+1} + u_{2n} u_{2n+1} + u_{2n+1} u_{2n+2} \\ &= u_{2n}^2 + u_{2n} u_{2n+1} + u_{2n+1} u_{2n+2} \\ &= u_{2n}(u_{2n} + u_{2n+1}) + u_{2n+1} u_{2n+2} \\ &= u_{2n} u_{2n+2} + u_{2n+1} u_{2n+2} \\ &= (u_{2n} + u_{2n+1}) u_{2n+2} \\ &= u_{2n+2}^2. \end{aligned}$$

Thus, S is inductive and it follows that S consists of all positive integers; so the given formula holds for all $n \geq 1$. ▮

Proof of (10): Our objective is to prove

$$u_{m+n+1} = u_m u_n + u_{m+1} u_{n+1} \qquad (*)$$

for all $m \geq 0$, $n \geq 0$. Since both m and n vary there is a "double infinity" of choices for the pair of indices m and n, so some kind of "double induction" is required.

With each pair of integers $m \geq 0$, $n \geq 0$ let us associate the point (m, n) in the X–Y plane. Plotting all such integral points, we have the accompanying picture.

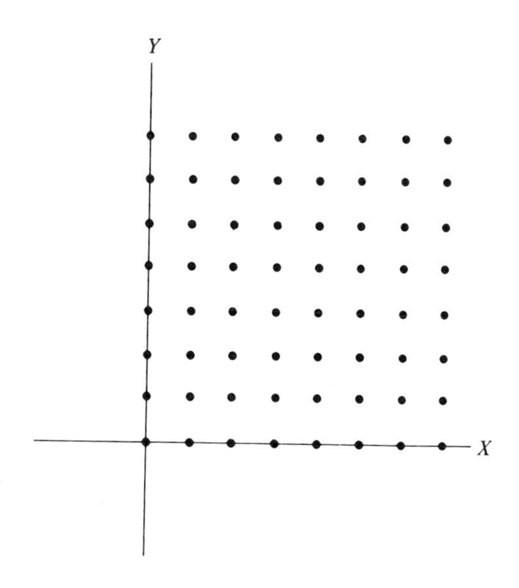

For each $n \geq 0$, let $\pi(n)$ be the statement

the relation $(*)$ holds for this n and all $m \geq 0$.

In other words, $\pi(n)$ asserts that $(*)$ holds for all integral points in row n of the picture (where the rows are naturally numbered, from the bottom upwards, by 0, 1, 2, 3, . . .). The idea of our proof, then, is to apply induction for the rows of the picture.

First of all, $\pi(0)$ is true. To see this, one notes that for $n = 0$ and all $m \geq 0$, the relation $(*)$ reads

$$u_{m+1} = u_m u_0 + u_{m+1} u_1$$

and this is true since $u_0 = 0$, $u_1 = 1$.

Furthermore, $\pi(1)$ is true. In fact, for $n = 1$ and all $m \geq 0$, the relation $(*)$ takes the form

$$u_{m+2} = u_m u_1 + u_{m+1} u_2$$

and this is true since $u_1 = 1$, $u_2 = 1$.

Finally, suppose inductively that $\pi(k)$ is true for $k = 0, 1, 2, \ldots, n$ (where $n \geq 2$); we must show that $\pi(n+1)$ is true. In other words, assuming $(*)$ is true for rows $0, 1, 2, \ldots, n$, we have to show it is also true for row $n + 1$. Now, we have

$$\begin{aligned} u_{m+(n+1)+1} &= u_{m+(n+1)} + u_{m+(n+1)-1} \\ &= u_{m+n+1} + u_{m+(n-1)+1} \\ &= (u_m u_n + u_{m+1} u_{n+1}) + (u_m u_{n-1} + u_{m+1} u_n) \\ &= u_m(u_n + u_{n-1}) + u_{m+1}(u_{n+1} + u_n) \\ &= u_m u_{n+1} + u_{m+1} u_{n+2} \end{aligned}$$

so indeed $\pi(n+1)$ is true. Thus, $\pi(n)$ is true for all n, and the proof is complete. ▮

This proof, it should be noted, uses a slight—but obviously permissible—modification of version (3) of mathematical induction; namely, instead of starting from $n = 1$, we start from $n = 0$.

2-3-20 / *PROBLEMS*

1. If $\{D, P\}$ is an ordered domain, show that
   (*i*) $a < b \Leftrightarrow a + c < b + c$.
   (*ii*) $a - c < a - d \Leftrightarrow d < c$.
   (*iii*) $a < b \Rightarrow a \neq b$.
   (*iv*) If $c > 0$, then $ac < bc \Leftrightarrow a < b$.
   (*v*) If $a < 0$ and $b < 0$, then $ab > 0$.

2. In an ordered domain $\{D, P\}$ show that
   (*i*) If $a < b$ and $c < d$, then $a + c < b + d$.
   (*ii*) If $0 < a < b$ and $0 < c < d$, then $ac < bd$.

3. Suppose $\{D, P\}$ is an ordered domain,
   (*i*) Prove that if $a < b$, then $a^3 < b^3$.
   (*ii*) What about the converse?
   (*iii*) Generalize (*i*) and (*ii*) for an arbitrary odd power. How about even powers?

4. Prove the following properties of $\leq$ in an ordered domain $\{D, P\}$:
(*i*) $a \geq 0, b \geq 0 \Rightarrow a + b \geq 0$.
(*ii*) $a \geq 0, b \geq 0 \Rightarrow ab \geq 0$.
(*iii*) $a \leq b, b \leq c \Rightarrow a \leq c$.
(*iv*) $a \leq b \Leftrightarrow a + c \leq b + c$.
(*v*) If $c \geq 0$, then $a \leq b \Rightarrow ac \leq bc$.
(*vi*) If $c \leq 0$, then $a \leq b \Rightarrow cb \leq ca$.
(*vii*) $a \leq b \Leftrightarrow -b \leq -a$.
(*viii*) $a \leq 0, b \leq 0 \Rightarrow ab \geq 0$.
(*ix*) $a \leq 0, b \geq 0 \Rightarrow ab \leq 0$.
(*x*) $a^2 \geq 0$ for all $a \in D$.
(*xi*) $a \leq b, c \leq d \Rightarrow a + c \leq b + d$.
(*xii*) $0 \leq a \leq b,\ 0 \leq c \leq d \Rightarrow 0 \leq ac \leq bd$.

5. In the ordered domain $\{D, P\}$, prove:
(*i*) $a \geq 0, b > 0 \Rightarrow a + b > 0$.
(*ii*) $a \geq 0, b > 0 \Rightarrow ab \geq 0$.
(*iii*) $a \leq b, b < c \Rightarrow a < c$.
(*iv*) If $c > 0$, then $a \leq b \Leftrightarrow ac \leq bc$.
(*v*) If $c \geq 0$, then $a < b \Rightarrow ac \leq bc$.
(*vi*) If $c < 0$, then $a \leq b \Leftrightarrow bc \leq ac$.
(*vii*) If $c \leq 0$, then $a < b \Rightarrow bc \leq ac$.
(*viii*) $a \leq 0, b < 0 \Rightarrow ab \geq 0$.
(*ix*) $a \leq 0, b > 0 \Rightarrow ab \leq 0$.
(*x*) $a < 0, b \geq 0 \Rightarrow ab \leq 0$.
(*xi*) $a \leq b, c < d \Rightarrow a + c < b + d$.
(*xii*) $0 \leq a \leq b, 0 \leq c < d \Rightarrow 0 \leq ac \leq bd$.
(*xiii*) $0 < a \leq b, 0 \leq c < d \Rightarrow 0 \leq ac < bd$.

6. Show that the equation $x^2 + e = 0$ cannot be solved in an ordered domain $\{D, P\}$.

7. For a, b in an ordered domain show that

$$a^2 + b^2 \geq 2ab$$

(where $2ab$ means $ab + ab$). When does equality hold?

8. If $a \leq b$ and $b \leq a$ in an ordered domain $\{D, P\}$ show that $a = b$. What happens if $b < a$?

9. For any a, b in the ordered domain $\{D, P\}$, prove:

$$\big| |a| - |b| \big| \leq |a - b| \leq |a| + |b|.$$

10. For any a in the ordered domain $\{D, P\}$, is it true that

$$-|a| = \min\{a, -a\}$$

where, of course, "min" means the **minimum** (that is, the smaller) of the two elements?

11. For elements a, b, c in an ordered domain $\{D, P\}$,

(*i*) $\min\{a, b\} + \max\{a, b\} = a + b$,
(*ii*) $\min\{\max\{a, b\}, c\} = \max\{\min\{a, c\}, \min\{b, c\}\}$,
(*iii*) $\max\{\min\{a, b\}, c\} = \min\{\max\{a, c\}, \max\{b, c\}\}$,
(*iv*) $\max\{a, b, c\} + \min\{a + b, a + c, b + c\} = a + b + c$,
(*v*) $\min\{\max\{a, b\}, \max\{a, c\}, \max\{b, c\}\}$
$= \max\{\min\{a, b\}, \min\{a, c\}, \min\{b, c\}\}$.

12. Can you make $D = \mathbf{Z}[\sqrt{3}]$ into an ordered domain? How about $D = \mathbf{Z}[i]$?

13. Prove that

$$a^2 - ab + b^2 \geq 0$$

for all a, b in the ordered domain $\{D, P\}$.

14. Use mathematical induction to prove each of the following formulas for all integers greater than or equal to 1.

(*i*) $\displaystyle\sum_{i=1}^{n} i = 1 + 2 + \cdots + n = \frac{n(n+1)}{2}$,

(*ii*) $\displaystyle\sum_{i=1}^{n} \frac{i(i+1)}{2} = 1 + 3 + 6 + \cdots + \frac{n(n+1)}{2} = \frac{n(n+1)(n+2)}{6}$,

(*iii*) $\displaystyle\sum_{i=1}^{n} \frac{1}{i(i+1)} = \frac{1}{2} + \frac{1}{6} + \frac{1}{12} + \cdots + \frac{1}{n(n+1)} = \frac{n}{n+1}$,

(*iv*) $\displaystyle\sum_{i=1}^{n} (2i-1)^2 = 1^2 + 3^2 + 5^2 + \cdots + (2n-1)^2 = \frac{4n^3 - n}{3}$,

(*v*) $\displaystyle\sum_{i=1}^{n} i^3 = 1^3 + 2^3 + 3^3 + \cdots + n^3 = \left\{\frac{n(n+1)}{2}\right\}^2$.

15. Find formulas for each of the following sums, and then prove their validity by induction

(*i*) $\displaystyle\sum_{i=1}^{n} 2i = 2 + 4 + 6 + \cdots + (2n)$,

(*ii*) $\displaystyle\sum_{i=1}^{n} (2i-1) = 1 + 3 + 5 + \cdots + (2n-1)$,

$n \geq 1$.

16. For $a, d, r \neq 1$ in **R**, what is the sum of the first n terms of
(*i*) the arithmetic progression: $a, a + d, a + 2d, \ldots, a + (n - 1)d, \ldots,$
(*ii*) the geometric progression: $a, ar, ar^2, \ldots, ar^{n-1}, \ldots.$
Prove your assertions by induction.

17. List several places in Chapter I where some form of mathematical induction or well ordering is used, explicitly or implicitly, in a proof.

18. Prove the parts of 2-3-19 which were not proved in the text.

19. Use induction to prove that 7 divides $5^{2n+1} + 2^{2n+1}$ for all $n \geq 0$.

20. For any $n \geq 1$ prove inductively that a set with n elements has exactly 2^n subsets (among which are included the empty set and the set itself).

21. Suppose $\{D, P\}$ is a well-ordered domain.
(*i*) We know that P has a smallest element e; does P have a biggest element?
(*ii*) Does every subset of D have a smallest element?
(*iii*) Is there a subset of D which is not inductive?
Justify your answers.

22. In a well-ordered domain $\{D, P\}$, we have:
(*i*) $a \geq b \Leftrightarrow a > b - e$.
(*ii*) If S is an inductive subset of D which has a smallest element s, then

$$S = \{a \in D \mid a \geq s\} = \{a \in D \mid a > s - e\}.$$

In virtue of this result, we are justified in starting a proof by mathematical induction from any integer n, when the situation calls for it, rather than solely from $n = 1$ (as was stated in 2-3-17).

23. Induction is used, when feasible, to prove an infinite number of statements. What is the idea underlying the inductive approach?

24. Suppose we have three pegs on one of which there are piled n circular disks—where each disk is smaller than the one on which it lies. Disks may be moved from one peg to another under the following restrictions:
(*i*) Only one disk may be moved at a time.
(*ii*) A disk may not be placed on top of a smaller disk.
Prove that under these rules, it is possible to transfer the entire pile of disks to another peg; moreover, this transfer can be accomplished in $2^n - 1$ moves.
(*iii*) If the peg to which all the disks are to be moved is specified in advance, what should the first move be?

2-4. Computation Rules

The informal phrase "and so on" occurs often in mathematical arguments to signify that a proof may be completed by repeating, as many times as required, the steps of the proof which have already been described. In fact, "and so on" is usually a cover up for some kind of mathematical induction. In this section we shall illustrate such uses of induction by showing, with a certain amount of care, how the basic rules for computation which we normally take for granted have meaning and validity in any ring.

2-4-1. Generalized Associative Law for Multiplication. Suppose $n \geq 3$, and let $a_1, a_2, \ldots, a_n$ be arbitrary elements of the ring R; then the meaning of

$$a_1 a_2 a_3 \cdots a_n$$

is unambiguous. In other words, all ways of inserting parentheses in this expression, and then carrying out the required multiplications, give the same result.

Proof: The meaning of this result is as follows. According to the definition of a ring, we know how to multiply two elements of R, but the definition says nothing about multiplying more than two elements. In particular, the symbol $a_1 a_2 a_3 \cdots a_n$ has no meaning unless we insert parentheses which prescribe the order in which one should perform a sequence of multiplications (where it is understood that at each step one multiplies exactly two elements of R) until all the elements $a_1, a_2, \ldots, a_n$ are used up. The assertion here is that no matter how the parentheses are placed the end result is always the same. For example, when $n = 6$, this result implies the equality of

$$((((a_1 a_2)a_3)a_4)a_5)a_6 \qquad \text{and} \qquad (a_1 a_2)((a_3 a_4)(a_5 a_6)).$$

Turning to the proof, we consider first the case $n = 3$. There are exactly two possible ways to insert parentheses—namely, $(a_1 a_2)a_3$ or $a_1(a_2 a_3)$—and according to the associative law both of these determine the same element of R. As was noted long ago [see 2-1-2, part (5)], this common value is denoted by $a_1 a_2 a_3$, and there is no ambiguity.

To deal with the general case, it is convenient to assign, at the start, a specific meaning to the symbol $a_1 a_2 a_3 \cdots a_n$. More precisely, let us take

$$a_1 a_2 a_3 \cdots a_n = ((\cdots(((a_1 a_2)a_3)a_4)\cdots)a_{n-1})a_n.$$

In other words, we start with $a_1 a_2$, then compute $(a_1 a_2)a_3$, and keep throwing in the remaining a_i's on the right, one at a time. Of course, this definition—as indicated by the use of three dots, involves an induction or recursion. In more detail, the definition of $a_1a_2a_3 \cdots a_n$ for $n \geq 3$ is given by

$$a_1a_2a_3 = (a_1a_2)a_3,$$
$$a_1a_2a_3a_4 = (a_1a_2a_3)a_4,$$

and then, inductively

$$a_1a_2a_3 \cdots a_{n-1}a_n = (a_1a_2a_3 \cdots a_{n-1})a_n. \qquad (\#)$$

We must show that any way of inserting parentheses gives this value.

Thus having observed the validity of the generalized associative law for $n = 3$, we consider $n \geq 4$ and suppose inductively that the generalized associative law holds for all $m < n$ (that is, for $m = 3, 4, \ldots, n-1$). Consider an arbitrary, but fixed, choice of parentheses for the multiplication of $a_1, a_2, \ldots, a_n$. At the last step in the required sequence of multiplications we have, in virtue of the induction hypothesis (which says that in any product of less than n terms the result is unambiguous and there is no need for parentheses),

$$(a_1 \cdots a_s)(a_{s+1} \cdots a_n) \qquad (*)$$

for some s with $1 \leq s \leq n-1$. If $s = n-1$, the product $(*)$ takes the form $(a_1 \cdots a_{n-1})(a_n)$ which, according to our definition, is equal to $a_1a_2 \cdots a_n$—so the proof in this case is finished. If $s < n-1$, then the expression $(a_{s+1} \cdots a_n)$ of $(*)$ includes two or more terms a_i and we may rewrite $(*)$ as

$$\begin{aligned}(a_1 \cdots a_s)(a_{s+1} \cdots a_n) &= (a_1 \cdots a_s)((a_{s+1} \cdots a_{n-1})(a_n)) \\ &= ((a_1 \cdots a_s)(a_{s+1} \cdots a_{n-1}))(a_n) \\ &= (a_1 \cdots a_{n-1})(a_n) \\ &= a_1 \cdots a_n.\end{aligned}$$

For completeness, we list the justifications for the various equalities above: first, by application of the definition (#) to the term $a_{s+1} \cdots a_n$; second, by the ordinary associative law; third, by the induction hypothesis; fourth, by the definition (#). This completes the proof. ∎

In exactly the same way, we have the generalized associative law for addition—after all, the preceding proof can obviously be applied to any binary operation which satisfies the associative law. Thus, the expression

$$a_1 + a_2 + \cdots + a_n$$

has an unambiguous meaning. Even more, because addition is commutative we have the following more general result.

2-4-2. Generalized Commutative–Associative Law for Addition. If $n \geq 2$ and $a_1, a_2, \ldots, a_n$ are elements of the ring R, then all sums of these n elements, in any order, are equal.

Proof: According to the generalized associative law for addition the sum

$$a_1 + a_2 + \cdots + a_n$$

is well defined (the result being independent of how the parentheses are inserted). By the same token, if we permute, or rearrange, the elements $a_1, a_2, \ldots, a_n$ and add them in this new order, then here too the sum has unambiguous meaning. (For example, when $n = 6$, one possible rearrangement of $a_1 + a_2 + a_3 + a_4 + a_5 + a_6$ is $a_5 + a_3 + a_1 + a_4 + a_6 + a_2$.) We shall show, by induction on n, that the sum of any such rearrangement is always equal to $a_1 + a_2 + \cdots + a_n$.

For $n = 2$, there is only one possible rearrangement of $a_1 + a_2$—namely $a_2 + a_1$; and these are equal by virtue of the commutative law for addition. Now, suppose inductively that our generalized commutative–associative law holds for $2, 3, \ldots, n - 1$ and consider the sum of any rearrangement of $a_1, a_2, \ldots, a_n$. In this sum, a_n may appear as the last term, or as the first term, or as an in-between term. Thus, depending on the location of a_n, the sum takes one of the forms

(*i*) $(\cdots) + (a_n)$,
(*ii*) $(a_n) + (\cdots)$,
(*iii*) $(\cdots) + (a_n) + (\cdots)$.

(Of course, the generalized associative law for addition permits us to insert parentheses in these ways.) Now, applying the commutative law, (*ii*) becomes $(\cdots) + (a_n)$; and applying the commutative law and the generalized associative law to (*iii*) leads to $(\cdots) + (a_n)$. Consequently, in all three cases, the sum is equal to

$$(\cdots) + (a_n).$$

But this $(\cdots)$ represents the sum of some rearrangement of $a_1, a_2, \ldots, a_{n-1}$; so, by the induction hypothesis, it is equal to $a_1 + \cdots + a_{n-1}$. We obtain, therefore,

$$(\cdots) + (a_n) = (a_1 + a_2 + \cdots + a_{n-1}) + a_n$$

which, according to the definition, equals $a_1 + \cdots + a_{n-1} + a_n$. This completes the proof. ▌

The idea of the preceding proof is quite clear, in spite of the imperfections caused by our lack of a suitable notation for the sum of the terms of a permutation or rearrangement of $a_1, a_2, \ldots, a_n$. Such a notation will be introduced in Chapter IV.

At this point, it becomes convenient to write sums and products in more compact form, so we introduce the standard Σ and Π notations. More precisely, we write

$$\sum_{i=1}^{n} a_i = a_1 + a_2 + \cdots + a_n,$$

$$\prod_{i=1}^{n} a_i = a_1 a_2 \cdots a_n,$$

and, in particular, in virtue of what has gone before we have

$$\sum_{i=1}^{n} a_i = \left(\sum_{i=1}^{n-1} a_i\right) + a_n, \qquad \prod_{i=1}^{n} a_i = \left(\prod_{i=1}^{n-1} a_i\right) a_n.$$

2-4-3. Generalized Distributive Laws. If $n \geq 2$ and $a, b_1, \ldots, b_n$ are elements of the ring R, then

$$a\left(\sum_{i=1}^{n} b_i\right) = \sum_{i=1}^{n} (ab_i) \qquad \text{and} \qquad \left(\sum_{i=1}^{n} b_i\right) a = \sum_{i=1}^{n} (b_i a).$$

Proof: The first of these is a compact form for the assertion

$$a(b_1 + b_2 + \cdots + b_n) = ab_1 + ab_2 + \cdots + ab_n \tag{$*$}$$

and it is an easy consequence of things we already know. In fact, for $n = 2$ the assertion is surely true—it is precisely the distributive law axiom for a ring. Then, suppose inductively that this generalized distributive law $(*)$ holds for $n - 1$. We have, therefore,

$$\begin{aligned} a\left(\sum_{i=1}^{n} b_i\right) &= a\left(\sum_{i=1}^{n-1} b_i + b_n\right) \\ &= a\left(\sum_{i=1}^{n-1} b_i\right) + ab_n \\ &= \sum_{i=1}^{n-1} (ab_i) + ab_n \\ &= \sum_{i=1}^{n} (ab_i) \end{aligned}$$

so that $(*)$ holds for n also. Thus, by mathematical induction, this generalized distributive law holds for all $n \geq 2$. ▌

In similar fashion, the other generalized distributive law holds for all $n \geq 2$. It says

$$(b_1 + b_2 + \cdots + b_n)a = b_1 a + b_2 a + \cdots + b_n a.$$

2-4-4. Proposition. If $a_1, a_2, \ldots, a_m, b_1, b_2, \ldots, b_n$ are elements of the ring R, then

$$\left(\sum_{i=1}^{m} a_i\right)\left(\sum_{j=1}^{n} b_j\right) = \sum_{\substack{i=1,\ldots,m\\ j=1,\ldots,n}} a_i b_j .$$

Proof: The left-hand side of this equation is the product.

$$(a_1 + a_2 + \cdots + a_m)(b_1 + b_2 + \cdots + b_n)$$

while the right-hand side is the sum of all the terms in the rectangular array (with m rows and n columns)

$$\begin{matrix} a_1 b_1 & a_1 b_2 & \cdots & a_1 b_n \\ a_2 b_1 & a_2 b_2 & \cdots & a_2 b_n \\ \vdots & \vdots & & \vdots \\ a_m b_1 & a_m b_2 & \cdots & a_m b_n . \end{matrix}$$

According to the generalized commutative–associative law, 2-4-2, the order in which the terms of the rectangular array are added does not matter, and this is in accord with the notation

$$\sum_{\substack{i=1,\ldots,m\\ j=1,\ldots,n}} (a_i b_j),$$

which does not specify in what order the terms are to be added. The proof we are about to give amounts to adding one row at a time. Namely, in virtue of the two generalized distributive laws, (both of which are used)

$$\begin{aligned} \left(\sum_{i=1}^{m} a_i\right)\left(\sum_{j=1}^{n} b_j\right) &= (a_1 + a_2 + \cdots + a_m)\left(\sum_{j=1}^{n} b_j\right) \\ &= a_1\left(\sum_{j=1}^{n} b_j\right) + a_2\left(\sum_{j=1}^{n} b_j\right) + \cdots + a_m\left(\sum_{j=1}^{n} b_j\right) \\ &= \sum_{j=1}^{n} (a_1 b_j) + \sum_{j=1}^{n} (a_2 b_j) + \cdots + \sum_{j=1}^{n} (a_m b_j) \\ &= \sum_{\substack{i=1,\ldots,m\\ j=1,\ldots,n}} (a_i b_j). \end{aligned}$$

Another way to write the details of this proof is

$$\begin{aligned} \left(\sum_{i=1}^{m} a_i\right)\left(\sum_{j=1}^{n} b_j\right) &= \sum_{i=1}^{m}\left(a_i\left(\sum_{j=1}^{n} b_j\right)\right) \\ &= \sum_{i=1}^{m}\left(\sum_{j=1}^{n} (a_i b_j)\right) \\ &= \sum_{\substack{i=1,\ldots,m\\ j=1,\ldots,n}} (a_i b_j). \end{aligned}$$

2-4-5. Remark. Consider the product $\Pi_{i=1}^{n} a_i$ of n elements of the ring R. In the special case where all the a_i are equal—say, $a = a_1 = a_2 = \cdots = a_n$—it is customary to write a^n for $\Pi_{i=1}^{n} a_i$. So according to the meaning of the product, Π, we have

$$a^1 = a, \quad a^2 = (a^1)(a), \ldots, a^n = (a^{n-1})a, \ldots.$$

As a matter of fact, precisely these rules may be used to give a direct, recursive, definition of a^n for all $n \geq 1$ (in other words, each power of a, after the first, is obtained by multiplying the preceding power of a, on the right, by a). Of course, one should still think of a^n as the product of n copies of a.

From the generalized associative law for multiplication we obtain the "familiar" rules

$$\begin{aligned} &(i) \quad (a^m)(a^n) = a^{m+n}, \\ &(ii) \quad (a^m)^n = a^{mn}, \end{aligned} \qquad \text{for all } m, n \geq 1.$$

In fact, a^{m+n} equals $\underbrace{aa \cdots a}_{m+n}$, where there are $m + n$ terms, and by 2-4-1, we may insert parentheses to obtain

$$\underbrace{aa \cdots a}_{m+n} = \underbrace{(a \cdots a)}_{m}\underbrace{(a \cdots a)}_{n}$$

where the first parenthesis contains m terms (each of which is a) and the second parenthesis contains n terms; and the right side surely equals $(a^m)(a^n)$. This proves (i).

The proof of (ii) consists of grouping the mn copies of a (which make up a^{mn}), via parentheses, into n groups each of which has m copies of a. In detail,

$$\begin{aligned} a^{mn} &= \underbrace{aa \cdots aa}_{mn} \\ &= \overbrace{\underbrace{(a \cdots a)}_{m}\,\underbrace{(a \cdots a)}_{m} \cdots \underbrace{(a \cdots a)}_{m}}^{n} \\ &= \underbrace{(a^m)(a^m) \cdots (a^m)}_{n} \\ &= (a^m)^n. \end{aligned}$$

Note that a^0 is not defined as a rule. For it if were defined we would still want property (i) to be satisfied; so in particular a^0 would satisfy

$$a^0 a^n = a^{0+n} = a^n \qquad \text{and} \qquad a^n a^0 = a^{n+0} = a^n.$$

Thus, a^0 would have to behave somewhat like an identity for multiplication. But the ring R, in which we are working, need not have a multiplicative identity, and in such a situation there is no obvious way to define a^0. In addition, a^n need not be defined for negative n; since, if $n = -1$ and condition (*i*) still holds, we have

$$a^{-1}a^1 = a^0 = a^1a^{-1}$$

—so that a^{-1} cannot be defined unless a^0 is already defined.

2-4-6. Remark. Consider the sum $\Sigma_{i=1}^n a_i$ of n elements of the ring R. In the special case where all the a_i are equal—say, $a = a_1 = a_2 = \cdots = a_n$—it is customary to write na for $\Sigma_{i=1}^n a_i$. Note that here $n \in \mathbf{Z}$, $a \in R$, and $na \in R$, but na is *not* a product of two elements of R; however, we still refer to na as the product of a and (the integer) n.

According to the meaning of the sum Σ we have

$$1a = a, \quad 2a = (1a) + a, \ldots, na = (n-1)a + a.$$

Again, in analogy with the statement made for powers in 2-4-5, these rules could be used to define na recursively for all $n \geq 1$. Of course, na should still be thought of as the sum of n copies of a. The reason for using the notation na when adding (or the notation a^n when multiplying) rather than some other notation, is that it obeys the usual kind of rules. Thus, we have

$$\left.\begin{array}{rl} (i) & (m+n)a = ma + na, \\ (ii) & n(ma) = (nm)a, \\ (iii) & m(a+b) = ma + mb, \\ (iv) & (ma)b = m(ab) = a(mb), \\ (v) & (ma)(nb) = (mn)(ab), \end{array}\right. \qquad \text{for all } a, b \in R,\ m, n \geq 1.$$

The proofs of these rules are rather trivial. Since na is the analog for addition of a^n for multiplication, we see that $(m+n)a$ is the analog of a^{m+n}, and $ma + na$ is the analog of $(a^m)(a^n)$. Thus, rule (*i*) is the additive analog of $a^m a^n = a^{m+n}$—so it follows from the generalized associative law for addition. Similarly, rule (*ii*) is the additive analog of $(a^m)^n = a^{mn}$, and it follows from the generalized associative law for addition. Rule (*iii*) is an immediate consequence of the generalized commutative–associative law for addition, since

$$\begin{aligned} m(a+b) &= \underbrace{(a+b) + (a+b) + \cdots + (a+b)}_{m} \\ &= (\underbrace{a + a + \cdots + a}_{m}) + (\underbrace{b + b + \cdots + b}_{m}) \\ &= ma + mb. \end{aligned}$$

To prove (*iv*), let $a = a_1 = a_2 = \cdots = a_m$ and $b = b_1 = b_2 = \cdots = b_m$; so in virtue of 2-4-3,

$$(ma)b = \left(\sum_{i=1}^m a_i\right) b = \sum_{i=1}^m (a_i b) = m(ab)$$

and

$$a(mb) = a\left(\sum_{i=1}^{m} b_i\right) = \sum_{i=1}^{m} (ab_i) = m(ab).$$

Finally, to prove (v) one may take $a = a_1 = a_2 = \cdots = a_m$ and $b = b_1 = b_2 = \cdots = b_n$ and use 2-4-4.

2-4-7. Remark. Can we extend the definition of na so it is applicable for all $n \leq 0$, and in such a way that rules (i)–(v) then hold for all integers m and n? Suppose this has been done. Then putting $n = 0$ in (i), we have

$$ma = (m + 0)a = ma + 0a$$

which tells us that

$$0a = 0 \qquad \text{for all } a \in R. \tag{$*$}$$

(Note that here the 0 on the left side is in $\mathbf{Z}$, while the 0 on the right side is in R.) Furthermore, if $n < 0$ we may put $m = -n$ in rule (i) and obtain

$$(-n)a + na = ((-n) + n)(a) = 0a = 0$$

which tells us that

$$na = -((-n)a) \qquad \text{for all } n < 0 \text{ and } a \in R. \tag{$**$}$$

In virtue of these observations, it is only natural that we should define na for $n \leq 0$ according to the rules $(*)$ and $(**)$. In particular, in the case $n = -1$, we have

$$(-1)a = -((-(-1))a) = -a, \qquad a \in R.$$

More generally, the use of $(**)$ for the definition of na is permissible because, if $n < 0$, then $(-n) > 0$, so $(-n)a$ has meaning (as in 2-4-6), and then $(-n)a$ has an additive inverse $-((-n)a)$ in R. This is not the only way one could define na when $n < 0$. An alternate method would be to put $na = (-n)(-a)$; but this definition is essentially the same as $(**)$ because we have

$$na = -((-n)a) = (-n)(-a), \qquad n < 0, a \in R. \tag{$***$}$$

To check the validity of $(***)$, we note first that if $m > 0$, then

$$0 = m0 = m(a + (-a)) = ma + m(-a)$$

and consequently

$$-(ma) = m(-a), \qquad m > 0, a \in R.$$

Now, when $n < 0$ we may apply this for $m = -n > 0$ to obtain $-((-n)a) = (-n)a$—so $(***)$ does hold.

We know the meaning of na for every $n \in \mathbf{Z}$, and need to verify the rules (*i*)–(*v*) of 2-4-6 for all integers m and n. This is not hard. For example, consider rule (*iii*): If $m > 0$, its validity was proved in 2-4-6; if $m = 0$, (*iii*) is clearly valid; if $m < 0$ we have

$$\begin{aligned} m(a + b) &= (-m)(-(a + b)) \\ &= (-m)((-a) + (-b)) \\ &= (-m)(-a) + (-m)(-b) \\ &= ma + mb \end{aligned}$$

so (*iii*) holds in this case also. Similarly, rule (*iv*) holds for $m > 0$ and for $m = 0$; while if $m < 0$ we have

$$\begin{aligned} (ma)b &= ((-m)(-a))b \\ &= (-m)((-a)(b)) \\ &= (-m)(-(ab)) \\ &= m(ab) \end{aligned}$$

and in analogous fashion $a(mb) = m(ab)$—so rule (*iv*) holds for all $m \in \mathbf{Z}$.

The verification of rules (*i*), (*ii*), and (*v*) is straightforward but somewhat tedious because there are a number of cases to consider—depending on whether m and n are positive, zero, or negative. The details are left to the reader.

There is still another familiar rule for computation which merits a reasonably careful proof.

2-4-8. Binomial Theorem. Suppose R is a commutative ring with unity e; then for any $a, b \in R$ we have

$$(a + b)^n = \sum_{i=0}^{n} \binom{n}{i} a^{n-i} b^i, \qquad n \leq 1.$$

Proof: Before undertaking the proof some preparatory comments are in order. First of all, we need to discuss the meaning of the "binomial coefficients" $\binom{n}{i}$. One starts from the **factorial** symbol, which is defined as

$$0! = 1, \qquad 1! = 1,$$

and inductively, for any $n \geq 1$

$$(n + 1)! = (n + 1)(n!).$$

It follows that for any $n \geq 1$,

$$n! = (n)(n - 1)(n - 2) \cdots (2)(1)$$

—in other words, $n!$ is the product of all the positive integers from 1 to n. Now, for any $n \geq 1$ and any k satisfying $0 \leq k \leq n$, let us define the **binomial**

coefficient $\binom{n}{k}$ by

$$\binom{n}{k} = \frac{n!}{k!(n-k)!}, \qquad n \geq 1, \quad 0 \leq k \leq n.$$

For example, we have

$$\binom{5}{2} = \frac{5\cdot4\cdot3\cdot2\cdot1}{(2\cdot1)\cdot(3\cdot2\cdot1)} = 10, \qquad \binom{5}{5} = \frac{5\cdot4\cdot3\cdot2\cdot1}{5\cdot4\cdot3\cdot2\cdot1\cdot(0!)} = 1,$$

$$\binom{7}{3} = \frac{7\cdot6\cdot5\cdot4\cdot3\cdot2\cdot1}{(3\cdot2\cdot1)\cdot(4\cdot3\cdot2\cdot1)} = 35, \qquad \binom{5}{3} = 10,$$

$$\binom{7}{4} = 35.$$

More generally, one observes immediately that

(*i*) $\binom{n}{k} = \dfrac{n(n-1)\cdots(n-k+1)}{1\cdot2\cdot3\cdots k}$,

(*ii*) $\binom{n}{0} = \binom{n}{n} = 1$,

(*iii*) $\binom{n}{n-k} = \binom{n}{k}$.

All these facts are presumably familiar to the reader. Less familiar is the following useful relation between binomial coefficients

$$\binom{n+1}{k} = \binom{n}{k} + \binom{n}{k-1}, \qquad 0 < k < n+1. \tag{$\neq$}$$

Its proof consists of a straightforward computation—namely,

$$\begin{aligned}
\binom{n}{k} + \binom{n}{k-1} &= \frac{n!}{k!(n-k)!} + \frac{n!}{(k-1)!(n-k+1)!} \\
&= \frac{n!(n-k+1)}{k!(n-k)!(n-k+1)} + \frac{(n!)k}{(k-1)!(k)(n-k+1)!} \\
&= \frac{n!(n-k+1)+(n!)k}{(k!)(n-k+1)!} \\
&= \frac{(n!)(n-k+1+k)}{(k!)(n+1-k)!} \\
&= \frac{(n+1)!}{k!(n+1-k)!} \\
&= \binom{n+1}{k}
\end{aligned}$$

Incidentally, at this stage we do not yet know that $\binom{n}{k}$ is always an integer, but the formula (#) may be used to give an easy proof of this fact. More precisely, if $n = 1$, then the only possibilities for k are $k = 0$ and $k = 1$; since $\binom{1}{0} = 1$ and $\binom{1}{1} = 1$, we see that $\binom{n}{k}$ is an integer when $n = 1$ and k is any integer satisfying $0 \le k \le 1$. Now, suppose inductively that $\binom{n}{k}$ is an integer for a fixed n and all k satisfying $0 \le k \le n$. We must show that $\binom{n+1}{k}$ is an integer for each k satisfying $0 \le k \le n + 1$. For $k = 0$ or $n + 1$ we have $\binom{n+1}{0} = \binom{n+1}{n+1} = 1$, so $\binom{n+1}{k}$ is an integer in these cases. Furthermore, for k satisfying $0 < k < n + 1$, the relation (#) expresses $\binom{n+1}{k}$ as a sum of two terms, each of which is an integer by the induction hypothesis, so $\binom{n+1}{k}$ is an integer. We conclude that $\binom{n}{k}$ is always an integer.

Turning to the formula of the binomial theorem, 2-4-8, we may write out, or expand, the right side. Since $\binom{n}{0} = \binom{n}{n} = 1$ and $a^0 = b^0 = e \in R$, the result is

$$\sum_{i=0}^{n} \binom{n}{i} a^{n-i}b^i = \binom{n}{0} a^n b^0 + \binom{n}{1} a^{n-1}b^1 + \cdots + \binom{n}{n-1} a^1 b^{n-1} + \binom{n}{n} a^0 b^n$$

$$= a^n + \binom{n}{1} a^{n-1}b + \binom{n}{2} a^{n-2}b^2 + \cdots + \binom{n}{n-1} ab^{n-1} + b^n.$$

Thus, what we have called the binomial theorem is indeed the standard statement, known to all—except that we deal with it in the general context of a commutative ring with unity. Now, let us give a formal proof, by induction on n. For $n = 1$, we have

$$\sum_{i=0}^{1} \binom{1}{i} a^{1-i}b^i = \binom{1}{0} a^1 b^0 + \binom{1}{1} a^0 b^1 = a + b$$

so the binomial theorem holds in this case. Next, suppose inductively that the binomial theorem holds for n. We must prove the validity of the binomial theorem for $n + 1$. In other words, given

$$(a + b)^n = \sum_{i=0}^{n} \binom{n}{i} a^{n-i}b^i,$$

we must show that

$$(a + b)^{n+1} = \sum_{i=0}^{n+1} \binom{n+1}{i} a^{n+1-i}b^i.$$

This is not hard—in fact, we have

$$\begin{aligned}
(a+b)^{n+1} &= (a+b)^n(a+b) \\
&= \left(\sum_{i=0}^{n} \binom{n}{i} a^{n-i} b^i\right)(a+b) \\
&= \sum_{i=0}^{n} \binom{n}{i} a^{n-i+1} b^i + \sum_{i=0}^{n} \binom{n}{i} a^{n-i} b^{i+1} \\
&= \binom{n}{0} a^{n+1} b^0 + \sum_{i=1}^{n} \binom{n}{i} a^{n-i+1} b^i + \sum_{i=0}^{n-1} \binom{n}{i} a^{n-i} b^{i+1} + \binom{n}{n} a^0 b^{n+1} \\
&= a^{n+1} + \sum_{i=1}^{n} \binom{n}{i} a^{n+1-i} b^i + \sum_{i=1}^{n} \binom{n}{i-1} a^{n-(i-1)} b^{(i-1)+1} + b^{n+1} \\
&= \binom{n+1}{0} a^{n+1} b^0 + \sum_{i=1}^{n} \left[\binom{n}{i} + \binom{n}{i-1}\right] a^{n+1-i} b^i + \binom{n+1}{n+1} a^0 b^{n+1} \\
&= \sum_{i=0}^{n+1} \binom{n+1}{i} a^{n+1-i} b^i.
\end{aligned}$$

This completes the proof. ▮

If the commutative ring R does not have a unity for multiplication, then a^0 and b^0 are not defined, but the binomial theorem still holds with only trivial modifications of notation. The nice symmetrical expression $\sum_{i=0}^{n} \binom{n}{i} a^{n-i} b^i$ cannot be used because when $i = 0$ or $i = n$ we find terms containing b^0 and a^0, respectively. So the appropriate form for the binomial theorem is

$$(a+b)^n = a^n + \binom{n}{1} a^{n-1} b + \binom{n}{2} a^{n-2} b^2 + \cdots + \binom{n}{n-1} a b^{n-1} + b^n.$$

Naturally, the proof follows the same principles as before, and it may safely be left to the reader.

We conclude this section with an interesting application of the binomial theorem.

2-4-9. Theorem (Fermat: 1601–1665). If p is prime, then

$$n^p \equiv n \pmod{p} \qquad \text{for all } n \in \mathbf{Z}.$$

To put it another way

$$\alpha^p = \alpha \qquad \text{for every } \alpha \in \mathbf{Z}_p.$$

Proof: Consider the first assertion. It is clearly true for $n = 0$ and $n = 1$.

Suppose, inductively, that the assertion holds for the positive integer n; we must prove its validity for $n+1$. According to the binomial theorem,

$$(n+1)^p = n^p + \binom{p}{1} n^{p-1} + \cdots + \binom{p}{k} n^{p-k} + \cdots + \binom{p}{p-1} n + 1^p. \qquad (*)$$

Now, let us look at the binomial coefficients $\binom{p}{k}$ for $k = 1, 2, \ldots, p-1$. We know that

$$\binom{p}{k} = \frac{p(p-1)\cdots(p-k+1)}{1 \cdot 2 \cdots k}$$

is an integer. Since the prime p appears in the numerator and all the terms in the denominator are positive integers less than p, it follows that the integer $\binom{p}{k}$ is divisible by p. Consequently, all the terms on the right side of $(*)$, except for the two end terms, are divisible by p; so we have

$$(n+1)^p \equiv n^p + 1^p \equiv n + 1 \pmod{p}$$

(note the use of the induction hypothesis) and our assertion holds for $n+1$. Therefore, Fermat's theorem holds for all $n \geq 0$.

What if n is negative? If $n < 0$, then surely there exists an integer $m \geq 0$ such that $n \equiv m \pmod{p}$. (In fact, by applying the division algorithm to n and p, we see that m may be taken to satisfy $0 \leq m < p$.) Then, in virtue of the foregoing and the obvious fact (see 2-4-10,) that raising two congruent integers to the same power preserves the congruence, we have

$$n^p \equiv m^p \equiv m \equiv n \pmod{p}.$$

This completes the proof of the first assertion.

Finally, consider any element $\alpha \in \mathbf{Z}_p$. According to the meaning of residue class (mod p), α can be written in the form $\alpha = \lfloor n \rfloor_p$ for some $n \in \mathbf{Z}$. Because $n^p \equiv n \pmod{p}$, we have $\lfloor n^p \rfloor_p = \lfloor n \rfloor_p$, and therefore,

$$\alpha^p = (\lfloor n \rfloor_p)^p = \lfloor n^p \rfloor_p = \lfloor n \rfloor_p = \alpha.$$

This completes the proof. ∎

2-4-10 / *PROBLEMS*

1. Prove, in complete detail, that in any ring

$$a_3 + a_5 + a_1 + a_4 + a_2 = a_1 + a_2 + a_3 + a_4 + a_5.$$

How many such rearrangements of a_1, a_2, a_3, a_4, a_5 are there?

2. For elements of a commutative ring, state and prove the generalized commutative–associative law for multiplication.

3. We proved 2-4-4 by adding the various rows of a certain rectangular array with m rows and n columns. Prove this result in another way—namely, by what amounts to summing the columns of the rectangular array.

4. Supply the missing proofs for the rules discussed in 2-4-7.

5. Suppose a and b are integers with $a \equiv b(\text{mod } m)$. Show that for all $n \geq 1$ we have
$$a^n \equiv b^n \ (\text{mod } m)$$
and for all $n \in \mathbf{Z}$ we have
$$na \equiv nb \ (\text{mod } m).$$
Even more, if $f(x) = c_0 + c_1 x + \cdots + c_r x^r$ with $c_0, c_1, \ldots, c_r \in \mathbf{Z}$, then $f(a) \equiv f(b) \ (\text{mod} m)$.

6. If a belongs to the ring R and n is an integer greater than 0, is na equal to $(-n)(-a)$? Why?

7. In any ordered domain D, prove by induction:
   (*i*) $a_1^2 + a_2^2 + \cdots + a_n^2 \geq 0$, and $a_1^2 + a_2^2 + \cdots + a_n^2 = 0 \Leftrightarrow a_1 = a_2 = \cdots = a_n = 0$.
   (*ii*) If a is negative, then any odd positive power of a is also negative.
   (*iii*) If $a < b$ and $n > 0$ is odd, then $a^n < b^n$.

8. In an ordered domain, show that $a^9 = b^9$ implies $a = b$. Does this result apply for any positive odd power? How about even powers?

9. In 2-4-8 we used induction to prove that the binomial coefficient $\binom{n}{k}$ is always an integer. For each $\binom{n}{k}$ consider the point (n, k) in the plane. Use the collection of all such integral points to explain the idea on which we based the proof that $\binom{n}{k}$ is always an integer.

10. In the statement of the binomial theorem, why must R be commutative? What happens if R is not commutative? What is $(a + b)^n$ in this situation?

11. For a, b, c in the commutative ring R, what is $(a + b + c)^n$?

12. In the commutative ring R, compute
    (*i*) $(2a + 3b)^5$, (*ii*) $(2a - 3b)^7$.

13. An element $a \neq 0$ of the ring R is said to be **nilpotent** if there exists a positive integer n for which $a^n = 0$.
    (*i*) Show that in an integral domain there are no nilpotent elements.
    (*ii*) Can you find any nilpotent elements in $\mathcal{M}(\mathbf{Z}, 2)$?

14. Evaluate the following binomial coefficients

(*i*) $\binom{11}{4}$, (*ii*) $\binom{15}{5}$,

(*iii*) $\binom{999}{3}$, (*iv*) $\binom{1001}{998}$.

15. Prove, by induction on n, that a set with n elements has exactly 2^n subsets.

16. (*i*) A set with n elements has exactly $\binom{n}{k}$ subsets which consist of k elements (when, of course, $0 \leq k \leq n$).

(*ii*) Use the binomial expansion of $2^n = (1 + 1)^n$ in conjunction with part (*i*) to conclude that a set with n elements has exactly 2^n subsets.

17. Suppose D is a well-ordered domain and we are given elements $a > 0$ and b of D. Prove the existence of an integer n for which $na > b$. In other words, prove that a well-ordered domain satisfies what is commonly known as the **Archimedean property.**

18. Suppose R is a ring with unity e, and a is an element of R which has an inverse (denoted by a^{-1}) for multiplication. Define a^n for all $n \in \mathbf{Z}$. State the multiplicative analogs (where meaningful) of the five rules about na which were discussed in 2-4-6 and 2-4-7. Examine these multiplicative analogs and decide which ones are valid (and under what hypotheses); then prove the valid ones.

2-5. Characterization of the Integers

When should we consider two rings to be the "same"? Must they be fully identical? Several instances related to this kind of question have already come up, but they were glossed over at the time. For example, in 2-2-9 it was noted informally that the ring of Gaussian integers $\mathbf{Z}[i] = \{a + bi \mid a, b \in \mathbf{Z}\}$ (in which the operations are the standard ones for complex numbers) is "essentially the same" as the ring gotten by taking the set $\mathbf{Z} \times \mathbf{Z} = \{(a, b) \mid a, b \in \mathbf{Z}\}$ and defining the operations of addition and multiplication by

$$(a, b) + (c, d) = (a + c, b + d),$$
$$(a, b) \cdot (c, d) = (ac - bd, ad + bc).$$

The basic difference is in notation; nothing is really affected by writing (a, b) for $a + bi$, or vice versa.

Furthermore, in 2-2-13, $\mathcal{S}(X)$, the set of all subsets of a given set X, was made into a ring by setting up a 1–1 correspondence between its elements and the elements of the ring Map(X, $\mathbf{Z}_2$) and then transferring the ring operations bodily from Map(X, $\mathbf{Z}_2$) to $\mathcal{S}(X)$. Obviously, these two rings should be considered to be the "same."

What are the essential ingredients of sameness? Clearly, it depends on the nature of the objects under discussion and on which of their properties we choose to focus. (For example, during political campaigns one often hears the statement that two opposing candidates are the "same." This is surely a matter of definition; it depends on which characteristics or positions of the candidates are counted and which ones are ignored.) If we are dealing with rings, then, as indicated by the examples mentioned above, two rings may be considered to be the "same" even though they are different sets. On the other hand, we have seen (in 2-2-9, for example) that it is often possible to make a given set, such as $\mathbf{Z} \times \mathbf{Z}$, into a ring in several apparently different ways.

Thus, for a ring, it is not the underlying set alone which matters. In fact, we recall that when being fussy about the definition of a ring, we referred to it as a triple $\{R, +, \cdot\}$ consisting of a set R and two operations $+$ and $\cdot$ which satisfy certain axioms. In other words, there are three features which distinguish a ring—the set of its elements and the two operations. After all, to define a ring these are the three things which must be specified. Consequently, for the "sameness" of two rings, it is natural to require that we have, simultaneously, the sameness of the two sets, of the two operations $+$, and of the two operations $\cdot$. Now, let us turn to the clarification of what is meant by sameness for sets and for operations.

2-5-1. Remark. In the context of our discussion, it should occasion no surprise that two sets are considered to be the same when there is a 1–1 correspondence between them. This is a familiar notion, but it is useful for us to develop it carefully.

Consider two arbitrary sets X and Y. By a **function** or **mapping** f from X to Y (which we denote by $f\colon X \to Y$) we mean, as indicated earlier in 2-2-12, a rule which assigns to each element $x \in X$ a unique element $f(x)$ of Y. In other words, one may think of a function as a black box or machine: every time one throws in an element x of X out comes an element $f(x)$ of Y. We write $f\colon x \to f(x)$ or simply $x \to f(x)$, and refer to the element $f(x)$ as the **image** of x under the mapping f, or as the **value** of f at x. Note that a function f must be defined for every $x \in X$; if there is even a single x for which there is no assigned image $f(x)$, then f is not a function. Furthermore, to each $x \in X$ there is assigned exactly one element $f(x)$; if there appear to be several possibilities for $f(x)$, then, in order for f to be a function, exactly one of these possibilities must be specified as the value of $f(x)$.

Extending the notation used in 2-2-12, we denote the set of all mappings from X into Y by $\text{Map}(X, Y)$. In keeping with 2-2-12, two mappings from X into Y are equal when they take equal values for every element of X—symbolically: For $f, g \in \text{Map}(X, Y)$,

$$f = g \Leftrightarrow f(x) = g(x) \text{ for all } x \in X.$$

Suppose we are given a function $f: X \to Y$. Let $f(X)$ denote the set of all elements that arise as images of elements of X under f—in other words,

$$f(X) = \{f(x) \mid x \in X\}.$$

Thus, $f(X)$ is a subset of Y, and it is known as the **image** of X under f, or simply as the **image** of f. When there is no danger of confusion, we denote it by **im f**. As a rule $f(X) \neq Y$; but when $f(X) = Y$, that is, when every element of Y is the image under f of some element of X, the mapping $f: X \to Y$ is said to be **onto** or **surjective**. Thus, f is surjective if and only if given any $y \in Y$ there exists an $x \in X$ such that $f(x) = y$.

Continuing with an arbitrary function $f: X \to Y$, it may well be that two distinct elements of X have the same image—that is, $x_1 \neq x_2$, but $f(x_1) = f(x_2)$. When this is not the case, that is, when f has the property that distinct elements of X always have distinct images under f, then f is said to be **one to one** or **injective**. Thus, f is injective if and only if it satisfies the following condition for any $x_1, x_2 \in X$:

$$\text{If } x_1 \neq x_2, \text{ then } f(x_1) \neq f(x_2). \qquad (*)$$

Of course, this condition may be reformulated as follows:

$$\text{If } f(x_1) = f(x_2), \text{ then } x_1 = x_2. \qquad (**)$$

Note that according to our definition f is **not** injective if and only if there exist distinct elements of X which have the same image under f.

2-5-2. Examples. (1) Suppose $X = \mathbf{R}$, $Y = \mathbf{R}$ and let $f: \mathbf{R} \to \mathbf{R}$ be given by

$$f(x) = \sqrt{x}, \qquad x \in \mathbf{R}.$$

In other words, f is the rule which assigns to each real number x its square root $\sqrt{x}$. (Of course, by square root we always mean the positive square root.) Unfortunately, if x is negative, then it has no square root—meaning that there is no real number whose square is x. Thus, f is not defined for every $x \in \mathbf{R}$, so according to our definition, f is not a function.

(2) Suppose $X = \mathbf{C}$, $Y = \mathbf{C}$ and let $f: \mathbf{C} \to \mathbf{C}$ be given by

$$f(z) = \sqrt{z}, \qquad z \in \mathbf{C}.$$

In order for f to be a function we must, first of all, be certain that every complex number has a square root; in other words, given any $z = a + bi \in \mathbf{C}$ we must be certain of the existence of real numbers x and y such that

$$(x + yi)^2 = a + bi.$$

This amounts to solving the simultaneous equations

$$\begin{aligned} x^2 - y^2 &= a, \\ 2xy &= b \end{aligned} \qquad (\#)$$

for real numbers x and y. If we put

$$\text{sign } b = \begin{cases} + & \text{for } b \geq 0, \\ - & \text{for } b < 0, \end{cases}$$

and read this symbol as the "sign of b," then the reader may easily derive (or simply check) the fact that a solution of these equations is given by

$$x = \sqrt{\frac{a + \sqrt{a^2 + b^2}}{2}}, \qquad y = (\text{sign } b)\sqrt{\frac{-a + \sqrt{a^2 + b^2}}{2}}. \qquad (\#\#)$$

Note that all these square roots have meaning because the expressions inside them are always greater than or equal to 0. The need for sign b arises in connection with the second equation, $2xy = b$. Under our convention about always taking positive square root,

$$\sqrt{b^2} = |b| \qquad \text{and} \qquad (\text{sign } b)\sqrt{b^2} = b$$

and consequently the given values for x and y satisfy both of the equations (#).

We now know that every complex number has a square root. But if $x + yi$ is a square root of $a + bi$, then so is $-x - yi$. Moreover, for all we know there may be other square roots of $a + bi$. Thus, we are in some difficulty with regard to which square root of $z = a + bi$ to take as $f(z)$. One way out is to take $\sqrt{z}$ to be $x + yi$ where x and y are given by (# #). Another way is to make no choice for $\sqrt{z}$; in this case, f if not a function.

(3) Suppose $X = \mathbf{Z}$, $Y = \mathbf{Z}$ and let f: $\mathbf{Z} \to \mathbf{Z}$ be given by

$$f(x) = 2x, \qquad x \in \mathbf{Z}.$$

Surely, f is a function which maps each integer to twice that integer. In particular, $f(1) = 2$, $f(5) = 10$, $f(0) = 0$, $f(79) = 158$, $f(-1) = -2$, $f(-17) = -34$. The image of f is

$$f(\mathbf{Z}) = \{f(x) \mid x \in \mathbf{Z}\} = \{2x \mid x \in \mathbf{Z}\} = 2\,\mathbf{Z}$$

—the set of all even integers. Therefore, any odd integer is not in the image of f, and f is not surjective. On the other hand, f is injective; for if m and n are integers which have the same image under f, then $m = n$, because

$$f(m) = f(n) \Rightarrow 2m = 2n \Rightarrow m = n$$

—so criterion (**) for injectivity is satisfied.

(4) Consider the mapping f: $\mathbf{Z} \to \mathbf{Z}$ defined by

$$f(x) = x^2, \qquad x \in \mathbf{Z}.$$

Thus, f maps each integer to its square; it is surely a function in terms of our usage of this word. Since $f(-1) = f(+1) = 1$ and, in general, $f(-x) = f(x)$ for all $x \in \mathbf{Z}$, we see that f is not injective. Furthermore, the image of f,

$$f(\mathbf{Z}) = \{x^2 \mid x \in \mathbf{Z}\}$$

consists of all integers which are perfect squares; consequently, f is not surjective.

(5) Consider the "squaring" function f: $\mathbf{C} \to \mathbf{C}$; in other words

$$f(z) = z^2, \qquad z \in \mathbf{C}.$$

Is f surjective? The image of f is $\{z^2 \mid z \in \mathbf{C}\}$, the set of all squares of complex numbers. But we have seen in (2) above that every complex number has a square root; or, to put it another way, every complex number can be expressed as the square of some complex number. Therefore, f is surjective.

Is f injective? Obviously, the answer is no, because $f(-z) = f(z)$ for all $z \in \mathbf{C}$.

(6) Consider the mapping f: $\mathbf{Z} \to \mathbf{Z}$ defined by

$$f(x) = x + 2, \qquad x \in \mathbf{Z}.$$

Given any $n \in \mathbf{Z}$, we have $f(n-2) = n$, so f is surjective. Since $f(x) = f(y) \Rightarrow x + 2 = y + 2 \Rightarrow x = y$, we see that f is injective.

2-5-3. Remark. Suppose we have three sets X, Y, Z and two mappings f: $X \to Y$ and g: $Y \to Z$. Then we may define the **composite** (or **composition**) function

$$g \circ f\colon X \to Z$$

by taking

$$(g \circ f)(x) = g(f(x)) \qquad \text{for all } x \in X.$$

The name "composite" is used because we "compose" the two mappings g and f to obtain $g \circ f$; this means that $g \circ f$ amounts to applying the mapping f and then applying g to the result. It is convenient to describe this situation by the "picture"

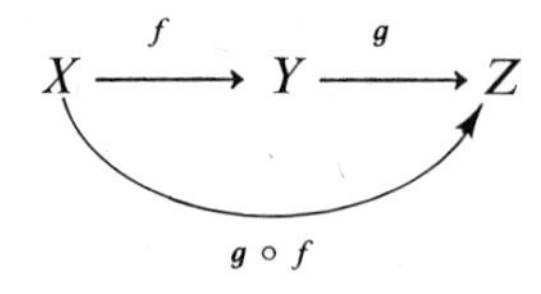

or by the phrase: If $f \in \text{Map}(X, Y)$ and $g \in \text{Map}(Y, Z)$, then we have $g \circ f \in \text{Map}(X, Z)$.

Two simple properties of the composite function $g \circ f$ are:

(*i*) If both f and g are surjective, so is $g \circ f$.
(*ii*) If both f and g are injective, so is $g \circ f$.

Proof: (*i*) Consider any $z \in Z$. Since $g\colon Y \to Z$ is onto, there exists $y \in Y$ such that $g(y) = z$. Then, because $f\colon X \to Y$ is onto there exists $x \in X$ such that $f(x) = y$. Consequently, $(g \circ f)(x) = z$—so $g \circ f$ is surjective.

(*ii*) Suppose x_1 and x_2 are distinct elements of X—so $x_1 \neq x_2$. Because f is injective, we have $f(x_1) \neq f(x_2)$; and then because g is injective, we have $g(f(x_1)) \neq g(f(x_2))$. Thus,

$$x_1 \neq x_2 \Rightarrow (g \circ f)(x_1) \neq (g \circ f)(x_2)$$

—so $g \circ f$ is injective. ▮

2-5-4. Remark. Suppose we are given a map $f\colon X \to Y$ which is both 1–1 and onto; in other words, f is both injective and surjective. Let us take any element $y \in Y$. Because f is surjective, there exists $x \in X$ for which $f(x) = y$. Moreover, this element x is uniquely determined by y. In fact, if $x' \in X$ is also and element for which $f(x') = y$, then, because f is injective, $f(x) = f(x')$ implies $x = x'$. Therefore, we may define a mapping

$$g\colon Y \to X$$

by putting

$$g(y) = x, \qquad y \in Y$$

where x is the element determined by y according to the procedure described above. In other words, g and f are connected to each other by the relation

$$x = g(y) \Leftrightarrow y = f(x).$$

Our situation may be described by the pictures or diagrams

$$X \xrightarrow{f} Y \xrightarrow{g} X, \qquad Y \xrightarrow{g} X \xrightarrow{f} Y$$

and both of these can be coalesced into the diagram

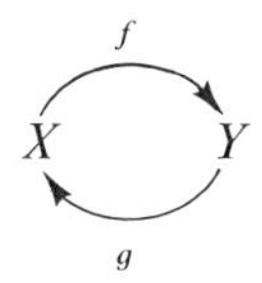

Now, let us examine the two composite maps

$$g \circ f\colon X \to X \qquad \text{and} \qquad f \circ g\colon Y \to Y.$$

For any $x \in X$, we have

$$(g \circ f)(x) = g(f(x)) = g(y) = x$$

which says that applying f (to x) and then following it by g takes us back to the element we started with (namely, to x). As is customary, let us denote the "identity map" of X (meaning the mapping which carries every element of X to itself) by $1: X \to X$—so

$$1(x) = x \qquad \text{for all } x \in X.$$

In virtue of this notation and the meaning of equality of mappings, we have proved that

$$g \circ f = 1 \quad \text{(the identity map of } X\text{).} \tag{$*$}$$

Similarly, for any $y \in Y$ we have

$$(f \circ g)(y) = f(g(y)) = f(x) = y$$

so that

$$f \circ g = 1 \quad \text{(the identity map of } Y\text{).} \tag{$**$}$$

Because properties ($*$) and ($**$) are satisfied, we say that the mapping g is an **inverse** of f. One often writes f^{-1} (and reads it as "f inverse") instead of g; so when f is injective and surjective we have $f \circ f^{-1} = 1, f^{-1} \circ f = 1$.

Incidentally, it should be noted that the inverse map $g: Y \to X$ constructed from our surjective and injective map $f: X \to Y$ is also surjective and injective. In fact, given any $x \in X$ we have, according to ($*$), $g(f(x)) = x$—so $f(x)$ is an element of Y whose image under g is x, and hence g is surjective. In addition, $g(y_1) = g(y_2)$ implies $f(g(y_1)) = f(g(y_2))$, which says that $y_1 = y_2$—hence g is injective.

Once we know that g is both surjective and injective, the foregoing discussion can be applied to g; in particular, g has an inverse, and the reader may easily convince himself that f is an inverse of g. (In this connection see Problem 11 at the end of this section.)

2-5-5. Definition. There is said to be a **1–1 correspondence** between the sets X and Y when there exists a mapping $f: X \to Y$ which is both injective and surjective. This mapping f is then called a **1–1 correspondence** between X and Y.

This mapping f need not be unique; once X and Y are in 1–1 correspondence, there may be many 1–1 correspondences between them.

Our definition is somewhat awkward because it distinguishes between X and Y in the sense that the mapping goes from X to Y. However, the definition could be stated, equivalently, in terms of a mapping from Y to X because, as seen in 2-5-4, there exists a mapping of $X \to Y$ which is surjective and injective if and only if there exists a mapping of $Y \to X$ which is surjective and injective. The important thing is that when the sets X and Y are in 1–1 correspondence the mapping describing this correspondence may be taken in either direction.

We have settled the meaning of "sameness" for sets, and now turn our attention to the meaning of sameness for two rings $\{R, +, \cdot\}$ and $\{R', +, \cdot\}$. As indicated earlier this requires, simultaneously, the sameness of the two sets R and R', the sameness of the two operations $+$, and the sameness of the two operations $\cdot$. (There is no great danger in using the same symbols $+$ and $\cdot$ for the operations in both rings.) Thus, we must have a 1–1 correspondence between the sets R and R' under which the operations in R and R' correspond to each other. In more detail, suppose we denote the 1–1 correspondence by

$$R \leftrightarrow R'$$

and write the correspondence between elements as

$$\begin{aligned} a &\leftrightarrow a', \qquad a \in R, \quad a' \in R', \\ b &\leftrightarrow b', \qquad b \in R, \quad b' \in R'. \end{aligned}$$

Then correspondence of the operations in R and R' means that the result of adding (or multiplying) two elements of R corresponds to the result of adding (or multiplying) the two corresponding elements of R'. Expressing this in symbols, we have

$$a + b \leftrightarrow a' + b' \qquad \text{and} \qquad ab \leftrightarrow a'b'. \tag{\#}$$

Now, according to our notation, given an element of R, the unique element corresponding to it is denoted by adjoining a prime ($'$). In particular, we have

$$a + b \leftrightarrow (a + b)' \qquad \text{and} \qquad ab \leftrightarrow (ab)'.$$

Consequently, because the correspondence is 1–1, (#) tells us that

$$(a + b)' = a' + b' \qquad \text{and} \qquad (ab)' = a'b'$$

for all $a, b \in R$. These last two relations express the sameness of the operations in R and R'.

From the point of view of algebra, it is more important to focus on the sameness, or preservation, of the operations than on the 1–1 correspondence between sets. This is the motif underlying the following definition.

2-5-6. Definition. Let $\{R, +, \cdot\}$ and $\{S, +, \cdot\}$ be rings. A mapping $\phi: R \to S$ is said to be a **homomorphism** of R into S when

$$\phi(a + b) = \phi(a) + \phi(b),$$

and

$$\phi(ab) = \phi(a)\phi(b),$$

for all $a, b \in R$. If, in addition, the mapping is both injective and surjective (that is, if it is 1–1 and onto), then ϕ is said to be an **isomorphism** of R onto S.

Thus, a homomorphism $\phi: R \to S$ is a mapping under which the ring operations are preserved. A homomorphism provides us with a tool for comparing two rings. Even more, we observe that the notion of an isomorphism ϕ of R onto S is the precise mathematical formulation of our vague notion of "sameness" for rings. In fact, when ϕ is an isomorphism, suppose we denote S by R' and for any $a \in R$ write $\phi(a) = a'$—then $a \leftrightarrow a'$ is a 1–1 correspondence between R and R' which expresses our intuitive notion of sameness for the rings R and R'.

2-5-7. Examples. (1) Consider the mapping ϕ: $\mathbf{Z} \to \mathbf{Z}$ defined by

$$\phi(n) = 2n, \qquad n \in \mathbf{Z}.$$

In 2-5-2, part (3) we observed that ϕ is injective but not surjective. Is ϕ a homomorphism? For any $m, n \in \mathbf{Z}$ we have

$$\phi(m + n) = 2(m + n) = 2m + 2n = \phi(m) + \phi(n)$$

which says that ϕ preserves addition. On the other hand, for any $m, n \in \mathbf{Z}$

$$\phi(mn) = 2mn \qquad \text{and} \qquad \phi(m)\phi(n) = (2m)(2n) = 4mn.$$

Since these are not always equal, ϕ does not preserve multiplication; therefore, ϕ is not a homomorphism.

We note in passing: If we modify the situation slightly and consider the map ϕ: $\mathbf{Z} \to 2\,\mathbf{Z}$ defined by $\phi(n) = 2n$, then ϕ is injective and surjective but not a homomorphism.

In the same vein, let the mapping ϕ: $\mathbf{Q} \to \mathbf{Q}$ be given by

$$\phi(\alpha) = 2\alpha, \qquad \alpha \in \mathbf{Q}$$

—or what is the same, since α can be written as m/n where $m, n \in \mathbf{Z}$ and $n \neq 0$, $\phi(m/n) = 2m/n$. Then ϕ is surjective, because given any $m/n \in \mathbf{Q}$ we have $\phi(m/2n) = m/n$; and ϕ is injective, because $m/n \neq m'/n'$ implies $\phi(m/n) \neq \phi(m'/n')$ since $2m/n \neq 2m'/n'$. In addition, ϕ is not a homomorphism—one checks easily that ϕ preserves addition, but not multiplication.

(2) Consider the mapping $\phi\colon \mathbf{Z} \to \mathbf{Z}$ given by

$$\phi(a) = a^2, \qquad a \in \mathbf{Z}.$$

As noted in 2-5-2, part (4) ϕ is neither surjective nor injective. Is it a homomorphism? Multiplication is preserved since for $a, b \in \mathbf{Z}$

$$\phi(ab) = (ab)^2 = a^2b^2 = \phi(a)\phi(b)$$

but addition is not preserved, since

$$\phi(a + b) = a^2 + 2ab + b^2 \quad \text{and} \quad \phi(a) + \phi(b) = a^2 + b^2$$

—hence ϕ is not a homomorphism.

(3) Suppose R and S are any rings. We may always define a mapping $\phi\colon R \to S$ by

$$\phi(a) = 0 \qquad \text{for all } a \in R.$$

Clearly, ϕ is a homomorphism. It is known as the **trivial homomorphism** (or as the **zero map**) and is not of much interest—no information about the rings R and S can be gleaned from the trivial homomorphism.

(4) Let us fix an integer $c \neq 0$ and define the mapping $\phi\colon \mathbf{Z} \to \mathbf{Z}$ by

$$\phi(x) = x + c, \qquad x \in \mathbf{Z}.$$

In particular, $\phi(0) = c$, $\phi(5) = 5 + c$, $\phi(-9) = -9 + c$; the action of ϕ amounts to adding c to any integer. Now, ϕ is injective because for $x, y \in \mathbf{Z}$

$$\phi(x) = \phi(y) \Rightarrow x + c = y + c \Rightarrow x = y$$

and ϕ is surjective because given any $y \in \mathbf{Z}$ we have

$$\phi(y - c) = y.$$

In addition, $\phi(x + y) = x + y + c$ and $\phi(x) + \phi(x) = x + y + 2c$ so (because $c \neq 0$) ϕ does not preserve addition; and ϕ does not preserve multiplication since $\phi(xy) = xy + c$ and $\phi(x)\phi(y) = xy + cx + cy + c^2$.

(5) Fix an integer $m > 1$; then we can always define a natural mapping $\phi\colon \mathbf{Z} \to \mathbf{Z}_m$ by

$$\phi(a) = \lfloor a \rfloor_m, \qquad a \in \mathbf{Z}.$$

In other words, ϕ maps each integer to its residue class modulo m, and it is often referred to as the **residue class map** mod m.

In the first place, ϕ is a homomorphism, because for any $a, b \in \mathbf{Z}$ we have

$$\begin{aligned} \phi(a + b) &= \lfloor a + b \rfloor_m = \lfloor a \rfloor_m + \lfloor b \rfloor_m = \phi(a) + \phi(b), \\ \phi(ab) &= \lfloor ab \rfloor_m = \lfloor a \rfloor_m \cdot \lfloor b \rfloor_m = \phi(a)\phi(b). \end{aligned}$$

Thus, ϕ is a homomorphism precisely because of the way in which addition and multiplication are defined in $\mathbf{Z}_m$.

Moreover, ϕ is surjective—in fact, given any $\alpha \in \mathbf{Z}_m$ we may write it in the form $\alpha = \lfloor a \rfloor_m$ with $a \in \mathbf{Z}$, and then $\phi(a) = \alpha$. However, ϕ is not injective—in fact, if $a' \equiv a \pmod{m}$ (that is, if a' is an element of the residue class $\lfloor a \rfloor_m$), then $\phi(a') = \phi(a)$ because $\lfloor a' \rfloor_m = \lfloor a \rfloor_m$.

(6) Let us fix integers $m > 1$, $n > 1$ and consider the rings $\mathbf{Z}_m$, $\mathbf{Z}_n$ and the direct sum $\mathbf{Z}_m \oplus \mathbf{Z}_n$. As the reader will recall, (see 2-2-11) the ring $\mathbf{Z}_m \oplus \mathbf{Z}_n$ consists of all ordered pairs (α, β) where $\alpha \in \mathbf{Z}_m$, $\beta \in \mathbf{Z}_n$, and the operations are componentwise. We define a mapping $\phi\colon \mathbf{Z} \to \mathbf{Z}_m \oplus \mathbf{Z}_n$ by putting

$$\phi(a) = (\lfloor a \rfloor_m, \lfloor a \rfloor_n), \qquad a \in \mathbf{Z}.$$

In particular, when $m = 5$, $n = 6$ the action of ϕ is of form $\phi(a) = (\lfloor a \rfloor_5, \lfloor a \rfloor_6)$ —so, for example,

$$\begin{aligned}
\phi(2) &= (\lfloor 2 \rfloor_5, \lfloor 2 \rfloor_6),\\
\phi(5) &= (\lfloor 5 \rfloor_5, \lfloor 5 \rfloor_6) = (\lfloor 0 \rfloor_5, \lfloor 5 \rfloor_6),\\
\phi(6) &= (\lfloor 6 \rfloor_5, \lfloor 6 \rfloor_6) = (\lfloor 1 \rfloor_5, \lfloor 0 \rfloor_6),\\
\phi(32) &= (\lfloor 32 \rfloor_5, \lfloor 32 \rfloor_6) = (\lfloor 2 \rfloor_5, \lfloor 2 \rfloor_6),\\
\phi(-10) &= (\lfloor -10 \rfloor_5, \lfloor -10 \rfloor_6) = (\lfloor 0 \rfloor_5, \lfloor 2 \rfloor_6).
\end{aligned}$$

Because of the way the operations are defined in residue class rings and in direct sums, the mapping $\phi\colon \mathbf{Z} \to \mathbf{Z}_m \oplus \mathbf{Z}_n$ is a homomorphism. In more detail, for $a, b \in \mathbf{Z}$ we have

$$\begin{aligned}
\phi(a+b) &= (\lfloor a+b \rfloor_m, \lfloor a+b \rfloor_n)\\
&= (\lfloor a \rfloor_m + \lfloor b \rfloor_m, \lfloor a \rfloor_n + \lfloor b \rfloor_n)\\
&= (\lfloor a \rfloor_m, \lfloor a \rfloor_n) + (\lfloor b \rfloor_m, \lfloor b \rfloor_n)\\
&= \phi(a) + \phi(b),
\end{aligned}$$

and

$$\begin{aligned}
\phi(ab) &= (\lfloor ab \rfloor_m, \lfloor ab \rfloor_n)\\
&= (\lfloor a \rfloor_m \lfloor b \rfloor_m, \lfloor a \rfloor_n \lfloor b \rfloor_n)\\
&= (\lfloor a \rfloor_m, \lfloor a \rfloor_n) \cdot (\lfloor b \rfloor_m, \lfloor b \rfloor_n)\\
&= \phi(a)\phi(b).
\end{aligned}$$

The mapping ϕ is not injective; for example, $\phi(0) = (\lfloor 0 \rfloor_m, \lfloor 0 \rfloor_n) = (0, 0)$, the zero element of $\mathbf{Z}_m \oplus \mathbf{Z}_n$, and also $\phi(mn) = (\lfloor mn \rfloor_m, \lfloor mn \rfloor_n) = (0, 0)$.

The question of whether ϕ is surjective is more complicated but extremely interesting. It asks: Given any element $(\alpha, \beta) \in \mathbf{Z}_m \oplus \mathbf{Z}_n$ which we may, of course, write in the form $(\lfloor a \rfloor_m, \lfloor b \rfloor_n)$, does there exist $c \in \mathbf{Z}$ for which $\phi(c) =$

$(\lfloor a \rfloor_m, \lfloor b \rfloor_n)$—that is, for which $(\lfloor c \rfloor_m, \lfloor c \rfloor_n) = (\lfloor a \rfloor_m, \lfloor b \rfloor_n)$? Thus, we seek $c \in \mathbf{Z}$ for which both

$$\lfloor c \rfloor_m = \lfloor a \rfloor_m \qquad \text{and} \qquad \lfloor c \rfloor_n = \lfloor b \rfloor_n$$

—or, translating to the language of congruences, we seek $c \in \mathbf{Z}$ such that

$$c \equiv a \pmod{m} \qquad \text{and} \qquad c \equiv b \pmod{n}.$$

Can this be done? Obviously, if $m = n$ and $a \not\equiv b \pmod{m}$, then no such c exists, and ϕ is not surjective. However, in the general case, when m, n, a, b are arbitrary, the tools required for settling the question of existence of c are not yet at our disposal. At this stage, therefore, we are unable to decide if, or when, ϕ is surjective.

(7) Fix integers $m_1 > 1$, $m_2 > 1$ and consider the direct sum ring $\mathbf{Z}_{m_1} \oplus \mathbf{Z}_{m_2}$. We may define mappings

$$f_1 \colon \mathbf{Z}_{m_1} \to \mathbf{Z}_{m_1} \oplus \mathbf{Z}_{m_2} \qquad \text{and} \qquad f_2 \colon \mathbf{Z}_{m_2} \to \mathbf{Z}_{m_1} \oplus \mathbf{Z}_{m_2}$$

by putting for $\alpha_1 \in \mathbf{Z}_{m_1}$, $\alpha_2 \in \mathbf{Z}_{m_2}$

$$f_1(\alpha_1) = (\alpha_1, 0) \qquad \text{and} \qquad f_2(\alpha_2) = (0, \alpha_2)$$

—or, writing this another way,

$$f_1(\lfloor a \rfloor_{m_1}) = (\lfloor a \rfloor_{m_1}, 0) \qquad \text{and} \qquad f_2(\lfloor a \rfloor_{m_2}) = (0, \lfloor a \rfloor_{m_2}).$$

Each of these maps is clearly an injective homomorphism, but not surjective.

In addition, we may define mappings

$$g_1 \colon \mathbf{Z}_{m_1} \oplus \mathbf{Z}_{m_2} \to \mathbf{Z}_{m_1} \qquad \text{and} \qquad g_2 \colon \mathbf{Z}_{m_1} \oplus \mathbf{Z}_{m_2} \to \mathbf{Z}_{m_2}$$

by putting

$$g_1(\alpha_1, \alpha_2) = \alpha_1 \qquad \text{and} \qquad g_2(\alpha_1, \alpha_2) = \alpha_2$$

or, what is the same thing,

$$g_1(\lfloor a \rfloor_{m_1}, \lfloor b \rfloor_{m_2}) = \lfloor a \rfloor_{m_1} \qquad \text{and} \qquad g_2(\lfloor a \rfloor_{m_1}, \lfloor b \rfloor_{m_2}) = \lfloor b \rfloor_{m_2}.$$

Each of these maps is clearly a surjective homomorphism, but not injective.

For $i = 1, 2$ one often refers to f_i as the **injection** on the ith coordinate and g_i as the **projection** on the ith coordinate.

Naturally, the notions of injection and projection may be extended to the case of a direct sum $\mathbf{Z}_{m_1} \oplus \mathbf{Z}_{m_2} \oplus \cdots \oplus \mathbf{Z}_{m_n}$ where $n \geq 2$. Even more, if $R_1, R_2, \ldots, R_n$ are any rings, we may associate with the direct sum $R_1 \oplus R_2 \oplus \cdots \oplus R_n$ injective maps $f_1, f_2, \ldots, f_n$ and projection maps $g_1, g_2, \ldots, g_n$. The details are left to the reader.

(8) Consider the domain of complex numbers $\mathbf{C}$. A generic complex number is of form $z = x + yi$ where x and y are real numbers. As is well

known, with each complex number $z = x + yi$ we may associate its **conjugate** $\bar{z} = x - yi$. Among the standard properties of conjugation (which the reader may easily verify if he is not familiar with them already) we have, for all z, z_1, z_2 in $\mathbf{C}$:

(*i*) $z\bar{z} = x^2 + y^2$,
(*ii*) $\bar{\bar{z}} = z$,
(*iii*) $\overline{z_1 + z_2} = \bar{z}_1 + \bar{z}_2$,
(*iv*) $\overline{z_1 z_2} = \bar{z}_1 \cdot \bar{z}_2$.

In words, the last two properties say that the conjugate of a sum is the sum of the conjugates and that the conjugate of a product is the product of the conjugates, respectively.

Now, consider the mapping $\phi: \mathbf{C} \to \mathbf{C}$ defined by

$$\phi(z) = \bar{z}, \qquad z \in \mathbf{C}.$$

Naturally, ϕ is known as the "conjugation map." Let us verify that ϕ is an isomorphism. It is surjective because, given any $z \in \mathbf{C}$ we have

$$\phi(\bar{z}) = \bar{\bar{z}} = z,$$

so z is the image of $\bar{z}$ under ϕ. It is injective because

$$\begin{aligned} \phi(z_1) = \phi(z_2) &\Rightarrow \bar{z}_1 = \bar{z}_2 \\ &\Rightarrow \bar{\bar{z}}_1 = \bar{\bar{z}}_2 \\ &\Rightarrow z_1 = z_2. \end{aligned}$$

Finally, ϕ is a homomorphism because for any $z_1, z_2 \in \mathbf{C}$

$$\begin{aligned} \phi(z_1 + z_2) &= \overline{z_1 + z_2} = \bar{z}_1 + \bar{z}_2 = \phi(z_1) + \phi(z_2), \\ \phi(z_1 z_2) &= \overline{z_1 z_2} = \bar{z}_1 \cdot \bar{z}_2 = \phi(z_1) \cdot \phi(z_2). \end{aligned}$$

(Thus the statement "ϕ is a homomorphism" is essentially another way of expressing the fact that the conjugate of a sum, or product, is the sum, or product of the conjugates.)

For an arbitrary ring R, any isomorphism $\phi: R \to R$ (that is, any isomorphism of R with itself) is said to be an **automorphism** of R. Thus, the foregoing discussion can be summarized by the statement: conjugation is an automorphism of $\mathbf{C}$.

(9) Let $R = \mathcal{M}(\mathbf{R}, 2)$, the ring of all 2×2 matrices with entries from $\mathbf{R}$. Let S denote the subset of R consisting of all elements of form

$$\begin{pmatrix} a & -b \\ b & a \end{pmatrix}, \qquad a, b \in \mathbf{R}.$$

For example,

$$\begin{pmatrix} \sqrt{2} & -3 \\ 3 & \sqrt{2} \end{pmatrix}, \qquad \begin{pmatrix} \sqrt{2} & 0 \\ 0 & \sqrt{2} \end{pmatrix}, \qquad \begin{pmatrix} 0 & \pi \\ -\pi & 0 \end{pmatrix}$$

are elements of S. It is easy to see that the nonempty set S is a subring of R. In fact, taking $\left(\begin{smallmatrix} c & -d \\ d & c \end{smallmatrix}\right)$, $c, d \in R$, which is also an element of S, it suffices, according to 2-2-7, to observe that

$$\begin{pmatrix} a & -b \\ b & a \end{pmatrix} - \begin{pmatrix} c & -d \\ d & c \end{pmatrix} = \begin{pmatrix} a-c & -(b-d) \\ b-d & a-c \end{pmatrix} \in S$$

and

$$\begin{pmatrix} a & -b \\ b & a \end{pmatrix} \cdot \begin{pmatrix} c & -d \\ d & c \end{pmatrix} = \begin{pmatrix} ac-bd & -(ad+bc) \\ ad+bc & ac-bd \end{pmatrix} \in S.$$

[This is identical with what was done in 2-2-10 for the subset of $\mathscr{M}(\mathbf{Z}, 2)$ denoted there by T.]

Now, let us define the mapping $\phi: \mathbf{C} \to S$ by

$$\phi(a+bi) = \begin{pmatrix} a & -b \\ b & a \end{pmatrix}, \qquad a, b \in \mathbf{R}.$$

The map ϕ is injective (since distinct complex numbers obviously map under ϕ to distinct matrices) and surjective (since from any given matrix belonging to S we can immediately produce the complex number whose image it is under ϕ). Moreover, ϕ is a homomorphism because

$$\begin{aligned} \phi((a+bi)+(c+di)) &= \phi((a+c)+(b+d)i) \\ &= \begin{pmatrix} a+c & -(b+d) \\ b+d & a+c \end{pmatrix} \end{aligned}$$

and

$$\begin{aligned} \phi(a+bi)+\phi(c+di) &= \begin{pmatrix} a & -b \\ b & a \end{pmatrix} + \begin{pmatrix} c & -d \\ d & c \end{pmatrix} \\ &= \begin{pmatrix} a+c & -(b+d) \\ b+d & a+c \end{pmatrix} \end{aligned}$$

while

$$\begin{aligned} \phi((a+bi)(c+di)) &= \phi((ac-bd)+(ad+bc)i) \\ &= \begin{pmatrix} ac-bd & -(ad+bc) \\ ad+bc & ac-bd \end{pmatrix} \end{aligned}$$

and

$$\begin{aligned} \phi(a+bi)\phi(c+di) &= \begin{pmatrix} a & -b \\ b & a \end{pmatrix} \begin{pmatrix} c & -d \\ d & c \end{pmatrix} \\ &= \begin{pmatrix} ac-bd & -(ad+bc) \\ ad+bc & ac-bd \end{pmatrix}. \end{aligned}$$

Consequently, ϕ is an isomorphism of **C** onto S, and S is just a disguised version of the complex numbers.

If we have an isomorphism between two rings, then they are essentially the same and, as indicated earlier, if we have a homomorphism of one ring into another, then their respective operations are related. Our next results describe how two rings are related under a homomorphism.

2-5-8. Proposition. Suppose R and S are rings and $\phi: R \to S$ is a homomorphism; then

(1) $\phi(0) = 0$,
(2) $\phi(-a) = -(\phi(a))$, $\quad a \in R$,
(3) $\phi(a - b) = \phi(a) - \phi(b)$, $\quad a, b \in R$,
(4) if R has an identity e and ϕ is surjective, then S has an identity—namely, $\phi(e)$.

Proof: (1) It should be noted that the 0 in $\phi(0)$ is the zero element of R and the 0 on the right side is the zero element of S; there is no serious danger of confusion. As for the proof, we have

$$\phi(0) = \phi(0 + 0) = \phi(0) + \phi(0),$$

which implies that $\phi(0)$ is the zero element of S.

(2) From the relations

$$0 = \phi(0) = \phi\big(a + (-a)\big) = \phi(a) + \phi(-a),$$

it follows that $\phi(-a)$ is the additive inverse of $\phi(a)$ in the ring S; in other words, $\phi(-a) = -(\phi(a))$. Of course, the "minus" sign has two meanings; in $\phi(-a)$ it denotes the additive inverse in R, while in $-(\phi(a))$ it denotes the additive inverse in S.

(3) For any $a, b \in R$ we have

$$\phi(a - b) = \phi\big(a + (-b)\big) = \phi(a) + \phi(-b) = \phi(a) - \phi(b).$$

(4) By hypothesis, $ea = a = ae$ for all $a \in R$. Now consider any element $s \in S$. We need to show that $\phi(e)s = s = s\phi(e)$. Because ϕ is surjective there exists an element $c \in R$ for which $\phi(c) = s$. Since $ec = c = ce$, application of ϕ gives $\phi(ec) = \phi(c) = \phi(ce)$ and

$$\phi(e)\phi(c) = \phi(c) = \phi(c)\,\phi(e),$$

which says that $\phi(e)s = s = s\phi(e)$. ∎

2-5-9. Proposition. Suppose $\phi: R \to S$ is a homomorphism of rings. Then:

(1) The image of ϕ, $\phi(R)$, is a subring of S.
(2) If we define the **kernel of** ϕ (and denote it as "ker ϕ") by

$$\ker \phi = \{a \in R \mid \phi(a) = 0\},$$

then ker ϕ is a subring of R.
(3) The map ϕ is injective $\Leftrightarrow$ $\ker \phi = (0)$.
(4) If ϕ is injective, then ϕ is an isomorphism of R onto $\phi(R)$.

Proof: (1) Consider any two elements of $\phi(R)$; they are of form

$$s_1 = \phi(a), \qquad s_2 = \phi(b), \qquad a, b \in R, \quad s_1, s_2 \in S.$$

Since $s_1 - s_2 = \phi(a) - \phi(b) = \phi(a - b) \in \phi(R)$ and $s_1 s_2 = \phi(a)\phi(b) = \phi(ab) \in \phi(R)$, we see that $\phi(R)$ is closed under subtraction and multiplication and, therefore, $\phi(R)$ is a subring of S.

(2) Suppose a and b are any elements of ker ϕ, so according to the definition of kernel we have $\phi(a) = 0$ and $\phi(b) = 0$. To prove that ker ϕ is a subring, it suffices to show that $a - b \in \ker \phi$ and $ab \in \ker \phi$; and the way to accomplish this is to apply ϕ to $a - b$ and ab. We have then

$$\phi(a - b) = \phi(a) - \phi(b) = 0,$$
$$\phi(ab) = \phi(a)\phi(b) = 0,$$

so $a - b$ and ab are in ker ϕ, and ker ϕ is a subring of R.

Actually a stronger statement holds about the kernel of a homomorphism, but we will not investigate it at this point.

(3) The element 0 of R always belongs to the kernel of ϕ, since $\phi(0) = 0$. Now, suppose $\ker \phi = (0)$—meaning that ker ϕ consists of the zero element alone. Then for $a, b \in R$, we have

$$\begin{aligned} \phi(a) = \phi(b) &\Rightarrow \phi(a - b) = \phi(a) - \phi(b) = 0 \\ &\Rightarrow a - b \in \ker \phi \\ &\Rightarrow a - b = 0 \\ &\Rightarrow a = b, \end{aligned}$$

so ϕ is 1–1. This proves the implication $\Leftarrow$.

To prove the implication $\Rightarrow$, suppose conversely that ϕ is 1–1. If $a \in \ker \phi$, the $\phi(a) = 0 = \phi(0)$, and by one-to-oneness $a = 0$. Thus, 0 is the sole element of the kernel or, what is the same thing, $\ker \phi = (0)$.

This result is quite useful. In order to decide if a homomorphism ϕ is injective, it suffices to determine which elements are the ones which map to 0 under ϕ; in particular, if $\phi(a) = 0$ implies $a = 0$, then $\ker \phi = (0)$ and the homomorphism is injective.

(4) According to part (1), $\phi(R)$ is a subring of S. Clearly, $\phi: R \to \phi(R)$ is still an injective homomorphism. But it is also surjective, so ϕ is an isomorphism of R onto $\phi(R)$. ▮

This last result is often convenient—for example, in connection with Example (9) of 2-5-7. More precisely, consider the mapping $\phi: \mathbf{C} \to \mathcal{M}(\mathbf{R}, 2)$ defined by

$$\phi(a + bi) = \begin{pmatrix} a & -b \\ b & a \end{pmatrix}.$$

As before, one verifies that ϕ is a homomorphism. Since $\left(\begin{smallmatrix} 0 & 0 \\ 0 & 0 \end{smallmatrix}\right)$ is clearly the zero element of $\mathcal{M}(\mathbf{R}, 2)$, it is immediate that $\ker \phi = (0)$—so ϕ is injective. Therefore, according to our last result, ϕ is an isomorphism of $\mathbf{C}$ onto $\phi(\mathbf{C})$. Obviously, $\phi(\mathbf{C})$ is precisely the set S defined in 2-5-7, part (9). In particular, there is no need to provide a separate proof that S is a subring of $\mathcal{M}(\mathbf{R}, 2)$, as was done there—this fact is automatic in virtue of part (1) of 2-5-9.

2-5-10. Proposition. Suppose R, S, T are rings. Suppose we are given mappings $\phi: R \to S$, $\psi: S \to T$ and consider the composite mapping $\psi \circ \phi: R \to T$ defined, as usual, by

$$(\psi \circ \phi)(a) = \psi(\phi(a)), \qquad a \in R.$$

Then:

(*i*) If both ϕ and ψ are 1–1, then so is $\psi \circ \phi$.
(*ii*) If both ϕ and ψ are onto, then so is $\psi \circ \phi$.
(*iii*) If both ϕ and ψ are homomorphisms, then so is $\psi \circ \phi$.
(*iv*) If both ϕ and ψ are isomorphisms, then so is $\psi \circ \phi$.
(*v*) If $\phi: R \to S$ is an isomorphism, so is $\phi^{-1}: S \to R$.

Proof: (*i*) and (*ii*) were proved in 2-5-3.

(*iii*) We must show that $\psi \circ \phi \in \mathrm{Map}(R, T)$ preserves both addition and multiplication. These properties do indeed hold because, for any $a, b \in R$

$$\begin{aligned}(\psi \circ \phi)(a + b) &= \psi[\phi(a + b)] \\ &= \psi[\phi(a) + \phi(b)] \\ &= \psi[\phi(a)] + \psi[\phi(b)] \\ &= (\psi \circ \phi)(a) + (\psi \circ \phi)(b)\end{aligned}$$

and

$$\begin{aligned}(\psi \circ \phi)(ab) &= \psi[\phi(ab)] \\ &= \psi[\phi(a)\phi(b)] \\ &= \psi[\phi(a)] \cdot \psi[\phi(b)] \\ &= [(\psi \circ \phi)(a)][(\psi \circ \phi)(b)]\end{aligned}$$

(*iv*) This is immediate from (*i*), (*ii*), and (*iii*).

(*v*) The basic facts about ϕ^{-1} were discussed in 2-5-4. Let us recall some of them explicitly. Given $s \in S$—because ϕ is surjective there exists $a \in R$ for which $\phi(a) = s$, and this element a is unique because ϕ is injective. We have then, by definition $\phi^{-1}(s) = a$. Furthermore, as proved in 2-5-4, $\phi^{-1} \circ \phi$: $R \to R$ is the identity map, $\phi \circ \phi^{-1}$: $S \to S$ is the identity map, and ϕ^{-1}: $S \to R$ is both surjective and injective.

To prove ϕ^{-1} is an isomorphism, it suffices to show that it is a homomorphism. For this, suppose $s_1, s_2 \in S$. We may write $\phi^{-1}(s_1) = a_1 \in R$, $\phi^{-1}(s_2) = a_2 \in R$—or what amounts to the same thing, $\phi(a_1) = s_1$, $\phi(a_2) = s_2$. Then

$$
\begin{aligned}
\phi^{-1}(s_1 + s_2) &= \phi^{-1}(\phi(a_1) + \phi(a_2)) && \text{(substitution)}\\
&= \phi^{-1}(\phi(a_1 + a_2)) && (\phi \text{ is a homomorphism})\\
&= (\phi^{-1} \circ \phi)(a_1 + a_2) && \text{(definition of composite)}\\
&= a_1 + a_2 && (\phi^{-1} \circ \phi = \text{identity})\\
&= \phi^{-1}(s_1) + \phi^{-1}(s_2) && \text{(substitution),}
\end{aligned}
$$

$$
\begin{aligned}
\phi^{-1}(s_1 s_2) &= \phi^{-1}[\phi(a_1)\phi(a_2)]\\
&= \phi^{-1}[\phi(a_1 a_2)]\\
&= (\phi^{-1} \circ \phi)(a_1 a_2)\\
&= a_1 a_2\\
&= \phi^{-1}(s_1)\phi^{-1}(s_2).
\end{aligned}
$$

This completes the proof. ▌

In virtue of the last result, there exists an isomorphism of R onto S if and only if there exists an isomorphism of S onto R; and in such a situation there is nothing lost in saying that the rings **R and S are isomorphic.**

We now have more than enough tools to prove a theorem which may be considered to be the focus or culmination of almost the entire development in this chapter.

2-5-11. Theorem. Let $\{D, P\}$ be a well ordered domain with identity e; then $\mathbf{Z}$ is order-isomorphic to D—more precisely, we can exhibit an isomorphism of $\mathbf{Z}$ onto D which preserves order.

Proof: Of course, both $\mathbf{Z}$ and D are rings, and we shall first produce an isomorphism ϕ of $\mathbf{Z}$ onto D. Then we shall prove that ϕ preserves order (after explaining the meaning of the phrases "order-isomorphism" and "preserves order").

Let us define a map $\phi\colon \mathbf{Z} \to D$ by

$$\phi(n) = ne, \qquad n \in \mathbf{Z}.$$

The meaning or definition of ne in D, and the properties of such symbols were discussed in 2-4-6 and 2-4-7. Using these properties, we see that for $m, n \in \mathbf{Z}$

$$\phi(m+n) = (m+n)e = me + ne = \phi(m) + \phi(n),$$
$$\phi(mn) = (mn)e = (me)(ne) = \phi(m)\phi(n).$$

Consequently, ϕ is a homomorphism.

Since $\{D, P\}$ is an ordered domain, we know from 2-3-6 that $e > 0$. Therefore, $2e = e + e > 0$, $3e = 2e + e > 0$, and inductively $me > 0$ for all integers $m > 0$. To put it another way, for every integer $m > 0$ we have $\phi(m) > 0$ in $\{D, P\}$. Consider the set

$$S = \{me \mid m > 0\}$$

which may also be written as

$$S = \{\phi(m) \mid m > 0\}.$$

In words, S is the image under ϕ of the set of all positive integers. Clearly, $S \subset P$ and $e \in S$. Moreover, the set S is inductive, since $me \in S$ implies $me + e = (m+1)e \in S$. Consequently, because $\{D, P\}$ is well ordered, 2-3-14 tells us that $S = P$.

Since $P = \{me \mid m > 0\}$, it now follows that

$$\begin{aligned} -P &= \{-(me) \mid m > 0\} = \{(-m)e \mid m > 0\} \\ &= \{me \mid m < 0\} = \{\phi(m) \mid m < 0\}. \end{aligned}$$

If we let $\mathbf{Z}_{>0}$ denote the set of all positive integers and $\mathbf{Z}_{<0}$ denote the set of all negative integers, the preceding remarks yield

$$\begin{aligned} \phi(\mathbf{Z}_{>0}) &= \{\phi(m) \mid m > 0\} = P, \\ \phi(\mathbf{Z}_{<0}) &= \{\phi(m) \mid m < 0\} = -P. \end{aligned} \tag{$*$}$$

In particular, for the homomorphism $\phi\colon \mathbf{Z} \to D$ we have

$$\begin{aligned} m > 0 &\Rightarrow \phi(m) > 0, \\ m = 0 &\Rightarrow \phi(m) = 0, \\ m < 0 &\Rightarrow \phi(m) < 0. \end{aligned} \tag{$**$}$$

In words, $(**)$ asserts that under ϕ, the image of a positive integer is a positive element of D, the image of zero is zero, and the image of a negative integer is a negative element of D; even more, $(*)$ asserts that under ϕ, the image of the set of all positive integers is the set of all positive elements of D, and the image of the set of all negative integers is the set of all negative elements of D.

With all the machinery in place, it is easy to show that ϕ is both injective and surjective. In fact, if $m \in \ker \phi$, then $\phi(m) = 0$, so according to (*) we have $m = 0$; hence, $\ker \phi = (0)$ and ϕ is injective. Furthermore, according to the definition of ordered domain, D is a disjoint union $D = P \cup \{0\} \cup -P$; so in virtue of (*) we have

$$D = \phi(\mathbf{Z}_{>0}) \cup \{\phi(0)\} \cup \phi(\mathbf{Z}_{<0})$$

which guarantees that every element of D is the image of some integer under ϕ—that is, ϕ is surjective. [The proof of surjectivity of ϕ could have been expressed more directly in the form: $\mathbf{Z} = \mathbf{Z}_{>0} \cup \{0\} \cup \mathbf{Z}_{<0}$ implies

$$\phi(\mathbf{Z}) = \phi(\mathbf{Z}_{>0}) \cup \{\phi(0)\} \cup \phi(\mathbf{Z}_{<0}) = P \cup \{0\} \cup -P = D].$$

Therefore, $\phi\colon \mathbf{Z} \to D$ is an isomorphism.

Thus, $\mathbf{Z}$ and D are essentially the same, as rings. But except for their ring structures both $\mathbf{Z}$ and D are well-ordered domains; so on each of them we have an order relation defined in terms of either the set of positive elements or the relation $<$ (see Section 2-3). In order for $\mathbf{Z}$ and D to be essentially the same, as well-ordered domains, it is natural to require that the isomorphism ϕ (or some other isomorphism between $\mathbf{Z}$ and D) should also preserve order. The "preservation of order" may be defined in terms of the set of positive elements, but it is more convenient and intuitive to do so in terms of the relation $<$. More precisely, an arbitrary mapping $\phi\colon \mathbf{Z} \to D$ is said to preserve order when, for any integers m and n

$$m < n \Rightarrow \phi(m) < \phi(n)$$

—in other words, m is less than n (in $\mathbf{Z}$) implies that $\phi(m)$ is less than $\phi(n)$ (in D). Then, by an **order-isomorphism** we mean an isomorphism which preserves order. Clearly, the notion of order-isomorphism is applicable for any two ordered, or well-ordered domains—and two such domains should be considered to be the same if they are order-isomorphic.

Returning to our isomorphism $\phi\colon n \to ne$ of $\mathbf{Z}$ onto D, it is immediate from (**) that ϕ preserves order—in fact,

$$\begin{aligned} m < n &\Rightarrow m - n < 0 \\ &\Rightarrow \phi(m - n) < 0 \\ &\Rightarrow \phi(m) - \phi(n) < 0 \\ &\Rightarrow \phi(m) < \phi(n). \end{aligned}$$

Thus, ϕ is indeed an order isomorphism, and the proof is complete. ▌

According to this result, there is really only one well-ordered domain, namely $\mathbf{Z}$! Any well-ordered domain D is an order-isomorphic "copy" of $\mathbf{Z}$; they can differ only in the names of their elements, but their *internal structures* are the same. Any statement about $\mathbf{Z}$ can be transferred to D, and conversely. The way to formulate this in mathematical language is to say that we have a **characterization of** $\mathbf{Z}$, or, more precisely that the axioms for a well-ordered domain **characterize** $\mathbf{Z}$. If these axioms hold, then we are dealing with some "version" of $\mathbf{Z}$.

Note that if D_1 and D_2 are any two well-ordered domains, then they are order-isomorphic. In fact, by 2-5-11, there exist order-preserving isomorphisms ϕ_1: $\mathbf{Z} \to D_1$ and ϕ_2: $\mathbf{Z} \to D_2$. It then follows easily that ϕ_1^{-1} : $D_1 \to \mathbf{Z}$ is an order preserving isomorphism, and so is

$$\phi_2 \circ \phi_1^{-1}: \quad D_1 \to D_2 .$$

Having developed a set of axioms which determine $\mathbf{Z}$ completely we could, in theory at least, go back and prove *all* the number-theoretic results of Chapter I in a formal rigorous way without knowing anything about $\mathbf{Z}$ other than the axioms for a well-ordered domain.

2-5-12 / *PROBLEMS*

1. Consider the sets $X = \{1, 2, 3\}$ and $Y = \{1, 2, 3, 4\}$.
(*i*) Define a mapping ϕ: $X \to Y$ by putting

$$\phi(1) = 3, \qquad \phi(2) = (1), \qquad \phi(3) = 4.$$

Is ϕ injective? Is it surjective?
(*ii*) Define a mapping ψ: $Y \to X$ by putting

$$\psi(1) = 2, \qquad \psi(2) = 3, \qquad \psi(3) = 1, \qquad \psi(4) = 3.$$

Is ψ injective? Is it surjective?
(*iii*) Compute the composite mappings $\psi \circ \phi \in \text{Map}(X, X)$ and $\phi \circ \psi \in \text{Map}(Y, Y)$. For each one, decide if it is injective and if it is surjective.

2. In each of the following cases is the given map ϕ: $\mathbf{Z} \to \mathbf{Z}$ injective? surjective? a homomorphism? an isomorphism?
(*i*) $\phi(x) = x + 5$, (*ii*) $\phi(x) = 3x - 2$,
(*iii*) $\phi(x) = |x|$, (*iv*) $\phi(x) = -x$,
(*v*) $\phi(x) = x^3 - x$.
Answer the same questions (if possible) when ϕ is the mapping of $\mathbf{Q} \to \mathbf{Q}$ defined by the given formulas, and also when ϕ is viewed as a mapping of $\mathbf{R} \to \mathbf{R}$.

3. Let $m = 3$. In each of the following cases, is the given map ϕ: $\mathbf{Z}_m \to \mathbf{Z}_m$ injective ? surjective? a homomorphism? an isomorphism?

(*i*) $\phi(\underline{\lvert x \rvert}_m) = 2\underline{\lvert x \rvert}_m$, (*ii*) $\phi(\underline{\lvert x \rvert}_m) = \underline{\lvert 2 \rvert}_m + \underline{\lvert x \rvert}_m$,
(*iii*) $\phi(\underline{\lvert x \rvert}_m) = -\underline{\lvert x \rvert}_m$, (*iv*) $\phi(\underline{\lvert x \rvert}_m) = (\underline{\lvert x \rvert}_m)^2$,
(*v*) $\phi(\underline{\lvert x \rvert}_m) = (\underline{\lvert x \rvert}_m)^3$.
Answer the same questions for $m = 5$ and for $m = 6$. Investigate these questions when m is an arbitrary integer greater than 1.

4. When there is a 1–1 correspondence between the sets X and Y, let us denote this by $X \leftrightarrow Y$. Show that the relation of 1–1 correspondence is reflexive, symmetric, and transitive. In more detail, for sets X, Y, Z
(*i*) $X \leftrightarrow X$,
(*ii*) if $X \leftrightarrow Y$, then $Y \leftrightarrow X$,
(*iii*) if $X \leftrightarrow Y$ and $Y \leftrightarrow Z$, then $X \leftrightarrow Z$.

5. Consider any sets X, Y, Z, W and any mappings $f: X \to Y$, $g: Y \to Z$, $h: Z \to W$; then we have mappings $h \circ (g \circ f)$ and $(h \circ g) \circ f$ in Map(X, W). Prove that

$$h \circ (g \circ f) = (h \circ g) \circ f.$$

Of course, this may be viewed as a sort of associative law for mappings.

6. Consider the three element set $X = \{1, 2, 3\}$.
(*i*) Find two mappings $f: X \to X$ and $g: X \to X$ such that $g \circ f \neq f \circ g$.
(*ii*) Can you do this in such a way that both f and g are 1–1 onto?
(*iii*) How many elements are there in Map (X, X)? How many of these are surjective maps?

7. (*i*) Exhibit two mappings $f: \mathbf{Z} \to \mathbf{Z}$ and $g: \mathbf{Z} \to \mathbf{Z}$ for which $g \circ f = 1$ (the identity map of $\mathbf{Z}$) and $f \circ g \neq 1$.
(*ii*) Can you do this in such a way that f is not surjective and g is not injective?

8. Suppose R and S are rings and consider a mapping $\phi: R \to S$. Now; ϕ may be injective or it may not; ϕ may be surjective or it may not; ϕ may be a homomorphism or it may not. Thus, there are eight "types" for ϕ. Exhibit a ϕ of each type; of course, you are free to choose R and S in each case.

9. Let us fix elements a, b in the ring R, and let us define mappings $\phi: R \to R$ and $\psi: R \to R$ by

$$\phi(x) = a + x, \qquad \psi(x) = bx.$$

Find the mappings $\psi \circ \phi$, $\phi \circ \psi$, $\phi \circ \phi$, $\psi \circ \psi$.

We now have a total of six mappings before us. Which ones are injective? surjective? homomorphisms? Can any of these properties be affected by imposing additional conditions on R (for example, such restrictions as: R is commutative; R has a unity; R is a domain, ... and so on) and selecting a and b judiciously?

10. Given sets X, Y and mappings $f: X \to Y$, $g: Y \to X$, prove that if $g \circ f = 1$, the identity map of X, then f is injective and g is surjective.

11. Suppose a mapping $f: X \to Y$ is given.
(*i*) Show that f is 1–1 and onto $\Leftrightarrow$ there exists a mapping $g: Y \to X$ for which $g \circ f = 1$ and $f \circ g = 1$. As indicated in 2-5-4, when such a mapping g exists we say that "f has an inverse" or that "g is an inverse of f."
(*ii*) When the above situation holds, we have:
(*a*) the inverse of f is unique.
(*b*) the mapping g has an inverse (as it too is 1–1 and onto), and its unique inverse is f.

12. Suppose $\phi: R_1 \to R_2$ is a surjective homomorphism (where R_1 and R_2 are rings). Show that if R_1 is commutative, then so is R_2.

13. The relation of isomorphism between rings is reflexive, symmetric and transitive. In more detail, when the rings R and R' are isomorphic, let us write $R \simeq R'$; then for rings R_1, R_2, R_3, prove
(*i*) $R_1 \simeq R_1$,
(*ii*) if $R_1 \simeq R_2$, then $R_2 \simeq R_1$,
(*iii*) if $R_1 \simeq R_2$ and $R_2 \simeq R_3$, then $R_1 \simeq R_3$.

14. If the rings R and S are isomorphic, and S is an integral domain, show that R is an integral domain.

15. Verify that the mapping $\phi: \mathbf{Z}_{24} \to \mathbf{Z}_6$ defined by

$$\phi(\lfloor a \rfloor_{24}) = \lfloor a \rfloor_6$$

is a homomorphism. Is it surjective? What is the kernel?

16. Suppose m and n are integers greater than 1 with $n \mid m$. Find a "natural" homomorphism $\phi: \mathbf{Z}_m \to \mathbf{Z}_n$. Is it surjective? What is the kernel? What happens if $n \nmid m$?

17. Suppose $\{R, +, \cdot\}$ is a ring with unity e. Let us define, for $a, b \in R$

$$a \perp b = a + b + e, \qquad a \circ b = a + b + ab.$$

Prove that $\{R, \perp, \circ\}$ is a ring. Does it have a unity? Show that the mapping $\phi: a \to a - e$ is an isomorphism of the ring $\{R, +, \cdot\}$ onto the ring $\{R, \perp, \circ\}$. What is the inverse isomorphism?

18. For any rings R_1 and R_2 the direct sum rings $R_1 \oplus R_2$ and $R_2 \oplus R_1$ are isomorphic. More generally, if $R_1, R_2, \ldots, R_n$ are rings and σ is a permutation of the set $\{1, 2, \ldots, n\}$, then the rings $R_1 \oplus R_2 \oplus \cdots \oplus R_n$ and $R_{\sigma 1} \oplus R_{\sigma 2} \oplus \ldots \oplus R_{\sigma n}$ are isomorphic.

19. If $\phi_1 : R \to R_1$ and $\phi_2 : R \to R_2$ are homomorphisms of rings, then the mapping $\phi : R \to R_1 \oplus R_2$ defined by

$$\phi(a) = (\phi_1(a), \phi_2(a)), \qquad a \in R$$

is also a homomorphism.

20. Suppose $\phi_1 : R_1 \to S_1$ and $\phi_2 : R_2 \to S_2$ are homomorphisms of rings. Define $\phi_1 \oplus \phi_2 : R_1 \oplus R_2 \to S_1 \oplus S_2$ by

$$(\phi_1 \oplus \phi_2)(a_1, a_2) = (\phi_1(a_1), \phi_2(a_2)), \qquad a_1 \in R_1, a_2 \in R_2$$

and show it is a homomorphism. Furthermore, if both ϕ_1 and ϕ_2 are surjective, then so is $\phi_1 \oplus \phi_2$; and if both ϕ_1 and ϕ_2 are injective, then so is $\phi_1 \oplus \phi_2$; consequently, if both ϕ_1 and ϕ_2 are isomorphisms, then so is $\phi_1 \oplus \phi_2$.

21. Suppose D_1, D_2, D_3 are ordered domains and $\phi_1 : D_1 \to D_2, \phi_2 : D_2 \to D_3$ are order-isomorphisms. Show that ϕ_1^{-1} and $\phi_2 \circ \phi_1$ are order-isomorphisms.

22. Are the rings $\mathbf{Z}[i] = \{a + bi \mid a, b \in \mathbf{Z}\}$ and $\mathbf{Z}[\sqrt{2}] = \{a + b\sqrt{2} \mid a, b \in \mathbf{Z}\}$ isomorphic?

23. (*i*) Consider the rings R_1, R_2 and their direct sum $R_1 \oplus R_2$. As in 2-5-7, part (7), define mappings

$$f_1 : R_1 \to R_1 \oplus R_2, \qquad f_2 : R_2 \to R_1 \oplus R_2,$$
$$g_1 : R_1 \oplus R_2 \to R_1, \qquad g_2 : R_1 \oplus R_2 \to R_2,$$

by putting

$$\begin{matrix} f_1(a_1) = (a_1, 0), & f_2(a_2) = (0, a_2), \\ g_1(a_1, a_2) = a_1, & g_2(a_1, a_2) = a_2, \end{matrix} \qquad a_1 \in R_1, \quad a_2 \in R_2.$$

Then, f_1 and f_2 are injective homomorphisms, and g_1 and g_2 are surjective homomorphisms.

By taking composites we obtain the following homomorphisms

$$\begin{aligned} &g_1 \circ f_1\colon\ R_1 \to R_1, && g_2 \circ f_1\colon R_1 \to R_2,\\ &g_1 \circ f_2\colon\ R_2 \to R_1, && g_2 \circ f_2\colon R_2 \to R_2,\\ &f_1 \circ g_1\colon\ R_1 \oplus R_2 \to R_1 \oplus R_2, && f_2 \circ g_2\colon R_1 \oplus R_2 \to R_1 \oplus R_2, \end{aligned}$$

and, in fact,

$$\begin{gathered} g_1 \circ f_1 = 1, \qquad g_2 \circ f_2 = 1,\\ g_1 \circ f_2 = 0, \qquad g_2 \circ f_1 = 0,\\ f_1 \circ g_1 + f_2 \circ g_2 = 1. \end{gathered}$$

[Here, as discussed in 2-2-12, 0 refers to the zero map, and $+$ denotes the sum of two functions in $\mathrm{Map}(R_1 \oplus R_2, R_1 \oplus R_2)$. Of course, the composition $\circ$ of two functions has no connection with the product $\cdot$ of two functions as discussed in 2-2-12.]

(*ii*) More generally, consider the direct sum of rings, $R_1 \oplus R_2 \oplus \cdots \oplus R_n$. For each $i = 1, 2, \ldots, n$ define mappings

$$f_i\colon R_i \to R_1 \oplus \cdots \oplus R_n, \qquad g_i\colon R_1 \oplus \cdots \oplus R_n \to R_i$$

by putting, for $a_1 \in R_1, a_2 \in R_2, \ldots, a_n \in R_n$

$$f_i(a_i) = (0, \ldots, 0, a_i, 0, \ldots, 0), \qquad g_i(a_1, a_2, \ldots, a_n) = a_i.$$

Then each f_i is an injective homomorphism, known as the **injection** on the ith coordinate; and each g_i is a surjective homomorphism, known as the **projection** on the ith coordinate.

For each pair of integers i, j with $1 \le i, j \le n$ we have a homomorphism

$$g_i \circ f_j\colon R_j \to R_i$$

—in fact,

$$g_i \circ f_j = \begin{cases} 1, & \text{if } i = j,\\ 0, & \text{if } i \neq j. \end{cases}$$

Furthermore, for each i we have a homomorphism

$$f_i \circ g_i\colon R_1 \oplus \cdots \oplus R_n \to R_1 \oplus \cdots \oplus R_n$$

and these satisfy the relation

$$f_1 \circ g_1 + f_2 \circ g_2 + \cdots + f_n \circ g_n = 1$$

(the identity map of $R_1 \oplus \cdots \oplus R_n$).

Miscellaneous Problems

1. In **Z**, let us write

$$a \top b = \max\{a, b\}, \qquad a \perp b = \min\{a, b\}.$$

Is $\{\mathbf{Z}, \top, \perp\}$ a ring? Which axioms are satisfied and which are not?

2. (*i*) Show that a nonempty subset S of **Z** that is closed under subtraction is a subring of **Z**; so there exists (see Problem 11 of 2-2-15) an integer $d \geq 0$ for which $S = d\mathbf{Z} = \{nd \mid n \in \mathbf{Z}\}$.

(*ii*) Show by example that a nonempty subset of **Z** that is closed under addition need not be a subring.

3. Consider the set $\mathscr{C}$ of all continuous real-valued functions on the closed interval [0, 1]. Show it is a commutative ring with unity. Is it an integral domain? If we write $X = [0, 1]$ how does $\mathscr{C}$ compare with the ring Map(X, **R**)? Show that the set of all differentiable functions on [0, 1] is a subring of $\mathscr{C}$.

4. Suppose f and g are two real-valued functions (on [0, 1], say) that have as many derivatives as desired. For any $n \geq 1$, find a formula for the nth derivative of the product fg; prove the formula by induction.

5. Which of the following subsets of Map(**R**, **R**) is a commutative ring with unity?

(*i*) $\{f \mid f(1) = 0\}$;
(*ii*) $\{f \mid f(1) \neq 0\}$;
(*iii*) $\{f \mid f(1) = f(0)\}$;
(*iv*) $\{f \mid |f(1)| \leq M\}$, where M is a fixed real number;
(*v*) $\{f \mid |f(x)| \leq M$ for all $x \in \mathbf{R}\}$, M fixed.

6. Show that in the definition of a ring, Axioms A3 and A4 may be replaced by the statement: For any $a, b \in R$, there exists an element $x \in R$ such that $a + x = b$.

7. Suppose $\{R, +, \cdot\}$ satisfies all the requirements for a ring with unity e except the commutative law for addition. By expanding $(a + b)(e + e)$, show that addition is commutative, so $\{R, +, \cdot\}$ is indeed a ring.

8. In an ordered domain, the cancellation law for multiplication can be proved from the other assumptions. In other words, show that if R is a commutative ring with unity in which the axioms for an ordering (as in 2-3-1) hold, then R is a domain.

9. Show that $\mathbf{Z}_p$—or, more generally, any finite integral domain—cannot be made into an ordered domain.

10. Suppose R satisfies all the axioms for a ring except the commutative law for addition. Prove that if there exists $c \in R$ for which $ca = cb \Rightarrow a = b$, then R is a ring.

11. Suppose the ring R has a unique element e such that $ea = a$ for all $a \in R$ (in other words, e is a unique left identity); show that e is an identity for multiplication.

12. If A, B, C are subsets of the universal set X, prove the following:

(*i*) $A = \varnothing \Leftrightarrow B = (A \cap B^c) \cup (A^c \cap B)$;

(*ii*) $A \cap B = A - (A - B) = B - (B - A) = (A \cup B) - (A + B)$;

(*iii*) $A - C \subset (A - B) \cup (B - C)$;

(*iv*) $(A \cup B) - (B \cap C) = (A - B) \cup (B - C)$;

(*v*) $(A - B) \cap C = A \cap C - B \cap C$;

(*vi*) $A \cap (B + C) = A \cap B + A \cap C$.

13. Suppose A, B, C are elements of $\mathscr{S}(X)$. Show that if $A \cup B = A \cup C$ and $A \cap B = A \cap C$ then $B = C$.

14. (*i*) If A and B are finite sets prove that

$$\#(A \cup B) = \#(A) + \#(B) - \#(A \cap B).$$

(*ii*) If A_1, A_2, A_3 are finite sets prove that

$$\begin{aligned} \#(A_1 \cup A_2 \cup A_3) = \#(A_1) + \#(A_2) + \#(A_3) \\ - \#(A_1 \cap A_2) - \#(A_1 \cap A_3) \\ - \#(A_2 \cap A_3) + \#(A_1 \cap A_2 \cap A_3). \end{aligned}$$

(*iii*) Generalize the preceding to the case of n finite sets $A_1, A_2, \ldots, A_n$.

15. (*i*) If A and B are finite sets then prove that

$$\#(A + B) = \#(A) + \#(B) - 2\#(A \cap B)$$

(*ii*) If A_1, A_2, A_3 are finite sets, find an expression for $\#(A_1 + A_2 + A_3)$.

(*iii*) Can you generalize the preceding to the case of n finite sets $A_1, A_2, \ldots, A_n$?

16. Suppose $\{A_\alpha \mid \alpha \in \mathfrak{A}\}$ is an arbitrary collection, finite or infinite, of subsets A_α of the set X. (Here, $\mathfrak{A}$ is simply the set of all indices.) Prove the generalized de Morgan laws:

$$\left(\bigcap_\alpha A_\alpha\right)^c = \bigcup_\alpha A_\alpha^c, \qquad \left(\bigcup_\alpha A_\alpha\right)^c = \bigcap_\alpha A_\alpha^c.$$

17. A ring R is said to be **Boolean** when $a^2 = a$ for every $a \in R$.
(*i*) Give an example of a Boolean ring.
(*ii*) Prove: If R is Boolean then $2a = 0$ for every $a \in R$.
(*iii*) Show that a Boolean ring is commutative.
(*iv*) Prove: If R is a Boolean ring with identity e then (excluding 0 and e) every element of R is a zero divisor.
(*v*) Prove: If R and S are Boolean rings, so is their direct sum $R \oplus S$. What about the converse?

18. In the commutative ring R, suppose a and b are nilpotent elements (see 2-4-10, Problem 13), and c is arbitrary. Show that $a + b$, $a - b$ and ca are nilpotent.

19. Give examples of a ring R and a subring S such that
(*i*) R and S have the same unity.
(*ii*) R and S have different unities.
(*iii*) R has unity but S does not.
(*iv*) S has unity but R does not.
(*v*) S is commutative but R is not.
(*vi*) R has unity and is not a domain, and S is a domain. Do R and S have the same unity?

20. Give examples of rings R and S and a surjective homomorphism $\phi: R \to S$ such that
(*i*) S is commutative but R is not.
(*ii*) S has a unity but R does not.
(*iii*) S is a domain but R is not.
(*iv*) R is a domain but S is not.

21. Suppose R is a ring and S is a set on which we have two operations $+$ and $\cdot$. Show that if $\phi: R \to S$ is a surjective map that preserves the operations (that is, it satisfies the requirements for a homomorphism) then S is a ring. One says: a homomorphic image of a ring is a ring.

22. Suppose ϕ is a 1–1 mapping of the set S onto the ring $R = \{R, +, \cdot\}$. Show that S becomes a ring, isomorphic to R, when its operations are defined by:

$$\begin{aligned} a \oplus b &= \phi^{-1}(\phi(a) + \phi(b)) \\ a \odot b &= \phi^{-1}(\phi(a) \cdot \phi(b)) \end{aligned} \qquad a, b \in S.$$

What if ϕ is given as a map of S onto R?

23. Suppose R is a ring (not necessarily commutative) with unity e. We say that the element $a \in R$ has a **right inverse** when there exists $b \in R$ for which $ab = e$, and that $a \in R$ has a **left inverse** when there exists $c \in R$ for which $ca = e$.

(*i*) If b is a right inverse of a, and c is a left inverse of a, show that $b = c$. We then have $ab = ba = e$, and b is said to be an **inverse** of a.

(*ii*) If a has an inverse, it is unique—denote it by a^{-1}. Show that a^{-1} also has an inverse, namely a.

(*iii*) Prove: If both a and b have inverses, then so does ab—in fact, $(ab)^{-1} = b^{-1}a^{-1}$.

(*iv*) Give an example of a ring R and two elements, $a, b \in R$ such that a and b have inverses but $(ab)^{-1} \neq a^{-1}b^{-1}$.

24. Consider $A = \begin{pmatrix} a & b \\ c & d \end{pmatrix}$ in $\mathscr{M}(\mathbf{Z}, 2)$; then prove the following:

(*i*) There exists $B \in \mathscr{M}(\mathbf{Z}, 2)$ for which $AB = I$ (that is, A has a right inverse) if and only if $ad - bc = \pm 1$.

(*ii*) There exists $B \in \mathscr{M}(\mathbf{Z}, 2)$ for which $AB = I$ if and only if there exists $C \in \mathscr{M}(\mathbf{Z}, 2)$ for which $CA = I$. That is, A has a right inverse if and only if A has a left inverse.

(*iii*) If $AB = I$ and $CA = I$ then $B = C$.

(*iv*) If $AB = I$ then $BA = I$ (that is, if A has a right inverse then it has an inverse) and conversely.

(*v*) If there exists B such that $AB = I$, then B is unique.

25. Consider $A = \begin{pmatrix} a & b \\ c & d \end{pmatrix}$ in $\mathscr{M}(\mathbf{Q}, 2)$ [or in $\mathscr{M}(\mathbf{R}, 2)$]. Then there exists $B \in \mathscr{M}(\mathbf{Q}, 2)$ [or in $\mathscr{M}(\mathbf{R}, 2)$] for which $AB = I \Leftrightarrow ad - bc \neq 0$. Discuss the remaining assertions of the preceding problem in this situation.

26. Let R be a ring with unity e. Suppose $a \in R$ has a right inverse. Then show that the following are equivalent:

(*i*) The right inverse of a is unique;

(*ii*) a has an inverse;

(*iii*) a is not a left zero-divisor (that is, $ab = 0 \Leftrightarrow b = 0$).

27. Consider a ring R. If m is the smallest positive integer such that $ma = 0$ for all $a \in R$, we say that R has **characteristic** m (in particular, R is said to have **finite** characteristic) and write char $R = m$. If no such m exists (so there is no $n > 0$ such that $na = 0$ for all $a \in R$) we say that R has characteristic 0.

Let D be an integral domain; then

(*i*) Prove: If there exist $n > 0$ in $\mathbf{Z}$ and $b \neq 0$ in D such that $nb = 0$ then $na = 0$ for all $a \in D$; so D has finite characteristic.

(*ii*) Prove: If D has finite characteristic then char D is the smallest positive integer d such that $db = 0$ for some $b \neq 0$ in D.

(*iii*) Prove: char D is 0 or a prime.

(*iv*) For an arbitrary $m > 0$, can you produce a ring R of characteristic m?

28. Consider a ring R without unity, and the product set $R \times \mathbf{Z}$—which we denote by $\bar{R}$. If we define operations in $\bar{R}$ by:

$$(a, m) + (b, n) = (a + b, m + n),$$
$$(a, m) \cdot (b, n) = (ab + na + mb, mn),$$

then $\bar{R}$ becomes a ring with unity (0, 1). What is the characteristic of $\bar{R}$? The mapping $a \to (a, 0)$ is an injective homomorphism of R into $\bar{R}$; thus $\bar{R}$ contains an isomorphic copy of R as a subring, and R has been "imbedded" in a ring with unity.

Explain the reasons underlying the choice of $\bar{R} = R \times \mathbf{Z}$ and the definition of its operations. Can you imbed the given ring R in a ring with unity whose characteristic is the same as the characteristic of R?

29. If R is a ring with unity of characteristic 0 show that R contains a subring that is isomorphic to $\mathbf{Z}$.

30. If a is an odd integer, prove that $a^2 \equiv 1 (\text{mod } 8)$, $a^4 \equiv 1 (\text{mod } 16)$ and, by induction,

$$a^{2^n} \equiv 1 \ (\text{mod } 2^{n+2}).$$

31. (*i*) Show that if p is prime then for any positive r and n,

$$a \equiv b \ (\text{mod } p^n) \Rightarrow a^{p^r} \equiv b^{p^r} \ (\text{mod } p^{n+r}).$$

(*ii*) Conversely, show that if the prime p is odd and $p \nmid a$ then

$$a^{p^r} \equiv b^{p^r} \ (\text{mod } p^{n+r}) \Rightarrow a \equiv b \ (\text{mod } p).$$

32. (*i*) Why is the relation $|a + b| \le |a| + |b|$ in an ordered domain known as the triangle inequality?

(*ii*) If $a_1, a_2, \ldots, a_n$ are elements of an ordered domain prove, by induction, that

$$\left| \sum_{i=1}^{n} a_i \right| \le \sum_{i=1}^{n} |a_i| .$$

(*iii*) Decide whether the following formula holds for elements of an ordered domain

$$(a_1 b_1 + a_2 b_2 + \cdots + a_n b_n)^2 \le (a_1{}^2 + \cdots + a_n{}^2)(b_1{}^2 + \cdots + b_n{}^2).$$

33. As usual, $[a, b]$ represents the closed interval $\{x \in \mathbf{R} \mid a \leq x \leq b\}$; prove that

$$x \longrightarrow \left(\frac{x-a}{b-a}\right) d + \left(\frac{x-b}{a-b}\right) c$$

is a 1–1 mapping of $[a, b]$ onto $[c, d]$.

34. Consider sets X, Y and a mapping $f\colon X \to Y$; then:

(*i*) Show that f is injective if and only if there exists $g\colon Y \to X$ such that $g \circ f = 1$.

(*ii*) Show that f is surjective if and only if there exists $h\colon Y \to X$ such that $f \circ h = 1$.

(*iii*) Show that if $g\colon Y \to X$ and $h\colon Y \to X$ are such that $g \circ f = 1$ and $f \circ h = 1$ then $g = h$.

(*iv*) Show that f is injective and surjective if and only if there exists $g\colon Y \to X$ such that $g \circ f = 1$ and $f \circ g = 1$.

35. Define a function f, recursively, on all positive integers as follows:

$$f(1) = 1, \qquad f(n+1) = 2f(n) + 1.$$

Find the value of $f(n)$. How is this related to Problem 24 of 2-3-20?

36. Use the fact that a set with n elements has exactly $\binom{n}{k}$ subsets with k elements to prove the relation

$$\binom{n+1}{k} = \binom{n}{k} + \binom{n}{k-1}, \qquad 0 < k < n+1$$

for binomial coefficients (which was mentioned in the proof of 2-4-8). How is this relation connected with Pascal's triangle?

37. Given a set M with m elements and a nonempty subset M_1.

(*i*) Show that the number of subsets of M with an odd number of elements is 2^{m-1}, and so is the number of subsets with an even number of elements.

(*ii*) Show that the number of subsets of M containing an odd number of elements from M_1 is 2^{m-1}.

38. Prove the following properties of binomial coefficients:

(*i*) $\displaystyle\sum_{k=0}^{n} \binom{n}{k} = 2^n,$ (*ii*) $\displaystyle\sum_{k=0}^{n} (-1)^k \binom{n}{k} = 0,$

(*iii*) $\displaystyle\sum_{k=1}^{n} k(-1)^k \binom{n}{k} = 0,$ (*iv*) $\displaystyle\sum_{k=r}^{n} (-1)^k \binom{k}{r} \binom{n}{k} = 0, \qquad n \geq r,$

(*v*) $\displaystyle\sum_{i=0}^{m} \binom{r}{i} \binom{s}{m-i} = \binom{r+s}{m}.$

39. Let $\sigma(n, k) = 1^k + 2^k + \cdots + n^k$. By induction on n (keeping k fixed) prove that

$$\binom{k+1}{1}\sigma(n, 1) + \binom{k+1}{2}\sigma(n, 2) + \cdots + \binom{k+1}{k}\sigma(n, k) = (n+1)^k - (n+1).$$

Use this formula to compute $\sigma(n, i)$ for $i = 1, 2, 3, 4$.

40. Suppose the three conditions of (2-3-4) are satisfied in the domain D—namely:

(i) $a > 0, b > 0 \Rightarrow a + b > 0$,

(ii) $a > 0, b > 0 \Rightarrow ab > 0$.

(iii) For any $a \in D$ exactly one of $a > 0$, $a = 0$, $a < 0$ holds.

Let $P = \{a \in D \mid a > 0\}$. Is $\{D, P\}$ then an ordered domain?

41. Can 2-3-16 be proved by considering the set

$$S = \{a \in P \mid \pi(b) \text{ is true for all } b < a\}?$$

42. Suppose $\{D, P\}$ is an ordered domain. We have seen in 2-3-14 that if $\{D, P\}$ is well ordered then mathematical induction holds. Prove the converse. To put it another way, show that the following conditions are equivalent in $\{D, P\}$:

(i) Any inductive set of positive elements that contains the identity e must be P itself.

(ii) Every nonempty subset of P has a smallest element.

43. Show that Condition (i) of Problem 42 may be replaced by our other formulations of mathematical induction—namely, those of 2-3-15 and 2-3-16.

44. In $\mathbf{Z}_6$ show that $R_1 = \{0, 2, 4\}$ and $R_2 = \{0, 3\}$ are subrings with $R_1 \cap R_2 = (0)$ and $R_1 + R_2 = \mathbf{Z}_6$. [Recall that $R_1 + R_2 = \{a_1 + a_2 \mid a_i \in R_i\}$.] One then says that $\mathbf{Z}_6$ is the **(internal) direct sum** of the rings R_1 and R_2. Prove that every element $a \in \mathbf{Z}_6$ can be written uniquely in the form $a = a_1 + a_2$, $a_1 \in R_1$, $a_2 \in R_2$.

Express $\mathbf{Z}_{10}$ and $\mathbf{Z}_{12}$ as internal direct sums of subrings. What about $\mathbf{Z}_m$ in general?

45. Can you find an injective homomorphism of $\mathbf{Z}_m$ into $\mathbf{Z}_n$ when $m \mid n$—in other words, does $\mathbf{Z}_n$ contain an isomorphic copy of $\mathbf{Z}_m$? What if $(m, n/m) = 1$?

46. Consider the rings

(*i*) $\mathbf{Z}_3$, (*ii*) $\mathbf{Z}_4$, (*iii*) $\mathbf{Z}_5$, (*iv*) $\mathbf{Z}_p$, (*v*) $\mathbf{Z}_m$.

In each case, find all subrings of the given ring. What are their characteristics? In each case, find as many automorphisms of the given ring as you can.

47. Consider the rings

(*i*) $\mathbf{Z}$, (*ii*) $\mathbf{Q}$, (*iii*) $\mathbf{R}$.

In each case show that the only automorphism of the given ring is the identity.

48. (*i*) If m is a square free integer show that $\mathbf{Z}[\sqrt{m}] = \{a + b\sqrt{m} \mid a, b \in \mathbf{Z}\}$ is an integral domain. How many automorphisms can you find?

(*ii*) If, in addition, $m \equiv 1 \pmod 4$ and we put $\omega = (1 + \sqrt{m})/2$ then $\mathbf{Z}[\omega] = \{a + b\omega \mid a, b \in \mathbf{Z}\}$ is an integral domain. How many automorphisms can you find?

49. Prove that if ϕ is an **endomorphism** of the ring R (meaning: a homomorphism of R into itself), then $S = \{a \in R \mid \phi(a) = a\}$ is a subring of R.

50. Show that all 2×2 matrices of form $\begin{pmatrix} a & -b \\ b & a \end{pmatrix}$ with $a, b \in \mathbf{Z}$ constitute an integral domain which is isomorphic to the domain of Gaussian integers, $\mathbf{Z}[i]$.

51. Suppose R is a ring with unity e and n is an integer greater than one. Consider $\mathcal{M}(R, n)$, the set of all $n \times n$ matrices with entries from R. We write a typical matrix of $\mathcal{M}(R, n)$ in the abbrieviated form (a_{ij})—which signifies that the entry of the matrix in the i, j place is the element a_{ij} of R. Define addition in $\mathcal{M}(R, n)$ by

$$(a_{ij}) + (b_{ij}) = (c_{ij}), \qquad \text{where} \quad c_{ij} = a_{ij} + b_{ij}, \quad i, j = 1, \ldots, n.$$

Define multiplication in $\mathcal{M}(R, n)$ by

$$(a_{ij}) \cdot (b_{ij}) = (c_{ij}), \qquad \text{where} \quad c_{ij} = \sum_{k=1}^{n} a_{ik} b_{kj}, \quad i, j = 1, \ldots, n.$$

Verify that $\mathcal{M}(R, n)$ is a ring with unity. The unity element I has e along the main diagonal and 0 elsewhere—

$$I = \begin{pmatrix} e & & & 0 \\ & e & & \\ & & \ddots & \\ 0 & & & e \end{pmatrix}.$$

Define also multiplication of a matrix by an element of R (called "multiplication by a scalar") by

$$c(a_{ij}) = (ca_{ij}), \qquad c \in R.$$

In other words, every entry of the matrix is multiplied by c.

Show that the following are subrings of $\mathscr{M}(R, n)$.

(*i*) $\{cI \mid c \in R\}$; these are known as the **scalar matrices**—each one is of form (a_{ij}), where $a_{11} = a_{22} = \cdots = a_{nn} = c$ and $a_{ij} = 0$ for $i \neq j$.

(*ii*) The set of **diagonal matrices;** these are of form (a_{ij}), where $a_{ij} = 0$ for $i \neq j$—in other words, all entries off the main diagonal are 0.

(*iii*) The **(upper) triangular matrices;** these are of form (a_{ij}) where $a_{ij} = 0$ for $i < j$—in other words, all entries "above" the main diagonal are 0.

(*iv*) The **strictly triangular matrices;** these are of form (a_{ij}) where $a_{ij} = 0$ for $i \leq j$—in other words, all entries above and on the main diagonal are 0.

Fix an integer r, $1 \leq r \leq n$. Is the set of matrices that are 0 outside the rth row—that is, $\{(a_{ij}) \mid a_{ij} = 0 \text{ when } i \neq j\}$—a subring? What about $\{(a_{ij}) \mid a_{ij} = 0 \text{ when } i \geq r\}$?

52. (*i*) Show that $\mathscr{M}(R, n)$ always has zero divisors; in more detail, exhibit $A, B \in \mathscr{M}(R, n)$ such that $AB = BA = 0$.

(*ii*) Show that if there exist $a, b \in R$ with $ab \neq 0$, then $\mathscr{M}(R, n)$ is not commutative.

53. Show that if $\phi\colon R \to R'$ is a homomorphism of rings, then the mapping $(a_{ij}) \to (\phi(a_{ij}))$ is a homomorphism of $\mathscr{M}(R, n) \to \mathscr{M}(R', n)$.

54. For each pair i, j with $1 \leq i, j \leq n$ let E_{ij} denote the matrix with e in the i, j place and 0 elsewhere. Then show that any matrix of $\mathscr{M}(R, n)$ has a unique expression of form

$$\sum_{i,j=1}^{n} a_{ij}E_{ij}, \qquad a_{ij} \in R.$$

Show also that the E_{ij}'s multiply according to the rule

$$E_{rs}E_{uv} = \begin{cases} 0 & \text{if } s \neq u, \\ E_{rv} & \text{if } s = u. \end{cases}$$

55. Consider the **real quaternions** Q. These are all the expressions $\alpha = a_0 + a_1 i + a_2 j + a_3 k$ where $a_0, a_1, a_2, a_3 \in \mathbf{R}$ and i, j, k are formal symbols. If also $\beta = b_0 + b_1 i + b_2 j + b_3 k \in Q$ then it is understood that

$$\alpha = \beta \Leftrightarrow a_0 = b_0, a_1 = b_1, a_2 = b_2, a_3 = b_3.$$

We define addition and multiplication of quaternions by:

$$\alpha + \beta = (a_0 + b_0) + (a_1 + b_1)i + (a_2 + b_2)j + (a_3 + b_3)k,$$

and

$$\alpha\beta = (a_0 b_0 - a_1 b_1 - a_2 b_2 - a_3 b_3) + (a_0 b_1 + a_1 b_0 + a_2 b_3 - a_3 b_2)i + (a_0 b_2 - a_1 b_3 + a_2 b_0 + a_3 b_1)j + (a_0 b_3 + a_1 b_2 - a_2 b_1 + a_3 b_0)k.$$

Multiplication is not really mysterious. One wants to have

$$i^2 = j^2 = k^2 = -1, \qquad ij = -ji = k, \quad jk = -kj = i, \quad ki = -ik = j,$$

and, assuming these rules, when α and β are multiplied according to the "usual rules" the result is the given expression for $\alpha\beta$.

Then Q is a noncommutative ring with zero-element $0 = 0 + 0i + 0j + 0k$ and unity $1 = 1 + 0i + 0j + 0k$. Moreover, every $\alpha \neq 0$ in Q has an inverse; in fact, writing $a_0{}^2 + a_1{}^2 + a_2{}^2 + a_3{}^2 = d$ (so $d \neq 0$) and $\beta = (a_0/d) - (a_1/d)i - (a_2/d)j - (a_3/d)k$, we have $\alpha\beta = \beta\alpha = 1$.

Prove that Q is isomorphic to the ring of quaternions consisting of all 2×2 matrices $\begin{pmatrix} \alpha & \beta \\ -\bar{\beta} & \bar{\alpha} \end{pmatrix}$ with $\alpha, \beta \in \mathbf{C}$ as defined in 2-2-15, Problem 29.

56. By an **arithmetic function** or a **number-theoretic function** we mean an element of Map($\mathbf{Z}_{>0}, \mathbf{Z}$), where $\mathbf{Z}_{>0}$ denotes the set of all positive integers. An arithmetic function f is said to be **multiplicative** if $f(mn) = f(m)f(n)$ whenever m and n are relatively prime. Then:

(*i*) If $f \in$ Map($\mathbf{Z}_{>0}, \mathbf{Z}$) is multiplicative and $g \in$ Map($\mathbf{Z}_{>0}, \mathbf{Z}$) is defined by

$$g(n) = \sum_{d \mid n} f(d),$$

then g is multiplicative.

(*ii*) Define $\mu \in$ Map($\mathbf{Z}_{>0}, \mathbf{Z}$) by

$$\mu(n) = \begin{cases} 1 & \text{if } n = 1, \\ 0 & \text{if some square divides } n, \\ (-1)^r & \text{if } n \text{ is the product of } r \text{ distinct primes.} \end{cases}$$

Then μ, which is known as the **Möbius function,** is multiplicative and it satisfies

$$\sum_{d \mid n} \mu(d) = \begin{cases} 1 & \text{if } n = 1, \\ 0 & \text{if } n > 1. \end{cases}$$

(*iii*) If $f \in$ Map($\mathbf{Z}_{>0}, \mathbf{Z}$) and g is defined by

$$g(n) = \sum_{d \mid n} f(d), \tag{$*$}$$

then

$$f(n) = \sum_{d \mid n} \mu(d) g\left(\frac{n}{d}\right) = \sum_{d \mid n} \mu\left(\frac{n}{d}\right) g(d). \qquad (**)$$

This is known as the **Möbius inversion formula.** Conversely, if $g \in \mathrm{Map}(\mathbf{Z}_{>0}, \mathbf{Z})$ is given and f is defined by $(**)$ then the formula $(*)$ holds.

(*iv*) If $g(n) = \sum_{d \mid n} f(d)$ and g is multiplicative, then so is f.

(*v*) How are the preceding results affected if we work in $\mathrm{Map}(\mathbf{Z}_{>0}, \mathbf{R})$?

57. Prove the following:

(*i*) $\sum_{d^2 \mid n} \mu(d) = |\mu(n)|$

(*ii*) If f is a multiplicative arithmetic function then

$$\sum_{d \mid n} \mu(d) f(d) = \prod_{p \mid n} (1 - f(p)), \qquad p \text{ prime}.$$

58. Let $D = \mathrm{Map}(\mathbf{Z}_{>0}, \mathbf{Z})$, the set of all arithmetic functions. Define addition $+$ and **convolution** $*$ in D by

$$(f + g)(n) = f(n) + g(n),$$

$$(f * g)(n) = \sum_{d \mid n} f(d) g\left(\frac{n}{d}\right).$$

Show that $\{D, +, *\}$ is an integral domain—the zero element, 0, is given by

$$0(n) = 0 \qquad \text{for all } n \in \mathbf{Z}_{>0},$$

and the identity for multiplication, e, is given by

$$e(n) = \begin{cases} 1 & \text{if } n = 1, \\ 0 & \text{if } n > 1. \end{cases}$$

CONGRUENCES AND POLYNOMIALS

The problem of solving equations of form

$$a_n x^n + a_{n-1}x^{n-1} + \cdots + a_1 x + a_0 = 0$$

is a familiar one. In high school, one is concerned primarily with the case where the "coefficients" $a_0, a_1, \ldots, a_n$ are real numbers, and one usually seeks a real value for the "unknown" x. Of course, the same kind of problem can be formulated for any commutative ring R; namely, if $a_0, a_1, \ldots, a_n$ all belong to R, is there a value in R for the unknown x for which the equation is satisfied? This question is, however, much too difficult, and we shall only attack small parts of it.

In this chapter, our approach will be on two distinct levels. On the abstract theoretical level we shall set up precise terminology about such things as polynomials, solutions or roots of polynomial equations, and factorization of polynomials, and we shall prove some basic results about them in the case of an arbitrary commutative ring with unity R. On the concrete practical level we shall deal with a specific kind of ring R, the ring of residue classes $\mathbf{Z}_m$, and learn the theory and practice of solving polynomial equations with coefficients in $\mathbf{Z}_m$.

3-1. Linear Congruences

Let us fix $m > 1$ and write $\mathbf{Z}_m = \{\alpha, \beta, \gamma, \delta, \ldots\}$. In order to study equations of form

$$\alpha_n x^n + \alpha_{n-1} x^{n-1} + \cdots + \alpha_1 x + \alpha_0 = 0, \qquad \alpha_0, \alpha_1, \ldots, \alpha_n \in \mathbf{Z}_m,$$

it is appropriate to begin with the simplest case, namely, when $n = 1$. Thus, we deal first with the equation $\alpha_1 x + \alpha_0 = 0$, which by a trivial change of notation we prefer to write as the **linear equation**

$$\alpha x = \beta, \qquad \alpha, \beta \in \mathbf{Z}_m, \quad \alpha \neq 0.$$

By a **solution** of this equation, we mean an element $\gamma \in \mathbf{Z}_m$ such that $\alpha\gamma = \beta$. In common language, one might then say that "γ is a value of x that satisfies the equation."

Equations of this type are unfamiliar, so let us try to develop a feeling for them by looking at a few examples.

3-1-1. Examples. (*i*) Consider $m = 7$, $\alpha = \lfloor 6 \rfloor_7 \in \mathbf{Z}_7$, $\beta = \lfloor 2 \rfloor_7 \in \mathbf{Z}_7$. We want to solve the equation

$$\lfloor 6 \rfloor_7 \cdot x = \lfloor 2 \rfloor_7.$$

A solution must come from $\mathbf{Z}_7$, so there are exactly seven possibilities: $\lfloor 0 \rfloor_7$, $\lfloor 1 \rfloor_7$, $\lfloor 2 \rfloor_7$, $\lfloor 3 \rfloor_7$, $\lfloor 4 \rfloor_7$, $\lfloor 5 \rfloor_7$, $\lfloor 6 \rfloor_7$. Using the rules for computation in $\mathbf{Z}_7$ (which were discussed in Section 1-7) we find, by trial and error, that $\gamma = \lfloor 5 \rfloor_7$ is a solution because $\lfloor 6 \rfloor_7 \cdot \lfloor 5 \rfloor_7 = \lfloor 2 \rfloor_7$, and in fact it is the only solution of the given equation.

(*ii*) Consider $m = 10$, $\alpha = \lfloor 6 \rfloor_{10} \in \mathbf{Z}_{10}$, $\beta = \lfloor 3 \rfloor_{10} \in \mathbf{Z}_{10}$; which means that our equation is

$$\lfloor 6 \rfloor_{10} \cdot x = \lfloor 3 \rfloor_{10}.$$

By trying each of the 10 elements of $\mathbf{Z}_{10}$, one discovers that this equation has no solutions.

(*iii*) Consider the equation

$$\lfloor 6 \rfloor_{15} \cdot x = \lfloor 9 \rfloor_{15},$$

which is the case $m = 15$, $\alpha = \lfloor 6 \rfloor_{15} \in \mathbf{Z}_{15}$, $\beta = \lfloor 9 \rfloor_{15} \in \mathbf{Z}_{15}$. There are 15 possible solutions, and by testing all the possibilities it is easy to see that there are exactly three solutions—namely, $\lfloor 4 \rfloor_{15}$, $\lfloor 9 \rfloor_{15}$, $\lfloor 14 \rfloor_{15}$.

(*iv*) Consider the equation

$$\lfloor 6 \rfloor_{15} \cdot x = \lfloor 8 \rfloor_{15}.$$

It is straightforward to verify that there are no solutions in $\mathbf{Z}_{15}$.

These examples indicate that peculiar or surprising things can happen. The linear equation $\alpha x = \beta$ in $\mathbf{Z}_m$ may have no solutions, or one solution, or several solutions. Furthermore, the trial-and-error method used to solve these equations is clearly unsatisfactory. It is feasible only when m is small. Surely, we need more powerful techniques.

Returning to our general linear equation $\alpha x = \beta$ in $\mathbf{Z}_m$, let us write $\alpha = \lfloor a \rfloor_m$, $\beta = \lfloor b \rfloor_m$ where $a, b \in \mathbf{Z}$. We may also write an arbitrary element $\gamma \in \mathbf{Z}$ in the form $\gamma = \lfloor c \rfloor_m$ where $c \in \mathbf{Z}$. (As a rule, elements of $\mathbf{Z}$ will be denoted by $a, b, c, d, \ldots$.) Now, by making use of elementary facts about congruence classes (from Section 1-7) we have

$$
\begin{aligned}
&\gamma = \lfloor c \rfloor_m \in \mathbf{Z}_m \text{ is a solution of } \alpha x = \beta \\
&\Leftrightarrow \alpha\gamma = \beta \\
&\Leftrightarrow \lfloor a \rfloor_m \lfloor c \rfloor_m = \lfloor b \rfloor_m \\
&\Leftrightarrow \lfloor ac \rfloor_m = \lfloor b \rfloor_m \\
&\Leftrightarrow ac \equiv b \pmod{m} \\
&\Leftrightarrow c \in \mathbf{Z} \text{ is a solution of the congruence } ax \equiv b \pmod{m}.
\end{aligned}
$$

Naturally, an equation of form $ax \equiv b \pmod{m}$ is known as a **linear congruence**, and by a **solution** of such an equation we mean (as has already been indicated) an integer c for which $ac \equiv b \pmod{m}$.

Note that the role played by the symbol x depends on the context; in $\alpha x = \beta$ the "values" for x are to come from $\mathbf{Z}_m$, while in $ax \equiv b \pmod{m}$ the "values" for x are to come from $\mathbf{Z}$. Of course, the important thing is to understand the connection or relation between the solutions of these two kinds of equations. This connection was proved above, so we merely state it formally.

3-1-2. Proposition. Suppose $\alpha, \beta \in \mathbf{Z}_m$ and $a, b \in \mathbf{Z}$ are *any* representatives for α and β, respectively—that is, $\alpha = \lfloor a \rfloor_m$, $\beta = \lfloor b \rfloor_m$. Then $\gamma = \lfloor c \rfloor_m \in \mathbf{Z}_m$ is a solution of the linear equation $\alpha x = \beta \Leftrightarrow c \in \mathbf{Z}$ is a solution of the linear congruence $ax \equiv b \pmod{m}$.

According to this result, in order to solve a linear equation in $\mathbf{Z}_m$ we need only solve a linear congruence (which involves integers rather than residue classes). For example, to solve the equations discussed in 3-1-1, we may "replace" them by the linear congruences $6x \equiv 2 \pmod{7}$, $6x \equiv 3 \pmod{10}$, $6x \equiv 9 \pmod{15}$, $6x \equiv 8 \pmod{15}$, respectively. Of course, other replacements are possible. For example, instead of replacing the linear equation $\lfloor 6 \rfloor_{15} x = \lfloor 9 \rfloor_{15}$ of case (*iii*) of 3-1-1 by $6x \equiv 9 \pmod{15}$, it could also be replaced by $21x \equiv 24 \pmod{15}$, or $-9x \equiv 69 \pmod{15}$, or $81x \equiv -21 \pmod{15}$, and so on—because, as stated in 3-1-2, any representatives a of $\lfloor 6 \rfloor_{15}$ and b of $\lfloor 9 \rfloor_{15}$ may be used. In this connection, it is useful to point out explicitly how the solutions for the many possible replacements of a linear equation $\alpha x = \beta$ in $\mathbf{Z}_m$ by congruences of form $ax \equiv b \pmod{m}$ are related; namely,

3-1-3. Proposition. Suppose a, a', b, b' are integers with $a \equiv a' \pmod{m}$ and $b \equiv b' \pmod{m}$. Then the solutions of the linear congruence $ax \equiv b \pmod{m}$ are identical with the solutions of the linear congruence $a'x \equiv b' \pmod{m}$.

Proof: The hypothesis says (in virtue of our knowledge of Section 1-7) that $\lfloor a \rfloor_m = \lfloor a' \rfloor_m$ and $\lfloor b \rfloor_m = \lfloor b' \rfloor_m$. Call these α and β, respectively. Consequently, by applying 3-1-2 twice, we have: $c \in \mathbf{Z}$ is a solution of $ax \equiv b \pmod{m}$ $\Leftrightarrow$ $\gamma = \lfloor c \rfloor_m \in \mathbf{Z}_m$ is a solution of $\alpha x = \beta$ $\Leftrightarrow$ $c \in \mathbf{Z}$ is a solution of $a'x \equiv b' \pmod{m}$. This does it. ▮

Incidentally, the reader should find it easy to prove this fact directly from the properties of congruence, without any recourse to residue classes.

This result, 3-1-3, tells us, for example, that to solve the congruence $3155x \equiv 7847 \pmod{18}$ is the same as solving the somewhat simpler congruence $5x \equiv -1 \pmod{18}$ [since $3155 \equiv 5 \pmod{18}$ and $7847 \equiv -1 \pmod{18}$].

Now, as the next stage of our discussion, we turn to the problem of solving the general linear congruence

$$ax \equiv b \pmod{m}, \qquad a, b \in \mathbf{Z}.$$

Consider an integer x_0; then

$$\begin{aligned} &x_0 \text{ is a solution of } ax \equiv b \pmod{m} \\ &\quad \Leftrightarrow ax_0 \equiv b \pmod{m} \\ &\quad \Leftrightarrow m \mid (b - ax_0) \\ &\quad \Leftrightarrow \text{there exists } y_0 \in \mathbf{Z} \text{ such that } my_0 = b - ax_0 \\ &\quad \Leftrightarrow \text{there exists } y_0 \in \mathbf{Z} \text{ such that } ax_0 + my_0 = b. \end{aligned}$$

This proves:

3-1-4. Proposition. The linear congruence $ax \equiv b \pmod{m}$ has a solution $x_0 \Leftrightarrow$ the linear diophantine equation $ax + my = b$ has a solution $\{x_0, y_0\}$.

This result "transforms" a linear congruence into a linear diophantine equation. Note that the same x_0 appears on both sides of the $\Leftrightarrow$ sign. In particular, because $\{x_0 = 5, y_0 = -4\}$ is a solution of $11x + 13y = 3$, it follows that $x_0 = 5$ is a solution of $11x \equiv 3 \pmod{13}$.

Because we learned long ago (in Section 1-6) how to solve linear diophantine equations completely, we are now in a position to settle all questions facing us. However, one preliminary remark is needed.

Suppose $c \in \mathbf{Z}$ is a solution of $ax \equiv b \pmod{m}$. If $c' \equiv c \pmod{m}$ (that is, if $c' \in \lfloor c \rfloor_m$), then $ac' \equiv ac \equiv b \pmod{m}$, so $c' \in \mathbf{Z}$ is also a solution. It is customary to group all the elements of $\lfloor c \rfloor_m$ together and say that $\lfloor c \rfloor_m$ is a solution (or "one solution") of $ax \equiv b \pmod{m}$. We shall often do so, even though it is not quite accurate. The object $\lfloor c \rfloor_m$, being an element of $\mathbf{Z}_m$, cannot possibly be a solution of $ax \equiv b \pmod{m}$; however, $\lfloor c \rfloor_m$ is indeed a solution of the corresponding equation $\alpha x = \beta$ with $\alpha = \lfloor a \rfloor_m$, $\beta = \lfloor b \rfloor_m$ in $\mathbf{Z}_m$. A phrase like "$\lfloor c \rfloor_m$ is a solution of $ax \equiv b \pmod{m}$" will be, for us, simply a convenient way to say that every integer belonging to $\lfloor c \rfloor_m$ is a solution.

3-1-5. Theorem. The linear congruence $ax \equiv b \pmod{m}$ has a solution $\Leftrightarrow$ (a, m) divides b.

Furthermore, if $x_0 \in \mathbf{Z}$ is any solution of $ax \equiv b \pmod{m}$ and we write $d = (a, m)$, $m = m'd$, then all the solutions are of form

$$\bar{x} = x_0 + tm', \qquad t = 0, \pm 1, \pm 2, \ldots.$$

This infinite collection of solutions breaks up into d distinct residue classes

$$\lfloor x_0 \rfloor_m, \quad \lfloor x_0 + m' \rfloor_m, \quad \lfloor x_0 + 2m' \rfloor_m, \ldots, \lfloor x_0 + (d-1)m' \rfloor_m,$$

so we say that there are exactly d solutions of $ax \equiv b \pmod{m}$. To put it another way, the equation $\alpha x = \beta$ where $\alpha = \lfloor a \rfloor_m$, $\beta = \lfloor b \rfloor_m$ has, in $\mathbf{Z}_m$, the d solutions:

$$\{\lfloor x_0 + sm' \rfloor_m \mid s = 0, 1, \ldots, d-1\}.$$

Proof: We have $m > 1$, and there is no harm in assuming $a \neq 0$, as the case $a = 0$ is surely not worth considering. According to 3-1-4 and 1-6-1,

$ax \equiv b \pmod{m}$ has a solution $\Leftrightarrow$ $ax + my = b$ has a solution $\Leftrightarrow$ (a, m) divides b. This proves the first part.

If x_0 is a solution of $ax \equiv b \pmod{m}$, then, by 3-1-4, there is associated with it a solution $\{x_0, y_0\}$ of the diophantine equation $ax + my = b$. So 1-6-1 (with an obvious modification of notation) tells us that all solutions of $ax + my = b$ are given by

$$\begin{aligned} \bar{x} &= x_0 + t\left(\frac{m}{d}\right), \\ \bar{y} &= y_0 - t\left(\frac{a}{d}\right). \end{aligned} \qquad t = 0, \pm 1, \pm 2, \ldots.$$

By another application of 3-1-4, it follows that all solutions of $ax \equiv b \pmod{m}$ are given by

$$\bar{x} = x_0 + tm', \qquad t = 0, \pm 1, \pm 2, \ldots.$$

This shows, in particular, that as soon as a single solution of the congruence $ax \equiv b \pmod{m}$ is known, we can immediately produce all solutions, and there are an infinite number of them.

Now, from the set of all solutions let us select d of them—namely, let us take

$$x_0, \quad x_0 + m', \quad x_0 + 2m', \ldots, x_0 + (d-1)m'.$$

We assert, first of all, that no two of these are congruent modulo m. To see this, consider any two of them

$$x_0 + im' \qquad \text{and} \qquad x_0 + jm'$$

where, of course, $0 \le i \le d-1$ and $0 \le j \le d-1$. Then

$$\begin{aligned} x_0 + im' \equiv x_0 + jm' \pmod{m} &\Rightarrow (i-j)m' \equiv 0 \pmod{m} \\ &\Rightarrow m \mid (i-j)m' \\ &\Rightarrow m'd \mid (i-j)m' \\ &\Rightarrow d \mid (i-j). \end{aligned}$$

Because of the constraints on i and j, this implies $i = j$—so, indeed, our d solutions are incongruent to each other modulo m. This also guarantees (by 1-7-11) that the d residue classes

$$\lfloor x_0 \rfloor_m, \quad \lfloor x_0 + m' \rfloor_m, \quad \lfloor x_0 + 2m' \rfloor_m, \ldots, \lfloor x_0 + (d-1)m' \rfloor_m$$

are distinct (and disjoint).

Furthermore, we assert that every solution is congruent (mod m) to one of our d selected solutions. To see this, consider any solution $x_0 + tm'$. By the division algorithm, we may write $t = qd + r$ with $0 \leq r < d$. Then

$$x_0 + tm' = x_0 + qm + rm' \equiv x_0 + rm' \pmod{m}$$

while $x_0 + rm'$ is clearly an element of our set $\{x_0 + sm' \mid s = 0, 1, \ldots, d-1\}$ of d selected solutions. We see, therefore, that any solution $x_0 + tm'$ belongs to one of the residue classes $\lfloor x_0 + sm' \rfloor_m$, $s = 0, 1, 2, \ldots, d-1$. On the other hand, if an integer w belongs to one of these d residue classes, say $w \in \lfloor x_0 + im' \rfloor_m$ where $0 \leq i \leq d-1$, then $w \equiv x_0 + im' \pmod{m}$ and it can be written as $w = x_0 + im' + km = x_0 + (i + kd)m'$, which is of form $x_0 + tm'$. It follows that the set of all solutions decomposes into the d distinct residue classes $\lfloor x_0 + sm' \rfloor_m$, $s = 0, 1, \ldots, d-1$. Symbolically, we have

$$\{x_0 + tm' \mid t \in \mathbf{Z}\} = \lfloor x_0 \rfloor_m \cup \lfloor x_0 + m' \rfloor_m \cup \cdots \cup \lfloor x_0 + (d-1)m' \rfloor_m$$

where the union on the right is a disjoint union.

Finally, according to 3-1-2, we obtain all the solutions of $\alpha x = \beta$, where $\alpha = \lfloor a \rfloor_m$, $\beta = \lfloor b \rfloor_m$, by taking the residue classes of all the solutions of $ax \equiv b \pmod{m}$. In view of the foregoing, taking the residue classes of all $x_0 + tm'$, $t \in \mathbf{Z}$ leaves us, precisely, with the d elements $\lfloor x_0 + sm' \rfloor_m$ of $\mathbf{Z}_m$. This completes the proof. ▮

3-1-6. Examples. We show how the preceding results apply to numerical examples.

(*i*) Consider the linear congruence

$$6x \equiv 2 \pmod{7}.$$

Here $m = 7$, $a = 6$, $b = 2$, and $d = (a, m) = (6, 7) = 1$, which surely divides $b = 2$. Therefore, a solution exists.

Obviously, $x_0 = 5$ is a solution. Since $m' = m/d = 7$, 3-1-5 tells us that all solutions are given by

$$\{x_0 + tm' \mid t \in \mathbf{Z}\} = \{5 + 7t \mid t \in \mathbf{Z}\}.$$

An explicit listing of all the solutions is

$$\ldots, -23, -16, -9, -2, 5, 12, 19, 26, 33, 40, \ldots.$$

Because $d = 1$, the set of all solutions consists of a single residue class, $\lfloor 5 \rfloor_7$.

As for the corresponding linear equation in $\mathbf{Z}_7$, $\lfloor 6 \rfloor_7 x = \lfloor 2 \rfloor_7$, we see that it has the single solution $\lfloor 5 \rfloor_7$ in $\mathbf{Z}_7$. Of course, this fact was already indicated in part (*i*) of 3-1-1.

(*ii*) Consider the linear equation

$$\lfloor 6 \rfloor_{10} x = \lfloor 3 \rfloor_{10}$$

in $\mathbf{Z}_{10}$. Corresponding to this, we may consider the linear congruence

$6x \equiv 3 \pmod{10}$. Here, $m = 10$, $a = 6$, $b = 3$. Since $(a, m) = 2$ does not divide $b = 3$, the congruence has no solution. Hence, in virtue of 3-1-2, the equation $\lfloor 6 \rfloor_{10}\, x = \lfloor 3 \rfloor_{10}$ has no solution in $\mathbf{Z}_{10}$—a fact that was noted in part (*ii*) of 3-1-1.

(*iii*) Consider the congruence

$$6x \equiv 9 \pmod{15}.$$

Here $m = 15$, $a = 6$, $b = 9$. Clearly, $d = (a, m) = (6, 15) = 3$, and this divides $b = 9$—so a solution exists. It is not hard to find a solution; for example, $x_0 = 4$. (Because the numbers are small and we need to find only one solution, it is probably most efficient to find it by trial and error.) Since $m' = m/d = 15/3 = 5$, all the solutions are of form

$$\{x_0 + tm' \mid t \in \mathbf{Z}\} = \{4 + 5t \mid t \in \mathbf{Z}\}.$$

In detail, the solutions are

$$\ldots, -11, -6, -1, 4, 9, 14, 19, 24, 29, \ldots.$$

As described in 3-1-5, the set of all solutions decomposes into $d = 3$ congruence classes—namely,

$$\lfloor 4 \rfloor_{15}, \quad \lfloor 9 \rfloor_{15}, \quad \lfloor 14 \rfloor_{15}$$

(because the solutions are $\lfloor x_0 \rfloor_{15}$, $\lfloor x_0 + m' \rfloor_{15}$, $\lfloor x_0 + 2m' \rfloor_{15}$). Note that if we start with the solution $x_0 = 9$, instead of $x_0 = 4$, we end up with all solutions given by three residue classes: $\lfloor 9 \rfloor_{15}$, $\lfloor 14 \rfloor_{15}$, $\lfloor 19 \rfloor_{15}$; and these are the same residue classes as above. Of course, the theorem says that we arrive at the same three residue classes no matter which solution x_0 is taken at the start.

In connection with part (*iii*) of 3-1-1, we now "understand" why the linear equation $\lfloor 6 \rfloor_{15}\, x = \lfloor 9 \rfloor_{15}$ in $\mathbf{Z}_{15}$ has exactly three solutions—namely, the three elements $\lfloor 4 \rfloor_{15}$, $\lfloor 9 \rfloor_{15}$, $\lfloor 14 \rfloor_{15}$ of $\mathbf{Z}_{15}$.

(*iv*) It is time we turned to less trivial examples. Consider the linear congruence

$$11{,}799x \equiv 8715 \pmod{1081}.$$

Because $11{,}799 \equiv 989 \pmod{1081}$ and $8715 \equiv 67 \pmod{1081}$, it follows from 3-1-3 that the solutions of this congruence are identical with those of

$$989x \equiv 67 \pmod{1081}.$$

Here $m = 1081$, $a = 989$, $b = 67$. A straightforward computation gives $(a, m) = (989, 1081) = 23$, which does not divide 67—so there is no solution.

(*v*) Consider the linear equation

$$\lfloor 791 \rfloor_{2023}\, x = \lfloor 1204 \rfloor_{2023} \qquad (*)$$

in $\mathbf{Z}_{2023}$. To solve this, we replace it with the associated linear congruence

$$791x \equiv 1204 \pmod{2023} \tag{**}$$

for which $m = 2023$, $a = 791$, $b = 1204$. Applying the Euclidean algorithm, we obtain $d = (a, m) = (791, 2023) = 7$. Since 7 divides $b = 1204$, a solution exists—in fact, there will be seven solutions.

We need to find a single solution x_0 of our congruence. Guess work is not practical here. Instead, we locate a solution of the linear diophantine equation

$$791x + 2023y = 1204. \tag{***}$$

From the Euclidean algorithm (which the reader has presumably computed already), the gcd $7 = (791, 2023)$ can be expressed as a linear combination

$$7 = 133 \cdot 791 - 52 \cdot 2023.$$

Therefore,

$$(172 \cdot 133)791 - (172 \cdot 52)2023 = 1204$$

and the pair $\{172 \cdot 133, -172 \cdot 52\}$ is a solution of the linear diophantine equation (***). Consequently, $172 \cdot 133 = 22{,}876$ is a solution of the linear congruence (**). Because $22{,}876 \equiv 623 \pmod{2023}$, we know that 623 is a solution of (**). We put $x_0 = 623$. (This choice of x_0 is clearly more convenient than taking $x_0 = 22{,}876$; of course, the choice of x_0 does not affect the end result.) Then $m' = m/d = 2023/7 = 289$ and all solutions of (**) are given by the seven residue classes

$$\{\lfloor 623 + s \cdot 289 \rfloor_{2023} \mid s = 0, 1, \ldots, 6\}.$$

By checking the arithmetic, one sees that

$$\lfloor 623 \rfloor_{2023},\quad \lfloor 912 \rfloor_{2023},\quad \lfloor 1201 \rfloor_{2023},\quad \lfloor 1490 \rfloor_{2023},$$
$$\lfloor 1779 \rfloor_{2023},\quad \lfloor 45 \rfloor_{2023},\quad \lfloor 334 \rfloor_{2023}$$

are the seven solutions of the linear equation (*) in $\mathbf{Z}_{2023}$.

It may be of interest to note how this last example was constructed. Given the desire to produce a linear congruence $ax \equiv b \pmod{m}$ with exactly seven solutions, it was necessary to have $d = (a, m) = 7$; the choices $a = 7 \cdot 113 = 791$, $m = 7 \cdot 289 = 2023$ clearly accomplished this. It remained to choose an integer b divisible by $(a, m) = 7$. The seven solutions would then be $\{\lfloor x_0 + 289s \rfloor_{2023} \mid s = 0, 1, \ldots, 6\}$ where x_0 is any integer which satisfies the congruence. In order to guarantee that the work involved in trying to find a single solution of $ax \equiv b \pmod{m}$ by trial and error would be prohibitive, the smallest positive integer solution would have to be sufficiently large. (In this way, anyone who tried to solve by testing $x_0 = 1, 2, 3, 4, \ldots$ could be expected to give up the effort.) The choice of $x_0 = 45$ accomplished this, and because

$$791 \cdot 45 = 35{,}595 \equiv 1204 \pmod{2023},$$

we were led to the choice of $b = 1204$.

Of course, when solving $791x \equiv 1204 \pmod{2023}$ we found one solution to be $x_0 = 623$. Our method of solution does not lead immediately to the smallest positive solution 45. This solution is not lost; it arises when we construct all seven solutions from the single solution $x_0 = 623$.

Now that a single linear congruence is under control (in the sense that we know how to solve it), it is natural to examine the problem of solving several simultaneous congruences.

3-1-7. Examples. (*i*) Consider the two congruences

$$x \equiv 5 \pmod 8, \qquad x \equiv 4 \pmod 9.$$

One way to solve these simultaneously is to note that all solutions of the first congruence are given by

$$x = 5 + 8t, \qquad t \in \mathbf{Z}. \tag{\#}$$

Substituting in the second congruence, we have $5 + 8t \equiv 4 \pmod 9$, or

$$8t \equiv -1 \pmod 9.$$

Since $(8, 9) = 1$, which divides -1, this congruence has a solution; clearly, $t_0 = 1$ is a solution, and then all solutions are of form $t = 1 + 9s$, $s \in \mathbf{Z}$. Substituting in (#), it follows that all solutions of our simultaneous congruences are given by

$$x = 5 + 8(1 + 9s) = 13 + 72s, \qquad s \in \mathbf{Z}.$$

So we have the infinite set of solutions

$$\ldots, 13, 85, 157, 229, 301, \ldots$$

and this set can be described simply as the residue class $\lfloor 13 \rfloor_{72}$.

(*ii*) Similarly, consider the simultaneous congruences

$$x \equiv 5 \pmod 8, \qquad x \equiv 4 \pmod{18}.$$

Again, the first has all its solutions of form $5 + 8t$, $t \in \mathbf{Z}$, and substituting in the second congruence gives

$$8t \equiv -1 \pmod{18}.$$

But this congruence has no solutions—so our simultaneous congruences have no solution.

The same procedure surely carries over to more than two simultaneous congruences, but things get somewhat tedious as the number of congruences increases. In such a situation we can often make use of the following crucial result.

3-1-8. Chinese Remainder Theorem. Suppose the integers $m_1, m_2, \ldots, m_n$, all of which are greater than 1, are relatively prime in pairs, and we put $m_1 m_2 \cdots m_n = m$. Then, for any choice of integers $a_1, a_2, \ldots, a_n$ the simultaneous congruences

$$\begin{aligned} x &\equiv a_1 \pmod{m_1}, \\ x &\equiv a_2 \pmod{m_2}, \\ &\vdots \\ x &\equiv a_n \pmod{m_n} \end{aligned}$$

have a solution. Moreover, if $x_0 \in \mathbf{Z}$ is a solution, then the elements of the residue class $\lfloor x_0 \rfloor_m$ constitute the set of all solutions.

Proof: Let us prove the last part first. Suppose x_0 is a solution of our simultaneous congruences and consider the residue class $\lfloor x_0 \rfloor_m$. We need to show that for $y_0 \in \mathbf{Z}$

$$y_0 \text{ is a solution} \Leftrightarrow y_0 \in \lfloor x_0 \rfloor_m .$$

Now, $y_0 \in \lfloor x_0 \rfloor_m \Rightarrow y_0 \equiv x_0 \pmod{m} \Rightarrow y_0 \equiv x_0 \pmod{m_i}$ for each $i = 1, \ldots, n$, and then y_0 is clearly a solution of all the congruences. Conversely, if y_0 is a solution of the simultaneous congruences, then, for each $i = 1, \ldots, n$ we have $y_0 \equiv a_i \pmod{m_i}$ and $x_0 \equiv a_i \pmod{m_i}$, so that $y_0 \equiv x_0 \pmod{m_i}$. In other words, $m_i \mid (y_0 - x_0)$ for $i = 1, \ldots, n$. By definition of least common multiple, it follows that

$$[m_1, m_2, \ldots, m_n] \mid (y_0 - x_0).$$

Because $m_1, \ldots, m_n$ are positive and relatively prime in pairs, one sees easily that

$$[m_1, m_2, \ldots, m_n] = m_1 m_2 \cdots m_n = m.$$

Consequently, $m \mid (y_0 - x_0)$, $y_0 \equiv x_0 \pmod{m}$, and $y_0 \in \lfloor x_0 \rfloor_m$. Thus, we have shown that the elements of $\lfloor x_0 \rfloor_m$ are indeed all the solutions.

As for existence of a solution, instead of following the method used in 3-1-7 which involves working through the congruences one at a time, we give a proof which treats all the congruences simultaneously and in a uniform manner. For each $j = 1, \ldots, n$ we put $m_j' = m/m_j$; in other words, m_j' is the product of all m_i with the factor m_j excluded—symbolically:

$$m_j' = \prod_{i \neq j} m_i, \qquad i, j = 1, 2, \ldots, n.$$

By hypothesis, $m_1, m_2, \ldots, m_n$ are relatively prime in pairs, and this may be expressed by the notation:

$$(m_i, m_j) = 1, \qquad \text{when } i \neq j.$$

Since any m_j is relatively prime to each of the other m_i, it follows from 1-3-10, part (*iv*) that it is relatively prime to their product—in symbols,

$$(m_j', m_j) = \left(\prod_{i \neq j} m_i, m_j\right) = 1, \qquad j = 1, \ldots, n.$$

Now, according to the theory of a single linear congruence, because $(m_j', m_j) = 1$ there exists, for each $j = 1, \ldots, n$, an integer b_j such that

$$m_j' b_j \equiv 1 \pmod{m_j}.$$

We observe further that

$$m_j' b_j \equiv 0 \pmod{m_i} \qquad \text{for } i \neq j$$

because $m_i \mid m_j'$ when $i \neq j$. Consequently, for each j, the integer $m_j' b_j a_j$ satisfies

$$m_j' b_j a_j \equiv a_j \pmod{m_j},$$
$$m_j' b_j a_j \equiv 0 \pmod{m_i} \quad \text{for } i \neq j.$$

By applying the rule for addition of congruences, we deduce that the integer

$$x_0 = m_1' b_1 a_1 + m_2' b_2 a_2 + \cdots + m_n' b_n a_n = \sum_{j=1}^{n} m_j' b_j a_j$$

satisfies $x_0 \equiv a_i \pmod{m_i}$ for all $i = 1, 2, \ldots, n$. This proves that a solution of our simultaneous congruences exists; moreover, in virtue of what we already know about the set of all solutions, the solution of our simultaneous linear congruences is often said to be **unique** (mod m). The proof is now complete. ▌

It is worthwhile to examine the idea underlying our proof of the Chinese remainder theorem. The simplest nontrivial system of congruences of the form considered in the Chinese remainder theorem is surely

$$x \equiv 1 \pmod{m_1},$$
$$x \equiv 0 \pmod{m_2},$$
$$x \equiv 0 \pmod{m_3},$$
$$\vdots$$
$$x \equiv 0 \pmod{m_n}.$$

Of course, there is no reason for specifying the first congruence as the one where 1 appears on the right-hand side. In fact, for any choice of j from $\{1, 2, \ldots, n\}$ we have a "simplest system" of the same type—namely,

$$\begin{aligned} x &\equiv 1 \pmod{m_j}, \\ x &\equiv 0 \pmod{m_i} \qquad \text{for each } i \neq j. \end{aligned}$$

We first showed that for each j this system has a solution—namely, $m_j'b_j$. Then from these $m_j'b_j$, $j = 1, \ldots, n$ we constructed a solution of the general system of congruences by taking an appropriate linear combination—namely, $a_1 m_1' b_1 + a_2 m_2' b_2 + \cdots + a_n m_n' b_n$.

3-1-9. Example. The proof of the Chinese remainder theorem provides a straightforward recipe for finding a solution of simultaneous congruences. To illustrate, let us solve the simultaneous congruences

$$\begin{aligned} x &\equiv 1 \pmod 2, \\ x &\equiv 2 \pmod 3, \\ x &\equiv 3 \pmod 5, \\ x &\equiv 5 \pmod 7. \end{aligned}$$

In this situation we have

$$\begin{aligned} m_1 &= 2, \qquad m_2 = 3, \qquad m_3 = 5, \qquad m_4 = 7, \\ a_1 &= 1, \qquad a_2 = 2, \qquad a_3 = 3, \qquad a_4 = 5. \end{aligned}$$

(Of course, a solution exists because m_1, m_2, m_3, m_4 are relatively prime in pairs.) Then $m = 2 \cdot 3 \cdot 5 \cdot 7 = 210$ and

$$m_1' = 105, \qquad m_2' = 70, \qquad m_3' = 42, \qquad m_4' = 30.$$

We need to find integers b_1, b_2, b_3, b_4 which satisfy the congruences $105b_1 \equiv 1 \pmod 2$, $70b_2 \equiv 1 \pmod 3$, $42b_3 \equiv 1 \pmod 5$, $30b_4 \equiv 1 \pmod 7$, respectively. This is easy, for by 3-1-3 these congruences reduce to $1b_1 \equiv 1 \pmod 2$, $1b_2 \equiv 1 \pmod 3$, $2b_3 \equiv 1 \pmod 5$, $2b_4 \equiv 1 \pmod 7$, respectively—so we may take

$$b_1 = 1, \qquad b_2 = 1, \qquad b_3 = -2, \qquad b_4 = 4.$$

(Note that any choices of the b's are permissible, so long as each one satisfies the appropriate congruence. For example, we could take $b_1 = -1$, $b_2 = -2$, $b_3 = 3$, $b_4 = 4$). Now, the theorem tells us that

$$\begin{aligned} x_0 &= m_1'b_1a_1 + m_2'b_2a_2 + m_3'b_3a_3 + m_4'b_4a_4 \\ &= (105)(1)(1) + (70)(1)(2) + (42)(-2)(3) + (30)(4)(5) \\ &= 593 \end{aligned}$$

is a solution. In fact, the set of all solutions is $\lfloor 593 \rfloor_{210} = \lfloor 173 \rfloor_{210}$, and 173 is the smallest positive solution.

It is important to observe that once we have performed the steps needed to solve this system of congruences, we can, almost immediately, write down the solution of any other system of congruences with the same m_1, m_2, m_3, m_4. For example, consider the system

$$x \equiv 0 \pmod{2},$$
$$x \equiv 1 \pmod{3},$$
$$x \equiv 4 \pmod{5},$$
$$x \equiv 2 \pmod{7}.$$

In this situation, we have, as before,

$$\begin{array}{llll} m_1 = 2, & m_2 = 3, & m_3 = 5, & m_4 = 7, \\ m_1' = 105, & m_2' = 70, & m_3' = 42, & m_4' = 30, \\ b_1 = 1, & b_2 = 1, & b_3 = -2, & b_4 = 4. \end{array}$$

Only the a_i's are new—namely,

$$a_1 = 0, \qquad a_2 = 1, \qquad a_3 = 4, \qquad a_4 = 2.$$

Thus, a solution is given by

$$(105)(1)(0) + (70)(1)(1) + (42)(-2)(4) + (30)(4)(2) = -26$$

and $\lfloor -26 \rfloor_{210}$ is the set of all solutions.

Of course, the same principle applies whenever we take a system of congruences for which the Chinese remainder theorem is applicable and change the a_i's but not the m_i's.

3-1-10. Remark. The Chinese remainder theorem may be interpreted in another way. Consider the system of congruences

$$x \equiv a_i \pmod{m_i}, \qquad i = 1, \ldots, n. \tag{$*$}$$

Looking at the ith congruence, we note that if x_0 is an integer, then, according to the definition of residue class, $x_0 \equiv a_i \pmod{m_i} \Leftrightarrow x_0 \in \lfloor a_i \rfloor_{m_i}$. Therefore, an integer x_0 is a solution of the system of simultaneous congruences $(*) \Leftrightarrow x_0$ belongs to every one of the residue classes $\lfloor a_i \rfloor_{m_i}$, $i = 1, \ldots, n$—that is, $\Leftrightarrow x_0$ belongs to the intersection of the residue classes $\lfloor a_i \rfloor_{m_i}$, $i = 1, \ldots, n$. (Here, of course, we are viewing a residue class $\lfloor a_i \rfloor_{m_i}$ as a set of integers.) This says that the set of all solutions of the system $(*)$ is precisely the intersection

$$\bigcap_{i=1}^{n} \lfloor a_i \rfloor_{m_i} = \lfloor a_1 \rfloor_{m_1} \cap \lfloor a_2 \rfloor_{m_2} \cap \cdots \cap \lfloor a_n \rfloor_{m_n}.$$

What is this intersection? In the case where the m_i are relatively prime in pairs, the Chinese remainder theorem tells us this intersection is nonempty; even more, the intersection is a residue class (mod $m_1 m_2 \cdots m_n$). For example, from this point of view, the result proved in 3-1-9 may be expressed in the form

$$\lfloor 1 \rfloor_2 \cap \lfloor 2 \rfloor_3 \cap \lfloor 3 \rfloor_5 \cap \lfloor 5 \rfloor_7 = \lfloor 173 \rfloor_{210}.$$

3-1-11. Examples. We conclude this section with several additional examples. These will indicate how we can go about solving any collection of simultaneous linear congruences.

(*i*) Consider the simultaneous congruences

$$x \equiv 6 \pmod{17},$$
$$x \equiv 17 \pmod{24},$$
$$x \equiv 13 \pmod{33}.$$

Because $m_1 = 17$, $m_2 = 24$, $m_3 = 33$ are not relatively prime in pairs, the Chinese remainder theorem does not apply. Instead, we must look carefully at the last two congruences, wherein lies the source of the difficulty. For an integer x_0, we have [since $24 = 3 \cdot 8$ and $(3, 8) = 1$]

x_0 is a solution of $x \equiv 17 \pmod{24}$
$\Leftrightarrow x_0 \equiv 17 \pmod{24}$
$\Leftrightarrow 24 \mid (x_0 - 17)$
$\Leftrightarrow 3 \mid (x_0 - 17)$ and $8 \mid (x_0 - 17)$
$\Leftrightarrow x_0 \equiv 17 \pmod 3$ and $x_0 \equiv 17 \pmod 8$
$\Leftrightarrow x_0 \equiv 2 \pmod 3$ and $x_0 \equiv 1 \pmod 8$
$\Leftrightarrow x_0$ is a solution of both $x \equiv 2 \pmod 3$ and $x \equiv 1 \pmod 8$.

Thus, the congruence $x \equiv 17 \pmod{24}$ may be replaced by the pair of congruences $x \equiv 2 \pmod 3$ and $x \equiv 1 \pmod 8$ without losing or gaining any solutions. In similar fashion, the congruence $x \equiv 13 \pmod{33}$ can be replaced by the pair of congruences $x \equiv 1 \pmod 3$ and $x \equiv 2 \pmod{11}$. Therefore, our system of three congruences may be replaced by the system

$$x \equiv 6 \pmod{17},$$
$$x \equiv 2 \pmod{3},$$
$$x \equiv 1 \pmod{8},$$
$$x \equiv 1 \pmod{3},$$
$$x \equiv 2 \pmod{11}.$$

The second and fourth of these congruences are clearly contradictory—after all, an integer cannot be congruent to 2 and also congruent to 1 (mod 3)—so our system of congruences has no solution. To put it another way, we have shown that

$$\lfloor 6 \rfloor_{17} \cap \lfloor 17 \rfloor_{24} \cap \lfloor 13 \rfloor_{33} = \emptyset,$$

(*ii*) Consider the three congruences

$$x \equiv 5 \pmod{18},$$
$$x \equiv -1 \pmod{24},$$
$$x \equiv 17 \pmod{33}.$$

Again, the Chinese remainder theorem does not apply, so by the procedure used above, we may "break down" each of these congruences and replace this system by the system

$$(1) \quad x \equiv 1 \pmod{2},$$
$$(2) \quad x \equiv 5 \pmod{9},$$
$$(3) \quad x \equiv -1 \pmod{3},$$
$$(4) \quad x \equiv -1 \pmod{8},$$
$$(5) \quad x \equiv 2 \pmod{3},$$
$$(6) \quad x \equiv 6 \pmod{11}.$$

(Of course, this new system has exactly the same solutions as the original system; this is what "replacement" is about.) The Chinese remainder theorem still does not apply. However, suppose we look at congruences (2), (3), and (5). If x_0 is a solution of (2), then, clearly, x_0 is a solution of both (3) and (5). [Note that (3) and (5) are really the same congruence.] It follows that congruences (3) and (5) may be discarded without affecting the solution set. Similarly, if x_0 is a solution of (4), then it is automatically a solution of (1), so congruence (1) may also be discarded. We are left, therefore, with the system

$$x \equiv 5 \pmod{9},$$
$$x \equiv -1 \pmod{8},$$
$$x \equiv 6 \pmod{11}.$$

Now, the Chinese remainder theorem does apply. It turns out (the computations being left to the reader) that this system has the unique solution $\lfloor 743 \rfloor_{792}$. Hence, the set $\lfloor 743 \rfloor_{792}$ consists of all integers which are solutions of our original system of linear congruences; this is often expressed as "743 is the unique solution (mod 792)."

From all this, we conclude that

$$\lfloor 5 \rfloor_{18} \cap \lfloor -1 \rfloor_{24} \cap \lfloor 17 \rfloor_{33} = \lfloor 1 \rfloor_2 \cap \lfloor 5 \rfloor_9 \cap \lfloor -1 \rfloor_3 \cap \lfloor -1 \rfloor_8 \cap \lfloor 2 \rfloor_3 \cap \lfloor 6 \rfloor_{11}$$
$$= \lfloor 5 \rfloor_9 \cap \lfloor -1 \rfloor_8 \cap \lfloor 6 \rfloor_{11}$$
$$= \lfloor 743 \rfloor_{792}.$$

(*iii*) Consider the pair of congruences

$$2x \equiv 5 \pmod 6,$$
$$5x \equiv 3 \pmod{11}.$$

The first of these has no solutions so, a fortiori, the simultaneous congruences have no solutions.

(*iv*) Consider the pair of congruences

$$\begin{aligned} 2x &\equiv 4 \pmod 6, \\ 5x &\equiv 3 \pmod{11}. \end{aligned} \tag{$*$}$$

It is easy to see that the first of these has the two solutions $\lfloor 2 \rfloor_6$ and $\lfloor 5 \rfloor_6$, and the second has the unique solution $\lfloor 5 \rfloor_{11}$. In other words, x_0 is a solution of the first congruence if and only if $x_0 \equiv 2 \pmod 6$ or $x_0 \equiv 5 \pmod 6$, and x_0 is a solution of the second congruence $\Leftrightarrow$ $x_0 \equiv 5 \pmod{11}$. Therefore, x_0 is a solution of the pair of congruences ($*$) if and only if x_0 is a solution of either the pair of congruences

$$\begin{aligned} x &\equiv 2 \pmod 6, \\ x &\equiv 5 \pmod{11}, \end{aligned} \tag{$**$}$$

or the pair of congruences

$$\begin{aligned} x &\equiv 5 \pmod 6, \\ x &\equiv 5 \pmod{11}. \end{aligned} \tag{$***$}$$

Thus, to solve ($*$), we need to solve both ($**$) and ($***$). This is easy. By the Chinese remainder theorem, ($**$) has the unique solution $\lfloor 38 \rfloor_{66}$ and ($***$) has the unique solution $\lfloor 5 \rfloor_{66}$—so the system ($*$) has the two solutions $\lfloor 5 \rfloor_{66}$ and $\lfloor 38 \rfloor_{66}$. In other words, the solutions of ($*$) are those integers x_0 which satisfy either $x_0 \equiv 5 \pmod{66}$ or $x_0 \equiv 38 \pmod{66}$.

The content of our solution of this problem may be expressed compactly as follows: The set of solutions of $2x \equiv 4 \pmod 6$ is the set $\lfloor 2 \rfloor_6 \cup \lfloor 5 \rfloor_6$, and the set of solutions of $5x \equiv 3 \pmod{11}$ is the set $\lfloor 5 \rfloor_{11}$, so the set of solutions of the simultaneous pair ($*$) is the set

$$(\lfloor 2 \rfloor_6 \cup \lfloor 5 \rfloor_6) \cap \lfloor 5 \rfloor_{11} = (\lfloor 2 \rfloor_6 \cap \lfloor 5 \rfloor_{11}) \cup (\lfloor 5 \rfloor_6 \cap \lfloor 5 \rfloor_{11})$$
$$= \lfloor 38 \rfloor_{66} \cup \lfloor 5 \rfloor_{66}.$$

(v) Consider the simultaneous congruences

$$4x \equiv 6 \pmod{10},$$
$$9x \equiv 15 \pmod{21}.$$

The first has the two solutions $\lfloor 4 \rfloor_{10}$, $\lfloor 9 \rfloor_{10}$, and the second has the three solutions $\lfloor 4 \rfloor_{21}$, $\lfloor 11 \rfloor_{21}$, $\lfloor 18 \rfloor_{21}$, so to solve the simultaneous congruences we must solve each of the following six pairs of congruences.

$$(1)\ \begin{cases} x \equiv 4 \pmod{10}, \\ x \equiv 4 \pmod{21}, \end{cases} \qquad (4)\ \begin{cases} x \equiv 9 \pmod{10}, \\ x \equiv 11 \pmod{21}, \end{cases}$$

$$(2)\ \begin{cases} x \equiv 9 \pmod{10}, \\ x \equiv 4 \pmod{21}, \end{cases} \qquad (5)\ \begin{cases} x \equiv 4 \pmod{10}, \\ x \equiv 18 \pmod{21}, \end{cases}$$

$$(3)\ \begin{cases} x \equiv 4 \pmod{10}, \\ x \equiv 11 \pmod{21}, \end{cases} \qquad (6)\ \begin{cases} x \equiv 9 \pmod{10}, \\ x \equiv 18 \pmod{21}. \end{cases}$$

It is not hard to solve each of these pairs—the results are, respectively, $\lfloor 4 \rfloor_{210}$, $\lfloor 109 \rfloor_{210}$, $\lfloor 74 \rfloor_{210}$, $\lfloor 179 \rfloor_{210}$, $\lfloor 144 \rfloor_{210}$, $\lfloor 39 \rfloor_{210}$. Thus, an integer x_0 is a solution if and only if x_0 is congruent to one of 4, 39, 74, 109, 144, 179 (mod 210).

There are six solutions (mod 210), and the compact way to describe the method for locating the solutions is as follows: The set of solutions of $4x \equiv 6 \pmod{10}$ is $\lfloor 4 \rfloor_{10} \cup \lfloor 9 \rfloor_{10}$, and the set of solutions of $9x \equiv 15 \pmod{21}$ is $\lfloor 4 \rfloor_{21} \cup \lfloor 11 \rfloor_{21} \cup \lfloor 18 \rfloor_{21}$—so the set of solutions of the simultaneous pair is

$$\begin{aligned}
&(\lfloor 4 \rfloor_{10} \cup \lfloor 9 \rfloor_{10}) \cap (\lfloor 4 \rfloor_{21} \cup \lfloor 11 \rfloor_{21} \cup \lfloor 18 \rfloor_{21}) \\
&\quad = (\lfloor 4 \rfloor_{10} \cap \lfloor 4 \rfloor_{21}) \cup (\lfloor 4 \rfloor_{10} \cap \lfloor 11 \rfloor_{21}) \cup (\lfloor 4 \rfloor_{10} \cap \lfloor 18 \rfloor_{21}) \\
&\qquad \cup (\lfloor 9 \rfloor_{10} \cap \lfloor 4 \rfloor_{21}) \cup (\lfloor 9 \rfloor_{10} \cap \lfloor 11 \rfloor_{21}) \cup (\lfloor 9 \rfloor_{10} \cap \lfloor 18 \rfloor_{21}) \\
&\quad = \lfloor 4 \rfloor_{210} \cup \lfloor 74 \rfloor_{210} \cup \lfloor 144 \rfloor_{210} \cup \lfloor 109 \rfloor_{210} \cup \lfloor 179 \rfloor_{210} \cup \lfloor 39 \rfloor_{210}.
\end{aligned}$$

3-1-12 / *PROBLEMS*

1. Without solving, determine how many solutions each of the following linear congruences has:

(i) $5x \equiv 89 \pmod{99}$, (ii) $9x \equiv 15 \pmod{21}$,
(iii) $28x \equiv 48 \pmod{1197}$.

2. Without solving, determine how many solutions each of the following linear equations has:

(i) $\lfloor 68 \rfloor_{289} x = \lfloor 72 \rfloor_{289}$, ($ii$) $\lfloor 5311 \rfloor_{7571} x = \lfloor 1243 \rfloor_{7571}$,
(iii) $\lfloor 120 \rfloor_{239} x = \lfloor 25 \rfloor_{239}$.

3. For $\alpha, \beta \in \mathbf{Z}_m$ find a necessary and sufficient condition that the linear equation $\alpha x = \beta$ have a unique solution in $\mathbf{Z}_m$.

4. Solve completely:

(*i*) $158x \equiv -22 \pmod{194}$, (*ii*) $194x \equiv -22 \pmod{158}$,
(*iii*) $84x \equiv 156 \pmod{605}$, (*iv*) $84x \equiv 156 \pmod{438}$.

5. Solve completely:

(*i*) $\lfloor 9 \rfloor_{24} x = \lfloor 6 \rfloor_{24}$, (*ii*) $\lfloor 12 \rfloor_{26} x = \lfloor 14 \rfloor_{26}$,
(*iii*) $\lfloor 1001 \rfloor_{4845} x = \lfloor 438 \rfloor_{4845}$, (*iv*) $\lfloor 165 \rfloor_{253} x = \lfloor 87 \rfloor_{253}$.

6. Produce a linear congruence which has exactly four solutions.

7. Consider the following collection of congruences:

$$
\begin{aligned}
&(1) \quad x \equiv 3 \pmod{7},\\
&(2) \quad x \equiv -2 \pmod{11},\\
&(3) \quad x \equiv 7 \pmod{12},\\
&(4) \quad x \equiv 8 \pmod{13},\\
&(5) \quad x \equiv 9 \pmod{14},\\
&(6) \quad x \equiv 1 \pmod{15},\\
&(7) \quad x \equiv -3 \pmod{17},\\
&(8) \quad x \equiv 5 \pmod{18}.
\end{aligned}
$$

Solve the system of simultaneous congruences for each of the following choices of congruences from this collection.

(*i*) (1) and (2), (*ii*) (2) and (3),
(*iii*) (1), (3), and (4), (*iv*) (1), (2), (4), and (7),
(*v*) (2), (4)–(7), (*vi*) (1), (2), (4), (7), and (8),
(*vii*) (1) and (5), (*viii*) (3) and (5),
(*ix*) (3) and (6), (*x*) (3) and (8),
(*xi*) (5) and (8), (*xii*) (6) and (8),
(*xiii*) (3), (5), and (6), (*xiv*) (5), (7), and (8).

8. Find the following sets:

(*i*) $\lfloor -2 \rfloor_{15} \cap \lfloor 3 \rfloor_{19}$, (*ii*) $\lfloor 4 \rfloor_{21} \cap \lfloor -4 \rfloor_{22} \cap \lfloor 7 \rfloor_{23}$,
(*iii*) $\lfloor 2 \rfloor_{3} \cap \lfloor 1 \rfloor_{5} \cap \lfloor -2 \rfloor_{7} \cap \lfloor 3 \rfloor_{11}$, (*iv*) $\lfloor 11 \rfloor_{30} \cap \lfloor 20 \rfloor_{51}$.

9. Evaluate $(\lfloor 5 \rfloor_{50} \cup \lfloor 30 \rfloor_{50}) \cap \lfloor 7 \rfloor_{23}$. Can you find a pair of simultaneous linear congruences for which this is the set of all solutions?

10. Viewing residue classes as sets of integers, prove

(*i*) $\lfloor 3 \rfloor_{21} \subset \lfloor 3 \rfloor_{7}$, (*ii*) $\lfloor 16 \rfloor_{24} \subset \lfloor 4 \rfloor_{6}$,
(*iii*) $\lfloor 1 \rfloor_{3} = \lfloor 1 \rfloor_{6} \cup \lfloor 4 \rfloor_{6}$, (*iv*) $\lfloor 2 \rfloor_{5} = \lfloor 2 \rfloor_{15} \cup \lfloor 7 \rfloor_{15} \cup \lfloor 12 \rfloor_{15}$,
(*v*) $\lfloor 4 \rfloor_{6} = \lfloor 4 \rfloor_{24} \cup \lfloor 10 \rfloor_{24} \cup \lfloor 16 \rfloor_{24} \cup \lfloor 22 \rfloor_{24}$.

Of course, the unions in (*iii*), (*iv*), (*v*) are disjoint.

11. Suppose m and n are integers greater than or equal to 2 and $m \mid n$; then for any $a \in \mathbf{Z}$, $\lfloor a \rfloor_n \subset \lfloor a \rfloor_m$—in fact, the residue class $\lfloor a \rfloor_m$ decomposes into the disjoint union

$$\lfloor a \rfloor_m = \lfloor a \rfloor_n \cup \lfloor a + m \rfloor_n \cup \lfloor a + 2m \rfloor_n \cup \cdots \lfloor a + (d-1)m \rfloor_n$$

where $n = md$.

12. Consider the linear congruence

$$ax \equiv b \pmod{m} \qquad (*)$$

and suppose $d = (a, m)$ divides b. Putting $a = da'$, $b = db'$, $m = dm'$, consider the linear congruence

$$a'x \equiv b' \pmod{m'}. \qquad (**)$$

Then $(*)$ and $(**)$ are equivalent in the sense that they have the same set of solutions. Prove this.

Now, $(*)$ has d solutions (mod m), while $(**)$ has a unique solution (mod m'). Explain what is going on.

13. Solve the simultaneous congruences

$$3x \equiv 5 \pmod{22},$$
$$11x \equiv 3 \pmod{28},$$
$$5x \equiv 89 \pmod{99}.$$

14. Solve the pair of congruences

$$9x \equiv 6 \pmod{24},$$
$$12x \equiv 14 \pmod{26}.$$

15. Use the method of 3-1-7 to do the various parts of Problem 7 above.

16. (Sun-Tse, 1st century AD). Find the smallest positive integer which leaves the remainders 2, 3, 2 upon division by 3, 5, 7, respectively.

17. (Brahmagupta, 7th century AD). When the eggs in a basket are removed 2, 3, 4, 5, 6 at a time there are left over 1, 2, 3, 4, 5, eggs, respectively. When they are removed 7 at a time, there is none left over. How many eggs are in the basket?

18. If $m = m_1 m_2$ with $(m_1, m_2) = 1$, show that the set of solutions of $ax \equiv b \pmod{m}$ is identical with the set of solutions of the pair

$$ax \equiv b \pmod{m_1}, \qquad ax \equiv b \pmod{m_2}.$$

19. Prove that the pair of congruences

$$x \equiv a \pmod{m},$$
$$x \equiv b \pmod{n}$$

has a solution if and only if (m, n) divides $(a - b)$. Moreover, in this situation, the solution is unique mod$[m, n]$—in more detail, if x_0 is a solution, then $\lfloor x_0 \rfloor_{[m,n]}$ is the set of all solutions.

20. Prove the Chinese remainder theorem by using the method illustrated in 3-1-7.

21. Suppose p is prime. Then for every $a \in \{1, 2, \ldots, p-1\}$ the congruence $ax \equiv 1 \pmod{p}$ has a unique solution—that is, there exists a unique $a' \in \{1, 2, \ldots, p-1\}$ for which $aa' \equiv 1 \pmod{p}$. Use this to prove that

$$(p-1)! \equiv -1 \pmod{p}.$$

This result is known as Wilson's theorem. It may also be expressed in the form: the product of all nonzero elements of $\mathbf{Z}_p$ is -1.

3-2. Units and Fields

One of the questions considered in the preceding section was that of solving the linear equation $\alpha x = \beta$ in $\mathbf{Z}_m$. The same question may also be formulated in a more general context—namely, here and throughout this chapter, we let $R = \{a, b, c, \ldots, u, v, \ldots\}$ denote a **commutative ring with unity** e, and make the following definition.

3-2-1. Definition. Given $a, b \in R$, $a \neq 0$, we say that a **divides** b in R (or a is a **factor** of b, or b is **divisible** by a, or b is a **multiple** of a) and write $a \mid b$ when there exists $c \in R$ for which $ac = b$. There is no requirement that c be unique. If no such element $c \in R$ exists, we say that a **does not divide** b in R, and write $a \nmid b$.

Thus, $a \mid b$ if and only if the linear equation $ax = b$ has a solution in R.

3-2-2. Remark. Of course, the definition of divisibility in R is the same as the one given long ago for divisibility in $\mathbf{Z}$. More precisely, we have extended or generalized the notion of divisibility from $\mathbf{Z}$ (which is a commutative ring with unity) to an arbitrary commutative ring with unity, R. Among the properties of divisibility in $\mathbf{Z}$ that are also valid in R, we have:

(*i*) $a \mid 0$ for every nonzero $a \in R$.
(*ii*) If $a \mid b$ and $b \mid c$, then $a \mid c$.
(*iii*) $(\pm e) \mid a$ for all $a \in R$; $(\pm a) \mid a$ for all $a \neq 0$.
(*iv*) $a \mid b \Leftrightarrow a \mid (-b) \Leftrightarrow (-a) \mid b \Leftrightarrow (-a) \mid (-b)$.
(*v*) If $a \mid b$ and $a \mid c$, then $a \mid (xb + yc)$ for all $x, y \in R$—in particular $a \mid (b + c)$ and $a \mid (b - c)$.

The proofs of these assertions are immediate; they are identical with the proofs in $\mathbf{Z}$, as given in 1-1-2.

On the other hand, many properties of divisibility in $\mathbf{Z}$ need not hold in R. For example, from $a \mid e$ in R we cannot conclude that $a = \pm e$; also, $a \mid b$ and $b \mid a$ together need not imply $a = \pm b$. (Illustrations of these statements will appear shortly.) Furthermore, there is no reason to expect the division algorithm, or anything like it, to hold in R.

Of course, the question of whether a divides b depends on the ring R in which they are viewed. For example, $2 \nmid 3$ in $\mathbf{Z}$; but since 2, 3, 3/2 are all perfectly good rational numbers with $(2)(3/2) = 3$, it follows that $2 \mid 3$ in $\mathbf{Q}$.

3-2-3. Examples. (*i*) The case $R = \mathbf{Z}$ is familiar. It was treated in detail in Chapter I. We have nothing to add here.

(*ii*) Suppose $R = \mathbf{Z}_7$, which is indeed a commutative ring with unity $\lfloor 1 \rfloor_7$. Because $\lfloor 3 \rfloor_7 \cdot \lfloor 2 \rfloor_7 = \lfloor 6 \rfloor_7$ and $\lfloor 6 \rfloor_7 \cdot \lfloor 4 \rfloor_7 = \lfloor 3 \rfloor_7$, the nonzero elements $\lfloor 3 \rfloor_7$ and $\lfloor 6 \rfloor_7$ divide each other in $\mathbf{Z}_7$ (that is, $\lfloor 3 \rfloor_7 \mid \lfloor 6 \rfloor_7$ and $\lfloor 6 \rfloor_7 \mid \lfloor 3 \rfloor_7$)—but note that $\lfloor 3 \rfloor_7 \neq \pm \lfloor 6 \rfloor_7$.

In addition, it is not hard to verify that for any $\alpha, \beta \in \mathbf{Z}_7$ with $\alpha \neq 0$, we have $\alpha \mid \beta$. One way to see this is directly from the multiplication table for $\mathbf{Z}_7$; the row associated with any $\alpha \neq 0$ includes every $\beta \in \mathbf{Z}_7$. Another, more sophisticated, way to see this is as follows. Write $\alpha = \lfloor a \rfloor_7$, $\beta = \lfloor b \rfloor_7$ where $a, b \in \mathbf{Z}$, and $7 \nmid a$ since $\alpha \neq 0$. Then, as seen in Section 3-1,

$$\begin{aligned} \alpha \mid \beta &\Leftrightarrow \text{the linear equation } \alpha x = \beta \text{ has a solution in } \mathbf{Z}_7 \\ &\Leftrightarrow \text{the linear congruence } ax \equiv b \pmod{7} \text{ has a solution.} \end{aligned}$$

But since $(a, 7) = 1$, this linear congruence does have a solution; hence, $\alpha \mid \beta$.

In particular, in $\mathbf{Z}_7$, $\lfloor 2 \rfloor_7 \mid \lfloor 1 \rfloor_7$, $\lfloor 3 \rfloor_7 \mid \lfloor 1 \rfloor_7$, $\lfloor 4 \rfloor_7 \mid \lfloor 1 \rfloor_7$, $\lfloor 5 \rfloor_7 \mid \lfloor 1 \rfloor_7$, $\lfloor 6 \rfloor_7 \mid \lfloor 1 \rfloor_7$.

(*iii*) Consider the commutative ring with unity, $\mathbf{Z}_{12}$. All kinds of interesting things can happen here. We list some facts about $\mathbf{Z}_{12}$—they can be verified easily by trial and error, or somewhat more efficiently by use of the multiplication table for $\mathbf{Z}_{12}$.

(1) $\lfloor 5 \rfloor_{12}$ divides every element of $\mathbf{Z}_{12}$,

(2) $\lfloor 5 \rfloor_{12}$ is divisible by $\lfloor 1 \rfloor_{12}$, $\lfloor 5 \rfloor_{12}$, $\lfloor 7 \rfloor_{12} = -\lfloor 5 \rfloor_{12}$, $\lfloor 11 \rfloor_{12} = -\lfloor 1 \rfloor_{12}$; no other elements of $\mathbf{Z}_{12}$ divide $\lfloor 5 \rfloor_{12}$.

(3) $\lfloor 6 \rfloor_{12}$ divides only itself and $\lfloor 0 \rfloor_{12}$,

(4) $\lfloor 6 \rfloor_{12}$ is divisible by all nonzero elements of $\mathbf{Z}_{12}$ except $\lfloor 4 \rfloor_{12}$ and $\lfloor 8 \rfloor_{12}$,

(5) The identity $\lfloor 1 \rfloor_{12}$ is divisible only by the four elements $\lfloor 1 \rfloor_{12}$, $\lfloor 5 \rfloor_{12}$, $\lfloor 7 \rfloor_{12}$, $\lfloor 11 \rfloor_{12}$,

(6) $\lfloor 3 \rfloor_{12}$ and $\lfloor 7 \rfloor_{12}$ divide each other, but $\lfloor 3 \rfloor_{12} \neq \pm \lfloor 7 \rfloor_{12}$,

(7) $\lfloor 4 \rfloor_{12}$ and $\lfloor 8 \rfloor_{12}$ divide each other (which is not surprising because $\lfloor 4 \rfloor_{12} = -\lfloor 8 \rfloor_{12}$),

(8) The result of "dividing" $\lfloor 8 \rfloor_{12}$ by $\lfloor 4 \rfloor_{12}$ is not unique, since

$$\lfloor 4 \rfloor_{12} \cdot \lfloor 2 \rfloor_{12} = \lfloor 4 \rfloor_{12} \cdot \lfloor 5 \rfloor_{12} = \lfloor 4 \rfloor_{12} \cdot \lfloor 8 \rfloor_{12} = \lfloor 4 \rfloor_{12} \cdot \lfloor 11 \rfloor_{12} = \lfloor 8 \rfloor_{12}.$$

It is instructive to prove several of these results in a more theoretical manner—namely, by the method introduced in (*ii*) above. We write elements of $\mathbf{Z}_{12}$ in the form $\alpha = \lfloor a \rfloor_{12}$, $\beta = \lfloor b \rfloor_{12}$, it being understood that $0 \leq a, b \leq 11$.

(1) To decide which elements of $\mathbf{Z}_{12}$ are divisible by $\lfloor 5 \rfloor_{12}$, we observe that $\alpha = \lfloor 5 \rfloor_{12}$ divides $\beta = \lfloor b \rfloor_{12} \Leftrightarrow$ the linear congruence $5x \equiv b \pmod{12}$ has a solution. But according to 3-1-5, this congruence has a solution for every b because $(5, 12) \mid b$. Thus, $\alpha = \lfloor 5 \rfloor_{12}$ divides every $\beta = \lfloor b \rfloor_{12}$ in $\mathbf{Z}_{12}$.

(2) To decide which elements of $\mathbf{Z}_{12}$ divide $\lfloor 5 \rfloor_{12}$, we observe that $\alpha = \lfloor a \rfloor_{12} \neq 0$ divides $\beta = \lfloor 5 \rfloor_{12} \Leftrightarrow ax \equiv 5 \pmod{12}$ has a solution $\Leftrightarrow (a, 12) \mid 5$ $\Leftrightarrow (a, 12) = 1$. Thus, the congruence has a solution only when a is prime to 12; and the divisors of $\lfloor 5 \rfloor_{12}$ are $\lfloor 1 \rfloor_{12}$, $\lfloor 5 \rfloor_{12}$, $\lfloor 7 \rfloor_{12}$, $\lfloor 11 \rfloor_{12}$.

(3) $\alpha = \lfloor 6 \rfloor_{12}$ divides $\beta = \lfloor b \rfloor_{12} \Leftrightarrow 6x \equiv b \pmod{12}$ has a solution $\Leftrightarrow$ $(6, 12) \mid b \Leftrightarrow 6 \mid b \Leftrightarrow b = 0$ or 6.

(4) $\alpha = \lfloor a \rfloor_{12} \neq 0$ divides $\beta = \lfloor 6 \rfloor_{12} \Leftrightarrow ax \equiv 6 \pmod{12}$ has a solution $\Leftrightarrow (a, 12) \mid 6 \Leftrightarrow a = 1, 2, 3, 5, 6, 7, 9, 10, 11$.

The remaining assertions may now safely be left to the reader.

(*iv*) Consider $\mathbf{Q}$, the domain of rational numbers (see 2-2-1). The rational numbers 37/79 and 47/97 divide each other, since

$$\left(\frac{37}{79}\right)\left(\frac{79 \cdot 47}{37 \cdot 97}\right) = \frac{47}{97} \quad \text{and} \quad \left(\frac{47}{97}\right)\left(\frac{97 \cdot 37}{47 \cdot 79}\right) = \frac{37}{79}.$$

More generally, any nonzero rational number a divides any rational number b. In fact, a is of form m_1/n_1 with $m_1, n_1 \in \mathbf{Z}$, $n_1 \neq 0$, and b is of form m_2/n_2 with $m_2, n_2 \in \mathbf{Z}$, $n_2 \neq 0$; furthermore, because $a \neq 0$ we have $m_1 \neq 0$. Then

$$\left(\frac{m_1}{n_1}\right)\left(\frac{n_1 \cdot m_2}{m_1 \cdot n_2}\right) = \frac{m_2}{n_2}$$

and $n_1m_2/m_1n_2 \in \mathbf{Q}$ because $n_1m_2 \in \mathbf{Z}$, $m_1n_2 \in \mathbf{Z}$, $m_1n_2 \neq 0$, so that $a \mid b$. In particular, any nonzero element $a \in \mathbf{Q}$ divides the identity element $1 = 1/1 = n/n$ of $\mathbf{Q}$.

(*v*) In the domain of real numbers, $\mathbf{R}$ (see 2-2-1), any nonzero element divides any real number. (This is a "familiar fact" from elementary school, but for us it is really an "assumption" about real numbers. It would require a massive amount of work to develop $\mathbf{R}$ in a careful, axiomatic way and then prove this statement about divisibility.) In particular, any nonzero real number divides the identity 1.

(*vi*) Consider the complex numbers, $\mathbf{C}$. This is a commutative ring with unity, $1 + 0i$; in fact, as seen in 2-2-1, $\mathbf{C}$ is an integral domain. Recall that an element $\alpha \in \mathbf{C}$ is of form $\alpha = a + bi$ where $a, b \in \mathbf{R}$; and $\alpha \neq 0$ when a and b are not both 0—in other words, $\alpha \neq 0 \Leftrightarrow a^2 + b^2 \neq 0$.

We assert that every complex number $\alpha \neq 0$ divides $1 = 1 + 0i$. To see this, note first that because $\alpha \neq 0$ division by $a^2 + b^2$, in $\mathbf{R}$, is permissible (according to part (*v*) above). Hence $a/(a^2 + b^2) \in \mathbf{R}$, $b/(a^2 + b^2) \in \mathbf{R}$, and

$$(a + bi)\left(\frac{a}{a^2 + b^2} - \frac{b}{a^2 + b^2}\, i\right) = 1 .$$

More generally, for any complex numbers $\alpha = a + bi \neq 0$ and $\beta = c + di$ we have $\alpha \mid \beta$. In fact, it is straightforward to verify that

$$(a + bi)\left(\frac{ac + bd}{a^2 + b^2} + \frac{ad - bc}{a^2 + b^2}\, i\right) = c + di .$$

This may look mysterious, but it is not. All we have done is solved the equation

$$(a + bi)(x + yi) = c + di$$

by multiplying both sides by $a - bi$.

(*vii*) Consider the integral domain $\mathbf{Z}[i]$, which was discussed in 2-2-4. An element of $\mathbf{Z}[i]$ is of form $a + bi$ with $a, b \in \mathbf{Z}$. Because $(1 + i)(1 - i) = 2$, we see that both $1 + i$ and $1 - i$ divide 2. We also have $(3 + 2i) \mid (-2 + 3i)$ and $(-2 + 3i) \mid (3 + 2i)$, since $-2 + 3i = (3 + 2i)i$ and $3 + 2i = (-2 + 3i)(-i)$—but $-2 + 3i \neq \pm(3 + 2i)$. On the other hand, $(2 + i) \nmid (3 + 2i)$. To see this, one tries to solve

$$(2 + i)(x + yi) = 3 + 2i.$$

Multiplying both sides by $2 - i$ gives

$$(5x) + (5y)i = 8 + i.$$

This equation has no solutions in integers x and y—so $(2 + i) \nmid (3 + 2i)$ in $\mathbf{Z}[i]$.

Clearly, the four elements ± 1, $\pm i$ are divisors of 1 in $\mathbf{Z}[i]$; it is not hard to see that there are no others.

(*viii*) Consider the integral domain (of 2-2-5)

$$\mathbf{Z}[\sqrt{2}] = \{a + b\sqrt{2} \mid a, b \in \mathbf{Z}\}.$$

To decide, for example, if $2 + 3\sqrt{2}$ divides $3 + 5\sqrt{2}$, we seek integers x, y such that

$$(2 + 3\sqrt{2})(x + y\sqrt{2}) = 3 + 5\sqrt{2}.$$

Multiplying both sides by $2 - 3\sqrt{2}$ yields

$$(-14)x + (-14y)\sqrt{2} = -24 + \sqrt{2}.$$

Since no integers x, y satisfy this equation, it follows that $(2 + 3\sqrt{2}) \nmid (3 + 5\sqrt{2})$ in $\mathbf{Z}[\sqrt{2}]$.

In similar fashion, we see that $(2 + 3\sqrt{2}) \mid (4 + 13\sqrt{2})$. In fact, from

$$(2 + 3\sqrt{2})(x + y\sqrt{2}) = 4 + 13\sqrt{2}$$

we obtain

$$(-14)x + (-14y)\sqrt{2} = -70 + 14\sqrt{2}.$$

Thus, $x = 5$, $y = -1$ and indeed $(2 + 3\sqrt{2})(5 - \sqrt{2}) = 4 + 13\sqrt{2}$.

Finally, because $(1 + \sqrt{2})(-1 + \sqrt{2}) = 1$ and $(3 + 2\sqrt{2})(3 - 2\sqrt{2}) = 1$, the elements $1 + \sqrt{2}$, $-1 + \sqrt{2}$, $3 + 2\sqrt{2}$, $3 - 2\sqrt{2}$ are divisors of 1. There are many other divisors of 1; can you find some of them?

3-2-4. Definition. Suppose R is a commutative ring with unity e; an element $a \in R$ is said to be a **unit** when $a \mid e$.

3-2-5. Remarks. (1) According to the definition, a is a unit of R if and only if the equation $ax = e$ has a solution in R—or, to put it another way, if and only if a has an inverse for multiplication.

(2) If a is a unit, then its inverse is unique; we denote it by a^{-1}.

Proof: Suppose both b and c are inverses of a, so $ab = e = ba$ and $ac = e = ca$. Then

$$b = be = b(ac) = (ba)c = ec = c.$$

(3) If a is a unit and $ab = ac$, then $b = c$; in other words, units can be "canceled."

Proof: Multiply both sides of $ab = ac$ by a^{-1}. (Note that if a is not a unit, then one is not permitted to cancel it. For example, in $\mathbf{Z}_{12}$ we have $\lfloor 4 \rfloor_{12} \cdot \lfloor 5 \rfloor_{12} = \lfloor 4 \rfloor_{12} \cdot \lfloor 8 \rfloor_{12}$, but $\lfloor 5 \rfloor_{12} \neq \lfloor 8 \rfloor_{12}$—$\lfloor 4 \rfloor_{12}$ may not be canceled because it is not a unit.)

(4) An element of R cannot be both a unit and a zero-divisor.

Proof: If a is a zero-divisor, there exists $c \neq 0$ in R such that $ac = 0$. If, in addition, a is a unit, then a^{-1} exists and

$$c = (a^{-1}a)c = a^{-1}(ac) = a^{-1}0 = 0$$

which contradicts $c \neq 0$.

(5) If a is a unit, then for any $b \in R$ the equation $ax = b$ has a unique solution; in other words, a unit divides any element of R, and the result of this "division" is unique.

Proof: Clearly, $x = a^{-1}b$ is a solution, and by cancellation [see (3) above] the solution is unique.

(6) e is a unit, and so is $-e$.

(7) If a is a unit, so is a^{-1}—and $(a^{-1})^{-1} = a$.

Proof: The relation $a^{-1}a = aa^{-1} = e$ says that a^{-1} is a unit whose unique inverse is a—hence, $(a^{-1})^{-1} = a$.

(8) If both a and b are units, then so is ab; that is, the product of units is a unit.

Proof: Clearly, $b^{-1}a^{-1}$ is the inverse of ab.

(9) For any $a \neq 0$ in R, we define

$$a^0 = e.$$

If a is a unit, we have $aa^{-1} = e$. From (8), we see that $a^2 = a \cdot a$ is a unit whose inverse is $(a^{-1})(a^{-1}) = (a^{-1})^2$. Therefore, a simple induction guarantees that for any positive integer m, a^m is a unit whose inverse is $(a^{-1})^m$; in other words,

$$(a^m)^{-1} = (a^{-1})^m, \qquad m > 0.$$

Now, if n is negative, then $-n > 0$, so replacing m by $-n$ in the above, we have $(a^{-n})^{-1} = (a^{-1})^{-n}$. Then, because we want exponents to behave properly, let us define a^n (that is, a to some negative exponent) by

$$(a^n) = (a^{-n})^{-1} = (a^{-1})^{-n}, \qquad n < 0.$$

It is now easy to verify the standard rules for exponents—namely, if a and b are units of R and m and n are *any* integers, then

$$\begin{aligned} &(i) \quad a^m a^n = a^{m+n}, \\ &(ii) \quad (a^m)^n = a^{mn}, \qquad m, n \in \mathbf{Z}. \\ &(iii) \quad (ab)^m = a^m b^m, \end{aligned}$$

The details are left to the reader. Of course, the multiplicative situation here is entirely analogous to the additive situation discussed in 2-4-6 and 2-4-7.

3-2-6. Examples. Pursuing the examples of 3-2-3 a bit further, we observe:

(*i*) The only units of $\mathbf{Z}$ are $+1$ and -1.

(*ii*) In $\mathbf{Z}_7$, the units are all six nonzero elements.

(*iii*) The units of $\mathbf{Z}_{12}$ are the four elements $\lfloor 1 \rfloor_{12}, \lfloor 5 \rfloor_{12}, \lfloor 7 \rfloor_{12}, \lfloor 11 \rfloor_{12}$; there are no others.

(*iv*) In $\mathbf{Q}$, in $\mathbf{R}$, and in $\mathbf{C}$, every nonzero element is a unit.

(*v*) We already know that ± 1, $\pm i$ are units of $\mathbf{Z}[i]$. Are there any others? Suppose $a + bi$ is another unit in $\mathbf{Z}[i]$. Then there is an element $c + di \in \mathbf{Z}[i]$ such that

$$(a + bi)(c + di) = 1.$$

By performing the multiplication we see that $ac - bd = 1$, $bc + ad = 0$. On the other hand, by multiplying both sides of the equation by $a - bi$, we obtain

$$(a^2 + b^2)c + (a^2 + b^2)di = a - bi.$$

Thus,

$$\begin{aligned} (a^2 + b^2)c &= a, \\ (a^2 + b^2)d &= -b. \end{aligned}$$

Upon multiplying the first of these by c, the second by d, and adding we have

$$(a^2 + b^2)(c^2 + d^2) = ac - bd = 1.$$

This is a contradiction because c and d are integers while $a^2 + b^2 \geq 2$ (because, by hypothesis, $a + bi \neq 0, \pm 1, \pm i$). Consequently, ± 1, $\pm i$ are the only units in $\mathbf{Z}[i]$.

(*vi*) We have already observed [in 3-2-3, part (*viii*)] that

$$(1 + \sqrt{2})(-1 + \sqrt{2}) = 1.$$

Therefore, $\alpha = 1 + \sqrt{2}$ is a unit of $\mathbf{Z}[\sqrt{2}]$, and so is its inverse $\alpha^{-1} = -1 + \sqrt{2}$ [which is often written as $1/(1 + \sqrt{2})$]. Then according to 3-2-5, part (8), $\alpha^2 = 3 + 2\sqrt{2}$ is a unit, and so is $(\alpha^{-1})^2 = 3 - 2\sqrt{2}$. [It may be noted in passing that we have already observed, in 3-2-3, part (*viii*), that $3 + 2\sqrt{2}$ and $3 - 2\sqrt{2}$ are units with $(3 + 2\sqrt{2})(3 - 2\sqrt{2}) = 1$.] By induction, α^n and $(\alpha^{-1})^n = \alpha^{-n}$ are units for every positive integer n. We may rephrase this as: α^n is a unit for every $n \in \mathbf{Z}$. Of course, $-\alpha^n$ is also a unit, so every element of the set

$$\{\pm\alpha^n = \pm(1 + \sqrt{2}^n) \,|\, n \in \mathbf{Z}\} \tag{$*$}$$

is a unit.

Note that $\alpha = 1 + \sqrt{2}$ is a real number greater than 2, and α^{-1} then lies between 0 and 1. Hence, the elements of the set (*) are distributed along the real number line as follows

$$\cdots < -\alpha^2 < -\alpha < -2 < -1 < -\alpha^{-1} < -\alpha^{-2} < \cdots < 0 <$$
$$\cdots < \alpha^{-3} < \alpha^{-2} < \alpha^{-1} < 1 < 2 < \alpha < \alpha^2 < \alpha^3 < \cdots.$$

The set (*) contains an infinite number of distinct elements, so the number of units of $\mathbf{Z}[\sqrt{2}]$ is surely infinite. However, it may be that there exist units of $\mathbf{Z}[\sqrt{2}]$ which are not members of the set (*). This is an interesting question (and the answer is known), but it is probably inappropriate to try to settle it here.

3-2-7. Definition. A commutative ring with unity in which every nonzero element has an inverse for multiplication is said to be a **field** (it will usually be denoted by F).

In other words, a commutative ring with unity is a field if and only if every nonzero element is a unit.

3-2-8. Examples. (*i*) In virtue of 3-2-6, $\mathbf{Z}_7$ is a field, and so are $\mathbf{Q}$, $\mathbf{R}$, and $\mathbf{C}$.

(*ii*) On the other hand, $\mathbf{Z}_{12}$ and $\mathbf{Z}[i]$ are not fields—each has nonzero elements which are not units.

(*iii*) What about the commutative ring with unity $\mathbf{Z}[\sqrt{2}]$? Is it a field? We have already seen that there are an infinite number of units—in fact, $\pm(1 + \sqrt{2})^n$ is a unit for every $n \in \mathbf{Z}$. However, we do not know if there are any nonzero elements which are not units. For this, let us consider a typical element—say $2 + 3\sqrt{2}$—and decide if it is a unit. If

$$(2 + 3\sqrt{2})(x + y\sqrt{2}) = 1,$$

then by the method of 3-2-3, part (*viii*),

$$(-14)x + (-14y)\sqrt{2} = 2 - 3\sqrt{2}.$$

Since this has no integer solutions x and y, $2 + 3\sqrt{2}$ is not a unit, so $\mathbf{Z}[\sqrt{2}]$ is not a field.

(*iv*) Consider

$$\mathbf{Q}[i] = \mathbf{Q}[\sqrt{-1}] = \{a + bi \mid a, b \in \mathbf{Q}\}.$$

The verification that this is a commutative ring with unity is straightforward; it goes just like the verification that $\mathbf{Z}[i]$ is a commutative ring with unity. We assert that $\mathbf{Q}[i]$ is a field. For this, we take an arbitrary nonzero element $a + bi$, and show it is a unit. We must solve

$$(a + bi)(x + yi) = 1$$

—where x and y are to come from $\mathbf{Q}$. Multiplying by the "conjugate" $a - bi$, we have

$$(a^2 + b^2)x + (a^2 + b^2)yi = a - bi.$$

Of course, $a^2 + b^2 \neq 0$, because not both a and b are 0. Hence, we can "divide" by $a^2 + b^2$ in the field $\mathbf{Q}$—so

$$x = \frac{a}{a^2 + b^2} \in \mathbf{Q}, \qquad y = \frac{-b}{a^2 + b^2} \in \mathbf{Q}$$

provide a solution; that is,

$$(a + bi)\left(\frac{a}{a^2 + b^2} - \frac{b}{a^2 + b^2}\, i\right) = 1.$$

and $\mathbf{Q}[i]$ is a field.

(*v*) It is easy to check that

$$\mathbf{Q}[\sqrt{-2}] = \{a + b\sqrt{-2} \mid a, b \in \mathbf{Q}\}$$

is a commutative ring with unity. Moreover, it is a field because the arbitrary nonzero element $a + b\sqrt{-2}$ has the inverse

$$\left(\frac{a}{a^2 + 2b^2}\right) - \left(\frac{b}{a^2 + 2b^2}\right)\sqrt{-2}.$$

3-2-9. Remarks. (1) There is a parallelism between addition and multiplication in the axioms for a field. More precisely, both addition and multiplication satisfy (*i*) closure for the operation, (*ii*) the associative law, (*iii*) the commutative law, (*iv*) existence of an identity for the operation, and (*v*) existence of an inverse (for addition, this applies to every element; while for multiplication, the zero element is excluded).

(2) In a field F, any equation of form $ax = b$, $a \neq 0$, $a, b \in F$ has a solution; in fact, $a^{-1}b$ is the unique solution. Thus, any nonzero element divides every

element of the field. Of course, as is customary in **Q**, **R**, and **C**, one often denotes the inverse a^{-1} by $1/a$ (where 1 is the unity of F) and then $a^{-1}b$ is also written in the form b/a.

(3) If F is a field, then it is an integral domain. To see this, note that if $a \neq 0$ is any element of F, then a is a unit; so according to 3-2-5, part (4) a is not a zero-divisor. Thus, F has no zero-divisors, and is therefore an integral domain.

The converse is false; for example, **Z** is an integral domain, but is not a field.

Now, let us exhibit some additional fields.

3-2-10. Theorem. For every prime p, $\mathbf{Z}_p$ is a field.

Proof: We denote the identity $\lfloor 1 \rfloor_p$ of $\mathbf{Z}_p$ by 1 (this should not cause serious confusion). Consider any nonzero element $\alpha \in \mathbf{Z}_p$, and write it in the form $\alpha = \lfloor a \rfloor_p$ for some choice of $a \in \mathbf{Z}$. Then

$$\begin{aligned} \alpha \text{ has an inverse } &\Leftrightarrow \alpha x = 1 \text{ has a solution in } \mathbf{Z}_p \\ &\Leftrightarrow ax \equiv 1 \pmod{p} \text{ has a solution in } \mathbf{Z} \\ &\Leftrightarrow (a, p) \text{ divides } 1 \\ &\Leftrightarrow (a, p) = 1. \end{aligned}$$

Since $\alpha \neq 0$ we know (from the meaning of residue class) that $p \nmid a$, so indeed $(a, p) = 1$. It follows that α has an inverse, and consequently, $\mathbf{Z}_p$ is a field. ▮

It is instructive to give another proof of this result. According to 2-1-19, $\mathbf{Z}_m$ is an integral domain $\Leftrightarrow$ m is prime. Hence, $\mathbf{Z}_p$ is an integral domain with a finite number of elements, and to complete the proof it suffices to show:

3-2-11. Theorem. A finite integral domain is a field.

Proof: Suppose D is an integral domain with n elements; denote it by

$$D = \{a_1, a_2, \ldots, a_n\}.$$

Of course, the elements $a_1, a_2, \ldots, a_n$ are distinct, and the zero element 0 and the unity element e appear in this set.

For any $a \neq 0$ in D, let us look at the elements $aa_1, aa_2, \ldots, aa_n$. If $aa_i = aa_j$, then, by cancellation (which is permissible in an integral domain), $a_i = a_j$ and $i = j$. Consequently, the n elements $aa_1, aa_2, \ldots, aa_n$ are distinct—so they fill up all of D; in other words,

$$D = \{aa_1, aa_2, \ldots, aa_n\}.$$

Now, e appears in this set, which implies the existence of an a_t for which $aa_t = e$. Thus, a has an inverse, and D is a field. ▌

In virtue of 3-2-10, for any prime p, every nonzero element of $\mathbf{Z}_p$ is a unit. What about the units of $\mathbf{Z}_m$ for arbitrary $m > 1$? If m is composite, we know that $\mathbf{Z}_m$ is not an integral domain—hence, it is not a field, and not every nonzero element is a unit.

3-2-12. Definition. For $m > 1$, we let $\phi(m)$ denote the number of units of $\mathbf{Z}_m$; ϕ is known as the **Euler ϕ-function** or as the **totient** function. If we use the notations $\mathbf{Z}_m{}^*$, for the set of all units of $\mathbf{Z}_m$ and $\#$ for the phrase "the number of elements in the set," then

$$\phi(m) = \#(\mathbf{Z}_m^*).$$

3-2-13. Remark. We reinterpret the Euler ϕ-function. Consider any $\alpha = \lfloor a \rfloor_m \in \mathbf{Z}_m$. By exactly the same argument used in the proof of 3-2-10, we see that

$$\alpha \text{ is a unit in } \mathbf{Z}_m \Leftrightarrow (a, m) = 1. \qquad (*)$$

In other words, α has an inverse in $\mathbf{Z}_m$ if and only if a and m are relatively prime.

What happens if we use a different representative a' for α? Clearly, for $\alpha = \lfloor a' \rfloor_m$ we have, as above,

$$\alpha \text{ is a unit in } \mathbf{Z}_m \Leftrightarrow (a', m) = 1.$$

One immediate conclusion to be drawn is: If $\lfloor a \rfloor_m = \lfloor a' \rfloor_m$, then $(a, m) = 1 \Leftrightarrow (a', m) = 1$. This is a weak form of the following general fact about congruences:

$$\text{If } \lfloor a \rfloor_m = \lfloor a' \rfloor_m \text{ [that is, if } a \equiv a' \pmod{m}] \text{ then } (a, m) = (a', m). \qquad (**)$$

The proof is easy. Let $d = (a, m)$ and $d' = (a', m)$. We may write $a' = a + tm$. It follows that $d \mid a'$, and then $d \mid d'$. Similarly, $d' \mid d$—so $d = d'$.

This fact permits us to speak of the greatest common divisor (α, m) when α is any element of $\mathbf{Z}_m$—namely, we choose any representative a of α and define

$$(\alpha, m) = (a, m).$$

According to $(**)$, this definition is independent of the choice of representative for α. Naturally, when $(\alpha, m) = 1$, we say that the residue class α is **relatively prime to** m.

Thus, α is a unit in $\mathbf{Z}_m \Leftrightarrow (\alpha, m) = 1$, and consequently $\phi(m)$ is the number of residue classes (mod m) which are relatively prime to m. The m

residue classes (mod m) may be listed as

$$\lfloor 1 \rfloor_m, \quad \lfloor 2 \rfloor_m, \quad \lfloor 3 \rfloor_m, \ldots, \lfloor m \rfloor_m.$$

To decide how many of these are relatively prime to m, we need only look at their respective representatives $1, 2, \ldots, m$. The conclusion is:

$\phi(m)$ is the number of positive integers
less than or equal to m that are relatively prime to m.

Incidentally, this formulation of the definition of $\phi(m)$ is the one that is found in all number theory books.

As illustrations, we observe that $\phi(6) = 2$ (of the integers 1, 2, 3, 4, 5, 6 only 1 and 5 are relatively prime to 6—or, to put it another way, $\lfloor 1 \rfloor_6$ and $\lfloor 5 \rfloor_6$ are the only units of $\mathbf{Z}_6$), $\phi(12) = 4$ (only 1, 5, 7, 11 are relatively prime to 12), $\phi(5) = 4$, $\phi(7) = 6$, $\phi(11) = 10$, $\phi(13) = 12$. In fact, for any prime p,

$$\phi(p) = p - 1$$

because $1, 2, 3, \ldots, p - 1$ are relatively prime to p, and p is not. More generally, we have:

3-2-14. Proposition. For any prime p and any $r \geq 1$

$$\phi(p^r) = p^r - p^{r-1} = p^{r-1}(p - 1).$$

Proof: Consider the set of p^r integers

$$P = \{1, 2, \ldots, p^r\}.$$

Because p is prime, an integer is relatively prime to p^r if and only if it is relatively prime to p—so $\phi(p^r)$ equals the number of elements of P which are relatively prime to p. Now, let us determine how many elements of P are not relatively prime to p. Because p is prime, an integer is not relatively prime to p if and only if it is divisible by p. Therefore, the elements of P which are not relatively prime to p are

$$p, 2p, 3p, \ldots, p^{r-1} \cdot p$$

and there are p^{r-1} such elements. Hence, the number of elements of P which are relatively prime to p is $p^r - p^{r-1}$, which proves the desired formula for $\phi(p^r)$. ∎

According to this result, we have, for example,

$$\phi(7^3) = 7^2(7 - 1) = 294 \qquad \text{and} \qquad \phi(5^5) = 5^4(5 - 1) = 2500.$$

We know how to compute the value of the Euler ϕ-function for prime powers p^r, and our objective is to find ϕ for any $m > 1$. To accomplish this, we make use of the fact that ϕ is "multiplicative"—by which is meant that the following property holds:

$$\text{if } (m, n) = 1, \text{ then } \phi(mn) = \phi(m)\phi(n).$$

The proof of this property will be given shortly; but for the moment, let us suppose it is true. Then, given any $m > 1$, we consider its unique factorization into prime powers

$$m = p_1^{r_1} p_2^{r_2} \cdots p_s^{r_s}$$

where $p_1, p_2, \ldots, p_s$ are distinct primes and the exponents $r_1, r_2, \ldots, r_s$ are all greater than or equal to 1. Clearly, $p_1^{r_1}$ and $p_2^{r_2} \cdots p_s^{r_s}$ are relatively prime, so according to the multiplicative property

$$\phi(m) = \phi(p_1^{r_1})\, \phi(p_2^{r_2} \cdots p_s^{r_s}).$$

This procedure may be repeated as many times as necessary; in the end (that is, inductively) we arrive at

$$\begin{aligned}
\phi(m) &= \phi(p_1^{r_1})\, \phi(p_2^{r_2}) \cdots \phi(p_s^{r_s}) \\
&= (p_1^{r_1} - p_1^{r_1 - 1})(p_2^{r_2} - p_2^{r_2 - 1}) \cdots (p_s^{r_s} - p_s^{r_s - 1}) \\
&= p_1^{r_1}\left(1 - \frac{1}{p_1}\right) p_2^{r_2}\left(1 - \frac{1}{p_2}\right) \cdots p_s^{r_s}\left(1 - \frac{1}{p_s}\right) \\
&= m\left(1 - \frac{1}{p_1}\right)\left(1 - \frac{1}{p_2}\right) \cdots \left(1 - \frac{1}{p_s}\right) \\
&= m \prod_{i=1}^{s} \left(1 - \frac{1}{p_i}\right).
\end{aligned}$$

Furthermore, because the p_i are precisely those primes which divide m, this may be written as

$$m \prod_{p \mid m} \left(1 - \frac{1}{p}\right).$$

We have proved:

3-2-15. Theorem. If m is any integer greater than 1, then

$$\phi(m) = m \prod_{p \mid m} \left(1 - \frac{1}{p}\right).$$

As an illustration of this result, suppose $m = 36{,}000 = 2^5 \cdot 3^2 \cdot 5^3$. Then

$$\phi(36{,}000) = 36{,}000(1 - \tfrac{1}{2})(1 - \tfrac{1}{3})(1 - \tfrac{1}{5})$$

which is somewhat awkward as we must clear of fractions. Instead, it is more efficient to make use of the general expression

$$\begin{aligned}\phi(m) &= (p_1^{r_1} - p_1^{r_1-1})(p_2^{r_2} - p_2^{r_2-1}) \cdots (p_s^{r_s} - p_s^{r_s-1}) \\ &= p_1^{r_1-1} p_2^{r_2-1} \cdots p_s^{r_s-1}(p_1 - 1)(p_2 - 1) \cdots (p_s - 1).\end{aligned}$$

Thus, we have

$$\begin{aligned}\phi(36{,}000) &= 2^4 \cdot 3^1 \cdot 5^2(2 - 1)(3 - 1)(5 - 1) \\ &= 9600.\end{aligned}$$

Now, let us turn to the missing part of the proof of 3-2-15. We must prove

3-2-16. Lemma. If m and n are relatively prime positive integers, then

$$\phi(mn) = \phi(m)\phi(n).$$

According to our definition of the Euler ϕ-function, it must be shown that

$$\#\,(\mathbf{Z}_{mn}^*) = \#\,(\mathbf{Z}_m^*)\ \#\,(\mathbf{Z}_n^*), \qquad (m, n) = 1.$$

We shall give a fairly sophisticated algebraic proof of this formula; it will be accomplished in a sequence of steps.

3-2-17. Theorem. If $m > 1$, $n > 1$, $(m, n) = 1$, then the ring $\mathbf{Z}_{mn}$ and the direct sum ring $\mathbf{Z}_m \oplus \mathbf{Z}_n$ are isomorphic; that is,

$$\mathbf{Z}_{mn} \approx \mathbf{Z}_m \oplus \mathbf{Z}_n$$

Proof: A typical element of $\mathbf{Z}_{mn}$ is of form $\lfloor a \rfloor_{mn}$ where $a \in \mathbf{Z}$. We also recall (see 2-2-11) that the elements of $\mathbf{Z}_m \oplus \mathbf{Z}_n$ are the ordered pairs (α, β) where $\alpha \in \mathbf{Z}_m$, $\beta \in \mathbf{Z}_n$, and that the operations of addition and multiplication in $\mathbf{Z}_m \oplus \mathbf{Z}_n$ are componentwise.

Let us define a mapping

$$f\colon \mathbf{Z}_{mn} \to \mathbf{Z}_m \oplus \mathbf{Z}_n$$

by putting

$$f\left(\lfloor a \rfloor_{mn}\right) = \left(\lfloor a \rfloor_m, \lfloor a \rfloor_n\right).$$

This definition seems to depend on the choice of the representative a of the residue class $\lfloor a \rfloor_{mn}$, so we must verify that f is well defined. In other words, we must show that if $\lfloor a' \rfloor_{mn} = \lfloor a \rfloor_{mn}$, then $f(\lfloor a' \rfloor_{mn}) = f(\lfloor a \rfloor_{mn})$; but this is immediate from

$$\begin{aligned} \lfloor a' \rfloor_{mn} = \lfloor a \rfloor_{mn} &\Rightarrow a' \equiv a \pmod{mn} \\ &\Rightarrow a' \equiv a \pmod{m} \quad \text{and} \quad a' \equiv a \pmod{n} \\ &\Rightarrow \lfloor a' \rfloor_m = \lfloor a \rfloor_m \quad \text{and} \quad \lfloor a' \rfloor_n = \lfloor a \rfloor_n \\ &\Rightarrow (\lfloor a' \rfloor_m, \lfloor a' \rfloor_n) = (\lfloor a \rfloor_m, \lfloor a \rfloor_n) \\ &\Rightarrow f(\lfloor a' \rfloor_{mn}) = f(\lfloor a \rfloor_{mn}). \end{aligned}$$

The verification that f is a homomorphism is straightforward. For any $\lfloor a \rfloor_{mn}, \lfloor b \rfloor_{mn} \in \mathbf{Z}_{mn}$, we have

$$\begin{aligned} f(\lfloor a \rfloor_{mn} + \lfloor b \rfloor_{mn}) &= f(\lfloor a + b \rfloor_{mn}) \\ &= (\lfloor a + b \rfloor_m, \lfloor a + b \rfloor_n) \end{aligned}$$

and

$$\begin{aligned} f(\lfloor a \rfloor_{mn}) + f(\lfloor b \rfloor_{mn}) &= (\lfloor a \rfloor_m, \lfloor a \rfloor_n) + (\lfloor b \rfloor_m, \lfloor b \rfloor_n) \\ &= (\lfloor a \rfloor_m + \lfloor b \rfloor_m, \lfloor a \rfloor_n + \lfloor b \rfloor_n) \\ &= (\lfloor a + b \rfloor_m, \lfloor a + b \rfloor_n). \end{aligned}$$

Thus, $f(\lfloor a \rfloor_{mn} + \lfloor b \rfloor_{mn}) = f(\lfloor a \rfloor_{mn}) + f(\lfloor b \rfloor_{mn})$ and f preserves addition. In similar fashion,

$$\begin{aligned} f(\lfloor a \rfloor_{mn} \cdot \lfloor b \rfloor_{mn}) &= f(\lfloor ab \rfloor_{mn}) \\ &= (\lfloor ab \rfloor_m, \lfloor ab \rfloor_n), \end{aligned}$$

and

$$\begin{aligned} f(\lfloor a \rfloor_{mn}) \cdot f(\lfloor b \rfloor_{mn}) &= (\lfloor a \rfloor_m, \lfloor a \rfloor_n) \cdot (\lfloor b \rfloor_m, \lfloor b \rfloor_n) \\ &= (\lfloor a \rfloor_m \cdot \lfloor b \rfloor_m, \lfloor a \rfloor_n \cdot \lfloor b \rfloor_n) \\ &= (\lfloor ab \rfloor_m, \lfloor ab \rfloor_n). \end{aligned}$$

Thus, $f(\lfloor a \rfloor_{mn} \cdot \lfloor b \rfloor_{mn}) = f(\lfloor a \rfloor_{mn}) \cdot f(\lfloor b \rfloor_{mn})$ and f preserves multiplication.

The next step is to show that f is surjective. Consider an arbitrary element of $\mathbf{Z}_m \oplus \mathbf{Z}_n$; it is an ordered pair (α, β) where $\alpha \in \mathbf{Z}_m$, $\beta \in \mathbf{Z}_n$, and, of course, we may write $\alpha = \lfloor a \rfloor_m$, $\beta = \lfloor b \rfloor_n$ where $a, b \in \mathbf{Z}$. The problem facing us is to exhibit the element $(\lfloor a \rfloor_m, \lfloor b \rfloor_n)$ of $\mathbf{Z}_m \oplus \mathbf{Z}_n$ as the image under f of some element $\lfloor c \rfloor_{mn}$ of $\mathbf{Z}_{mn}$—that is, we seek $\lfloor c \rfloor_{mn}$ such that

$$f(\lfloor c \rfloor_{mn}) = (\lfloor a \rfloor_m, \lfloor b \rfloor_n).$$

According to the definition of f, this means we are searching for $c \in \mathbf{Z}$ such that

$$(\lfloor c \rfloor_m, \lfloor c \rfloor_n) = (\lfloor a \rfloor_m, \lfloor b \rfloor_n) \in \mathbf{Z}_m \oplus \mathbf{Z}_n$$

—which translates to the requirement that both

$$\lfloor c \rfloor_m = \lfloor a \rfloor_m \qquad \text{and} \qquad \lfloor c \rfloor_n = \lfloor b \rfloor_n .$$

Putting this in the language of congruences, we want $c \in \mathbf{Z}$ for which

$$c \equiv a \pmod{m},$$
$$c \equiv b \pmod{n}.$$

But now, because m and n are relatively prime (this is the first time we have used the hypothesis $(m, n) = 1$), the Chinese remainder theorem, 3-1-8, guarantees the existence of $c \in \mathbf{Z}$ which satisfies these simultaneous congruences. This proves that f is surjective. (More precisely, the statement that f is surjective may be viewed as a restatement of the Chinese remainder theorem in the current circumstances.)

It remains only to show that f is injective (that is, one-to-one). One proof consists of the simple observation that f is a mapping of $\mathbf{Z}_{mn}$, which has mn elements, *onto* $\mathbf{Z}_m \oplus \mathbf{Z}_n$, which (because it consists of all ordered pairs (α, β) where $\alpha \in \mathbf{Z}_m$, $\beta \in \mathbf{Z}_n$) also has mn elements—so f must be one-to-one. Another, more algebraic proof, consists of showing that $\ker f = (0)$—namely,

$$
\begin{aligned}
\lfloor c \rfloor_{mn} \in \ker f &\Rightarrow f(\lfloor c \rfloor_{mn}) = 0 \in \mathbf{Z}_m \oplus \mathbf{Z}_n \\
&\Rightarrow (\lfloor c \rfloor_m, \lfloor c \rfloor_n) = (\lfloor 0 \rfloor_m, \lfloor 0 \rfloor_n) \\
&\Rightarrow \lfloor c \rfloor_m = \lfloor 0 \rfloor_m \text{ and } \lfloor c \rfloor_n = \lfloor 0 \rfloor_n \\
&\Rightarrow c \equiv 0 \pmod{m} \text{ and } c \equiv 0 \pmod{n} \\
&\Rightarrow m \mid c \text{ and } n \mid c \\
&\Rightarrow mn \mid c \qquad [\text{since} \quad (m, n) = 1] \\
&\Rightarrow c \equiv 0 \pmod{mn} \\
&\Rightarrow \lfloor c \rfloor_{mn} = \lfloor 0 \rfloor_{mn} = 0 \in \mathbf{Z}_{mn}.
\end{aligned}
$$

This completes the proof. ▮

Now, $\mathbf{Z}_{mn}$ is a commutative ring with unity $\lfloor 1 \rfloor_{mn}$ and $\mathbf{Z}_m \oplus \mathbf{Z}_n$ is a commutative ring with unity $(\lfloor 1 \rfloor_m, \lfloor 1 \rfloor_n)$. Because these rings are isomorphic, it is only natural to expect that they have the "same" units; in particular, these rings should have the same number of units. In other words, if the set of units of $\mathbf{Z}_m \oplus \mathbf{Z}_n$ is denoted by $(\mathbf{Z}_m \oplus \mathbf{Z}_n)^*$, we expect to have

$$\#(\mathbf{Z}_{mn}^*) = \#(\mathbf{Z}_m \oplus \mathbf{Z}_n)^* .$$

This relation is indeed true. It is an immediate consequence of the following general result.

3-2-18. Proposition. Suppose R is a commutative ring with unity e which is isomorphic to the commutative ring R with unity e'. Let $f: R \to R'$ be an isomorphism, and for $a \in R$ let us put $a' = f(a)$; then

(*i*) $f(e) = e'$,
(*ii*) there is a one-to-one correspondence between the set of units of R and the set of units of R'—namely, u is a unit of $R \Leftrightarrow u' = f(u)$ is a unit of R'.

Proof: Part (*i*) is immediate from 2-5-8. As for (*ii*)—u is a unit of $R \Rightarrow$ there exists $v \in R$ such that $uv = e \Rightarrow f(uv) = f(e) \Rightarrow f(u)f(v) = e' \Rightarrow u'v' = e' \Rightarrow u'$ is a unit of R'. Conversely, if u' is a unit of R, then there exists $v' \in R'$ such that $u'v' = e'$. Now, there exist unique elements $u, v \in R$ for which $u' = f(u)$, $v' = f(v)$. Hence, $f(uv) = f(u)f(v) = u'v' = e' = f(e)$—so $uv = e$ (because f is one-to-one) and u is indeed a unit of R. ▮

Our next task is to find the number of units of $\mathbf{Z}_m \oplus \mathbf{Z}_n$—that is, to compute $\#(\mathbf{Z}_m \oplus \mathbf{Z}_n)^*$—in terms of the units of $\mathbf{Z}_m$ and $\mathbf{Z}_n$. This is accomplished via our next result.

3-2-19. Proposition. Suppose R_1 is a commutative ring with unity e_1 and R_2 is a commutative ring with unity e_2. Then the direct sum $R_1 \oplus R_2$ is a commutative ring with unity (e_1, e_2). Moreover, for $u_1 \in R_1$, $u_2 \in R_2$, (u_1, u_2) is a unit of $R_1 \oplus R_2$ if and only if u_1 is a unit of R_1 and u_2 is a unit of R_2.

Proof: The first assertion is trivial. Because both R_1 and R_2 are commutative, $R_1 \oplus R_2$ is clearly a commutative ring; and for any $(a_1, a_2) \in R_1 \oplus R_2$

$$(e_1, e_2)(a_1, a_2) = (e_1 a_1, e_2 a_2) = (a_1, a_2)$$

so (e_1, e_2) is a unity element.

If (u_1, u_2) is a unit of $R_1 \oplus R_2$, then there exists $(v_1, v_2) \in R_1 \oplus R_2$ such that $(u_1, u_2)(v_1, v_2) = (e_1, e_2)$; therefore, $(u_1 v_1, u_2 v_2) = (e_1, e_2)$ and $u_1 v_1 = e_1$, $u_2 v_2 = e_2$—so u_1 is a unit of R_1 and u_2 is a unit of R_2.

Conversely, if u_1 is a unit of R_1 and u_2 is a unit of R_2, then there exist $v_1 \in R_1$, $v_2 \in R_2$ such that $u_1 v_1 = e_1$, $u_2 v_2 = e_2$. Consequently, $(u_1, u_2)(v_1, v_2) = (u_1 v_1, u_2 v_2) = (e_1, e_2)$ and (u_1, u_2) is a unit of $R_1 \oplus R_2$. ▮

3-2-20. Remark. Applying this for $\alpha \in \mathbf{Z}_m$, $\beta \in \mathbf{Z}_n$, we have: (α, β) is a unit of $\mathbf{Z}_m \oplus \mathbf{Z}_n \Leftrightarrow \alpha$ is a unit of $\mathbf{Z}_m$ and β is a unit of $\mathbf{Z}_n$. In other words,

$$(\alpha, \beta) \in (\mathbf{Z}_m \oplus \mathbf{Z}_n)^* \Leftrightarrow \alpha \in \mathbf{Z}_m^* \quad \text{and} \quad \beta \in \mathbf{Z}_n^*.$$

It follows that $(\mathbf{Z}_m \oplus \mathbf{Z}_n)^*$ has the same number of elements as the product set $\mathbf{Z}_m^* \times \mathbf{Z}_n^*$ (which consists of all ordered pairs (μ, ν) where $\mu \in \mathbf{Z}_m^*$, $\nu \in \mathbf{Z}_n^*$); symbolically,

$$\#(\mathbf{Z}_m \oplus \mathbf{Z}_n)^* = \#(\mathbf{Z}_m^* \times \mathbf{Z}_n^*),$$

But because a product set (of two sets) consists of **all** ordered pairs, it is clear that

$$\#(\mathbf{Z}_m^* \times \mathbf{Z}_n^*) = \#(\mathbf{Z}_m^*)\#(\mathbf{Z}_n^*).$$

Our preliminaries are now complete. To prove 3-2-16 we need only string them together—in fact,

$$\begin{aligned} \phi(mn) &= \#(\mathbf{Z}_{mn}^*) \\ &= \#(\mathbf{Z}_m \oplus \mathbf{Z}_n)^* \\ &= \#(\mathbf{Z}_m^* \times \mathbf{Z}_n^*) \\ &= \#(\mathbf{Z}_m^*)\#(\mathbf{Z}_n^*) \\ &= \phi(m)\phi(n). \end{aligned}$$

A more wordy version of this proof goes as follows: By definition, $\phi(mn)$ is the number of units of $\mathbf{Z}_{mn}$. In virtue of 3-2-17, the rings $\mathbf{Z}_{mn}$ and $\mathbf{Z}_m \oplus \mathbf{Z}_n$ are isomorphic, so according to 3-2-18 their units are in one-to-one correspondence. Hence, $\phi(mn)$ equals the number of units of $\mathbf{Z}_m \oplus \mathbf{Z}_n$—which number is denoted by $\#(\mathbf{Z}_m \oplus \mathbf{Z}_n)^*$. Now, 3-2-19 and 3-2-20 tell us that this number equals $\#(\mathbf{Z}_m^*)\#(\mathbf{Z}_n^*)$ (the product of the number of units of $\mathbf{Z}_m$ and the number of units of $\mathbf{Z}_n$) and this product equals $\phi(m)\phi(n)$. ▮

3-2-21. Examples. (1) If $m > 2$, then $\phi(m)$ is even.

Proof: If m is a power of 2, $m = 2^r$, then $r \geq 2$ because $m > 2$. Hence, $\phi(m) = \phi(2^r) = 2^{r-1}$, which is even. If m is not a power of 2 there exists an odd prime p that divides m, and we may write

$$m = p^r m' \qquad \text{where } (p, m') = 1, r \geq 1.$$

In other words, p^r is the power of p which appears in the factorization of m. Then

$$\phi(m) = \phi(m')\phi(p^r) = \phi(m')p^{r-1}(p-1).$$

Since $p - 1$ is even, so is $\phi(m)$.

(2) Find all m for which $\phi(m) = 14$.

Solution: The only possible prime factors of m are 2, 3, 5, 7, 11, 13; for if p^r is the power of a prime $p \geq 17$ which appears in the factorization of m, then

$$\phi(m) \geq \phi(p^r) = p^{r-1}(p-1) \geq 16.$$

Consequently, the prime-power factorization of m must look like

$$m = 2^r 3^s 5^t 7^u 11^v 13^w, \qquad r, s, t, u, v, w \geq 0.$$

Now, $\phi(13^w) = 13^{w-1}(13-1) = 13^{w-1}(12)$, which is not a factor of $\phi(m) = 14 = 2 \cdot 7$; hence, 13 is not a factor of m. Similarly $\phi(11^v) = 11^{v-1}(10)$, $\phi(7^u) = 7^{u-1}(6)$, $\phi(5^t) = 5^{t-1}(4)$—so 11, 7, 5 are not factors of m. Thus, m is reduced to the form $m = 2^r 3^s$. Furthermore, the form of $\phi(m) = 2 \cdot 7$ requires $r \leq 2$ and $s \leq 1$. One checks the few remaining possibilities and learns that there are no integers m for which $\phi(m) = 14$.

3-2-22. Exercise. The crucial step in the evaluation of $\phi(m)$ was the proof that ϕ is multiplicative. This was accomplished in 3-2-16, 3-2-17, 3-2-18, 3-2-19, and 3-2-20. Our discussion here will lead to a more elementary and more standard (but less incisive) proof that ϕ is multiplicative.

(1) A set of integers is called a **complete residue system** mod m if it contains exactly one element from each residue class mod m; we shall say that such a set is a $C(m)$.

Clearly, the set $\{0, 1, 2, 3, 4\}$ is a $C(5)$, $\{-3, -2, -1, 0, 1, 2, 3\}$ is a $C(7)$, and $\{-12, -4, 4, 13, 22, 82, 91\}$ is a $C(7)$.

(2) For a set S of integers, the following are equivalent:

(*i*) S is a $C(m)$.
(*ii*) Every integer is congruent (mod m) to exactly one element of S.
(*iii*) S has m elements, no two of which are congruent (mod m).
(*iv*) The elements of the set $\{\lfloor s \rfloor_m \mid s \in S\}$ are distinct, and this set equals $\mathbf{Z}_m$.

(3) Suppose $S = \{s_1, s_2, \ldots, s_m\}$ is a $C(m)$; then

(*i*) For every integer r, the set

$$S + r = \{s_1 + r, s_2 + r, \ldots, s_m + r\}$$

is a $C(m)$.

(*ii*) If q is relatively prime to m, the set

$$qS = \{qs_1, qs_2, \ldots, qs_m\}$$

is a $C(m)$.

(*iii*) If r is any integer and q is relatively prime to m, the set

$$qS + r = \{qs_1 + r, qs_2 + r, \ldots, qs_m + r\}$$

is a $C(m)$.

(4) A set of integers is called a **reduced residue system** mod m if it contains exactly one element from each residue class (mod m) which is prime to m, and no other elements; we shall say that such a set is an $R(m)$.

Clearly, $\{1, 2, 3, 4\}$ is an $R(5)$, $\{-3, -2, -1, +1, +2, +3\}$ is an $R(7)$, and $\{-12, 41, -29, 11\}$ is an $R(12)$.

(5) For a set S of integers, the following are equivalent:

(*i*) S is an $R(m)$.
(*ii*) Every integer a with $(a, m) = 1$ is congruent (mod m) to exactly one element of S.
(*iii*) S has $\phi(m)$ elements, all of which are relatively prime to m, and no two of which are congruent (mod m).
(*iv*) The elements of the set $\{\lfloor s \rfloor_m \mid s \in S\}$ are distinct, and this set equals $\mathbf{Z}_m{}^*$.

(6) Suppose $S = \{s_1, s_2, \ldots, s_{\phi(m)}\}$ is an $R(m)$; then

(*i*) If q is relatively prime to m, the set

$$qS = \{qs_1, qs_2, \ldots, qs_{\phi(m)}\}$$

is an $R(m)$.
(*ii*) If $\{t_1, t_2, \ldots, t_{\phi(m)}\}$ is also an $R(m)$, then

$$\prod_{i=1}^{\phi(m)} s_i \equiv \prod_{i=1}^{\phi(m)} t_i \pmod{m}.$$

(7) If a is any integer relatively prime to m, and $\{s_1, s_2, \ldots, s_{\phi(m)}\}$ is an $R(m)$, then

$$\prod_{i=1}^{\phi(m)} s_i \equiv \prod_{i=1}^{\phi(m)} (as_i) = a^{\phi(m)} \prod_{i=1}^{\phi(m)} s_i \pmod{m}$$

and it follows that

$$a^{\phi(m)} \equiv 1 \pmod{m}.$$

This is a famous result due to Euler (1707–1783).

In particular, if m is a prime p and a is an integer not divisible by p, then

$$a^{p-1} \equiv 1 \pmod{p}$$

and, in addition

$$a^p \equiv a \pmod{p}$$

[a result discovered by Fermat, which was proved earlier in 2-4-9].

(8) Suppose $(m, n) = 1$, $S = \{s_1, s_2, \ldots, s_{\phi(m)}\}$ is an $R(m)$, and $T = \{t_1, t_2, \ldots, t_{\phi(n)}\}$ is an $R(n)$. Consider the set

$$nS + mT = \{ns_i + mt_j \,|\, i = 1, \ldots, \phi(m), j = 1, \ldots, \phi(n)\}$$

which may also be written as

$$nS + mT = \{ns + nt \,|\, s \in S, t \in T\}.$$

(*i*) No two elements of $nS + mT$ are congruent (mod mn). In particular, the elements of $nS + mT$ are distinct and $\#(nS + mT) = \phi(m)\phi(n)$.

(*ii*) Every element of $nS + mT$ is relatively prime to m, and to n also—hence, it is prime to mn.

(*iii*) If a is an integer prime to mn, then a is congruent (mod mn) to some element of $nS + mT$.

Putting (*i*), (*ii*), (*iii*) together, we see that $nS + mT$ is an $R(mn)$, so it has $\phi(mn)$ elements. Consequently,

$$\phi(mn) = \phi(m)\phi(n).$$

3-2-23 / *PROBLEMS*

1. Are the following statements true or false? Justify your answers.

(*i*) $\lfloor 8 \rfloor_{24} \mid \lfloor 15 \rfloor_{24}$ in $\mathbf{Z}_{24}$,

(*ii*) $\lfloor 8 \rfloor_{29} \mid \lfloor 15 \rfloor_{29}$ in $\mathbf{Z}_{29}$,

(*iii*) $(3 - 2i) \mid (2 - 23i)$ in $\mathbf{Z}[i]$,

(*iv*) $(3 - 2\sqrt{2}) \mid (2 - 23\sqrt{2})$ in $\mathbf{Z}[\sqrt{2}]$,

(*v*) $(3 + 4\sqrt{2}) \mid (7 - 2\sqrt{2})$ in $\mathbf{Z}[\sqrt{2}]$,

(*vi*) $(3 + 4\sqrt{2}) \mid (7 - 2\sqrt{2})$ in $\mathbf{Q}[\sqrt{2}]$,

(*vii*) $(2 + \sqrt{6}) \mid (3 - 2\sqrt{6})$ in $\mathbf{Z}[\sqrt{6}]$,

(*viii*) $(2 + \sqrt{6}) \mid (3 - 2\sqrt{6})$ in $\mathbf{R}$.

2. Find all units of

(*i*) $\mathbf{Z}_6$, (*ii*) $\mathbf{Z}_{15}$, (*iii*) $\mathbf{Z}_{24}$.

3. Find all units of

(*i*) $\mathbf{Z}[\sqrt{-2}]$, (*ii*) $\mathbf{Z}[\sqrt{-5}]$, (*iii*) $\mathbf{Q}[\sqrt{-5}]$.

4. We know that $\mathbf{Z}_{13}$ is a field; find the multiplicative inverse of every nonzero element.

5. If m is composite, show that $\mathbf{Z}_m$ is not a field.

6. (*i*) Give an example of a nonzero element of a commutative ring with unity which is neither a unit nor a zero-divisor.
(*ii*) In $\mathbf{Z}_{15}$, verify that every nonzero element is either a unit or a zero-divisor.
(*iii*) Prove that a nonzero element of any residue class ring $\mathbf{Z}_m$ is either a unit or a zero-divisor (but not both).

7. Prove that $\mathbf{Q}[\sqrt{2}]$ is a field. What about $\mathbf{Q}[\sqrt{m}]$, where m is any square-free integer, positive or negative?

8. In an integral domain D, show that both $a \mid b$ and $b \mid a$ if and only if $a = bu$ where u is a unit of D.

9. Find as many units as you can in

(*i*) $\mathbf{Z}[\sqrt{3}]$, (*ii*) $\mathbf{Z}[\sqrt{6}]$,
(*iii*) $\mathbf{Z}[\sqrt{7}]$, (*iv*) $\mathbf{Z}[\sqrt{10}]$.

10. Which of the following are units of $\mathbf{Z}_{851}$?

(*i*) $\lfloor 28 \rfloor_{851}$, (*ii*) $\lfloor 115 \rfloor_{851}$, (*iii*) $\lfloor 111 \rfloor_{851}$, (*iv*) $\lfloor 100 \rfloor_{851}$.

For each one that is a unit, find its inverse.

11. Prove the rules for exponents listed in 3-2-5, part (9) for the units $a, b \in R$. To what extent are they true if a and b are not units?

12. We have seen in 3-2-5, part (8) that the product of two units is a unit. Make a multiplication table for the units of:
(*i*) $\mathbf{Z}_5$, (*ii*) $\mathbf{Z}_6$, (*iii*) $\mathbf{Z}_7$, (*iv*) $\mathbf{Z}_{10}$,
(*v*) $\mathbf{Z}_{11}$, (*vi*) $\mathbf{Z}_{12}$, (*vii*) $\mathbf{Z}_{14}$, (*viii*) $\mathbf{Z}_{15}$.

13. Evaluate $\phi(m)$ when m is
(*i*) 11^{15}, (*ii*) $2^3 \cdot 7^4 \cdot 13^2 \cdot 17 \cdot 67^3$, (*iii*) 18,900,
(*iv*) 16,807, (*v*) 305,453.

14. Show that $\phi(m) = \phi(2m) \Leftrightarrow m$ is odd.

15. Find the smallest positive even integer a for which $\phi(m) = a$ has no solution (that is, for which no such m exists).

16. Derive or justify the expression

$$\phi(m) = \prod_{p \mid m} (p^{\nu_p(m)} - p^{\nu_p(m)-1}).$$

17. There are 10 integers m for which $\phi(m) = 24 = 2^3 \cdot 3$; find them.

18. (*i*) If A and B are finite sets, then $\#(A \times B) = (\#A)(\#B)$.
(*ii*) If $A_2 \subset A_1$ and $B_2 \subset B_1$ are all finite sets, then

$$\begin{aligned}&\#[(A_1 - A_2) \times (B_1 - B_2)] \\ &\quad = (\#A_1)(\#B_1) - (\#A_1)(\#B_2) - (\#A_2)(\#B_1) + (\#A_2)(\#B_2).\end{aligned}$$

19. (*i*) If $(m, n) = 1$, then for any $a \in \mathbf{Z}$

$$\lfloor a \rfloor_m \cap \lfloor a \rfloor_n = \lfloor a \rfloor_{mn}, \qquad \text{(as sets)}.$$

(*ii*) What happens if m and n are not relatively prime?

20. For integers $m_1, m_2, \ldots, m_r$, all greater than 1, show that the map

$$a \longrightarrow (\lfloor a \rfloor_{m_1}, \lfloor a \rfloor_{m_2}, \ldots, \lfloor a \rfloor_{m_r})$$

of

$$\mathbf{Z} \longrightarrow \mathbf{Z}_{m_1} \oplus \mathbf{Z}_{m_2} \oplus \cdots \oplus \mathbf{Z}_{m_r}$$

is a homomorphism. What is its kernel? When is it surjective?

21. (*i*) In $\mathbf{Z}_7$, compute α^6 for each $\alpha \neq 0$.
(*ii*) In $\mathbf{Z}_{12}$, compute α^4 for $\alpha = \lfloor 1 \rfloor_{12}, \lfloor 5 \rfloor_{12}, \lfloor 7 \rfloor_{12}, \lfloor 11 \rfloor_{12}$.
(*iii*) For each $\alpha \neq 0$ in $\mathbf{Z}_{13}$ compute α^{12}.
Why do the answers come out as they do?

22. Is the congruence

$$11^{7 \cdot 17 \cdot 37} \equiv 11 \pmod{7 \cdot 17 \cdot 37}$$

valid? How is this related to the result of Fermat, 2-4-9?

23. Suppose R is a $C(m)$, S is a $C(n)$, and $(m, n) = 1$. Show that $nR + mS$ is a $C(mn)$.

24. For each of the following values of m compute $\Sigma_{d|m}\, \phi(d)$; that is, compute the sum of all $\phi(d)$ as d runs over all positive divisors of m. (Note that 1 is always counted as a divisor of m, and so is m.)
(*i*) 24, (*ii*) 27, (*iii*) 49, (*iv*) 50, (*v*) 60.

25. (*i*) Prove that if p is prime and $r \geq 1$, then

$$\sum_{i=0}^{r} \phi(p^i) = p^r.$$

(*ii*) For arbitrary $m > 1$ prove that

$$\sum_{d \mid m} \phi(d) = m$$

by using an induction based on writing $m = p^r n$ where p is prime and $(p, n) = 1$.

(*iii*) Give an alternate proof of (*ii*) by showing that for each divisor d of m, the number of elements from the set $\{1, 2, \ldots, m\}$ which have d as their greatest common divisor with m is precisely $\phi(m/d)$.

26. Is a subring of a field a field? Is it a domain? Explain.

27. (*i*) Suppose $f: R \to S$ is an isomorphism of rings. If the element $a \in R$ has an inverse for multiplication, so does $f(a) \in S$—and $f(a^{-1}) = (f(a))^{-1}$. Does this assertion apply when $f: R \to S$ is a surjective homomorphism?

(*ii*) Prove: If R and S are isomorphic rings, then R is a domain $\Leftrightarrow$ S is a domain, and R is a field $\Leftrightarrow$ S is a field.

28. Let α be a formal symbol, and consider

$$\mathbf{Z}_2[\alpha] = \{a + b\alpha \mid a, b \in \mathbf{Z}_2\}.$$

Thus, $\mathbf{Z}_2[\alpha]$ consists of the four elements $0 = 0 + 0\alpha$, $1 = 1 + 0\alpha$, $\alpha = 0 + 1\,\alpha$, and $1 + \alpha$. Define addition and multiplication in $\mathbf{Z}_2[\alpha]$ just as for polynomials except that we write $\alpha^2 = 1 + \alpha$. Make addition and multiplication tables for $\mathbf{Z}_2[\alpha]$, and show that it is a field with four elements.

3-3. Polynomials and Polynomial Functions

Before we discuss solving expressions like

$$a_0 + a_1 x + a_2 x^2 + \cdots + a_n x^n = 0$$

in certain special situations, it is appropriate to examine in some detail what is meant by such an expression. This requires a good deal of fussing, and is not very exciting—but it needs to be done.

3-3-1. Notation. Throughout this section we suppose $\{R, \perp, \cdot\}$ is a commutative ring with unity. The elements of R will be denoted by $a, b, c, \ldots$. In addition, henceforth, we shall denote the unity element of R by 1 instead of e; this will be a small convenience and should cause no difficulty. Note the choice of $\perp$ to denote addition in R. The more usual symbol $+$ is reserved temporarily for something else.

Let x be a symbol which has no connection of any kind with R; in particular, x is not an element of R. Such a formal symbol x is called an **indeterminate** over R or a **transcendental** over R. Consider all "formal" symbols or expressions of form

$$a_0 x^0 + a_1 x^1 + a_2 x^2 + \cdots + a_n x^n \qquad (*)$$

where n is an integer greater than or equal to 0 and $a_0, a_1, a_2, \ldots, a_n$ are elements of R. Any expression of form (∗) is called a **polynomial in x with coefficients in R**, or in abbreviated terminology, a **polynomial in x over R**. [Of course, it is more common to write polynomials in the form

$$a_0 + a_1 x + a_2 x^2 + \cdots + a_n x^n$$

and we shall eventually do so; however, at this stage, there are certain small technical advantages in using the notation of (∗).] The set of all such polynomials, for all choices of $n \geq 0$ and all choices of $a_0, a_1, \ldots, a_n$ in R, is denoted by $R[x]$. Our immediate objective is to show how $R[x]$ can be made into a ring—but first a number of preliminary remarks are in order.

Note first that the notation for a polynomial is somewhat deceptive. The symbol + has nothing to do with addition; at this point, it is merely a symbol that provides separation between parts of a polynomial—any other symbol could serve just as well. The symbols $x^0, x^1, x^2, \ldots, x^n$ which should be read as x upper zero, x upper 1, x upper 2, ..., x upper n, have nothing to do with powers of x (after all, since x is not in R, multiplying x by itself has no meaning), they are just formal symbols. Furthermore, pieces of a polynomial like $a_0 x^0, a_1 x^1, \ldots, a_n x^n$ have no connection with multiplication (after all, we have no way of multiplying an element $a_i \in R$ and a formal symbol x^i which has no connection with R), we simply prefer to write things in this way.

One may naturally ask: if the symbols are not what they seem to be, why not put the notation in another, less suggestive, form? The answer is that, in the end, all our symbols will behave exactly as the notation leads us to expect.

3-3-2. Conventions. By a **term** of the polynomial

$$a_0 x^0 + a_1 x^1 + a_2 x^2 + \cdots + a_n x^n \qquad (*)$$

we mean any one of the symbols $a_0 x^0, a_1 x^1, a_2 x^2, \ldots, a_n x^n$. As indicated earlier, the elements $a_0, a_1, a_2, \ldots, a_n$ of R should be referred to as the **coefficients** of the polynomial, or of the respective terms. The term $a_n x^n$ is called the **leading term** and its coefficient a_n is known as the **leading coefficient** of the polynomial. It is customary (but not essential) to arrange the notation so that $a_n \neq 0$. If $a_n = 1$, the polynomial is said to be **monic**.

According to the definition, an element of $\mathbf{Z}[x]$—that is, a polynomial in x over $\mathbf{Z}$—should look something like $2x^0 + 3x^1 + 7x^2 + 5x^3 + 4x^4$. On the other hand, in high school and elsewhere, we often came upon polynomials that looked like $3x^6 - 2x^4 + x^2 - 1$, or $5x^7$, or $x^8 - x^2 - 2x^3 + 3x^4$. These differ from our formal definition of polynomial in several ways—for example, the use of the − sign, the appearance of an unadorned element of $\mathbf{Z}$ (without an x^i), writing a term like x^2 without a coefficient, putting terms in the wrong order, or omitting some terms. The apparent discrepancies will be ironed out eventually.

A typical element of $R[x]$ will be denoted by $f(x)$—that is,

$$f(x) = a_0 x^0 + a_1 x^1 + a_2 x^2 + \cdots + a_n x^n$$

and this expression may also be written compactly as

$$f(x) = \sum_{i=0}^{n} a_i x^i .$$

Note that $f(x)$ is **not** a function; it represents a polynomial and nothing else. Also, it is not quite appropriate to refer to Σ as "sum" or "summation," since + does not represent addition; it is probably best at this point to read the symbol Σ as "sigma."

Now, suppose $g(x) \in R[x]$ is another polynomial—say

$$g(x) = b_0 x^0 + b_1 x^1 + b_2 x^2 + \cdots + b_m x^m = \sum_{i=0}^{m} b_i x^i .$$

We shall say that the polynomials $f(x)$ and $g(x)$ are **equal**, and write $f(x) = g(x)$, when $a_i = b_i$ for all i. Strictly speaking this definition has meaning only when $m = n$; but when $m \neq n$, say $m < n$, it is to be understood that we are taking $b_{m+1} = b_{m+2} = \cdots = b_n = 0$. Since we have permitted the introduction of terms with 0 coefficient without affecting the polynomial, it is only natural that we should also permit the dropping of terms with 0 coefficient (even if they are in the middle) without affecting the polynomial. Thus, for example, the polynomials $a_0 x^0 + a_3 x^3 + a_5 x^5$ and $a_0 x^0 + 0x^1 + 0x^2 + a_3 x^3 + 0x^4 + a_5 x^5$ are considered to be equal. In general, any missing term is considered to have coefficient 0. Note that any term $a_i x^i$ of a polynomial thus becomes a polynomial (that is, an element of $R[x]$) in its own right—namely,

$$a_i x^i = 0x^0 + 0x^1 + \cdots + 0x^{i-1} + a_i x^i .$$

Once we have agreed about throwing in or throwing out terms with coefficient 0, why not go all the way? Instead of writing a generic polynomial as

$$f(x) = a_0 x^0 + a_1 x^1 + \cdots + a_n x^n = \sum_{i=0}^{n} a_i x^i \qquad \text{for some } n \geq 0,$$

let us write it in the form

$$f(x) = \sum_{i=0}^{\infty} a_i x^i$$

with the understanding that almost all a_i equal 0 (that is, only a finite number of the coefficients a_i are not equal to 0). Consequently, if $g(x) = \sum_{i=0}^{\infty} b_i x^i$ is also a polynomial in x over R, then our definition of equality for polynomials takes the form

$$f(x) = g(x) \Leftrightarrow a_i = b_i, \text{ for } i = 0, 1, 2, \ldots.$$

Now, let us turn to addition and multiplication of polynomials. From our high school experience, this is a familiar matter. For example, using the standard high school notation: If $f(x)$ and $g(x)$ are the polynomials with integer coefficients

$$f(x) = 2 + 3x^2 + x^3, \qquad g(x) = 1 - x + 2x^4,$$

then these may be rewritten as

$$\begin{aligned} f(x) &= 2 + 0x + 3x^2 + 1x^3 + 0x^4, \\ g(x) &= 1 + (-1)x + 0x^2 + 0x^3 + 2x^4, \end{aligned}$$

and then

$$\begin{aligned} f(x) + g(x) &= (2 + 1) + (0 + (-1))x + (3 + 0)x^2 + (1 + 0)x^3 + (0 + 2)x^4 \\ &= 3 - x + 3x^2 + x^3 + 2x^4 \end{aligned}$$

and

$$\begin{aligned} f(x) \cdot g(x) = {} & [(2)(1)] + [(2)(-1) + (0)(1)]x + [(2)(0) + (0)(-1) + (3)(1)]x^2 \\ & + [(2)(0) + (0)(0) + (3)(-1) + (1)(1)]x^3 \\ & + [(2)(2) + (0)(0) + (3)(0) + (1)(-1) + (0)(1)]x^4 \\ & + [(0)(2) + (3)(0) + (1)(0) + (0)(-1)]x^5 \\ & + [(3)(2) + (1)(0) + (0)(0)]x^6 \\ & + [(1)(2) + (0)(0)]x^7 + [(0)(2)]x^8 \\ = {} & 2 - 2x + 3x^2 - 2x^3 + 3x^4 + 6x^6 + 2x^7 \end{aligned}$$

It is precisely these familiar rules for addition and multiplication of polynomials which we carry over to $R[x]$.

In detail: Suppose $\{R, \perp, \cdot\}$ is a commutative ring with unity and consider any polynomials

$$f(x) = \sum_0^\infty a_i x^i, \quad g(x) = \sum_0^\infty b_i x^i$$

in $R[x]$. We define their sum $f(x) \oplus g(x)$ by

$$\left(\sum_0^\infty a_i x^i\right) \oplus \left(\sum_0^\infty b_i x^i\right) = \sum_0^\infty (a_i \perp b_i) x^i.$$

In other words, as expected, we have put $f(x) \oplus g(x) = \Sigma_0^\infty c_i x^i$ where $c_i = a_i \perp b_i$ for each $i = 0, 1, 2, \ldots$. (Note that since a_i, b_i come from R, their sum is denoted by $a_i \perp b_i$.) Of course, $c_i = a_i \perp b_i = 0$ for almost all i, because a_i equals 0 for almost all i and so does b_i; thus $\Sigma_0^\infty (a_i \perp b_i)x^i$ has only a finite number of nonzero coefficients, and it is indeed a polynomial in x over R.

The definition of the product $f(x) \odot g(x)$ is somewhat more complicated than the definition of the sum $f(x) \oplus g(x)$; but again, as for $\oplus$, it is entirely in keeping with our high school procedures—namely,

$$\left(\sum_0^\infty a_i x^i\right) \odot \left(\sum_0^\infty b_i x^i\right) = \sum_0^\infty c_i x^i$$

where $c_0 = a_0 \cdot b_0$, $c_1 = a_0 \cdot b_1 \perp a_1 \cdot b_0$, $c_2 = a_0 \cdot b_2 \perp a_1 \cdot b_1 \perp a_2 \cdot b_0$, $c_3 = a_0 \cdot b_3 \perp a_1 \cdot b_2 \perp a_2 \cdot b_1 \perp a_3 \cdot b_0, \ldots$, and in general, for any $i = 0, 1, 2, 3, \ldots$

$$c_i = a_0 \cdot b_i \perp a_1 \cdot b_{i-1} \perp \cdots \perp a_{i-1} \cdot b_1 \perp a_i \cdot b_0 = \sum_{r=0}^{i} a_r \cdot b_{i-r}.$$

Clearly, c_i is the sum of all possible products $a_r \cdot b_s$ where $r + s = i$, so it may be written as

$$c_i = \sum_{r+s=i} a_r \cdot b_s, \qquad 0 \leq r, s \leq i.$$

(Here, Σ denotes summation in R and, of course, $\cdot$ denotes multiplication in R. We have used the symbols $\oplus$ and $\odot$ for addition and multiplication in $R[x]$ because the symbols $\perp$, $\cdot$, $+$ already have meaning, and we prefer to maintain the distinction between the new operations and the old ones.) We must verify that $f(x) \odot g(x) = \Sigma_0^\infty c_i x^i$ is really an element of $R[x]$—that is, that $c_i = 0$ for almost all i. To do this, let a_n be the last nonzero coefficient of $f(x)$ and let b_m be the last nonzero coefficient of $g(x)$. Thus, $a_i = 0$ for all $i > n$, and $b_i = 0$ for all $i > m$; so we may write

$$f(x) = \sum_0^n a_i x^i, \qquad g(x) = \sum_0^m b_i x^i.$$

Consequently, in virtue of our high school experience, we surely expect c_i to equal 0 for every $i > n + m$. In fact, when $i > n + m$ is fixed, each of the terms $a_r \cdot b_s$ which appear in $c_i = \Sigma_{r+s=i} a_r \cdot b_s$ must be 0. To see this, note that if $i > n + m$ and $r + s = i$ with $0 \leq r, s \leq i$, then either $r > n$ or $s > m$ or both—for if $r \leq n$ and $s \leq m$, then $i = r + s \leq n + m$—so, by the way n and m are chosen, in each term $a_r \cdot b_s$ either $a_r = 0$ or $b_s = 0$ or both. Hence, $c_i = 0$ for $i > n + m$ and $f(x) \odot g(x)$ is an element of $R[x]$ which may also be written in the form

$$f(x) \odot g(x) = \sum_0^{n+m} c_i x^i.$$

It is worth noting, once more, that the multiplication we have defined in $R[x]$ is entirely consistent with the way the reader has always multiplied polynomials, as illustrated above for $f(x) = 2 + 3x^2 + x^3$ and $g(x) = 1 - x + 2x^4$ in $\mathbf{Z}[x]$.

3-3-3. Theorem. If R is a commutative ring with unity, then $R[x]$ is a commutative ring with unity which contains an isomorphic copy of R. More precisely, there is a "natural" injective homomorphism of the ring $\{R, \perp, \cdot\}$ into the ring $\{R[x], \oplus, \odot\}$.

Proof: First of all, we need to verify that $\{R[x], \oplus, \odot\}$ is a commutative ring with unity. Closure has already been proved for both $\oplus$ and $\odot$. The associative law for addition holds: For if $f(x) = \Sigma_0^\infty a_i x^i$, $g(x) = \Sigma_0^\infty b_i x^i$, $h(x) = \Sigma_0^\infty c_i x^i$ are any elements of $R[x]$, then, in virtue of the associative law for $\perp$ in R, it is easy to see that both polynomials $(f(x) \oplus g(x)) \oplus h(x)$ and $f(x) \oplus (g(x) \oplus h(x))$ are equal to $\Sigma_0^\infty (a_i \perp b_i \perp c_i)x^i$. The polynomial $\Sigma_0^\infty 0x^i = 0x^0 + 0x^1 + 0x^2 + \cdots$ is clearly an identity for addition—we denote it by 0. In addition, the polynomial $\Sigma_0^\infty (-a_i)x^i$ is surely an additive inverse for $f(x) = \Sigma_0^\infty a_i x^i$—so we shall write $-(f(x))$ for $\Sigma_0^\infty (-a_i)x^i$. As for the commutative law for addition, for any i the coefficient of x^i in $f(x) \oplus g(x)$ is $a_i \perp b_i$, which is equal to $b_i \perp a_i$, the coefficient of x^i in $g(x) \oplus f(x)$.

Turning to the requirements for $\odot$ in $R[x]$, we note first that the polynomial

$$1x^0 = 1x^0 + 0x^1 + 0x^2 + 0x^3 + \cdots$$

is obviously an identity element for multiplication. The commutative law for $\odot$ holds because, for each i, the coefficient of x^i in $f(x) \odot g(x)$ is

$$a_0 \cdot b_i \perp a_1 \cdot b_{i-1} \perp \cdots \perp a_{i-1} \cdot b_1 \perp a_i \cdot b_0$$

while the coefficient of x^i in $g(x) \odot f(x)$ is

$$b_0 \cdot a_i \perp b_1 \cdot a_{i-1} \perp \cdots \perp b_{i-1} \cdot a_1 \perp b_i \cdot a_0$$

—and these are equal because both addition and multiplication in R are commutative. To prove the validity of the associative law for $\odot$, we must show that, for each $i \geq 0$, the coefficient of x^i in $(f(x) \odot g(x)) \odot h(x)$ is equal to the coefficient of x^i in $f(x) \odot (g(x) \odot h(x))$. By being careful about subscripts, it may be shown that in each case the coefficient of x^i turns out to be

$$\sum_{r+s+t=i} a_r \cdot b_s \cdot c_t, \qquad 0 \leq r, s, t \leq i$$

(that is, the sum of all products of form $a_r \cdot b_s \cdot c_t$ where the sum of the subscripts is i). The details are left to the reader.

Finally, it is only necessary to check one distributive law, because $\odot$ is commutative. This is a straightforward matter which may safely be left to the reader. The proof that $\{R[x], \oplus, \odot\}$ is a commutative ring with unity is now complete, so we turn to the "imbedding" or "injection" of R into $R[x]$.

Let us define the mapping $\phi\colon R \to R[x]$ by

$$\phi(a) = ax^0, \qquad a \in R.$$

In other words, the mapping ϕ takes the element $a \in R$ to the polynomial $ax^0 = ax^0 + 0x^1 + 0x^2 + \cdots$ in $R[x]$. The verification that ϕ is a homomorphism is easy; in fact, for $a, b \in R$ we have

$$\phi(a \perp b) = (a \perp b)x^0 = (ax^0) \oplus (bx^0) = \phi(a) \oplus \phi(b),$$
$$\phi(a \cdot b) = (a \cdot b)x^0 = (ax^0) \odot (bx^0) = \phi(a) \odot \phi(b).$$

(Note the distinction which must be made between the operations $\perp$ and $\cdot$ in R as opposed to $\oplus$ and $\odot$ in $R[x]$.) Since ϕ is a homomorphism, we know (from 2-5-8) that $\phi(0) = 0$ or, what is the same thing, $0 \in \ker \phi$. Furthermore, if $a \in \ker \phi$, then $\phi(a) = 0$, which says that $ax^0 = ax^0 + 0x^1 + 0x^2 + \cdots$ is the zero polynomial—hence, $a = 0$. Consequently, $\ker \phi = (0)$ and according to 2-5-9, ϕ is injective. Thus, ϕ is a ("natural") injective homomorphism of the ring $\{R, \perp, \cdot\}$ into the ring $\{R[x], \oplus, \odot\}$.

Now, making use of parts (1) and (4) of 2-5-9, the image

$$\phi(R) = \{ax^0 \mid a \in R\}$$

of the homomorphism ϕ is a subring of $\{R[x], \oplus, \odot\}$ and, because ϕ is injective, ϕ is an isomorphism of R onto $\phi(R)$. In particular, $\phi(R)$ (or better, $\{\phi(R), \oplus, \odot\}$) is an isomorphic copy of R which is contained in $R[x]$, and it is also customary to say that ϕ is an "imbedding" of R into $R[x]$. ∎

3-3-4. Notation. Having proved that $\{R[x], \oplus, \odot\}$ is a ring, let us sketch how, and with what justification, the notation may be adjusted to make it conform with the standard notation.

For $i \geq 0$, let us write $1x^i$ as x^i. Furthermore, let us denote x^1 by x. Each x^i may be viewed as representing the term $1x^i$ in some polynomial, and also as the polynomial $0x^0 + 0x^1 + \cdots + 1x^i + 0x^{i+1} + \cdots$. In particular, x may be viewed as an element of $R[x]$, namely, $x = 0x^0 + 1x^1$. Once x belongs to the ring $R[x]$, we may multiply it by itself—thus

$$x \odot x = (0x^0 + 1x^1) \odot (0x^0 + 1x^1) = 0x^0 + 0x^1 + 1x^2 = 1x^2 = x^2$$

and, more generally, for all $i \geq 2$ we have, by induction

$$\begin{aligned} \underbrace{x \odot x \odot \cdots \odot x}_{i \text{ terms}} &= (0x^0 + 1x^1) \odot (0x^0 + 1x^1) \odot \cdots \odot (0x^0 + 1x^1) \\ &= 0x^0 + 0x^1 + \cdots + 0x^{i-1} + 1x^i \\ &= 1x^i \\ &= x^i. \end{aligned}$$

According to this, x^i is what it looks like! More precisely, the result of multiplying i copies of x in the ring $R[x]$ is x^i, which means x to the power i; on the other hand, heretofore in this section we have been using x^i solely as a formal symbol, called "x upper i." In virtue of the above, the formal symbol x^i is to be interpreted as the ith power of x for all $i \geq 1$.

We observe next that the symbol x^0 has two possible meanings. On the one hand, it is x upper zero which is the same as the element $1x^0$ of $R[x]$. On the other hand, x^0 can refer to raising the element x of $R[x]$ to the zero power. Since in a ring with identity the zero power of any nonzero element is the identity, we see that x^0 (that is, x to the power zero) equals $1x^0$, the identity for multiplication in the ring $R[x]$. Thus, the two interpretations of x^0 give the same result, and our notations are still compatible.

We have already proved that the ring R is isomorphic to the subring $\phi(R) = \{ax^0 \mid a \in R\}$ of $R[x]$, so let us replace the elements of $\phi(R)$ by the corresponding elements of R. Thus, we denote the polynomial ax^0 by a, and in writing polynomials we simply omit the x^0. An arbitrary polynomial now takes the familiar form

$$a_0 + a_1 x + a_2 x^2 + \cdots + a_n x^n. \tag{$*$}$$

If a is any element of R, then a also represents the polynomial $a + 0x + 0x^2 + \cdots$ in $R[x]$. Consequently, R may be considered as a subset of $R[x]$—that is, $R \subset R[x]$.

Now, consider any element $a_i \in R$ and any power x^i. Both of these are elements of $R[x]$, so it is of interest to find their product (with respect of $\odot$, of course). We have

$$\begin{aligned} a_i \odot x^i = a_i x^0 \odot 1x^i &= (a_i x^0 + 0x^1 + \cdots) \odot (0x^0 + 0x^1 + \cdots + 1x^i) \\ &= 0x^0 + 0x^1 + \cdots + (a_i \cdot 1)x^i \\ &= a_i x^i \end{aligned}$$

which tells us that the polynomial $a_i x^i$ in $R[x]$ (or the term $a_i x^i$ of some polynomial) is exactly what it looks like—it represents the product in $R[x]$ of the two elements a_i and x^i of $R[x]$.

Given two polynomials (or rather, terms) $a_i x^i$ and $a_j x^j$ where $i < j$, let us compute their sum (with respect to $\oplus$, of course). We have

$$\begin{aligned} a_i x^i \oplus a_j x^j &= (0 + 0x + \cdots + a_i x^i + \cdots) \oplus (0 + 0x + \cdots + a_j x^j + \cdots) \\ &= 0 + 0x + \cdots + a_i x^i + \cdots + a_j x^j + \cdots \\ &= a_i x^i + a_j x^j \end{aligned}$$

—so it follows inductively that for any polynomial $(*)$ of $R[x]$ we have

$$a_0 + a_1 x + a_2 x^2 + a_n x^n = a_0 \oplus a_1 x \oplus a_2 x^2 \oplus \cdots \oplus a_n x^n.$$

Thus, the formal symbol + may be replaced by $\oplus$ (the symbol for addition in $R[x]$) whenever it appears, and our polynomial is built up from the element x and the elements $a_1, \ldots, a_n$ of R (all of which belong to $R[x]$) by repeated addition, $\oplus$, and multiplication, $\odot$. Furthermore, because $\oplus$ is a commutative operation, the order in which the terms of a polynomial $a_0 \oplus a_1 x \oplus \cdots \oplus a_n x^n$ (which is also the same as $a_0 \oplus a_1 \odot x \oplus \cdots \oplus a_n \odot x^n$) appear does not matter.

It remains to examine the imbedding of R in $R[x]$. We have agreed to use the symbol a to represent the polynomial $ax^0 = ax^0 + 0x + \cdots$ (such a polynomial is known as a **constant** polynomial, and, in general, the term a_0 of the polynomial $a_0 + a_1 x + a_2 x^2 + \cdots + a_n x^n$ is known as the **constant term**); the justification being the fact that the rings $\{R, \perp, \cdot\}$ and $\{\phi(R), \oplus, \odot\}$ (which is a subring of $\{R[x], \oplus, \odot\}$) are isomorphic. Any elements $a, b \in R$ may be viewed as $a = ax^0$, $b = bx^0$ in $R[x]$, so adding them with respect to the operation $\oplus$ in $R[x]$, we have

$$a \oplus b = ax^0 \oplus bx^0 = (a \perp b)x^0 = a \perp b$$

This means that, on elements of R, the operations $\oplus$ and $\perp$ are the "same." Similarly, the operations $\odot$ and $\cdot$ are the same on elements of R, since

$$a \odot b = ax^0 \odot bx^0 = (a \cdot b)x^0 = a \cdot b.$$

Therefore, the operations $\perp$ and $\cdot$ may be replaced by $\oplus$ and $\odot$, respectively, whenever they appear, and everything remains compatible. In particular, the ring $\{R, \perp, \cdot\}$ is now written as $\{R, \oplus, \odot\}$, and it is clearly a subring of $\{R[x], \oplus, \odot\}$. The symbols $\perp$, $\cdot$, and + have been dropped completely!

If one insists on having the notation conform exactly with the familiar notation, it is only necessary to replace $\oplus$ by the symbol + and $\odot$ by the symbol $\cdot$; thereby concluding with the ring $\{R[x], +, \cdot\}$ and the subring $\{R, +, \cdot\}$. In any event, the $-$ sign represents the usual additive inverse in a ring; so, for example, the term $-(a_i x^i)$ represents $(-a_i)x^i$, and $a_0 - a_1 x - a_2 x^2 + a_3 x^3$ is another way of writing $a_0 + (-a_1)x + (-a_2)x^2 + a_3 x^3$. This completes our discussion of the notation in a polynomial ring.

Given any polynomial $f(x) = a_0 + a_1 x + \cdots + a_n x^n \in R[x]$ with $a_n \neq 0$ we define its **degree** [which is written as $\deg f(x)$] to be n. In other words, for a polynomial $f(x) = \Sigma_{i=0}^{\infty} a_i x^i$, its degree is the highest power or exponent of x which has a nonzero coefficient. (For example, the polynomial $2 - 3x + x^2 - 6x^4 \in \mathbf{Z}[x]$ is of degree 4; however, if this same polynomial is viewed as a polynomial in $\mathbf{Z}_3[x]$, then its degree is 2.) It is understood that a constant polynomial—meaning one of form $f(x) = a$—has degree 0, provided $a \neq 0$. This leaves the zero polynomial $f(x) = 0 = 0 + 0x + 0x^2 + \cdots$; we define its degree to be $-\infty$, where it is taken for granted that $-\infty$ is less than any integer, and $-\infty$ plus any integer equals $-\infty$.

3-3-5. Proposition. Suppose any two polynomials $f(x)$ and $g(x)$ in $R[x]$ are given; then

(*i*) $\deg(f(x) \pm g(x)) \leq \max\{\deg f(x), \deg g(x)\}$.
Moreover, if $\deg f(x) \neq \deg g(x)$, then equality holds.

(*ii*) $\deg(f(x) \cdot g(x)) \leq \deg f(x) + \deg g(x)$.
Moreover, if the leading coefficient (by which we mean the last nonzero coefficient) of just one of these polynomials is not a zero-divisor, then equality holds.

Proof: This is really a familiar result from high school, except for some slight complications which may be caused by the nature of the ring R.

If either $f(x) = 0$ or $g(x) = 0$ or both, it is easy to see that both (*i*) and (*ii*) hold. The details, which depend on our conventions about $-\infty$, may safely be left to the reader.

Suppose, therefore, that $f(x) \neq 0$ and $g(x) \neq 0$; so we have $f(x) = a_0 + a_1 x + \cdots + a_n x^n$ with $a_n \neq 0$ and $g(x) = b_0 + b_1 x + \cdots + b_m x^m$ with $b_m \neq 0$. In particular, $\deg f(x) = n \geq 0$ and $\deg g(x) = m \geq 0$. If $\deg f(x) \neq \deg g(x)$, there is nothing lost in assuming that $\deg f(x) < \deg g(x)$; thus $n < m$ and the leading term of $f(x) \pm g(x)$ is $\pm b_m x^m$—so

$$\deg(f(x) \pm g(x)) = m = \max\{\deg f(x), \deg g(x)\}$$

which proves the second part of (*i*). If, however, $\deg f(x) = \deg g(x)$, then $m = n$ and

$$f(x) \pm g(x) = (a_0 \pm b_0) + (a_1 \pm b_1)x + \cdots + (a_m \pm b_m)x^m.$$

Unfortunately, it may well be that $a_m \pm b_m = 0$, so the only conclusion we can draw is

$$\deg(f(x) \pm g(x)) \leq m = \max\{\deg f(x), \deg g(x)\}.$$

This completes the proof of (*i*).

As for (*ii*), surely $f(x) \cdot g(x)$ has no term after $a_n b_m x^{n+m}$ whose coefficient is not equal to 0. Now, it may well be that $a_n b_m = 0$, so the only conclusion we can draw is

$$\deg(f(x) \cdot g(x)) \leq n + m = \deg f(x) + \deg g(x).$$

However, $a_n b_m = 0$ cannot occur if either of the factors is not a zero-divisor, so in this case $\deg(f(x) \cdot g(x)) = n + m$ and equality holds. This completes the proof of (*ii*). ▮

To illustrate this result, consider $f(x) = 1 - x + 6x^3$, $g(x) = 2 + x + 18x^3$, $h(x) = 1 + 12x^2$. When these are viewed as polynomials in $\mathbf{Z}[x]$, we have

$$\begin{array}{ccc} \deg f(x) = 3, & \deg g(x) = 3, & \deg h(x) = 2, \\ \deg(f(x) + g(x)) = 3, & \deg(g(x) + h(x)) = 3, & \deg(f(x) + h(x)) = 3, \\ \deg(f(x) \cdot g(x)) = 6, & \deg(g(x) \cdot h(x)) = 5, & \deg(f(x) \cdot h(x)) = 5. \end{array}$$

On the other hand, these polynomials can also be viewed in $\mathbf{Z}_{24}[x]$. The degrees of $f(x)$, $g(x)$, $h(x)$ do not change, but in $\mathbf{Z}_{24}[x]$ we have

$$\begin{aligned} f(x) + g(x) &= 3, \\ g(x) + h(x) &= 3 + x + 12x^2 + 18x^3, \\ f(x) + h(x) &= 2 - x + 12x^2 + 6x^3, \\ f(x) \cdot g(x) &= 2 - x - x^2 + 6x^3 - 12x^4 + 12x^6, \\ g(x) \cdot h(x) &= 2 + x + 6x^3, \\ f(x) \cdot h(x) &= 1 - x + 12x^2 - 6x^3, \end{aligned}$$

so that

$$\begin{array}{ccc} \deg(f(x) + g(x)) = 0, & \deg(g(x) + h(x)) = 3, & \deg(f(x) + h(x)) = 3, \\ \deg(f(x) \cdot g(x)) = 6, & \deg(g(x) \cdot h(x)) = 3, & \deg(f(x) \cdot h(x)) = 3. \end{array}$$

3-3-6. Proposition. Suppose R is a commutative ring with unity. Then

$$R[x] \text{ is an integral domain} \Leftrightarrow R \text{ is an integral domain.}$$

Furthermore, when R is an integral domain D, the units of $D[x]$ are precisely the units of D.

Proof: Suppose R is an integral domain and $f(x)$ is a zero-divisor in $R[x]$. Thus, $f(x) \neq 0$, and there exists $g(x) \neq 0$ in $R[x]$ such that

$$f(x) \cdot g(x) = 0.$$

Therefore, $\deg(f(x) \cdot g(x)) = \deg 0 = -\infty$. On the other hand, $\deg f(x) \geq 0$, $\deg g(x) \geq 0$; and according to 3-3-5, $\deg(f(x) \cdot g(x)) = \deg f(x) + \deg g(x) \geq 0$, since by the hypothesis on R the leading coefficient of $f(x)$ is not a zero-divisor. This is a contradiction; so there cannot be any zero-divisors $f(x)$ in $R[x]$, and $R[x]$ is an integral domain. This proves $\Leftarrow$.

As for $\Rightarrow$; if $R[x]$ is an integral domain, then R is a commutative ring with unity which has no zero-divisors because it is contained in $R[x]$—so R is an integral domain.

Finally, suppose D is an integral domain. If $a \in D$ is a unit, there exists $b \in D$ such that $ab = 1$. But a, b, 1 are also elements of $D[x]$, and 1 is the unity of $D[x]$ also, so a is a unit of $D[x]$. Conversely, suppose $f(x) \in D[x]$ is a unit. Then there exists $g(x) \in D[x]$ such that

$$f(x) \cdot g(x) = 1.$$

Of course, $f(x) \neq 0$, $g(x) \neq 0$, so that $\deg f(x) \geq 0$, $\deg g(x) \geq 0$. Now, $\deg(1) = 0$, and $\deg(f(x) \cdot g(x)) = \deg f(x) + \deg g(x)$ because the leading coefficients of $f(x)$ and $g(x)$ are not equal to 0 and they come from the integral domain D. [In general, the fact that the degree of a product of two polynomials is equal to the sum of their degrees is familiar from high school. Its validity there is a consequence of the fact that we always dealt with polynomials whose coefficients came from $\mathbf{Z}$, $\mathbf{Q}$, $\mathbf{R}$, or $\mathbf{C}$—all of which are integral domains. Only when the coefficients come from a ring which is not an integral domain is it possible that $\deg(f(x) \cdot g(x)) \neq \deg f(x) + \deg g(x)$.] Therefore,

$$\deg f(x) + \deg g(x) = 0,$$

so that both $f(x)$ and $g(x)$ have degree 0. Consequently, these polynomials must be of form $f(x) = a$, $g(x) = b$ where a and b are nonzero elements of D. In particular, $f(x)$ and $g(x)$ are elements of D whose product is 1, so $f(x)$ is a unit of D. This completes the proof that the units of $D[x]$ are the same as the units of D. ▌

In virtue of this result, it is clear that $\mathbf{Z}_{12}[x]$ is not an integral domain because $\mathbf{Z}_{12}$ is not an integral domain. On the other hand, $\mathbf{Z}[x]$ is an integral domain (because $\mathbf{Z}$ is), and its only units are ± 1. Furthermore, $\mathbf{Q}[x]$ is an integral domain (because $\mathbf{Q}$ is an integral domain) and its units are the nonzero elements of $\mathbf{Q}$ (because these are the units of $\mathbf{Q}$). More generally, because any field is an integral domain whose units are all its nonzero elements, we have:

3-3-7. Corollary. If F is a field, then $F[x]$ is an integral domain whose set of units is F^*, the set of all nonzero elements of F.

So far, we have discussed polynomials; now let us turn to polynomial functions.

3-3-8. Remark. Suppose we are given a polynomial

$$f(x) = a_0 + a_1 x + \cdots + a_n x^n = \sum_0^\infty a_i x^i$$

in $R[x]$. For any $c \in R$ consider the expression

$$a_0 + a_1 c + \cdots + a_n c^n.$$

This involves multiplying and adding elements of R; so it represents an element of R, which we write as $f(c)$. For obvious reasons, this procedure may be called **substitution** of c for x. Note that only elements of R may be substituted for x.

There is nothing strange about substitution, it is part of our everyday experience. For example, if $f(x) = 2 - x + x^3 \in \mathbf{Z}[x]$, then $f(1) = 2, f(2) = 8$, $f(0) = 2, f(-1) = 2$, and so on.

Now, suppose that, in addition to $f(x) = \Sigma_0^\infty a_i x^i \in R[x]$ we have the polynomial

$$g(x) = b_0 + b_1 x + \cdots + b_m x^m = \sum_0^\infty b_i x^i$$

in $R[x]$. Let us consider their sum and their product

$$s(x) = f(x) + g(x), \qquad t(x) = f(x) \cdot g(x).$$

Expressing these explicitly, we have

$$\begin{aligned} s(x) &= (a_0 + b_0) + (a_1 + b_1)x + \cdots + (a_i + b_i)x^i + \cdots \\ &= \sum_0^\infty (a_i + b_i)x^i \end{aligned}$$

and

$$\begin{aligned} t(x) &= a_0 b_0 + (a_1 b_0 + a_0 b_1)x + \cdots + (a_i b_0 + a_{i-1} b_1 + \cdots + a_0 b_i)x^i + \cdots \\ &= \sum_{i=0}^\infty \left(\sum_{r+s=i} a_r b_s \right) x^i. \end{aligned}$$

What happens if c is substituted for x in $s(x)$ and $t(x)$?

If we consider the concrete situation, $f(x) = 2 - x + x^3 \in \mathbf{Z}[x]$, $g(x) = 1 + x + x^2 \in \mathbf{Z}[x]$, then

$$\begin{aligned} s(x) &= f(x) + g(x) = 3 + x^2 + x^3, \\ t(x) &= f(x) \cdot g(x) = 2 + x + x^2 + x^4 + x^5 \end{aligned}$$

—and, for example, $f(1) = 2, g(1) = 3, s(1) = 5, t(1) = 6$, so $s(1) = f(1) + g(1)$ and $t(1) = f(1) \cdot g(1)$; $f(0) = 2$, $g(0) = 1$, $s(0) = 3$, $t(0) = 2$, so $s(0) = f(0) + g(0)$ and $t(0) = f(0) \cdot g(0)$; $f(-1) = 2$, $g(-1) = 1$, $s(-1) = 3$, $t(-1) = 2$, so $s(-1) = f(-1) + g(-1)$ and $t(-1) = f(-1) \cdot g(-1)$. More generally, for any $c \in \mathbf{Z}$, we have $f(c) = 2 - c + c^3$, $g(c) = 1 + c + c^2$, $s(c) = 3 + c^2 + c^3$, $t(c) = 2 + c + c^2 + c^4 + c^5$—so as expected

$$s(c) = f(c) + g(c) \qquad \text{and} \qquad t(c) = f(c) \cdot g(c).$$

Returning to the general situation, it is not hard to see that the same rules always hold. In more detail, for any $c \in R$, a trivial computation in R gives

$$\begin{aligned} f(c) + g(c) &= (a_0 + a_1 c + a_2 c^2 + \cdots + a_n c^n) \\ &\quad + (b_0 + b_1 c + b_2 c^2 + \cdots + b_m c^m) \\ &= (a_0 + b_0) + (a_1 + b_1)c + \cdots + (a_i + b_i)c^i + \cdots \\ &= s(c). \end{aligned}$$

The compact way to write this proof is

$$\begin{aligned} f(c) + g(c) &= \left(\sum_0^\infty a_i c^i\right) + \left(\sum_0^\infty b_i c^i\right) \\ &= \sum_0^\infty (a_i + b_i)c^i \\ &= s(c) \end{aligned}$$

and this is valid because Σ_0^∞ is really a finite sum whenever it appears.

In the same spirit, we have in the general case

$$\begin{aligned} f(c) \cdot g(c) &= \left(\sum_0^\infty a_i c^i\right) \cdot \left(\sum_0^\infty b_i c^i\right) \\ &= \sum_{i=0}^\infty \left(\sum_{r+s=i} a_r b_s\right) c^i \\ &= t(c). \end{aligned}$$

The reader who feels uncomfortable with infinite summation symbols may use the explicit finite expressions for $f(c)$, $g(c)$, $t(c)$ to verify that $t(c) = f(c) \cdot g(c)$. The key to the matter is that computations with "polynomial expressions in c" go just like the same computations with the corresponding polynomials in x.

3-3-9. Remark. If $f(x) \in R[x]$ is given, then for any $c \in R$ we can compute $f(c) \in R$. Thus, we have a function or mapping

$$c \to f(c), \qquad c \in R$$

of $R \to R$ which is determined by $f(x)$. For want of a better notation, let us denote this mapping of $R \to R$, which carries c to $f(c)$ for every $c \in R$, by $\bar{f}$. In other words, we have defined $\bar{f} \in \mathrm{Map}(R, R)$ by the rule

$$\bar{f}(c) = f(c), \qquad c \in R.$$

Such a function $\bar{f}$: $R \to R$ which arises from some polynomial $f(x) \in R[x]$ is known as a **polynomial function** over R. Clearly, polynomials and polynomial functions are different things.

The notion of a polynomial function and the notation for it are probably confusing—after all a new and somewhat strange level of abstraction has been introduced—so let us do a concrete example. Suppose $R = \mathbf{Z}_4$ and we consider the polynomial $f(x) = 2 - 3x + x^3 \in \mathbf{Z}_4[x]$. What is $\bar{f}$? It is the mapping of $\mathbf{Z}_4 \to \mathbf{Z}_4$ [that is, the element of Map($\mathbf{Z}_4$, $\mathbf{Z}_4$)] which maps $c \to f(c) = 2 - 3c + c^3$ for every $c \in \mathbf{Z}_4$; in other words,

$$\bar{f}(c) = 2 - 3c + c^3, \qquad c \in \mathbf{Z}_4.$$

Of course, $\bar{f}$, like any other element of Map($\mathbf{Z}_4$, $\mathbf{Z}_4$), is specified by what it does to every element of $\mathbf{Z}_4$; thus, letting 0, 1, 2, 3 represent the four elements $\lfloor 0 \rfloor_4, \lfloor 1 \rfloor_4, \lfloor 2 \rfloor_4, \lfloor 3 \rfloor_4$ of $\mathbf{Z}_4$, the mapping $\bar{f} \in$ Map($\mathbf{Z}_4$, $\mathbf{Z}_4$) is given by

$$\bar{f}(0) = 2, \qquad \bar{f}(1) = 0, \qquad \bar{f}(2) = 0, \qquad \bar{f}(3) = 0.$$

3-3-10. Remark. It is natural to ask whether polynomials and polynomial functions are not really different versions of the same reality. Is it true that distinct (that is, different) polynomials $f(x)$ and $g(x)$ always lead to distinct polynomial functions $\bar{f}$ and $\bar{g}$? Or, can it be that $f(x) \neq g(x)$ but yet $\bar{f} = \bar{g}$? In our past experience, we invariably considered polynomials and polynomial functions as the same thing; in fact, one never distinguished between them in any way. However, the following simple example shows why this assumption is unwarranted in general. (In Section 3-5, we shall see why this assumption can be justified in the situations dealt with in high school.)

Let $R = \mathbf{Z}_p$, where p is prime, and consider the polynomials

$$f(x) = x^p, \qquad g(x) = x$$

in $\mathbf{Z}_p[x]$. Obviously, these are different polynomials—$f(x) \neq g(x)$. On the other hand, we assert that $\bar{f} = \bar{g}$. To prove that the elements $\bar{f}, \bar{g} \in$ Map($\mathbf{Z}_p$, $\mathbf{Z}_p$) are the same, we must show that $\bar{f}(\alpha) = \bar{g}(\alpha)$ for every $\alpha \in \mathbf{Z}_p$. But for $\alpha \in \mathbf{Z}_p$, we have

$$\bar{f}(\alpha) = f(\alpha) = \alpha^p, \qquad \bar{g}(\alpha) = g(\alpha) = \alpha$$

and these are equal in virtue of Fermat's theorem, 2-4-9. Thus, distinct polynomials can indeed lead to the same polynomial function.

There is more to be said about the connection between polynomials and polynomial functions. To any polynomial $f(x) \in R[x]$ there is associated the mapping $\bar{f}: R \to R$ [that is, the element $\bar{f} \in$ Map(R, R)] that is given by $\bar{f}(c) = f(c)$ for all $c \in R$. Thus, we can define a mapping

$$\Lambda: R[x] \to \text{Map}(R, R)$$

by putting

$$\Lambda(f(x)) = \bar{f}, \qquad f(x) \in R[x].$$

In other words, Λ is the mapping which carries each polynomial $f(x)$ to its associated polynomial function $\bar{f}$. Now, in 2-2-12 we saw that, for any set X, Map(X, R) becomes a ring when the operations are defined in an appropriate manner. In particular, in the case $X = R$, Map(R, R) becomes a ring in which the operations take the following form: If ϕ and ψ are elements of Map(R, R), then $\phi + \psi$ and $\phi \cdot \psi$ are the elements of Map(R, R) defined by

$$\begin{aligned}(\phi + \psi)(c) &= \phi(c) + \psi(c),\\ (\phi \cdot \psi)(c) &= \phi(c) \cdot \psi(c),\end{aligned} \qquad c \in R.$$

3-3-11. Proposition. The mapping

$$\Lambda: R[x] \to \text{Map}(R, R)$$

given by

$$\Lambda(f(x)) = \bar{f}, \qquad f(x) \in R[x]$$

is a homomorphism of rings.

Proof: Given any $f(x)$, $g(x) \in R[x]$, let us write $s(x) = f(x) + g(x)$ and $t(x) = f(x) \cdot g(x)$. We need to show that

$$\begin{aligned}\Lambda(f(x) + g(x)) &= \Lambda(f(x)) + \Lambda(g(x)),\\ \Lambda(f(x)g(x)) &= \Lambda(f(x)) \cdot \Lambda(g(x)).\end{aligned}$$

In connection with the first of these, we observe that

$$\Lambda(f(x) + g(x)) = \Lambda(s(x)) = \bar{s} \qquad \text{and} \qquad \Lambda(f(x)) + \Lambda(g(x)) = \bar{f} + \bar{g}.$$

It remains to verify that $\bar{s} = \bar{f} + \bar{g}$, and because we are dealing with elements of Map(R, R) this requires verification that $\bar{s}(c) = (\bar{f} + \bar{g})(c)$ for every $c \in R$. But this is not hard.

$$\begin{aligned}\bar{s}(c) &= s(c) && \text{(by definition of } \bar{s}\text{)},\\ &= f(c) + g(c) && \text{(by 3-3-8)},\\ &= \bar{f}(c) + \bar{g}(c) && \text{(by definition of } \bar{f} \text{ and } \bar{g}\text{)},\\ &= (\bar{f} + \bar{g})(c) && \text{(by definition of + in Map}(R, R)).\end{aligned}$$

In similar fashion, Λ preserves multiplication; in fact, $\Lambda(f(x)g(x)) = \Lambda(t(x)) = \bar{t}$, $\Lambda(f(x)) \cdot \Lambda(g(x)) = \bar{f} \cdot \bar{g}$, and $\bar{t} = \bar{f} \cdot \bar{g}$ because, for every $c \in R$, we have $\bar{t}(c) = t(c) = f(c) \cdot g(c) = \bar{f}(c) \cdot \bar{g}(c) = (\bar{f} \cdot \bar{g})(c)$. This completes the proof. ▌

There are two kinds of problems about polynomials around which our concern will center for the rest of this chapter: solving them and factoring them. We conclude this section with some preliminary comments about these questions.

3-3-12. Definition. The element $c \in R$ is said to be a **root** of the polynomial $f(x) = a_0 + a_1 x + \cdots + a_n x^n \in R[x]$ when $f(c) = 0$.

Other terminologies used for this notion are: c is a **solution** of the equation $f(x) = a_0 + a_1 x + \cdots + a_n x^n = 0$; c is a place where the polynomial $f(x)$ (or the polynomial function $\bar{f}$) takes the value 0—or simply, c is a **zero** of the polynomial $f(x)$ (or of the polynomial function $\bar{f}$). One may also take a "geometric" view. In fact, the set

$$G(f) = \{(r, f(r)) \mid r \in R\}$$

is obviously a set of points in the direct product $R \times R$. In analogy with the familiar situation in $\mathbf{R} \times \mathbf{R}$ (which is just another way of denoting the Euclidean plane), we may call $G(f)$ the **graph** of $f(x)$—and then a root of $f(x)$ is a value of the unknown (that is, indeterminate) x where the graph crosses the horizontal axis.

It is important to emphasize the ring R from which the coefficients of $f(x)$ come—for according to our definition, the roots of $f(x)$ can only come from R. For example, consider the polynomial

$$f(x) = 3x^3 - 4x^2 + 2x + 3.$$

If we view $f(x) \in \mathbf{Z}[x]$, then $f(3) = 54$, so 3 is not a root of $f(x)$. On the other hand, if we view $f(x) \in \mathbf{Z}_6[x]$, then $f(3) = 0$ and 3 is a root of $f(x)$—more precisely, here $f(x)$ is really $\lfloor 3 \rfloor_6 x^3 - \lfloor 4 \rfloor_6 x^2 + \lfloor 2 \rfloor_6 x + \lfloor 3 \rfloor_6$ (and we are keeping the notation simple by dropping $\lfloor \;\; \rfloor_6$ throughout) so that $f(\lfloor 3 \rfloor_6) = \lfloor 54 \rfloor_6 = \lfloor 0 \rfloor_6$, and this is written as $f(3) = 0$.

When R is a commutative ring with unity, so is $R[x]$, and therefore (as indicated in 3-2-1) it is meaningful to talk about divisibility or factorization in $R[x]$. The connection between a root of the polynomial $f(x)$ and its factorization is given by:

3-3-13. Factor Theorem. Given $f(x) \in R[x]$; then $c \in R$ is a root of $f(x) \Leftrightarrow (x - c)$ divides $f(x)$.

Proof: Let us prove $\Leftarrow$ first. If $(x - c)$ divides $f(x)$, then there exists $g(x) \in R[x]$ such that

$$f(x) = (x - c)g(x).$$

By the properties of substitution, 3-3-8, we have $f(c) = (c - c)g(c) = 0$, so c is indeed a root of $f(x)$.

As for the implication $\Rightarrow$, let us write

$$f(x) = a_0 + a_1 x + a_2 x^2 + \cdots + a_n x^n.$$

Since c is a root of $f(x)$, we have

$$f(c) = a_0 + a_1 c + a_2 c^2 + \cdots + a_n c^n = 0.$$

Because everything in sight belongs to $R[x]$, we may then write

$$f(x) = f(x) - f(c) = a_1(x - c) + a_2(x^2 - c^2) + \cdots + a_n(x^n - c^n). \qquad (*)$$

Now, in $R[x]$, we clearly have $x^2 - c^2 = (x - c)(x + c)$, $(x^3 - c^3) = (x - c)(x^2 + cx + c^2)$, and for arbitrary $i \geq 2$

$$(x^i - c^i) = (x - c)(x^{i-1} + cx^{i-2} + \cdots + c^{i-2}x + c^{i-1}).$$

Thus, $(x - c)$ is a factor of each term on the right side of $(*)$, so (in virtue of the generalized distributive law in $R[x]$) $(x - c)$ is a factor of $f(x)$—or, in other words, $(x - c)$ divides $f(x)$. This completes the proof. ▌

3-3-14. Example. Consider the polynomial $f(x) = x^2 - 1 \in \mathbf{Z}_{15}[x]$. Its roots, if any, must come from $\mathbf{Z}_{15}$—so the only possibilities are 0, 1, 2, ..., 14, where according to our convention each such integer a represents the element $\lfloor a \rfloor_{15} \in \mathbf{Z}_{15}$. When each of these 15 possibilities is tested, we find that 1, 4, 11, 14 (meaning $\lfloor 1 \rfloor_{15}$, $\lfloor 4 \rfloor_{15}$, $\lfloor 11 \rfloor_{15}$, $\lfloor 14 \rfloor_{15}$) are the roots of $f(x)$. This is surprising perhaps, since we normally expect a polynomial of degree 2 to have at most 2 roots. However, having seen in Section 3-1 that a linear equation $\alpha x = \beta$ in $\mathbf{Z}_m$ may have more than one root, our surprise here should not be very great.

Since 1, 4, 11, 14 are roots, 3-3-13 tells us that $(x - 1)$, $(x - 4)$, $(x - 11)$, $(x - 14)$ all divide $x^2 - 1$ in $\mathbf{Z}_{15}[x]$. In fact, we have in $\mathbf{Z}_{15}[x]$

$$x^2 - 1 = (x - 1)(x - 14) = (x - 4)(x - 11).$$

Here we have two distinct factorizations, neither of which can be factored further. Thus, factorization in $\mathbf{Z}_{15}[x]$ need not be unique.

Incidentally, in $\mathbf{Z}_{15}$, $14 = -1$ and $11 = -4$, so we also have the factorizations

$$x^2 - 1 = (x - 1)(x + 1) = (x - 4)(x + 4)$$

—but these factorizations are really the same as the preceding ones.

3-3-15 / *PROBLEMS*

1. (*i*) Find the sum, difference, and product of the polynomials $f(x) = 3 + 2x - x^2 - 3x^3$ and $g(x) = 3x^2 + 2x^3 + x - 1$ in $\mathbf{Z}[x]$.
(*ii*) Perform the same operations when $f(x)$ and $g(x)$ are viewed in
(*a*) $\mathbf{Z}_4[x]$, (*b*) $\mathbf{Z}_5[x]$, (*c*) $\mathbf{Z}_6[x]$, (*d*) $\mathbf{Z}_7[x]$, (*e*) $\mathbf{Z}_{11}[x]$.

2. With the usual understanding that R is a commutative ring with unity, provide the missing details in the proof (see 3-3-3) that $R[x]$ is a commutative ring with unity.

3. In connection with 3-3-4, explain carefully why $1 = 1x^0$ is the unity of $R[x]$.

4. Suppose $c \in R$ and $f(x) = a_0 + a_1x + \cdots + a_nx^n \in R[x]$. How do we know that

$$cf(x) = (ca_0) + (ca_1)x + \cdots + (ca_n)x^n?$$

5. In $R[x]$, what can be said about xa for $a \in R$? More generally, given $a, b \in R$ what is the meaning, if any, of ax^ibx^j?

6. When R is a commutative ring with unity, which of the following are subrings of $R[x]$?
(*i*) all polynomials with constant term equal to 0,
(*ii*) all polynomials with constant term equal to 1,
(*iii*) all polynomials of degree less than or equal to 7,
(*iv*) all polynomials of degree equal to 7,
(*v*) all polynomials in which the even powers of x all appear with coefficient 0 (x^0 is considered to be an even power),
(*vi*) all polynomials in which all the odd powers of x appear with coefficient 0,
(*vii*) all polynomials for which the coefficients of both x and x^2 are 0.

7. Which of the following belong to $\mathbf{Z}[x]$?
(*i*) $1/x^2$, (*ii*) x^{-1}, (*iii*) $\sqrt{x}$.

8. Is it true that in any polynomial ring $R[x]$ we have

$$(x - 1)(x^{n-1} + x^{n-2} + \cdots + x + 1) = x^n - 1?$$

9. How many polynomials of degree 4 are there in $\mathbf{Z}_3[x]$? How many of degree 5? How many of degree less than or equal to 5?

10. For integers $m \geq 0$, $n \geq 2$, how many polynomials are there in $\mathbf{Z}_n[x]$ of degree exactly m? How many of degree less than or equal to m?

11. Since $\mathbf{Z}_6$ is not an integral domain, we know that $\mathbf{Z}_6[x]$ is not an integral domain. Which of the following elements of $\mathbf{Z}_6[x]$ is a zero-divisor? Justify your answers.
(i) $2x$, *(ii)* $x + 5$, *(iii)* $x + 2$, *(iv)* $2x + 4$,
(v) $5x - 1$, *(vi)* $2x + 3$, *(vii)* $3x - 2$.

12. Can you find polynomials $f(x)$ and $g(x)$ in $\mathbf{Z}_{12}[x]$ such that $\deg f(x) = \deg g(x) = 5$, $\deg(f(x) + g(x)) = 3$, and $\deg(f(x)g(x)) = 8$?

13. Prove: If x and y are indeterminates over R, then the rings $R[x]$ and $R[y]$ are isomorphic. Even more, if both R and R' are commutative rings with unity which are isomorphic, x is an indeterminate over R, and y is an indeterminate over R', then the rings $R[x]$ and $R'[y]$ are isomorphic.

14. *(i)* Show that the polynomials $f_1(x) = x, f_2(x) = x^3, f_3(x) = x^6 + 2x^2 + x$, $f_4(x) = x^9 + 8x^3 + x$ in $\mathbf{Z}_3[x]$ all determine the same polynomial function.
(ii) What happens if they are viewed as polynomials in $\mathbf{Z}_5[x]$?

15. *(i)* Consider the polynomial $f(x) = x^3 - x + 2 \in \mathbf{Z}_3[x]$. Describe the polynomial function $\bar{f}$: $\mathbf{Z}_3 \to \mathbf{Z}_3$ that it determines concretely—that is, by exhibiting $\bar{f}(0), \bar{f}(1), \bar{f}(2)$. Find two other polynomials in $\mathbf{Z}_3[x]$ which determine the same polynomial function.
(ii) Answer the same questions as above when $\mathbf{Z}_3$ is replaced by
(a) $\mathbf{Z}_5$, *(b)* $\mathbf{Z}_7$, *(c)* $\mathbf{Z}_{11}$.

16. Can you find a polynomial $f(x) \in \mathbf{Z}_3[x]$ whose polynomial function $\bar{f} \in \mathrm{Map}(\mathbf{Z}_3, \mathbf{Z}_3)$ is given by

$$\bar{f}(0) = 1, \qquad \bar{f}(1) = 2, \qquad \bar{f}(2) = 1?$$

17. *(i)* Show that $\mathrm{Map}(\mathbf{Z}_3, \mathbf{Z}_3)$ has 27 elements, as does the set $\mathbf{Z}_3^2[x]$ of all polynomials in $\mathbf{Z}_3[x]$ of degree less than or equal to 2.
(ii) Consider the mapping $\Lambda: f(x) \to \bar{f}$ of

$$\mathbf{Z}_3^2[x] \longrightarrow \mathrm{Map}(\mathbf{Z}_3, \mathbf{Z}_3).$$

Is it one-to-one and onto?

18. Find the roots of $f(x) = 3 + 2x - x^2 - 3x^3$ when it is viewed as a polynomial over
(i) $\mathbf{Z}_4$, *(ii)* $\mathbf{Z}_5$, *(iii)* $\mathbf{Z}_6$, *(iv)* $\mathbf{Z}_7$, *(v)* $\mathbf{Z}_{11}$.
Do the same for $g(x) = 3x^2 + 2x^3 + x - 1$.

19. Does $(x - 3)$ divide $x^4 + x^3 + x + 4$ when they are viewed as polynomials over
(i) $\mathbf{Z}$, *(ii)* $\mathbf{Q}$, *(iii)* $\mathbf{Z}_3$, *(iv)* $\mathbf{Z}_4$,
(v) $\mathbf{Z}_5$, *(vi)* $\mathbf{Z}_6$, *(vii)* $\mathbf{Z}_7$, *(viii)* $\mathbf{Z}_{11}$?

20. Fix any $c \in R$ where, as usual, R is a commutative ring with unity. Define a function

$$E_c \colon R[x] \to R$$

by putting

$$E_c(f(x)) = f(c), \qquad f(x) \in R[x].$$

Show that E_c, which is called the **evaluation function** at c, is a surjective homomorphism. What is the kernel of E_c? When is $E_c = E_{c'}$?

21. (*i*) Suppose $D = P \cup \{0\} \cup -P$ is an ordered domain. Show that $D[x]$ becomes an ordered domain when we take as the set of positive elements those polynomials whose leading coefficient belongs to P.

(*ii*) In this way, $\mathbf{Z}[x]$ becomes an ordered domain which contains a least positive element—namely, 1. However, $\mathbf{Z}[x]$ is not well ordered.

(*iii*) Show that $D[x]$ also becomes an ordered domain when an element $f(x) = a_0 + a_1 x + \cdots + a_n x^n \in D[x]$ is defined to be positive when the first nonzero coefficient (starting from the left) is in P (that is, is positive). Thus, in virtue of part (*i*), the same domain $D[x]$ can be made into an ordered domain in more than one way (that is, with different sets of positive elements).

22. Show that $\mathbf{Z}_3[2x] = \mathbf{Z}_3[x]$; in other words, every polynomial in $2x$ (really, $\lfloor 2 \rfloor_3 x$, of course) over $\mathbf{Z}_3$ can be expressed as a polynomial in x over $\mathbf{Z}_3$, and conversely.

23. (*i*) Suppose R is a commutative ring with unity and c is a unit of R; then $R[x] = R[cx]$.

(*ii*) If, in addition d is any element of R, then $R[x] = R[cx + d]$.

(*iii*) Is $R[x^2]$ equal to $R[x]$?

24. If $f(x) = a_0 + a_1 x + a_2 x^2 + \cdots + a_n x^n \in R[x]$, let us call

$$f'(x) = a_1 + 2a_2 x^1 + 3a_3 x^2 + \cdots + na_n x^{n-1} \in R[x]$$

the **derivative** of $f(x)$. In compact notation

$$f(x) = \sum_{i=0}^{n} a_i x^i, \qquad f'(x) = \sum_{i=1}^{n} i a_i x^{i-1}.$$

Prove the usual rules for derivatives—namely.

(*i*) $(f(x) + g(x))' = f'(x) + g'(x)$,

(*ii*) $(cf(x))' = cf'(x)$, $\qquad c \in R$

(*iii*) $(f(x)g(x))' = f(x)g'(x) + f'(x)g(x)$,

(*iv*) $[(f(x))^n]' = n(f(x))^{n-1} f'(x)$,

25. If R is a commutative ring with unity, let $R[[x]]$ denote the set of all "formal power series" in x over R. In other words, the elements of $R[[x]]$ are of form

$$a_0 + a_1 x + a_2 x^2 + \cdots = \sum_{i=0}^{\infty} a_i x^i$$

where there are no restrictions on the elements a_i of R. Defining addition and multiplication in $R[[x]]$ just like for ordinary polynomials, show that $R[[x]]$ becomes a commutative ring with unity which contains $R[x]$ as a subring. Which are the units of $R[[x]]$? If R is an integral domain D, show that $D[[x]]$ is an integral domain.

3-4. Factorization in $F[x]$

Consider a field F and the polynomial ring $F[x]$. According to 3-3-7, $F[x]$ is an integral domain whose set of units is F^*, the set of all nonzero elements of F. In other words, a polynomial $u(x) \in F[x]$ is a unit $\Leftrightarrow$ $u(x)$ is a constant polynomial of form $u(x) = u$ where $u \in F^*$. [Since $\deg(0) = -\infty$, we may also say that $u(x)$ is a unit $\Leftrightarrow$ $\deg(u(x)) = 0$.] Our main objective in this section is to study questions of divisibility and factorization in $F[x]$. We shall see that as far as these questions are concerned $F[x]$ behaves very much like **Z**—except that the units complicate matters somewhat.

Let us examine the role played by units in questions of divisibility. As an illustration, consider the polynomials

$$f(x) = (3 - \sqrt{2})x^2 + (1 - \sqrt{2})x + (1 + \sqrt{2})$$

and

$$g(x) = (1 + 2\sqrt{2})x^2 - x + (3 + 2\sqrt{2})$$

in $\mathbf{R}[x]$. It is easy to see that, although these polynomials appear to be unrelated, they actually divide each other. In fact, if we take $u(x) = u = 1 + \sqrt{2}$ and $v(x) = v = -1 + \sqrt{2}$ then $g(x) = f(x)u$ [which says that $f(x) \mid g(x)$] and $f(x) = g(x)v$ [which says that $g(x) \mid f(x)$]. Of course, $u = 1 + \sqrt{2}$ and $v = -1 + \sqrt{2}$ are nonzero elements of **R**, so they are units of $\mathbf{R}[x]$; even more, u and v are inverses of each other—

$$u^{-1} = \frac{1}{1 + \sqrt{2}} = -1 + \sqrt{2} = v, \qquad v^{-1} = \frac{1}{-1 + \sqrt{2}} = 1 + \sqrt{2} = u.$$

This situation is a manifestation of the following general result.

3-4-1. Proposition. Suppose $f(x)$ and $g(x)$ are nonzero polynomials in $F[x]$; then the following conditions are equivalent:

(1) $f(x)$ and $g(x)$ divide each other,
(2) there exists a unit $u(x) = u$ of $F[x]$ such that $g(x) = f(x)u$,
(3) there exists a unit $v(x) = v$ of $F[x]$ such that $f(x) = g(x)v$.

Proof: (1) $\Rightarrow$ (2): Since $f(x) \mid g(x)$ there exists $u(x) \in F[x]$ for which $g(x) = f(x)u(x)$, and since $g(x) \mid f(x)$ there exists $v(x) \in F[x]$ for which $f(x) = g(x)v(x)$. Consequently,

$$f(x) = f(x)u(x)v(x).$$

Because $F[x]$ is an integral domain, the cancellation law applies—so the nonzero element $f(x)$ may be canceled to give

$$1 = u(x)v(x).$$

Thus, $u(x)$ is a unit of $F[x]$; so $u(x) = u$, an element of F^*, and $g(x) = f(x)u$.

(2) $\Rightarrow$ (3): Given $g(x) = f(x)u$ with $u = u(x)$ a unit of $F[x]$. Since $u \in F^*$ there exists $v \in F^*$ such that $uv = 1$. Thus, $v = v(x)$ is a unit of $F[x]$ and

$$g(x)v = f(x)uv = f(x).$$

(3) $\Rightarrow$ (1): Given $f(x) = g(x)v$ with $v = v(x)$ a unit of $F[x]$, we see that $g(x) \mid f(x)$. Upon taking $u = u(x)$ for which $vu = 1$, we obtain $f(x)u = g(x)$ so that $f(x) \mid g(x)$. This completes the proof. ▌

3-4-2. Remark. When condition (2) is satisfied—that is, when $g(x) = f(x)u$ with $u = u(x)$ a unit of $F[x]$—it is customary to say that $g(x)$ is an **associate** of $f(x)$, and to denote this by $g(x) \sim f(x)$. With this notation, the assertion of 3-4-1 takes the form: $g(x) \sim f(x) \Leftrightarrow f(x) \sim g(x) \Leftrightarrow f(x)$ and $g(x)$ divide each other.

It is easy to see that $\sim$ satisfies the following properties:

(*i*) $f(x) \sim f(x)$, [take $u = u(x) = 1$].
(*ii*) If $g(x) \sim f(x)$, then $f(x) \sim g(x)$.
(*iii*) If $f(x) \sim g(x)$ and $g(x) \sim h(x)$, then $f(x) \sim h(x)$.

Of course, (*iii*) is immediate because if $f(x)$ and $g(x)$ divide each other and also $g(x)$ and $h(x)$ divide each other, then $f(x)$ and $h(x)$ divide each other.

These three properties, which we have already encountered on a number of occasions, are known as the reflexive, symmetric, and transitive properties (or laws), respectively. They occur within a very general framework which we shall now discuss, even though it is not essential to our needs.

3-4-3. Digression. Consider an arbitrary set $X = \{a, b, c, \ldots, x, y, z, \ldots\}$ and suppose we are given a relation ρ on X. We do not wish to be overly fancy or precise about the meaning of a "relation." Suffice it to say that for any pair of elements a, b, in X (in the given order) either a is related to b with respect to the relation ρ—in which case we write $a\rho b$—or else a is not related to b with respect to ρ—in which case we write $a\not\rho b$. In other words, for any given pair of elements a, b of X the statement $a\rho b$ is either true or it is false (but not both!).

Let us list some examples of the notion of relation.

(1) Let X be the set $\mathbf{R}$ of all real numbers; for $a, b \in \mathbf{R}$, $a\rho b$ is taken to mean $a < b$.
(2) Let X be the set of all triangles in the plane; for $a, b \in X$, let $a\rho b$ have the meaning that the triangle a has the same area as the triangle b.
(3) Let X be an integral domain D; for $a, b \in D$, $a\rho b$ means a divides b.
(4) Let X be a polynomial ring $F[x]$; for $f(x), g(x) \in F[x]$ we define $f(x)\ \rho\ g(x)$ to mean $\deg f(x) = \deg g(x)$.
(5) Let $X = \mathbf{Z}$ and ρ be congruence (mod m). In other words, when $m > 1$ is fixed and $a, b \in \mathbf{Z}$, we write $a\rho b$ to signify that $a \equiv b(\text{mod } m)$.
(6) Let X be any ring R and fix a subring S; for $a, b \in R$ take $a\rho b$ to mean $a - b \in S$.

The most interesting relations are those which satisfy the reflexive, symmetric, and transitive properties. In more detail, any such relation ρ is known as an **equivalence relation**, and it satisfies

(*i*) $a\rho a$ for all $a \in X$.
(*ii*) If $a\rho b$, then $b\rho a$.
(*iii*) If $a\rho b$ and $b\rho c$, then $a\rho c$.

Of the examples listed above, it is clear that (2), (4), (5), (6) are equivalence relations while (1) and (3) are not.

The importance of equivalence relations derives from the following facts. If ρ is an equivalence relation on the set X, then [in analogy with what was done in 1-7-8 for congruence (mod m)], for every $a \in X$ we write

$$\lfloor a \rfloor_\rho = \{b \in X \mid b\ \rho\ a\}$$

and call it the **equivalence class** of a (with respect to ρ). By the same kinds of arguments used in 1-7-10, keeping in mind that $b \in \lfloor a \rfloor_\rho \Leftrightarrow b\ \rho\ a$, it follows that the equivalence classes satisfy the properties:

(*i*) $a \in \lfloor a \rfloor_\rho$ for all $a \in X$,

(*ii*) $b \in \lfloor a \rfloor_\rho \Leftrightarrow a \in \lfloor b \rfloor_\rho$,

(*iii*) $b \in \lfloor a \rfloor_\rho \Rightarrow \lfloor b \rfloor_\rho = \lfloor a \rfloor_\rho$,

(*iv*) $\lfloor a \rfloor_\rho \cap \lfloor b \rfloor_\rho \neq \varnothing \Rightarrow \lfloor a \rfloor_\rho = \lfloor b \rfloor_\rho$.

Consequently, each element of X belongs to some equivalence class, and any two equivalence classes are disjoint or identical. We conclude that the set of equivalence classes with respect to ρ provides a disjoint decomposition (or **partition**) of X.

Returning to $F[x]$, we note that in virtue of 3-4-2 and 3-4-3 the relation $\rho = \sim$ (where both ρ and $\sim$ signify "is an associate of") is an equivalence relation. In truth, as may be seen from 3-4-1 and 3-4-2, the relation $\sim$ is defined only for nonzero polynomials (that is, on the set $F[x] - \{0\}$, obtained by removing the zero polynomial from $F[x]$); however, we shall be careless and say that $\sim$ is an equivalence relation on $F[x]$, instead of insisting on $F[x] - \{0\}$. Of course, any equivalence class is of form

$$\lfloor f(x) \rfloor_{\sim} = \{g(x) \in F[x] \mid g(x) \sim f(x)\}$$

where $f(x) \neq 0$, [so $g(x) \in \lfloor f(x) \rfloor_{\sim} \Leftrightarrow g(x) \sim f(x) \Leftrightarrow g(x) = f(x)$ times a unit] and the equivalence classes provide a disjoint decomposition of $F[x] - \{0\}$.

Our next result indicates that divisibility is really a question about equivalence classes, but we prefer to express it in less abstract terms.

3-4-4. Proposition. Given nonzero polynomials $f(x)$, $g(x) \in F[x]$, then, $f(x) \mid g(x) \Leftrightarrow$ any associate of $f(x)$ divides any associate of $g(x)$. More precisely, if $f'(x)$ is an associate of $f(x)$ and $g'(x)$ is an associate of $g(x)$: then

$$f(x) \mid g(x) \Leftrightarrow f'(x) \mid g'(x).$$

Proof: By hypothesis, we have $f'(x) = f(x)u$, $g'(x) = g(x)v$ where $u = u(x)$ and $v = v(x)$ are units of $F[x]$ (that is, $u, v \in F^*$). Both u^{-1} and v^{-1} exist and are units, so we also have $f(x) = f'(x)u^{-1}$, $g(x) = g'(x)v^{-1}$. Now: $f(x) \mid g(x) \Rightarrow$ there exists $h(x) \in F[x]$ with $g(x) = f(x)h(x) \Rightarrow g'(x) = g(x)v = f(x)h(x)v = (f(x)u)(u^{-1}h(x)v) = f'(x)(h(x)u^{-1}v)$, which says that $f'(x) \mid g'(x)$. The proof in the other direction goes exactly the same way. ▮

In the situation considered above, any element of $\lfloor f(x) \rfloor_{\sim}$ divides any element of $\lfloor g(x) \rfloor_{\sim}$, so it would be meaningful to say that $\lfloor f(x) \rfloor_{\sim}$ divides $\lfloor g(x) \rfloor_{\sim}$. It is simpler, however, to work with a specific representative (chosen according to the next result) of each equivalence class to settle questions of divisibility.

3-4-5. Proposition. Every nonzero polynomial is an associate of a unique monic polynomial. In other words, every equivalence class of $F[x]$ contains exactly one monic polynomial.

Proof: Write the given nonzero polynomial explicitly as $f(x) = a_0 + a_1x + a_2x^2 + \cdots + a_nx^n$ with $a_n \neq 0$. Since $a_n \in F^*$, it has an inverse $a_n^{-1} \in F^*$, and a_n^{-1} is a unit of $F[x]$. (In familiar situations such as F equals **Q** or **R** we would write $1/a_n$ for a_n^{-1}.) Then the polynomial

$$\begin{aligned} f'(x) &= a_n^{-1}f(x) = f(x)a_n^{-1} \\ &= (a_0 a_n^{-1}) + (a_1a_n^{-1})x + \cdots + (a_{n-1}a_n^{-1})x^{n-1} + x^n \end{aligned}$$

is monic and it is an associate of $f(x)$.

As for uniqueness, suppose $f''(x)$ is another monic associate of $f(x)$. Then $f''(x)$ is an associate of $f'(x)$ and we have

$$f''(x) = f'(x)u$$

for some $u \in F^*$. Of course, both $f'(x)$ and $f''(x)$ have the same degree, n—and by comparing the leading coefficients of $f''(x)$ and $f'(x)u$, we see that $u = 1$. Thus, there is indeed a unique monic associate of $f(x)$.

To prove the second version (which is really just a rephrasing of what has already been proved) we observe that every equivalence class is of form $\lfloor f(x) \rfloor_\sim$ with $f(x) \neq 0$. As above, let $f'(x)$ be the unique monic associate of $f(x)$. In general, for $g(x) \in F[x]$, we have:

$g(x)$ is a monic polynomial in $\lfloor f(x) \rfloor_\sim \Leftrightarrow g(x)$ is a monic associate *of* $f(x)$.

It follows that $f'(x)$ is the unique monic polynomial in $\lfloor f(x) \rfloor_\sim$, and the proof is complete. ∎

So far, in this section, we have been dealing with peripheral or preliminary results. The heart of the matter for us is to study factorization in $F[x]$ and to show that it is "essentially unique." The problem of factorization is not a strange one. In Chapter I, we made a detailed study of the integers **Z**, and proved that every nonzero integer has an essentially unique factorization into primes. Moreover (as the reader may convince himself by a careful examination of the structure of the discussion in sections 1-2, 1-3, 1-4), in the final analysis, the key result, on which the entire edifice of uniqueness of factorization rests, turns out to be the division algorithm. We shall see that the same kind of situation occurs in $F[x]$—namely, there is a division algorithm in $F[x]$, and one eventually derives uniqueness of factorization from it.

3-4-6. Division Algorithm. Given polynomials $f(x) \neq 0$ and $g(x)$ in $F[x]$, there exist unique polynomials $q(x)$ and $r(x)$ in $F[x]$ such that

$$g(x) = q(x)f(x) + r(x) \qquad \text{where} \quad \deg r(x) < \deg f(x).$$

Proof: Note that in view of our convention about the degree of the zero polynomial being $-\infty$ $(\deg(0) = -\infty)$, the condition $\deg r(x) < \deg f(x)$ is really another way of saying that: either $r(x) = 0$, or else, when $r(x) \neq 0$ its degree is less than the degree of $f(x)$. Naturally, $q(x)$ and $r(x)$ are known as the **quotient** and **remainder**, respectively.

The proof is not hard but somewhat formal, so let us indicate how it goes by a concrete example. Consider $f(x) = 3x^2 - x + 1$ and $g(x) = 2x^4 + 5x^3 - x^2 + 2$ in $\mathbf{R}[x]$. If we multiply $f(x)$ by $\frac{2}{3}x^2$, the result is

$$(\tfrac{2}{3}x^2)f(x) = 2x^4 - \tfrac{2}{3}x^3 + \tfrac{2}{3}x^2$$

which enables us to eliminate the leading term, $2x^4$, of $g(x)$ by subtracting $(\frac{2}{3}x^2)f(x)$—we have then

$$g(x) - (\tfrac{2}{3}x^2)f(x) = \tfrac{17}{3}x^3 - \tfrac{5}{3}x^2 + 2$$

or, what is the same,

$$g(x) = (\tfrac{2}{3}x^2)f(x) + (\tfrac{17}{3}x^3 - \tfrac{5}{3}x^2 + 2).$$

(We observe in passing that this cannot be done when $f(x)$ and $g(x)$ are viewed in $\mathbf{Z}[x]$, since $\frac{2}{3}$ is not an element of $\mathbf{Z}$; however, because $\mathbf{R}$ is a field, the steps we have performed are permissible; similarly, because $\mathbf{Q}$ is a field, the same procedure is applicable when $f(x)$ and $g(x)$ are viewed in $\mathbf{Q}[x]$.)

If we put $g'(x) = \frac{17}{3}x^3 - \frac{5}{3}x^2 + 2$, then

$$g(x) = (\tfrac{2}{3}x^2)f(x) + g'(x) \qquad (*)$$

and the same process can be applied to $f(x)$ and $g'(x)$. More precisely $(\frac{17}{9}x)f(x)$ has the same leading coefficient as $g'(x)$—and upon subtraction we obtain

$$g'(x) - (\tfrac{17}{9}x)f(x) = \tfrac{2}{9}x^2 - \tfrac{17}{9}x + 2.$$

Denoting the right side by $g''(x)$, we have

$$g'(x) = (\tfrac{17}{9}x)f(x) + g''(x)$$

and this may be substituted in $(*)$ to give

$$g(x) = (\tfrac{2}{3}x^2 + \tfrac{17}{9}x)f(x) + g''(x). \qquad (**)$$

Obviously, $\deg g'(x) > \deg g''(x)$, and each time our process is repeated the degree of the "left-over term" (meaning $g'(x), g''(x), \ldots$) diminishes—so eventually one obtains a left-over term of degree less than $\deg f(x) = 2$. In fact, here we have

$$g''(x) - \tfrac{2}{27}f(x) = -\tfrac{49}{27}x + \tfrac{52}{27}$$

—so calling the right side $g'''(x)$ yields

$$g''(x) = \tfrac{2}{27}f(x) + g'''(x)$$

and then

$$g(x) = (\tfrac{2}{3}x^2 + \tfrac{17}{9}x + \tfrac{2}{27})f(x) + g'''(x). \qquad (***)$$

Because $\deg g'''(x) = 1 < 2 = \deg f(x)$, we put $r(x) = g'''(x)$ and $q(x) = \frac{2}{3}x^2 + \frac{17}{9}x + \frac{2}{27}$, thus getting

$$g(x) = q(x)f(x) + r(x), \qquad \deg r(x) < \deg f(x).$$

Hence, we have shown explicitly how the division algorithm is valid for the given polynomials $f(x)$ and $g(x)$. Of course, the familiar process of "long division" is designed as a compact, efficient method for carrying out the computations used above, and thus finding $q(x)$ and $r(x)$—in fact,

$$
\begin{array}{r|rrrrr}
 & & & \frac{2}{3}x^2 & +\ \frac{17}{9}x & +\ \frac{2}{27} \\
\cline{2-6}
3x^2 - x + 1 & 2x^4 & +\ 5x^3 & -\ x^2 & +\ 0x & +\ 2 \\
 & 2x^4 & -\ \frac{2}{3}x^3 & +\ \frac{2}{3}x^2 & \vdots & \vdots \\
\cline{2-4}
 & & \frac{17}{3}x^3 & -\ \frac{5}{3}x^2 & +\ 0x & \vdots \\
 & & \frac{17}{3}x^3 & -\ \frac{17}{9}x^2 & +\ \frac{17}{9}x & \vdots \\
\cline{3-5}
 & & & \frac{2}{9}x^2 & -\ \frac{17}{9}x & +\ 2 \\
 & & & \frac{2}{9}x^2 & -\ \frac{2}{27}x & +\ \frac{2}{27} \\
\cline{4-6}
 & & & & -\ \frac{49}{27}x & +\ \frac{52}{27}
\end{array}
$$

Now, let us give a fairly precise formal proof of the division algorithm in its full generality; we do the existence part first—namely, given $f(x) \neq 0$ and $g(x)$ in $F[x]$ we show that there exist $q(x), r(x) \in F[x]$ such that

$$g(x) = q(x)f(x) + r(x), \qquad \deg r(x) < \deg f(x). \qquad (*)$$

If $\deg f(x) = 0$, then $f(x) = a_0$ with $a_0 \neq 0$, so upon putting $q(x) = a_0^{-1}g(x)$ and $r(x) = 0$ we see that $(*)$ holds. Therefore, we may assume, henceforth, that $\deg f(x) \geq 1$.

Suppose next that $\deg g(x) < \deg f(x)$ (this always includes the case where $g(x) = 0$); if we then put $q(x) = 0$ and $r(x) = g(x)$, then $(*)$ clearly holds. Therefore, we may assume, henceforth, that $\deg g(x) \geq \deg f(x) \geq 1$.

Let us write $n = \deg g(x)$, $m = \deg f(x)$—so $n \geq m \geq 1$—and proceed by induction on n. For each $n \geq 1$, let $\pi(n)$ be the following statement:

$$\pi(n): \quad \begin{cases} \text{if } f(x) \neq 0 \text{ and } \deg g(x) = n, \text{ then there exist} \\ q(x), r(x) \in F[x] \text{ such that } (*) \text{ holds.} \end{cases}$$

It is not hard to verify that $\pi(1)$ is true. In fact, if $n = 1$, then $m = 1$, so $g(x)$ and $f(x)$ take the form

$$g(x) = b_0 + b_1x, \qquad b_1 \neq 0,$$
$$f(x) = a_0 + a_1x, \qquad a_1 \neq 0.$$

Putting $q(x) = b_1 a_1^{-1}$ and $r(x) = b_0 - b_1 a_1^{-1} a_0$, we see immediately that $(*)$ holds. Hence, $\pi(1)$ is true.

Now, suppose inductively that $\pi(k)$ is true for $k = 1, 2, \ldots, n-1$, and consider the polynomial $g(x)$ of degree n. We may write

$$g(x) = b_0 + b_1 x + \cdots + b_n x^n, \qquad b_n \neq 0,$$
$$f(x) = a_0 + a_1 x + \cdots + a_m x^m, \qquad a_m \neq 0,$$

where, as usual, $n \geq m \geq 1$. Clearly, $(b_n a_m^{-1} x^{n-m}) f(x)$ is a polynomial whose leading term is $b_n x^n$, so upon subtracting it from $g(x)$ we obtain a polynomial

$$g'(x) = g(x) - (b_n a_m^{-1} x^{n-m}) f(x)$$

whose degree is less than or equal to $n - 1$. If $g'(x) = 0$, then taking $q(x) = b_n a_m^{-1} x^{n-m}$ and $r(x) = 0$ gives the validity of $(*)$. If $\deg g'(x) = 0$, then we may take $q(x) = b_n a_m^{-1} x^{n-m}$, $r(x) = g'(x)$; and $(*)$ holds because $\deg f(x) \geq 1$. It remains to consider the (most common) case where $\deg g'(x) \geq 1$ (and $\deg g'(x)$ is still less than or equal to $n - 1$). By the induction hypotheses, there exist $q'(x), r'(x) \in F[x]$ such that

$$g'(x) = q'(x) f(x) + r'(x), \qquad \deg r'(x) < \deg f(x).$$

Therefore,

$$g(x) = [b_n a_m^{-1} x^{n-m} + q'(x)] f(x) + r'(x),$$

so if we put $q(x) = b_n a_m^{-1} x^{n-m} + q'(x)$ and $r(x) = r'(x)$, then $(*)$ holds for $g(x)$. Thus, $\pi(n)$ is also true. Because $\pi(n)$ is now true for every $n = 1, 2, \ldots$, the existence proof for the division algorithm is complete.

To prove uniqueness for the division algorithm, we suppose (just as was done in **Z**; see 1-2-1) that there are two versions—say

$$g(x) = q_1(x) f(x) + r_1(x), \qquad \deg r_1(x) < \deg f(x),$$
$$g(x) = q_2(x) f(x) + r_2(x), \qquad \deg r_2(x) < \deg f(x).$$

Consequently,

$$(q_1(x) - q_2(x)) f(x) = r_2(x) - r_1(x). \qquad (\#)$$

Applying the rules for degrees of polynomials (see 3-3-5) we have

$$\deg(r_2(x) - r_1(x)) < \deg f(x).$$

On the other hand, since $f(x) \neq 0$, its leading coefficient, which is a nonzero element of F, is surely not a zero-divisor, so we have

$$\deg((q_1(x) - q_2(x)) f(x)) = \deg(q_1(x) - q_2(x)) + \deg f(x).$$

Therefore,

$$\deg(q_1(x) - q_2(x)) + \deg f(x) < \deg f(x)$$

and

$$\deg(q_1(x) - q_2(x)) < 0.$$

This tells us that $q_1(x) - q_2(x) = 0$, and then from (#), $r_2(x) - r_1(x) = 0$. Thus, $q_1(x) = q_2(x)$, $r_1(x) = r_2(x)$, and the division algorithm is unique. ▮

Once the division algorithm has been proved, the discussion of factorization in $F[x]$ proceeds along a familiar path; it is completely parallel to the discussion of factorization in **Z**, as given in Chapter I. It would be perfectly feasible, at this point, to challenge the reader to provide the details leading to uniqueness of factorization into primes in $F[x]$. We shall, however, take a less risky approach—proofs will be sketched briefly, but definitions and statements of results will be given carefully.

3-4-7. Euclidean Algorithm. Given $f(x), g(x) \in F[x]$ with $f(x) \neq 0$, we have

$$\begin{aligned}
g(x) &= q_1(x)f(x) + r_1(x), & \deg r_1(x) &< \deg f(x),\\
f(x) &= q_2(x)r_1(x) + r_2(x), & \deg r_2(x) &< \deg r_1(x),\\
r_1(x) &= q_3(x)r_2(x) + r_3(x), & \deg r_3(x) &< \deg r_2(x),\\
&\ \vdots & &\ \vdots\\
r_{i-2}(x) &= q_i(x)r_{i-1}(x) + r_i(x), & \deg r_i(x) &< \deg r_{i-1}(x),\\
&\ \vdots & &\ \vdots\\
r_{n-2}(x) &= q_n(x)r_{n-1}(x) + r_n(x), & \deg r_n(x) &< \deg r_{n-1}(x),\\
r_{n-1}(x) &= q_{n+1}(x)r_n(x) + 0.
\end{aligned}$$

Proof: The only point at issue is that by repeating the division algorithm, as indicated above, one reaches a remainder of 0 after a finite number of steps. Suppose this is false; then for each $i = 1, 2, 3, \ldots$ the remainder $r_i(x)$ is nonzero and hence has degree greater than or equal to 0. Thus, we have an infinite sequence of polynomials $r_1(x), r_2(x), \ldots$ for which

$$\deg r_1(x) > \deg r_2(x) > \cdots > \deg r_i(x) > \deg r_{i+1}(x) > \cdots \geq 0.$$

This is clearly impossible. ▮

The notation is arranged so $r_n(x)$ denotes the last nonzero remainder. All the preceding remainders $r_1(x), r_2(x), \ldots, r_{n-1}(x)$ are also nonzero. Note, incidentally, that if we ever arrive at a remainder $r_i(x)$ of degree 0—that is, $r_i(x) = c \neq 0$ in F—then this $r_i(x)$ ends up being $r_n(x)$ because $r_i(x) = c$, being a unit of $F[x]$, surely divides $r_{i-1}(x)$, so the next remainder is 0.

3-4-8. Examples. (1) Consider the polynomials $f(x) = 12x^3 + 16x^2 - 3x$, $g(x) = 12x^4 + 4x^3 - 13x^2 + 14x + 3$ in $\mathbf{R}[x]$. Their Euclidean algorithm takes the form

$$\begin{aligned} 12x^4 + 4x^3 - 13x^2 + 14x + 3 & \\ = (x - 1)(12x^3 + 16x^2 - 3x) &+ (6x^2 + 11x + 3), \\ 12x^3 + 16x^2 - 3x = (2x - 1)(6x^2 + 11x + 3) &+ (2x + 3), \\ 6x^2 + 11x + 3 = (3x + 1)(2x + 3), & \end{aligned}$$

and it is obtained by use of long division. In the notation of 3-4-7, we have: $q_1(x) = (x - 1)$, $r_1(x) = 6x^2 + 11x + 3$, $q_2(x) = (2x - 1)$, $r_2(x) = (2x + 3)$, $q_3(x) = (3x + 1)$.

(2) Consider $f(x) = 2x^3 + 5x^2 + 5x + 3$ and $g(x) = 4x^3 + 3x - 3$ as polynomials in $\mathbf{Z}_7[x]$. By long division, and keeping in mind that all coefficients come from $\mathbf{Z}_7$, we obtain for the Euclidean algorithm

$$\begin{aligned} 4x^3 + 3x - 3 &= 2(2x^3 + 5x^2 + 5x + 3) + (4x^2 - 2), \\ 2x^3 + 5x^2 + 5x + 3 &= (4x + 3)(4x^2 - 2) + (6x + 2), \\ 4x^2 - 2 &= (3x - 1)(6x + 2). \end{aligned}$$

In particular, $6x + 2$ (which is the same as $-x - 5$) is the last nonzero remainder.

3-4-9. Definition. Let nonzero polynomials $f(x), g(x) \in F[x]$ be given. A polynomial $d(x) \in F[x]$ which satisfies

(*i*) $d(x) \mid f(x)$ and $d(x) \mid g(x)$

is said to be a **common divisor** of $f(x)$ and $g(x)$. If, in addition, $d(x)$ satisfies the conditions,

(*ii*) if $c(x) \mid f(x)$ and $c(x) \mid g(x)$, then $c(x) \mid d(x)$,
(*iii*) $d(x)$ is monic,

then $d(x)$ is said to be a **greatest common divisor** of $f(x)$ and $g(x)$.

3-4-10. Proposition. Given nonzero polynomials $f(x), g(x) \in F[x]$, then:

(*i*) Their greatest common divisor exists and is unique—we denote it by $d(x) = (f(x), g(x))$.
(*ii*) We can compute $d(x)$ via the Euclidean algorithm; more precisely, if $r_n(x)$ is the last nonzero remainder in the Euclidean algorithm, then its unique monic associate (which is obtained by multiplying $r_n(x)$ by the inverse of its leading coefficient) is $d(x)$.

(*iii*) We can express $d(x)$ as a linear combination of $f(x)$ and $g(x)$; that is, we can find $s(x), t(x) \in F[x]$ for which

$$d(x) = s(x)f(x) + t(x)g(x).$$

(*iv*) The gcd $d(x)$ is the monic polynomial of smallest degree which can be expressed as a linear combination of $f(x)$ and $g(x)$.

Proof: (*i*) *uniqueness*. Suppose that both $d(x)$ and $d'(x)$ are gcd's. Then they must divide each other—$d(x) \mid d'(x)$ and $d'(x) \mid d(x)$. Because $d(x)$ and $d'(x)$ are both monic, we must have $d(x) = d'(x)$. This shows that if a gcd exists, then it is unique.

existence. Let $d(x)$ be the unique monic associate of $r_n(x)$, the last nonzero remainder of the Euclidean algorithm of $f(x)$ and $g(x)$—so if $d \neq 0$ is the leading coefficient of $r_n(x)$, then $d(x) = d^{-1}r_n(x)$. Thus, $d(x) \mid r_n(x)$ and working upward through the Euclidean algorithm we obtain, in turn,

$$d(x) \mid r_{n-1}(x), d(x) \mid r_{n-2}(x), \ldots, d(x) \mid f(x), d(x) \mid g(x)$$

—so $d(x)$ is a common divisor of $f(x)$ and $g(x)$.

On the other hand, if $c(x) \mid f(x)$ and $c(x) \mid g(x)$, then working downward through the Euclidean algorithm we obtain in turn,

$$c(x) \mid r_1(x), c(x) \mid r_2(x), \ldots, c(x) \mid r_n(x)$$

—so $c(x) \mid d(x)$. Hence, $d(x)$ satisfies all the requirements for the gcd of $f(x)$ and $g(x)$.

This proves (*i*) and (*ii*). Of course, it is essentially the same proof as was given for 1-3-2.

(*iii*) By working either upward or downward through the Euclidean algorithm (as was done in 1-3-6) we see that $r_n(x)$ can be expressed as a linear combination of $f(x)$ and $g(x)$—say,

$$r_n(x) = s'(x)f(x) + t'(x)g(x).$$

Then if, as before, d denotes the leading coefficient of $r_n(x)$, we have

$$d(x) = d^{-1}r_n(x) = (d^{-1}s'(x))f(x) + (d^{-1}t'(x))g(x)$$

and putting $s(x) = d^{-1}s'(x)$, $t(x) = d^{-1}t'(x)$ does it.

(*iv*) Suppose $d'(x)$ is a monic polynomial of degree less than or equal to $\deg d(x)$ which can be expressed as a linear combination of $f(x)$ and $g(x)$—say

$$d'(x) = p(x)f(x) + q(x)g(x).$$

Then $d(x) \mid d'(x)$ and $\deg d(x) \leq \deg d'(x)$. Therefore, $\deg d(x) = \deg d'(x)$; and because $d(x) \mid d'(x)$ with both of them monic, we conclude that $d(x) = d'(x)$. This completes the proof. ▌

3-4-11. Remarks. (1) In $\mathbf{Z}$, we have the notion of order and the consequent notion of "size" of an integer. In general, there is no ordering on a polynomial ring $F[x]$; however, the notion of degree of a polynomial can often be used to give a rough measure of size. For example, in the Euclidean algorithm, the degree provides a measure of the size of the remainders, and it is the key ingredient for guaranteeing that the Euclidean algorithm has only a finite number of steps.

(2) What is the significance of the requirement that $d(x)$ be monic in the definition of gcd (see 3-4-9)? Suppose our definition of gcd requires only that conditions (*i*) and (*ii*) hold—that is, that $d(x)$ is a common divisor of $f(x)$ and $g(x)$ which is divisible by every common divisor. Then any two polynomials which satisfy these conditions divide each other; so they are associates, and belong to the same equivalence class with respect to $\sim$ (the equivalence relation on $F[x]$ which we discussed earlier in this section). The only way to assure uniqueness of the gcd is to specify, somehow, a choice of a single element of this equivalence class; clearly, taking the unique monic one is a convenient way to do this.

(3) Incidentally, the requirement that the gcd be monic plays exactly the same role as the requirement, in $\mathbf{Z}$, that the gcd d of two nonzero integers a and b be positive (see 1-3-1). In more detail, suppose the definition of gcd in $\mathbf{Z}$ only calls for a common divisor of a and b which is divisible by every common divisor. Then any two integers d_1, d_2 which fulfill these conditions for gcd must divide each other. As in 3-4-1, this implies that d_2 equals d_1 times a unit of $\mathbf{Z}$; so, as in 3-4-2, we may say that d_2 is an associate of d_1 and write $d_2 \sim d_1$. Now, $\sim$ is clearly an equivalence relation on $\mathbf{Z}$ [really, on $\mathbf{Z} - (0)$], and the equivalence class $\lfloor d_1 \rfloor_\sim$ determined by d_1 consists of all integers of form d_1 times a unit. But there are only two units in $\mathbf{Z}$—namely, ± 1—so the equivalence class $\lfloor d_1 \rfloor_\sim$ consists of the two elements d_1, $-d_1$ (and, in general, any equivalence class has exactly two elements). In particular, $d_2 = \pm d_1$. To obtain uniqueness for the gcd, we must specify an element of the equivalence class

$$\lfloor d_1 \rfloor_\sim = \{d_1, -d_1\}$$

and the best way to do this is to choose the positive element of the pair.

(4) Consider the concrete example of 3-4-8. If

$$f(x) = 12x^3 + 16x^2 - 3x, \qquad g(x) = 12x^4 + 4x^3 - 13x^2 + 14x + 3$$

in $\mathbf{R}[x]$, then the last nonzero remainder in the Euclidean algorithm is $2x + 3$—so the gcd is $d(x) = x + \frac{3}{2}$. To express $2x + 3$ as a linear combination of $f(x)$ and $g(x)$, we have

$$\begin{aligned} 2x + 3 &= f(x) - (2x - 1)(6x^2 + 11x + 3) \\ &= f(x) - (2x - 1)[g(x) - (x - 1)f(x)] \\ &= (2x^2 - 3x + 2)f(x) + (1 - 2x)g(x). \end{aligned}$$

Consequently,

$$d(x) = x + \tfrac{3}{2} = (x^2 - \tfrac{3}{2}x + 1)f(x) + (\tfrac{1}{2} - x)g(x).$$

In similar fashion, consider

$$f(x) = 2x^3 + 5x^2 + 5x + 3, \qquad g(x) = 4x^3 + 3x - 3$$

in $\mathbf{Z}_7[x]$. The last nonzero remainder of the Euclidean algorithm is $6x + 2$, which is expressed as a linear combination of $f(x)$ and $g(x)$ by

$$\begin{aligned} 6x + 2 &= f(x) - (4x + 3)(4x^2 - 2) \\ &= f(x) - (4x + 3)\big(g(x) - 2f(x)\big) \\ &= (8x + 7)f(x) + (-4x - 3)g(x) \\ &= xf(x) + (3x + 4)g(x). \end{aligned}$$

Then, 6^{-1} (the inverse of 6 in $\mathbf{Z}_7$) equals 6, since $6 \cdot 6 = 1$ in $\mathbf{Z}_7$; so $d(x)$, the gcd, equals $x + 5$, and

$$d(x) = 6^{-1}(6x + 2) = x + 5 = (6x)f(x) + (4x + 3)g(x).$$

(5) In virtue of 3-4-10, the gcd of $f(x)$ and $g(x)$ is the monic polynomial of the largest degree which is a common divisor. As a matter of fact, it is not hard to see that this property may be taken as the definition of the gcd of $f(x)$ and $g(x)$. However, the definition we have given in 3-4-9—as the monic common divisor which is divisible by every common divisor—is the more appropriate one for generalization to other rings in which one may wish to consider gcds.

(6) If $d(x) = \big(f(x), g(x)\big) = 1$, then $f(x)$ and $g(x)$ are said to be **relatively prime**. In particular, if the last nonzero remainder $r_n(x)$ of the Euclidean algorithm for $f(x)$ and $g(x)$ is a constant—$r_n(x) = d \neq 0$, $d \in F$—then $d(x) = d^{-1}r_n(x) = 1$, so $f(x)$ and $g(x)$ are relatively prime.

(7) Our discussion of gcd may also be extended to situations where the Euclidean algorithm provides no information; for example, when $f(x) \mid g(x)$ or when one of the polynomials $f(x)$, $g(x)$ is 0. The treatment of such situations parallels the remarks made in 1-3-11 and 1-3-12.

3-4-12. Proposition. If $f(x) \mid g(x)h(x)$ while $f(x)$ and $g(x)$ are relatively prime, then $f(x) \mid h(x)$.

Proof: Since $\big(f(x),\ g(x)\big) = 1$ there exist $s(x), t(x) \in F[x]$ with

$$1 = s(x)f(x) + t(x)g(x).$$

Therefore,

$$h(x) = s(x)f(x)h(x) + t(x)g(x)h(x)$$

and both terms on the right side are divisible by $f(x)$. Hence, $f(x) \mid h(x)$. ▮

3-4-13. Definition. Suppose $p(x) \in F[x]$ is a nonzero polynomial which is not a unit. Then $p(x)$ is said to be **prime** or **irreducible** in $F[x]$ (or over F) when the only divisors of $p(x)$ are the units and the associates of $p(x)$.

According to the definition, only polynomials of degree greater than or equal to 1 are candidates for being primes. Now, any polynomial $f(x)$ of degree greater than or equal to 1 is automatically divisible by all the units (that is, by every element of F^*) and also by all its associates (that is, by all elements of form $f(x)u$ with u a unit, or to put it another way, by every element of $\underline{|f(x)|}_{\sim}$). The primes are those polynomials which have no divisors except the automatic ones. Our definition of irreducibility, or primeness, is equivalent to another very common formulation as given in the next result:

3-4-14. Proposition. Suppose $p(x) \in F[x]$ is a polynomial of degree $n \geq 1$; then

$p(x)$ is prime $\Leftrightarrow$ $p(x)$ cannot be expressed as a product of two polynomials of degree greater than or equal to 1

or, what is equivalent,

$p(x)$ is not prime $\Leftrightarrow$ $p(x)$ can be expressed as a product of two polynomials of degree greater than or equal to 1.

Proof: We shall prove the second version after recalling a few basic facts about polynomials in $F[x]$—namely:

(i) $\deg(f(x)g(x)) = \deg f(x) + \deg g(x)$,
(ii) $f(x)$ is a unit if and only if $\deg f(x) = 0$,
(iii) if two polynomials are associates they have the same degree.

Suppose $p(x)$ is not prime; so there exists a divisor $f(x)$ of $p(x)$ which is not a unit or an associate—and we may write

$$p(x) = f(x)g(x).$$

Since $f(x)$ is not a unit, $\deg f(x) \geq 1$. If $\deg g(x) = 0$, then $g(x)$ is a unit and $f(x)$ is an associate of $p(x)$—a contradiction. Hence $\deg g(x) \geq 1$, and we have proved the implication $\Rightarrow$.

Conversely, suppose $p(x)$ can be expressed as a product $p(x) = f(x)g(x)$ of two polynomials of degree greater than or equal to 1. We have $\deg f(x) \geq 1$, so $f(x)$ is not a unit. Since $\deg g(x) \geq 1$ and $\deg f(x) + \deg g(x) = \deg p(x) = n$, we have $\deg f(x) < n$—so $f(x)$ is not an associate of $p(x)$. Thus, $p(x)$, being divisible by $f(x)$ which is neither a unit nor an associate, is not prime. This proves the implication $\Leftarrow$. ▮

Naturally, a polynomial of degree 1 or more that is not irreducible (that is, it has a "nontrivial" factorization) is said to be **reducible** or **composite**.

Obviously, any polynomial of degree 1 is irreducible. The question of irreducibility for polynomials of degree 2 or more is complicated, and we shall eventually have a few things to say about it—but here we stick to the main theme.

3-4-15. Proposition. Suppose $p(x), f_1(x), f_2(x), \ldots, f_n(x)$ are polynomials of degree 1 or more over F. If $p(x)$ is prime and

$$p(x) \mid f_1(x)f_2(x) \cdots f_n(x),$$

then $p(x)$ divides at least one of the $f_i(x)$.

Proof: If $p(x)$ divides $f_1(x)$, we are finished. If $p(x) \nmid f_1(x)$, then because $p(x)$ is prime it is easy to see that $p(x)$ and $f_1(x)$ are relatively prime, $(p(x), f_1(x)) = 1$. Application of 3-4-12 now yields

$$p(x) \mid f_2(x)f_3(x) \cdots f_n(x)$$

—so, proceeding inductively, we obtain the desired result. ▮

3-4-16. Theorem. Let $f(x) \in F[x]$ be a polynomial of degree 1 or more. Then $f(x)$ can be expressed uniquely (up to order) as a product of a nonzero constant (that is, an element of F^*) and a finite number of monic irreducible polynomials over F.

Proof: *uniqueness*. Suppose $f(x)$ has two factorizations

$$f(x) = ap_1(x)p_2(x) \cdots p_r(x) = bq_1(x)q_2(x) \cdots q_s(x)$$

where $p_1(x), \ldots, p_r(x), q_1(x), \ldots, q_s(x)$ are monic and irreducible, and a, b are nonzero elements of F. Since a product of monic polynomials is monic, we see that a is the leading coefficient of $f(x)$, and so is b. Hence, $a = b$, and by cancellation we have

$$p_1(x)p_2(x) \cdots p_r(x) = q_1(x)q_2(x) \cdots q_s(x).$$

Now, $p_1(x)$ divides the right side, so according to 3-4-15 it divides at least one of the $q_i(x)$—and we may reindex so that $p_1(x) \mid q_1(x)$. Since both of these are monic and prime, it follows that they are equal. Then by cancellation,

$$p_2(x) \cdots p_r(x) = q_2(x) \cdots q_s(x)$$

and an induction takes care of the rest.

existence. Suppose the desired factorization does not exist for some polynomial of degree greater than or equal to 1. Let $f(x)$ then be a polynomial of minimal degree greater than or equal to 1 which does not factor into the product of a nonzero constant and a finite number of monic irreducible polynomials. Then $f(x)$ is not irreducible, so it factors into $f(x) = g(x)h(x)$, both of which have degree greater than or equal to 1. Since both $g(x)$ and $h(x)$ have degree less than $\deg f(x)$, minimality tells us that both $g(x)$ and $h(x)$ factor into the product of a nonzero constant and a finite number of monic irreducible polynomials. Putting these together provides $f(x)$ with a factorization of the same form—contradiction. ▌

Our development of unique factorization in $F[x]$ is substantially the same as the development of unique factorization in **Z** that was given in Chapter I, especially in Section 1-4. But there are some insubstantial alterations. According to the definition of a prime in $F[x]$, if $p(x)$ is prime, then so is any associate $p(x)u$, $u \in F^*$ (for example, in $\mathbf{R}[x]$, $2x + 3$, $x + (3/2)$, $\pi x + (3\pi/2)$ are associates of each other and each of them is a prime). Since nothing is said about this, the presumption is that these are distinct primes. But this is inappropriate, since all polynomials in the equivalence class $\lfloor p(x) \rfloor_\sim$ are interchangeable in questions of divisibility and factorization. Thus, when we want to have uniqueness of prime factorization, it is necessary to pick a single element from $\lfloor p(x) \rfloor_\sim$. This is the reason why the irreducible polynomials in 3-4-16 are required to be monic.

On the other hand, in **Z**, instead of permitting either element of an equivalence class (such as $\lfloor 7 \rfloor_\sim = \{7, -7\} = \lfloor -7 \rfloor_\sim$) to be considered as a prime, we always ignored the negative element and required that a prime be positive. The analogous choice for $F[x]$ would be to permit only monic polynomials to be irreducible. Of course, the reader who wishes can adjust the definitions of a prime in **Z** and in $F[x]$ to make them correspond more exactly, and then carry out the two treatments in exactly parallel fashion. In any case, the important objective is always to arrive at unique factorization into primes.

The reader should also note that the prime factorization of a polynomial includes a nonzero constant (in other words, a unit of $F[x]$). This corresponds to the fact that the prime factorization of an integer n includes a term ± 1 (that is, a unit of **Z**).

The problem of determining the prime factorization of a concrete polynomial or of just deciding if a given polynomial is irreducible is, to put it mildly, far from trivial. Much depends on the structure of the field F. For example, let us look at a simple polynomial of degree 2, $x^2 + 1$ in $F[x]$. If it is reducible over F (that is, if it factors in $F[x]$), its factorization must be of form

$$x^2 + 1 = (x - a)(x - b), \qquad a, b \in F.$$

So its reducibility will depend on whether the polynomial $x^2 + 1$ has a root in F. Thus:

(*i*) $x^2 + 1$ is irreducible in $\mathbf{R}[x]$, because there is no real number whose square is -1.

(*ii*) $x^2 + 1$ is reducible in $\mathbf{C}[x]$—in fact, for $i = \sqrt{-1}$,

$$x^2 + 1 = (x - i)(x + i).$$

(*iii*) $x^2 + 1$ is reducible in $\mathbf{Z}_2[x]$—in fact, here

$$x^2 + 1 = (x + 1)^2.$$

(*iv*) $x^2 + 1$ is irreducible over $\mathbf{Z}_7$ because, by trial and error, one sees that $x^2 + 1$ has no root in $\mathbf{Z}_7$.

(*v*) $x^2 + 1$ is reducible over $\mathbf{Z}_{13}$—in fact, in $\mathbf{Z}_{13}[x]$

$$(x^2 + 1) = (x - 5)(x - 8).$$

We shall return to such questions later.

3-4-17 / *PROBLEMS*

1. Find all units of the polynomial rings:
(*i*) $\mathbf{Q}[x]$, (*ii*) $\mathbf{R}[x]$, (*iii*) $\mathbf{Z}_2[x]$,
(*iv*) $\mathbf{Z}_3[x]$, (*v*) $\mathbf{Z}_7[x]$, (*vi*) $\mathbf{Z}_{11}[x]$,
(*vii*) $\mathbf{Z}_p[x]$, when p is a prime.

2. List all the associates of $2x^2 - x + 1$ in
(*i*) $\mathbf{Z}_3[x]$, (*ii*) $\mathbf{Z}_5[x]$, (*iii*) $\mathbf{Z}_7[x]$, (*iv*) $\mathbf{Z}_{11}[x]$.

3. Consider the polynomials

$$f_1(x) = 3x^2 - 4x + 2, \qquad f_2(x) = 4x^2 + 3x + 1,$$
$$f_3(x) = 2x^2 - 2x + 4, \qquad f_4(x) = 4x^2 + 2x - 1.$$

Which of these are associates in:
(*i*) $\mathbf{Z}_5[x]$, (*ii*) $\mathbf{Z}_7[x]$, (*iii*) $\mathbf{Z}_{11}[x]$?

4. Consider the polynomials $f(x) = x^4$, $g(x) = x^6$, $h(x) = x^3$ in $\mathbf{Z}_7[x]$, and let $\bar{f}, \bar{g}, \bar{h}$ denote the associated polynomial functions in Map($\mathbf{Z}_7$, $\mathbf{Z}_7$). Is it true that $\bar{f} \cdot \bar{g} = \bar{h}$?

5. Which of the following is an equivalence relation on the given set?

(*i*) X is the set of all triangles in the plane; ρ means "has the same perimeter as."

(*ii*) X is the set of all points in the plane; ρ means "is at distance less than or equal to 2 from."

(*iii*) X is the set of all living human beings; ρ means "is the father of."

(*iv*) X is the set of all living human beings; ρ means "is a brother of."

(*v*) X is the set of all living human beings; ρ means "has an ancester in common with."

(*vi*) $X = F[x]$ and for a fixed $f(x)$ of $\deg \geq 1$, $g(x)\ \rho\ h(x)$ means "$g(x) - h(x)$ is divisible by $f(x)$ (where $g(x)$ and $h(x)$ are arbitrary elements of $F[x]$)."

(*vii*) X is the collection of all sets; ρ means "can be put in 1–1 correspondence with."

(*viii*) $X = \mathbf{Z}$; $m\ \rho\ n$ means "$m - n$ is even."

(*ix*) $X = \mathbf{Z}$; $m\ \rho\ n$ means "$m - n$ is odd."

For those that are not equivalence relations, list the requirements that are not satisfied.

6. For each of the following, describe a set X and a relation ρ on it such that ρ is:

(*i*) reflexive, but neither symmetric nor transitive,
(*ii*) symmetric, but neither reflexive nor transitive,
(*iii*) transitive, but neither reflexive nor symmetric,
(*iv*) reflexive and symmetric, but not transitive,
(*v*) reflexive and transitive, but not symmetric,
(*vi*) symmetric and transitive, but not reflexive.

7. Consider the polynomial $f(x) = 3x^2 - 4x + 2$. Find the unique monic polynomial belonging to the equivalence class $\lfloor f(x) \rfloor_{\sim}$ when $f(x)$ is viewed in

(*i*) $\mathbf{Q}[x]$, (*ii*) $\mathbf{Z}_5[x]$, (*iii*) $\mathbf{Z}_7[x]$, (*iv*) $\mathbf{Z}_{11}[x]$.

8. (*a*) Compute the division algorithm in $\mathbf{Q}[x]$ for each of the following pairs of polynomials:

(*i*) $f(x) = x - 3$, $g(x) = x^3 + x^2 - 13x + 3$,
(*ii*) $f(x) = x + 2$, $g(x) = 2x^5 + 4x^3 - 7x^2 + 5x + 3$,
(*iii*) $f(x) = 2x - 1$, $g(x) = x^4 - 3x^2 + x + 5$,
(*iv*) $f(x) = x^2 + 2$, $g(x) = 3x^7 - 4$,
(*v*) $f(x) = 3x^2 - 2$, $g(x) = 2x^6 - x + 5$.

(*b*) Do the same thing when $f(x)$ and $g(x)$ are viewed in $\mathbf{R}[x]$.

9. Perform the division algorithm for each pair of polynomials in Problem 8 when they are viewed in

(*i*) $\mathbf{Z}_2[x]$, (*ii*) $\mathbf{Z}_3[x]$, (*iii*) $\mathbf{Z}_5[x]$, (*iv*) $\mathbf{Z}_7[x]$.

10. Compute the division algorithm for $f(x) = 4x^3 - 2x^2 + 1$ and $g(x) = 4x^5 + 2x^4 + 2x^3 - 2x^2 + x - 2$ when they are viewed in the polynomial ring
(*i*) $\mathbf{Q}[x]$, (*ii*) $\mathbf{R}[x]$, (*iii*) $\mathbf{Z}_3[x]$,
(*iv*) $\mathbf{Z}_5[x]$, (*v*) $\mathbf{Z}_7[x]$.

11. Suppose R is a commutative ring with unity and $f(x) \in R[x]$ is a polynomial of degree greater than or equal to 1 whose leading coefficient is a unit of R. Show that we have a division algorithm for $f(x)$ and any $g(x) \in R[x]$. Is it unique?

12. List as many places as you can where an equivalence relation has come up in this book.

13. Suppose $f(x), g(x) \in F[x]$ are both monic. Is it true that $f(x) + g(x)$, $f(x) - g(x)$, and $f(x) \cdot g(x)$ are monic?

14. For each pair of polynomials in $\mathbf{Q}[x]$ find their gcd and express it as a linear combination of them.
(*i*) $f(x) = 2x^3 - 4x^2 + x - 2$, $g(x) = x^3 - x^2 - x - 2$,
(*ii*) $f(x) = x^4 + x^3 + x^2 + x + 1$, $g(x) = x^3 - 1$,
(*iii*) $f(x) = x^2 + x + 1$, $g(x) = x^4 + x^2 + 1$,
(*iv*) $f(x) = x^3 - 1$, $g(x) = x^5 - x^4 + x^3 - x^2 + x - 1$.

15. For each pair of polynomials in the appropriate polynomial ring $F[x]$, find the gcd and express it as a linear combination of them.
(*i*) $f(x) = x^2 + x + 1$, $g(x) = x^4 + x + 1$ in $\mathbf{Z}_2[x]$,
(*ii*) $f(x) = x^2 + 1$, $g(x) = x^5 + 1$ in $\mathbf{Z}_2[x]$,
(*iii*) $f(x) = x^2 - x + 4$, $g(x) = x^3 + 2x^2 + 3x + 2$ in $\mathbf{Z}_3[x]$,
(*iv*) $f(x) = x^2 + 4$, $g(x) = x^3 + 2x^2 + 3x + 2$ in $\mathbf{Z}_5[x]$.

16. Exhibit two polynomials of degree 3 in $\mathbf{R}[x]$ such that the last nonzero remainder of their Euclidean algorithm is $x - 2$.

17. Suppose F is a field and K is a field which contains F (for example, $F = \mathbf{Q}$ and $K = \mathbf{R}$). Consider polynomials $f(x), g(x) \in F[x]$. Show that their gcd when they are viewed as polynomials in $K[x]$ is the same as their gcd when they are viewed as polynomials in $F[x]$.

18. Suppose that for nonzero polynomials $f(x), g(x) \in F[x]$ we define their greatest common divisor to be the monic polynomial of biggest degree which is a common divisor of $f(x)$ and $g(x)$. Show that this definition of gcd is equivalent to the one we have given in 3-4-9.

19. Show that our results about gcd are also valid when $f(x) \mid g(x)$ or when one of the polynomials $f(x), g(x)$ is 0. Things here go as indicated in 1-3-11 and 1-3-12.

20. (*i*) Suppose $p(x), f(x) \in F[x]$ are both of degree greater than or equal to 1. If $p(x)$ is prime and $p(x) \nmid f(x)$, show that $(p(x), f(x)) = 1$.
(*ii*) If $p(x)$ and $q(x)$ are both monic and prime with $p(x) \mid q(x)$, then $p(x) = q(x)$.

21. Given $f(x), g(x), h(x) \in F[x]$, we have:
(*i*) $(f(x), g(x)) = 1$ if and only if there exist $s(x), t(x) \in F[x]$ for which $s(x)f(x) + t(x)g(x) = 1$.
(*ii*) If $f(x) \mid h(x)$, $g(x) \mid h(x)$ and $(f(x), g(x)) = 1$, then $f(x)g(x) \mid h(x)$.
(*iii*) If $(f(x), g(x)) = 1$ and $(f(x), h(x)) = 1$, then $(f(x), g(x)h(x)) = 1$.
(*iv*) If $(f(x), g(x)) = 1$ and $h(x) \mid f(x)$, then $(h(x), g(x)) = 1$.
(*v*) If $(f(x), g(x)h(x)) = 1$, then $(f(x), g(x)) = (f(x), h(x)) = 1$.
(*vi*) If $(f(x), g(x)) = 1$, then $(f(x)h(x), g(x)) = (h(x), g(x))$.

22. How should one account for the fact that in $\mathbf{Z}_{15}[x]$, $x^2 - 1 = (x-1)(x-14) = (x-4)(x-11)$ has two distinct factorizations into primes?

23. Decide if the polynomial $x^2 + x + 1$ is irreducible over each of the following fields; if it is reducible, factor into irreducible polynomials.
(*i*) $\mathbf{Q}$, (*ii*) $\mathbf{R}$, (*iii*) $\mathbf{C}$, (*iv*) $\mathbf{Z}_2$,
(*v*) $\mathbf{Z}_3$, (*vi*) $\mathbf{Z}_5$, (*vii*) $\mathbf{Z}_7$, (*viii*) $\mathbf{Z}_{11}$.

24. Do Problem 23 for each of the following polynomials:
(*i*) $x^2 + 2$, (*ii*) $x^2 - 2$, (*iii*) $x^3 - 1$,
(*iv*) $x^3 + 1$, (*v*) $x^2 + 3x - 2$, (*vi*) $x^3 - 2$.

25. Suppose $f(x)$ is a monic polynomial of degree 2 or 3 over the field F. Show that $f(x)$ is irreducible over $F \Leftrightarrow f(x)$ has no roots in F. Why does this statement not carry over when $\deg f(x) = 4$?

26. Consider the polynomial $f(x) = x^7 - x \in \mathbf{Z}_7[x]$. Find all its roots. Can you find its factorization into primes?

27. Consider the polynomial $f(x) = x^2 + x + 1 \in \mathbf{Z}_5[x]$ and compute $g(x) = (f(x))^5 = (x^2 + x + 1)^5$. Compute the associated polynomial functions $\bar{f}, \bar{g} \in \operatorname{Map}(\mathbf{Z}_5, \mathbf{Z}_5)$ (that is, specify their values for each element of $\mathbf{Z}_5$) and show that $\bar{f} = \bar{g}$.

28. Suppose R and S are commutative rings with unity, and $\phi: R \to S$ is a homomorphism. Define a mapping

$$\phi^{\#}: R[x] \to S[x]$$

as follows: If $f(x) = \Sigma\, a_i x^i \in R[x]$, then $\phi^{\#}(f(x))$ [which may be written as $f^{\#}(x)$] equals $\Sigma\, \phi(a_i)x^i$. Show that $\phi^{\#}$ is a homomorphism; if ϕ is injective, so is $\phi^{\#}$; if ϕ is surjective, so is $\phi^{\#}$; if ϕ is an isomorphism, so is $\phi^{\#}$.

29. For any $m \geq 2$, let ϕ: $\mathbf{Z} \to \mathbf{Z}_m$ be the canonical map $\phi(a) = \lfloor a \rfloor_m$, so (with the notation of Problem 28) $\phi^{\#}$: $\mathbf{Z}[x] \to \mathbf{Z}_m[x]$ is a surjective homomorphism. Prove that ker $\phi^{\#}$ consists of those polynomials all of whose coefficients are divisible by m.

30. By making use of the mapping $\phi^{\#}$: $\mathbf{Z}[x] \to \mathbf{Z}_p[x]$ prove that if $f(x)$, $g(x)$, $h(x) \in \mathbf{Z}[x]$ with $f(x) = g(x)h(x)$ and every coefficient of $f(x)$ is divisible by the prime p, then at least one of $g(x)$ or $h(x)$ has the property that all its coefficients are divisible by p.

31. (*i*) Does $f(x) = x^2 + 3$ divide $g(x) = x^5 + x^3 + x^2 - 9$ in $\mathbf{Z}[x]$?
(*ii*) Find all $m \geq 2$ such that $f(x) \mid g(x)$ in $\mathbf{Z}_m[x]$.
(*iii*) Do the same for $f(x) = x^2 + 2$, $g(x) = x^6 + 5x^5 + 5x^4 + 10x^3 + 8x^2 + 4$.

32. For each monic irreducible polynomial $p(x) \in F[x]$ define $v_{p(x)}(f(x))$ for any $f(x) \neq 0 \in F[x]$ as the exponent to which $p(x)$ appears in the prime factorization of $f(x)$. State and prove results concerning the $v_{p(x)}$'s analogous to those discussed in Section 1-5 for the v_p's.

33. Define "least common multiple" in $F[x]$. Discuss its properties, and prove them.

3-5. Roots of Polynomials

Throughout this section, F will denote an arbitrary field, D an integral domain, and R a commutative ring with unity. To keep the discussion consistent, our general results, which deal mainly with roots of polynomials, will be stated in $F[x]$; but many of them are valid in $D[x]$, and some even hold in $R[x]$. These results are generalizations of facts that are more or less familiar from high school—where instead of working over a field F, we invariably considered only polynomials with coefficients coming from the fields **Q**, **R**, or **C**. From these general considerations, we shall pass to some remarks for specific choices of F—namely, **Q**, **R**, **C**, and (because of its number-theoretic importance) $\mathbf{Z}_p$.

3-5-1. Remainder Theorem. Consider an arbitrary polynomial $f(x) \in F[x]$; then for any $c \in F$ the remainder upon division of $f(x)$ by $(x - c)$ is precisely $f(c)$.

Proof: According to the division algorithm, when $(x - c)$ is divided into $f(x)$ we obtain, uniquely, a quotient $q(x)$ and a remainder $r(x)$ with $\deg r(x) < \deg(x - c) = 1$. Thus, $r(x)$ is a constant—$r(x) = r \in F$—and we have

$$f(x) = q(x)(x - c) + r.$$

Now, substitution gives

$$f(c) = q(c)(c - c) + r = r$$

which completes the proof. ▮

The remainder theorem may be rephrased as asserting the existence of a unique $q(x) \in F[x]$ such that

$$f(x) = q(x)(x - c) + f(c).$$

So as a corollary we have: $(x - c) \mid f(x) \Leftrightarrow$ there exists $q(x) \in F[x]$ for which $f(x) = q(x)(x - c) \Leftrightarrow f(c) = 0 \Leftrightarrow c$ is a root of $f(x)$. This proves:

3-5-2. Factor Theorem. For $c \in F$ and $f(x) \in F[x]$ we have

$$(x - c) \text{ divides } f(x) \Leftrightarrow c \text{ is a root of } f(x).$$

3-5-3. Remark. A stronger form of the factor theorem has already been proved. In fact, in 3-3-13, we proved it for $f(x) \in R[x]$, $c \in R$, by another method. It is worth noting that the method of proof used here in 3-5-1 and 3-5-2 can be carried over to $R[x]$. In more detail, even though there may not be a division algorithm for every pair of polynomials in $R[x]$, we can derive a division algorithm for $(x - c) \in R[x]$ and $f(x) \in R[x]$ precisely because $(x - c)$ is monic. In other words, given $c \in R$ and $f(x) \in R[x]$ there exist unique $q(x) \in R[x]$ and $r \in R$ such that

$$f(x) = q(x)(x - c) + r \tag{$*$}$$

and from this, the remainder theorem and the factor theorem follow as above. Incidentally, the reader can easily prove $(*)$ informally by actually doing the long division; because $(x - c)$ is monic there is nothing to stop the process until one arrives at a remainder with no x's—that is, a constant. Or else, one can prove $(*)$ formally by rewriting the proof of 3-4-6 (including uniqueness) practically word for word; everything goes through because the leading coefficient of $(x - c)$ is 1.

3-5-4. Theorem. Suppose $f(x) = a_0 + a_1 x + \cdots + a_n x^n \in F[x]$ is a polynomial of degree $n \geq 1$, and suppose we have n distinct elements $c_1, c_2, \ldots, c_n$ of F, all of which are roots of $f(x)$; then $f(x)$ has the factorization

$$f(x) = a_n(x - c_1)(x - c_2) \cdots (x - c_n).$$

Proof: The idea of the proof is simple. Since c_1 is a root of $f(x)$ there exists $f_1(x) \in F[x]$ for which

$$f(x) = (x - c_1) f_1(x).$$

Then $c_2, \ldots, c_n$ are roots of $f_1(x)$, because they are distinct from c_1. In particular, we can write $f_1(x) = (x - c_2)f_2(x)$, so that

$$f(x) = (x - c_1)(x - c_2)f_2(x), \qquad f_2(x) \in F[x].$$

This process continues until all the c's are exhausted, at which stage the expression looks like

$$f(x) = (x - c_1)(x - c_2) \cdots (x - c_n)f_n(x), \qquad f_n(x) \in F[x].$$

But now, by comparing degrees and leading coefficients, it follows that $f_n(x) = a_n$.

Rather than being content with the preceding sketch of a proof, let us give a rigorous, formal proof by induction on n.

If $n = 1$, then our polynomial is of form $f(x) = a_0 + a_1 x$, $a_1 \neq 0$, and we have a root c_1 of $f(x)$. Consequently, $0 = f(c_1) = a_0 + a_1 c_1$—so that $c_1 = -a_1^{-1}a_0$ and

$$f(x) = a_1(x + a_1^{-1}a_0) = a_1(x - c).$$

Thus, the theorem holds for $n = 1$.

Now suppose inductively that the theorem holds for n (that is, for any polynomial of degree n and n of its roots) and suppose we have $n + 1$ distinct roots $c_1, \ldots, c_n, c_{n+1}$ (in F) of a polynomial $f(x) = a_0 + a_1 x + \cdots + a_n x^n + a_{n+1}x^{n+1} \in F[x]$ of degree $n + 1$. Since c_{n+1} is a root, there exists $f_1(x) \in F[x]$ such that

$$f(x) = f_1(x)(x - c_{n+1}). \tag{\#}$$

For each $i = 1, 2, \ldots, n$ we then have

$$0 = f(c_i) = f_1(c_i)(c_i - c_{n+1}).$$

But $c_i - c_{n+1} \neq 0$ for $i = 1, \ldots, n$, because $c_1, c_2, \ldots, c_n, c_{n+1}$ are distinct; so that $f_1(c_i) = 0$. Consequently, we have n distinct roots $c_1, c_2, \ldots, c_n$ of $f_1(x)$, and from (#), $f_1(x)$ is clearly a polynomial of degree n whose leading coefficient is a_{n+1}. By the induction hypothesis, we know that

$$f_1(x) = a_{n+1}(x - c_1)(x - c_2) \cdots (x - c_n)$$

and hence

$$f(x) = a_{n+1}(x - c_1)(x - c_2) \cdots (x - c_n)(x - c_{n+1}).$$

So the theorem holds for $n + 1$. This completes the proof. ▌

This result has diverse and important consequences of which the most familiar is the following.

3-5-5. Corollary. A polynomial of degree $n \geq 1$ over the field F can have at most n distinct roots in F.

Proof: By contradiction. Suppose $f(x) = a_0 + a_1x + \cdots + a_nx^n \in F[x]$ is a polynomial of degree n which has more than n distinct roots. In particular, $f(x)$ surely has $n + 1$ distinct roots in F—call them $c_1, \ldots, c_n, c_{n+1}$. According to the theorem applied to $f(x)$ and the n distinct roots $c_1, c_2, \ldots, c_n$ [observe that the hypothesis of 3-5-4 does not require $c_1, \ldots, c_n$ to be all the roots of $f(x)$], $f(x)$ has the factorization

$$f(x) = a_n(x - c_1)(x - c_2) \cdots (x - c_n).$$

But then, since c_{n+1} is a root,

$$0 = f(c_{n+1}) = a_n(c_{n+1} - c_1)(c_{n+1} - c_2) \cdots (c_{n+1} - c_n).$$

Because we are in a field, at least one of the factors on the right-hand side must be 0; however, by our hypotheses, $a_n \neq 0$ and $c_1, c_2, \ldots, c_n, c_{n+1}$ are distinct—contradiction. ▮

3-5-6. Corollary. If $n \geq 0$ and $f(x) \in F[x]$ is a polynomial of degree less than or equal to n which has $n + 1$ (or more) distinct roots in F, then $f(x) = 0$—that is, $f(x)$ is the zero polynomial.

Proof: Suppose $\deg f(x) = k \leq n$. If $k = 0$, then, according to the hypothesis $f(x)$ has at least one root in F; but this is impossible because a polynomial of degree 0 is a nonzero constant, and thus has no roots. If $k \geq 1$, then, by hypothesis, $f(x)$ has $n + 1 \geq k + 1$ distinct roots in F, which contradicts 3-5-5. The only possibility remaining is $k = \deg f(x) = -\infty$—in other words, $f(x) = 0$. ▮

3-5-7. Corollary. Suppose $n \geq 0$ and $f(x), g(x) \in F[x]$ are both of degree less than or equal to n. If their values agree at $n + 1$ (or more) elements of F, then $f(x) = g(x)$.

Proof: The essential assertion here is: If $c_1, c_2, \ldots, c_{n+1}$ are distinct elements of F, and

$$f(c_i) = g(c_i), \qquad i = 1, 2, \ldots, n + 1,$$

then $f(x) = g(x)$. But this is easy. In fact, $\phi(x) = f(x) - g(x)$ is a polynomial of degree less than or equal to n which has $n + 1$ distinct roots in F, since for $i = 1, \ldots, n + 1$

$$\phi(c_i) = f(c_i) - g(c_i) = 0.$$

According to 3-5-6, $\phi(x) = 0$; so $f(x) = g(x)$. ▮

3-5-8. Corollary. Suppose the field F has an infinite number of elements. Then distinct polynomials over F give rise to distinct polynomial functions; in other words, the homomorphism of rings

$$\Lambda: F[x] \to \mathrm{Map}(F, F)$$

treated in 3-3-11 is injective.

Proof: From 3-3-11, we know that Λ is a homomorphism; so to prove Λ is injective, it suffices to show $\ker \Lambda = (0)$.

Suppose $f(x) \in \ker \Lambda$. Then its image $\bar{f}$ under Λ is 0—$\bar{f} = \Lambda(f(x)) = 0 \in \mathrm{Map}(F, F)$. Now an element of $\mathrm{Map}(F, F)$ equals 0 when it is the mapping which takes the value 0 for every element of F; hence, $\bar{f}(c) = 0$ for every $c \in F$. According to the definition of the polynomial function $\bar{f}$, we have

$$f(c) = \bar{f}(c) = 0 \qquad \text{for every } c \in F.$$

Thus, every element of F is a root of $f(x)$ and—this is the crucial step—because F is infinite, $f(x)$ has an infinite number of distinct roots in F.

However for the given polynomial $f(x) \in \ker \Lambda$, there surely exists an integer $n \geq 0$ for which $\deg f(x) \leq n$. Since $f(x)$ has more than $n + 1$ distinct roots, 3-5-6 guarantees that $f(x) = 0$. ▮

3-5-9. Remarks. (1) It has already been observed that 3-5-1 and 3-5-2 hold when F is replaced by D, or even by R. Here we note that all the subsequent results—namely, 3-5-4 to 3-5-8—are also valid over any integral domain D and, in particular, over $\mathbf{Z}$. It is a straightforward matter for the reader to check that in this situation our proofs of these results go through, word for word.

However, these results do not carry over for general R. For example, over $R = \mathbf{Z}_{15}$ (which is not an integral domain) the polynomial $x^2 - 1$ is of degree 2, but has the four roots 1, 4, 11, 14 in $\mathbf{Z}_{15}$; so, in particular, 3-5-5 is not valid over $\mathbf{Z}_{15}$.

(2) In high school, the polynomials considered always came from $\mathbf{Z}[x]$, $\mathbf{Q}[x]$, $\mathbf{R}[x]$, or $\mathbf{C}[x]$. Because $\mathbf{Z}$, $\mathbf{Q}$, $\mathbf{R}$, $\mathbf{C}$ all have an infinite number of elements, 3-5-8 tells us that in these cases distinct polynomials give rise to distinct polynomial functions. This fact justifies the customary carelessness in high school, where the distinction between polynomials and polynomial functions is glossed over. It is imprecise to do so, but it does not lead to seriously wrong results.

(3) When F has only a finite number of elements, the proof of 3-5-8 breaks down, and, in fact, the result is false (more on this in 3-5-15). For example, we have already seen in 3-3-10 that when $F = \mathbf{Z}_p$, the distinct

polynomials $f(x) = x^p$ and $g(x) = x$ give rise to the same polynomial function —that is, $\bar{f} = \bar{g}$, or put another way, $\Lambda(f(x)) = \Lambda(g(x))$. Of course, the polynomial

$$\phi(x) = x^p - x \in \mathbf{Z}_p[x]$$

then satisfies

$$\phi(\alpha) = 0 \qquad \text{for every } \alpha \in \mathbf{Z}_p$$

—so the polynomial function associated with $\phi(x)$ is $\Lambda(\phi(x)) = \bar{\phi} = 0$. Thus, $\phi(x)$ is a nonzero polynomial in ker Λ—so Λ is not injective.

Let us digress a bit from our main theme of studying roots of polynomials to discuss an interesting problem in which the previous results have an application. We work in the Euclidean plane $\mathbf{R} \times \mathbf{R}$ and consider a "curve-fitting" problem. Given a finite set of points in the plane, we wish to find a curve (if one exists) that passes through all of them; moreover, we want the curve to be the graph of some polynomial.

This problem is not unfamiliar. For example, suppose we are given two points (2, 3) and (5, 7) in the plane. Of course, there is a unique straight line which passes through these points. In fact, one sees easily that this straight line is the one whose equation is

$$y = \tfrac{4}{3}x + \tfrac{1}{3}.$$

In other words, if we let $f(x)$ denote the polynomial $\frac{4}{3}x + \frac{1}{3}$ of degree 1 in $\mathbf{R}[x]$, then its graph $G(f) = \{(a, f(a)) \mid a \in \mathbf{R}\}$ (as introduced in 3-3-12) is the unique straight line that passes through the points (2, 3) and (5, 7).

More generally, if we are given the points (a_1, b_1), (a_2, b_2), with $a_1 \neq a_2$, in the plane, then the reader may check that the straight line passing through these points has the equation

$$y = \frac{b_2 - b_1}{a_2 - a_1}x + \frac{a_2 b_1 - a_1 b_2}{a_2 - a_1}.$$

In other words, the graph of the linear polynomial

$$f(x) = \frac{b_2 - b_1}{a_2 - a_1}x + \frac{a_2 b_1 - a_1 b_2}{a_2 - a_1} \in \mathbf{R}[x]$$

is the straight line passing through the points (a_1, b_1) and (a_2, b_2). Note the need for $a_1 \neq a_2$; if $a_1 = a_2$, then the straight line in question is vertical and, in particular, it is impossible to find a polynomial $f(x) \in \mathbf{R}[x]$ whose graph passes through both points—after all, the graph of a polynomial includes exactly one point for each value of x.

What happens if we are given three points?—say, $(-1, 6)$, $(0, 1)$, $(2, 3)$. Here, one might guess that there is a "quadratic equation" of form

$$y = a + bx + cx^2$$

whose graph passes through the three given points. If so, we may substitute the coordinates of the points in this equation, to obtain

$$\begin{aligned} 6 &= a - b + c, \\ 1 &= a, \\ 3 &= a + 2b + 4c. \end{aligned}$$

Thus, $a = 1$, $b = -3$, $c = 2$; so the graph of the polynomial

$$f(x) = 1 - 3x + 2x^2 = 2x^2 - 3x + 1$$

passes through $(-1, 6)$, $(0, 1)$, and $(2, 3)$.

In general, if three points in the plane are given, we would substitute their coordinates in the equation $y = a + bx + cx^2$ thus obtaining three linear equations in the three unknowns a, b, c. Presumably these three equations can be solved (and we finally have the equation of a parabola; or a straight line when $c = 0$, which occurs when the three points are collinear), but how do we know?

Similarly, when four points in the plane are given, we can substitute their coordinates in the equation

$$y = f(x) = a + bx + cx^2 + dx^3,$$

thus getting four linear equations in the four unknowns a, b, c, d. Hopefully, these equations have a solution, and we finally have a cubic polynomial whose graph passes through the four given points.

Surely, the same method (known as the method of **undetermined coefficients**) can be extended to five, six, or more points; but it becomes increasingly harder to solve the simultaneous linear equations as the number of unknowns increases—and we cannot be certain that a solution of these equations exists, or whether a solution is unique. Fortunately, there is an elegant method, due to Lagrange (1736–1813), that settles explicitly all questions related to the problem of finding a polynomial whose graph passes through a finite number of given points.

3-5-10. Lagrange Interpolation. Suppose $n \geq 1$ and $a_1, a_2, \ldots, a_n$ are distinct elements of the field F. Then, given any elements $b_1, b_2, \ldots, b_n$ of F there exists a unique polynomial $f(x) \in F[x]$, of degree less than n, such that

$$f(a_i) = b_i, \qquad i = 1, 2, \ldots, n.$$

To put it another way, we can exhibit explicitly a unique polynomial $f(x)$ of degree less than n whose graph (in $F \times F$) passes through the n points $(a_1, b_1), (a_2, b_2), \ldots, (a_n, b_n)$ of $F \times F$.

Proof: For each $j = 1, 2, \ldots, n$ let us put

$$g_j(x) = \prod_{i \neq j} (x - a_i).$$

Thus, each $g_j(x)$ is the product of linear terms $(x - a_i) \in F[x]$, one for each $i \neq j$—so clearly $g_j(x)$ is a polynomial of degree $n - 1$ in $F[x]$. Since $a_1, a_2, \ldots, a_n$ are distinct, we observe that for each j,

$$g_j(a_j) = \prod_{i \neq j} (a_j - a_i) \neq 0.$$

On the other hand, when $i \neq j$, one of the terms in the product used to compute $g_j(a_i)$ is $(a_i - a_i)$—so

$$g_j(a_i) = 0, \qquad \text{for } i \neq j.$$

Now, for each $j = 1, 2, \ldots, n$ we put

$$f_j(x) = \frac{1}{g_j(a_j)}\, g_j(x).$$

All we have done is to multiply the polynomial $g_j(x)$ by the nonzero field element $1/g_j(a_j)$ [and this is valid because $g_j(a_j) \neq 0$]. Thus, each $f_j(x)$ is a polynomial of degree $n - 1$ in $F[x]$, and the values taken by $f_j(x)$ at $a_1, a_2, \ldots, a_n$ are

$$f_j(a_j) = 1,$$
$$f_j(a_i) = 0, \qquad \text{for } i \neq j.$$

Then,

$$f(x) = b_1 f_1(x) + b_2 f_2(x) + \cdots + b_n f_n(x)$$

[note the analogy of this construction of $f(x)$ with the way we proved the Chinese remainder theorem 3-1-8] is a polynomial of degree less than or equal to $n - 1$ in $F[x]$, and it satisfies

$$f(a_i) = b_i, \qquad i = 1, 2, \ldots, n.$$

As for uniqueness, if $f'(x) \in F[x]$ is also a polynomial of degree less than or equal to $n - 1$ such that $f'(a_i) = b_i$ for $i = 1, \ldots, n$, then we have two polynomials $f(x), f'(x)$ of degree less than or equal to $n - 1$ whose values agree at n elements of F—so, according to 3-5-7, $f(x) = g(x)$. ▮

3-5-11. Example. Consider the five points

$$(-1, 10), \quad (1, 2), \quad (2, 1), \quad (3, 18), \quad (0, 3)$$

in the Euclidean plane, $\mathbf{R} \times \mathbf{R}$. Let us find the unique polynomial of degree less than or equal to 4 (in $\mathbf{R}[x]$) that passes through these five points. This will be done by the method used in the proof of Lagrange interpolation.

We have, in the notation of 3-5-10, $n = 5$ and

$$\begin{array}{lllll} a_1 = -1, & a_2 = 1, & a_3 = 2, & a_4 = 3, & a_5 = 0, \\ b_1 = 10, & b_2 = 2, & b_3 = 1, & b_4 = 18, & b_5 = 3. \end{array}$$

Therefore,

$$g_1(x) = (x-1)(x-2)(x-3)(x-0)$$

—since $g_1(x)$ is, by definition, $(x - a_2)(x - a_3)(x - a_4)(x - a_5)$—and

$$\begin{aligned} g_2(x) &= (x+1)(x-2)(x-3)(x-0), \\ g_3(x) &= (x+1)(x-1)(x-3)(x-0), \\ g_4(x) &= (x+1)(x-1)(x-2)(x-0), \\ g_5(x) &= (x+1)(x-1)(x-2)(x-3). \end{aligned}$$

For each $j = 1, \ldots, 5$ we compute $g_j(a_j)$—thus obtaining

$$\begin{aligned} g_1(-1) &= (-2)(-3)(-4)(-1) = 24, \\ g_2(1) &= (2)(-1)(-2)(1) = 4, \\ g_3(2) &= (3)(1)(-1)(2) = -6, \\ g_4(3) &= (4)(2)(1)(3) = 24, \\ g_5(0) &= (1)(-1)(-2)(-3) = -6. \end{aligned}$$

Next, we compute

$$f_j(x) = \frac{1}{g_j(a_j)} \cdot g_j(x) \qquad \text{for } j = 1, \ldots, 5;$$

upon multiplying out the terms in $g_j(x)$, these turn out to be

$$\begin{aligned} f_1(x) &= \tfrac{1}{24}(x^4 - 6x^3 + 11x^2 - 6x), \\ f_2(x) &= \tfrac{1}{4}(x^4 - 4x^3 + x^2 + 6x), \\ f_3(x) &= -\tfrac{1}{6}(x^4 - 3x^3 - x^2 + 3x), \\ f_4(x) &= \tfrac{1}{24}(x^4 - 2x^3 - x^2 + 2x), \\ f_5(x) &= -\tfrac{1}{6}(x^4 - 5x^3 + 5x^2 + 5x - 6). \end{aligned}$$

If we now put

$$f(x) = b_1 f_1(x) + b_2 f_2(x) + b_3 f_3(x) + b_4 f_4(x) + b_5 f_5(x),$$

then a straightforward computation gives

$$\begin{aligned} f(x) &= 10f_1(x) + 2f_2(x) + 1f_3(x) + 18f_4(x) + 3f_5(x) \\ &= x^4 - 3x^3 + 2x^2 - x + 3. \end{aligned}$$

This is the desired polynomial satisfying

$$f(-1) = 10, \qquad f(1) = 2, \qquad f(2) = 1, \qquad f(3) = 18, \qquad f(0) = 3.$$

Returning to the central question of roots of polynomials, let us apply the previous results to polynomials over a finite field $\mathbf{Z}_p$.

3-5-12. Proposition. If p is prime and we denote the elements of the finite field $\mathbf{Z}_p$ by $0, 1, 2, \ldots, p-1$, then we have in $\mathbf{Z}_p[x]$ the factorizations

$$\begin{aligned} x^p - x &= x(x-1)(x-2)\cdots(x-(p-1)), \\ x^{p-1} - 1 &= (x-1)(x-2)\cdots(x-(p-1)). \end{aligned}$$

Moreover, for any $\alpha \in \mathbf{Z}_p$

$$\alpha^p = \alpha$$

and for any $\alpha \neq 0$ in $\mathbf{Z}_p$

$$\alpha^{p-1} = 1.$$

Proof: Every element $\alpha \in \mathbf{Z}_p = \{0, 1, \ldots, p-1\}$ is a root of the polynomial $f(x) = x^p - x$ (since $\alpha^p = \alpha$); hence, $f(x) = x^p - x \in \mathbf{Z}_p[x]$ is a monic polynomial of degree $p \geq 1$ and the p distinct elements $0, 1, 2, \ldots, p-1$ of $\mathbf{Z}_p$ are roots. So, according to 3-5-4, we have the factorization

$$x^p - x = x(x-1)(x-2)\cdots(x-(p-1)) \qquad \text{in } \mathbf{Z}_p[x].$$

Furthermore, $x^p - x = x(x^{p-1} - 1)$, and because $\mathbf{Z}_p[x]$ is an integral domain x may be canceled; thus leaving us with

$$x^{p-1} - 1 = (x-1)(x-2)\cdots(x-(p-1)) \qquad \text{in } \mathbf{Z}_p[x]. \quad \blacksquare$$

This result has a very useful number-theoretic consequence, which is usually ascribed to Wilson (1741–1793), although it is not clear who really discovered it.

3-5-13. Wilson's Theorem. If p is prime, then

$$(p-1)! \equiv -1 \pmod{p}.$$

Even more, for any $n > 1$

$$(n-1)! \equiv -1 \pmod{n} \Leftrightarrow n \text{ is prime.}$$

Proof: Look at the factorization

$$x^{p-1} - 1 = (x-1)(x-2)\cdots(x-(p-1))$$

in $\mathbf{Z}_p[x]$. In particular, the constant terms of the two sides of this equation must be equal. Hence,

$$-1 = (-1)(-2)\cdots(-(p-1)).$$

This equality is in $\mathbf{Z}_p$, so for ordinary integers it translates to

$$-1 \equiv (-1)(-2)\cdots(-(p-1)) \pmod{p}.$$

The right-hand side has $p-1$ terms, so we have

$$-1 \equiv (-1)^{p-1}(p-1)! \pmod{p}.$$

If p is an odd prime, then $(-1)^{p-1} = 1$ and we obtain $(p-1)! \equiv -1 \pmod{p}$. If $p = 2$, then surely $(p-1)! \equiv -1 \pmod{p}$, since $1 \equiv -1 \pmod{2}$. This proves the first part.

To prove the converse, let $(n-1)! \equiv -1 \pmod{n}$ be given, and suppose n is not prime. Then there exists a prime p which divides n, and $p < n$. Since $p \mid n$, we have

$$(n-1)! \equiv -1 \pmod{p}$$

and because $p < n$, p must be one of the terms in $(n-1)!$. Thus, $p \mid (n-1)!$ and $(n-1)! \equiv 0 \pmod{p}$. The conclusion is

$$0 \equiv -1 \pmod{p}$$

—a contradiction. Hence, n must be prime. ▌

3-5-14. Proposition. If a polynomial $f(x) \in \mathbf{Z}_p[x]$ is given, there exists a unique polynomial $r(x)$ of degree less than p in $\mathbf{Z}_p[x]$ which determines the same polynomial function as $f(x)$—that is, $\bar{r} = \bar{f} \in \mathrm{Map}(\mathbf{Z}_p, \mathbf{Z}_p)$. In particular, $r(x)$ has the same roots as $f(x)$.

Proof: *existence.* If $\deg f(x) < p$, which includes the case $f(x) = 0$, then we may take $r(x) = f(x)$; so suppose $\deg f(x) \geq p$. Applying the division algorithm to $x^p - x$ and $f(x)$, we obtain $q(x), r(x) \in \mathbf{Z}_p[x]$ for which

$$f(x) = q(x)(x^p - x) + r(x), \qquad \deg r(x) < \deg(x^p - x) = p.$$

For each $c \in \mathbf{Z}_p$, we have $c^p = c$, so that

$$\begin{aligned} \bar{f}(c) &= f(c) \\ &= q(c)(c^p - c) + r(c) \\ &= r(c) \\ &= \bar{r}(c). \end{aligned}$$

Thus, the polynomial functions $\bar{f}$ and $\bar{r}$ are equal; or, to put it another way, the polynomials $f(x)$ and $r(x)$ take the same value at each $c \in \mathbf{Z}_p$; that is, $f(c) = r(c)$ for each $c \in \mathbf{Z}_p$. In particular, $f(x)$ and $r(x)$ surely have the same roots. This show that $r(x)$ is a polynomial which has the required properties.

uniqueness. Suppose $s(x) \in \mathbf{Z}_p[x]$ is also a polynomial of degree less than p with $\bar{s} = \bar{f}$. Then $r(x)$ and $s(x)$ are both of degree less than p, and since $r(c) = s(c)$ for every $c \in \mathbf{Z}_p$ their values agree at the p distinct elements of $\mathbf{Z}_p$. According to 3-5-7, $r(x) = s(x)$. ▮

To illustrate this result, let us work in $\mathbf{Z}_7[x]$. If $f(x) = x^5 + 3x^2 - x + 4$, then, because its degree is less than 7, $r(x) = x^5 + 3x^2 - x + 4$. If $f(x) = x^7 - x$, then $r(x) = 0$, since $f(x) = 1 \cdot (x^7 - x) + 0$. If $f(x) = x^8$, then $r(x) = x^2$, because $x^8 = x(x^7 - x) + x^2$. If $f(x) = x^7$, then $r(x) = x$, because $x^7 = 1(x^7 - x) + x$. If $f(x) = 3x^9 + x^8 - x^7 - 3x^3 - x^2 + x + 5$, then $r(x) = 5$, since by long division

$$3x^9 + x^8 - x^7 - 3x^3 - x^2 + x + 5 = (3x^2 + x - 1)(x^7 - x) + 5.$$

In particular, the polynomial function determined by $f(x) = 3x^9 + x^8 - x^7 - 3x^3 - x^2 + x + 5$ takes the value 5 for every $c \in \mathbf{Z}_7$. The reader may check (in case there is any doubt) that $\bar{f} = \bar{r}$ in each of the foregoing examples, by comparing $f(c)$ and $r(c)$ for each $c \in \mathbf{Z}_7$.

In the preceding examples, we found $r(x)$ by taking the remainder when $f(x)$ is divided by $x^p - x$. Another way to obtain $r(x)$ is to start with $f(x)$, and in any power of x greater than or equal to 7 replace x^p by x; and then keep going until only powers of x less than the 7th remain—for example, $x^{10} - 2x^9 + 3x^8 - x^7 = x^7 \cdot x^3 - 2x^2 \cdot x^7 + 3x \cdot x^7 - x^7 \to x \cdot x^3 - 2x^2 \cdot x + 3x \cdot x - x = x^4 - 2x^3 + 3x^2 - x$.

3-5-15. Exercise. We have seen in 3-5-8 that when the field F is infinite, the homomorphism

$$\Lambda : F[x] \to \mathrm{Map}(F, F)$$

is injective. On the other hand, when F is the finite field $\mathbf{Z}_p$, the homomorphism

$$\Lambda\colon \mathbf{Z}_p[x] \to \mathrm{Map}(\mathbf{Z}_p, \mathbf{Z}_p)$$

is not injective—for the nonzero polynomial $x^p - x$ is clearly in the kernel of Λ.

Even more, if $f(x) \in \mathbf{Z}_p[x]$, then, with the notation of 3-5-14,

$$r(x) = 0 \Leftrightarrow (x^p - x) \mid f(x)$$

and it follows that

$$\bar{f} = 0 \in \mathrm{Map}(\mathbf{Z}_p, \mathbf{Z}_p) \Leftrightarrow (x^p - x) \mid f(x).$$

Consequently, ker Λ consists of all multiples of $x^p - x$. (Note that the polynomial 0 is also a multiple of $x^p - x$.) In particular, for $f(x), g(x) \in \mathbf{Z}_p[x]$, we have

$$\bar{f} = \bar{g} \Leftrightarrow (x^p - x) \mid (f(x) - g(x)).$$

We have seen in 3-5-5 that a polynomial of degree n over a field can have at most n distinct roots in the field. When the field in question is the finite field $\mathbf{Z}_p$ we can characterize those polynomials of degree n that have *exactly* n distinct roots.

3-5-16. Proposition. If $f(x) \in \mathbf{Z}_p[x]$ is of degree $n \geq 1$, then $f(x)$ has exactly n distinct roots in $\mathbf{Z}_p \Leftrightarrow f(x) \mid (x^p - x)$.

Proof: Let $g(x)$ be the unique monic associate of $f(x)$. Then $\deg g(x) = n$, $g(x)$ has precisely the same roots as $f(x)$, and $g(x) \mid (x^p - x)$ if and only if $f(x) \mid (x^p - x)$. Therefore, it suffices to prove our assertion for $g(x)$.

If $g(x)$ has the n distinct roots $c_1, c_2, \ldots, c_n$ in $\mathbf{Z}_p$, then in virtue of 3-5-4,

$$g(x) = (x - c_1)(x - c_2) \cdots (x - c_n).$$

Now, $c_1, c_2, \ldots, c_n$ are distinct elements from the set $\mathbf{Z}_p = \{0, 1, \ldots, p-1\}$ so $n \leq p$ and, even more, each $(x - c_i)$ appears exactly once on the right-hand side of

$$x^p - x = x(x-1)(x-2) \cdots (x - (p-1)).$$

Consequently, $g(x) \mid (x^p - x)$.

Conversely, if $g(x) \mid (x^p - x)$, then there exists $h(x) \in \mathbf{Z}_p[x]$ for which

$$g(x)h(x) = x^p - x = x(x-1)(x-2) \cdots (x - (p-1)). \qquad (*)$$

Now, $g(x)$ and $h(x)$ may be factored into primes of $\mathbf{Z}_p[x]$. But the prime factorization of $g(x)h(x)$ is given by the right side of (*). Since $g(x)$ is monic of degree n, $h(x)$ is clearly monic of degree $p - n$. Thus, by uniqueness of the prime factorization, $g(x)$ is the product of n of the linear factors on the right side of (*) [and $h(x)$ is the product of the remaining $p - n$ terms on the right side of (*)]; hence, it follows that $g(x)$ has n distinct roots. ▌

3-5-17. Examples. Let us illustrate the preceding result by examining several polynomials in $\mathbf{Z}_{11}[x]$ and deciding if they have the maximum possible number of distinct roots in $\mathbf{Z}_p$ (namely, n, when the polynomial has degree n).

(1) The polynomial

$$f(x) = x^{12} - 3x^7 + 4x^2 + 2 \in \mathbf{Z}_{11}[x]$$

cannot have 12 distinct roots in $\mathbf{Z}_{11}$, since $\mathbf{Z}_{11}$ has only 11 distinct elements. Of course, $f(x)$ does not divide $x^{11} - x$—its degree is too high—so again $f(x)$ does not have 12 distinct roots. How many distinct roots does $f(x)$ have? At this stage, we can do little more than test all the elements 0, 1, 2, ..., 10 of $\mathbf{Z}_p$ to see if they are roots. Note that since

$$f(x) = x(x^{11} - x) + (-3x^7 + 5x^2 + 2)$$

the roots of $f(x)$ are the same as those of $r(x) = -3x^7 + 5x^2 + 2$.

(2) The polynomial

$$f(x) = x^9 + x^8 + x^7 + x^6 + x^5 + x^4 + x^3 + x^2 + x + 1 \in \mathbf{Z}_{11}[x],$$

being of degree 9, can have at most nine distinct roots in $\mathbf{Z}_{11}$. But $f(x)$ divides $x^{11} - x$; in fact,

$$x^{11} - x = (x^2 - x)f(x) = x(x - 1)f(x)$$

—so $f(x)$ has nine distinct roots; and obviously the roots are 2, 3, 4, 5, 6, 7, 8, 9, 10, since from the factorization of $x^{11} - x$ into 11 linear terms it follows that

$$f(x) = (x - 2)(x - 3)(x - 4)(x - 5)(x - 6)(x - 7)(x - 8)(x - 9)(x - 10).$$

The reader may multiply this out to check it, and also for practice computing in $\mathbf{Z}_{11}[x]$.

(3) The polynomial

$$f(x) = x^2 + x + 3 \in \mathbf{Z}_{11}[x]$$

has at most two distinct roots in $\mathbf{Z}_{11}$. By long division, it is not hard to see that $f(x)$ does not divide $x^{11} - x$; so $f(x)$ does not have two distinct roots. If we test all 11 elements of $\mathbf{Z}_{11}$, it turns out that 5 is a root. [Once 5 has been found to be a root, there is no need to test for other roots because we have already shown that $f(x)$ does not have two distinct roots.] Then $f(x)$ factors in $\mathbf{Z}_{11}[x]$ as

$$f(x) = (x - 5)(x - 5) = (x - 5)^2.$$

Thus, 5 should be considered as a "double" root of $f(x)$—there are really two roots, but they are equal.

(4) Consider the polynomial

$$f(x) = 3x^2 - x + 3 \in \mathbf{Z}_{11}[x].$$

Its unique monic associate is (since 4 is the multiplicative inverse of 3 in $\mathbf{Z}_{11}$)

$$g(x) = 4f(x) = x^2 - 4x + 1 \in \mathbf{Z}_{11}[x].$$

Of course, the roots of $g(x)$ are the same as those of $f(x)$. By long division, one sees that $g(x) \mid (x^{11} - x)$, so $g(x)$ has two distinct roots in $\mathbf{Z}_{11}$. In fact, by trial and error, $g(x)$ has the two roots 7, 8 and

$$g(x) = (x - 7)(x - 8).$$

Naturally, $f(x)$ also has the two roots 7, 8 and

$$f(x) = 3g(x) = 3(x - 7)(x - 8).$$

(5) The polynomial

$$f(x) = x^2 - 4x - 4 \in \mathbf{Z}_{11}[x]$$

does not divide $x^{11} - x$, so it does not have two distinct roots. In fact, by trial and error, $f(x)$ has no roots in $\mathbf{Z}_{11}$. Thus, $f(x)$ has no linear factors; and because $f(x)$ has degree 2, we conclude that it is irreducible.

(6) Consider the polynomial

$$f(x) = x^3 - 5x^2 + 5x - 9 \in \mathbf{Z}_{11}[x].$$

It has at most three distinct roots in $\mathbf{Z}_{11}$. Because $f(x) \nmid (x^{11} - x)$ (which is left to the reader to check), $f(x)$ does not have three distinct roots. How many roots does $f(x)$ have? Clearly, 0 and 1 are not roots of $f(x)$, but 2 is a root. Therefore, $(x - 2)$ divides $f(x)$; in fact, we have

$$f(x) = (x - 2)(x^2 - 3x - 1) \in \mathbf{Z}_{11}[x].$$

Now, it is easy to see that $x^2 - 3x - 1$ has no roots in $\mathbf{Z}_{11}$ (so it is irreducible). Consequently, $f(x)$ has the single root, 2.

According to 3-5-12 or 3-5-16, $x^{p-1} - 1$ has $p - 1$ distinct roots in $\mathbf{Z}_p$. However, a somewhat stronger result holds:

3-5-18. Proposition. If p is prime and $d \mid (p-1)$, $d > 0$, then the polynomial $x^d - 1$ has exactly d distinct roots in $\mathbf{Z}_p$.

Proof: Let us write $p - 1 = dt$ and make use of 3-5-16. Thus, $x^d - 1$ divides $x^p - 1$, since

$$\begin{aligned} x^p - x &= x[(x^d)^t - 1] \\ &= x(x^d - 1)[(x^d)^{t-1} + (x^d)^{t-2} + \cdots + x^d + 1]. \end{aligned}$$

This completes the proof. ▮

In virtue of this result, the polynomial $x^9 - 1 \in \mathbf{Z}_{127}[x]$ has exactly nine roots in $\mathbf{Z}_{127}$ (for $p = 127$ is prime and $d = 9$ divides $p - 1 = 126$)—but we do nothing about trying to locate them. By the same reasoning $x^2 - 1$ has two distinct roots in $\mathbf{Z}_{127}$, and $x^7 - 1$ has seven distinct roots in $\mathbf{Z}_{127}$.

Now that we have learned a few facts about polynomials over $\mathbf{Z}_p$, let us have a look at polynomials of the type familiar to us from high school algebra—namely, those with coefficients from the fields $\mathbf{Q}$, $\mathbf{R}$, or $\mathbf{C}$. Of course, the same polynomial may be viewed over different fields, and its roots and factorization depend on the choice of the field in question. For example, consider the polynomial

$$x^2 + 1.$$

Viewing it in $\mathbf{Q}[x]$—clearly, it has no roots in $\mathbf{Q}$ (since there is no rational number whose square is -1) and thus $x^2 + 1$ is irreducible over $\mathbf{Q}$. Viewing it in $\mathbf{R}[x]$—there are no roots in $\mathbf{R}$ and $x^2 + 1$ is irreducible over $\mathbf{R}$. Viewing it in $\mathbf{C}[x]$—clearly, $\pm i$ are roots in $\mathbf{C}$ (in fact, the reason why one enlarges the reals, $\mathbf{R}$, to form the complexes, $\mathbf{C}$, is precisely to have a field in which $x^2 + 1$ has a root) and we have the factorization

$$x^2 + 1 = (x - i)(x + i)$$

over $\mathbf{C}$.

As another example, consider the polynomial

$$x^2 - 2.$$

As a polynomial in $\mathbf{Q}[x]$, it has no roots since there is no rational number whose square is 2. (This is the familiar statement that "the $\sqrt{2}$ is irrational," as observed in Miscellaneous Problem 20 of Chapter I.) Thus, $x^2 - 2$ is irreducible over $\mathbf{Q}$. As a polynomial in $\mathbf{R}[x]$, $x^2 - 2$ clearly has the roots $\pm\sqrt{2}$ (note that $\sqrt{2}$ is a perfectly good real number; it is the number associated with the hypotenuse of a right triangle both of whose legs have length 1), and we have the factorization

$$x^2 - 2 = (x - \sqrt{2})(x + \sqrt{2})$$

over **R**. Of course, the story over **C** now goes exactly like the story over **R**—$x^2 - 2$ has the roots $\pm\sqrt{2}$ in **C**, and its factorization in **C**$[x]$ is $(x - \sqrt{2})(x + \sqrt{2})$.

We begin our discussion of high school-type polynomials by considering polynomials over **C**. The key fact here is a theorem of immense importance whose proof is unfortunately far beyond the scope of this book—namely:

3-5-19. Fundamental Theorem of Algebra. Any polynomial $f(x) \in$ **C**$[x]$ of degree 1 or more has a root in **C**.

It may be remarked that every known proof of the fundamental theorem of algebra, and there are many of them, makes use of nonalgebraic properties of **C**. Note that although the theorem asserts the existence of a root for every polynomial in **C**$[x]$ it says nothing about how to find such a root. Nevertheless, the mere fact of existence of a single root has important consequences. Thus, suppose $c_1 \in$ **C** is the root of $f(x)$ specified by the theorem. Then $(x - c_1)$ divides $f(x)$; so there exists a polynomial $f_1(x) \in$ **C**$[x]$ of degree $n - 1$ such that

$$f(x) = (x - c_1)f_1(x).$$

If $n - 1 > 1$, then the fundamental theorem of algebra applies to $f_1(x)$; so there exists $c_2 \in$ **C** which is a root of $f_1(x)$, and we can then write $f_1(x) = (x - c_2)f_2(x)$ where $f_2(x) \in$ **C**$[x]$ is of degree $n - 2$—and hence

$$f(x) = (x - c_1)(x - c_2)f_2(x).$$

If $\deg f_2(x) = n - 2$ is still greater than 1, the process may be repeated. This procedure keeps going (the reader may set up a formal induction if he wishes) until one arrives at:

3-5-20. Theorem. Any polynomial $f(x) = a_0 + a_1 x + \cdots + a_n x^n \in$ **C**$[x]$ of degree $n \geq 1$ factors into n linear factors; more precisely, there exist $c_1, c_2, \ldots, c_n \in$ **C** such that

$$f(x) = a_n(x - c_1)(x - c_2) \cdots (x - c_n). \qquad (*)$$

Another way to express this result is: A polynomial $f(x) \in$ **C**$[x]$ of degree $n \geq 1$ has n roots in **C**. But note that these n roots $c_1, c_2, \ldots, c_n$ need not be distinct; this depends on whether or not the factors $(x - c_i)$, $i = 1, \ldots, n$ are distinct.

Now, according to 3-4-16, since $\mathbf{C}$ is a field, our $f(x) \in \mathbf{C}[x]$ can be expressed uniquely (up to order) as a product of a nonzero constant and a finite number of monic irreducible polynomials over $\mathbf{C}$. But any linear term of form $x - c_i$ is monic and irreducible, so

$$f(x) = a_n(x - c_1)(x - c_2) \cdots (x - c_n)$$

is the "factorization into primes"! In particular, it follows that a polynomial $f(x)$ of degree 2 or more cannot be irreducible—its reducibility is displayed by its factorization into linear factors. We conclude:

3-5-21. Corollary. A polynomial in $\mathbf{C}[x]$ is prime (that is, irreducible) if and only if it is of degree 1.

According to this result

$$\{x - c \mid c \in \mathbf{C}\}$$

is the set of all monic irreducible polynomials over $\mathbf{C}$. Furthermore, these primes of $\mathbf{C}[x]$ are distinct in the sense that if $c \neq c'$, then the primes $x - c$ and $x - c'$ are not associates. (In particular, the number of primes in $\mathbf{C}[x]$ is infinite.) On the other hand, an arbitrary irreducible polynomial, being of form $ax + b$, $a, b \in \mathbf{C}$, $a \neq 0$, is an associate of one of these—namely of $a^{-1}(ax + b) = x - (-b/a)$.

3-5-22. Remark. In virtue of the preceding discussion, we know that any quadratic polynomial

$$f(x) = ax^2 + bx + c \in \mathbf{C}[x]$$

has two roots in $\mathbf{C}$ (that is, $f(x)$ factors into the product of a and two monic polynomials of degree 1). These roots may be found explicitly by the method which is often referred to as "completing the square"—it goes as follows:

$$\begin{aligned}
z_0 \in \mathbf{C} \text{ is a root of } f(x) &\Leftrightarrow f(z_0) = az_0^2 + bz_0 + c = 0 \\
&\Leftrightarrow 4a^2z_0^2 + 4abz_0 + 4ac = 0 \qquad (\text{since } a \neq 0) \\
&\Leftrightarrow 4a^2z_0^2 + 4abz_0 = -4ac \\
&\Leftrightarrow 4a^2z_0^2 + 4abz_0 + b^2 = b^2 - 4ac \\
&\Leftrightarrow (2az_0 + b)^2 = b^2 - 4ac \\
&\Leftrightarrow 2az_0 + b = \pm\sqrt{b^2 - 4ac} \\
&\Leftrightarrow z_0 = \frac{-b \pm \sqrt{b^2 - 4ac}}{2a}.
\end{aligned}$$

Of course, this result is known as the "*quadratic formula*"; it says that the two roots of $ax^2 + bx + c$ are

$$\frac{-b + \sqrt{b^2 - 4ac}}{2a}, \qquad \frac{-b - \sqrt{b^2 - 4ac}}{2a}$$

and its factorization is

$$ax^2 + bx + c = a\left[x - \left(\frac{-b + \sqrt{b^2 - 4ac}}{2a}\right)\right]\left[x - \left(\frac{-b - \sqrt{b^2 - 4ac}}{2a}\right)\right]$$

However, one matter of concern remains—does the square root of the complex number $b^2 - 4ac$ really exist, and can we find it? More generally, given any complex number $s + ti$ (s and t are real) do there exist any complex numbers whose square is $s + ti$, and can we find them? This amounts to finding the roots of the polynomial

$$x^2 - (s + ti)$$

over **C**. According to 3-5-20, this polynomial has two roots in **C**. If $u + vi$ (*with* $u, v \in \mathbf{R}$) is a root, then

$$(u + vi)^2 = s + ti \tag{*}$$

and clearly $-(u + vi)$ is also a root. Thus, $u + vi$ and $-(u + vi)$ are the two roots, and the factorization is

$$x^2 - (s + ti) = [x - (u + vi)][x + (u + vi)].$$

The notation $\pm\sqrt{s + ti}$ refers to the two roots of $x^2 - (s + ti)$. It does not matter which element of the pair $\{u + vi, -u - vi\}$ we take to be $+\sqrt{s + ti}$ and which one to be $-\sqrt{s + ti}$. [Since **C** cannot be ordered (see 2-3-2) there is no satisfactory notion of positivity in **C**, so we are in no position to speak of the "positive square root" of $s + ti$.] Of course, when $s + ti$ is a positive real number—that is, when $t = 0$ and $s > 0$—we still take $\sqrt{s + ti} = \sqrt{s}$ to mean the positive square root—that is, the positive real number whose square is s.

What about finding the two square roots of $s + ti$ explicitly? From (*) we have the two equations

$$u^2 - v^2 = s,$$
$$2uv = t.$$

If $t = 0$ (that is, when $s + ti$ is real), then—if $s > 0$, we take $v = 0$ and $u = \sqrt{s}$ (the usual positive square root of the positive real number s), and if $s < 0$, we take $u = 0$ and $v = (\sqrt{-s})$. Of course, the case $t = 0$, $s = 0$ is trivial as $u = v = 0$ will do.

Now, if $t \neq 0$, then $u \neq 0$ and $v \neq 0$ (since $2uv = t$), so we may substitute $v = t/2u$ in the first equation. The end result is that one solution $u + vi$ of $x^2 - (s + ti)$ is given by

$$u = \sqrt{\frac{s + \sqrt{s^2 + t^2}}{2}}, \qquad v = \frac{t}{\sqrt{2(s + \sqrt{s^2 + t^2})}} \tag{\#}$$

and the other one is then $-u - vi$. Of course, all the square roots that occur here are positive square roots of positive real numbers (for when $t \neq 0$, $s^2 + t^2 > 0$ and $s + \sqrt{s^2 + t^2} > 0$.

As an example of all this, consider the polynomial

$$f(x) = x^2 + x + (1 + i) \in \mathbf{C}[x].$$

By the quadratic formula, the two roots are

$$\frac{-1 \pm \sqrt{-3 - 4i}}{2}.$$

As for $\pm\sqrt{-3 - 4i}$, these are the two roots $u + vi$, $-(u + vi)$ of $x^2 + 3 + 4i$. The reader may compute u and v from scratch, or simply substitute in (#)—thus obtaining

$$u = 1, \qquad v = -2.$$

Hence, the roots of $f(x)$ are

$$\frac{-1 \pm (1 - 2i)}{2}$$

—that is, $-i$ and $-1 + i$.

Next, let us consider any polynomial $f(x) \in \mathbf{R}[x]$ of degree 1 or more and with leading coefficient $a_n \neq 0$. From the foregoing, we know that when $f(x)$ is viewed in $\mathbf{C}[x]$ it factors in the form

$$f(x) = a_n(x - c_1)(x - c_2) \cdots (x - c_n), \qquad c_1, \ldots, c_n \in \mathbf{C}.$$

Some of the roots $c_1, \ldots, c_n$ may be nonreal complex numbers and others may be real. Concerning the nonreal roots, we have:

3-5-23. Proposition. If $f(x) = a_0 + a_1 x + \cdots + a_n x^n \in \mathbf{R}[x]$ is a polynomial of degree $n \geq 1$ and $c \in \mathbf{C}$ is a nonreal root of $f(x)$, then its complex conjugate $\bar{c}$ is also a root of $f(x)$.

Proof: This is the familiar statement that the complex (nonreal) roots of a real polynomial occur in conjugate pairs. The proof is easy; it makes use of such basic properties of conjugation in $\mathbf{C}$ as: the conjugate of a sum is the sum of conjugates, the conjugate of a product is the product of the conjugates, a real number is its own conjugate—namely,

$$\begin{aligned}
c \text{ is a root of } f(x) &\Rightarrow a_0 + a_1 c + a_2 c^2 + \cdots + a_n c^n = 0 \\
&\Rightarrow \overline{a_0 + a_1 c + a_2 c^2 + \cdots + a_n c^n} = \bar{0} \\
&\Rightarrow \overline{a_0} + \overline{a_1 c} + \overline{a_2 c^2} + \cdots + \overline{a_n c^n} = 0 \\
&\Rightarrow \bar{a}_0 + \bar{a}_1 \bar{c} + \bar{a}_2 \bar{c}^2 + \cdots + \bar{a}_n \bar{c}^n = 0 \\
&\Rightarrow a_0 + a_1 \bar{c} + a_2 \bar{c}^2 + \cdots + a_n \bar{c}^n = 0 \\
&\Rightarrow f(\bar{c}) = 0 \\
&\Rightarrow \bar{c} \text{ is a root of } f(x). \quad \blacksquare
\end{aligned}$$

Consequently, if $c \in \mathbf{C}$ is a nonreal complex root of

$$f(x) = a_n(x - c_1)(x - c_2) \cdots (x - c_n),$$

say $c = s + ti$, $s, t \in \mathbf{R}$, $t \neq 0$, then so is $\bar{c} = s - ti$, and both $(x - c)$ and $(x - \bar{c})$ appear among the factors of $f(x)$. Taking both of these, we have

$$(x - c)(x - \bar{c}) = x^2 - (2s)x + (s^2 + t^2).$$

Since $2s$ and $s^2 + t^2$ are real numbers, the right side is a quadratic polynomial in $\mathbf{R}[x]$. Furthermore, it is irreducible over $\mathbf{R}$—for it has no roots in $\mathbf{R}$, the two roots c and $\bar{c}$ being nonreal complex numbers.

In this way, each real root of $f(x)$ determines a linear factor (in $\mathbf{R}[x]$) of $f(x)$, and each pair of conjugate complex roots determines an irreducible quadratic factor (in $\mathbf{R}[x]$) of $f(x)$. We have proved:

3-5-24. Theorem. Any polynomial $f(x) = a_0 + a_1 x + \cdots + a_n x^n \in \mathbf{R}[x]$ of degree $n \geq 1$ factors into a product of a_n, monic linear polynomials in $\mathbf{R}[x]$, and monic irreducible quadratic polynomials in $\mathbf{R}[x]$.

Obviously, this factorization of $f(x)$ is its "unique factorization into primes" in $\mathbf{R}[x]$, whose existence and uniqueness were guaranteed in 3-4-16. We also note that according to this result a polynomial of degree 3 or more in $\mathbf{R}[x]$ factors into linear and/or quadratic terms, so it cannot be irreducible. Even more, we have:

3-5-25. Proposition. A polynomial $f(x) \in \mathbf{R}[x]$ is irreducible $\Leftrightarrow$ $f(x)$ is of two types:

(*i*) $f(x)$ is linear [that is, $\deg f(x) = 1$],
(*ii*) $f(x)$ is quadratic [that is, $\deg f(x) = 2$] with negative discriminant.

Proof: We have already observed that a polynomial of degree 3 or more cannot be irreducible in $\mathbf{R}[x]$ and, of course, any polynomial of degree 1 is irreducible. It remains, therefore, to decide under what circumstances a polynomial of degree 2,

$$f(x) = ax^2 + bx + c \in \mathbf{R}[x], \qquad a \neq 0,$$

is irreducible. The criterion will be based on the **discriminant** of $f(x)$, which is defined to be $b^2 - 4ac$ (which is an element of $\mathbf{R}$).

The only way $f(x)$ can factor in $\mathbf{R}[x]$ is as the product of two linear polynomials. Therefore, $f(x)$ is irreducible in $\mathbf{R}[x]$ $\Leftrightarrow$ it has no real root. Applying the method of completing the square (as in 3-5-22) to $f(x) = ax^2 + bx + c$ we see that its roots (about which we can only be certain, in general, that they are in **C**) are

$$\frac{-b \pm \sqrt{b^2 - 4ac}}{2a}, \qquad a, b, c \in \mathbf{R}.$$

Thus, there are no real roots of $f(x)$ $\Leftrightarrow$ the discriminant $b^2 - 4ac$ is less than 0. This completes the proof. ∎

In virtue of this result, the polynomials $x - \sqrt{2}$, $\sqrt{3}\,x - \pi$, $7x + 9$ in $\mathbf{R}[x]$ are irreducible because they are linear. The quadratic polynomials $x^2 + 7x - 9$, $x^2 - \pi x - \sqrt{2}$, $\sqrt{3}\,x^2 + (2 + \sqrt{3})x + \sqrt{2}$ in $\mathbf{R}[x]$ are reducible over **R** because in each case the discriminant $b^2 - 4ac$ is greater than 0. On the other hand, each of the quadratic polynomials $x^2 - 5x + 7$, $x^2 + \pi x + 2\sqrt{7}$, $\sqrt{3}\,x^2 - (1 + \sqrt{3})x + \sqrt{2}$ in $\mathbf{R}[x]$ is irreducible over **R** because the discriminant $b^2 - 4ac$ is less than 0.

We conclude our remarks about polynomials with real coefficients by noting a simple consequence of 3-5-24.

3-5-26. Corollary. A polynomial $f(x) \in \mathbf{R}[x]$ of odd degree n has at least one real root.

Proof: Because n is odd, the factorization of $f(x)$ into primes cannot consist solely of quadratic terms. There must be at least one linear factor; hence, there is at least one real root. ∎

When we turn to polynomials in $\mathbf{Q}[x]$, the results are of a different type. For polynomials in $\mathbf{C}[x]$, the form of their factorization into primes has been determined completely, but we have little information about finding the roots explicitly. [Actually, we have seen how to find the roots of any $f(x) \in \mathbf{C}[x]$ of degree 2; in addition it is well known that if $f(x) \in \mathbf{C}[x]$ is of degree 3 or 4, then its roots can be found explicitly—in fact, the roots can be expressed in terms of "radicals." It is only when $\deg f(x) \geq 5$ that there is no general method which always locates the roots.] Furthermore, these same statements apply for polynomials in $\mathbf{R}[x]$. On the other hand, for arbitrary $f(x) \in \mathbf{Q}[x]$ there is not much we can say about its prime factorization, but it turns out to be relatively easy to find all its roots in **Q**.

To see this, consider $f(x) = a_0 + a_1 x + \cdots + a_n x^n \in \mathbf{Q}[x]$. Since each a_i is in **Q** it can be expressed as the ratio of two integers. If we multiply $f(x)$ by the least common multiple of the denominators of all the a_i—call it m—the result $g(x) = mf(x)$ is a polynomial whose coefficients are integers and which has the same roots as $f(x)$. Thus, when we want to discuss the roots of $f(x) \in \mathbf{Q}[x]$ there is no harm in assuming that $f(x)$ has coefficients in $\mathbf{Z}$—

$$f(x) = a_0 + a_1 x + \cdots + a_n x^n \in \mathbf{Z}[x].$$

In addition, if the gcd of $a_1, a_2, \ldots, a_n$ is greater than 1, then it may be factored out, leaving a polynomial in $\mathbf{Z}[x]$ for which the gcd of all its coefficients is 1, and whose roots are the same as those of $f(x)$. Thus, we may assume, if necessary, that $f(x) \in \mathbf{Z}[x]$ is **primitive**—by which is meant that $(a_0, a_1, \ldots, a_n) = 1$, or what is the same, the coefficients of $f(x)$ are relatively prime.

For example, to find the rational roots of

$$\tfrac{1}{3}x^5 - \tfrac{3}{2}x^4 + \tfrac{5}{8}x^3 + \tfrac{7}{12}x^2 - 2x + \tfrac{5}{6},$$

we may work with the polynomial

$$8x^5 - 36x^4 + 15x^3 + 14x^2 - 48x + 20,$$

and to find the rational roots of

$$54x^5 - 21x^3 + 102x^2 - 9x + 24,$$

we may work with the primitive polynomial

$$18x^5 - 7x^3 + 34x^2 - 3x + 8.$$

3-5-27. Proposition. Consider a polynomial

$$f(x) = a_0 + a_1 x + \cdots + a_n x^n \in \mathbf{Z}[x]$$

of degree $n \geq 1$. If r/s is a rational root in lowest terms [meaning that r and s are integers with $(r, s) = 1$], then $r \mid a_0$ and $s \mid a_n$.

Proof: Since r/s is a root, we have

$$a_0 + a_1\left(\frac{r}{s}\right) + \cdots + a_{n-1}\left(\frac{r}{s}\right)^{n-1} + a_n\left(\frac{r}{s}\right)^n = 0$$

and multiplying by s^n gives

$$a_0 s^n + a_1 r s^{n-1} + \cdots + a_{n-1} r^{n-1} s + a_n r^n = 0. \qquad (*)$$

Except for the first term, all the terms on the left side are obviously divisible by r. Since 0 is divisible by r, it follows that the first term $a_0 s^n$ is divisible by r. Because $(r, s) = 1$ we know that $(r, s^n) = 1$, and therefore $r \mid a_0$.

By applying the same technique to $(*)$, we also obtain $s \mid a_n$. ▮

According to this result, when a polynomial with integer coefficients is given then there are only a finite number of rationals that can possibly be roots —namely, those of form r/s where $(r, s) = 1$, r divides the constant term of

$f(x)$, and s divides the leading coefficient of $f(x)$ (and, of course, we can insist that s be positive, but r may be negative). For example, for the polynomial

$$f(x) = 6x^5 + 19x^4 + 31x^3 + 17x^2 - 7x - 6 \in \mathbf{Z}[x]$$

the only possible rational roots are: $\pm 1, \pm 2, \pm 3, \pm 6, \pm \frac{1}{2}, \pm \frac{3}{2}, \pm \frac{1}{3}, \pm \frac{2}{3}, \pm \frac{1}{6}$ (since $a_0 = -6$, $a_5 = 6$). By testing these, we find all the rational roots of $f(x)$; they are: $-1, \frac{1}{2}, \frac{2}{3}$, and the factorization of $f(x)$ is

$$(2x - 1)(3x + 2)(x + 1)(x^2 + 2x + 3).$$

Note that this is the prime factorization of $f(x) \in \mathbf{Q}[x]$ (after all, $x^2 + 2x + 3$ is irreducible in $\mathbf{R}[x]$, so it is surely irreducible in $\mathbf{Q}[x]$).

The problem of deciding whether an arbitrary polynomial in $\mathbf{Q}[x]$ is irreducible is essentially unsettled. However, there is one result, due to Eisenstein (1823–1852), that enables us to recognize, or produce, many irreducible polynomials over $\mathbf{Q}$.

3-5-28. Eisenstein's Criterion. Suppose

$$f(x) = a_0 + a_1 x + \cdots + a_n x^n \in \mathbf{Z}[x]$$

is a polynomial of degree $n \geq 1$, and suppose there exists a prime p for which

(*i*) $p \mid a_i$ for $i = 0, 1, \ldots, n - 1$,
(*ii*) $p \nmid a_n$,
(*iii*) $p^2 \nmid a_0$.

Then, $f(x)$ is irreducible over $\mathbf{Q}$.

Proof: We merely sketch the proof (which goes in several steps) and leavc the details as an exercise for the reader.

(1) Show that $f(x)$ is irreducible over $\mathbf{Z}$. In fact, suppose

$$f(x) = (b_0 + b_1 x + \cdots + b_r x^r)(c_0 + c_1 x + \cdots + c_s x^s)$$

where all the coefficients are integers, $r \geq 1$, $s \geq 1$, $r + s = n$. Since $p \mid a_0$, $p^2 \nmid a_0$, and $a_0 = b_0 c_0$ exactly one of b_0, c_0 is divisible by p. Say, $p \mid b_0$ and $p \nmid c_0$. Now, $p \nmid b_r$, since p does not divide $a_n = b_r c_s$. Let $t \leq r < n$ be the smallest positive integer for which $p \nmid b_t$. It follows that $p \nmid a_t$—contradiction. It remains to show, therefore, that if a polynomial is irreducible in $\mathbf{Z}[x]$, then it is irreducible in $\mathbf{Q}[x]$.

(2) The product of two primitive polynomials (in $\mathbf{Z}[x]$) is primitive. For this suppose the product is not primitive, so there exists a prime p which divides all its coefficients. Now apply 3-4-17, Problem 30 to obtain a contradiction.

(3) Prove **Gauss' lemma** which says: If $f(x) \in \mathbf{Z}[x]$ of degree 1 or more factors over $\mathbf{Q}$, then it factors over $\mathbf{Z}$—in other words, if there exist $g(x)$, $h(x) \in \mathbf{Q}[x]$ for which $f(x) = g(x)h(x)$, then there exist $g'(x), h'(x) \in \mathbf{Z}[x]$ for which $f(x) = g'(x)h'(x)$.

To do this, there is no loss of generality in assuming, at the start, that $f(x)$ is primitive. Beginning from $f(x) = g(x)h(x)$ one may clear of fractions and then factor out the gcd of the coefficients for each of the polynomials on the right side. We finally have an equation of form

$$sf(x) = rg'(x)h'(x)$$

where $r, s \in \mathbf{Z}$, $g'(x), h'(x) \in \mathbf{Z}[x]$, and $g'(x), h'(x)$ are both primitive. Then $r = s$, which completes the proof. ▌

According to this result, the polynomials

$$8x^5 + 36x^4 + 15x^3 - 48x + 21 \qquad \text{and} \qquad 3x^5 - 18x + 34x^2 - 8x + 42$$

are irreducible over $\mathbf{Q}$—the first one satisfies the Eisenstein criterion for $p = 3$ and the second one for $p = 2$. The reader can now surely construct irreducible polynomials over $\mathbf{Q}$ at will.

Incidentally, $x^2 - 2$ satisfies the Eisenstein criterion for $p = 2$, so it is irreducible over $\mathbf{Q}$. Hence $x^2 - 2$ has no root in $\mathbf{Q}$; or, to put it another way, there is no rational number (that is, element of $\mathbf{Q}$) whose square is 2. We have proved $\sqrt{2}$ is irrational.

Furthermore, for any prime p and any $n \geq 2$ the polynomial

$$x^n - p$$

is irreducible over $\mathbf{Q}$ (since it is Eisenstein). Therefore, for any integer $n \geq 2$ there exists an irreducible polynomial over $\mathbf{Q}$ of degree n—in fact, there exist an infinite number of irreducible ones of degree n, because there are an infinite number of choices for the prime p.

3-5-29 / *PROBLEMS*

1. Use Lagrange interpolation to find a polynomial $f(x) \in \mathbf{R}[x]$ for which
   (*i*) $f(2) = 5, f(3) = 8,$ (*ii*) $f(-1) = 7, f(1) = 7,$
   (*iii*) $f(\sqrt{2}) = \pi, f(\sqrt{3}) = \pi^2.$
   What is really going on here?

2. Use Lagrange interpolation to find $f(x) \in \mathbf{Q}[x]$ of degree less than or equal to 2 for which
   (*i*) $f(0) = 1, f(1) = 2, f(2) = 4,$
   (*ii*) $f(-2) = 0, f(2) = 3, f(4) = 6,$
   (*iii*) $f(-1) = 2, f(1) = 1, f(2) = 3.$

3. Do each part of Problem 2 when the field $\mathbf{Q}$ is replaced by the field
   (*i*) $\mathbf{R}$, (*ii*) $\mathbf{Z}_5$, (*iii*) $\mathbf{Z}_7$, (*iv*) $\mathbf{Z}_{11}$.

4. We know that there exists a unique polynomial

$$f(x) = a + bx + cx^2 + dx^3 \in \mathbf{R}[x]$$

   of degree less than or equal to 3 whose graph passes through the four points

$$(-1, 5), \quad (0, 1), \quad (1, 5), \quad (2, 19).$$

   Find a, b, c, d by the method of undetermined coefficients. Then find $f(x)$ in another way—namely, by use of Lagrange interpolation.

5. Why does Lagrange interpolation not work when the field F is replaced by an integral domain D?

6. Find the unique polynomial $f(x)$ of degree less than or equal to 3 in $\mathbf{Z}_5[x]$ for which

$$f(1) = 4, \qquad f(2) = 2, \qquad f(3) = 1, \qquad f(4) = 1.$$

7. Do Problem 6 when $\mathbf{Z}_5$ is replaced by
   (*i*) $\mathbf{Z}_7$, (*ii*) $\mathbf{Z}_{11}$.

8. What is the remainder upon division of $3x^5 + x^4 + 2x^3 + 4x^2 + 5x + 6$ by $x - 3$ when these polynomials are viewed in
   (*i*) $\mathbf{Z}[x]$, (*ii*) $\mathbf{Q}[x]$, (*iii*) $\mathbf{Z}_2[x]$, (*iv*) $\mathbf{Z}_3[x]$,
   (*v*) $\mathbf{Z}_5[x]$, (*vi*) $\mathbf{Z}_6[x]$, (*vii*) $\mathbf{Z}_7[x]$, (*viii*) $\mathbf{Z}_{11}[x]$?

9. Give a proof of 3-5-4 based on the least common multiple of $x - c_1$, $x - c_2, \ldots, x - c_n$ in $F[x]$.

10. (*i*) Suppose $f(x), g(x) \in F[x]$ are of degrees m and n, respectively, and the number of elements in the field F is greater than $\max\{m, n\}$. If $f(c) = g(c)$ for all $c \in F$, then $f(x) = g(x)$.
    (*ii*) Is this result valid if F is replaced by an integral domain D?

11. In the text, we proved 3-5-4, 3-5-6, 3-5-7 over a field F. When F is replaced by R, an arbitrary commutative ring with unity, each of these results fails. For each result, give an example of a ring R for which it is false.

12. Consider the polynomials
    (*a*) $x^3 + x + 3$, (*b*) $x^4 - 5x^3 + x^2 - 3x + 2$, (*c*) $x^7 - x$,
    (*d*) $x^4 + 3x^2 + 2$, (*e*) $x^4 - 1$, (*f*) $x^4 + 1$.
    For each one find all its roots when it is viewed as a polynomial over the field:
    (*i*) $\mathbf{Z}_2$, (*ii*) $\mathbf{Z}_3$, (*iii*) $\mathbf{Z}_5$,
    (*iv*) $\mathbf{Z}_7$, (*v*) $\mathbf{Z}_{11}$, (*vi*) $\mathbf{Z}_{13}$.

    In each case (there are 36 cases: six polynomials and six fields) give the prime factorization.

13. Decide if each of the following polynomials is irreducible over

(*a*) $\mathbf{Q}$, (*b*) $\mathbf{R}$, (*c*) $\mathbf{C}$

and, if not, factor it when possible.

(*i*) $x^2 - 13$, (*ii*) $x^2 + 13$,
(*iii*) $x^2 - 5x + 6$, (*iv*) $x^2 + x + 1$,
(*v*) $x^2 + 2x - 2$, (*vi*) $x^3 - 2$,
(*vii*) $x^3 + 2$, (*viii*) $x^3 + x + 1$,
(*ix*) $x^3 + x^2 + 1$, (*x*) $x^3 + 2x^2 - 3x - 1$,
(*xi*) $x^3 + x + 2$, (*xii*) $x^3 + 2x^2 - 5x - 6$.

14. Treat the polynomials of Problem 13 in the same way over the fields:

(*i*) $\mathbf{Z}_2$, (*ii*) $\mathbf{Z}_3$, (*iii*) $\mathbf{Z}_5$,
(*iv*) $\mathbf{Z}_7$, (*v*) $\mathbf{Z}_{11}$, (*vi*) $\mathbf{Z}_{17}$.

15. (*i*) Find all irreducible polynomials of degree less than or equal to 3 in $\mathbf{Z}_3[x]$.

(*ii*) Find all irreducible polynomials of degree less than or equal to 5 over the field $\mathbf{Z}_2$. There are 10 of them.

16. The number of monic polynomials of degree 2 over the field $\mathbf{Z}_p$ is p^2; show that $(p^2 - p)/2$ of them are irreducible.

17. Find the square roots of the following complex numbers:

(*i*) i, (*ii*) $-i$, (*iii*) $3 - 4i$, (*iv*) $5 + 12i$.

18. Solve the following quadratic polynomials over $\mathbf{C}$:

(*i*) $x^2 + ix + (1 - i)$, (*ii*) $2x^2 - \sqrt{3}\,x + (2 + i)$,
(*iii*) $x^2 + (1 + i)x + (3 + 4i)$, (*iv*) $(3 - i)x^2 + (1 + 2i)x - (2 + 3i)$.

19. Find all rational roots of the following polynomials:

(*i*) $x^4 + x^3 + x^2 + x + 1$,
(*ii*) $x^5 + x^4 + x^3 + x^2 + x + 1$,
(*iii*) $x^6 + x^5 + x^4 + x^3 + x^2 + x + 1$,
(*iv*) $x^{50} - x^{20} + x^{10} - 1$,
(*v*) $x^{100} - x^{50} + 1$,
(*vi*) $9x^4 + 6x^3 + 19x^2 + 12x + 2$,
(*vii*) $x^{11} + 2x^9 - 2$,
(*viii*) $x^m + 3x^{m-1} - 3$, $m > 1$,
(*ix*) $4x^3 + 3x^2 + 6x + 12$,
(*x*) $5x^3 - x^2 - 20x + 4$.

20. Find the prime factorizations of the following polynomials over $\mathbf{C}$ and over $\mathbf{R}$:

(*i*) $x^3 + 1$, (*ii*) $x^4 + 1$, (*iii*) $x^5 + 1$.

21. (*i*) Find a polynomial of degree 4 in $\mathbf{Z}_{13}[x]$ which has four distinct roots.
(*ii*) Does $x^4 - 7x^3 - 7x - 1 \in \mathbf{Z}_{13}[x]$ have four distinct roots?
(*iii*) What about $3x^4 - x^3 + 2x^2 + 5x - 3 \in \mathbf{Z}_{13}[x]$?

22. Which of the following polynomials are irreducible over $\mathbf{Q}$?
(*i*) $x^5 + 3x^3 + 6x + 3$, (*ii*) $x^5 - 3x^3 + 3x^2 - 9$,
(*iii*) $x^5 + 6x^3 - 3x + 15$, (*iv*) $x^7 - 91$.

23. (*i*) For any $a \in F$, $f(x)$ is irreducible over $F \Leftrightarrow f(x + a)$ is irreducible over F.
(*ii*) Show that $x^2 + 1$ is irreducible over $\mathbf{Q}$ by making use of the Eisenstein criterion.

24. Use Wilson's theorem to prove that if p is a prime $\equiv 1 \pmod 4$, then the congruence

$$x^2 + 1 \equiv 0 \pmod p$$

has a solution—that is, there is an element of $\mathbf{Z}_p$ whose square is -1.

25. Suppose both of the polynomials $f(x), g(x) \in \mathbf{Q}[x]$ have the real number α as a root. Show that $f(x)$ and $g(x)$ are not relatively prime.

26. Suppose $f(x) \in F[x]$ is of degree greater than or equal to 1; then, $f(x)$ is irreducible if and only if it satisfies the following property: if $f(x) \mid g(x)h(x)$ and $f(x) \nmid g(x)$, then $f(x) \mid h(x)$. (That is, if $f(x)$ divides a product, then it must divide at least one of the factors.)

27. The polynomial $f(x) = x^3 - x - 1$ is irreducible over $\mathbf{Z}_2$. Take a formal symbol α and make believe it is a root of $f(x)$—so $\alpha^3 - \alpha - 1 = 0$ or $\alpha^3 = \alpha + 1$. Let $\mathbf{Z}_2[\alpha]$ denote the set of all formal expressions

$$a_0 + a_1\alpha + a_2\alpha^2, \qquad a_0, a_1, a_2 \in \mathbf{Z}_2.$$

Define addition in $\mathbf{Z}_2[\alpha]$ by

$$(a_0 + a_1\alpha + a_2\alpha^2) + (b_0 + b_1\alpha + b_2\alpha^2)$$
$$= (a_0 + b_0) + (a_1 + b_1)\alpha + (a_2 + b_2)\alpha^2.$$

Define multiplication in $\mathbf{Z}_2[\alpha]$ by multiplying out as for polynomials and replacing higher powers of α than the second by combination of lower powers—that is, $\alpha^3 = \alpha + 1$, $\alpha^4 = \alpha^2 + \alpha$.

Show that $\mathbf{Z}_2[\alpha]$ is a commutative ring with unity, having eight elements, and containing $\mathbf{Z}_2$.

Even more, $\mathbf{Z}_2[\alpha]$ is a field. One way to prove this is to simply look at the multiplication table for $\mathbf{Z}_2[\alpha]$. Another, more sophisticated way, is as follows: Given $a_0 + a_1\alpha + a_2\alpha^2 \neq 0$ in $\mathbf{Z}_2[\alpha]$, write $g(x) = a_0 + a_1x$

$+ a_2 x^2 \in \mathbf{Z}_2[x]$. Since $f(x)$ is prime, clearly $(g(x), f(x)) = 1$, so there exist $u(x), v(x) \in \mathbf{Z}_2[x] \subset (\mathbf{Z}_2[\alpha])[x]$ such that

$$g(x)u(x) + f(x)v(x) = 1.$$

Then, substituting α, we obtain $g(\alpha)u(\alpha) = 1$, so $u(\alpha)$, which is in $\mathbf{Z}_2[\alpha]$, is a multiplicative inverse of $g(\alpha) = a_0 + a_1\alpha + a_2\alpha^2$.

3-6. Solving Polynomials in $\mathbf{Z}_m[x]$

In this section we finally learn how to go about solving an arbitrary concrete equation of form

$$\alpha_n x^n + \alpha_{n-1}x^{n-1} + \cdots + \alpha_1 x + \alpha_0 = 0, \qquad \alpha_0, \alpha_1, \ldots, \alpha_n \in \mathbf{Z}_m$$

—or, as we find it more convenient to write,

$$\alpha_0 + \alpha_1 x + \cdots + a_n x^n = 0, \qquad \alpha_0, \alpha_1, \ldots, \alpha_n \in \mathbf{Z}_m. \tag{$*$}$$

A small part of this problem was treated, with complete success, in Section 3-1—namely, the case where the equation is linear (that is, where $n = 1$). Here we place no restrictions on the positive integer n. Of course, by a **solution** of the equation $(*)$ we mean an element $\gamma \in \mathbf{Z}_m$ for which $\alpha_0 + \alpha_1\gamma + \cdots + \alpha_n\gamma^n = 0$ (or, as it is commonly expressed: γ satisfies the equation); and then the phrase "solving the equation $(*)$" means finding all the solutions of $(*)$.

Equivalently, if we consider the polynomial

$$\begin{aligned} f(x) &= \alpha_n x^n + \alpha_{n-1}x^{n-1} + \cdots + \alpha_1 x + \alpha_0 \\ &= \alpha_0 + \alpha_1 x + \cdots + \alpha_n x^n \qquad\qquad (**) \\ &= \sum_{i=0}^{n} \alpha_i x^i \end{aligned}$$

in $\mathbf{Z}_m[x]$, then a solution of $(*)$ is the same as a root of the polynomial $f(x)$—so solving the equation $(*)$ amounts to finding all the roots (in $\mathbf{Z}_m$) of the polynomial $f(x)$.

Strictly speaking, the problem of finding all the roots of $f(x)$ can be settled in a finite number of steps; in fact, there are m choices for $\gamma \in \mathbf{Z}_m$ and for each of them we simply evaluate $f(\gamma)$ to decide if it is a root. But this is not what we have in mind; there is no mathematics involved, and anyway m may be very large. Instead, we develop a method which, in most cases, reduces considerably the amount of work involved. To do this, it is convenient to reformulate our problem in terms of integers (rather than residue classes) and congruences. Thus, for each $i = 1, \ldots, n$, let a_i be an integer for which $\lfloor a_i \rfloor_m = \alpha_i$, and consider the congruence

$$a_0 + a_1 x + \cdots + a_n x^n \equiv 0 \pmod{m}. \tag{$***$}$$

By a **solution** of this congruence we mean, as expected, an integer c such that

$$a_0 + a_1 c + \cdots + a_n c^n \equiv 0 \pmod{m}.$$

In parallel with the remarks in Section 3-1 we have the following facts about the solutions of the congruence (***) and their connection with the solutions of the equation (*):

3-6-1. Facts. (1) $\gamma = \lfloor c \rfloor_m \in \mathbf{Z}_m$ is a solution of the equation (*) $\Leftrightarrow$ $c \in \mathbf{Z}$ is a solution of the congruence (***). This is immediate because

$$\begin{aligned} &\alpha_0 + \alpha_1 \gamma + \cdots + \alpha_n \gamma^n = 0 \\ &\Leftrightarrow \lfloor a_0 \rfloor_m + \lfloor a_1 \rfloor_m (\lfloor c \rfloor_m) + \cdots + \lfloor a_n \rfloor_m (\lfloor c \rfloor_m)^n = \lfloor 0 \rfloor_m \\ &\Leftrightarrow \lfloor a_0 + a_1 c + \cdots + a_n c^n \rfloor_m = \lfloor 0 \rfloor_m \\ &\Leftrightarrow a_0 + a_1 c + \cdots + a_n c^n \equiv 0 \pmod{m}. \end{aligned}$$

(2) If $a_i' \equiv a_i \pmod{m}$ for $i = 1, \ldots, n$, then the solutions of the congruence $a_0 + a_1 x + \cdots + a_n x^n \equiv 0 \pmod{m}$ are identical with the solutions of the congruence $a_0' + a_1' x + \cdots + a_n' x^n \equiv 0 \pmod{m}$. This is a consequence of (1), and it also follows immediately from the fact that for $c \in \mathbf{Z}$, $a_0 + a_1 c + \cdots + a_n c^n \equiv a_0' + a_1' c + \cdots + a_n' c^n \pmod{m}$.

(3) In solving the equation (*) via the congruence (***), the choice of the representatives a_i for the various α_i, $i = 1, \ldots, n$, does not matter. This is clear from parts (1) and (2).

(4) If $c \in \mathbf{Z}$ is a solution of the congruence (***), then so is every element of $\lfloor c \rfloor_m$—since if $c' \equiv c \pmod{m}$, then $a_0 + a_1(c') + \cdots + a_n(c')^n \equiv a_0 + a_1 c + \cdots + a_n c^n \equiv 0 \pmod{m}$ [in fact, for any polynomial $g(x) \in \mathbf{Z}[x]$, $c' \equiv c \pmod{m}$ implies $g(c') \equiv g(c) \pmod{m}$]. Because of this, one often views all the elements of a residue class, or the residue class itself, as a single solution of the congruence (***)—especially when one is interested in counting the solutions. Of course, these residue classes are precisely the solutions of the equation (*).

With the preliminaries completed, let us turn to the problem at hand—expressed in the language of congruences—namely, to solve any equation of form

$$a_0 + a_1 x + \cdots + a_n x^n \equiv 0 \pmod{m}. \qquad (***)$$

The first case to consider is obviously when m is a prime p. In this situation, our problem amounts to finding the roots of the polynomial $a_0 + a_1 x + \cdots + a_n x^n$ over the field $\mathbf{Z}_p$ (the polynomial is really $\lfloor a_0 \rfloor_p + \lfloor a_1 \rfloor_p x + \cdots + \lfloor a_n \rfloor_p x^n \in \mathbf{Z}_p[x]$, but our use of the less cumbersome notation should cause no serious confusion). As seen in Section 3-5 we have some information about the roots of this polynomial. However, we have no general recipe for finding the roots, other than testing each element of $\mathbf{Z}_p$, and will have to content ourselves with this state of affairs. Of course, in practice we shall deal with small primes p, in which case the roots can be located easily.

As for the general case, where the modulus m in (***) is arbitrary, we shall see that solving it can be reduced to the case of a prime modulus p. As an initial step in this direction, let us examine a few specific examples where the modulus is a prime power p^r.

3-6-2. Examples. (*i*) Consider the congruence

$$x^3 - x + 4 \equiv 0 \pmod{7^2}. \tag{\#}$$

Instead of solving it by testing each of the 49 possibilities 0, 1, ..., 48, we begin with the observation that if the integer c is a solution, then it is surely a solution of the congruence

$$x^3 - x + 4 \equiv 0 \pmod{7}. \tag{\#\#}$$

The basic idea then is to find the solutions of (# #) and then to search for the solutions of (#) among them.

By trial and error one sees that (# #) has exactly one solution, 3 (or better, $\lfloor 3 \rfloor_7$). [We note in passing that the existence of a unique solution reflects the fact that, over the field $\mathbf{Z}_7$, the polynomial $x^3 - x + 4$ factors into $(x-3)(x^2+3x+1)$, and x^2+3x+1 is irreducible over $\mathbf{Z}_7$.] Consequently, if $c \in \mathbf{Z}$ is a solution of (#), then it must be an element of $\lfloor 3 \rfloor_7 = \{3 + 7t \mid t \in \mathbf{Z}\}$—so we seek an integer (or integers) t for which

$$c = 3 + 7t, \qquad t \in \mathbf{Z}$$

is a solution of (#). This requires

$$(3+7t)^3 - (3+7t) + 4 \equiv 0 \pmod{7^2}$$

which becomes

$$28 + 182t + 441t^2 + 343t^3 \equiv 0 \pmod{7^2}.$$

Reducing the coefficients modulo $7^2 = 49$, we have

$$28 + 35t \equiv 0 \pmod{7^2}$$

which is equivalent to

$$4 + 5t \equiv 0 \pmod{7}.$$

This linear congruence has the unique solution $\lfloor 2 \rfloor_7$; in other words, the set of all integer solutions is

$$t = 2 + 7u, \qquad u \in \mathbf{Z}.$$

Therefore,

$$c = 3 + 7(2 + 7u) = 17 + 7^2 u$$

for any and all $u \in \mathbf{Z}$, and we conclude that $\lfloor 17 \rfloor_{7^2}$ [or, as it is often written, 17 (mod 49)] is the unique solution of (#).

(*ii*) Consider the congruence

$$x^2 - x + 2 \equiv 0 \pmod{7^2}. \qquad (\mp)$$

Following the above technique, if $c \in \mathbf{Z}$ is a solution, then it must be a solution of

$$x^2 - x + 2 \equiv 0 \pmod{7}. \qquad (\mp\mp)$$

By trial-and-error, this has the unique solution 4—meaning $\lfloor 4 \rfloor_7$. [One normally expects the polynomial of second degree $x^2 - x + 2$ to have two roots over $\mathbf{Z}_7$; but $x^2 - x + 2$ factors over $\mathbf{Z}_7$ into $(x - 4)(x - 4)$, so 4 is a double root.] Thus, c must be of form $c = 4 + 7t$ and it satisfies

$$(4 + 7t)^2 - (4 + 7t) + 2 \equiv 0 \pmod{7^2}.$$

This reduces to

$$14 \equiv 0 \pmod{7^2}$$

which is impossible. The conclusion is: The congruence $x^2 - x + 2 \equiv 0 \pmod{7^2}$ has no solutions.

(*iii*) To solve the equation

$$f(x) = x^2 + x + 1 \equiv 0 \pmod{7^2} \qquad (*)$$

we look first at

$$f(x) = x^2 + x + 1 \equiv 0 \pmod{7}. \qquad (**)$$

This has the two solutions $\lfloor 2 \rfloor_7$ and $\lfloor 4 \rfloor_7$, corresponding to the fact that over $\mathbf{Z}_7$ we have the factorization $x^2 + x + 1 = (x - 2)(x - 4)$. Thus, if $c \in \mathbf{Z}$ is a solution of $(*)$, then it belongs to either $\lfloor 2 \rfloor_7$ or $\lfloor 4 \rfloor_7$; in other words, c has to be of form

$$c = 2 + 7t \qquad \text{or} \qquad c = 4 + 7t$$

for some choice of the integer t. Substituting the case $c = 2 + 7t$ in $(*)$ leads to

$$7 + 35t + 49t^2 \equiv 0 \pmod{7^2},$$

which reduces to

$$1 + 5t \equiv 0 \pmod{7}.$$

This has the unique solution $\lfloor 4 \rfloor_7$, so the integer t is of form $4 + 7u$, $u \in \mathbf{Z}$, and consequently $c = 30 + 7^2 u$. Thus, $\lfloor 30 \rfloor_{49}$ is a solution of $(*)$.

In similar fashion, substituting the case $c = 4 + 7t$ in $(*)$ yields

$$3 + 2t \equiv 0 \pmod 7,$$

which has the unique solution $\lfloor 2 \rfloor_7 = \{2 + 7u \mid u \in \mathbf{Z}\}$. It follows that $\lfloor 18 \rfloor_{49}$ is a solution of $(*)$.

The conclusion is: $x^2 + x + 1 \equiv 0 \pmod{7^2}$ has the two solutions $\lfloor 30 \rfloor_{7^2}$, $\lfloor 18 \rfloor_{7^2}$.

(*iv*) Suppose we wish to solve

$$x^2 + x + 1 \equiv 0 \pmod{7^3}. \qquad (***)$$

The same kind of approach is clearly valid. If $c \in \mathbf{Z}$ is a solution, then it is surely a solution of

$$x^2 + x + 1 \equiv 0 \pmod{7^2}.$$

As seen in part (*iii*), this congruence has the two solutions $\lfloor 30 \rfloor_{7^2}$ and $\lfloor 18 \rfloor_{7^2}$, so c has to be of the form

$$c = 30 + 7^2 t \qquad \text{or} \qquad c = 18 + 7^2 t$$

for appropriate choices of $t \in \mathbf{Z}$.

In the first case, we have then

$$(30 + 7^2 t)^2 + (30 + 7^2 t) + 1 \equiv 0 \pmod{7^3}$$

which leads to

$$931 + (61)(49)t \equiv 0 \pmod{7^3},$$
$$19 + 61t \equiv 0 \pmod 7,$$
$$5 + 5t \equiv 0 \pmod 7.$$

This has the unique solution $\lfloor 6 \rfloor_7$, and then $(***)$ has the solution

$$\lfloor 30 + 7^2 \cdot 6 \rfloor_{7^3} = \lfloor 324 \rfloor_{7^3}.$$

(This solution can also be written as $\lfloor -19 \rfloor_{7^3}$; it arises in this form when we take $t = -1$ instead of $t = 6$.)

In the second case, we have

$$(18 + 7^2 t)^2 + (18 + 7^2 t) + 1 \equiv 0 \pmod{7^3}$$

which becomes

$$343 + 37 \cdot 7^2 t + 7^4 t^2 \equiv 0 \pmod{7^3}$$

and ends as

$$2t \equiv 0 \pmod 7.$$

The only solution here is $\lfloor 0 \rfloor_7$, so for $t = 0$ we obtain $c = 18$ and the solution $\lfloor 18 \rfloor_{7^3}$ of (***).

We have shown: The congruence $x^2 + x + 1 \equiv 0 \pmod{7^3}$ has the two solutions $\lfloor -19 \rfloor_{7^3}$, $\lfloor 18 \rfloor_{7^3}$.

(v) To solve

$$x^2 - x - 12 \equiv 0 \pmod{7^2}$$

we consider first the congruence

$$x^2 - x - 12 \equiv 0 \pmod 7$$

which is the same as

$$x^2 - x + 2 \equiv 0 \pmod 7.$$

As noted in part (ii) this has the unique solution $\lfloor 4 \rfloor_7$. Therefore, any integer solution of the original congruence is of form $c = 4 + 7t$, and it satisfies

$$(4 + 7t)^2 - (4 + 7t) - 12 \equiv 0 \pmod{7^2}.$$

In other words,

$$49t^2 + 49t \equiv 0 \pmod{7^2}$$

so that

$$0t \equiv 0 \pmod 7,$$

a congruence that is true for any choice of t. To put it another way, the validity of this congruence is independent of t, and it has the seven solutions $t \equiv 0, 1, 2, 3, 4, 5, 6 \pmod 7$. It follows that $x^2 - x - 12 \equiv 0 \pmod{7^2}$ has seven solutions—namely, $\lfloor 4 \rfloor_{7^2}$, $\lfloor 11 \rfloor_{7^2}$, $\lfloor 18 \rfloor_{7^2}$, $\lfloor 25 \rfloor_{7^2}$, $\lfloor 32 \rfloor_{7^2}$, $\lfloor 39 \rfloor_{7^2}$, $\lfloor 46 \rfloor_{7^2}$.

From the preceding examples it would appear that the way to solve an arbitrary polynomial congruence modulo a prime power p^r is to solve it first (mod p), then use these solutions to solve the congruence (mod p^2), and proceed upward in this way one step at a time through the congruences modulo increasing powers of p until one reaches p^r. Indeed, this is the way things go, but before analyzing and simplifying the passage from the solutions (mod p^s) to the solutions (mod p^{s+1}), it is necessary to make a preliminary comment.

3-6-3. Remark. Given any polynomial with integer coefficients

$$f(x) = a_0 + a_1 x + \cdots + a_n x^n = \sum_{i=0}^{n} a_i x^i \in \mathbf{Z}[x],$$

we associate with it another polynomial with integer coefficients

$$f'(x) = a_1 + 2a_2 x + \cdots + na_n x^{n-1} = \sum_{i=1}^{n} ia_i x^{i-1} \in \mathbf{Z}[x],$$

and call $f'(x)$ the **derivative** of $f(x)$. For example, if $f(x) = 2 + 3x - x^2 + 5x^3 + x^6 - 2x^7$, then $f'(x) = 3 - 2x + 15x^2 + 6x^5 - 14x^6$.

The reader who is familiar with calculus may verify that this "formal derivative" satisfies the usual properties of a derivative (as indicated in 3-3-15, Problem 24), but this is irrelevant here. The application of the derivative which we require arises as follows. For any $h \in \mathbf{Z}$ consider the polynomial

$$f(x + h) = a_0 + a_1(x + h) + a_2(x + h)^2 + \cdots + a_n(x + h)^n \in \mathbf{Z}[x].$$

If we expand, and combine all terms in which h does not appear, the result is

$$a_0 + a_1 x + \cdots + a_n x^n \text{ which equals } f(x).$$

Combining all the terms of the expansion in which h appears to the first power, we have

$$h(a_1 + 2a_2 x + 3a_3 x^2 + \cdots + na_n x^{n-1}) \text{ which equals } hf'(x).$$

The remaining terms all include h to a power greater than or equal to 2; so their sum may be put in the form

$$h^2 \cdot (\text{some element of } \mathbf{Z}[x]).$$

Thus, we have

$$f(x + h) = f(x) + hf'(x) + h^2 \cdot (\text{an element of } \mathbf{Z}[x])$$

and for any $c \in \mathbf{Z}$ this yields

$$f(c + h) = f(c) + hf'(c) + h^2 \cdot (\text{an integer}).$$

We are now in a position to analyze successfully the connections between the solutions of a congruence modulo a prime power p^s and its solutions modulo the next highest prime power p^{s+1}.

3-6-4. Proposition. Given a polynomial $f(x) = a_0 + a_1 x + \cdots + a_n x^n \in \mathbf{Z}[x]$, a prime p, and an integer $s \geq 1$, we have:

(*i*) If $c \in \mathbf{Z}$ is a solution of the congruence

$$f(x) \equiv 0 \pmod{p^{s+1}},$$

then c is a solution of

$$f(x) \equiv 0 \pmod{p^i}$$

for each $i = 1, 2, \ldots, s$.

(*ii*) If $c_s \in \mathbf{Z}$ is a solution of the congruence

$$f(x) \equiv 0 \pmod{p^s} \tag{$*$}$$

and we put

$$c_{s+1} = c_s + tp^s, \qquad t \in \mathbf{Z},$$

then c_{s+1} is a solution of the congruence

$$f(x) \equiv 0 \pmod{p^{s+1}} \tag{$**$}$$

if and only if the integer t satisfies the relation

$$f'(c_s)t \equiv -\frac{f(c_s)}{p^s} \pmod{p} \tag{$\#$}$$

(*iii*) If $\lfloor c_s \rfloor_{p^s}$ is a solution of ($*$), then the number of distinct solutions $\lfloor c_{s+1} \rfloor_{p^{s+1}}$ of ($**$) which arise from it [that is, for which $c_{s+1} \equiv c_s \pmod{p^s}$] is

$$\begin{array}{ll} 0 & \text{when } f'(c_s) \equiv 0 \pmod{p} \text{ and } \dfrac{f(c_s)}{p^s} \not\equiv 0 \pmod{p}, \\ 1 & \text{when } f'(c_s) \not\equiv 0 \pmod{p}, \\ p & \text{when } f'(c_s) \equiv 0 \pmod{p} \text{ and } \dfrac{f(c_s)}{p^s} \equiv 0 \pmod{p}. \end{array}$$

Proof: (*i*) This is trivial; $f(c) \equiv 0 \pmod{p^{s+1}}$ implies $f(c) \equiv 0 \pmod{p^i}$ for every $i < s + 1$.

(*ii*) By hypothesis, $f(c_s) \equiv 0 \pmod{p^s}$, so $p^s \mid f(c_s)$ and $f(c_s)/p^s$, which appears on the right side of ($\#$), is indeed an integer. Then

$$\begin{aligned} & c_{s+1} = c_s + tp^s \text{ is a solution of } (**) \\ \Leftrightarrow\ & f(c_s + tp^s) \equiv 0 \pmod{p^{s+1}} \\ \Leftrightarrow\ & f(c_s) + (tp^s)f'(c_s) + (tp^s)^2 \cdot (\text{an integer}) \equiv 0 \pmod{p^{s+1}} \\ \Leftrightarrow\ & f(c_s) + tp^s f'(c_s) \equiv 0 \pmod{p^{s+1}} \qquad (\text{as } 2s \geq s + 1) \\ \Leftrightarrow\ & p^s\left(\frac{f(c_s)}{p^s} + tf'(c_s)\right) \equiv 0 \pmod{p^{s+1}} \\ \Leftrightarrow\ & \frac{f(c_s)}{p^s} + tf'(c_s) \equiv 0 \pmod{p} \\ \Leftrightarrow\ & f'(c_s)t \equiv -\frac{f(c_s)}{p^s} \pmod{p}, \end{aligned}$$

so (*ii*) is proved.

(*iii*) If c_s (more precisely, $\lfloor c_s \rfloor_{p^s}$) is a solution of $f(x) \equiv 0 \pmod{p^s}$, then a solution c_{s+1} (more precisely, $\lfloor c_{s+1} \rfloor_{p^{s+1}}$) of $f(x) \equiv 0 \pmod{p^{s+1}}$ is said to "*arise from* c_s (or, from $\lfloor c_s \rfloor_{p^s}$)" when c_{s+1} is of form $c_s + tp^s$ for some $t \in \mathbf{Z}$—or, equivalently, when $c_{s+1} \equiv c_s \pmod{p^s}$. Thus, the phrase "arise from" is meant to indicate the passage [described in part (ii)] from a solution of (*) to a solution of (**). Our purpose here is to see how many solutions of (**) we obtain in this way. Of course, because counting is involved, it is understood that we are dealing with residue classes as solutions.

Now, if both $\lfloor c_s + t_1 p^s \rfloor_{p^{s+1}}$ and $\lfloor c_s + t_2 p^s \rfloor_{p^{s+1}}$ are solutions of (**) which arise from $\lfloor c_s \rfloor_{p^s}$, then clearly

$$\lfloor c_s + t_1 p^s \rfloor_{p^{s+1}} = \lfloor c_s + t_2 p^s \rfloor_{p^{s+1}} \Leftrightarrow t_1 \equiv t_2 \pmod{p}$$

—so they are distinct solutions if and only if $t_1 \not\equiv t_2 \pmod{p}$. Therefore, in virtue of part (*ii*), the number of distinct solutions of $f(x) \equiv 0 \pmod{p^{s+1}}$ arising from $\lfloor c_s \rfloor_{p^s}$ is precisely the number of distinct solutions of the congruence

$$f'(c_s)t \equiv -\frac{f(c_s)}{p^s} \pmod{p}. \qquad (\#)$$

But this is a linear congruence modulo the prime p, and the theory of a linear congruence was settled completely in Section 3-1. Accordingly, it is easy to see that (#) has either 0, 1, or p solutions, and the conditions under which these three possibilities occur are those stated in (*iii*). This completes the proof. ▌

3-6-5. Remark. Basing ourselves on this result, let us summarize the method for solving a polynomial congruence

$$f(x) \equiv 0 \pmod{p^r}.$$

We start by solving

$$f(x) \equiv 0 \pmod{p}$$

by trial and error, as a rule. In keeping with the notation of 3-6-4, denote a typical such solution by c_1. The next step involves solving

$$f(x) \equiv 0 \pmod{p^2}.$$

To find a typical solution $c_2 = c_1 + tp$ of this congruence, we solve the linear congruence

$$f'(c_1)t \equiv -\frac{f(c_1)}{p} \pmod{p}$$

[which is the congruence (#) in the case $s = 1$] for t. Of course, this must be done for each choice of c_1.

In the same way, c_2 is used to find a typical solution $c_3 = c_2 + tp^2$ of the congruence

$$f(x) \equiv 0 \pmod{p^3}$$

—where t must satisfy

$$f'(c_2)t \equiv -\frac{f(c_2)}{p^2} \pmod{p}.$$

This procedure is repeated to solve $f(x) \equiv 0$ modulo $p^4, p^5, \ldots, p^r$. At each stage c_i leads either 0, 1, or p choices for c_{i+1}. We finally have solutions c_r of the original congruence. Incidentally, it is clear that

$$c_2 \equiv c_1 \pmod{p},\ c_3 \equiv c_2 \pmod{p^2},\ \ldots,\ c_{s+1} \equiv c_s \pmod{p^s},\ \ldots$$

—so if we start from a specific c_1 and finally have with c_r, then

$$c_1 \equiv c_2 \equiv \cdots \equiv c_r \pmod{p}.$$

How do we know that we have located all solutions of $f(x) \equiv 0 \pmod{p^r}$ in this way? This is not hard. Suppose c (more precisely, $\lfloor c \rfloor_{p^r}$) is any solution of $f(x) \equiv 0 \pmod{p^r}$; we must show that c occurs as the end result of some chain of solutions $c_1, c_2, \ldots, c_r = c$. Now, c (and $\lfloor c \rfloor_p$) is a solution of $f(x) \equiv 0 \pmod{p}$, so c may be chosen as c_1. To find c_2 when $c_1 = c$, we solve

$$f'(c)t \equiv -\frac{f(c)}{p} \pmod{p}.$$

Since $p^r \mid f(c)$, the right-hand side is congruent to 0 (mod p), so one solution for t is $t = 0$, and we then obtain $c_2 = c$. Proceeding in this fashion (and using the fact that $f(c)/p^s \equiv 0 \pmod{p}$ for every $s < r$) we have a chain of solutions

$$c_1 = c,\ c_2 = c,\ c_3 = c,\ \ldots,\ c_s = c,\ \ldots,\ c_r = c.$$

This does it.

It is of some interest to distinguish one situation in which the method described above goes very smoothly.

3-6-6. Corollary. Suppose c_1 is a solution of $f(x) \equiv 0 \pmod{p}$ with $f'(c_1) \not\equiv 0 \pmod{p}$; then the congruence $f(x) \equiv 0 \pmod{p^r}$ has a solution $\lfloor c_r \rfloor_{p^r}$ which is the uniques olution arising from $\lfloor c_1 \rfloor_p$ via a chain of solutions $c_1, c_2, \ldots, c_r$.

Proof: It suffices to guarantee that a chain of solutions $c_1, c_2, \ldots, c_r$ exists, and that the chain $\lfloor c_1 \rfloor_p, \lfloor c_2 \rfloor_{p^2}, \ldots, \lfloor c_r \rfloor_{p^r}$ is unique. This is done inductively.

Starting from c_1, we find $c_2 = c_1 + tp$ by solving

$$f'(c_1)t \equiv -\frac{f(c_1)}{p} \pmod{p}.$$

Because $f'(c_1) \not\equiv 0 \pmod{p}$, 3-6-4, part (*iii*) tells us that a solution $\lfloor c_2 \rfloor_{p^2}$ arising from $\lfloor c_1 \rfloor_p$ exists and is unique. Next, we must locate $c_3 = c_2 + tp^2$ by solving

$$f'(c_2)t \equiv -\frac{f(c_2)}{p^2} \pmod{p}.$$

Since $c_1 \equiv c_2 \pmod{p}$, we have $f'(c_1) \equiv f'(c_2) \pmod{p}$, so this congruence may be put in the form

$$f'(c_1)t \equiv -\frac{f(c_2)}{p^2} \pmod{p}.$$

Hence, a solution $\lfloor c_3 \rfloor_{p^3}$ arising from $\lfloor c_2 \rfloor_{p^2}$ exists and is unique.

This state of affairs recurs as we move along the chain; since $c_s \equiv c_1 \pmod{p}$ and $f'(c_s) \equiv f'(c_1) \pmod{p}$ for each $s < r$, to find $c_{s+1} = c_s + tp^s$ from c_s we need to solve

$$f'(c_1)t \equiv -\frac{f(c_s)}{p^s} \pmod{p} \qquad (\#\#)$$

—so, clearly, $\lfloor c_{s+1} \rfloor_{p^{s+1}}$ arising from $\lfloor c_s \rfloor_{p^s}$ exists and is unique. Thus, $\lfloor c_r \rfloor_{p^r}$ exists and is unique. The details are left to the reader. ▌

Note: In treating numerical examples with $f'(c_1) \not\equiv 0 \pmod{p}$, the foregoing says, in particular, that at each stage we may solve

$$f'(c_1)t \equiv -\frac{f(c_s)}{p^s} \pmod{p}, \qquad s = 1, 2, \ldots, r-1$$

instead of the customary congruence

$$f'(c_s)t \equiv -\frac{f(c_s)}{p^s} \pmod{p}.$$

This reduces somewhat the work involved in arriving at c_r.

3-6-7. Examples. Let us illustrate the general technique discussed in 3-6-4, 3-6-5, 3-6-6 by using it to solve the congruences that were examined in 3-6-2.

(*i*) To solve

$$f(x) = x^3 - x + 4 \equiv 0 \pmod{7^2}, \qquad (p = 7)$$

we note first, as before, that $c_1 = 3$ is a solution of

$$f(x) = x^3 - x + 4 \equiv 0 \pmod 7.$$

In fact, $\lfloor 3 \rfloor_7$ is the only solution.

Then $f'(x) = 3x^2 - 1$, so the equation

$$f'(c_1)t \equiv -\frac{f(c_1)}{p} \pmod p$$

used to find $c_2 = c_1 + tp = 3 + 7t$ becomes [since $f(3) = 28$, and $f'(3) = 26$]

$$26t \equiv -4 \pmod 7.$$

Because $f'(c_1) = 26 \not\equiv 0 \pmod 7$, there is a unique solution $\lfloor c_2 \rfloor_{7^2}$. In fact, solving $5t \equiv -4 \pmod 7$ [which is a simpler version of $26t \equiv -4 \pmod 7$] yields $t = 2$, so $c_2 = 3 + 7 \cdot 2 = 17$; thus $\lfloor 17 \rfloor_{7^2}$ is the unique solution of the original congruence.

(*ii*) To solve

$$f(x) = x^2 - x + 2 \equiv 0 \pmod{7^2}, \qquad (p = 7)$$

we observe that

$$f(x) = x^2 - x + 2 \equiv 0 \pmod 7$$

has the unique solution $\lfloor c_1 \rfloor_7 = \lfloor 4 \rfloor_7$. Then $f'(x) = 2x - 1$, $f(c_1) = f(4) = 14$, $f'(c_1) = f'(4) = 7$, and the equation

$$f'(c_1)t \equiv -\frac{f(c_1)}{p} \pmod p$$

via which we find $c_2 = c_1 + tp = 4 + 7t$ takes the form

$$7t \equiv -2 \pmod 7 \qquad \text{or} \qquad 0t \equiv -2 \pmod 7.$$

This has no solution, so c_2 does not exist—or equivalently, according to the criterion of 3-6-4, part (*iii*), there are no solutions $\lfloor c_2 \rfloor_7{}^2$ arising from $\lfloor c_1 \rfloor_7$. Thus the original congruence has no solution.

(*iii*) To solve

$$f(x) = x^2 + x + 1 \equiv 0 \pmod{7^2}$$

we start from the two solutions

$$c_1 = 2, \qquad c_1 = 4$$

of

$$f(x) = x^2 + x + 1 \equiv 0 \pmod 7.$$

Each of these solutions must be pursued to the next stage.

If $c_1 = 2$, then $f(c_1) = f(2) = 7$ and $f'(c_1) = f'(2) = 5$, so that $c_2 = c_1 + tp = 2 + 7t$ must satisfy

$$5t \equiv -\tfrac{7}{7} \pmod 7.$$

Now $5t \equiv -1 \pmod 7$ has the solution $t = 4$, so $c_2 = 2 + 7 \cdot 4 = 30$ is a solution of the original congruence. According to 3-6-4, part (*iii*), since $f'(2) \not\equiv 0 \pmod 7$, $\lfloor 30 \rfloor_{7^2}$ is the unique solution of the original which arises from $\lfloor c_1 \rfloor_7 = \lfloor 2 \rfloor_7$.

If $c_1 = 4$, then $f(c_1) = f(4) = 21$ and $f'(c_1) = f'(4) = 9$ so $c_2 = c_1 + tp = 4 + 7t$ has to satisfy

$$9t \equiv -\tfrac{21}{7} \pmod 7 \qquad \text{or} \qquad 2t \equiv -3 \pmod 7.$$

Taking $t = 2$ we obtain $c_2 = 18$—and $\lfloor 18 \rfloor_{7^2}$ is the unique solution of the original arising from $\lfloor c_1 \rfloor_7 = \lfloor 4 \rfloor_7$.

Thus, $x^2 + x + 1 \equiv 0 \pmod{49}$ has exactly two solutions: $\lfloor 30 \rfloor_{7^2}$, $\lfloor 18 \rfloor_{7^2}$.

(*iv*) To solve

$$f(x) = x^2 + x + 1 \equiv 0 \pmod{7^3}$$

we need to continue the two chains of solutions

$$c_1 = 2, \quad c_2 = 30 \qquad\qquad c_1 = 4, \quad c_2 = 18$$

found above one step further. Recalling from above that $f'(c_1) \not\equiv 0 \pmod p$ $(p = 7)$, and making use of 3-6-6, we see that $c_3 = c_2 + tp^2$ may be obtained, in each case, by solving

$$f'(c_1)t \equiv -\frac{f(c_2)}{p^2} \pmod p.$$

In the case, $c_1 = 2$, $c_2 = 30$ we have

$$f'(2)t \equiv -\frac{f(30)}{7^2} \pmod 7,$$

which becomes

$$5t \equiv -\frac{931}{49} = -19 \equiv 2 \pmod 7.$$

Taking $t = 6$ gives $c_3 = 324$; so $\lfloor 324 \rfloor_{7^3}$ is the unique solution of $f(x) \equiv 0 \pmod{7^3}$ which arises from $\lfloor c_1 \rfloor_7 = \lfloor 2 \rfloor_7$.

In the case, $c_1 = 4$, $c_2 = 18$ we have

$$9t \equiv -\frac{343}{49} \pmod 7 \qquad \text{or} \qquad 2t \equiv 0 \pmod 7.$$

Taking $t = 0$ gives $c_3 = 18$; so $\lfloor 18 \rfloor_{7^3}$ is the unique solution of $f(x) \equiv 0 \pmod{7^3}$ which arises from $\lfloor c_1 \rfloor_7 = \lfloor 4 \rfloor_7$.

Thus, $x^2 + x + 1 \equiv 0 \pmod{7^3}$ has exactly two solutions: $\lfloor 324 \rfloor_{7^3}$, $\lfloor 18 \rfloor_{7^3}$.

(v) It may be safely left for the reader to solve

$$f(x) = x^2 - x - 12 \equiv 0 \pmod{7^2}$$

and find its seven solutions.

3-6-8. Remark. Having seen how the problem of solving a polynomial congruence modulo any prime power p^r reduces essentially to that of solving the congruence modulo the prime p, let us now show how the problem of solving a polynomial congruence

$$f(x) \equiv 0 \pmod{m}, \qquad (+)$$

where m is arbitrary, reduces to solving the congruence modulo certain prime powers.

Suppose the prime-power factorization of m is $m = p_1^{r_1} p_2^{r_2} \cdots p_n^{r_n}$ and for convenience let us put $m_1 = p^{r_1}$, $m_2 = p_2^{r_2}, \ldots, m_n = p_n^{r_n}$. Thus, $m = m_1 \cdots m_n$ and, because $m_1, m_2, \ldots, m_n$ are relatively prime in pairs, $[m_1, m_2, \ldots, m_n] = m_1 m_2 \cdots m_n$. Consider the system of congruences

$$\begin{aligned} f(x) &\equiv 0 \pmod{m_1}, \\ f(x) &\equiv 0 \pmod{m_2}, \\ &\vdots \\ f(x) &\equiv 0 \pmod{m_i}, \\ &\vdots \\ f(x) &\equiv 0 \pmod{m_n}. \end{aligned} \qquad (++)$$

We say that an integer c is a **solution** of this system when it is a solution of each of the congruences—that is, when $f(c) \equiv 0 \pmod{m_i}$ for $i = 1, 2, \ldots, n$. Clearly, the integer c is a solution of the congruence $(+)$ if and only if it is a solution of the system $(++)$ [since m divides $f(c) \Leftrightarrow m_i$ divides $f(c)$ for $i = 1, \ldots, n$].

Now, we can find solutions (if any exist) of $(++)$ as follows. For each $i = 1, \ldots, n$ we solve the ith congruence $f(x) \equiv 0 \pmod{m_i}$ by the procedure of 3-6-5. Suppose $c^{(1)}, c^{(2)}, \ldots, c^{(n)}$ are solutions of the individual congruences —more precisely, $c^{(i)}$ is a solution of $f(x) \equiv 0 \pmod{m_i}$ for each $i = 1, 2, \ldots, n$. By the Chinese remainder theorem, we can find an integer c satisfying

$$c \equiv c^{(i)} \pmod{m_i}, \qquad i = 1, 2, \ldots, n.$$

Then for each $i = 1, 2, \ldots, n$

$$f(c) \equiv f(c^{(i)}) \equiv 0 \pmod{m_i}$$

so c is a solution of the system $(++)$ and of the original congruence $f(x) \equiv 0 \pmod{m}$.

How many solutions are there? This involves counting elements of residue class rings. Let S_m denote the set of elements of $\mathbf{Z}_m$ which are solutions of $f(x) \equiv 0 \pmod{m}$; so $\#(S_m)$ is the number of solutions of $f(x) \equiv 0 \pmod{m}$—let us write this as $\#(m)$. Similarly, for each i, S_{m_i} is the set of elements of $\mathbf{Z}_{m_i}$ which are solutions of $f(x) \equiv 0 \pmod{m_i}$—so $\#(m_i) = \#(S_{m_i})$ is the number of solutions of $f(x) \equiv 0 \pmod{m_i}$. We produce a 1–1 correspondence between S_m and the product set $S_{m_1} \times S_{m_2} \times \cdots \times S_{m_n}$. If $\lfloor c \rfloor_m \in S_m$, then, for each i, $\lfloor c \rfloor_{m_i}$ is a solution of the ith congruence—that is, $\lfloor c \rfloor_{m_i} \in S_{m_i}$. Thus,

$$\lfloor c \rfloor_m \to (\lfloor c \rfloor_{m_1}, \lfloor c \rfloor_{m_2}, \ldots, \lfloor c \rfloor_{m_n})$$

is a mapping of $S_m \to S_{m_1} \times S_{m_2} \times \cdots \times S_{m_n}$. On the other hand, an element of the product set is of form

$$(\lfloor c^{(1)} \rfloor_{m_1}, \lfloor c^{(2)} \rfloor_{m_2}, \ldots, \lfloor c^{(n)} \rfloor_{m_n}),$$

where for each i, $c^{(i)}$ is a solution of the ith congruence. Then for an integer c satisfying $c \equiv c^{(i)} \pmod{m_i}$, $i = 1, \ldots, n$ (whose existence is guaranteed by the Chinese remainder theorem), we have

$$(\lfloor c \rfloor_{m_1}, \lfloor c \rfloor_{m_2}, \ldots, \lfloor c \rfloor_{m_n}) = (\lfloor c^{(1)} \rfloor_{m_1}, \lfloor c^{(2)} \rfloor_{m_2}, \ldots, \lfloor c^{(n)} \rfloor_{m_n})$$

and (again by the Chinese remainder theorem) $\lfloor c \rfloor_m$ is uniquely determined by the given element of the product set. Thus,

$$(\lfloor c^{(1)} \rfloor_{m_1}, \lfloor c^{(2)} \rfloor_{m_2}, \ldots, \lfloor c^{(n)} \rfloor_{m_n}) \to \lfloor c \rfloor_m$$

is a mapping of $S_{m_1} \times S_{m_2} \times \cdots \times S_{m_n} \to S_m$.

It is easy to see that the two mappings just defined are inverses of each other, and provide a one-to-one correspondence between the sets S_m and $S_{m_1} \times S_{m_2} \times \cdots \times S_{m_n}$. In particular,

$$\#(S_m) = \#(S_{m_1} \times S_{m_2} \times \cdots \times S_{m_n})$$

and

$$\#(m) = \prod_{i=1}^{n} \#(m_i).$$

Observing that in the derivation of this formula the form of the m_i's does not enter—only the fact that they are relatively prime in pairs—we can summarize the discussion as follows.

3-6-9. Proposition. Suppose $m = m_1 m_2 \cdots m_n$ where $m_1, \ldots, m_n$ are relatively prime in pairs. Then we can find the solutions of

$$f(x) \equiv 0 \pmod{m}$$

from the solutions of the individual congruences of the system

$$\begin{aligned} f(x) &\equiv 0 \pmod{m_1}, \\ f(x) &\equiv 0 \pmod{m_2}, \\ &\vdots \\ f(x) &\equiv 0 \pmod{m_n}, \end{aligned}$$

via the Chinese remainder theorem. The number of solutions of $f(x) \equiv 0 \pmod{m}$ is given by

$$\#(m) = \prod_{i=1}^{n} \#(m_i).$$

In particular, $\#(m) = 0 \Leftrightarrow$ some $\#(m_i) = 0$; in other words, $f(x) \equiv 0 \pmod{m}$ has no solution $\Leftrightarrow$ one (or more) of the congruences $f(x) \equiv 0 \pmod{m_i}$ has no solution.

3-6-10. Example. We solve

$$f(x) = x^3 + 90x^2 + 83x + 99 \equiv 0 \pmod{3^2 \cdot 5^2 \cdot 7}.$$

As per 3-6-8 and 3-6-9, we first solve each of the congruences

$$\begin{aligned} f(x) &\equiv 0 \pmod{3^2}, \\ f(x) &\equiv 0 \pmod{5^2}, \\ f(x) &\equiv 0 \pmod{7}. \end{aligned}$$

These are, respectively, the congruences

$$\begin{aligned} &\text{(I)} && x^3 + 2x \equiv 0 \pmod{3^2}, \\ &\text{(II)} && x^3 + 15x^2 + 8x - 1 \equiv 0 \pmod{5^2}, \\ &\text{(III)} && x^3 - x^2 - x + 1 \equiv 0 \pmod{7}. \end{aligned}$$

To solve (I), we start with

$$x^3 + 2x \equiv 0 \pmod{3},$$

which has the three solutions 0, 1, 2 (mod 3); in fact, in $\mathbf{Z}_3[x]$ we have

$$x^3 + 2x = x(x-1)(x-2).$$

By the method of 3-6-4 and 3-6-5 these solutions lead, respectively, to the solutions 0, 4, 5 of (I). Thus, $x^3 + 2x \equiv 0 \pmod{3^2}$ has exactly three solutions: $\lfloor 0 \rfloor_{3^2}, \lfloor 4 \rfloor_{3^2}, \lfloor 5 \rfloor_{3^2}$.

To solve (II), we start with

$$x^3 + 15x^2 + 8x - 1 \equiv 0 \pmod{5},$$

which is $x^3 + 3x - 1 \equiv 0 \pmod 5$. This has the two solutions 3, 4 (mod 5); in fact, 3 is a double solution (or root), because in $\mathbf{Z}_5[x]$, we have

$$x^3 + 15x^2 + 8x - 1 = x^3 + 3x - 1 = (x - 3)^2(x - 4).$$

By the method of 3-6-4 and 3-6-5, there is no solution of (II) arising from the solution 3. On the other hand, the solution 4 leads, by this method, to the solution 19 of (II). Thus, $x^3 + 15x^2 + 8x - 1 \equiv 0 \pmod{5^2}$ has exactly one solution: $\lfloor 19 \rfloor_{5^2}$.

To solve (III) is easy. It has the two solutions 1, 6 (mod 7); in fact, in $\mathbf{Z}_7[x]$ we have

$$x^3 - x^2 - x + 1 = (x - 1)^2(x - 6).$$

Thus, $x^3 - x^2 - x + 1 \equiv 0 \pmod 7$ has the two solutions: $\lfloor 1 \rfloor_7$, $\lfloor 6 \rfloor_7 = \lfloor -1 \rfloor_7$.

Returning to the original congruence, it has $3 \cdot 1 \cdot 2 = 6$ solutions. They are to be found, by the Chinese remainder theorem, from the six congruences of form

$$x \equiv a_1 \pmod 9,$$
$$x \equiv a_2 \pmod{25},$$
$$x \equiv a_3 \pmod 7,$$

where $a_1 = 0, 4$, or 5, $a_2 = 19$, $a_3 = 1$ or -1. The six solutions turn out to be

$$\lfloor 419 \rfloor_{1575}, \quad \lfloor 594 \rfloor_{1575}, \quad \lfloor 769 \rfloor_{1575}, \quad \lfloor 869 \rfloor_{1575}, \quad \lfloor 1044 \rfloor_{1575}, \quad \lfloor 1219 \rfloor_{1575}.$$

3-6-11 / *PROBLEMS*

1. Use the elementary technique illustrated in 3-6-2 to solve:
(*i*) $x^3 - x + 4 \equiv 0 \pmod{7^3}$,
(*ii*) $x^3 - x + 4 \equiv 0 \pmod{7^4}$,
(*iii*) $x^2 + x + 1 \equiv 0 \pmod{7^4}$,
(*iv*) $x^2 - x - 12 \equiv 0 \pmod{7^3}$.

2. Do each part of Problem 1 by the general method.

3. By solving $f(x) = x^2 - 5x - 15 \equiv 0 \pmod{3^3}$ show that it has exactly two solutions: $\lfloor 15 \rfloor_{27}$, $\lfloor 17 \rfloor_{27}$.

4. Let $f(x)$ be any one of the polynomials
(*i*) $x^3 - x + 3$, (*v*) $x^3 + 2x - 5$,
(*ii*) $x^3 - 2x + 3$, (*vi*) $x^3 - 2x^2 + 1$,
(*iii*) $x^3 + 2x - 3$, (*vii*) $x^2 + x + 7$,
(*iv*) $x^3 + x^2 - 4$, (*viii*) $x^4 + x + 1$.
Solve the congruence $f(x) \equiv 0 \pmod m$ for each of the following choices of m:
(*a*) 3, (*b*) 9, (*c*) 27, (*d*) 81.

5. Do Problem 4 for m equal to
(*a*) 2, (*b*) 4, (*c*) 8, (*d*) 16, (*e*) 32, (*f*) 64.

6. Do Problem 4 for the following values of m:
(*a*) 5, (*b*) 25, (*c*) 125.

7. Do Problem 4 for m equal to
(*a*) 7, (*b*) 49, (*c*) 343.

8. Do Problem 4 when m equals
(*a*) 15, (*b*) 45, (*c*) 75, (*d*) 135.

9. Do Problem 4 when m equals
(*a*) 21, (*b*) 63, (*c*) 147, (*d*) 189.

10. Do Problem 4 for m equal to
(*a*) 35, (*b*) 175.

11. Do Problem 4 when m is
(*a*) 210, (*b*) 280, (*c*) 336, (*d*) 490.

12. Consider $f(x) = x^3 - 4x^2 - 11x + 30$ viewed as a polynomial in $\mathbf{Z}_m[x]$ when m equals
(*i*) 419, (*ii*) 463, (*iii*) 91, (*iv*) 143.
In each case, find all the roots in $\mathbf{Z}_m$. (Incidentally, 419 and 463 are prime.)

13. Without actually finding the solutions, decide how many solutions there are to each of the following congruences.
(*i*) $x^3 - x + 1 \equiv 0 \pmod{3^{10} \cdot 13^5}$,
(*ii*) $x^3 - x + 1 \equiv 0 \pmod{5^7 \cdot 7^5}$,
(*iii*) $x^3 - x + 1 \equiv 0 \pmod{7^8 \cdot 11^4}$,
(*iv*) $x^3 + 5x - 3 \equiv 0 \pmod{3^{10} \cdot 5^5}$,
(*v*) $x^3 + 2x - 3 \equiv 0 \pmod{3^5 \cdot 5^2}$.

14. Construct, if possible, a polynomial of degree 3 for which the number of solutions of $f(x) \equiv 0 \pmod{5^2}$ is:
(*i*) 0, (*ii*) 1, (*iii*) 5, (*iv*) 2, (*v*) 3, (*vi*) 4.
Which additional integers can represent the number of solutions?

15. If $f(x)$ is of degree 3 and p is prime, what are the possibilities for the number of solutions of $f(x) \equiv 0 \pmod{p^2}$.

16. Suppose c_1 is a solution of $f(x) \equiv 0 \pmod{p}$ with $f'(c_1) \not\equiv 0 \pmod{p}$. Prove that if c is a solution of $f(x) \equiv 0 \pmod{p^r}$ with $c \equiv c_1 \pmod{p}$, then $\lfloor c \rfloor_{p^r}$ is the solution of $f(x) \equiv 0 \pmod{p^r}$ arising uniquely from $\lfloor c_1 \rfloor_p$.

17. Suppose $f(x) \in \mathbf{Z}[x]$ is a polynomial of degree n. For any $h \in \mathbf{Z}$, show that

$$f(x+h) = f(x) + f'(x)h + \frac{1}{2!}f''(x)h^2 + \cdots + \frac{1}{n!}f^{(n)}(x)h^n$$

where $f''(x)$ is the derivative of $f'(x)$ and, recursively, $f^{(i)}(x)$ is the derivative of $f^{(i-1)}(x)$. Furthermore, each term $f^{(i)}(x)h^i/i!$ belongs to $\mathbf{Z}[x]$.

The reader with calculus experience should recognize this as a form of Taylor's theorem.

3-7. Quadratic Reciprocity

In the last section, we studied the problem of solving a polynomial congruence

$$f(x) = a_0 + a_1x + \cdots + a_n x^n \equiv 0 \pmod{m}, \qquad a_0, a_1, \ldots, a_n \in \mathbf{Z}.$$

Equivalently, this is the problem of finding the roots in $\mathbf{Z}_m$ of the polynomial

$$f(x) = a_0 + a_1x + \cdots + a_n x^n \in \mathbf{Z}_m[x].$$

(Strictly speaking, this latter polynomial should be written as

$$\lfloor a_0 \rfloor_m + \lfloor a_1 \rfloor_m x + \cdots + \lfloor a_n \rfloor_m x^n \in \mathbf{Z}_m[x]$$

but we prefer to let an integer represent itself or its residue class, depending on the context. This has been done many times, so the danger of confusion should not be great.) It was seen that this reduces to the problem of solving the congruence modulo a prime. More precisely, once we know how to find all solutions of

$$f(x) = a_0 + a_1x + \cdots + a_n x^n \equiv 0 \pmod{p}$$

when p is prime then, by the recipe of 3-6-5, we can find all solutions of

$$f(x) = a_0 + a_1x + \cdots + a_n x^n \equiv 0 \pmod{p^r}$$

for any choice of $r \geq 1$. Once this has been done for every prime power p^r which appears in the prime factorization of m, we can make use of the Chinese remainder theorem, as in 3-6-8, to obtain all the solutions of the original congruence mod m.

Our concern turns, therefore, to solving the congruence

$$f(x) = a_0 + a_1x + \cdots + a_n x^n \equiv 0 \pmod{p}, \qquad p \text{ prime}$$

or, equivalently, to finding the roots in the field $\mathbf{Z}_p$ of the polynomial

$$f(x) = a_0 + a_1x + \cdots + a_n x^n \in \mathbf{Z}_p[x].$$

Unfortunately, this is too hard in general (though for reasonably small p, it is feasible to simply test every element of $\mathbf{Z}_p$ and thus find all the roots). This is not surprising; after all, in the familiar situation of polynomials with real coefficients we have no general method which always locates the real roots.

Consequently, just as is done in the study of polynomials with real coefficients, it is natural to restrict the degree n of $f(x)$. When $n = 1$, we are dealing with a linear congruence

$$a_0 + a_1 x \equiv 0 \pmod{p}$$

and, in virtue of our work in Section 3-1, we know everything about solving it. When $n = 2$, which means that the polynomial is "quadratic," we are dealing (after a trivial change of notation) with a congruence of form

$$f(x) = ax^2 + bx + c \equiv 0 \pmod{p} \qquad (*)$$

or, equivalently, with the polynomial

$$f(x) = ax^2 + bx + c \in \mathbf{Z}_p[x]. \qquad (**)$$

Again, we emphasize that a, b, c are viewed, interchangeably, as integers or as elements of $\mathbf{Z}_p$. In particular, since $f(x) \in \mathbf{Z}_p[x]$ has degree 2, $a \neq 0 \in \mathbf{Z}_p$; or what is the same thing $p \nmid a$, when a is viewed in $\mathbf{Z}$.

If $p = 2$, the only possible roots of $(**)$ [or solutions of $(*)$] are 0 and 1—so it is trivial to find the roots. Therefore, throughout this section, we shall always assume that

$$p \text{ is an odd prime.}$$

3-7-1. Remark. As a first step, let us transform and simplify the problem of finding the roots of

$$f(x) = ax^2 + bx + c \in \mathbf{Z}_p[x]. \qquad (*)$$

The idea is to use the standard technique of "completing the square," as was done in 3-5-22. Note first that $a \neq 0$ in $\mathbf{Z}_p$, by hypothesis, and $4 \neq 0$ in $\mathbf{Z}_p$, because p is an odd prime. Therefore, $4a \neq 0$ in the field $\mathbf{Z}_p$. Now, for $x_0 \in \mathbf{Z}_p$, we have

$$\begin{aligned} x_0 \text{ is a root of } (*) &\Leftrightarrow ax_0^2 + bx_0 + c = 0 \\ &\Leftrightarrow 4a(ax_0^2 + bx_0) = -4ac \\ &\Leftrightarrow 4a^2x_0^2 + 4abx_0 + b^2 = b^2 - 4ac \\ &\Leftrightarrow (2ax_0 + b)^2 = b^2 - 4ac. \end{aligned}$$

At this point, if we were dealing with real or complex numbers, the next step would be to take square roots and finally obtain

$$x_0 = \frac{-b \pm \sqrt{b^2 - 4ac}}{2a}$$

—but in $\mathbf{Z}_p$, the meaning of " $\pm\sqrt{b^2 - 4ac}$" is far from clear. Therefore, in order to be careful, let us consider the polynomial in a new indeterminate y,

$$y^2 - (b^2 - 4ac) \in \mathbf{Z}_p[y] \qquad (\#)$$

or, what is the same thing, the equation

$$y^2 = b^2 - 4ac$$

over $\mathbf{Z}_p$. If x_0 is a root of $(*)$, then obviously $y_0 = 2ax_0 + b \in \mathbf{Z}_p$ is a root of $(\#)$. Conversely, suppose $y_0 \in \mathbf{Z}_p$ is a root of $(\#)$. Then, because $2a \neq 0$ in $\mathbf{Z}_p$, the linear equation (over $\mathbf{Z}_p$)

$$2ax + b = y_0$$

has a unique solution $x_0 \in \mathbf{Z}_p$. In fact, this solution is

$$x_0 = (2a)^{-1}(y_0 - b)$$

where $(2a)^{-1}$ denotes the multiplicative inverse of $2a$ in the field $\mathbf{Z}_p$. Thus, $(2ax +_0 b)^2 = {y_0}^2 = b^2 - 4ac$, and from our equivalences, x_0 is a root of $(*)$. We have proved:

$$(*) \text{ has the root } x_0 \Leftrightarrow (\#) \text{ has the root } y_0 = 2ax_0 + b.$$

All this means that the problem of finding the roots of any quadratic polynomial $ax^2 + bx + c \in \mathbf{Z}_p[x]$ reduces to the problem of finding the roots of a simpler quadratic polynomial $y^2 - (b^2 - 4ac)$. Therefore, with a trivial change of notation, it suffices for us to consider the problem of finding the roots of polynomials of form

$$x^2 - a \in \mathbf{Z}_p[x] \qquad (a \neq 0 \text{ in } \mathbf{Z}_p).$$

(Any root is then entitled to be called a "square root of a" in $\mathbf{Z}_p$, and a may be said to be a "square" in $\mathbf{Z}_p$. Of course, there are at most two distinct square roots of a, because $x^2 - a$ is a polynomial of degree 2 over the field $\mathbf{Z}_p$.) Equivalently, our problem may be restated as that of solving the congruence

$$x^2 \equiv a \pmod{p}, \qquad (a \in \mathbf{Z}, p \nmid a).$$

3-7-2. Definition. Suppose p is an odd prime and a is an integer with $p \nmid a$. We say that a is a **quadratic residue** (mod p) when the congruence $x^2 \equiv a \pmod{p}$ has a solution; if the congruence has no solution, a is called a **quadratic nonresidue** (mod p).

It is extremely convenient to introduce a symbol or notation due to Legendre (1752–1833)—namely,

$$\left(\frac{a}{p}\right) = \begin{cases} +1, & \text{when } a \text{ is a quadratic residue (mod } p), \\ -1, & \text{when } a \text{ is a quadratic nonresidue (mod } p). \end{cases}$$

One reads $\left(\frac{a}{p}\right)$ as "the Legendre symbol of a over p."

There are several equivalent formulations for these definitions; in fact,

$$\begin{aligned} \left(\frac{a}{p}\right) = 1 &\Leftrightarrow a \text{ is a quadratic residue (mod } p) \\ &\Leftrightarrow x^2 \equiv a \pmod{p} \text{ has a solution} \\ &\Leftrightarrow \text{the polynomial } x^2 - a \in \mathbf{Z}_p[x] \text{ has a root in } \mathbf{Z}_p \\ &\Leftrightarrow a \left(\text{more precisely, } \lfloor a \rfloor_p\right) \text{ is a square in } \mathbf{Z}_p \end{aligned}$$

and

$$\begin{aligned} \left(\frac{a}{p}\right) = -1 &\Leftrightarrow a \text{ is a quadratic nonresidue (mod } p) \\ &\Leftrightarrow x^2 \equiv a \pmod{p} \text{ has no solution} \\ &\Leftrightarrow \text{the polynomial } x^2 - a \in \mathbf{Z}_p[x] \text{ has no root in } \mathbf{Z}_p \\ &\Leftrightarrow a \left(\text{more precisely, } \lfloor a \rfloor_p\right) \text{ is not a square in } \mathbf{Z}_p. \end{aligned}$$

3-7-3. Examples. (1) Let $p = 7$. The congruence $x^2 \equiv 2 \pmod{7}$ has a solution—in fact, 3 is a solution, and so is 4—so $\left(\frac{2}{7}\right) = 1$. On the other hand, $\left(\frac{3}{7}\right) = -1$ because the congruence $x^2 \equiv 3 \pmod{7}$ has no solution, or equivalently because the polynomial $x^2 - 3 \in \mathbf{Z}_7[x]$ has no root in $\mathbf{Z}_7$. In similar fashion, $\left(\frac{1}{7}\right) = 1$, $\left(\frac{4}{7}\right) = 1$, $\left(\frac{5}{7}\right) = -1$, $\left(\frac{6}{7}\right) = -1$, $\left(\frac{-1}{7}\right) = -1$, $\left(\frac{-2}{7}\right) = -1$, and so on.

(2) Let $p = 11$. In the field $\mathbf{Z}_{11}$, we have

$$\begin{array}{lllll} 1^2 = 1, & 2^2 = 4, & 3^2 = 9, & 4^2 = 5, & 5^2 = 3, \\ 10^2 = 1, & 9^2 = 4, & 8^2 = 9, & 7^2 = 5, & 6^2 = 3. \end{array}$$

Therefore, 1, 3, 4, 5, 9 are quadratic residues (mod 11), and 2, 6, 7, 8, 10 are quadratic nonresidues (mod 11). In other words, $\left(\frac{a}{11}\right) = 1$ for $a = 1, 3, 4, 5, 9$ and $\left(\frac{a}{11}\right) = -1$ for 2, 6, 7, 8, 10.

What about $\left(\frac{-1}{11}\right)$? In $\mathbf{Z}_{11}$, $-1 = 10$, which is not a square in $\mathbf{Z}_{11}$—so $x^2 + 1 \in \mathbf{Z}_{11}[x]$ has no root in $\mathbf{Z}_{11}$, and $\left(\frac{-1}{11}\right) = -1$.

Similarly, in $\mathbf{Z}_{11}$, $9 = -2 = 20 = -13 \cdots$ so they all have the same Legendre symbol—that is, $\left(\frac{-2}{11}\right) = \left(\frac{20}{11}\right) = \left(\frac{-13}{11}\right) = \left(\frac{9}{11}\right) = 1$.

(3) Consider $p = 13$. Then $(\frac{a}{13}) = 1$ for $a = 1, 3, 4, 9, 10, 12$ and $(\frac{a}{13}) = -1$ for $a = 2, 5, 6, 7, 8, 11$—since, in $\mathbf{Z}_{13}$,

$$\begin{array}{cccccc} 1^2 = 1, & 2^2 = 4, & 3^2 = 9, & 4^2 = 3, & 5^2 = 12, & 6^2 = 10, \\ 12^2 = 1, & 11^2 = 4, & 10^2 = 9, & 9^2 = 3, & 8^2 = 12, & 7^2 = 10. \end{array}$$

Thus, for example, $x^2 \equiv 10 \pmod{13}$ has a solution, and $x^2 \equiv 8 \pmod{13}$ does not have a solution.

Note that for any $c = 1, 2, \ldots, 12$, viewed as an element of $\mathbf{Z}_{13}$, we have

$$c^2 = (-c)^2 = (13 - c)^2 \text{ in } \mathbf{Z}_{13}.$$

Therefore, if $c \in \mathbf{Z}$ is a solution of $x^2 \equiv a \pmod{13}$, then so is $13 - c$. Even more, these facts carry over to any p—namely, since

$$c^2 = (-c)^2 = (p - c)^2 \text{ in } \mathbf{Z}_p$$

or, equivalently (when we work in $\mathbf{Z}$) since

$$c^2 \equiv (-c)^2 \equiv (p - c)^2 \pmod{p},$$

it follows that if $c \in \mathbf{Z}_p$ is a root of $x^2 - a \in \mathbf{Z}_p[x]$, then so is $p - c = -c$ in $\mathbf{Z}_p$, or equivalently, (when we work in $\mathbf{Z}$) if $c \in \mathbf{Z}$ is a solution of $x^2 \equiv a \pmod{p}$, then so is $p - c$.

Our objective is to compute the Legendre symbol $(\frac{a}{p})$ in all meaningful cases, and thus to settle the question of when the congruence $x^2 \equiv a \pmod{p}$ has a solution (or, equivalently to determine which elements of any finite field $\mathbf{Z}_p$ are squares). We begin with some elementary properties of the Legendre symbol.

3-7-4. Proposition. Suppose a and b are integers which are relatively prime to the odd prime p; then

(*i*) $\left(\dfrac{a^2}{p}\right) = 1, \left(\dfrac{1}{p}\right) = 1,$

(*ii*) $a \equiv b \pmod{p} \Rightarrow \left(\dfrac{a}{p}\right) = \left(\dfrac{b}{p}\right),$

(*iii*) $\left(\dfrac{a}{p}\right) \equiv a^{(p-1)/2} \pmod{p},$

(*iv*) $\left(\dfrac{ab}{p}\right) = \left(\dfrac{a}{p}\right)\left(\dfrac{b}{p}\right),$

(*v*) $\left(\dfrac{-1}{p}\right) = (-1)^{(p-1)/2}.$

Proof: (*i*) The congruence $x^2 \equiv a^2 \pmod p$ obviously has a solution—a itself will do—so $\left(\frac{a^2}{p}\right) = 1$; and then $\left(\frac{1}{p}\right) = \left(\frac{1^2}{p}\right) = 1$.

(*ii*) Since $a \equiv b \pmod p$, $c \in \mathbf{Z}$ is a solution of $x^2 \equiv a \pmod p$ if and only if it is a solution of $x^2 \equiv b \pmod p$. Thus, there are only two possibilities: either both $x^2 \equiv a \pmod p$ and $x^2 \equiv b \pmod p$ have a solution, or neither one has a solution. It follows that $\left(\frac{a}{p}\right) = \left(\frac{b}{p}\right)$.

(*iii*) If $\left(\frac{a}{p}\right) = 1$, then there exists $x_0 \in \mathbf{Z}$ with $x_0^2 \equiv a \pmod p$. Raising both sides of the congruence to the power $(p-1)/2$, gives

$$a^{(p-1)/2} \equiv (x_0{}^2)^{(p-1)/2} = x_0^{p-1} \equiv 1 \pmod p.$$

[The congruence $x_0^{p-1} \equiv 1 \pmod p$ is valid in virtue of 3-2-22, part (7)—or by making use of 3-5-12.] Thus: If $\left(\frac{a}{p}\right) = 1$, then $\left(\frac{a}{p}\right) \equiv a^{(p-1)/2} \pmod p$.

On the other hand, suppose $\left(\frac{a}{p}\right) = -1$, so $x^2 \equiv a \pmod p$ has no solutions. By the theory of linear congruences, for each $r \in \{1, 2, \ldots, p-1\}$ there exists an $r' \in \{1, 2, \ldots, p-1\}$ with

$$rr' \equiv a \pmod p$$

[that is, r' is a solution of $rx \equiv a \pmod p$]. Moreover, $r' \neq r$; because if $r' = r$, then $r^2 \equiv a \pmod p$, which cannot be since $x^2 \equiv a \pmod p$ has no solutions. Thus, the set $\{1, 2, \ldots, p-1\}$, with $p-1$ elements, breaks up into pairs $\{r, r'\}$, and there are $(p-1)/2$ such pairs. By Wilson's theorem,3-5-13, we have

$$-1 \equiv (p-1)! \equiv \underbrace{\prod}_{\text{all pairs}} (rr') \equiv a^{(p-1)/2} \pmod p.$$

Thus: If $\left(\frac{a}{p}\right) = -1$, then $\left(\frac{a}{p}\right) \equiv a^{(p-1)/2} \pmod p$.

The proof of our result, with which the name Euler is usually associated, is now complete.

(*iv*) In virtue of (*iii*), we have

$$\left(\frac{a}{p}\right)\left(\frac{b}{p}\right) \equiv a^{(p-1)/2} b^{(p-1)/2} = (ab)^{(p-1)/2} \equiv \left(\frac{ab}{p}\right) \pmod p.$$

Now $\left(\frac{a}{p}\right)\left(\frac{b}{p}\right)$ is $+1$ or -1, and so is $\left(\frac{ab}{p}\right)$; if $\left(\frac{a}{p}\right)\left(\frac{b}{p}\right) \neq \left(\frac{ab}{p}\right)$, we have $1 \equiv -1 \pmod p$ or $2 \equiv 0 \pmod p$—which is impossible because p is an odd prime. Hence, $\left(\frac{a}{p}\right)\left(\frac{b}{p}\right) = \left(\frac{ab}{p}\right)$.

(*v*) According to (*iii*), $\left(\frac{-1}{p}\right) \equiv (-1)^{(p-1)/2} \pmod p$. Again, because each side is $+1$ or -1, it follows that they are equal. ∎

It is customary to rephrase part (*v*) in view of the fact that

$$(-1)^{(p-1)/2} = 1 \Leftrightarrow \frac{p-1}{2} \text{ is even} \Leftrightarrow \frac{p-1}{2} \text{ is of form } 2n$$

$$\Leftrightarrow p-1 \text{ is of form } 4n \Leftrightarrow p \text{ is of form } 4n+1$$

$$\Leftrightarrow p \equiv 1 \pmod 4$$

and

$$(-1)^{(p-1)/2} = -1 \Leftrightarrow \frac{p-1}{2} \text{ is odd} \Leftrightarrow \frac{p-1}{2} \text{ is of form } 2n+1$$

$$\Leftrightarrow p-1 \text{ is of form } 4n+2 \Leftrightarrow p \text{ is of form } 4n+3$$

$$\Leftrightarrow p \equiv 3 \pmod 4.$$

We have proved:

3-7-5. Corollary. Suppose p is an odd prime; then

$$\left(\frac{-1}{p}\right) = (-1)^{(p-1)/2} = \begin{cases} 1 & \text{if } p \equiv 1 \pmod 4, \\ -1 & \text{if } p \equiv 3 \pmod 4. \end{cases}$$

We may illustrate the utility of this corollary by computing $\left(\frac{-1}{p}\right)$ for each of the primes $p = 41, 47, 67, 179, 197, 601$ and interpreting the result in various ways: $\left(\frac{-1}{41}\right) = 1$ since $41 \equiv 1 \pmod 4$, so the congruence $x^2 \equiv -1 \pmod{41}$ has a solution; $\left(\frac{-1}{47}\right) = -1$ since $47 \equiv 3 \pmod 4$, so the congruence $x^2 \equiv -1 \pmod{47}$ has no solution; $\left(\frac{-1}{67}\right) = -1$ because $67 \equiv 3 \pmod 4$, so the polynomial $x^2 + 1$ is irreducible over $\mathbf{Z}_{67}$; $\left(\frac{-1}{179}\right) = -1$ because $179 \equiv 3 \pmod 4$, so the element -1 is not a square in $\mathbf{Z}_{179}$; $\left(\frac{-1}{197}\right) = 1$ because $197 \equiv 1 \pmod 4$, so the element -1 is a square in $\mathbf{Z}_{197}$; $\left(\frac{-1}{601}\right) = 1$ because $601 \equiv 1 \pmod 4$, so the polynomial $x^2 + 1 \in \mathbf{Z}_{601}[x]$ has two roots in $\mathbf{Z}_{601}$—that is, $x^2 + 1$ is reducible over $\mathbf{Z}_{601}$.

Our next result provides a way, somewhat better than trial and error, to compute the Legendre symbol $\left(\frac{a}{p}\right)$. However, its importance lies not in itself, but rather in the crucial role it plays in our derivation of the fundamental theorems about the Legendre symbol.

3-7-6. Gauss' Lemma. Suppose p is an odd prime and $p \nmid a$. Consider the set

$$S = \left\{1a, 2a, 3a, \ldots, \left(\frac{p-1}{2}\right)a\right\}.$$

Let n denote the number of least positive residues modulo p of the elements of S which are greater than $p/2$. Then

$$\left(\frac{a}{p}\right) = (-1)^n.$$

Proof: First, let us clarify the definition of n and show, via some concrete examples, how Gauss' lemma enables us to compute $\left(\frac{a}{p}\right)$. For each element of S take the smallest positive integer in its residue class (mod p); in other words, for each $i = 1, 2, \ldots, (p-1)/2$ apply the division algorithm for p and ia—the remainder is the smallest positive integer in $\lfloor ia \rfloor_p$, and is referred to as the "least positive residue" (mod p) of ia. [Clearly, $p \nmid ia$ for $i = 1, \ldots, (p-1)/2$, so we never have residue 0.] Thus, for each $i = 1, \ldots, (p-1)/2$ we obtain an integer from $1, 2, \ldots, p-1$. Counting how many of these are greater than $p/2$ determines n.

For example: For $p = 13$ and $a = 3$ we have $(p-1)/2 = 6$, so here

$$S = \{3, 6, 9, 12, 15, 18\}$$

and the least positive residues (mod 13) of these integers are

$$\{3, 6, 9, 12, 2, 5\}.$$

Two of these are greater than $13/2 = p/2$; so $n = 2$, and Gauss' lemma asserts that $\left(\frac{3}{13}\right) = (-1)^2 = 1$.

Similarly, when $p = 13$, $a = 5$ we have $S = \{5, 10, 15, 20, 25, 30\}$; so the least positive residues (mod 13) are $\{5, 10, 2, 7, 12, 6\}$. Three of these are greater than $\frac{13}{2}$, so $n = 3$ and $\left(\frac{5}{13}\right) = (-1)^3 = -1$.

We now turn to the proof of Gauss' lemma. Consider the least positive residues (mod p) of the elements of S. These $(p-1)/2$ elements are distinct, because no two of the $(p-1)/2$ elements of S are congruent (mod p). Let $r_1, r_2, \ldots, r_n$ be the least positive residues which are greater than $p/2$, and let $s_1, s_2, \ldots, s_m$ be the least positive residues which are less than $p/2$. Of course, no least positive residue is equal to $p/2$, because $p/2$ is not an integer. In particular, $n + m = (p-1)/2$.

The elements $r_1, r_2, \ldots, r_n, s_1, s_2, \ldots, s_m$ are distinct. Thus, $p - r_1, p - r_2, \ldots, p - r_n$ are surely distinct, and $0 < p - r_i < p/2$ for each $i = 1, \ldots, n$ because $p/2 < r_i < p$. Consequently, all the elements

$$p - r_1, p - r_2, \ldots, p - r_n, s_1, s_2, \ldots, s_m \tag{$*$}$$

are greater than 0 and less than $p/2$ [that is, they are greater than or equal to 1 and less than or equal to $(p-1)/2$]. We assert that they are distinct. For this, it suffices to show that no $p - r_i$ can equal any s_j. In fact, for any r_i and s_j there exist integers x_0, y_0 with $1 \le x_0, y_0 \le (p-1)/2$ for which

$$r_i \equiv x_0 a \pmod{p} \qquad \text{and} \qquad s_j \equiv y_0 a \pmod{p}$$

—so if $p - r_i = s_j$, then

$$p - x_0 a \equiv y_0 a \pmod{p}.$$

And then $(x_0 + y_0)a \equiv 0 \pmod p$ which implies $p \mid (x_0 + y_0)$. But this is impossible because $1 < x_0 + y_0 \le p - 1$. The conclusion is: The set $\{p - r_1, \ldots, p - r_n, s_1, \ldots, s_m\}$ with $m + n = (p-1)/2$ elements is precisely the set of integers $\{1, 2, \ldots, (p-1)/2\}$. Therefore, [with congruences (mod p)]

$$\begin{aligned} \left(\frac{p-1}{2}\right)! &= \prod_{i=1}^{n}(p - r_i)\prod_{j=1}^{m} s_j \\ &\equiv (-1)^n \prod_{i=1}^{n} r_i \prod_{j=1}^{m} s_j \\ &\equiv (-1)^n a^{(p-1)/2}\left(\frac{p-1}{2}\right)! \end{aligned}$$

Since $((p-1)/2)!$ is prime to p, we may cancel and obtain

$$1 \equiv (-1)^n a^{(p-1)/2} \pmod p.$$

From this we conclude

$$\left(\frac{a}{p}\right) = a^{(p-1)/2} \equiv (-1)^n \pmod p$$

and because $\left(\frac{a}{p}\right)$ and $(-1)^n$ are $+1$ or -1, equality holds. This completes the proof. ▮

In preparation for our next result, which is rather technical in nature, we have need for a simple definition. For any real number α, let us denote the largest integer less than or equal to α by $[\alpha]$. For example, $[\ \ ]$, which is known as the "greatest integer" function, takes the values: $[\frac{7}{2}] = 3$, $[5] = 5$, $[12.7] = 12$, $[\pi] = 3$, $[-3] = -3$, $[-\pi] = -4$, $[-12.7] = -13$—and, in general, $[\alpha] \le \alpha < [\alpha] + 1$.

We shall require only one fact about $[\ \ ]$. Namely—suppose

$$b = qa + r, \qquad 0 \le r < a$$

is the division algorithm for a and b. Then $b/a = q + (r/a)$, and since $0 \le r/a < 1$, we have $[b/a] = q$. Thus, the division algorithm always takes the form

$$b = \left[\frac{b}{a}\right]a + r, \qquad 0 \le r < a.$$

3-7-7. Lemma. Suppose p is an odd prime and $p \nmid a$. As in Gauss' lemma, let n denote the number of least positive residues (mod p) of the elements of $S = \{1a, 2a, \ldots, ((p-1)/2)a\}$ which are greater than $p/2$. If we put

$$t = \sum_{i=1}^{(p-1)/2}\left[\frac{ia}{p}\right],$$

then

$$(a-1)\frac{p^2-1}{8} \equiv t - n \pmod 2.$$

Proof: As in the proof of 3-7-6, let $r_1, r_2, \ldots, r_n$ be the least positive residues (mod p) of S which are greater than $p/2$, and $s_1, s_2, \ldots, s_m$ be the least positive residues (mod p) of S which are less than $p/2$. We have seen that

$$\{p - r_1, \ldots, p - r_n, s_1, \ldots, s_m\} = \left\{1, 2, \ldots, \frac{p-1}{2}\right\}$$

and consequently adding all the terms in each set yields

$$\sum_{i=1}^{(p-1)/2} i = \sum_{i=1}^{n} (p - r_i) + \sum_{i=1}^{m} s_i.$$

The left-hand side is

$$1 + 2 + \cdots + \frac{p-1}{2} = \frac{\left(\frac{p-1}{2}\right)\left(\frac{p-1}{2}+1\right)}{2} = \frac{p^2-1}{8},$$

while the right-hand side is $np - \Sigma_{i=1}^{n} r_i + \Sigma_{i=1}^{m} s_i$—so

$$\frac{p^2-1}{8} = np - \sum_{i=1}^{n} r_i + \sum_{i=1}^{m} s_i. \qquad (*)$$

Next, we observe that the division algorithm for p and ia takes the form

$$ia = \left[\frac{ia}{p}\right] p + \text{rem}_i$$

where the remainder terms rem_i run over the set $\{r_1, \ldots, r_n, s_1, \ldots, s_m\}$ as i runs from 1 to $(p-1)/2$. Therefore,

$$\begin{aligned} a\left(\frac{p^2-1}{8}\right) = a\left(\sum_{i=1}^{(p-1)/2} i\right) &= \sum_{i=1}^{(p-1)/2} ia \\ &= \sum_{i=1}^{(p-1)/2} \left(\left[\frac{ia}{p}\right] p + \text{rem}_i\right) \\ &= p \sum_{i=1}^{(p-1)/2} \left[\frac{ia}{p}\right] + \sum_{i=1}^{(p-1)/2} \text{rem}_i \\ &= pt + \sum_{i=1}^{n} r_i + \sum_{i=1}^{m} s_i \qquad (**) \end{aligned}$$

Subtracting (*) from (**) gives

$$(a-1)\left(\frac{p^2-1}{8}\right)=p(t-n)+2\sum_{i=1}^{n} r_i.$$

Viewing this equation modulo 2 we obtain [since p is odd, and hence congruent to 1 (mod 2)],

$$(a-1)\left(\frac{p^2-1}{8}\right)\equiv t-n \pmod 2.$$

This completes the proof. ▮

3-7-8. Theorem. If p is an odd prime, then

$$\left(\frac{2}{p}\right)=(-1)^{(p^2-1)/8}=\begin{cases}1, & \text{if } p\equiv \pm 1 \pmod 8,\\ -1, & \text{if } p\equiv \pm 3 \pmod 8.\end{cases}$$

Proof: Putting $a=2$ in 3-7-7, we have

$$\frac{p^2-1}{8}\equiv t-n \pmod 2.$$

Furthermore, when $a=2$,

$$t=\sum_{i=1}^{(p-1)/2}\left[\frac{2i}{p}\right]=\left[\frac{2}{p}\right]+\left[\frac{4}{p}\right]+\cdots+\left[\frac{p-1}{p}\right].$$

Each of these terms is 0 because each fraction is less than 1. Hence, $t=0$. Of course, $-n\equiv n \pmod 2$—so making use of Gauss' lemma, we have

$$\left(\frac{2}{p}\right)=(-1)^n=(-1)^{(p^2-1)/8}.$$

The value of the Legendre symbol $\left(\frac{2}{p}\right)$ depends, therefore, on whether $(p^2-1)/8$ is odd or even. Now, any odd prime p is one of the forms $8n+1$, $8n+3$, $8n+5$, $8n+7$—or to put it another way, p is of form $8n+1$, $8n+3$, $8n-3$, or $8n-1$. If p is of form $8n\pm 1$ [that is, $p\equiv \pm 1 \pmod 8$], then

$$\frac{p^2-1}{8}=\frac{(8n\pm 1)^2-1}{8}=\frac{64n^2\pm 16n}{8}=8n^2\pm 2n$$

which is even. On the other hand, if p is of form $8n\pm 3$ [that is, $p\equiv \pm 3 \pmod 8$],

then

$$\frac{p^2-1}{8}=\frac{(8n\pm 3)^2-1}{8}=\frac{64n^2\pm 48n+8}{8}=8n^2\pm 6n+1$$

which is odd. This completes the proof. ▌

As illustrations of this result, we compute $\left(\frac{2}{p}\right)$ for $p=41, 47, 227, 229$; namely, $\left(\frac{2}{41}\right)=1$ because $41\equiv 1 \pmod 8$, $\left(\frac{2}{47}\right)=1$ because $47\equiv -1 \pmod 8$ $\left(\frac{2}{227}\right)=-1$ because $227\equiv 3 \pmod 8$, $\left(\frac{2}{229}\right)=-1$ because $229\equiv 5\equiv -3 \pmod 8$.

3-7-9. Remark. Thus far in our investigation of the value of the Legendre symbol $\left(\frac{a}{p}\right)$ our attention has focused on $\left(\frac{-1}{p}\right)$ and $\left(\frac{2}{p}\right)$. Why this emphasis? The explanation is fairly obvious. If a is any nonzero integer, then its prime factorization is of form

$$a=(\pm 1)2^{r_0}\prod_{i=1}^{s} q_i^{r_i}, \qquad r_0\geq 0, \quad r_i>0, \quad i=1,\ldots,s.$$

The sign of a determines the choice of $+1$ or -1. The exponent r_0 is greater than or equal to 0, rather than just greater than 0, because $r_0=0$ occurs when $2\nmid a$. The exponents r_i are all greater than 0 because we include only the odd primes q_i which divide a; in particular, because $p\nmid a$, the q_i are odd primes different from p. The prime 2 has been distinguished from the other primes; this is due to a fact of life—the prime 2 behaves differently from the others.

According to 3-7-4, the Legendre symbol is multiplicative—so

$$\left(\frac{a}{p}\right)=\left(\frac{\pm 1}{p}\right)\left(\frac{2}{p}\right)^{r_0}\left(\frac{q_1}{p}\right)^{r_1}\cdots\left(\frac{q_s}{p}\right)^{r_s}=\left(\frac{\pm 1}{p}\right)\left(\frac{2}{p}\right)^{r_0}\prod_{i=1}^{s}\left(\frac{q_i}{p}\right)^{r_i}.$$

Thus, to compute $\left(\frac{a}{p}\right)$ in general, it suffices to know $\left(\frac{-1}{p}\right)$ [there is no need to bother with $\left(\frac{1}{p}\right)$—it is always equal to 1; and anyway when a is positive, the factor $+1$ should be ignored], $\left(\frac{2}{p}\right)$, and $\left(\frac{q}{p}\right)$, where q is an odd prime different from p. Having already determined $\left(\frac{-1}{p}\right)$ and $\left(\frac{2}{p}\right)$, it remains to consider $\left(\frac{q}{p}\right)$. For this we need one more result.

3-7-10. Lemma. If p and q are distinct odd primes, then

$$\left(\frac{p-1}{2}\right)\left(\frac{q-1}{2}\right)=\sum_{i=1}^{(p-1)/2}\left[\frac{qi}{p}\right]+\sum_{i=1}^{(q-1)/2}\left[\frac{pi}{q}\right].$$

Proof: We give an elegant geometric proof, due to Eisenstein, for this apparently formidable formula.

In the Euclidean plane, **R × R**, consider the rectangle with vertices at (0, 0), $(p/2, 0)$, $(0, q/2)$, $(p/2, q/2)$, shown in the accompanying figure.

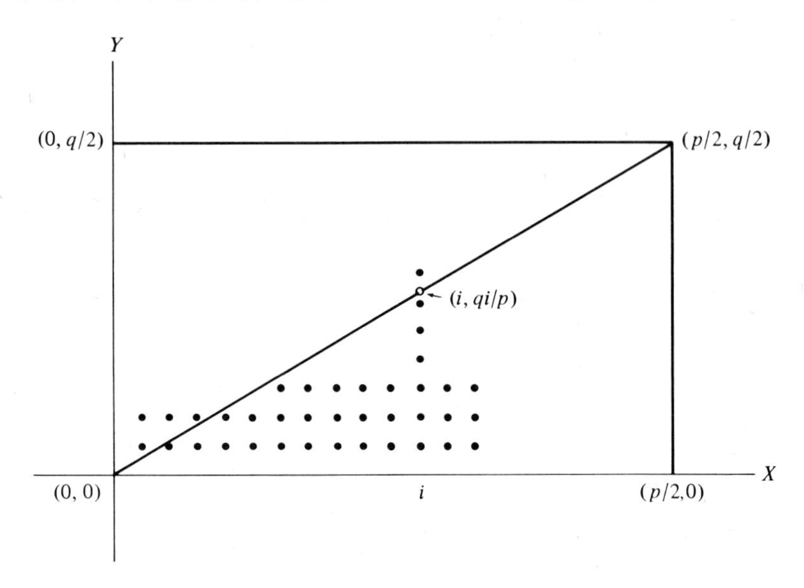

Let us count the number of **lattice points** (meaning points both of whose coordinates are integers) inside this rectangle (but not on the edges). Since both p and q are odd, the set of lattice points clearly consists of all points (m, n) where m and n are integers satisfying $1 \leq m \leq (p-1)/2$, $1 \leq n \leq (q-1)/2$. The number of such points is obviously $((p-1)/2)((q-1)/2)$.

We can also count these lattice points in another way. Consider the diagonal from (0, 0) to $(p/2, q/2)$. The points on it (and inside the rectangle) are those points on the straight line $y = qx/p$ whose coordinates satisfy $0 < x < p/2$, $0 < y < q/2$. None of our lattice points is on this diagonal—in fact, if x is any integer i with $0 < i < p/2$, the corresponding point on the diagonal is $(i, qi/p)$, and qi/p cannot be an integer because p and q are distinct primes and $i < p$. It suffices, therefore, to count the lattice points inside each of the triangles.

Consider the "lower triangle" with vertices (0, 0), $(p/2, 0)$, $(p/2, q/2)$. The way to count its lattice points is as follows: For each $i = 1, 2, \ldots, (p-1)/2$ consider the vertical line segment, inside the triangle, at $x = i$—this segment runs from $(i, 0)$ to $(i, qi/p)$, with endpoints excluded—and count the lattice points on it. This amounts to counting the number of integers greater than 0 and less than qi/p. Since qi/p is not an integer, this number is precisely $[qi/p]$. Hence, the number of lattice points inside the lower triangle is

$$\sum_{i=1}^{(p-1)/2} \left[\frac{qi}{p}\right].$$

In similar fashion, to count the lattice points inside the "upper triangle" with vertices (0, 0), $(0, q/2)$, $(p/2, q/2)$, one uses, for each $i = 1, \ldots, (q-1)/2$, the horizontal segment at $y = i$—which runs from $(0, i)$ to $(pi/q, i)$—and ends up with

$$\sum_{i=1}^{(q-1)/2} \left[\frac{pi}{q}\right]$$

lattice points inside the triangle. The proof is now complete. ▮

3-7-11. Quadratic Reciprocity Law. If p and q are distinct odd primes, then

$$\left(\frac{p}{q}\right)\left(\frac{q}{p}\right) = (-1)^{((p-1)/2)((q-1)/2)}$$

$$= \begin{cases} +1, & \text{if } p \text{ or } q \text{ is} \equiv 1 \pmod 4 \\ -1, & \text{if both } p \text{ and } q \text{ are} \equiv 3 \pmod 4 \end{cases}$$

Proof: Apply 3-7-7 for p and $a = q$. Since q is odd, $2 \mid (q-1)$; so the result is $0 \equiv t - n \pmod 2$, or $t \equiv n \pmod 2$. In virtue of Gauss' lemma,

$$\left(\frac{q}{p}\right) = (-1)^n = (-1)^t, \qquad \text{where } t = \sum_{i=1}^{(p-1)/2} \left[\frac{qi}{p}\right].$$

Now, by interchanging the roles of p and q, we obtain

$$\left(\frac{p}{q}\right) = (-1)^{t'}, \qquad \text{where } t' = \sum_{i=1}^{(q-1)/2} \left[\frac{pi}{q}\right].$$

Therefore, in virtue of 3-7-10,

$$\left(\frac{p}{q}\right)\left(\frac{q}{p}\right) = (-1)^{\sum_{i=1}^{(p-1)/2}[pi/q] + \sum_{i=1}^{(q-1)/2}[qi/p]} = (-1)^{((p-1)/2)((q-1)/2)}.$$

It remains to settle when $(\frac{p}{q})(\frac{q}{p})$ is $+1$ or -1, and this obviously depends on whether $((p-1)/2)((q-1)/2)$ is even or odd. Since $((p-1)/2)((q-1)/2)$ is odd $\Leftrightarrow$ both $(p-1)/2$ and $(q-1)/2$ are odd $\Leftrightarrow$ both $p \equiv 3 \pmod 4$ and $q \equiv 3 \pmod 4$, we see that $(\frac{p}{q})(\frac{q}{p}) = -1$ if both p and q are $\equiv 3 \pmod 4$, and $(\frac{p}{q})(\frac{q}{p}) = 1$ in all other cases (that is, if $p \equiv 1 \pmod 4$ or $q \equiv 1 \pmod 4$, or both). This completes the proof. ▮

The first complete proof of the reciprocity law was given by Gauss, who considered the theorem so important that he gave eight proofs altogether. So many proofs have been given over the years that their number is in doubt;

it has been asserted that there are more than 150 of them. For this reason, most books on elementary number theory give substantially the same proof (due primarily to Eisenstein), and so have we.

Because the square of any Legendre symbol is 1, we may multiply through by $\left(\frac{p}{q}\right)$ in the quadratic reciprocity law and arrive at the following formulation. If p and q are distinct odd primes, then

$$\left(\frac{q}{p}\right) = \begin{cases} -\left(\frac{p}{q}\right), & \text{if } p \equiv q \equiv 3 \pmod 4, \\ +\left(\frac{p}{q}\right), & \text{otherwise.} \end{cases}$$

For example, $\left(\frac{97}{79}\right) = \left(\frac{79}{97}\right)$, because $97 \equiv 1 \bmod 4)$; $\left(\frac{59}{23}\right) = -\left(\frac{23}{59}\right)$ because $23 \equiv 3 \pmod 4$ and $58 \equiv 3 \pmod 4$.

In virtue of the reciprocity law, coupled with 3-7-5 and 3-7-8 [which enable us to evaluate $\left(\frac{-1}{p}\right)$ and $\left(\frac{2}{p}\right)$—and which are sometimes called "complements" or "supplements" of the reciprocity law] we can compute $\left(\frac{a}{p}\right)$ in any concrete case, as will be seen in the examples which follow.

The reason for the name "quadratic reciprocity law" is that it displays a reciprocity between the problems of existence of solutions for

$$x^2 \equiv q \pmod p \qquad \text{and} \qquad x^2 \equiv p \pmod q.$$

More precisely, if $p \equiv q \equiv 3 \pmod 4$, then $\left(\frac{q}{p}\right) = -\left(\frac{p}{q}\right)$, so $\left(\frac{q}{p}\right) = 1 \Leftrightarrow \left(\frac{p}{q}\right) = -1$, and $\left(\frac{q}{p}\right) = -1 \Leftrightarrow \left(\frac{p}{q}\right) = 1$; hence, in this situation

$$x^2 \equiv q \pmod p \text{ has a solution} \Leftrightarrow x^2 \equiv q \pmod p \text{ has no solution}$$

and

$$x^2 \equiv q \pmod p \text{ has no solution} \Leftrightarrow x^2 \equiv p \pmod q \text{ has a solution.}$$

On the other hand, if p or q is congruent to 1 (mod 4), then $\left(\frac{q}{p}\right) = \left(\frac{p}{q}\right)$, so that $\left(\frac{q}{p}\right) = 1 \Leftrightarrow \left(\frac{p}{q}\right) = 1$ and $\left(\frac{q}{p}\right) = -1 \Leftrightarrow \left(\frac{p}{q}\right) = -1$; hence, in this situation

$$x^2 \equiv q \pmod p \text{ has a solution} \Leftrightarrow x^2 \equiv p \pmod q \text{ has a solution}$$

and

$$x^2 \equiv q \pmod p \text{ has no solution} \Leftrightarrow x^2 \equiv p \pmod q \text{ has no solution.}$$

We are in no position to discuss the immense significance of the quadratic reciprocity law in more advanced branches of mathematics. All we can do is give some examples to illustrate the kinds of questions which can now be answered.

3-7-12. Examples. (1) What is the value of $(\frac{-23}{41})$? This involves a straightforward computation

$$\begin{aligned}
\left(\frac{-23}{41}\right) &= \left(\frac{-1}{41}\right)\left(\frac{23}{41}\right) && \text{[3-7-4, part } (iv)], \\
&= \left(\frac{23}{41}\right) && [\text{3-7-5, } 41 \equiv 1 \pmod 4], \\
&= \left(\frac{41}{23}\right) && [\text{3-7-11, } 41 \equiv 1 \pmod 4], \\
&= \left(\frac{18}{23}\right) && \text{[3-7-4, part } (ii)], \\
&= \left(\frac{3^2}{23}\right)\left(\frac{2}{23}\right) && \text{[3-7-4, part } (iv)], \\
&= \left(\frac{2}{23}\right) && \text{[3-7-4, part } (i)], \\
&= 1 && [\text{3-7-8, } 23 \equiv -1 \pmod 8].
\end{aligned}$$

In particular, the congruence $x^2 \equiv -23 \pmod{41}$ has a solution.

(2) Is -59 a square in $\mathbf{Z}_{131}$? This depends on the value of $(\frac{-59}{131})$, so we compute

$$\begin{aligned}
\left(\frac{-59}{131}\right) &= \left(\frac{-1}{131}\right)\left(\frac{59}{131}\right) && \text{[3-7-4]}, \\
&= (-1)\left(\frac{59}{131}\right) && [131 \equiv 3 \pmod 4], \\
&= (-1)(-1)\left(\frac{131}{59}\right) && [59 \equiv 3 \pmod 4,\ 131 \equiv 3 \pmod 4], \\
&= \left(\frac{13}{59}\right) && \text{[3-7-4]}, \\
&= \left(\frac{59}{13}\right) && [13 \equiv 1 \pmod 4], \\
&= \left(\frac{7}{13}\right) && \text{[3-7-4]}, \\
&= \left(\frac{13}{7}\right) && [13 \equiv 1 \pmod 4],
\end{aligned}$$

$$= \left(\frac{6}{7}\right) \qquad [3\text{-}7\text{-}4],$$

$$= \left(\frac{2}{7}\right)\left(\frac{3}{7}\right) \qquad [3\text{-}7\text{-}4],$$

$$= \left(\frac{3}{7}\right) \qquad [7 \equiv -1 \pmod 8],$$

$$= -1 \qquad \text{(trial and error)}.$$

Therefore, -59 is not a square in $\mathbf{Z}_{131}$.

Another way to do this computation is to start from $-59 \equiv 72 \pmod{131}$ —so

$$\left(\frac{-59}{131}\right) = \left(\frac{72}{131}\right) = \left(\frac{2^3 \cdot 3^2}{131}\right)$$

$$= \left(\frac{2^2}{131}\right)\left(\frac{2}{131}\right)\left(\frac{3^2}{131}\right)$$

$$= \left(\frac{2}{131}\right) = -1 \qquad [\text{as } 131 \equiv 3 \pmod 8].$$

(3) Is the polynomial $x^2 + 189$ irreducible over $\mathbf{Z}_{491}$? We compute—starting with the observation that 491 is prime, and $189 = 3^3 \cdot 7$—

$$\left(\frac{-189}{491}\right) = \left(\frac{-1}{491}\right)\left(\frac{3^2}{491}\right)\left(\frac{3}{491}\right)\left(\frac{7}{491}\right)$$

$$= (-1)\left(\frac{3}{491}\right)\left(\frac{7}{491}\right)$$

$$= \left(\frac{491}{3}\right)\left(\frac{7}{491}\right) = \left(\frac{2}{3}\right)\left(\frac{7}{491}\right)$$

$$= (-1)\left(\frac{7}{491}\right) = \left(\frac{491}{7}\right) = \left(\frac{1}{7}\right)$$

$$= 1.$$

[Among the facts used in this computation are: $491 \equiv 3 \pmod 4$, $(\frac{2}{3}) = -1$ because $x^2 \equiv 2 \pmod 3$ has no solution, $7 \equiv 3 \pmod 4$.] This shows that $x^2 + 189 \in \mathbf{Z}_{491}[x]$ has a root in $\mathbf{Z}_{491}$, so it is not irreducible. Of course, our computation here (as is true, in general, when the Legendre symbol is computed) only settles the existence question for a root; it provides no information about locating a root.

(4) For which odd prime p is 3 a quadratic residue? We must determine all p for which $\left(\frac{3}{p}\right) = 1$. [Of course, the case $p = 3$ is excluded since quadratic residue and Legendre symbol $\left(\frac{a}{p}\right)$ are defined only when $p \nmid a$—see 3-7-2.] By quadratic reciprocity,

$$\left(\frac{3}{p}\right)\left(\frac{p}{3}\right) = (-1)^{((3-1)/2)((p-1)/2)} = (-1)^{(p-1)/2}$$

or, upon multiplying both sides by $\left(\frac{p}{3}\right)$

$$\left(\frac{3}{p}\right) = (-1)^{(p-1)/2}\left(\frac{p}{3}\right).$$

Now, $\left(\frac{3}{p}\right) = 1$ in exactly two situations:

$$(i)\ \ (-1)^{(p-1)/2} = \left(\frac{p}{3}\right) = 1, \qquad (ii)\ \ (-1)^{(p-1)/2} = \left(\frac{p}{3}\right) = -1.$$

Recalling that

$$(-1)^{(p-1)/2} = \begin{cases} 1, & \text{if } p \equiv 1 \pmod 4, \\ -1, & \text{if } p \equiv 3 \pmod 4, \end{cases}$$

and noting that for an odd prime $p \neq 3$ we have clearly

$$\left(\frac{p}{3}\right) = \begin{cases} 1, & \text{if } p \equiv 1 \pmod 3, \\ -1, & \text{if } p \equiv 2 \pmod 3, \end{cases}$$

so that the two situations in question are:

(*i*) $p \equiv 1 \pmod 4$ and $p \equiv 1 \pmod 3$,
(*ii*) $p \equiv 3 \pmod 4$ and $p \equiv 2 \pmod 3$.

Solving the simultaneous congruences by inspection (it is not worth the effort to use the Chinese remainder theorem) we have

(*i*) $p \equiv 1 \pmod{12}$, (*ii*) $p \equiv 11 \equiv -1 \pmod{12}$.

This shows

$$\left(\frac{3}{p}\right) = 1 \Leftrightarrow p \equiv \pm 1 \pmod{12}.$$

(5) For which primes p is the polynomial $x^2 + 3$ reducible over $\mathbf{Z}_p$?

This is another way of asking when $x^2 + 3$ has a root in $\mathbf{Z}_p$, or when does the congruence $x^2 \equiv -3 \pmod p$ have a solution. For $p = 2$ and $p = 3$ the congruence surely has a solution—so suppose $p > 3$. We must determine the primes p for which $\left(\frac{-3}{p}\right) = 1$. Now,

$$\begin{aligned}\left(\frac{-3}{p}\right) &= \left(\frac{-1}{p}\right)\left(\frac{3}{p}\right)\\ &= (-1)^{(p-1)/2}\left(\frac{3}{p}\right)\\ &= (-1)^{(p-1)/2}(-1)^{((p-1)/2)((3-1)/2)}\left(\frac{p}{3}\right)\\ &= (-1)^{(p-1)/2}(-1)^{(p-1)/2}\left(\frac{p}{3}\right)\\ &= \left(\frac{p}{3}\right)\end{aligned}$$

and: $\left(\frac{p}{3}\right) = 1$ for $p \equiv 1 \pmod 3$, $\left(\frac{p}{3}\right) = -1$ for $p \equiv 2 \pmod 3$. Thus, except for the primes 2 and 3, $\left(\frac{-3}{p}\right) = 1 \Leftrightarrow p \equiv 1 \pmod 3$ (that is: $\left(\frac{-3}{p}\right) = 1 \Leftrightarrow p$ is of form $3n + 1$).

(6) When is -2 a quadratic residue $(\text{mod } p)$? We note first that the Legendre symbol $\left(\frac{-2}{p}\right) = \left(\frac{-1}{p}\right)\left(\frac{2}{p}\right)$ equals 1 under two sets of circumstances

$$(i)\quad \left(\frac{-1}{p}\right) = \left(\frac{2}{p}\right) = 1, \qquad (ii)\quad \left(\frac{-1}{p}\right) = \left(\frac{2}{p}\right) = -1.$$

In virtue of 3-7-5 and 3-7-8 these become:

(*i*) $p \equiv 1 \pmod 4$ and $p \equiv \pm 1 \pmod 8$,
(*ii*) $p \equiv 3 \pmod 4$ and $p \equiv \pm 3 \pmod 8$.

Since $p \equiv 1 \pmod 4$ and $p \equiv -1 \pmod 8$ are impossible simultaneously, (*i*) becomes $p \equiv 1 \pmod 8$. In similar fashion, (*ii*) becomes $p \equiv 3 \pmod 8$. Thus, -2 is a quadratic residue $(\text{mod } p) \Leftrightarrow p \equiv 1$ or $3 \pmod 8$.

(7) What is the value of $\left(\frac{5}{p}\right)$ (p an odd prime, not equal to 5)? By quadratic reciprocity,

$$\left(\frac{5}{p}\right)\left(\frac{p}{5}\right) = (-1)^{((5-1)/2)((p-1)/2)} = 1.$$

Therefore,

$$\left(\frac{5}{p}\right) = \left(\frac{p}{5}\right) = \begin{cases} 1 & \text{if } p \equiv \pm 1 \pmod 5,\\ -1 & \text{if } p \equiv \pm 2 \pmod 5.\end{cases}$$

There is another kind of question for which the techniques of this section may be applied. Long ago, in 1-4-5, we proved that the number of primes of form $4n + 3$ is infinite—but at that time we were unable to prove the infinitude of primes of form $4n + 1$. With the tools now at our disposal, this difficulty is easy to overcome.

3-7-13. Proposition. The number of primes of form $4n + 1$ is infinite.

Proof: Suppose there are only a finite number of such primes; denote them by

$$p_1 < p_2 < \cdots < p_m.$$

Consider the integer

$$N = (2p_1 p_2 \cdots p_m)^2 + 1.$$

It is of form $4n + 1$, so there surely exists an odd prime p which divides N. The fact that p divides $(2p_1 p_2 \cdots p_m)^2 + 1$ may be restated as

$$(2p_1 p_2 \cdots p_m)^2 \equiv -1 \pmod p.$$

In other words, the congruence $x^2 \equiv -1 \pmod p$ has a solution, and therefore $\left(\frac{-1}{p}\right) = 1$. But $\left(\frac{-1}{p}\right) = 1 \Rightarrow p \equiv 1 \pmod 4 \Rightarrow p$ is of form $4n + 1 \Rightarrow p$ is one of the p_i. This is a contradiction because no p_i divides N. ▮

3-7-14. Exercise. The Legendre symbol $\left(\frac{a}{p}\right)$ was defined only when p is an odd prime with $p \nmid a$. We extend the Legendre symbol to other "denominators," as was done by Jacobi (1804–1851).

(1) Let a and b be relatively prime integers with b odd and positive. Thus, the prime factorization of b looks like $b = p_1 p_2 \cdots p_r$, where the p_i are odd primes which need not be distinct. Define the **Jacobi symbol** $\left(\frac{a}{b}\right)$ by

$$\left(\frac{a}{b}\right) = \prod_{i=1}^{r} \left(\frac{a}{p_i}\right)$$

where $\left(\frac{a}{p_i}\right)$ is the Legendre symbol. For example, if $a = 13$, $b = 17{,}325 = 3 \cdot 3 \cdot 5 \cdot 5 \cdot 7 \cdot 11$, then $(a, b) = 1$ and

$$\left(\frac{13}{17{,}325}\right) = \left(\frac{13}{3}\right)\left(\frac{13}{3}\right)\left(\frac{13}{5}\right)\left(\frac{13}{5}\right)\left(\frac{13}{7}\right)\left(\frac{13}{11}\right).$$

Of course, the Jacobi symbol $\left(\frac{a}{b}\right)$ is always ± 1.

When b is an odd prime p the Legendre symbol $\left(\frac{a}{p}\right)$ and the Jacobi symbol $\left(\frac{a}{p}\right)$ are indistinguishable; but no harm can come of it since they are equal—that is, they take the same values. The Jacobi symbol is said to be an "extension" of the Legendre symbol, in the sense that it is defined for more cases, and when both are defined they are equal.

(2) With a and $b = p_1 p_2 \cdots p_r$ as above, we say that a is a quadratic residue (mod b) when the congruence $x^2 \equiv a \pmod{b}$ has a solution. [Obviously, this definition can be made even when b is not odd; it is customary to talk about quadratic residues (mod b) so long as b is positive and $(a, b) = 1$.] We have then: a is a quadratic residue (mod b) $\Rightarrow a$ is a quadratic residue modulo each $p_i \Rightarrow$ each $\left(\frac{a}{p_i}\right) = 1 \Rightarrow \left(\frac{a}{b}\right) = 1$. However, the converse is false; for example,

$$\left(\frac{2}{33}\right) = \left(\frac{2}{3}\right)\left(\frac{2}{11}\right) = (-1)(-1) = 1$$

but since $x^2 \equiv 2 \pmod{3}$ has no solutions, it is immediate that 2 is not a quadratic residue (mod 33). Of course, $\left(\frac{a}{b}\right) = -1$ implies a is not a quadratic residue (mod b).

(3) The Jacobi symbol has the following elementary properties. If b and b' are positive odd integers and $(aa', bb') = 1$, then

(i) $a \equiv a' \pmod{b} \Rightarrow \left(\frac{a}{b}\right) = \left(\frac{a'}{b}\right)$,

(ii) $\left(\frac{a}{bb'}\right) = \left(\frac{a}{b}\right)\left(\frac{a}{b'}\right)$,

(iii) $\left(\frac{aa'}{b}\right) = \left(\frac{a}{b}\right)\left(\frac{a'}{b}\right)$,

(iv) $\left(\frac{a^2}{b}\right) = \left(\frac{a}{b^2}\right) = 1$,

(v) $\left(\frac{a^2 a'}{b^2 b'}\right) = \left(\frac{a'}{b'}\right)$.

(4) Furthermore, the quadratic reciprocity law and its complements carry over to the Jacobi symbol—namely, if a and b are positive odd integers with $(a, b) = 1$, then

(i) $\left(\frac{-1}{b}\right) = (-1)^{(b-1)/2}$,

(ii) $\left(\frac{2}{b}\right) = (-1)^{(b^2-1)/8}$,

(iii) $\left(\frac{a}{b}\right)\left(\frac{b}{a}\right) = (-1)^{((a-1)/2)((b-1)/2)}$.

The proofs make use of the elementary facts that for odd integers u and v we have

$$\frac{u-1}{2}+\frac{v-1}{2}\equiv\frac{uv-1}{2}\pmod 2$$

and

$$\frac{u^2-1}{8}+\frac{v^2-1}{8}\equiv\frac{(uv)^2-1}{8}\pmod 2$$

—and then, inductively, these statements extend to n odd integers $u_1, u_2, \ldots u_n$, giving

$$\sum_{i=1}^{n}\frac{u_i-1}{2}\equiv\frac{(\prod u_i)-1}{2}\pmod 2,$$

$$\sum_{i=1}^{n}\frac{u_i^2-1}{8}\equiv\frac{(\prod u_i)^2-1}{8}\pmod 2.$$

(5) Note: If we had defined the Jacobi symbol by

$$\left(\frac{a}{b}\right)=\begin{cases}1, & \text{if } a \text{ is a quadratic residue} \pmod b,\\ -1, & \text{if } a \text{ is not a quadratic residue} \pmod b,\end{cases}$$

then the reciprocity law would not hold. For example, if $a = 5$, $b = 21$, then $(-1)^{((a-1)/2)((b-1)/2)} = 1$, $\left(\frac{b}{a}\right) = \left(\frac{21}{5}\right) = 1$ because $x^2 \equiv 21 \equiv 1 \pmod 5$ has a solution, and $\left(\frac{a}{b}\right) = \left(\frac{5}{21}\right) = -1$ because $x^2 \equiv 5 \pmod{21}$ has no solution (in fact, it suffices to observe that $x^2 \equiv 5 \pmod 3$ has no solution).

(6) The computation of a Jacobi symbol is straightforward; for example, $a = 231$ and $b = 1105$ are relatively prime odd integers, and

$$\left(\frac{231}{1105}\right)=\left(\frac{1105}{231}\right)=\left(\frac{181}{231}\right)=\left(\frac{231}{181}\right)=\left(\frac{50}{181}\right)=\left(\frac{5^2}{181}\right)\left(\frac{2}{181}\right)=\left(\frac{2}{181}\right)=-1.$$

3-7-15 / *PROBLEMS*

Throughout p is understood to be an odd prime with $p \nmid a$.

1. List all the quadratic residues and all the quadratic nonresidues for each of the following primes:
(*i*) 17, (*ii*) 19, (*iii*) 23, (*iv*) 29, (*v*) 31.

2. Find all solutions of

$$x^2 \equiv a \pmod{13}$$

for each of the following values of a:
(*i*) 1, (*ii*) 3, (*iii*) 4, (*iv*) 5,
(*v*) 8, (*vi*) 9, (*vii*) 10, (*viii*) 12.

3. Do Problem 2 for

$$x^2 \equiv a \pmod{13^2}.$$

4. Prove that the congruence $x^2 \equiv a \pmod{p}$ has no roots, or else it has two distinct roots. In the latter case, how are the two roots related?

5. Prove the reduction step of 3-7-1 by working with congruences (rather than over $\mathbf{Z}_p$).

6. We have seen that $a \equiv b \pmod{p} \Rightarrow \left(\frac{a}{p}\right) = \left(\frac{b}{p}\right)$. Prove this by viewing a and b in $\mathbf{Z}_p$.

7. Use Gauss' lemma, 3-7-6 (and nothing else) to compute the Legendre symbol in the 24 cases given by

$$a = \pm 2,\ \pm 3,\ 5,\ 6, \qquad p = 11,\ 17,\ 19,\ 23.$$

8. Use our general results to evaluate $\left(\frac{a}{p}\right)$ for each of the 25 cases:

$$a = -1,\ \pm 2,\ \pm 3, \qquad p = 7,\ 11,\ 13,\ 17,\ 19.$$

9. Show that we always have $a^{(p-1)/2} \equiv \pm 1 \pmod{p}$ (where, as usual, p is an odd prime and $p \nmid a$).

10. Prove that of the nonzero elements of $\mathbf{Z}_p$, half are squares in $\mathbf{Z}_p$ and half are not squares. In other words, half of the integers $1, 2, \ldots, p-1$ are quadratic residues (mod p) and half are quadratic nonresidues.

11. (*i*) Show that: If both a and b are quadratic residues (mod p), then so is ab.
(*ii*) If both a and b are quadratic nonresidues (mod p), then ab is a quadratic residue.
(*iii*) What happens if one is a quadratic residue and the other is a quadratic nonresidue?

12. Decide if the congruence $x^2 \equiv a \pmod{p}$ has a root for each of the 36 cases given by:

$$a = 5,\ -7,\ 11,\ -11,\ 13,\ -13, \qquad p = 97,\ 101,\ 103,\ 617,\ 619,\ 911.$$

1.3 Prove that $\Sigma_{a=1}^{p-1} \left(\frac{a}{p}\right) = 0$.

14. Decide if the following congruences have solutions—and how many?

(i) $x^2 \equiv 2 \pmod{47}$, *(ii)* $x^2 \equiv -2 \pmod{47}$,
(iii) $x^2 \equiv 2 \pmod{67}$, *(iv)* $x^2 \equiv -2 \pmod{67}$,
(v) $x^2 \equiv 2 \pmod{94}$, *(vi)* $x^2 \equiv -2 \pmod{94}$,
(vii) $x^2 \equiv 2 \pmod{134}$, *(viii)* $x^2 \equiv -2 \pmod{134}$,
(ix) $x^2 \equiv 2 \pmod{268}$, *(x)* $x^2 \equiv -2 \pmod{268}$.

15. For any $r > 1$ prove that $x^2 \equiv a \pmod{p^r}$ has a solution $\Leftrightarrow x^2 \equiv a \pmod{p}$ has a solution. In such a situation, how many solutions does $x^2 \equiv a \pmod{p^r}$ have?

16. *(i)* Suppose p and q are distinct odd primes with $\left(\frac{a}{p}\right) = \left(\frac{a}{q}\right) = -1$. Show that $x^2 \equiv a \pmod{pq}$ has no solutions.
(ii) What happens if $\left(\frac{a}{p}\right) = 1$ and $\left(\frac{a}{q}\right) = -1$?

17. Compute the Legendre symbols:

(i) $\left(\frac{2}{97}\right)$, *(ii)* $\left(\frac{19}{23}\right)$, *(iii)* $\left(\frac{-29}{67}\right)$, *(iv)* $\left(\frac{-38}{29}\right)$,

(v) $\left(\frac{-23}{59}\right)$, *(vi)* $\left(\frac{-79}{97}\right)$, *(vii)* $\left(\frac{94}{67}\right)$, *(viii)* $\left(\frac{135}{229}\right)$,

(ix) $\left(\frac{-135}{229}\right)$, *(x)* $\left(\frac{2\cdot 3\cdot 5\cdot 7\cdot 11}{127}\right)$, *(xi)* $\left(\frac{630}{503}\right)$, *(xii)* $\left(\frac{-999}{73}\right)$.

18. *(i)* Suppose p and q are distinct odd primes, both of which are congruent to $3 \pmod{5}$. Show that if $x^2 \equiv q \pmod{p}$ has no solutions, then $x^2 \equiv p \pmod{q}$ has two solutions.
(ii) What if both p and q are congruent to $1 \pmod{4}$?

19. Starting from Wilson's theorem, prove that

$$\left(\left(\frac{p-1}{2}\right)!\right)^2 (-1)^{(p-1)/2} \equiv -1 \pmod{p}.$$

20. In $\mathbf{Z}_p$, the product of all nonzero elements which are squares is

$$(-1)^{(p+1)/2} = \begin{cases} 1, & \text{if } p \equiv 3 \pmod{4}, \\ -1, & \text{if } p \equiv 1 \pmod{4}. \end{cases}$$

Restate this result in terms of quadratic residues and congruences.

21. Find all primes for which

$$x^2 \equiv 11 \pmod{p}$$

has a solution. (Excluding $p = 2$ or 11, there are 10 classes of such primes.)

22. Find all primes p for which

$$(i)\ \left(\frac{6}{p}\right) = 1, \qquad (ii)\ \left(\frac{10}{p}\right) = 1, \qquad (iii)\ \left(\frac{14}{p}\right) = 1.$$

23. Find all odd primes for which both 2 and 3 are quadratic residues.

24. Find all primes p for which both

$$\left(\frac{5}{p}\right) = -1 \quad \text{and} \quad \left(\frac{11}{p}\right) = +1.$$

25. (*i*) Prove that the number of primes of form $3n - 1$ is infinite.
(*ii*) Prove that the number of primes of form $3n + 1$ is infinite.

26. Prove that the congruence

$$x^2 \equiv -3 \pmod{7^2 \cdot 19^2 \cdot 23}$$

has no solutions; but that the congruence

$$x^2 \equiv -3 \pmod{7^2 \cdot 19^2 \cdot 31}$$

has exactly eight solutions (do not try to find them).

27. Suppose $(a, b) = 1$ where b is odd and positive. Show that the congruence $x^2 \equiv a \pmod{b}$ has a solution $\Leftrightarrow$ $\left(\frac{a}{p}\right) = 1$ for every prime p which divides b. In such a situation, how many solutions does the congruence have?

28. For which values of c does the congruence

$$3x^2 - 2x + c \equiv 0 \pmod{7}$$

have a solution? How many solutions does it then have?

29. (*i*) Does $x^2 \equiv 30 \pmod{91}$ have a solution? How about $x^2 \equiv 44 \pmod{91}$?
(*ii*) For how many a's does the congruence

$$x^2 \equiv a \pmod{91}$$

have a solution? Find at least 10 of them.

30. Without trying to solve, decide how many solutions there are to the congruence

$$37x^2 + 348x + 311 \equiv 0 \pmod{617}$$

(617 is prime).

31. If for the congruence $ax^2 + bx + c \equiv 0 \pmod{p}$ we let $\Delta = b^2 - 4ac$ denote the discriminant, then it has

$$\text{no solutions} \Leftrightarrow \left(\frac{\Delta}{p}\right) = -1,$$

$$\text{exactly one solution} \Leftrightarrow p \mid \Delta,$$

$$\text{two distinct solutions} \Leftrightarrow \left(\frac{\Delta}{p}\right) = 1.$$

Give a concrete example of each type.

32. (*i*) If both a and b are quadratic residues (mod p), then $ax^2 \equiv b \pmod{p}$ has a solution.
(*ii*) What if both a and b are nonresidues?
(*iii*) What if one is a residue and the other a nonresidue? Give a concrete example of each type, and exhibit a solution when one exists.

33. Suppose q is the smallest positive integer which is a quadratic nonresidue (mod p); prove that q is prime. Furthermore, $q, 2q, \ldots, (q-1)q$ are all less than p—hence, $q < \sqrt{p+1}$.

Miscellaneous Problems

1. Solve the following pairs of simultaneous congruences or equations:

(*i*) $5x + 12y \equiv 15 \pmod{22}$;
$13x - 7y \equiv 19 \pmod{22}$.

(*ii*) $8x \equiv 4 \pmod{14}$;
$5x \equiv 3 \pmod{11}$.

(*iii*) $\lfloor 4 \rfloor_{17} x - \lfloor 9 \rfloor_{17} y = \lfloor 16 \rfloor_{17}$;
$\lfloor 8 \rfloor_{17} x + \lfloor 5 \rfloor_{17} y = \lfloor 3 \rfloor_{17}$.

(*iv*) $\lfloor 4 \rfloor_{18} x - \lfloor 9 \rfloor_{18} y = \lfloor 16 \rfloor_{18}$;
$\lfloor 8 \rfloor_{18} x + \lfloor 5 \rfloor_{18} y = \lfloor 3 \rfloor_{18}$.

(*v*) $36x \equiv 27 \pmod{45}$;
$35x \equiv 27 \pmod{45}$.

2. By using the method of 3-1-7, prove the following generalization of the Chinese remainder theorem. The system of linear congruences $x \equiv a_i \pmod{m_i}$, $i = 1, \ldots, n$ has a solution $\Leftrightarrow$ $(m_j, m_k) \mid (a_j - a_k)$ for all $j \neq k$, $j, k = 1, \ldots, n$. Furthermore, the solution is then unique mod $[m_1, \ldots, m_n]$—that is, if $x_0 \in \mathbf{Z}$ is a solution then $\lfloor x_0 \rfloor_{[m_1, \ldots, m_n]}$ is the set of all solutions.

3. Suppose $m_1, m_2, \ldots, m_n$ are relatively prime in pairs, and consider the system of linear congruences,

$$a_i x \equiv b_i \pmod{m_i}, \qquad i = 1, \ldots, n.$$

Find a necessary and sufficient condition that this system have a solution. How many solutions are there? What happens if $m_1, m_2, \ldots, m_n$ are not relatively prime in pairs?

4. Prove that the number of fractions a/b (with a, b both positive integers) in lowest terms, of value 1 or less, and with denominator m or less, is $\phi(1) + \phi(2) + \cdots + \phi(m)$.

5. Prove that for $m > 1$, the sum of the integers from the set $\{1, 2, \ldots, m\}$ which are relatively prime to m is $\frac{1}{2}(m\phi(m))$.

6. If the positive integer a is fixed, prove that the equation $\phi(m) = a$ has at most a finite number of solutions for m.

7. Suppose $d \mid m$ and $d \neq m$; show that

$$m - \phi(m) > d - \phi(d).$$

8. Prove that if m and n are positive integers with $(m, n) = d$ then

$$\phi(mn) = \frac{d\phi(m)\phi(n)}{\phi(d)}.$$

9. Show that the Euler ϕ-function satisfies

$$\phi(n) = \sum_{d|n} \mu(d)\left(\frac{n}{d}\right) = \sum_{d|n} \mu\left(\frac{n}{d}\right)d,$$

where μ is the Möbius function (see Chapter II, Miscellaneous Problem 56)

10. Suppose f is a function defined on all positive integers n, with $f(n) > 0$. Prove that

$$g(n) = \prod_{d|n} f(d) \Rightarrow f(n) = \prod_{d|n} g(d)^{\mu(n/d)}.$$

11. Consider polynomials $f(x), g(x) \in R[x]$, where R is a commutative ring with unity, and the composite polynomial $g(f(x)) \in R[x]$. [For example, if $f(x) = 3x^3 - 1$, $g(x) = x^2 + x + 1$ in $\mathbf{Z}[x]$, then $g(f(x)) = (f(x))^2 + f(x) + 1 = (3x^3 - 1)^2 + (3x^3 - 1) + 1 = 9x^6 - 3x^3 + 1$.] Denote the derivative by $'$ (see 3-3-15, Problem 24). Show that

$$[g(f(x))]' = [g'(f(x))] \cdot f'(x).$$

12. The element c of the field F is said to be a root of **multiplicity** $r \geqslant 1$ of the polynomial $f(x) \in F[x]$ when $(x - c)^r$ is the power of $(x - c)$, which appears in the prime factorization of $f(x)$—or, what is the same thing, when

$$(x - c)^r \mid f(x), \qquad \text{but} \qquad (x - c)^{r+1} \nmid f(x).$$

When $r = 1$, c is said to be a **simple root**, and when $r > 1$, c is said to be a **multiple root.**

(*i*) $c \in F$ is a multiple root of $f(x) \Leftrightarrow c$ is a root of $(f(x), f'(x))$, the gcd of $f(x)$ and its derivative $f'(x)$.

(*ii*) Suppose c is a root of $f(x)$, then c is a multiple root of $f(x)$ $\Leftrightarrow$ c is a root of $f'(x)$.

13. For a polynomial $f(x) \in \mathbf{Z}_p[x]$, show that $f'(x) = 0 \Leftrightarrow f(x)$ is a polynomial in x^p. [That is, $f(x)$ is of form $a_0 + a_1 x^p + a_2 x^{2p} + \cdots + a_r x^{rp}$; one also denotes this by $f(x) = g(x^p)$, where $g(x) \in \mathbf{Z}_p[x]$.]

14. Let R be an arbitrary ring; it need not be commutative or have a unity. Show that $R[x]$ (by which is meant the set of all expressions $a_0 + a_1 x + \cdots + a_n x^n$ with coefficients in R) is a ring with respect to the standard operations.
What about $R[[x]]$, the set of all formal power series (see 3-3-15, Problem 25), when the ring R is arbitrary?

15. How many elements are there in the ring $R[x]$? What is the characteristic of this ring? What is its center?

16. Suppose R is a commutative ring with unity, then so is $R[x]$. Taking another indeterminate y, we can form the polynomial ring $(R[x])[y]$ in y over $R[x]$—denote it also by $R[x, y]$. Generalize this to form a commutative ring with unity $R[x_1, x_2, \ldots, x_n]$ for each $n \geq 1$. What are its units? What if R is not commutative?

17. Suppose R is a commutative ring with unity. If a is a unit in R and b is an arbitrary element of R, define the mapping $\phi : R[x] \to R[x]$ by

$$\phi(f(x)) = f(ax + b), \qquad f(x) \in R[x].$$

Show that ϕ is an automorphism of the ring $R[x]$.

18. We describe a somewhat more formal method for defining a polynomial ring than the one given in Section 3-3.

(*i*) Suppose the ring R with unity 1, is given. Let S denote the set of all infinite sequences

$$\alpha = (a_0, a_1, a_2, a_3, \ldots)$$

of elements a_i from R such that only a finite number of the a_i are not equal to 0 (in other words, almost all a_i equal 0). If $\beta = (b_0, b_1, b_2, b_3, \ldots)$ is also an element of S define the sum of α and β by

$$\alpha + \beta = (a_0 + b_0, a_1 + b_1, a_2 + b_2, \ldots)$$

and their product by

$$\alpha\beta = (c_0, c_1, c_2, c_3, \ldots),$$

where, for $i = 0, 1, 2, \ldots,$

$$c_i = a_0 b_i + a_1 b_{i-1} + \cdots + a_i b_0 = \sum_{r=0}^{i} a_r b_{i-r} = \sum_{r+s=i} a_r b_s .$$

Then $\alpha + \beta$ and $\alpha\beta$ are elements of S, and S is a ring with unity $(1, 0, 0, 0, \ldots)$.

(*ii*) Let R_0 denote the subset of S consisting of all elements of form $(a, 0, 0, 0, \ldots, 0, \ldots)$, $a \in R$. Then R_0 is a subring of S which is isomorphic to R under the mapping

$$a \longrightarrow (a, 0, 0, 0, \ldots, 0, \ldots), \qquad a \in R.$$

We identify R and R_0—and denote $(a, 0, 0, \ldots, 0, \ldots)$ also by a. Thus, R is viewed as a subring of S.

(*iii*) Let us denote $(0, 1, 0, 0, \ldots, 0, \ldots)$ by x. So $x \in S$. Then we have $x^2 = (0, 0, 1, 0, 0, \ldots, 0, \ldots)$ and, inductively,

$$x^i = (0, 0, \ldots, 0, 1, 0, \ldots, 0, \ldots),$$

where the 1 is in the $(i + 1)$ place. For any $a_i \in R \subset S$ we have

$$a_i x^i = (0, \ldots, 0, a_i, 0, \ldots),$$

and consequently

$$(a_0, a_1, a_2, a_3, \ldots) = a_0 + a_1 x + a_2 x^2 + \cdots = \sum_i a_i x^i.$$

Now, denote S by $R[x]$; show that $R[x]$ has all the properties we want it to have (that is, as in Section 3-3).

19. Let F be a field. For $a \in F$, $b \in F^*$ let us denote $ab^{-1} = b^{-1}a$ by a/b; in particular, $b^{-1} = 1/b$ and $0b^{-1} = 0/b = 0$. Verify the following for a, $c \in F$, $b, d \in F^*$.

(*i*) $\dfrac{a}{b} = \dfrac{ad}{bd}$. (*ii*) $\dfrac{a}{b} = \dfrac{c}{d} \Leftrightarrow ad = bc$.

(*iii*) $\left(\dfrac{a}{b}\right) \cdot \left(\dfrac{c}{d}\right) = \dfrac{ac}{ad}$. (*iv*) $\dfrac{a}{b} + \dfrac{c}{d} = \dfrac{ad + bc}{bd}$.

(*v*) $\left(\dfrac{a}{b}\right)^n = \dfrac{a^n}{b^n}$, $n > 0$. What happens if n is negative?

20. Suppose an integral domain D is contained in the field F; we say that F is a **quotient field** of D when every element of F can be expressed in the form $a/b = ab^{-1}$ with a, $b \in D$. We show that an arbitrary integral domain D has a quotient field (details are left to the reader).

Let D' denote the set of nonzero elements of D. Consider the product set

$$D \times D' = \{(a, b) \mid a \in D, b \in D'\}.$$

Of course, equality for elements of $D \times D'$ is defined by

$$(a, b) = (c, d) \quad \Leftrightarrow \quad a = c, \quad b = d.$$

Define operations on $D \times D'$ as follows:

$$(a, b) + (c, d) = (ad + bc, bd), \qquad (a, b) \cdot (c, d) = (ac, bd).$$

Clearly, $D \times D'$ is closed under these operations.

We want (a, b) to behave like the "fraction" a/b, but also need to take into account the expectation that a fraction can be written in more than one way. To do this, define a relation $\equiv$ on $D \times D'$ by

$$(a, b) \equiv (c, d) \quad \Leftrightarrow \quad ad = bc.$$

Then $\equiv$ is an equivalence relation; denote the set of equivalence classes by $\overline{D \times D'}$, and the equivalence class to which (a, b) belongs by $\lfloor (a, b) \rfloor$. Note that

$$\lfloor (a, b) \rfloor = \lfloor (c, d) \rfloor \quad \Leftrightarrow \quad ad = bc.$$

If we put

$$\lfloor (a, b) \rfloor + \lfloor (c, d) \rfloor = \lfloor (ad + bc, bd) \rfloor,$$
$$\lfloor (a, b) \rfloor \cdot \lfloor (c, d) \rfloor = \lfloor (ac, bd) \rfloor,$$

then $+$ and $\cdot$ are well-defined operations under which $\overline{D \times D'}$ becomes a field (call it F)—the zero element is $\lfloor (0, 1) \rfloor$; the additive inverse of $\lfloor (a, b) \rfloor$ is $\lfloor (-a, b) \rfloor$; the identity for multiplication is $\lfloor (1, 1) \rfloor = \lfloor (a, a) \rfloor$, $a \neq 0$.

Define a mapping of $D \to \overline{D \times D'} = F$ by

$$a \longrightarrow \lfloor (a, 1) \rfloor.$$

It is an injective homomorphism. Hence, the field F contains an isomorphic copy of D. We identify D with its image, so $D \subset F$, and $\lfloor (a, 1) \rfloor$ is simply a. For $b \in D'$, $\lfloor (1, b) \rfloor$ is then $b^{-1} = 1/b$, so every element $\lfloor (a, b) \rfloor$ of F is of form ab^{-1}, $a \in D$, $b \in D'$. Thus F is indeed a quotient field of D.

In particular, in the case when $D = \mathbf{Z}$ we have constructed the rational field, $\mathbf{Q}$.

21. The quotient field of an integral domain D is unique. More precisely, suppose F and F' are quotient fields of D, then there exists an isomorphism ψ of F onto F', which is the identity on D [that is, $\psi(a) = a$ for all $a \in D$].

More generally, suppose the field F_i is a quotient field of the integral domain D_i, $i = 1, 2$. If ϕ is an isomorphism of $D_1 \to D_2$ then there exists an isomorphism ψ: $F_1 \to F_2$, which is an extension of ϕ [that is, $\psi(a) = \phi(a)$ for all $a \in D_1$]. Prove these statements.

22. If D is an ordered domain then so is its quotient field F (meaning that F satisfies the order requirements of an ordered domain) when the set of positive elements is taken as $\{a/b \mid a,\ b \in D,\ ab > 0\}$. In particular show that $\mathbf{Q}$ is an ordered field.

23. (*i*) Let α and β be distinct elements of the ordered field F. Show that there exists an element of F between them; even more, there are an infinite number of elements between them.

(*ii*) Prove that the positive elements of an ordered field are never well ordered.

24. If the ordered domain D is "archimedean" (see 2-4-10, Problem 17) then so is its quotient field.

25. Fix a polynomial $g(x) \in F[x]$ of degree 1 or more. Show that an arbitrary element $f(x)$ of $F[x]$ can be expressed uniquely in the form

$$f(x) = r_0(x) + r_1(x)g(x) + r_2(x)g(x)^2 + \cdots + r_m(x)g(x)^m,$$

where $r_m(x) \neq 0$ and $\deg(r_i(x)) < \deg(g(x))$ for $i = 0, 1, \ldots, m$. In analogy with the situation in $\mathbf{Z}$, one may refer to this expression as the expansion of $f(x)$ in "base $g(x)$."

26. Consider the polynomial ring $F[x]$ over a field F. It is an integral domain, so it has a quotient field, called the field of **rational forms (or functions)** over F and denoted by $F(x)$. The elements of $F(x)$ are "fractions"—in fact,

$$F(x) = \left\{ \frac{f(x)}{g(x)} \,\middle|\, f(x),\ g(x) \in F[x],\ g(x) \neq 0 \right\}.$$

Given an element

$$\frac{f(x)}{g(x)} \in F(x),$$

let

$$g(x) = p_1(x)^{n_1} p_2(x)^{n_2} \cdots p_r(x)^{n_r} = \prod_{i=1}^{r} p_i(x)^{n_i}$$

be the factorization of $g(x)$ into distinct prime (that is, irreducible) polynomials $p_1(x), \ldots p_r(x)$ in $F[x]$. Then $f(x)/g(x)$ can be written in the form

$$\frac{f(x)}{g(x)} = h(x) + \sum_{j=1}^{n_1} \frac{h_{1,j}(x)}{p_1(x)^j} + \sum_{j=1}^{n_2} \frac{h_{2,j}(x)}{p_2(x)^j} + \cdots + \sum_{j=1}^{n_r} \frac{h_{r,j}(x)}{p_r(x)^j},$$

where $h(x)$ and all $h_{i,j}(x)$ are in $F[x]$ and

$$\deg(h_{i,j}(x)) < \deg(p_i(x)) \begin{cases} i = 1, 2, \ldots, r, \\ j = 1, 2, \ldots, n_i, \end{cases}$$

Moreover, this expression for $f(x)/g(x)$ is unique.

In the case where F is the real field $\mathbf{R}$ we know (see 3-5-25) that an irreducible polynomial must be of degree 1 or 2; so the numerators in the expression for $f(x)/g(x)$ are of degree 0 (that is, constants) or 1. This expression plays a key role in calculus when showing how to integrate an arbitrary rational function.

27. Suppose a and b are elements of the domain D; then for relatively prime integers m and n, we have

$$a^m = b^m, \quad a^n = b^n \quad \Rightarrow \quad a = b.$$

28. Can you prove 3-5-5 by using unique factorization into primes?

29. Give examples of the following:
(*i*) Two units whose sum is a unit.
(*ii*) Two units whose sum is not a unit.
(*iii*) Two zero-divisors whose sum is a zero-divisor.
(*iv*) Two zero-divisors whose sum is not a zero-divisor.

30. Prove the following statement: In $F[x]$, the polynomial $f(x) = a_0 + a_1 x + \cdots + a_n x^n$ of degree n is irreducible $\Leftrightarrow$ the "reverse" polynomial $f^*(x) = a_n + a_{n-1}x + \cdots + a_1 x^{n-1} + a_0 x^n$ is irreducible.

31. For any prime p, consider the **cyclotomic polynomial**

$$f(x) = x^{p-1} + x^{p-2} + \cdots + x + 1 \in \mathbf{Z}[x].$$

It satisfies $(x - 1)f(x) = x^p - 1$. Substitute $y + 1$ for x and use the Eisenstein criterion to show that $f(x)$ is irreducible over $\mathbf{Q}$. Justify all steps carefully.

32. Find the gcd of

$$f(x) = x^5 + x^3 + 3x^2 - 2x + 6, \qquad g(x) = x^6 + 3x^4 + x^3 + 5x^2 + 2x + 6,$$

and express it as a linear combination of $f(x)$ and $g(x)$ when they are viewed over the field

(*i*) $\mathbf{Q}$, (*ii*) $\mathbf{Z}_2$, (*iii*) $\mathbf{Z}_3$, (*iv*) $\mathbf{Z}_5$, (*v*) $\mathbf{Z}_7$, (*vi*) $\mathbf{Z}_{11}$.

Can you find the prime factorizations of these polynomials over the given fields?

33. Over which fields $\mathbf{Z}_p$ does $x^2 + x + 1$ divide $x^5 + x + 1$?

34. (*i*) For each $n > 1$ consider the polynomials

$$60x^n - 143, \quad \text{and} \quad 60x^n + 91x + 143.$$

Show that neither one has a rational root.

(*ii*) Find all $n > 1$ and positive integers $a < 500$ for which the polynomial $60x^n - a$ has a rational root.

35. Determine all rational numbers for which the polynomial $f(x) = 7x^2 - 5x$ takes an integer value.

36. If $g(x) \mid f(x)$ in $\mathbf{Z}[x]$, show that $g(c) \mid f(c)$ for every $c \in \mathbf{Z}$. Given $f(x) \in \mathbf{Z}[x]$ of degree n make use of this fact and Lagrange interpolation to construct an algorithm which, for each $r \leq n$, determines (in a finite number of steps) all polynomials of degree r in $\mathbf{Z}[x]$ which divide $f(x)$.

37. If p is prime, make use of Lagrange interpolation to show that every element of Map($\mathbf{Z}_p$, $\mathbf{Z}_p$) is a polynomial function over $\mathbf{Z}_p$. In other words, the homomorphism of rings

$$\Lambda: \mathbf{Z}_p[x] \longrightarrow \text{Map}(\mathbf{Z}_p, \mathbf{Z}_p)$$

is surjective. What is the kernel of Λ? Show that ker Λ consists of all polynomials in $\mathbf{Z}_p[x]$ which are divisible by $x^p - x$. Furthermore, Λ provides a 1–1 correspondence between the set of all polynomials in $\mathbf{Z}_p[x]$ of degree less than p and Map($\mathbf{Z}_p$, $\mathbf{Z}_p$).

38. Show that the polynomial functions over any infinite field F (or integral domain D) form an integral domain.

39. Let F be a finite field with n elements—say, $F = \{a_1, a_2, \ldots, a_p\}$ and put

$$\phi(x) = \prod_{i=1}^{n} (x - a_i) \in F[x].$$

[For example, if $F = \mathbf{Z}_p$, then $\phi(x) = x^p - x$.] If $f(x), g(x) \in F(x)$ determine the same polynomial function—that is, if $\bar{f} = \bar{g}$—then $\phi(x)$ divides $f(x) - g(x)$. What about the converse? What is the kernel of the homomorphism $\Lambda: F[x] \to \text{Map}(F, F)$?

40. For the field F consider the domains
(*i*) $F[x]$, (*ii*) $F(x)$, (*iii*) $F[[x]]$ (see 3-3-15, Problem 25). In each case, decide if the map described by $f(x) \to f(-x)$ is an automorphism.

41. Define a mapping of $\mathbf{Z}[x] \to \mathbf{Z}_p[x]$ by:

$$f(x) = \sum_{i=0}^{n} a_i x^i \to \bar{f}(x) = \sum_{i=0}^{n} \lfloor a_i \rfloor_p x^i.$$

Show it to be an epimorphism. What is the kernel? Making use of the map $\Lambda: \mathbf{Z}_p[x] \to \text{Map}(\mathbf{Z}_p, \mathbf{Z}_p)$ show that

$$\Lambda\left((\overline{f(x)})^p\right) = \Lambda\left(\overline{(f(x))^p}\right) = \Lambda\left(\overline{f(x)}\right).$$

42. Show that the greatest integer function [ ] satisfies the following properties; for $\alpha, \beta \in \mathbf{R}$ and $n \in \mathbf{Z}$

(*i*) $[\alpha] \leq \alpha \leq [\alpha] + 1$,

(*ii*) $\alpha - 1 \leq [\alpha] \leq \alpha$,

(*iii*) $0 \leq \alpha - [\alpha] < 1$,

(*iv*) $[\alpha + n] = [\alpha] + n$,

(*v*) $[\alpha] + [\beta] \leq [\alpha + \beta] \leq [\alpha] + [\beta] + 1$,

(*vi*) $|\alpha - [\alpha] - \frac{1}{2}| \leq \frac{1}{2}$,

(*vii*) $[\alpha] + [-\alpha] = \begin{cases} 0 & \text{if } \alpha \in \mathbf{Z}, \\ -1 & \text{otherwise,} \end{cases}$

(*viii*) $[2\alpha] - 2[\alpha] = \begin{cases} 1 & \text{if } [2\alpha] \text{ is odd,} \\ 0 & \text{if } [2\alpha] \text{ is even,} \end{cases}$

(*ix*) $\left[\frac{[\alpha]}{n}\right] = \left[\frac{\alpha}{n}\right]$, n positive,

(*x*) $[\alpha + \frac{1}{2}]$is the nearest integer to α. If two integers are equally close to α, $[\alpha + \frac{1}{2}]$ is the bigger one.

(*xi*) $[2\alpha] + [2\beta] \geq [\alpha] + [\beta] + [\alpha + \beta]$

43. If m and n are positive integers with $(m, n) = 1$ show that

$$\sum_{i=1}^{n-1} \left[\frac{mi}{n}\right] = \frac{(m-1)(n-1)}{2}.$$

Furthermore, show that if $(m, n) = d$ then

$$\sum_{i=1}^{n-1} \left[\frac{mi}{n}\right] = \frac{(m-1)(n-1)}{2} + \frac{d-1}{2}.$$

44. Denoting, as usual, the number of positive divisors of the integer i by $\tau(i)$, show that

$$\sum_{i=1}^{n} \tau(i) = \sum_{d=1}^{n} \left[\frac{n}{d}\right].$$

45. For any positive integer n and any $\alpha \in \mathbf{R}$, show that

$$\sum_{i=1}^{n-1} \left[\alpha + \frac{i}{n}\right] = [n\alpha].$$

46. Consider the real number α; then

(*i*) $\alpha \in \mathbf{Q} \Leftrightarrow$ there exists $m \in \mathbf{Z}$ for which $[m\alpha] = m\alpha$.

(*ii*) $\alpha \in \mathbf{Q} \Leftrightarrow$ there exists $m \in \mathbf{Z}$ for which $[m!\alpha] = m!\alpha$

(*iii*) We know that e, the natural base of logarithms, is given by

$$e = \sum_{n=0}^{\infty} \frac{1}{n!} = 1 + 1 + \frac{1}{2!} + \ldots.$$

For any positive m show that

$$[m!e] = m!\left(\sum_{n=0}^{m} \frac{1}{n!}\right) < m!e$$

Hence, e is irrational.

47. (*i*) Making use of the v_p notation from Section 1-5, show that the exponent to which the prime p appears in the factorization of $n!$ is

$$v_p(n!) = \sum_{i=1}^{\infty} \left[\frac{n}{p^i}\right].$$

(*ii*) Use part (*ix*) of Problem 42 to simplify the computation of $v_7(50{,}000!)$.

48. If the expansion of n in base p is $n = b_0 + b_1 p + \cdots + b_s p^s$ then

$$v_p(n!) = \frac{n - (b_0 + b_1 + \cdots + b_s).}{p - 1}$$

49. If $(a, m) = 1$ then, according to 3-2-22,$a^{\phi(m)} \equiv 1 (\text{mod } m)$. If s is the smallest integer for which $a^s \equiv 1 (\text{mod } m)$, show that $s \mid \phi(m)$.

50. (*i*) If $(m, 133) = (n, 133) = 1$, then $133 \mid (m^{18} - n^{18})$.
(*ii*) For every $n > 0$, we have $101010 \mid (n^{37} - n)$.

51. Suppose m is odd; then:
(*i*) The sum of the elements in any complete residue system mod m (see 3-2-22) is $\equiv 0 (\text{mod } m)$.
(*ii*) The sum of the elements in any reduced residue system mod m is $\equiv 0 (\text{mod } m)$.

52. For an odd prime p, show that,

$$\prod_{i=1}^{(p-1)/2} (2i)^2 = 2^2 \cdot 4^2 \cdots (p-1)^2 \equiv (-1)^{(p+1)/2} \quad (\text{mod } p).$$

53. For every $p > 3$, show that the sum of all squares in $\mathbf{Z}_p$ is 0; in other words, the sum of all the quadratic residues is congruent to 0 (mod p).

54. If $p \nmid a$, $p \nmid m$ and the integer n is arbitrary prove that

$$\sum_{a=0}^{p-1} \left(\frac{ma + n}{p}\right) = 0,$$

it being understood that $\left(\frac{0}{p}\right) = 0$.

55. If $p \geq 7$ there are consecutive quadratic residues (mod p) and consecutive quadratic nonresidues (mod p)—in other words, there exist integers $1 \leq a, b \leq p - 1$ for which

$$\left(\frac{a}{p}\right) = \left(\frac{a+1}{p}\right) = 1, \qquad \left(\frac{b}{p}\right) = \left(\frac{b+1}{p}\right) = -1.$$

56. Suppose the r consecutive integers a, $a+1, \dots, a+r-1$ have the same quadratic character (mod p), while $a-1$ and $a+r$ do not have this quadratic character—that is,

$$\left(\frac{a-1}{p}\right) \neq \left(\frac{a}{p}\right) = \left(\frac{a+1}{p}\right) = \cdots = \left(\frac{a+r-1}{p}\right) \neq \left(\frac{a+r}{p}\right).$$

Then $r \leq a$.

57. Suppose $a > \sqrt{p}$ and $(\frac{a}{p}) = -1$; then there is no sequence of a consecutive quadratic residues (mod p), nor is there a sequence of a consecutive quadratic nonresidues (mod p).

58. Consider the sequence $1, 2, \dots, p-1$ where p is an odd prime. Let

$A =$ the number of pairs a, $a+1$ with $\left(\frac{a}{p}\right) = 1$ and $\left(\frac{a+1}{p}\right) = 1$,

$B =$ the number of pairs a, $a+1$ with $\left(\frac{a}{p}\right) = -1$ and $\left(\frac{a+1}{p}\right) = -1$,

$C =$ the number of pairs a, $a+1$ with $\left(\frac{a}{p}\right) = 1$ and $\left(\frac{a+1}{p}\right) = -1$,

$D =$ the number of pairs a, $a+1$ with $\left(\frac{a}{p}\right) = -1$ and $\left(\frac{a+1}{p}\right) = 1$.

(i) Prove that

$$\sum_{a=1}^{p-2}\left\{\left(\frac{a}{p}\right) + \left(\frac{a+1}{p}\right)\right\} = 2A - 2B, \qquad \sum_{a=1}^{p-2}\left\{\left(\frac{a}{p}\right) - \left(\frac{a+1}{p}\right)\right\} = 2C - 2D$$

(ii) For each $r = 1, 2, \dots, p-1$ let

$$f(r) = \sum_{a=1}^{p}\left(\frac{a}{p}\right)\left(\frac{a+r}{p}\right).$$

Then $f(r) = f(1)$, and by evaluating $\sum_{r=1}^{p-1} f(r)$ it follows that $f(1) = -1$. Consequently,

$$A + B - C - D = -1.$$

(iii) Show that

$$A = \frac{p-5}{4}, \quad B = C = D = \frac{p-1}{4} \qquad \text{when } p \equiv 1 \pmod 4$$

$$A = B = D = \frac{p-3}{4}, \quad C = \frac{p+1}{4} \qquad \text{when } p \equiv 3 \pmod 4.$$

59. Suppose a is odd; show that:

(*i*) $x^2 \equiv a \pmod 2$ has exactly one solution.

(*ii*) $x^2 \equiv a \pmod{2^2}$ has a solution $\Leftrightarrow$ $a \equiv 1 \pmod 4$; and in this case there are exactly two solutions.

(*iii*) $x^2 \equiv a \pmod{2^3}$ has a solution $\Leftrightarrow$ $a \equiv 1 \pmod 8$; and in this case there are exactly four solutions.

(*iv*) If $s \geq 3$ and $x^2 \equiv a \pmod{2^s}$ has a solution c_s then $x^2 \equiv a \pmod{2^{s+1}}$ has a solution of form $c_{n+1} = c_s + t2^{s-1}$.

(*v*) For any $n \geq 3$, $x^2 \equiv a \pmod{2^n}$ has a solution $\Leftrightarrow$ $a \equiv 1 \pmod 8$; and in this case there are exactly four solutions.

60. Suppose $(a, m) = 1$ and we write the prime factorization of m in the form

$$m = 2^{r_0} \prod_{i=1}^{t} p^{r_i} \qquad r_0 \geq 0, \quad r_i > 0 \quad \text{for} \quad i = 1, \ldots, t.$$

Find a necessary and sufficient condition for the congruence $x^2 \equiv a \pmod m$ to have a solution. Show that if a solution exists then the number of solutions is

$$\begin{aligned} &2^t && \text{when } r_0 \leq 1,\\ &2^{t+1} && \text{when } r_0 = 2,\\ &2^{t+2} && \text{when } r_0 \geq 3. \end{aligned}$$

61. Consider the ring of quaternions, Q (see Miscellaneous Problem 55 of Chapter II).

(*i*) Show that it is a division ring (by which is meant that it satisfies all the requirements for a field except the commutative law for multiplication).

(*ii*) Find the center of Q.

(*iii*) Show that the equation $x^2 = -1$ has an infinite number of solutions in Q.

62. Consider the domain $\mathbf{Z}[\sqrt{-6}] = \{a + b\sqrt{-6} \mid a, b \in \mathbf{Z}\}$. For each element $\alpha = a + b\sqrt{-6}$ of $\mathbf{Z}[\sqrt{-6}]$ call $\bar{\alpha} = a - b\sqrt{-6}$ the **conjugate** of α, and define the **norm** of α by

$$N(\alpha) = \alpha\bar{\alpha} = a^2 + 6b^2.$$

Prove the following:

(*i*) $N(1) = 1$; $N(\alpha\beta) = N(\alpha)N(\beta)$ for all $a, \beta \in \mathbf{Z}[\sqrt{-6}]$.

(*ii*) α is a unit $\Leftrightarrow$ $N(\alpha) = 1$; so $\mathbf{Z}[\sqrt{-6}]$ has only the two units ± 1.

(*iii*) $2, 3, 5, \sqrt{-6}, 2+\sqrt{6}$ are prime elements of $\mathbf{Z}[\sqrt{-6}]$. Of course, an element π is said to be **prime** when it cannot be expressed in the form $\pi = \alpha\beta$ where neither α nor β is a unit of $\mathbf{Z}[\sqrt{-6}]$.

(*iv*) If $\alpha \neq 0$ is not a unit in $\mathbf{Z}[\sqrt{-6}]$ then it can be written as a finite product of primes.

(*v*) In $\mathbf{Z}[\sqrt{-6}]$, factorization into primes need not be unique—since, for example,
$6 = 2 \cdot 3 = \sqrt{-6} \cdot \sqrt{-6}$ and $10 = 2 \cdot 5 = (2+\sqrt{-6})(2-\sqrt{-6})$.

63. For $\alpha = a + b\sqrt{-5}$ in the domain $\mathbf{Z}[\sqrt{-5}]$ write $\bar{\alpha} = a - b\sqrt{-5}$ and $N(\alpha) = \alpha\bar{\alpha} = a^2 + 5b^2$. Then:

(*i*) $N(\alpha\beta) = N(\alpha)N(\beta)$ for all $\alpha, \beta \in \mathbf{Z}[\sqrt{-5}]$.

(*ii*) The only units of $\mathbf{Z}[\sqrt{-5}]$ are ± 1.

(*iii*) Factorization into primes exists for elements of $\mathbf{Z}[\sqrt{-5}]$.

(*iv*) The following are primes: $2, 3, 7, 1 \pm \sqrt{-5}, 2 \pm \sqrt{-5}, 3 \pm \sqrt{-5}, 1 \pm 2\sqrt{-5}, 2 \pm 3\sqrt{-5}$.

(*v*) Factorization into primes need not be unique—for example,

$$\begin{aligned} 9 &= 3 \cdot 3 = (2+\sqrt{-5})(2-\sqrt{-5}), \\ 6 &= 2 \cdot 3 = (1+\sqrt{-5})(1-\sqrt{-5}), \\ 14 &= 2 \cdot 7 = (3+\sqrt{-5})(3-\sqrt{-5}), \\ 21 &= 3 \cdot 7 = (1+2\sqrt{-5})(1-2\sqrt{-5}), \\ 49 &= 7 \cdot 7 = (2+3\sqrt{-5})(2-3\sqrt{-5}). \end{aligned}$$

64. Consider the domain $\mathbf{Z}[\sqrt{5}] = \{a + b\sqrt{5} \mid a, b \in \mathbf{Z}\}$. For $\alpha = a + b\sqrt{5}$ put $\bar{\alpha} = a - b\sqrt{5}$ and $N(\alpha) = a^2 - 5b^2$. Then:

(*i*) $2+\sqrt{5}, 9+4\sqrt{5}, 2-\sqrt{5}, 9-4\sqrt{5}$ are units of $\mathbf{Z}[\sqrt{5}]$.

(*ii*) For every $n \in \mathbf{Z}$, $\pm(2+\sqrt{5})^n$ is a unit.

(*iii*) The elements $3+2\sqrt{5}$, $-4+\sqrt{5}$, $-13+6\sqrt{5}$ are associates in $\mathbf{Z}[\sqrt{5}]$.

(*iv*) α is a unit $\Leftrightarrow$ $N(\alpha) = \pm 1$.

(*v*) If $N(\alpha)$ is a prime in $\mathbf{Z}$ then α is a prime of $\mathbf{Z}[\sqrt{5}]$.

(*vi*) The elements $3+2\sqrt{5}$, $-4+\sqrt{5}$, $-13+6\sqrt{5}$ are primes of $\mathbf{Z}[\sqrt{5}]$ (they should really be viewed as the same prime) and so are $\pm 2, 3 \pm \sqrt{5}$.

(*vii*) In $\mathbf{Z}[\sqrt{5}]$ factorization into primes exists, but is not unique [witness $4 = 2 \cdot 2 = (3+\sqrt{5})(3-\sqrt{5})$].

65. Let

$$\omega = \frac{1 + \sqrt{5}}{2}$$

and consider the integral domain $\mathbf{Z}[\omega] = \{a + b\omega \mid a, b \in \mathbf{Z}\}$. It contains the domain $\mathbf{Z}[\sqrt{5}]$. Let

$$\bar{\omega} = \frac{1 - \sqrt{5}}{2},$$

and for $\alpha = a + b\omega$ put $\bar{\alpha} = a + b\bar{\omega}$. Define $N(\alpha) = \alpha\bar{\alpha} = a^2 + ab - b^2 \in \mathbf{Z}$.

(*i*) The map $\alpha \to \bar{\alpha}$ is an automorphism of $\mathbf{Z}[\omega]$.

(*ii*) If $\alpha \in \mathbf{Z}[\sqrt{5}] \subset \mathbf{Z}[\omega]$ then $N(\alpha)$ coincides with the norm of α defined in the preceding problem.

(*iii*) $N(\alpha\beta) = N(\alpha)N(\beta)$ for all $\alpha, \beta \in \mathbf{Z}[\omega]$.

(*iv*) α is a unit $\Leftrightarrow$ $N(\alpha) = \pm 1$.

(*v*) Every unit of $\mathbf{Z}[\sqrt{5}]$ is a unit of $\mathbf{Z}[\omega]$. The converse is false—in fact, ω is a unit of $\mathbf{Z}[\omega]$, and so is $\pm\omega^n$ for every $n \in \mathbf{Z}$.

(*vi*) If $N(\alpha)$ is prime (in $\mathbf{Z}$) then α is prime (in $\mathbf{Z}[\omega]$).

(*vii*) Factorization into primes exists in $\mathbf{Z}[\omega]$.

(*viii*) Are 2, $3 \pm \sqrt{5}$ primes of $\mathbf{Z}[\omega]$?

(*ix*) Is factorization into primes unique in $\mathbf{Z}[\omega]$?

66. For $\alpha = a + bi$ in the domain of Gaussian integers $\mathbf{Z}[i]$, put $N(\alpha) = (a + bi)(a - bi) = a^2 + b^2$. Then:

(*i*) $N(\alpha\beta) = N(\alpha)N(\beta)$.

(*ii*) There are four units: ± 1, $\pm i$.

(*iii*) $N(\alpha)$ prime in $\mathbf{Z} \Rightarrow \alpha$ prime in $\mathbf{Z}[i]$.

(*iv*) $2 = (1 + i)(1 - i)$; this is a factorization of 2 into primes $(1 + i)$ and $(1 - i)$, which are associates.

(*v*) 3 is prime in $\mathbf{Z}[i]$; so is 7. More generally, any prime $p \equiv 3 \pmod 4$ is a prime of $\mathbf{Z}[i]$.

(*vi*) We have the factorization into primes $5 = (2 + i)(2 - i)$. How does it compare with the prime factorization $5 = (1 - 2i)(1 + 2i)$?

(*vii*) 13 factors into primes; in fact,

$$13 = (3 + 2i)(3 - 2i) = (2 + 3i)(2 - 3i).$$

Compare these factorizations.

(*viii*) Any prime $p \equiv 1 \pmod 4$ factors as the product of two distinct primes of $\mathbf{Z}[i]$.

(*ix*) Factorization into primes holds in $\mathbf{Z}[i]$.

(*x*) What about uniqueness?

67. (*i*) The polynomial $f(x) = x^2 + 1$ is irreducible over $\mathbf{Z}_3$. Extend the procedure of 3-2-23, Problem 28 and 3-5-29, Problem 27, by using an α which satisfies $\alpha^2 + 1 = 0$, to construct a field, $\mathbf{Z}_3[\alpha]$, with 9 elements.

(*ii*) Use the irreducible polynomial $f(x) = x^3 - x^2 + 1$ over $\mathbf{Z}_3$ to construct a field with 27 elements.

(*iii*) Consider an arbitrary field F and an irreducible polynomial $f(x) = a_0 + a_1x + \cdots + a_nx^n$ over F. Let ξ be a formal symbol which is taken to satisfy $f(\xi) = a_0 + a_1\xi + \cdots + a_n\xi^n = 0$. Generalizing the above, show how

$$F[\xi] = \{c_0 + c_1\xi + \cdots + c_{n-1}\xi^{n-1} \mid c_0, c_1, \ldots, c_{n-1} \in F\}$$

can be made into a field which contains F.

IV

GROUPS

In the preceding chapters, we studied algebraic systems with two operations—namely, the particular objects $\mathbf{Z}$ and $\mathbf{Z}_m$, and the general objects known as "rings." In this chapter, we turn to the study of logically simpler algebraic systems, known as "groups"—they are algebraic objects with a single operation.

It is a matter of pedagogical taste whether one does groups or rings first. Most algebra books follow the rather natural path of increasing complexity (that is, of more and more axioms), so they do groups first. Our approach, on the other hand, has been based on the idea of moving from the familiar to the unfamiliar and from the concrete to the abstract; thus, we were led to study the integers first, then rings, and then groups. Of course, when everything is said and done, one ends with the same body of knowledge, no matter what order is used for organizing the material.

The contents of this chapter are almost entirely pure algebra. As the facts about groups are developed, we shall often observe analogies with facts we already know about rings. Although groups are of interest in themselves and for various applications, our interest stems primarily from the number-theoretic connections. These connections involve $\mathbf{Z}_m^*$ (the set of units of $\mathbf{Z}_m$) which is a group whose structure we try to determine.

4-1. Basic Facts and Examples

Suppose G is a nonempty set on which we have a binary operation $\cdot$. We recall (see 2-1-2) what this means—namely, we are given a mapping from $G \times G \to G$ which assigns to each ordered pair $(a, b) \in G \times G$ an element $a \cdot b \in G$. The notation $\cdot$ for the operation is fairly common but, of course, any other notation is equally valid. In practice, we shall usually drop the $\cdot$ and write ab instead of $a \cdot b$.

In virtue of our past experience, the kinds of properties of this operation which interest us are:

(1) closure (this is really redundant, because closure is included in the definition of an operation),
(2) the associative law,
(3) the existence of an identity,
(4) the existence of inverses,
(5) the commutative law.

These properties appear as ingredients in our first result (and in all subsequent results too), which combines the basic definition of a group with consideration of equivalent formulations for its axioms.

4-1-1. Theorem. Suppose $\{G, \cdot\}$ is a **semigroup**; by this we mean that $\cdot$ is an operation on the nonempty set G with the properties

(1) G is closed under the operation,
(2) the associative law holds.

Then the following conditions are equivalent, and when any one of them holds, $\{G, \cdot\}$ is said to be a **group.**

I: $\begin{cases} (i) & G \text{ has an identity } e\text{; that is, } ea = ae = a \text{ for every } a \in G. \\ (ii) & \text{Every element of } G \text{ has an inverse; that is, given any } a \in G \\ & \text{there exists } a' \in G \text{ for which } a'a = aa' = e. \end{cases}$

II: $\begin{cases} \text{For any choice of } a, b \in G \text{ the equations } ax = b \text{ and } ya = b \text{ have} \\ \text{solutions in } G \text{ for } x \text{ and } y. \end{cases}$

III: $\begin{cases} (i) & G \text{ has a left identity } e\text{; that is, } ea = a \text{ for every } a \in G. \\ (ii) & \text{Every element of } G \text{ has a left inverse; that is, given any } a \in G \text{ there exists } a' \in G \text{ with } a'a = e. \end{cases}$

III′: $\begin{cases} (i) & G \text{ has a right identity } e\text{; that is, } ae = a \text{ for every } a \in G. \\ (ii) & \text{Every element of } G \text{ has a right inverse; that is, given any } a \in G \text{ there exists } a' \in G \text{ with } aa' = e. \end{cases}$

In addition, if the commutative law (that is, $ab = ba$ for all $a, b \in G$) holds in a group, or a semigroup, we say it is **abelian**—after Abel (1802–1829).

Proof: We need only prove

$$\text{I} \Rightarrow \text{II} \Rightarrow \text{III} \Rightarrow \text{I},$$

as the proof of I $\Rightarrow$ II $\Rightarrow$ III′ $\Rightarrow$ I will then go in the same way.

I $\Rightarrow$ II: Given $a, b \in G$ there exists, by I(*ii*), an inverse a' of a—so $a'a = aa' = e$. Then $x = a'b$ and $y = ba'$ are solutions of $ax = b$ and $ya = b$, respectively, since [making use of the associative law and I(*i*)] we have

$$a(a'b) = (aa')b = eb = b \quad \text{and} \quad (ba')a = b(a'a) = be = b.$$

Note how this proof requires e to be an identity on both right and left, and a' to be an inverse of a on both right and left.

II $\Rightarrow$ III: Fix an $a \in G$. Applying II for $b = a$, there is a solution in G of the equation $ya = a$; call it e—so $ea = a$. Now, e is a left identity for this a, but this does not say that the same e is a left identity for every element of G. Thus, we would like to show that for any $b \in G$ we have $eb = b$. For this, let c be a solution of $ax = b$ (its existence is guaranteed by II). So $ac = b$ and (using the associative law)

$$eb = e(ac) = (ea)c = ac = b.$$

Hence, e is indeed a left identity for every element of G. Furthermore, given $a \in G$ there exists, according to II, an $a' \in G$ such that $a'a = e$; so a has a left inverse.

Note how this proof requires both parts of II; that is, the solvability of both $ax = b$ and $ya = b$.

III $\Rightarrow$ I: We have, by hypothesis, $ea = a$ for every $a \in G$ and the existence of $a' \in G$ for which $a'a = e$. We want to prove that $ae = a$ for every $a \in G$ (so e is an identity on both sides) and $aa' = e$ (so a' is an inverse on both sides). For this, in virtue of III(*ii*), let a'' be a left inverse of a'—so $a''a' = e$. Then using the associative law and the fact that e is a left identity,

$$\begin{aligned} aa' &= e(aa') = (a''a')(aa') = a''[a'(aa')] \\ &= a''[(a'a)a'] = a''(ea') = a''a' = e, \end{aligned}$$

so a' is also a right inverse of a. Now, e is also a right identity, since

$$ae = a(a'a) = (aa')a = ea = a.$$

This completes the proof. ▌

We shall usually be somewhat careless and refer to a group G instead of $\{G, \cdot\}$. The operation $\cdot$, which is obviously called "multiplication," will be taken for granted; it may be commutative or it may not. On occasion, the symbol $+$ will be used for the operation (in which case, one naturally calls the operation "addition"). It is a strong convention in this subject that when $+$ is used, the operation is understood to be commutative. This is entirely consistent with our experience with rings, where addition is always commutative.

4-1-2. Examples. (1) The integers under addition, $\{\mathbf{Z}, +\}$, are clearly an abelian group; 0 is an identity element for this operation, and $-a$ is an inverse for $a \in \mathbf{Z}$.

(2) The integers under multiplication, $\{\mathbf{Z}, \cdot\}$, satisfy closure, associativity, the commutative law, and have an identity element 1—so it is an abelian semi group with identity. But it is not a group because not every element has an inverse.

(3) For any integer $m > 1$, consider $\{\mathbf{Z}_m, +\}$, the integers modulo m under addition. This is an abelian group; $0 = \lfloor 0 \rfloor_m$, is an identity element, and $\lfloor -a \rfloor_m$ is an inverse for $\lfloor a \rfloor_m \in \mathbf{Z}_m$.

(4) The integers modulo m under multiplication, $\{\mathbf{Z}_m, \cdot\}$, is an abelian (or simply, commutative) semigroup with identity $\lfloor 1 \rfloor_m$. It is not a group, because not every element has an inverse.

(5) More generally, consider any ring $\{R, +, \cdot\}$. If we recall the axioms for addition, they assert precisely that $\{R, +\}$ is an abelian group; it is known as "the additive group of the ring." What about $\{R, \cdot\}$, the set R under the operation of multiplication? In virtue of the axioms concerning multiplication in a ring, closure and associativity are satisfied—so $\{R, \cdot\}$ is a semigroup. This semigroup has an identity if and only if the ring $\{R, +, \cdot\}$ has a unity (that is, identity) for multiplication. However, $\{R, \cdot\}$ can never be a group, for even if an identity exists the element 0 can never have an inverse under the operation of multiplication.

There are many additional examples of groups to be given, but we defer them for a moment to discuss the elementary facts about groups (all of which are really known to us from the work on rings and fields).

4-1-3. Proposition. In any group $G = \{G, \cdot\}$ we have:

(1) The generalized associative law holds; that is, for any $a_1, a_2, \ldots, a_n \in G$ the meaning of $a_1 a_2 \cdots a_n$ is unambiguous—it does not matter how parentheses are inserted.
(2) The identity e is unique.
(3) The inverse of any $a \in G$ is unique; we denote it by a^{-1}.
(4) The cancellation laws hold; more precisely, left cancellation says: $ab = ac \Rightarrow b = c$, and right cancellation says: $ba = ca \Rightarrow b = c$.
(5) The solution of each of the equations $ax = b$ and $ya = b$ is unique.
(6) For every $a \in G$, $(a^{-1})^{-1} = a$.
(7) The inverse of a product is the product of the inverses in the reverse order; that is,

$$(ab)^{-1} = b^{-1}a^{-1} \qquad \text{for all } a, b \in G.$$

(8) The standard definitions of powers and their properties apply; thus for any $a \in G$ and $n \in \mathbf{Z}$,

$$a^0 = e, \qquad a^{-n} = (a^{-1})^n, \qquad a^m a^n = a^{m+n}, \qquad (a^m)^n = a^{mn}.$$

Proof: The details could safely be left to the reader, but for the record we comment briefly on each statement.

(1) This goes exactly like the proof of the generalized associative law for multiplication in a ring, 2-4-1. Since only the associative law is used in the proof, this assertion is obviously valid in any semigroup.

(2) If $e' \in G$ is also an identity, then $e = e'e = e'$.

(3) If both a' and a'' are inverses of a, then

$$a' = a'e = a'(aa'') = (a'a)a'' = ea'' = a''.$$

(4) If $ab = ac$, then $b = (a^{-1}a)b = a^{-1}(ab) = a^{-1}(ac) = (a^{-1}a)c = c$. Similarly, $ba = ca \Rightarrow b = c$.

(5) If both c_1 and c_2 are solutions of $ax = b$, then $ac_1 = b = ac_2$, so by cancellation, $c_1 = c_2$. Thus $a^{-1}b$ is the unique solution of $ax = b$. Similarly ba^{-1} is the unique solution of $ya = b$.

(6) Since $(a^{-1})^{-1}(a^{-1}) = e = aa^{-1}$, cancellation gives $(a^{-1})^{-1} = a$.

(7) By definition, $(ab)(ab)^{-1} = e$. On the other hand,

$$\begin{aligned}(ab)(b^{-1}a^{-1}) &= ((ab)b^{-1})(a^{-1}) = (a(bb^{-1}))a^{-1} \\ &= (ae)a^{-1} = aa^{-1} = e.\end{aligned}$$

So, by cancellation, $(ab)^{-1} = b^{-1}a^{-1}$.

(8) Most of the preceding assertions were proved in 3-2-5 when we were discussing the units of a commutative ring with unity. Even though our group

need not be commutative, the story about exponents still goes as in 3-2-5, part (9), because we are concerned here solely with the powers of a single element and these always commute with each other. ▮

4-1-4. Remark. If the group G is also abelian, then we have additional properties such as the generalized commutative–associative law (see 2-4-2), and $(ab)^n = a^n b^n$ for all $a, b \in G$, $n \in \mathbf{Z}$. The verifications should cause the reader no difficulty.

When the abelian group G is written additively (that is, with operation $+$) some of the properties of 4-1-3 take the following forms: the identity 0 is unique; the inverse of any $a \in G$ is unique, we denote it by $-a$; the cancellation law (because of commutativity, only one cancellation law is needed) holds—that is, $a + b = a + c \Rightarrow b = c$; the equation $a + x = b$ has a unique solution, namely, $b - a$; $-(-a) = a$ for every $a \in G$; $-(a + b) = -b - a = -a - b$; for $m, n \in \mathbf{Z}$, $a, b \in G$, we have $0a = 0$, $(-n)a = n(-a)$, $(m + n)a = ma + na$, $m(na) = (mn)a$, $m(a + b) = ma + mb$. The reader may well find these statements rather boring—after all, we proved them in a ring (more precisely, for the additive group of an arbitrary ring) in Sections 2-1 and 2-4.

We have seen in 4-1-3 that the cancellation laws hold in a group, and that they play a key role in proving various properties. The question naturally arises if the cancellation laws can be used as replacements for the decisive group axioms. More precisely: given a semigroup in which both cancellation laws hold, is it a group? The answer is in the negative, as may be seen from a simple example. Consider $\{2\mathbf{Z}, \cdot\}$, the even integers under multiplication. Closure, associativity, commutativity, and cancellation are clear—so this is a commutative semigroup with both cancellation laws. But it is not a group; there is no identity, and of course inverses cannot exist.

On the other hand, as our next result indicates, there is one situation in which the cancellation laws guarantee that a semigroup is a group.

4-1-5. Proposition. Suppose G is a finite semigroup. If both cancellation laws hold, then G is a group.

Proof: Suppose G consists of n elements, and write

$$G = \{c_1, c_2, \ldots, c_n\}.$$

Thus, $c_1, c_2, \ldots, c_n$ are distinct and every element of G appears among them. Now, let any $a, b \in G$ be given. Consider $ac_1, ac_2, \ldots, ac_n$; these n elements are distinct since, by left cancellation, $ac_i = ac_j \Rightarrow c_i = c_j \Rightarrow i = j$. Because G has n elements, we must have

$$G = \{ac_1, ac_2, \ldots, ac_n\}.$$

Since the element b appears in this set, we conclude that the equation $ax = b$ has a solution in G.

In similar fashion, using right cancellation, the equation $ya = b$ has a solution in G. Thus, Axiom II of 4-1-1 is satisfied and G is a group.

It is worthwhile to give a slightly more abstract formulation of this proof. For any $a \in G$, consider the mapping

$$\phi_a: G \to G$$

defined by

$$\phi_a(x) = ax, \qquad x \in G.$$

In other words, ϕ_a amounts to multiplication by a, on the left. By left cancellation, ϕ_a is one-to-one [that is, $\phi_a(x_1) = \phi_a(x_2) \Rightarrow ax_1 = ax_2 \Rightarrow x_1 = x_2$]. Because G is finite, the map ϕ_a must be onto G. Therefore, any $b \in G$ is in the image of ϕ_a—so the equation $ax = b$ has a solution. Similarly, to show that $ya = b$ has a solution, one works with right multiplication by a; the notation might look like $\psi_a: G \to G$ where $\psi_a(y) = ya$, $y \in G$. ▌

We turn next to additional examples of groups.

4-1-6. Example. Consider $\{\mathbf{R}, \cdot\}$, the reals under multiplication. Obviously this is a commutative semigroup with identity 1; however, as in 4-1-2, part (5), 0 cannot have an inverse, so this is not a group.

What happens if we throw out 0, and consider the set of all nonzero reals? Denoting it by $\mathbf{R}^* = \mathbf{R} - (0)$, we note that in $\{\mathbf{R}^*, \cdot\}$ we have closure (since the product of two nonzero real numbers is not zero), the associative law, the commutative law, the identity, 1, and every $a \in \mathbf{R}^*$ has an inverse for multiplication, namely $a^{-1} = 1/a \in \mathbf{R}^*$. Thus, $\{\mathbf{R}^*, \cdot\}$ is a commutative group.

More generally, consider any field $F = \{F, +, \cdot\}$. As before, $\{F, \cdot\}$ is not a group because 0 has no inverse. If, as usual, we let F^* denote the set of all nonzero elements of F, then F^* is closed under multiplication because in a field (which is an integral domain) the product of two nonzero elements is not zero. Moreover, according to the definition of a field, 3-2-7, every element of F^* is a unit—that is, has an inverse for multiplication. It follows that $F^* = \{F^*, \cdot\}$ is a commutative group; it is known as "the multiplicative group of the field F." Of course, F^* may also be characterized as the set or (group) of units of F.

In particular, if p is prime, then $\mathbf{Z}_p$ is a field and in virtue of the foregoing, $\mathbf{Z}_p^*$, the set of all nonzero elements (or equivalently, the set of units) of $\mathbf{Z}_p$ is a group with $p - 1$ elements.

4-1-7. Examples. For any $m > 1$, consider the commutative ring with unity $\mathbf{Z}_m = \{\mathbf{Z}_m, +, \cdot\}$. As noted in 4-1-2, $\{\mathbf{Z}_m, +\}$ is an abelian group, but

$\{\mathbf{Z}_m, \cdot\}$ is not a group. In order to extract a multiplicative group from $\mathbf{Z}_m$, we must obviously include only elements which have an inverse under multiplication. Therefore, let us consider $\mathbf{Z}_m^*$, the set of all units of $\mathbf{Z}_m$ (that is, all elements of $\mathbf{Z}_m$ which have a multiplicative inverse). It is straightforward to verify that $\{\mathbf{Z}_m^*, \cdot\}$ is an abelian group with $\phi(m)$ elements, but there is no additional effort involved in treating this in a more general context.

Thus, let $R = \{R, +, \cdot\}$ be any commutative ring with unity, and denote the set of all units of R by R^*. (This is consistent with the earlier notation when R is a field F. There F^* is defined as the set of all nonzero elements—which is the same as the set of all units.) Consider $\{R^*, \cdot\}$. Clearly, e (the identity) belongs to R^*, and the associative law for multiplication holds for elements of R^*. Furthermore, according to 3-2-5, parts (8) and (7), the product of units is a unit—so R^* is closed under multiplication—and any unit has an inverse which is also a unit—so every element of R^* has an inverse in R^*. Thus, $\{R^*, \cdot\}$ is an abelian group; it is known as the **group of units** of the ring R.

Incidentally, the notion of "group of units" carries over to an arbitrary ring R with unity e, even if it is not commutative. More precisely, we have $ea = ae = a$ for all $a \in R$, and we say that $a \in R$ is a unit when there exists $a' \in R$ with $aa' = a'a = e$. The properties of units then go as in 3-2-5, and it follows easily that the set of all units (which is once more denoted by R^*) is a group under multiplication. In more detail: If $a, b \in R^*$, then $(ab)(b'a') = e$ and $(b'a')(ab) = e$, so $ab \in R^*$ and R^* is closed under multiplication; associativity is clear; obviously $e \in R^*$; finally, if $a \in R^*$, then $aa' = a'a = e$ which also says that a' is a unit—so a has an inverse in R^*.

4-1-8. Example. Consider the field of complex numbers $\mathbf{C}$. Every element of $\mathbf{C}$ is uniquely of form $z = x + yi$ where $x, y \in \mathbf{R}$, and we have the notion of *absolute value* of a complex number—it is defined by

$$|z| = \sqrt{x^2 + y^2}.$$

Thus, if we view the complex numbers, geometrically, as the plane $\mathbf{R} \times \mathbf{R}$, then $|z|$ represents the distance from the origin $(0, 0)$ to the point (x, y).

Among the significant properties of absolute value one finds:

(*i*) $|z| = \sqrt{z\bar{z}}$ where $\bar{z} = x - yi$ is the complex conjugate.
(*ii*) $|z| \geq 0$; $|z| = 0 \Leftrightarrow z = 0$.
(*iii*) $|z_1 z_2| = |z_1|\,|z_2|$, $z_1, \quad z_2 \in \mathbf{C}$.
(*iv*) $|z_1 + z_2| \leq |z_1| + |z_2|$, $\quad z_1, z_2 \in \mathbf{C}$.

The verification of these facts is entirely mechanical—except for (*iv*), which is known as the **triangle inequality**—and the details may be left to the reader. We shall make heavy use of (*iii*), which says that the absolute value is multiplicative.

Let us look at

$$W = \{z \in \mathbf{C} \mid |z| = 1\}.$$

In words, W consists of all complex numbers of absolute value 1. Geometrically, W consists of all points whose distance from the origin is 1—that is, W is the circle of radius 1 with center at the origin. This explains why W is known as the **unit circle**.

We show W is a group under multiplication. If $z_1, z_2 \in W$, then $|z_1 z_2| = |z_1|\,|z_2| = 1 \cdot 1 = 1$, so $z_1 z_2 \in W$ and we have closure. The associative law is trivial because we are dealing with the product of complex numbers. The element $1 = 1 + 0i$ belongs to W, so there is an identity. Finally, consider $z \in W$; clearly $z \neq 0$ so, because $\mathbf{C}$ is a field, z has a multiplicative inverse—call it z^{-1} (we are not interested here in what z^{-1} looks like). From $1 = zz^{-1}$, we have

$$1 = |1| = |zz^{-1}| = |z|\,|z^{-1}| = |z^{-1}|.$$

Thus, $z^{-1} \in W$, and z has an inverse in W. Since multiplication of complex numbers is commutative, we have proved that W is an abelian group (often called the **circle group**) under multiplication.

4-1-9. Example. In ordinary 3-space consider a plane on which an equilateral triangle has been drawn, and suppose the vertices are labeled as in the accompanying diagram. Now, imagine a cardboard copy of this triangle

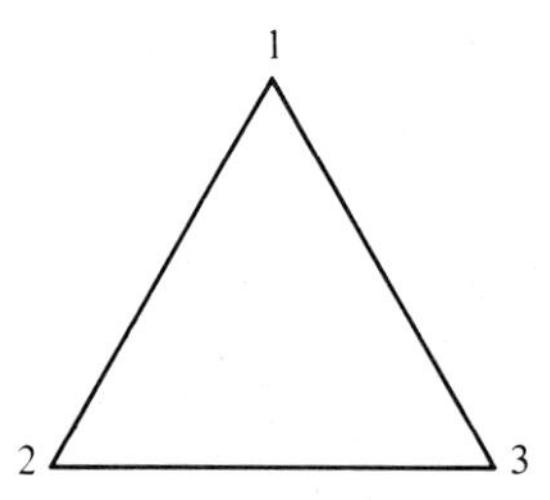

placed over it. By a **rigid motion** of the triangle we shall mean any operation of removing the cardboard from the plane, moving it around in space (we even permit the cardboard to pass through the plane—that is, our plane is conceptual rather than physical) and then placing it once more on the triangular area in the plane. The end result of a rigid motion finds the cardboard in the same place as before except that its vertices may not be where they were originally. For example, the vertex of the cardboard which was at 1 may end up at either 1, 2, or 3. (It should be understood that the labels 1, 2, 3 are considered part of the plane, and they never move!) Obviously, a rigid motion is described completely when we know how the vertices are moved. It now follows that there are exactly six rigid motions of the equilateral triangle; in fact, taking the vertex of the cardboard at 1 there are three choices as to where it can end up, and then there are two choices as to where the vertex at 2 ends up.

Let us introduce a notation for a rigid motion σ. If, for example, σ moves the vertices in the following way

$$1 \to 3, \qquad 2 \to 2, \qquad 3 \to 1,$$

then we shall write

$$\sigma = \begin{pmatrix} 1 & 2 & 3 \\ 3 & 2 & 1 \end{pmatrix}.$$

In other words, underneath each vertex we list the vertex at which it ends under σ. The six rigid motions of the equilateral triangle may therefore be denoted as follows.

$$e = \begin{pmatrix} 1 & 2 & 3 \\ 1 & 2 & 3 \end{pmatrix}, \qquad \sigma_1 = \begin{pmatrix} 1 & 2 & 3 \\ 2 & 3 & 1 \end{pmatrix}, \qquad \sigma_2 = \begin{pmatrix} 1 & 2 & 3 \\ 3 & 1 & 2 \end{pmatrix},$$

$$\tau_1 = \begin{pmatrix} 1 & 2 & 3 \\ 1 & 3 & 2 \end{pmatrix}, \qquad \tau_2 = \begin{pmatrix} 1 & 2 & 3 \\ 3 & 2 & 1 \end{pmatrix}, \qquad \tau_3 = \begin{pmatrix} 1 & 2 & 3 \\ 2 & 1 & 3 \end{pmatrix}.$$

Note that e is an "identity motion" in the sense that under it each vertex ends up where it started.

There is another kind of motion of our triangle which is of interest. By a **symmetry** of the equilateral triangle we mean a rotation of the cardboard triangle about some axis in such a way that it ends in the same place as before —but, of course, the vertices may have moved. Obviously, any symmetry is a rigid motion. To describe symmetries of our triangle consider the accompany-

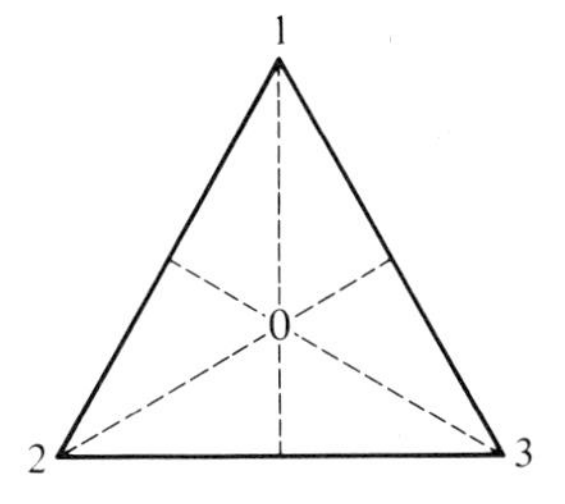

ing picture where the dotted lines are altitudes (or, equivalently, perpendicular bisectors, medians, angle bisectors). Consider the symmetry obtained by taking as axis the line through 0 perpendicular to the plane of the triangle and rotating about this axis by 120° in a counterclockwise direction. The action of this symmetry on the vertices is clearly that of σ_1. Similarly, by rotating by 240° in a counterclockwise direction about this axis, we have the motion σ_2. Furthermore, the symmetries which involve flipping the triangle over (that is, rotating by 180°) with respect to the altitudes through points 1, 2, or 3, are clearly τ_1, τ_2, and τ_3, respectively. Finally, e should be viewed as a trivial symmetry under which nothing moves.

Thus we see that the equilateral triangle has exactly six distinct symmetries—so for the equilateral triangle, the set of symmetries is the same as the set of rigid motions. If we denote this set by $G = \{e, \sigma_1, \sigma_2, \tau_1, \tau_2, \tau_3\}$, then G becomes a group when multiplication is defined as the composition of mappings. In more detail, for any $\sigma, \tau \in G$, $\sigma\tau$ is taken to mean $\sigma \circ \tau$—that is, apply the rigid motion τ and then apply σ to the result.

Now, G is closed under multiplication because the composition of two rigid motions is obviously a rigid motion. (If the elements of G are viewed as symmetries, it is not obvious that the composition of two symmetries is a symmetry. But we now know this, precisely because G is closed.) It is also instructive to verify closure concretely by constructing the multiplication table—it takes the form:

$\circ$	e	σ_1	σ_2	τ_1	τ_2	τ_3
e	e	σ_1	σ_2	τ_1	τ_2	τ_3
σ_1	σ_1	σ_2	e	τ_3	τ_1	τ_2
σ_2	σ_2	e	σ_1	τ_2	τ_3	τ_1
τ_1	τ_1	τ_2	τ_3	e	σ_1	σ_2
τ_2	τ_2	τ_3	τ_1	σ_2	e	σ_1
τ_3	τ_3	τ_1	τ_2	σ_1	σ_2	e

The reader is advised to derive the table on his own, to see how things go. For example, to compute

$$\tau_2\sigma_2 = \tau_2 \circ \sigma_2 = \begin{pmatrix} 1 & 2 & 3 \\ 3 & 2 & 1 \end{pmatrix} \circ \begin{pmatrix} 1 & 2 & 3 \\ 3 & 1 & 2 \end{pmatrix}$$

we note that σ_2 maps $1 \to 3$ and then τ_2 maps $3 \to 1$, so $\tau_2\sigma_2$ maps $1 \to 1$. Similarly, $\sigma_2\colon 2 \to 1$ and $\tau_2\colon 1 \to 3$, so $\tau_2\sigma_2\colon 2 \to 3$; and finally $\sigma_2\colon 3 \to 2$ and $\tau_2\colon 2 \to 2$, so $\tau_2\sigma_2\colon 3 \to 2$. Thus,

$$\tau_2\sigma_2 = \begin{pmatrix} 1 & 2 & 3 \\ 1 & 3 & 2 \end{pmatrix} = \tau_1.$$

The other products go in exactly the same way.

The associative law can be verified by brute force with the aid of the table;

but surely no one will actually undertake to check all possible cases. Instead we note that because we are dealing with mappings the associative law for the operation of composition is automatic—in fact, for σ, τ, $\rho \in G$ both $\sigma \circ (\tau \circ \rho)$ and $(\sigma \circ \tau) \circ \rho$ are given by the rule: first apply ρ, then apply τ, and then σ; so $\sigma \circ (\tau \circ \rho) = (\sigma \circ \tau) \circ \rho$. (This fact has already been observed in 2-5-12, Problem 5, and it will be discussed in more detail in 4-1-11.)

Clearly, e is a left identity, and from the table every element of G has a left inverse (this is so because e appears in every column of the table). Of course, the existence of a left inverse is also obvious from the geometry—any rigid motion can be followed by another one that returns all vertices to the original position. Hence, G is a group, known sometimes as the "group of the equilateral triangle." It is not commutative—for example, $\tau_1\sigma_2 = \tau_3$ while $\sigma_2\tau_1 = \tau_2$.

4-1-10. Example. Consider the set of all rigid motions of the square as shown in the accompanying figure. Any rigid motion is completely determined

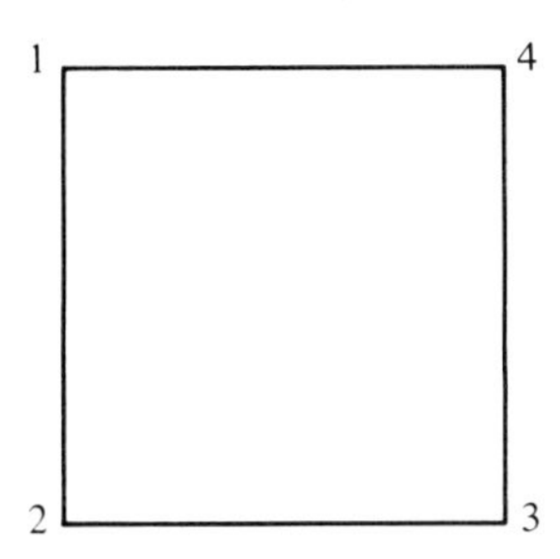

by what happens to the vertices at 1 and 2. Since the vertex at 1 can go to any one of four places and the vertex at 2 can then finish in either of two places, these are a total of eight such rigid motions. They may be denoted by

$$e = \begin{pmatrix} 1 & 2 & 3 & 4 \\ 1 & 2 & 3 & 4 \end{pmatrix}, \qquad \sigma_1 = \begin{pmatrix} 1 & 2 & 3 & 4 \\ 2 & 3 & 4 & 1 \end{pmatrix},$$

$$\sigma_2 = \begin{pmatrix} 1 & 2 & 3 & 4 \\ 3 & 4 & 1 & 2 \end{pmatrix}, \qquad \sigma_3 = \begin{pmatrix} 1 & 2 & 3 & 4 \\ 4 & 1 & 2 & 3 \end{pmatrix},$$

$$\tau_1 = \begin{pmatrix} 1 & 2 & 3 & 4 \\ 4 & 3 & 2 & 1 \end{pmatrix}, \qquad \tau_2 = \begin{pmatrix} 1 & 2 & 3 & 4 \\ 2 & 1 & 4 & 3 \end{pmatrix},$$

$$\rho_1 = \begin{pmatrix} 1 & 2 & 3 & 4 \\ 1 & 4 & 3 & 2 \end{pmatrix}, \qquad \rho_2 = \begin{pmatrix} 1 & 2 & 3 & 4 \\ 3 & 2 & 1 & 4 \end{pmatrix}.$$

It is tedious but straightforward to derive the multiplication table:

$\circ$	e	σ_1	σ_2	σ_3	τ_1	τ_2	ρ_1	ρ_2
e	e	σ_1	σ_2	σ_3	τ_1	τ_2	ρ_1	ρ_2
σ_1	σ_1	σ_2	σ_3	e	ρ_1	ρ_2	τ_2	τ_1
σ_2	σ_2	σ_3	e	σ_1	τ_2	τ_1	ρ_2	ρ_1
σ_3	σ_3	e	σ_1	σ_2	ρ_2	ρ_1	τ_1	τ_2
τ_1	τ_1	ρ_2	τ_2	ρ_1	e	σ_2	σ_3	σ_1
τ_2	τ_2	ρ_1	τ_1	ρ_2	σ_2	e	σ_1	σ_3
ρ_1	ρ_1	τ_1	ρ_2	τ_2	σ_1	σ_3	e	σ_2
ρ_2	ρ_2	τ_2	ρ_1	τ_1	σ_3	σ_1	σ_2	e

By using the same techniques as in 4-1-9, one sees that these eight rigid motions form a nonabelian group. It is known as the **group of the square** and also as the **octic group.**

What about the symmetries of the square? Naturally, since any symmetry is a rigid motion there are at most eight of them. Now consider the accompanying picture, where the horizontal and vertical lines through 0 pass

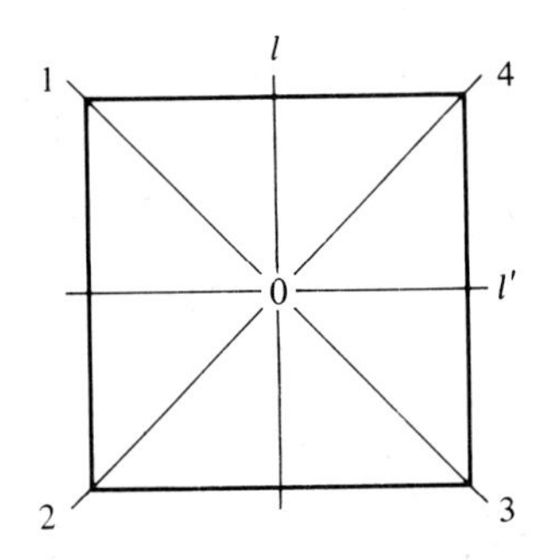

through the midpoints of the opposite sides. Except for the identity symmetry e, we have the following additional symmetries:

(*i*) Rotation by 90°, counterclockwise, about the axis through 0 perpendicular to the plane of the square; this is σ_1.

(*ii*) Rotation by 180°, counterclockwise, about the same axis; this is σ_2.

(*iii*) Rotation by 270°, counterclockwise (or by 90°, clockwise), about the same axis; this is σ_3.

(*iv*) Rotation by 180° (that is, flipping the square over) with respect to the vertical axis l; this is τ_1.

(*v*) Rotation by 180° (that is, flipping the square over) with respect to the horizontal axis l'; this is τ_2.

(*vi*) Rotation by 180° (that is, flipping) about the diagonal passing through 1 and 3; this is ρ_1.

(*vii*) Rotation by 180° (that is flipping) about the diagonal passing through 2 and 4; this is ρ_2.

Consequently, there are eight symmetries and they constitute the same group as the rigid motions of the square.

4-1-11. Example. Consider an arbitrary set $X = \{x, y, z, \ldots\}$ and let S_X denote the set of all permutations of X. Intuitively, a permutation of X is a rearrangement of shuffling of the elements of X; formally, the appropriate way to formulate this notion is to define a **permutation** as a mapping of $X \to X$ which is 1–1 and onto. Thus, we may write

$$S_X = \{\sigma \in \mathrm{Map}(X, X) \mid \sigma \text{ is 1–1 and onto}\}.$$

Let us define the product of two permutations to be their composition. In other words, for $\sigma, \tau \in S_X$ we take their product to be $\sigma \circ \tau$. According to the definition of composition (see 2-5-3), $\sigma \circ \tau \colon X \to X$ is defined by

$$(\sigma \circ \tau)(x) = \sigma(\tau(x)), \qquad \text{for all } x \in X.$$

Furthermore, as was shown in 2-5-3, $\sigma \circ \tau$ is surjective because both σ and τ are surjective, and $\sigma \circ \tau$ is injective because both σ and τ are injective. Therefore, $\sigma \circ \tau$ is a permutation of X—that is, $\sigma \circ \tau \in S_X$, and S_X is closed under the operation of composition, $\circ$.

It is now easy to see that $\{S_X, \circ\}$ is, in fact, a group. The associative law holds—for $\sigma, \tau, \rho \in S_X$ we have $\sigma \circ (\tau \circ \rho) = (\sigma \circ \tau) \circ \rho$ because for every $x \in X$, $[\sigma \circ (\tau \circ \rho)](x)$ and $[(\sigma \circ \tau) \circ \rho](x)$ both equal $\sigma(\tau(\rho(x)))$. The identity map $1 \colon X \to X$ defined by $1(x) = x$ for every $x \in X$ is an identity for S_X, because surely $1 \circ \sigma = \sigma \circ 1 = \sigma$ for every $\sigma \in S_X$. Finally, given $\sigma \in S_X$, because $\sigma \colon X \to X$ is 1–1 and onto there exists (according to 2-5-4) a mapping $\tau \colon X \to X$, called the "inverse map" of σ, such that $\sigma \circ \tau = 1$ and $\tau \circ \sigma = 1$, and in addition τ is 1–1 and onto. Thus, τ is an element of S_X and it is indeed an inverse of σ under the operation $\circ$. We have verified that $\{S_X, \circ\}$ is a group called the **group of permutations** (or **symmetries**) of the set X. Of course, we write σ^{-1} (instead of τ) for the inverse of σ; after all, this is the way we normally write inverses in a multiplicative group.

As a special case of the preceding, suppose the set X is finite—say, X has n elements. Since the structure of the group S_X is not affected by the nature of elements of X, there is nothing lost in taking

$$X = \{1, 2, \ldots, n\}.$$

Then we write S_n instead of S_X and refer to S_n as the **symmetric group on** n **letters.** An element $\sigma \in S_n$ is a permutation of the n integers $1, 2, \ldots, n$ and it is customary to describe σ completely (as was done in 4-1-9 and 4-1-10) by writing each of the n integers with its image under σ below it. For example, if $\sigma \in S_5$ is the mapping for which:

$$1 \to 4, \qquad 2 \to 1, \qquad 3 \to 3, \qquad 4 \to 5, \qquad 5 \to 2$$

then we write

$$\sigma = \begin{pmatrix} 1 & 2 & 3 & 4 & 5 \\ 4 & 1 & 3 & 5 & 2 \end{pmatrix}.$$

Obviously, there would be no harm in permuting the columns and writing σ as

$$\begin{pmatrix} 3 & 5 & 2 & 1 & 4 \\ 3 & 2 & 1 & 4 & 5 \end{pmatrix} \quad \text{or} \quad \begin{pmatrix} 2 & 5 & 3 & 1 & 4 \\ 1 & 2 & 3 & 4 & 5 \end{pmatrix}.$$

The identity element of S_5 is

$$1 = \begin{pmatrix} 1 & 2 & 3 & 4 & 5 \\ 1 & 2 & 3 & 4 & 5 \end{pmatrix}$$

and the inverse of σ is easily seen to be

$$\sigma^{-1} = \begin{pmatrix} 4 & 1 & 3 & 5 & 2 \\ 1 & 2 & 3 & 4 & 5 \end{pmatrix} = \begin{pmatrix} 1 & 2 & 3 & 4 & 5 \\ 2 & 5 & 3 & 1 & 4 \end{pmatrix}$$

—that is, one merely interchanges the two rows of σ. That this is the inverse depends on the fact that we already know how to multiply elements of S_5—for example, if

$$\tau = \begin{pmatrix} 1 & 2 & 3 & 4 & 5 \\ 3 & 5 & 4 & 2 & 1 \end{pmatrix},$$

then

$$\sigma \circ \tau = \begin{pmatrix} 1 & 2 & 3 & 4 & 5 \\ 3 & 2 & 5 & 1 & 4 \end{pmatrix}, \qquad \tau \circ \sigma = \begin{pmatrix} 1 & 2 & 3 & 4 & 5 \\ 2 & 3 & 4 & 1 & 5 \end{pmatrix},$$

$$\sigma^2 = \begin{pmatrix} 1 & 2 & 3 & 4 & 5 \\ 5 & 4 & 3 & 2 & 1 \end{pmatrix}, \qquad \tau^2 = \begin{pmatrix} 1 & 2 & 3 & 4 & 5 \\ 4 & 1 & 2 & 5 & 3 \end{pmatrix},$$

and, in particular,

$$\sigma \circ \sigma^{-1} = \sigma^{-1} \circ \sigma = \begin{pmatrix} 1 & 2 & 3 & 4 & 5 \\ 1 & 2 & 3 & 4 & 5 \end{pmatrix} = 1.$$

More generally, for arbitrary n, $\sigma \in S_n$ is written in the form

$$\sigma = \begin{pmatrix} 1 & 2 & 3 & \cdots & n \\ \sigma 1 & \sigma 2 & \sigma 3 & \cdots & \sigma n \end{pmatrix}.$$

The identity element of the group S_n is

$$1 = \begin{pmatrix} 1 & 2 & 3 & \cdots & n \\ 1 & 2 & 3 & \cdots & n \end{pmatrix}.$$

If, in addition, $\tau \in S_n$, then

$$\tau = \begin{pmatrix} 1 & 2 & 3 & \cdots & n \\ \tau 1 & \tau 2 & \tau 3 & \cdots & \tau n \end{pmatrix},$$

and we have

$$\sigma \circ \tau = \begin{pmatrix} 1 & 2 & 3 & \cdots & n \\ \sigma\tau 1 & \sigma\tau 2 & \sigma\tau 3 & \cdots & \sigma\tau n \end{pmatrix}, \qquad \tau \circ \sigma = \begin{pmatrix} 1 & 2 & 3 & \cdots & nn \\ \tau\sigma 1 & \tau\sigma 2 & \tau\sigma 3 & \cdots & \tau\sigma n \end{pmatrix}$$

As for the inverse, clearly

$$\sigma^{-1} = \begin{pmatrix} \sigma 1 & \sigma 2 & \sigma 3 & \cdots & \sigma n \\ 1 & 2 & 3 & \cdots & n \end{pmatrix}.$$

Note that in the expression for σ^{-1} the top row consists of all the elements $1, 2, \ldots, n$ except that they are not in the usual order—but the expression obviously denotes a mapping which is 1–1 and onto.

The number of elements in the group S_n is clearly $n!$; in fact, to determine a permutation σ there are n choices for $\sigma 1$, and then $n - 1$ choices for $\sigma 2$, and so on—so there are indeed $n!$ choices for $\sigma \in S_n$. In particular, S_3 has $3! = 6$ elements. Because the rigid motions of the equilateral triangle (see 4-1-9) can be viewed as permutations of the set $\{1, 2, 3\}$—that is, as elements of S_3—and there are six of them, it follows (because the operation for both rigid motions and elements of S_3 is the same: namely, composition of mappings) that the group of rigid motions of the equilateral triangle is S_3.

4-1-12 / *PROBLEMS*

1. In any group G, show
 (*i*) $ab = b \Rightarrow a = e$.
 (*ii*) $a^2 = a \Rightarrow a = e$.
 (*iii*) A left inverse of a given element is also a right inverse, and conversely.

2. Let elements a, b, c in the group G be given. Then the equation

$$axbcx = cabx$$

has a unique solution for x.

3. For elements a, b in the abelian group G, prove

$$(ab)^n = a^n b^n \qquad \text{for all } n \in \mathbf{Z}.$$

Of course, the group G does not have to be abelian—it is only necessary that $ab = ba$.

4. If $a^2 = e$ for all a in the group G, then G is abelian.

5. Suppose G is a group; show it is abelian $\Leftrightarrow$ $(ab)^2 = a^2 b^2$ for all $a, b \in G$.

6. In each case, decide which of the following properties of a group are satisfied: closure, associative law, identity, inverses, commutative law, cancellation laws. In particular, which of these are groups? semigroups?
 (*i*) $\{\mathbf{Z}, \circ\}$ where $a \circ b = 0$,
 (*ii*) $\{\mathbf{Z}, \circ\}$ where $a \circ b = a - b$,
 (*iii*) $\{\mathbf{Z}, \circ\}$ where $a \circ b = a + b + 1$,
 (*iv*) $\{\mathbf{Z}, \circ\}$ where $a \circ b = a$,

(*v*) $\{\mathbf{Z}, \circ\}$ where $a \circ b = a + b + ab$,
(*vi*) the set of all even integers under addition,
(*vii*) the set of all odd integers under addition,
(*viii*) the set of all even integers under multiplication,
(*ix*) the set of all odd integers under multiplication,
(*x*) the set of all positive rationals under addition,
(*xi*) the set of all positive rationals under multiplication,
(*xii*) the set of all rationals a satisfying $0 < a < 1$ under multiplication,
(*xiii*) $\{G, \circ\}$ where G is the set of all rationals with the single element -1 excluded, and the operation is $a \circ b = a + b + ab$,
(*xiv*) $\{\mathbf{Q}, \circ\}$ where $a \circ b = \max\{a, b\}$,
(*xv*) $\{\mathbf{R}, \circ\}$ where $a \circ b = (ab)/2$,
(*xvi*) $\{\mathbf{R}, \circ\}$ where $a \circ b = |a| + |b|$,
(*xvii*) $\{\mathbf{R}, \circ\}$ where $a \circ b = [a + b]$ (greatest integer function),
(*xviii*) $\{\mathbf{R}, \circ\}$ where $a \circ b = \sqrt{a^2 + b^2}$,
(*xix*) all irrational real numbers under addition,
(*xx*) all irrational real numbers under multiplication,
(*xxi*) $\{\mathbf{C}, \circ\}$ where $\alpha \circ \beta = |\alpha|\,|\beta|$,
(*xxii*) the elements 1, 5, 7, 11 of $\mathbf{Z}_{12}$ under multiplication,
(*xxiii*) the elements 0, 2, 5, 7 of $\mathbf{Z}_{10}$ under addition,
(*xxiv*) the elements 1, 3, 9 of $\mathbf{Z}_{10}$ under multiplication,
(*xxv*) The elements 3, 6, 9 of $\mathbf{Z}_{12}$ under addition,
(*xxvi*) $G = \{0, 1, 2, \ldots, m - 1\}$ and the operation $\circ$ is given by

$$a \circ b = a + b, \qquad \text{if } a + b < m,$$

$$a \circ b = r, \qquad \text{if } a + b \geq m \text{ and } r = (a + b) - m,$$

(*xxvii*) for a fixed prime p, $G = \{n/p^r \mid n \in \mathbf{Z}, r \geq 0\}$ and the operation is addition,
(*xxviii*) $G = \{a + b\sqrt{2} \mid a, b \in \mathbf{Q}, \text{ not both } 0\}$ under multiplication,
(*xxix*) $G = \{a + b\sqrt[3]{2} \mid a, b \in \mathbf{Z}\}$ under addition,
(*xxx*) $G = \{a + b\sqrt[3]{2} \mid a, b \in \mathbf{Q}, \text{ not both } 0\}$ under multiplication,
(*xxxi*) for an arbitrary nonempty set X, $G = \mathrm{Map}(X, X)$ and the operation is composition of mappings.
(*xxxii*) $G = \{(a, b) \in \mathbf{R} \times \mathbf{R} \mid \text{not both } 0\}$ and the operation $\circ$ is given by

$$(a, b) \circ (c, d) = (ac - bd, ad + bc),$$

(*xxxiii*) the nonzero elements of an integral domain D under multiplication,
(*xxxiv*) Take R as the underlying set of a ring $\{R, +, \cdot\}$ with operation $\circ$ given by $a \circ b = a + b + ab$,

7. (*i*) Do the four matrices

$$e = \begin{pmatrix} 1 & 0 \\ 0 & 1 \end{pmatrix}, \quad a = \begin{pmatrix} -1 & 0 \\ 0 & 1 \end{pmatrix}, \quad b = \begin{pmatrix} 1 & 0 \\ 0 & -1 \end{pmatrix}, \quad c = \begin{pmatrix} -1 & 0 \\ 0 & -1 \end{pmatrix}$$

in $\mathscr{M}(\mathbf{Z}, 2)$ constitute a group under multiplication? Make a table.

(*ii*) Do the same thing for the four matrices

$$e = \begin{pmatrix} 1 & 0 \\ 0 & 1 \end{pmatrix}, \quad a = \begin{pmatrix} 0 & 1 \\ 1 & 0 \end{pmatrix}, \quad b = \begin{pmatrix} 0 & -1 \\ -1 & 0 \end{pmatrix}, \quad c = \begin{pmatrix} -1 & 0 \\ 0 & -1 \end{pmatrix}$$

8. Derive the multiplication table for the rigid motions of the square, as given in 4-1-10. Verify that these rigid motions form a group. Compare S_4 (the symmetric group on four letters) and the group of the square.

9. How is the fact, 4-1-5, that a finite semigroup with cancellation is a group related to the fact, 3-2-11, that a finite integral domain is a field? How are the proofs related?

10. Why does the proof of 4-1-5 not go through when G is infinite?

11. By defining an operation appropriately, can you make the following sets into a group? If so make a table for each case where you have a group.

(*i*) $\{0, 1\}$, (*ii*) $\{-1, +1\}$, (*iii*) $\{a, b\}$,
(*iv*) $\{0, 1, 2\}$, (*v*) $\{0, +1, -1\}$, (*vi*) $\{a, b, c\}$.

12. On the set $G = \{e, a, b, c\}$ define an operation by the table

	e	a	b	c
e	e	a	b	c
a	a	e	c	b
b	b	b	a	e
c	c	c	e	a

Is this a group?

13. Show that if $G = \{e, a, b\}$ is a 3-element group with identity e, then its multiplication table must take the form:

	e	a	b
e	e	a	b
a	a	b	e
b	b	e	a

14. Show that if $G = \{e, a, b, c\}$ is a 4-element group with identity e, then (except for possible renaming of the elements) its multiplication table must take one of the two forms:

	e	a	b	c
e	e	a	b	c
a	a	b	c	e
b	b	c	e	a
c	c	e	a	b

	e	a	b	c
e	e	a	b	c
a	a	e	c	b
b	b	c	e	a
c	c	b	a	e

15. For the following elements of S_7

$$\sigma = \begin{pmatrix} 1 & 2 & 3 & 4 & 5 & 6 & 7 \\ 3 & 1 & 6 & 7 & 2 & 4 & 5 \end{pmatrix}, \qquad \tau = \begin{pmatrix} 1 & 2 & 3 & 4 & 5 & 6 & 7 \\ 6 & 5 & 2 & 1 & 4 & 7 & 3 \end{pmatrix},$$

$$\rho = \begin{pmatrix} 1 & 2 & 3 & 4 & 5 & 6 & 7 \\ 4 & 6 & 3 & 7 & 5 & 2 & 1 \end{pmatrix}$$

compute

(*i*) σ^7,	(*ii*) τ^7,	(*iii*) ρ^7,
(*iv*) σ^{-1}	(*v*) τ^{-1},	(*vi*) ρ^{-1},
(*vii*) $\sigma \circ \tau$,	(*viii*) $\tau \circ \sigma$,	(*ix*) $\rho \circ \sigma \circ \tau$,
(*x*) $\tau \circ \sigma \circ \tau^{-1}$	(*xi*) $\rho \circ \sigma \circ \rho^{-1}$,	(*xii*) $\rho^2 \circ \sigma \circ \rho^{-2}$,
(*xiii*) σ^{-7},	(*xiv*) τ^{-7},	(*xv*) ρ^{-7},
(*xvi*) $\sigma^3 \circ \rho^2 \circ \tau^{-2}$,	(*xvii*) $\tau^2 \circ \rho^2 \circ \tau^2 \circ \rho^2$,	(*xviii*) $\sigma \circ \tau \circ \sigma^{-1} \circ \tau^{-1}$.

16. Show that the rigid motions of the rectangle (which is not a square) (such as the one shown in the accompanying diagram) form a group.

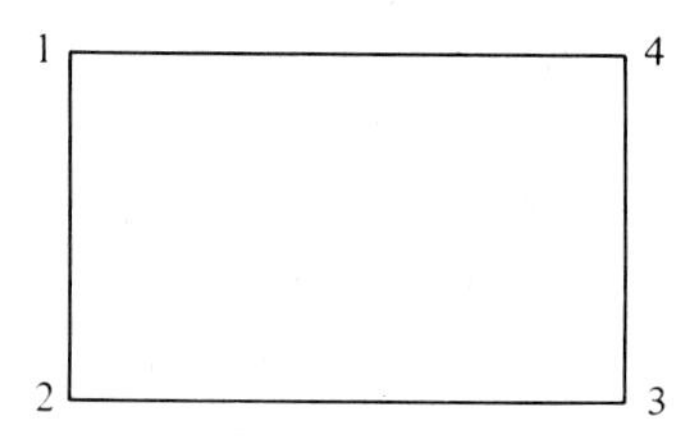

Make a table. How is this group (which is known as the **four group** or as **Klein's four group**) related to the group of the square and to S_4? What about the symmetries of this rectangle?

17. The regular pentagon (as the one shown in the accompanying figure) has 10 rigid motions.

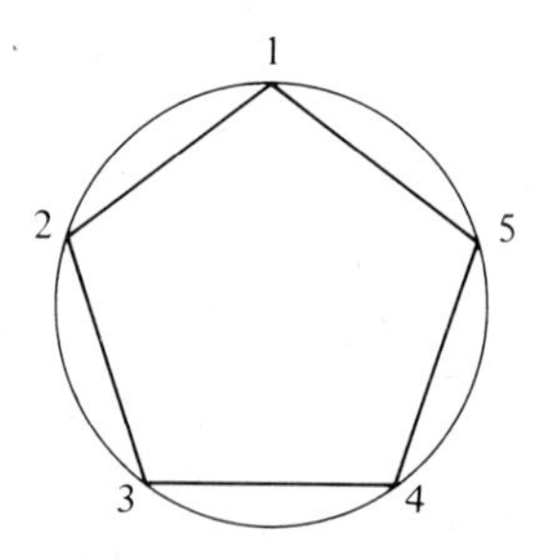

List them, and show that they form a group. If you have plenty of time and energy make a table. Compare this group with S_5. What about the symmetries of the regular pentagon?

18. Treat the 12 rigid motions of the regular hexagon in the manner of Problem 17.

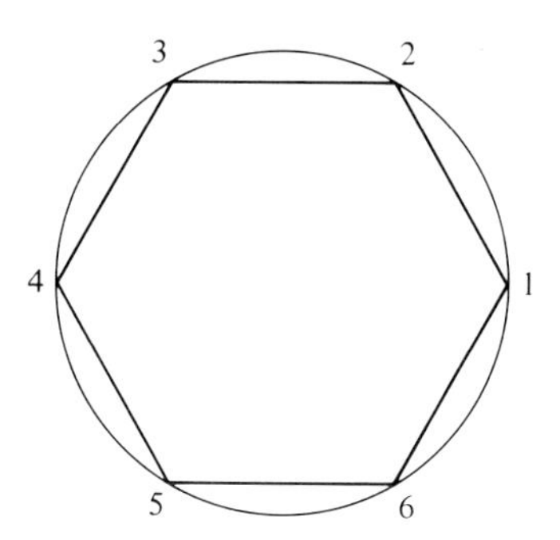

19. Give an example of two elements a, b in a group, with $(ab)^{-1} \neq a^{-1}b^{-1}$.

20. Prove: If a group has an even number of elements, then there exists an element not equal to e which is its own inverse.

21. Prove: If G is a group, then:
 (*i*) $\{a^{-1} \mid a \in G\} = G$—that is, the set of all inverses of elements of G is G itself.
 (*ii*) The mapping $\phi: G \to G$ which maps $a \to a^{-1}$ is 1–1 and onto. Moreover, ϕ is a homomorphism $\Leftrightarrow$ G is abelian.

22. If X is an arbitrary set, we know from 2-2-3 that $\mathscr{S}(X)$, the set of all subsets of X, can be made into a commutative ring with unity. What is the group of units of this ring?

23. Consider $\{\mathscr{S}(X), \circ\}$ where $\circ$ is given by
 (*i*) $A \circ B = A \cup B$,
 (*ii*) $A \circ B = A \cap B$,
 (*iii*) $A \circ B = A - B = \{x \in A \mid x \notin B\}$,
 (*iv*) $A \circ B = A + B = A \cup B - A \cap B$.
 In which cases do we have a group? a semigroup?

24. If X is a set with more than two elements, show that S_X, the group of permutations, is not abelian.

25. Consider the rings:
 (*i*) $\mathbf{Z}_5$, (*ii*) $\mathbf{Z} \oplus \mathbf{Z}$, (*iii*) $\mathbf{Z}_3 \oplus \mathbf{Z}_3$,
 (*iv*) $\mathbf{Z}_4 \oplus \mathbf{Z}_4$, (*v*) $\mathbf{Z}_3 \oplus \mathbf{Z}_4$, (*vi*) $\mathbf{Z}_3 \oplus \mathbf{Z}$,
 In each case, find the group of units and make a multiplication table for it.

26. If G_1 and G_2 are multiplicative groups, the product set $G_1 \times G_2 = \{(a_1, a_2) \mid a_1 \in G_1, a_2 \in G_2\}$ can be made into a group by defining multiplication componentwise—that is,

$$(a_1, a_2)(b_1, b_2) = (a_1 a_1, b_2 b_2).$$

The result is called the **direct product** $G_1 \times G_2$.

Generalize the direct product to any number of groups. When is the resulting group commutative?

27. The reals under addition, $\{\mathbf{R}, +\}$, are a group, and so are the positive reals under multiplication $\{\mathbf{R}_{>0}, \cdot\}$. Are these groups related?

28. Show, by example, that if $\{G, \cdot\}$ is a semigroup in which (*i*) there is a left identity e, and (*ii*) every element has a right inverse, then $\{G, \cdot\}$ need not be a group.

29. Prove that the following sets of complex numbers are groups under multiplication:

(*i*) $W_4 = \{1, i, -1, -i\}$; these are the four 4th roots of unity.

(*ii*) $W_3 = \left\{1, \dfrac{-1}{2} + \dfrac{\sqrt{3}}{2} i = \dfrac{-1 + \sqrt{-3}}{2}, \dfrac{-1}{2} - \dfrac{\sqrt{3}}{2} i = \dfrac{-1 - \sqrt{-3}}{2}\right\}$;

these are the three cube roots of unity.

(*iii*) $W_6 = \left\{1, \dfrac{1}{2} + \dfrac{\sqrt{3}}{2} i, \dfrac{-1}{2} + \dfrac{\sqrt{3}}{2} i, -1, \dfrac{-1}{2} - \dfrac{\sqrt{3}}{2} i, \dfrac{-1}{2} + \dfrac{\sqrt{3}}{2} i\right\}$

$= \left\{ \pm 1, \dfrac{\pm 1 \pm \sqrt{-3}}{2}\right\}$; these are the six 6th roots of unity.

(*iv*) $W_8 = \left\{ \pm 1, \pm i, \dfrac{1}{\sqrt{2}} (\pm 1 \pm i)\right\}$;

these are the eight 8th roots of unity.

(*v*) For any $n > 1$, $W_n = \{z \in \mathbf{C} \mid z^n = 1\}$; this is the set of all nth roots of unity.

30. (*i*) In $\mathcal{M}(\mathbf{R}, 2)$, the ring of all 2×2 matrices with entries from R, exhibit a concrete matrix of form

$$\begin{pmatrix} a & b \\ c & d \end{pmatrix} \qquad \text{with } a \neq 0, b \neq 0, c \neq 0, d \neq 0$$

which has an inverse. Find its inverse. Can you determine all matrices in $\mathcal{M}(\mathbf{R}, 2)$ which have an inverse? In other words, can you find the group of units of the ring $\mathcal{M}(\mathbf{R}, 2)$?

(*ii*) Do the same thing for the ring $\mathcal{M}(\mathbf{Z}, 2)$.

4-2. Subgroups and Cosets

4-2-1. Definition. A nonempty subset H of the group G is said to be a **subgroup** of G when, with respect to the operation induced from G, H becomes a group in its own right. In other words, H is a subgroup of $G = \{G, \cdot\}$ when $\{H, \cdot\}$ is itself a group.

The definition of a subgroup is quite natural; it is essentially parallel to the definition of subring of a ring as given in 2-2-6. More generally, for any kind of algebraic system, the notion of a "subsystem" would be defined in the same fashion.

To decide if a given subset of a group is indeed a subgroup we have the criteria given by our next result. These criteria are analogous to those for a subring, as discussed in 2-2-7.

4-2-2. Proposition. Suppose H is a nonempty subset of the group $G = \{G, \cdot\}$; then the following are equivalent:

(*i*) H is a subroup of G.
(*ii*) $a, b \in H \Rightarrow ab \in H$ and $a^{-1} \in H$; that is, H is closed under both multiplication and taking of inverses.
(*iii*) $a, b \in H \Rightarrow ab^{-1} \in H$.

Proof: $(i) \Rightarrow (ii)$: Suppose H is a subgroup of G, then, obviously, $a, b \in H \Rightarrow ab \in H$; however, because a^{-1} denotes the inverse of a as an element of the group G, it is not immediate that $a \in H \Rightarrow a^{-1} \in H$.

To see this, it surely suffices to prove that for any subgroup H of G we have:

(1) Its identity (denote it temporarily by e') is the same as the identity e of G.

(2) For $a \in H$, its inverse in the group H (denote it temporarily by a') is the same as its inverse a^{-1} in the group G.

Now, from

$$e'e' = e' = ee'$$

we obtain by cancellation in G, $e' = e$—so (1) holds. Then

$$aa' = e' = e = aa^{-1}$$

so by cancellation in G, $a' = a^{-1}$—so (2) holds. This completes the proof that $(i) \Rightarrow (ii)$.

$(ii) \Rightarrow (iii)$: By hypothesis, H is closed under multiplication and taking of inverses, Therefore,

$$a, b \in H \Rightarrow a, b^{-1} \in H \Rightarrow ab^{-1} \in H$$

so (iii) holds.

$(iii) \Rightarrow (i)$: For any two elements $a, b \in H$ we know, by hypothesis, that $ab^{-1} \in H$. Apply this first to the special case where the elements a and b of H are equal; the conclusion is, $aa^{-1} \in H$—that is, the identity e belongs to H, or to put it another way, H has an identity. Next apply it to $e \in H$ and any $a \in H$; the conclusion is, $ea^{-1} \in H$—that is, the inverse of an element of H belongs to H, or to put it another way, every element of H has an inverse in H. Furthermore, for $a, b \in H$ we now have $a, b^{-1} \in H$ and consequently $a(b^{-1})^{-1} \in H$; thus $ab \in H$, and we conclude that H is closed under multiplication. Since the associative law is automatic for elements of H, we have proved that H is a group—hence, a subgroup of G. ▌

The preceding result is stated for a multiplicative group. When the group operation is addition, $+$, the criteria that H be a subgroup take the equivalent forms

$$a, b \in H \Rightarrow a + b \in H \text{ and } -a \in H,$$

or

$$a, b \in H \Rightarrow a - b \in H; \text{ that is, } H \text{ is closed under subtraction.}$$

In a rather special situation (which will occur frequently in our discussion) it is possible to give an even simpler criterion for a subgroup—namely,

4-2-3. Proposition If the finite subset H of the group $G = \{G, \cdot\}$ is closed under the operation, then H is a subgroup of G.

Proof: Since H is closed under the operation, $H = \{H, \cdot\}$ is clearly a semigroup. Moreover, the cancellation laws hold in G, so they surely hold in H. Thus, H is a finite (in particular, finite implies nonempty) semigroup that satisfies both cancellation laws. Hence, according to 4-1-5, $\{H, \cdot\}$ is a group—that is, H is a subgroup of G.

There is also something to be gained by giving a more pedestrian proof of this result. Since H is closed under the operation, it suffices to show $a \in H \Rightarrow a^{-1} \in H$. We do this in a concrete fashion which indicates how one can actually compute a^{-1}.

Suppose $a \in H$. If $a = e$, then obviously $a^{-1} \in H$, and we are finished. So suppose $a \neq e$. Now, consider the sequence of powers of a

$$a, a^2, a^3, a^4, \ldots. \qquad (*)$$

Because H is closed under multiplication all these elements belong to H. But H has only a finite number of elements, so not all these powers of a can be distinct. Thus, there exist distinct positive integers i and j (say $i < j$) for which $a^i = a^j$. (We have just used an obvious but extremely important principle known as the **pigeon-hole principle**. It may be stated formally as follows: If more than n objects are placed in, or distributed among, n boxes—or pigeon holes—then there exists a box that contains two, or more, objects.) Then, multiplying, in G, by a^{-i} we obtain

$$a^{j-i} = e, \qquad j - i \geq 1.$$

Consequently, e appears in the sequence $(*)$ of powers of a. Let m be the smallest positive power of a for which

$$a^m = e.$$

Since $a \neq e$, we know that $m > 1$; thus $m - 1 \geq 1$ and $a^{m-1} \in H$. From

$$a \cdot a^{m-1} = a^m = e$$

we conclude that $a^{-1} = a^{m-1} \in H$—so H is a subgroup.

(In particular, one way to locate the inverse of a is to compute the sequence of powers of a; upon arriving at the first power of a which equals e, one takes the preceding power of a.) ▮

If G is any group, then G itself is a subgroup and so is (e) (the set, or subgroup, consisting of the identity alone). These subgroups provide no additional information about the group, and they are said to be **trivial**. We shall now turn to a number of examples of nontrivial subgroups.

4-2-4. Example. Consider $\{\mathbf{Z}, +\}$, the group of integers under addition. The set $2\mathbf{Z}$ of all even integers is obviously a subgroup. More generally, for any $m \geq 0$, $m\mathbf{Z}$ is a subgroup of the additive group of integers, since it is closed under subtraction—that is, if $ma_1, ma_2 \in m\mathbf{Z}$, then $ma_1 - ma_2 = m(a_1 - a_2) \in m\mathbf{Z}$.

Furthermore, we can show that any subgroup of $\{\mathbf{Z}, +\}$ is of form $m\mathbf{Z}$. To do this, consider an arbitrary subgroup H. If $H = (0)$, then for $m = 0$ we have $H = m\mathbf{Z}$; so suppose $H \neq (0)$. There exists an element $a \neq 0$ in H. If a is negative, then its inverse $-a$ is positive and belongs to H. Thus H contains a positive integer. Let m be the smallest positive integer belonging to H. Because H is a subgroup, $2m = m + m$ belongs to H, and so do $3m = 2m + m$, $4m = 3m + m, \ldots$; of course, the inverses $-m, -2m, -3m, -4m, \ldots$ also belong to H. In particular,

$$m\mathbf{Z} \subset H, \qquad m \geq 1.$$

If $m = 1$, then $m\mathbf{Z} = \mathbf{Z}$, and we have $H = \mathbf{Z} = 1\mathbf{Z}$. Suppose, therefore, that $m > 1$, and let a be an arbitrary element of H. Applying the division algorithm to a and m, there exist integers q and r for which

$$a = qm + r, \qquad 0 \leq r < m.$$

Now, $qm = mq \in m\mathbf{Z} \subset H$, so because H is a group

$$r = a - qm \in H.$$

Because m is the smallest positive integer belonging to H, it follows that $r = 0$. Thus, $a = qm$—so any element a of H belongs to $m\mathbf{Z}$ and we have $H \subset m\mathbf{Z}$. The conclusion is

$$H = m\mathbf{Z}.$$

We have proved: The set of all subgroups of $\{\mathbf{Z}, +\}$ is given by $\{m\mathbf{Z} \mid m \geq 0\}$.

Incidentally, it may be noted that the subgroups $m\mathbf{Z}$ are all distinct—in other words, for m_1, m_2, both greater than or equal to 0, we have

$$m_1\mathbf{Z} = m_2\mathbf{Z} \Leftrightarrow m_1 = m_2.$$

The details are left to the reader.

4-2-5. Example. Consider $\{\mathbf{Z}_{12}, +\}$, the additive group of the ring $\mathbf{Z}_{12}$. Is the set $S = \{0, 3, 6, 9\}$ a subgroup? According to 4-2-3, because S is finite, it suffices to check if S is closed under addition. This is a straightforward matter; the work may be organized into an addition table for the elements of S. It takes the form:

+	0	3	6	9
0	0	3	6	9
3	3	6	9	0
6	6	9	0	3
9	9	0	3	6

All the entries in the table are from S—in other words, S is closed under addition, and S is indeed a subgroup.

In the same way, the sets $\{0, 6\}$, $\{0, 4, 8\}$, and $\{0, 2, 4, 6, 8, 10\}$ are subgroups of $\{\mathbf{Z}_{12}, +\}$.

On the other hand, the set $T = \{0, 3, 5, 8, 11\}$ is not a subgroup because it is not closed under addition. In particular, not all entries of the addition table for T belong to T—for example, $5 + 8 = 1$, which is not in T.

As a matter of fact, it is not hard to see that except for the trivial subgroups (0) and $\mathbf{Z}_{12}$, we have already listed all the subgroups of $\{\mathbf{Z}_{12}, +\}$. (One way to do this is to imitate the method used in 4-2-4 to determine all subgroups of the additive group of integers.) More generally, in the next section we shall learn, among other things, how to find all subgroups of $\{\mathbf{Z}_m, +\}$ for any $m > 1$.

We may also consider $\{\mathbf{Z}_{12}^*, \cdot\}$, the group of units of the ring $\mathbf{Z}_{12}$. This is a group whose underlying set consists of the four elements 1, 5, 7, 11 of $\mathbf{Z}_{12}$. It is clear that $\{1, 5\}$ is a subgroup, as are $\{1, 7\}$ and $\{1, 11\}$—and, except for the trivial subgroups (1) and $\mathbf{Z}_{12}^*$, there are no others.

In the same spirit, consider $\{\mathbf{Z}_{13}, +\}$, the additive group of the ring $\mathbf{Z}_{13}$. Is the subset $S = \{0, 3, 6, 9, 12\}$ a subgroup? Because S is finite, only closure needs to be checked. Since $6 + 9 = 2$ in $\mathbf{Z}_{13}$, we see that closure fails for S, so it is not a subgroup. As a matter of fact, the group $\{\mathbf{Z}_{13}, +\}$ has no subgroups except the trivial ones. The reader may convince himself of this fact directly, but should he prefer to wait then he will see it fall out as a consequence of a general result later in this section (see 4-2-17).

We can also study $\{\mathbf{Z}_{13}^*, \cdot\}$ the group of units of the ring $\mathbf{Z}_{13}$. This is a group whose underlying set may be taken as the twelve elements 1, 2, 3, 4, ..., 10, 11, 12 of $\mathbf{Z}_{13}$. Is the subset $\{1, 5, 8, 12\}$ a subgroup? Constructing the multiplication table for this set, we have

$\cdot$	1	5	8	12
1	1	5	8	12
5	5	12	1	8
8	8	1	12	5
12	12	8	5	1

so this finite set is closed under multiplication and is, therefore, a subgroup of $\{\mathbf{Z}_{13}^*, \cdot\}$.

It is straightforward to check that among the nontrivial subgroups of $\{\mathbf{Z}_{13}^*, \cdot\}$ we also have

$$\{1, 12\}, \qquad \{1, 3, 9\}, \qquad \{1, 3, 4, 9, 10, 12\}.$$

In fact, it is not hard to see that we have already listed all the nontrivial subgroups of $\{\mathbf{Z}_{13}^*, \cdot\}$. This fact will also appear as an immediate consequence of general considerations in the next section. More generally, we shall study the structure of the group $\{\mathbf{Z}_m^*, \cdot\}$, for any $m > 1$, in Section 4-6.

4-2-6. Example. Consider the symmetric group on three letters, S_3, which may also be viewed as the group of rigid motions (or symmetries) of the equilateral triangle—see 4-1-9 and 4-1-11. The six elements of S_3 may be denoted as

$$e = \begin{pmatrix} 1 & 2 & 3 \\ 1 & 2 & 3 \end{pmatrix}, \qquad \sigma_1 = \begin{pmatrix} 1 & 2 & 3 \\ 2 & 3 & 1 \end{pmatrix}, \qquad \sigma_2 = \begin{pmatrix} 1 & 2 & 3 \\ 3 & 1 & 2 \end{pmatrix},$$

$$\tau = \begin{pmatrix} 1 & 2 & 3 \\ 1 & 3 & 2 \end{pmatrix}, \qquad \tau_2 = \begin{pmatrix} 1 & 2 & 3 \\ 3 & 2 & 1 \end{pmatrix}, \qquad \tau_3 = \begin{pmatrix} 1 & 2 & 3 \\ 2 & 1 & 3 \end{pmatrix}.$$

Naturally, the multiplication table for this group is the one already given in 4-1-9.

Obviously, the following are subgroups of $G = S_3$:

$$\{e\}, \quad H_1 = \{e, \tau_1\}, \quad H_2 = \{e, \tau_2\}, \quad H_3 = \{e, \tau_3\}, \quad A = \{e, \sigma_1, \sigma_2\}, \quad S_3.$$

Furthermore, there are no other subgroups; this may be seen in the most elementary manner. In more detail, suppose H is a subgroup. If $H \neq (e)$, then H contains one (or more) of the elements $\tau_1, \tau_2, \tau_3, \sigma_1, \sigma_2$. Taking powers of this element, H surely contains one of the subgroups H_1, H_2, H_3, A. If H is not one of these subgroups, choose an element of H not in the subgroup, then by repeatedly taking powers of this element and multiplying these with elements of the subgroup, it follows that $H = S_3$. Thus, there are no additional subgroups.

Incidentally, each of the nontrivial subgroups has a geometric interpretation in terms of rigid motions of the equilateral triangle. In particular, the subgroup A consists precisely of the three rotations about the axis perpendicular to the plane of the equilateral triangle and passing through its "center"; while, for each $i = 1, 2, 3$, the subgroup H_i consists of all rigid motions which leave the vertex i fixed.

Let us consider next the octic group (otherwise known as the group of the square) which was described in 4-1-10. It consists of the eight rigid motions of the square, (shown in the accompanying figure),

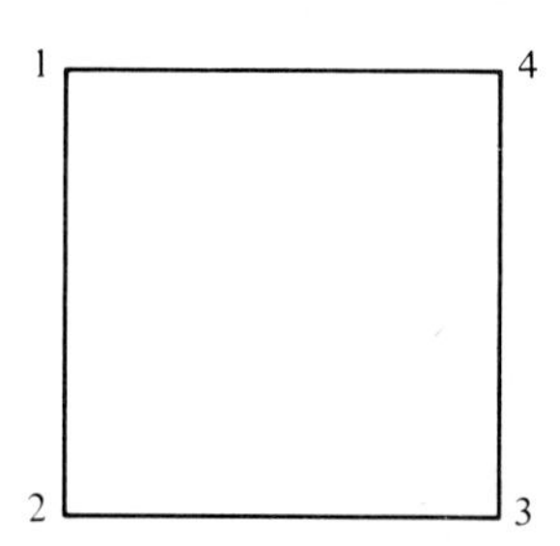

and its elements are denoted by

$$e = \begin{pmatrix} 1 & 2 & 3 & 4 \\ 1 & 2 & 3 & 4 \end{pmatrix}, \quad \sigma_1 = \begin{pmatrix} 1 & 2 & 3 & 4 \\ 2 & 3 & 4 & 1 \end{pmatrix}, \quad \sigma_2 = \begin{pmatrix} 1 & 2 & 3 & 4 \\ 3 & 4 & 1 & 2 \end{pmatrix},$$

$$\sigma_3 = \begin{pmatrix} 1 & 2 & 3 & 4 \\ 4 & 1 & 2 & 3 \end{pmatrix}, \quad \tau_1 = \begin{pmatrix} 1 & 2 & 3 & 4 \\ 4 & 3 & 2 & 1 \end{pmatrix}, \quad \tau_2 = \begin{pmatrix} 1 & 2 & 3 & 4 \\ 2 & 1 & 4 & 3 \end{pmatrix},$$

$$\rho_1 = \begin{pmatrix} 1 & 2 & 3 & 4 \\ 1 & 4 & 3 & 2 \end{pmatrix}, \quad \rho_2 = \begin{pmatrix} 1 & 2 & 3 & 4 \\ 3 & 2 & 1 & 4 \end{pmatrix}.$$

Obviously, the octic group is a subgroup of S_4.

From the multiplication table for the octic group (see 4-1-10) it is straightforward to check that each of the following finite subsets of the octic group is closed under multiplication:

$$\{e, \sigma_2\}, \quad \{e, \tau_1\}, \quad \{e, \tau_2\}, \quad \{e, \rho_1\}, \quad \{e, \rho_2\},$$
$$\{e, \sigma_1, \sigma_2, \sigma_3\}, \quad \{e, \rho_1, \rho_2, \sigma_2\}, \quad \{e, \tau_1, \tau_2, \pi_2\}$$

so it is a subgroup of the octic group. We shall leave it for the reader to investigate the geometric interpretation of these subgroups and to decide if there are any other nontrivial subgroups of the octic group.

4-2-7. Example. Let G be an arbitrary multiplicative group and consider the subset

$$\mathfrak{Z} = \{a \in G \mid ax = xa \text{ for all } x \in G\}.$$

In words, $\mathfrak{Z}$ consists of those elements of G which "commute" with every element of G. We show $\mathfrak{Z}$, which is known as the **center** of G, is a subgroup.

First of all, $\mathfrak{Z} \neq \emptyset$—in fact, $e \in \mathfrak{Z}$ since $ex = x = xe$ for all $x \in G$. It remains to verify that $\mathfrak{Z}$ is closed under multiplication and taking of inverses—but these are easy. If $a, b \in \mathfrak{Z}$, then $ax = xa$ and $bx = xb$ for all $x \in G$ and, therefore,

$$(ab)x = a(bx) = a(xb) = (ax)b = (xa)b = x(ab)$$

for all $x \in G$—so, $ab \in \mathfrak{Z}$. Furthermore, if $a \in \mathfrak{Z}$, then for every $x \in G$ we have

$$\begin{aligned} ax = xa &\Rightarrow a^{-1}(ax) = a^{-1}(xa) \\ &\Rightarrow x = a^{-1}xa \\ &\Rightarrow xa^{-1} = (a^{-1}xa)a^{-1} \\ &\Rightarrow xa^{-1} = (a^{-1}xa)a^{-1} \\ &\Rightarrow xa^{-1} = a^{-1}x \end{aligned}$$

which says that $a^{-1} \in \mathfrak{Z}$. Thus, $\mathfrak{Z}$ is indeed a subgroup.

Obviously, if the group G is commutative, then its center $\mathfrak{Z}$ equals G, and conversely—that is,

$$G \text{ is abelian} \Leftrightarrow \mathfrak{Z} = G.$$

4-2-8. Example. As noted in 4-1-8,

$$W = \{z \in \mathbf{C} \mid |z| = 1\}$$

is an abelian group under multiplication. It is known as the "circle group" because its elements are all complex numbers which lie on the "unit circle." Obviously, W is a subgroup of $\mathbf{C}^*$, the multiplicative group of units (or nonzero elements) of the field $\mathbf{C}$.

Now, for each positive integer n, consider the set

$$W_n = \{z \in \mathbf{C} \mid z^n = 1\}$$

—in words, W_n consists of all complex numbers whose nth power is 1, so we refer to W_n as the set of "nth roots of unity." The set W_n is nonempty, since the element 1 surely belongs to W_n. Even more, the set W_n is finite; in fact, for $z \in \mathbf{C}$

$$z \in W_n \Leftrightarrow z^n - 1 = 0$$
$$\Leftrightarrow z \text{ is a root of the polynomial } f(x) = x^n - 1 \in \mathbf{C}[x]$$

—but, according to 3-5-5, a polynomial of degree n over a field can have at most n distinct roots in the field—so $\#(W_n) \leq n$.

Of still greater interest is the fact that W_n is a subgroup of W. In the first place, $W_n \subset W$; in fact,

$$z \in W_n \Rightarrow z^n = 1 \Rightarrow |z^n| = |1|$$
$$\Rightarrow |z|^n = 1$$

and, since $|z|$ is a nonnegative real number, this implies $z \in W$. Then, because W_n is a nonempty finite subset of W, to prove it is a subgroup of W it suffices to verify that W_n is closed under multiplication. But this is trivial:

$$z_1, z_2 \in W_n \Rightarrow z_1{}^n = 1,\ z_2{}^n = 1$$
$$\Rightarrow (z_1 z_2)^n = 1$$
$$\Rightarrow z_1 z_2 \in W_n.$$

In particular, we have shown that W_n is a group with at most n elements.

The precise number of elements in W_n (that is, the number of nth roots of unity) is not of critical importance for us, but we digress to show the reader with some trigonometric experience why this number is n.

Consider an arbitrary point (that is, complex number) on the unit circle W. Drawing the radius (whose length is 1) from the origin to this point, and with the angle θ as indicated in the accompanying diagram, the coordinates of the point (in the X–Y plane) are clearly $(\cos\theta, \sin\theta)$. (This is, essentially, the definition of $\sin\theta$ and $\cos\theta$ for an arbitrary angle θ.)

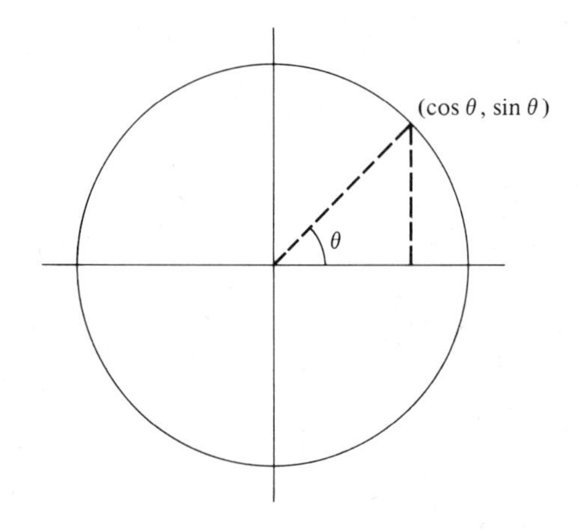

In other words, we are dealing with the complex number

$$\cos\theta + i\sin\theta.$$

Conversely, any complex number of form $\cos\theta + i\sin\theta$ is on the unit circle (because its distance from the origin is $\sqrt{\cos^2\theta + \sin^2\theta} = 1$). If we have two elements $z_1, z_2 \in W$ and write them in the form

$$z_1 = \cos\theta_1 + i\sin\theta_1, \qquad z_2 = \cos\theta_2 + i\sin\theta_2,$$

then, in virtue of standard trigonometric identities, the computation of the product of these complex numbers yields

$$z_1z_2 = \cos(\theta_1 + \theta_2) + i\sin(\theta_1 + \theta_2) \tag{*}$$

In other words, multiplication of two elements of the group W amounts to adding their angles—as illustrated in the accompanying picture. Of course,

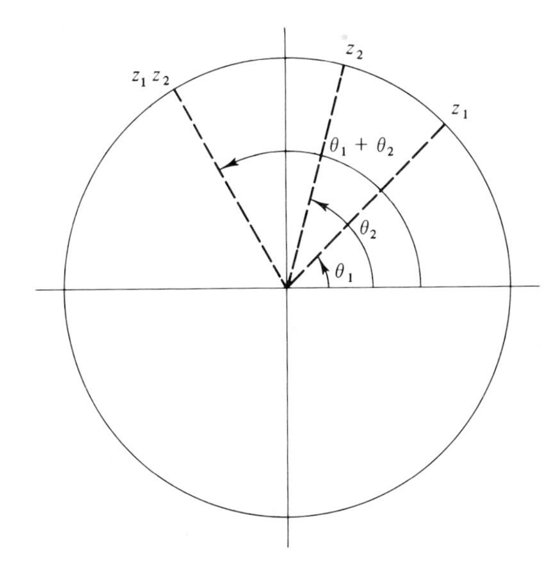

the identity is $1 = 1 + 0i = \cos 0 + i\sin 0$; that is, the identity is the complex number associated with the angle $\theta = 0$. Also, for an arbitrary element $z = \cos\theta + i\sin\theta$ of W its inverse z^{-1}, is clearly

$$\cos(-\theta) + i\sin(-\theta)$$

since

$$(\cos\theta + i\sin\theta)(\cos(-\theta) + i\sin(-\theta)) = \cos(\theta - \theta) + i\sin(\theta - \theta) = 1.$$

Now, let us fix an integer $n \geq 1$ and analyze the group W_n of nth roots of unity. We consider the angle $\alpha = 360/n$, or in radian measure $\alpha = 2\pi/n$, and put

$$\zeta = \cos\alpha + i\sin\alpha.$$

Of course, $\zeta \in W$. According to (∗), we have

$$\zeta^2 = \cos(2\alpha) + i\sin(2\alpha)$$

and inductively, for any $r \geq 1$,

$$\zeta^r = \cos(r\alpha) + i\sin(r\alpha).$$

In addition, $\zeta^0 = 1 = \cos(0\alpha) + i\sin(0\alpha)$. Obviously, each of the complex numbers

$$1 = \zeta^0, \zeta, \zeta^2, \ldots, \zeta^{n-1}$$

lies on the unit circle. Since

$$\begin{aligned} \zeta^n &= \cos(n\alpha) + i\sin(n\alpha) \\ &= \cos(2\pi) + i\sin(2\pi) \\ &= 1, \end{aligned}$$

ζ is an nth root of unity; hence, any power of ζ is also an nth root of unity and, in particular, $1, \zeta, \ldots, \zeta^{n-1}$ all belong to W_n. Moreover, these n complex numbers are distinct—for they obviously represent the n equally spaced points on the unit circle, going counterclockwise and starting from $(1, 0)$.

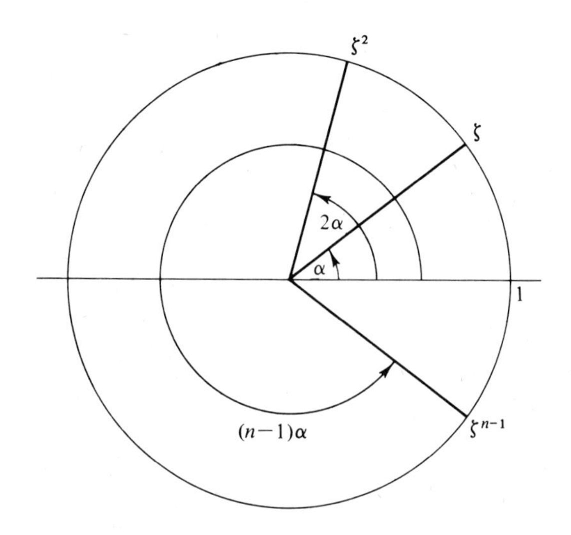

Thus, the group W_n has n elements and all of them are powers of the element $\zeta = \cos\alpha + i \sin\alpha$, where $\alpha = 2\pi/n$—namely,

$$W_n = \{1 = \zeta^0, \zeta, \zeta^2, \ldots, \zeta^{n-1}\} = \{\zeta, \zeta^2, \ldots, \zeta^{n-1}, \zeta^n = 1\}.$$

4-2-9. Proposition. The intersection of two subgroups of a given group is a subgroup. More generally, if $\{H_\alpha\}$ is any collection of subgroups of the group G, then their intersection, $\bigcap_\alpha H_\alpha$, is a subgroup of G.

Proof: For convenience, let us write $H = \bigcap_\alpha H_\alpha$. Because the identity e of G belongs to every subgroup of G, and, in particular, to every H_α it follows that $e \in H$—so H is nonempty. To show H is a subgroup, we verify the condition given in 4-2-2, part (*iii*); namely,

$$\begin{aligned} a, b \in H &\Rightarrow a, b \in H_\alpha \text{ for every } \alpha \\ &\Rightarrow ab^{-1} \in H_\alpha \text{ for every } \alpha \text{ (because } H_\alpha \text{ is a subgroup)} \\ &\Rightarrow ab^{-1} \in \bigcap_\alpha H_\alpha = H. \quad \blacksquare \end{aligned}$$

4-2-10. Example. Consider the group S_X, introduced in 4-1-11, of all permutations of a given set X. Fix an element y of X, and let

$$H_y = \{\sigma \in S_X \mid \sigma y = y\}.$$

In words, H_y is the set of all permutations of X (that is, mappings of $X \to X$ which are 1–1 and onto) which map the element y to itself (that is, which keep y fixed). Is H_α a subgroup of S_X?

In the first place, H_y is nonempty—the identity element 1 of S_X (which is the identity map $1: X \to X$) obviously belongs to H_y. In addition, H_y is closed under the operation; for if $\sigma, \tau \in H_y$, then

$$(\sigma \circ \tau)y = \sigma(\tau(y)) = \sigma y = y$$

so $\sigma \circ \tau \in H_y$. Finally, H_y is closed under taking of inverses; for if $\sigma \in H_y$, then $\sigma y = y$, and applying σ^{-1} gives $\sigma^{-1}(\sigma y) = \sigma^{-1}(y)$, so $\sigma^{-1} y = y$ and $\sigma^{-1} \in H_y$. Thus, H_y is a subgroup of S_X.

More generally, suppose $Y = \{y_1, y_2, \ldots, y_n\}$ is a finite subset of X, and let H_y denote the set of all permutations $\sigma \in S_X$ which leave Y pointwise fixed (that is, $\sigma y_i = y_i$ for $i = 1, 2, \ldots, n$). Then H_y is a subgroup of S_X, as the reader can easily verify in a straightforward manner. Another, more interesting, way to see this is to consider the intersection of subgroups

$$H_{y_1} \cap H_{y_2} \cap \cdots \cap H_{y_n}.$$

According to 4-2-9, this intersection is a subgroup of S_X. Moreover, it is immediate from the definitions (as the reader may convince himself) that

$$H_{y_1} \cap H_{y_2} \cap \cdots \cap H_{y_n} = H_Y.$$

Obviously, the same situation applies when Y is an arbitrary subset of X.

Having studied several examples of subgroups, our next objective is to see how a group can be decomposed with respect to a subgroup.

4-2-11. Definition. Suppose H is a subgroup of the group G. For any $a \in G$, let us write

$$aH = \{ah \mid h \in H\}$$

(in words, aH is the subset of G consisting of all elements of form ah, as h runs over H) and refer to aH as the **left coset** of H in G determined by a.

Similarly, we may write

$$Ha = \{ha \mid h \in H\}$$

and refer to it as the **right coset** of H in G determined by a.

No one will be surprised at the assertion that there is a complete parallelism between results about left cosets and results about right cosets. Because of this, we shall deal almost exclusively with left cosets and refer to them (somewhat carelessly) simply as **cosets**. It will usually be left to the reader to convince himself that each statement made about cosets (meaning: left cosets) has an analog for right cosets.

The definition of a coset was stated for the situation where the group operation is multiplication. When the group operation is addition, a coset should obviously be written in the form

$$a + H = \{a + h \mid h \in H\}.$$

4-2-12. Examples. (1) Consider the group $G = S_3 = \{e, \sigma_1, \sigma_2, \tau_1, \tau_2, \tau_3\}$ (with notation as in 4-1-9, 4-1-11, 4-2-6) and the subgroup $H_1 = \{e, \tau_1\}$. Each element of G determines a (left) coset of H_1 in G. Making use of the multiplication table for S_3, and dropping the $\circ$ symbol for multiplication, the cosets are seen to be

$$\begin{aligned}
eH_1 &= \{ee, e\tau_1\} = \{e, \tau_1\}, \\
\sigma_1 H_1 &= \{\sigma_1 e, \sigma_1\tau_1\} = \{\sigma_1, \tau_3\}, \\
\sigma_2 H_1 &= \{\sigma_2 e, \sigma_2\tau_1\} = \{\sigma_2, \tau_2\}, \\
\tau_1 H_1 &= \{\tau_1 e, \tau_1\tau_1\} = \{\tau_1, e\}, \\
\tau_2 H_1 &= \{\tau_2 e, \tau_2\tau_1\} = \{\tau_2, \sigma_2\}, \\
\tau_3 H_1 &= \{\tau_3 e, \tau_3\tau_1\} = \{\tau_3, \sigma_1\}.
\end{aligned}$$

Note that each coset contains two elements (because the subgroup H_1 has two elements) and that some of the cosets are equal (as sets)—namely,

$$eH_1 = \tau_1 H_1 = H_1, \qquad \sigma_1 H_1 = \tau_3 H_1, \qquad \sigma_2 H_1 = \tau_2 H_1$$

(The order in which the elements of a set are listed does not matter.)

The right cosets of H_1 in G are as follows.

$$\begin{aligned} H_1 e &= \{ee, \tau_1 e\} = \{e, \tau_1\}, \\ H_1\sigma_1 &= \{e\sigma_1, \tau_1\sigma_1\} = \{\sigma_1, \tau_2\}, \\ H_1\sigma_2 &= \{e\sigma_2, \tau_1\sigma_2\} = \{\sigma_2, \tau_3\}, \\ H_1\tau_1 &= \{e\tau_2, \tau_1\tau_1\} = \{\tau_1, e\}, \\ H_1\tau_2 &= \{e\tau_2, \tau_1\tau_2\} = \{\tau_2, \sigma_1\}, \\ H_1\tau_3 &= \{e\tau_3, \tau_1\tau_3\} = \{\tau_3, \sigma_2\}, \end{aligned}$$

and we have

$$H_1 e = H_1\tau_1 = H_1, \qquad H_1\sigma_1 = H_1\tau_2, \qquad H_1\sigma_2 = H_1\tau_3.$$

Note that the right cosets are not the same as the left cosets.

If we take the subgroup $H_2 = \{e, \tau_2\}$, then its left cosets in $G = S_3$ are

$$\begin{aligned} eH_2 &= \{e, \tau_2\}, \qquad & \sigma_1 H_2 &= \{\sigma_1, \tau_1\}, \qquad & \sigma_2 H_2 &= \{\sigma_2, \tau_3\}, \\ \tau_1 H_2 &= \{\tau_1, \sigma_1\}, \qquad & \tau_2 H_2 &= \{\tau_2, e\}, \qquad & \tau_3 H_2 &= \{\tau_3, \sigma_3\}, \end{aligned}$$

and its right cosets in $G = S_3$ are

$$\begin{aligned} H_2 e &= \{e, \tau_2\}, \qquad & H_2\sigma_1 &= \{\sigma_1, \tau_3\}, \qquad & H_2\sigma_2 &= \{\sigma_2, \tau_1\}, \\ H_2\tau_1 &= \{\tau_1, \sigma_2\}, \qquad & H_2\tau_2 &= \{\tau_2, e\}, \qquad & H_2\tau_3 &= \{\tau_3, \sigma_1\}. \end{aligned}$$

Here too the left and right cosets are not the same.

Finally, let us consider the subgroup $A = \{e, \sigma_1, \sigma_2\}$ of $G = S_3$. The left cosets of A are

$$\begin{aligned} eA &= \{e, \sigma_1, \sigma_2\}, \qquad & \sigma_1 A &= \{\sigma_1, \sigma_2, e\}, \qquad & \sigma_2 A &= \{\sigma_2, e, \sigma_1\}, \\ \tau_1 A &= \{\tau_1, \tau_2, \tau_3\}, \qquad & \tau_2 A &= \{\tau_2, \tau_3, \tau_1\}, \qquad & \tau_3 A &= \{\tau_3, \tau_1, \tau_2\}, \end{aligned}$$

and the right cosets of A are

$$\begin{aligned} Ae &= \{e, \sigma_1, \sigma_2\}, \qquad & A\sigma_1 &= \{\sigma_1, \sigma_2, e\}, \qquad & A\sigma_2 &= \{\sigma_2, e, \sigma_1\}, \\ A\tau_1 &= \{\tau_1, \tau_3, \tau_2\}. \qquad & A\tau_1 &= \{\tau_2, \tau_1, \tau_3\}, \qquad & A\tau_3 &= \{\tau_3, \tau_2, \tau_1\}. \end{aligned}$$

In this situation, the left and right cosets are the same—in the sense that every left coset is a right coset and conversely; in fact,

$$\begin{aligned} eA = \sigma_1 A = \sigma_2 A = Ae = A\sigma_1 = A\sigma_2 = A, \\ \tau_1 A = \tau_2 A = \tau_3 A = A\tau_1 = A\tau_2 = A\tau_3. \end{aligned}$$

(2) Consider the multiplicative group $G = \mathbf{Z}_{13}^* = \{1, 2, \ldots, 11, 12\}$ and the subgroup $H = \{1, 5, 8, 12\}$ which was mentioned in 4-2-5. Because G is commutative, there is no distinction between left and right cosets. Some typical cosets of H in G are

$$8H = \{8 \cdot 1, 8 \cdot 5, 8 \cdot 8, 8 \cdot 12\} = \{8, 1, 12, 5\},$$
$$7H = \{7 \cdot 1, 7 \cdot 5, 7 \cdot 8, 7 \cdot 12\} = \{7, 9, 4, 6\},$$
$$3H = \{3 \cdot 1, 3 \cdot 5, 3 \cdot 8, 3 \cdot 12\} = \{3, 2, 11, 10\}.$$

(3) Consider $G = \mathbf{Z}$, the group of integers under addition, and the subgroup $H = 7\mathbf{Z}$. Because the operation is addition, the coset of H in G determined by the element 5, say, takes the form

$$5 + 7\mathbf{Z} = \{5 + 7t \mid t \in \mathbf{Z}\}.$$

But, lo and behold, this coset is precisely what we have heretofore referred to as the residue (or congruence) class $\lfloor 5 \rfloor_7$. Obviously, then, for any $a \in \mathbf{Z}$, the coset it determines is

$$a + 7\mathbf{Z} = \{a + 7t \mid t \in \mathbf{Z}\} = \lfloor a \rfloor_7 .$$

More generally, as seen in 4-2-4, an arbitrary nontrivial subgroup of $\mathbf{Z}$ is of form $m\mathbf{Z}$ for some integer $m > 1$. Then, for any $a \in \mathbf{Z}$, the coset of $m\mathbf{Z}$ in $\mathbf{Z}$ which it determines is

$$a + m\mathbf{Z} = \{a + mt \mid t \in \mathbf{Z}\} = \lfloor a \rfloor_m .$$

This shows, in particular, that residue classes (mod m) are really very special cases of cosets—namely, a residue class (mod m) is a coset when we are dealing with the additive group $\mathbf{Z}$ and the subgroup $m\mathbf{Z}$.

In virtue of this, it is natural to inquire if the properties of residue classes (mod m) (see 1-7-10 and 1-7-12) have analogs for cosets in general. They do—in fact, we have

4-2-13. Properties. Suppose H is a subgroup of the group G; then for any $a, b \in G$ we have:

(1) H itself is a coset (of H in G).

Proof: Since $eH = \{eh \mid h \in H\} = \{h \mid h \in H\} = H$, the coset determined by e is H itself. ▮

(2) $a \in aH$. In words, every element of G belongs to the coset which it determines; in particular, every element of G belongs to some coset, and the cosets cover G.

Proof: Since $e \in H$, we have $a = ae \in aH$. ▮

(3) $b \in aH \Leftrightarrow a \in bH$. In words, b belongs to the coset determined by a if and only if a belongs to the coset determined by b.

Proof: Because the roles of a and b are interchangeable, it suffices to prove the implication $\Rightarrow$. If $b \in aH$, then $b = ah$ for some $h \in H$. Since H is a subgroup, we have $h^{-1} \in H$—so $a = bh^{-1} \in bH$. ▌

(4) $b \in aH \Rightarrow bH = aH$. In other words, if an element belongs to a coset, then it determines that coset.

Proof: First, let us show: $b \in aH \Rightarrow bH \subset aH$. In fact, given $b \in aH$, it is of form $b = ah'$ for some $h' \in H$. Then any element of bH is of form $bh = ah'h$ with $h \in H$. But $h'h \in H$ because H is a group—so $bh = ah'h \in aH$, and indeed $bh \subset aH$.

In virtue of (3) above, and the foregoing (with the roles of a and b interchanged) we have

$$b \in aH \Rightarrow a \in bH \Rightarrow aH \subset bH.$$

Combining this with the foregoing gives $b \in aH \Rightarrow bH = aH$. ▌

(5) $aH \cap bH \neq \varnothing \Rightarrow aH = bH$. In other words, if two cosets meet they are the same coset—so two cosets are disjoint sets or they are identical.

Proof: If $c \in aH \cap bH$, then $cH = aH$ and $cH = bH$. ▌

An immediate consequence of these properties of cosets is the following:

4-2-14. Proposition. If H is any subgroup of the group G, then G decomposes as a union of disjoint cosets of H in G. In particular, an element of G belongs to exactly one coset.

Proof: There is nothing to prove; we merely point out that if the cosets aH and bH are equal, $aH = bH$, then they are counted as a single coset—this permits us to talk about the union of "disjoint cosets." ▌

4-2-15. Remark. The preceding discussion of cosets can be understood better in terms of the general framework of an equivalence relation (see 3-4-3). As a point of departure, we observe a simple criterion for the elements $a, b \in G$ to belong to the same coset (that is, for aH to equal bH)—namely, by making use of 4-2-13, part (3),

$$aH = bH \Leftrightarrow a \in bH \Leftrightarrow b^{-1}a \in H.$$

Now, let us define the relation $a \equiv b \pmod{H}$ (which one reads in the obvious way) on G by

$$a \equiv b \pmod{H} \Leftrightarrow b^{-1}a \in H.$$

This is an equivalence relation. In fact: $a \equiv a \,(\text{mod } H)$, since $a^{-1}a = e \in H$; $a \equiv b \,(\text{mod } H) \Rightarrow b \equiv a \,(\text{mod } H)$; since $b^{-1}a \in H \Rightarrow a^{-1}b = (b^{-1}a)^{-1} \in H$; $a \equiv b \,(\text{mod } H)$ and $b \equiv c \,(\text{mod } H)$ imply $a \equiv c \,(\text{mod } H)$, since $b^{-1}a \in H$ and $c^{-1}b \in H$ imply $c^{-1}a = (c^{-1}b)(b^{-1}a) \in H$.

According to 3-4-3, the equivalence classes with respect to this equivalence relation provide a disjoint decomposition of G. Denoting the equivalence class to which a belongs by $\lfloor a \rfloor_H$, we recall that it is defined by

$$\lfloor a \rfloor_H = \{b \in G \mid b \equiv a \,(\text{mod } H)\}.$$

What is this subset $\lfloor a \rfloor_H$ of G? Since

$$\begin{aligned} b \in \lfloor a \rfloor_H &\Leftrightarrow b \equiv a \,(\text{mod } H) \\ &\Leftrightarrow a^{-1}b \in H \\ &\Leftrightarrow b \in aH, \end{aligned}$$

we see that $\lfloor a \rfloor_H = aH$. Thus, cosets are equivalence classes, and the coset decomposition of G with respect to the subgroup H is really the decomposition of G into equivalence classes under the equivalence relation $a \equiv b \,(\text{mod } H)$.

4-2-16. Theorem. (Lagrange). Suppose H is a subgroup of the group G. Denote the number of elements (which may be ∞) of G by $\#(G)$, and call it the **order** of G. Denote the number of left cosets (which may be ∞) of H in G by $(G : H)$, and call it the **index** of H in G.

If the group G is finite, then

$$\#(G) = (G : H)\, \#(H).$$

In particular, if G is finite the order of any subgroup divides the order of G.

Proof: We observe that any left coset, aH, of H in G has the same number of elements as H. In fact, the mapping $h \to ah$ of

$$H \to aH$$

is onto (by definition of aH) and 1–1 (since, by cancellation, $ah_1 = ah_2$ implies $h_1 = h_2$); so $\#(aH) = \#(H)$.

Therefore, G is a union of disjoint cosets (there are $(G : H)$ of them), each of which consists of $\#(H)$ elements—so $\#(G) = (G : H)\, \#(H)$.

We have proved: For any subgroup H of the finite group G, the order of G equals the order of H times the index of H in G. ▌

4-2-17. Corollary. If the group G has a prime number of elements, then it has no subgroups except the trivial ones.

Proof: Suppose $\#(G) = p$. If H is a subgroup of G, then $\#(H)$ divides $\#(G)$, so $\#(H)$ is 1 or p. Consequently, the only possibilities are $H = (e)$ or $H = G$. ▮

This tells us, for example, that the group $\{\mathbf{Z}_{13}, +\}$ (which has 13 elements) has no nontrivial subgroups—an assertion which was made in 4-2-5.

4-2-18 / *PROBLEMS*

1. Suppose H is a subgroup of G and F is a nonempty subset of H; show that F is a subgroup of $H \Leftrightarrow F$ is a subgroup of G.

2. In the additive group of integers, $\{\mathbf{Z}, +\}$, which of the following subsets are subgroups:
(*i*) $6\mathbf{Z} \cap 12\mathbf{Z}$, (*ii*) $6\mathbf{Z} \cup 12\mathbf{Z}$,
(*iii*) $6\mathbf{Z} \cap 10\mathbf{Z}$, (*iv*) $6\mathbf{Z} \cup 10\mathbf{Z}$,
(*v*) all integers divisible by 3 or 5,
(*vi*) all integers divisible by 3 and 5.
For those that are subgroups, express them in the form $m\mathbf{Z}$. (According to 4-2-4, we know that every subgroup of $\{\mathbf{Z}, +\}$ is of form $m\mathbf{Z}$.)

3. Verify that the following are subgroups of $\{\mathbf{Z}_{15}, +\}$:
(*i*) $\{0, 5, 10\}$, (*ii*) $\{0, 3, 6, 9, 12\}$.
Find other subgroups, if you can.

4. The group $\{\mathbf{Z}_{15}^*, \cdot\}$ consists of the eight elements 1, 2, 4, 7, 8, 11, 13, 14. Which of the following subsets are subgroups?
(*i*) $\{1, 2, 4, 8\}$, (*ii*) $\{1, 2, 4, 8, 11, 13\}$,
(*iii*) $\{1, 4, 7, 13\}$, (*iv*) $\{1, 7, 14\}$.
Find additional subgroups, if you can.

5. (*i*) Show that if $\{R, +, \cdot\}$ is a ring and S is a subring, then S is a subgroup of $\{R, +\}$, the additive group of R.
(*ii*) On the other hand, a subgroup S of $\{R, +\}$ need not be a subring. Prove this by exhibiting an example.
(*iii*) Give an example of a ring $\{R, +, \cdot\}$ for which every subgroup of $\{R, +$ is$\}$ also a subring.

6. Show that in 4-2-3 the hypothesis of finiteness for the subset H cannot be dropped. More precisely, give an example of an infinite subset H of a group G such that H is closed under the operation but is not a subgroup.

7. Find as many subgroups of $\{\mathbf{Z}_m, +\}$ as you can for each of the following values of m:
(*i*) 5, (*ii*) 6, (*iii*) 7,
(*iv*) 10, (*v*) 14, (*vi*) 18.
Have you found them all?

8. Do Problem 7 for the groups $\{\mathbf{Z}_m^*, \cdot\}$.

9. If m is a nonnegative integer, then

$$m\mathbf{Z} = (0) \Leftrightarrow m = 0 \quad \text{and} \quad m\mathbf{Z} = \mathbf{Z} \Leftrightarrow m = 1.$$

More generally, if m and n are nonnegative integers, then

$$m\mathbf{Z} \subset n\mathbf{Z} \Leftrightarrow n \mid m$$

(we permit $0 \mid 0$) and consequently

$$m\mathbf{Z} = n\mathbf{Z} \Leftrightarrow m = n.$$

Thus, the subgroups $m\mathbf{Z}$ ($m \geq 0$) are all distinct.

10. The group S_4 has 24 elements; find a subgroup with the following number of elements
(*i*) 2, (*ii*) 3, (*iii*) 4, (*iv*) 6, (*v*) 8, (*vi*) 12.

11. Prove: If H is a subgroup of G, then

$$H = \{x \in G \mid xH \subset H\}.$$

12. Suppose $H_1 \subset H_2 \subset H_3 \subset \cdots$ is an ascending sequence (which may be finite of infinite) of subgroups of a given group G. If we put

$$H = \bigcup_i H_i,$$

then H is a subgroup of G.

13. In a group G, the union of two subgroups is a subgroup if and only if one of them is contained in the other.

14. Suppose we are given an arbitrary set X and an abelian group $\{G, +\}$. Show how Map(X, G) can be made into an abelian group. What happens if the operation in G is multiplication? and if G is not abelian?

15. For an arbitrary element a in the group G, define a mapping $L_a: G \to G$ by putting

$$L_a(x) = ax, \qquad \text{for all } x \in G.$$

Then the map L_a, which is known as "left multiplication by a," is 1–1 and onto; in other words, L_a is a permutation of the set G—$L_a \in S_G$. In the same way, "right multiplication by a" is also an element of S_G.

16. For each element a in the group $\{\mathbf{Z}_7^*, \cdot\}$, find its inverse; express it as a power of a, if you can. Do the same thing for the multiplicative group:
(*i*) $\mathbf{Z}_8^*$, (*ii*) $\mathbf{Z}_{10}^*$, (*iii*) $\mathbf{Z}_{11}^*$, (*iv*) $\mathbf{Z}_{12}^*$,
(*v*) $\mathbf{Z}_{13}^*$, (*vi*) $\mathbf{Z}_{14}^*$, (*vii*) $\mathbf{Z}_{15}^*$, (*viii*) $\mathbf{Z}_{17}^*$,

17. For which elements of $\{\mathbf{Z}_{13}^*, \cdot\}$ can you find a nontrivial subgroup containing them?

18. (*i*) Show that in the group of the square (see 4-1-10) the element σ_2 belongs to the center, and τ_2 does not. Find the center.
(*ii*) The center of S_3 is (e).
(*iii*) Find the center of S_4. What about the center of S_X for an arbitrary set X?

19. Consider an arbitrary set X and a nonempty subset Y. Is $\{\sigma \in S_X \mid \sigma Y = Y\}$ a subgroup of S_X? Is $\{\sigma \in S_X \mid \sigma Y \subset Y\}$ a subgroup of S_X?

20. (*i*) Prove: For W_n, the group of complex nth roots of unity (see 4-2-8), we have (with $m \geq 1$, $n \geq 1$)

$$W_m \subset W_n \Leftrightarrow n \mid m.$$

(*ii*) Show that the set

$$W_\infty = \bigcup_{n=1}^{\infty} W_n$$

is a subgroup of the unit circle W. It is known as the group of all roots of unity.

21. List all left and right cosets for the subgroup $H_3 = \{e, \tau_3\}$ of S_3. [Notation as in 4-2-12, part (1)].

22. List all cosets of H in $\{\mathbf{Z}_{12}, +\}$ when H is the subgroup:
(*i*) $\{0, 4, 8\}$, (*ii*) $\{0, 3, 6, 9\}$, (*iii*) $\{0, 2, 4, 6, 8, 10\}$.

23. List all cosets of H in $\{\mathbf{Z}_{13}^*, \cdot\}$ when H is the subgroup:
(*i*) $\{1, 12\}$, (*ii*) $\{1, 3, 9\}$, (*iii*) $\{1, 3, 4, 9, 10, 12\}$.

24. Let G be the group of the square, as discussed in 4-2-6. Find all cosets (left and right) of H in G when H is the subgroup
(*i*) $\{e, \sigma_2\}$, (*ii*) $\{e, \tau_2\}$, (*iii*) $\{e, \rho_1\}$,
(*iv*) $\{e, \rho_1, \rho_2, \sigma_2\}$, (*v*) $\{e, \tau_1, \tau_2, \sigma_2\}$.

25. Consider the subgroup H_y of S_X (notation as in 4-2-10). For $\rho \in S_X$, characterize the left coset to which ρ belongs; more precisely, show that

$$\rho H_y = \{\tau \in S_X \mid \tau y = \rho y\}.$$

Can you say anything about the right coset $H_y \rho$?

26. Let Y be any subset of X and consider the subgroup

$$H_Y = \bigcap_{y \in Y} H_y$$

of S_X. Characterize the left coset to which $\rho \in S_X$ belongs.

27. Suppose H is a subgroup of G; show that

(*i*) the analogs of the properties of left cosets, which were listed in 4-2-13, hold for the right cosets of H in G,

(*ii*) G decomposes into a disjoint union of right cosets of H in G,

(*iii*) Let us define a relation on G by

$$a \approxeq b \pmod{H} \Leftrightarrow ba^{-1} \in H$$

(or equivalently $\Leftrightarrow ab^{-1} \in H$). Then this is an equivalence relation on G and its equivalence classes are the right cosets of H in G.

28. Compare the meanings of $a \equiv b \pmod{H}$ and $a \approxeq b \pmod{H}$ when the group G is additive. Do so, in particular, when G is the group $\{\mathbf{Z}, +\}$ and H is the subgroup $m\mathbf{Z}$.

29. Suppose H is a subgroup of G. If G is finite, then any right coset has the same number of elements as any left coset, and the number of right cosets is equal to the number of left cosets. More generally, for arbitrary G, finite or infinite.

(*i*) there is a 1–1 correspondence between the elements of any two left (or right) cosets.

(*ii*) there is a 1–1 correspondence between the elements of any left coset and any right coset,

(*iii*) there is a 1–1 correspondence between the set of left cosets and the set of right cosets. (Note: $aH \leftrightarrow Ha$ will not do.)

30. What happens to Lagrange's theorem, 4-2-16, when G is infinite—that is, when $\#(G) = \infty$?

31. Use Lagrange's theorem to help verify that

(*i*) The subgroups of $\{\mathbf{Z}_{12}, +\}$ listed in 4-2-5 are indeed all subgroups,

(*ii*) the subgroups of $\{\mathbf{Z}_{13}^*, \cdot\}$ listed in 4-2-5 are indeed all subgroups,

(*iii*) the subgroups of the octic group listed in 4-2-6 are indeed all subgroups. Is the four group (see 4-1-12, Problem 16) one of these subgroups?

32. Compare the groups $\{\mathbf{Z}_{12}, +\}$ and $\{\mathbf{Z}_{13}^*, \cdot\}$.

33. Use Lagrange's theorem to help find all subgroups of the group of rigid motions of the

(*i*) regular pentagon (see 4-1-12, Problem 17),

(*ii*) regular hexagon (see 4-1-12, Problem 18).

It is instructive to interpret the subgroups geometrically.

4-3. Cyclic Groups

Consider an element a of the multiplicative group G, and let H be any subgroup of G which contains a. Since $a \in H$, we have $a^2 \in H$ and, inductively, $a^n \in H$ for every $n \geq 1$. Furthermore, $a^0 = e \in H$, and because a subgroup is closed under taking of inverses, $a^{-n} = (a^n)^{-1} \in H$ for every $n \geq 1$. Thus, all powers of a (positive, negative, or zero) belong to H—in symbols: $\{a^n \mid n \in \mathbf{Z}\} \subset H$.

Let us introduce the notation

$$[a] = \{a^n \mid n \in \mathbf{Z}\}.$$

Then $[a]$ (call it "bracket of a" momentarily) is a subgroup of G—in fact, by the rules for exponents, given a^i and a^j in $[a]$ (any two elements of $[a]$ are of this form), then $a^i \cdot a^j = a^{i+j} \in [a]$ and $a^{-i} \in [a]$. We see immediately that:

4-3-1. Proposition. For any $a \in G$, $[a] = \{a^n \mid n \in \mathbf{Z}\}$ is a subgroup of G. Moreover,

(*i*) $[a]$ is contained in any subgroup of G that contains a,
(*ii*) $[a]$ is the intersection of all subgroups of G that contain a.

Proof: (*i*) has already been done. In virtue of (*i*) and the fact that $[a]$ is a subgroup of G which contains a, it follows that (*ii*) holds. ▌

One sometimes summarizes this result by saying: $[a]$ is the unique smallest subgroup of G that contains a.

If the operation in the group G is addition, then instead of powers of a we are concerned with multiples of a—namely,

$$a, \quad a + a = 2a, \quad a + 2a = 3a, \ldots, \; -a = (-1)a, \quad -(2a) = (-2)a, \ldots.$$

In other words, in this situation we have

$$[a] = \{na \mid n \in \mathbf{Z}\}.$$

Our discussion will deal almost exclusively with multiplicative groups. The reader will have no difficulty translating the results to additive groups.

4-3-2. Definition. If there exists an element a in the group G for which $G = [a]$, then G is said to be a **cyclic group** and a is said to be a **generator** of G. We also say that G is a cyclic group **generated by** a.

More generally, for any $a \in G$, $[a]$ is known as the **cyclic group—or subgroup—generated by** a (because, when we view the subgroup $[a]$ of G as a group in its own right, it is clearly cyclic and a is a generator).

4-3-3. Examples. (1) Consider the additive group of integers, $\{\mathbf{Z}, +\}$. What is [1]? Because we are in an additive situation, we know that

$$[1] = \{n \cdot 1 \mid n \in \mathbf{Z}\}.$$

In simple-minded fashion we obtain this by adding 1 to itself over and over (including zero times) and taking inverses (that is, negatives). In any case, we obviously have

$$[1] = \mathbf{Z},$$

so $\{\mathbf{Z}, +\}$ is a cyclic group, and 1 is a generator. If we compute $[-1]$ in the same way, we see that $[-1] = \mathbf{Z}$—so -1 is also a generator of $\{\mathbf{Z}, +\}$.

What about [2]? It consists of 2, 4, 6, 8, ..., and also 0, -2, -4, -6, -8, ... More precisely, we know that

$$[2] = \{n \cdot 2 \mid n \in \mathbf{Z}\} = 2\mathbf{Z}$$

and also

$$[-2] = \{(n)(-2) \mid n \in \mathbf{Z}\} = (-2)\mathbf{Z} = 2\mathbf{Z}.$$

More generally, for any $m \in \mathbf{Z}$

$$[m] = \{n \cdot m \mid n \in \mathbf{Z}\} = m\mathbf{Z}$$

and also

$$[-m] = [m].$$

Because $m\mathbf{Z} = \mathbf{Z} \Leftrightarrow m = \pm 1$ this tells us, in particular, that $+1$ and -1 are the only generators of $\{\mathbf{Z}, +\}$. In addition, this result tells us that every subgroup of the cyclic group $\{\mathbf{Z}, +\}$ is cyclic—in fact, as seen in 4-2-4, any subgroup is of form $m\mathbf{Z}$ with $m \geq 0$, and from above, $m\mathbf{Z} = [m]$, which is a cyclic group generated by m.

(2) Consider the additive group of reals $\{\mathbf{R}, +\}$. Let α be any nonzero element of $\mathbf{R}$. Then the cyclic subgroup generated by α is

$$[\alpha] = \{n\alpha \mid n \in \mathbf{Z}\}.$$

This cannot equal $\mathbf{R}$; for example, there is no integer n for which $n\alpha = \frac{1}{2}\alpha$, so the real number $\frac{1}{2}\alpha$ does not belong to $[\alpha]$. Therefore, $\{\mathbf{R}, +\}$ is an example of an infinite group which is not cyclic.

(3) Consider the additive group $\{\mathbf{Z}_6, +\}$, whose elements are 0, 1, 2, 3, 4, 5. What is [1]? By repeated addition of 1 to itself, we see that 1, 2, 3, 4, 5, $6 = 0$, ... all belong to [1]. Continuing the process of adding 1 beyond $6 = 0$, we have 1, 2, 3, 4, 5, $6 = 0$, 1, 2, ...; so the elements keep repeating in a "cyclical" fashion, and we have no new elements of [1]. Similarly, taking inverses gives $-5 = 1$, $-4 = 2$, $-3 = 3$, $-2 = 4$, $-1 = 5$, $-0 = 0$ as elements of [1]. In short, $[1] = \mathbf{Z}_6$; so $\{\mathbf{Z}_6, +\}$ is a cyclic group with generator 1.

Computing (in the same way) the cyclic subgroups of $\{\mathbf{Z}_6, +\}$ generated by each of the remaining elements we have

$$[0] = (0), \qquad [1] = \mathbf{Z}_6, \qquad [2] = \{0, 2, 4\},$$
$$[3] = \{0, 3\}, \qquad [4] = \{0, 2, 4\}, \qquad [5] = \mathbf{Z}_6.$$

In particular, 5 (which equals -1) is also a generator of $\mathbf{Z}_6$.

Next, let us consider the additive group $\{\mathbf{Z}_7, +\}$, whose elements are 0, 1, 2, 3, 4, 5, 6. Since [1] contains $1, 2, 3, 4, 5, 6, 7 = 0, \ldots,$ we have $[1] = \mathbf{Z}_7$—so $\{\mathbf{Z}_7, +\}$ is a cyclic group, and 1 is a generator. Similarly, [2] contains $2, 4, 6, 8 = 1, 3, 5, 7 = 0, \ldots$ which implies $[2] = \mathbf{Z}_7$. Even more, as is easily checked,

$$\mathbf{Z}_7 = [1] = [2] = [3] = [4] = [5] = [6],$$

so every nonzero element of $\mathbf{Z}_7$ is a generator of the cyclic group.

Pursuing this type of example to its logical conclusion: For any $m \geq 1$, the additive group $\{\mathbf{Z}_m, +\}$ (whose elements may be taken as $0, 1, 2, \ldots, m-1$) is cyclic since $[1] = \mathbf{Z}_m$. In particular, for any $m \geq 1$ there exists a cyclic group of order m—$\{\mathbf{Z}_m, +\}$ is such a group. The question of which elements of $\mathbf{Z}_m$ are generators will be settled later in this section; for the moment it is left as a challenge to the reader.

(4) Consider the group of rigid motions of the equilateral triangle—$S_3 = \{e, \sigma_1, \sigma_2, \tau_1, \tau_2, \tau_3\}$—which was introduced in 4-1-9. This is a finite group which is not cyclic. To see this, it suffices to check that for each $\sigma \in S_3$ we have $[\sigma] \neq S_3$.

Clearly, $[e] = (e)$. As for $[\sigma_1]$: it surely contains

$$\sigma_1, \sigma_1^2 = \sigma_2, \sigma_1^3 = \sigma_1\sigma_2 = e, \sigma_1^4 = \sigma_1, \sigma^5 = \sigma_2, \ldots$$

and, also

$$\sigma_1^0 = e, \quad \sigma_1^{-1} = \sigma_2, \quad \sigma_1^{-2} = \sigma_2^{-1} = \sigma_1, \quad \sigma_1^{-3} = e, \quad \sigma_1^{-4} = \sigma_2, \ldots.$$

Thus because of the way the terms repeat, we have

$$[\sigma_1] = \{\sigma_1^n \mid n \in \mathbf{Z}\} = \{e, \sigma_1, \sigma_2\}.$$

In similar fashion:

$$[\sigma_2] = \{e, \sigma_1, \sigma_2\} \qquad [\tau_1] = \{e, \tau_1\},$$
$$[\tau_2] = \{e, \tau_2\}, \qquad [\tau_3] = \{e, \tau_3\}$$

—so indeed S_3 is not cyclic.

(5) Consider $\{\mathbf{Z}_7^*, \cdot\}$; this is a multiplicative group consisting of the six elements 1, 2, 3, 4, 5, 6. Let us find

$$[2] = \{2^n \mid n \in \mathbf{Z}\}.$$

We have, in $\mathbf{Z}_7^*$,

$$2^1 = 2, \quad 2^2 = 4, \quad 2^3 = 1, \quad 2^4 = 2, \quad 2^5 = 4, \quad 2^6 = 1, \ldots$$

and using $2^{-1} = 2^2 = 4$ (which follows from $2^1 \cdot 2^2 = 2^3 = 1$) we have also

$$2^0 = 1, \quad 2^{-1} = 4, \quad 2^{-2} = (2^{-1})^2 = 2, \quad 2^{-3} = (2^{-1})^3 = 1, \quad 2^{-4} = 4, \ldots .$$

Consequently,

$$[2] = \{1, 2, 4\}.$$

In the same way, we find

$$[1] = \{1\}, \quad [3] = \{1, 2, 3, 4, 5, 6\} = \mathbf{Z}_7^*, \quad [4] = \{1, 2, 4\},$$
$$[5] = \{1, 2, 3, 4, 5, 6\} = \mathbf{Z}_7^*, \qquad [6] = \{1, 6\}.$$

In particular, $\mathbf{Z}_7^*$ is a cyclic group—it has two generators, 3 and 5.

The group $\{\mathbf{Z}_{17}^*, \cdot\}$, which consists of the 16 elements $1, 2, \ldots, 15, 16$, is also cyclic. The cyclic group generated by 2 is easily seen to be

$$[2] = \{1, 2, 4, 8, 9, 13, 15, 16\},$$

so 2 is not a generator of $\mathbf{Z}_{17}^*$. However, 3 is a generator, since the powers of 3 include

$$3, \quad 3^2 = 9, \quad 3^3 = 3 \cdot 9 = 10, \quad 3 = 3^4 \cdot 10 = 13, \quad 3^5 = 3 \cdot 13 = 5,$$
$$3^6 = 3 \cdot 5 = 15, \quad 3^7 = 3 \cdot 15 = 11, \quad 3^8 = 3 \cdot 11 = 16, \quad 3^9 = 3 \cdot 16 = 14,$$
$$3^{10} = 3 \cdot 14 = 8, \quad 3^{11} = 3 \cdot 8 = 7, \quad 3^{12} = 3 \cdot 7 = 4, \quad 3^{13} = 3 \cdot 4 = 12,$$
$$3^{14} = 3 \cdot 12 = 2, \quad 3^{15} = 3 \cdot 2 = 6, \quad 3^{16} = 3 \cdot 6 = 1, \quad 3^{17} = 3, \ldots .$$

The reader who does not mind doing arithmetic can check that 3, 5, 6, 7, 10, 11, 12, 14 are generators of $\mathbf{Z}_{17}^*$—in other words,

$$\mathbf{Z}_{17}^* = [3] = [5] = [6] = [7] = [10] = [11] = [12] = [14]$$

—whereas the remaining elements of $\mathbf{Z}_{17}^*$ are not generators.

More generally, later in this section we shall see that $\{\mathbf{Z}_p^*, \cdot\}$ is cyclic for any prime p, and also obtain some information about the generators of such a cyclic group.

On the other hand, $\{\mathbf{Z}_m^*, \cdot\}$ is not cyclic for all m [for example, $\mathbf{Z}_{12}^* = \{1, 5, 7, 11\}$ is not cyclic, since $[1] = \{1\}, [5] = \{1, 5\}, [7] = \{1, 7\}, [11] = \{1, 11\}$] and $\{\mathbf{Z}_m^*, \cdot\}$ may be cyclic even if m is not prime (for example, $\mathbf{Z}_{10}^* = \{1, 3, 7, 9\}$ is cyclic with generators 3 and 7). The full story as to when $\{\mathbf{Z}_m^*, \cdot\}$ is a cyclic group will unfold in Section 4-6.

(6) Consider the group of nth roots of unity, W_n, where n is any positive integer (see 4-2-8). This is a cyclic group of order n; in fact, as seen in 4-2-8, the element $\zeta = \cos\alpha + i \sin\alpha$, where $\alpha = 360/n$, is a generator.

(7) Any group G of prime order p is cyclic. In fact, suppose $a \in G$ is any element different from the identity. Then $[a]$ is a subgroup of G, so (by Lagrange) its order divides p. Since $[a]$ contains more than one element (both a and the identity belong to $[a]$), its order must be p; therefore, $[a] = G$ and G is cyclic. Our proof also shows that for a group with prime order, every element, except the identity, is a generator.

(8) Obviously, any cyclic group is abelian. Conversely, it is almost equally obvious that an abelian group need not be cyclic—for example, $\{\mathbf{Z}_{12}^*, \cdot\}$ is abelian but not cyclic.

4-3-4. Remark. Cyclic groups are characterized by the fact that they can be generated by a single element. We digress here to place a group, or a subgroup, of form $[a]$ in a more general context.

Let S be any nonempty subset of the group G; in particular, S may consist of the single element a. Consider all expressions obtained by writing a finite product whose individual terms are either elements of S or inverses of elements of S. Typical expressions of this type look as follows (with $a, b, c, d, \in S$):

$$a, \quad abcd, \quad ba^{-1}, \quad c^{-1}, \quad dabb^{-1}cda^{-1},$$
$$bbbaac^{-1}ccd^{-1}a^{-1}a^{-1}a^{-1}bac^{-1}, \quad baa^{-1}b^{-1}, \quad aaa, \quad c^{-1}, \quad c^{-1}.$$

The definition is worded in such a way that powers (other than 1 and -1) are not permitted, but we can certainly combine like terms and use powers without affecting anything.

Let $[S]$ denote the set of all elements of G which are represented by expressions (or "words") of the given type. Of course, the expression for an element of $[S]$ is not unique—for example, aa^{-1} and bb^{-1} both represent the element e (so, in particular, $e \in [S]$). Clearly, $[S]$ is closed under multiplication and taking of inverses; so $[S]$ is a subgroup of G, and is known as the subgroup of G **generated** by S. Note that if $S = \{a\}$ (that is, S consists of the single element a), then, according to the definition $[S] = \{a^n \mid n \in \mathbf{Z}\}$—so $[S]$ equals what we have earlier denoted by $[a]$, and the notations for $[a]$ and $[S]$ are compatible. (Strictly speaking, one should write $[\{a\}]$ instead of $[a]$.)

Among the properties of [ ], which the reader can easily verify, we have: $S \subset [S]$; $S \subset T \Rightarrow [S] \subset [T]$; if H is a subgroup which contains S, then $H \supset [S]$; $[S]$ is the intersection of all subgroups of G which contain S; $[S]$ is the unique smallest subgroup of G which contains S; S is a subgroup $\Leftrightarrow$ $[S] = S$.

As seen in the examples considered in 4-3-3, one way to find $[a]$ is to start taking powers of a. If and when we arrive at $a^m = e$, then, apparently, all the elements of $[a] = \{a^n \mid n \in \mathbf{Z}\}$ are already in hand. In particular, the set $\{a^n \mid n \in \mathbf{Z}\}$ need not be infinite, and distinct powers of a may be equal. Let us analyze the situation more carefully.

4-3-5. Definition. Let a be an element of the multiplicative group G. By the **order** of a we mean the smallest positive integer m for which $a^m = e$, and we then write ord $a = m$. (In particular, if $a = e$, then $m = 1$—so ord $e = 1$.) If no such m exists, then a is said to be of infinite order, and we write ord $a = \infty$.

If the group G is additive, then our definition becomes ord a is the smallest positive integer m for which $ma = 0$.

Let us illustrate this notion. In the group $\{\mathbf{Z}_{13}^*, \cdot\}$ we have: ord $2 = 12$ since the positive powers of 2 are 2^1, $2^2 = 4$, $2^3 = 8$, $2^4 = 3$, $2^5 = 2 \cdot 3 = 6$, $2^6 = 2 \cdot 6 = 12$, $2^7 = 2 \cdot 12 = 11$, $2^8 = 2 \cdot 11 = 9$, $2^9 = 2 \cdot 9 = 5$, $2^{10} = 2 \cdot 5 = 10$, $2^{11} = 2 \cdot 10 = 7$, $2^{12} = 2 \cdot 7 = 1$; ord $3 = 3$, since $3^1 = 3$, $3^2 = 9$, $3^3 = 1$; ord $4 = 6$, since $4^1 = 4$, $4^2 = 3$, $4^3 = 4 \cdot 3 = 12$, $4^4 = 4 \cdot 12 = 9$, $4^5 = 4 \cdot 9 = 10$, $4^6 = 4 \cdot 10 = 1$; ord $5 = 4$, since $5^1 = 5$, $5^2 = 12$, $5^3 = 5 \cdot 12 = 8$, $5^4 = 5 \cdot 8 = 1$.

In the additive group $\{\mathbf{Z}_{24}, +\}$ we have: ord $2 = 12$, since the positive multiples of 2 are $1 \cdot 2 = 2$, $2 \cdot 2 = 4$, $3 \cdot 2 = 6$, $4 \cdot 2 = 8$, $5 \cdot 2 = 10$, $6 \cdot 2 = 12$, $7 \cdot 2 = 14$, $8 \cdot 2 = 16$, $9 \cdot 2 = 18$, $10 \cdot 2 = 20$, $11 \cdot 2 = 22$, $12 \cdot 2 = 0$; ord $3 = 8$, since $1 \cdot 3 = 3$, $2 \cdot 3 = 6$, $3 \cdot 3 = 9$, $4 \cdot 3 = 12$, $5 \cdot 3 = 15$, $6 \cdot 3 = 18$, $7 \cdot 3 = 21$, $8 \cdot 3 = 0$; ord $4 = 6$, since $1 \cdot 4 = 4$, $2 \cdot 4 = 8$, $3 \cdot 4 = 12$, $4 \cdot 4 = 16$, $5 \cdot 4 = 20$, $6 \cdot 4 = 0$; ord $5 = 24$ since $1 \cdot 5 = 5$, $2 \cdot 5 = 10$, $3 \cdot 5 = 15$, $4 \cdot 5 = 20$, $5 \cdot 5 = 4 \cdot 5 + 5 = 20 + 5 = 1$, $6 \cdot 5 = 5 \cdot 5 + 5 = 6$, $7 \cdot 5 = 6 + 5 = 11$, $8 \cdot 5 = 16$, $9 \cdot 5 = 21$, $10 \cdot 5 = 2$, $11 \cdot 5 = 7$, $12 \cdot 5 = 12$, $13 \cdot 5 = 17$, $14 \cdot 5 = 22$, $15 \cdot 5 = 3$, $16 \cdot 5 = 8$, $17 \cdot 5 = 13$, $18 \cdot 5 = 18$, $19 \cdot 5 = 23$, $20 \cdot 5 = 4$, $21 \cdot 5 = 9$, $22 \cdot 5 = 14$, $23 \cdot 5 = 19$, $24 \cdot 5 = 0$.

4-3-6. Proposition. For an element a of the multiplicative group G, we have

$$\text{ord } a = \infty \Leftrightarrow \text{all powers } a^n,\ n \in \mathbf{Z}, \text{ are distinct}$$

and, when this is the case, ord $a = \#[a]$.

Proof: To prove $\Rightarrow$, suppose ord $a = \infty$. If all powers a^n are not distinct, then we have $a^i = a^j$ for some $i \neq j$—say, $i < j$. But then $a^{j-i} = e$ with $j - i > 0$, which contradicts the hypothesis that ord $a = \infty$.

Conversely, if all powers a^n are distinct then, in particular, $a^0 = e$ is not equal to a^m for any $m > 0$—so ord $a = \infty$. This proves $\Leftarrow$.

When the hypotheses hold, the group $[a] = \{a^n \mid n \in \mathbf{Z}\}$ has an infinite number of elements—that is, its order $\#[a]$ is ∞, which is equal to ord a. ▌

4-3-7. Proposition. For an element a of the multiplicative group G, we have

$$\text{ord } a < \infty \Leftrightarrow \text{not all } a^n,\ n \in \mathbf{Z}, \text{ are distinct.}$$

If this is the case, and ord $a = m$, then

(*i*) $a^n = e \Leftrightarrow m \mid n$,
(*ii*) $a^i = a^j \Leftrightarrow i \equiv j \pmod{m}$,
(*iii*) the elements $e = a^0, a^1, a^2, \ldots, a^{m-1}$ are distinct,
(*iv*) $[a] = \{e, a, a^2, \ldots, a^{m-1}\}$,
(*v*) ord $a = \#[a]$.

Proof: The first assertion is equivalent to the assertion of 4-3-6. In fact, taking "negatives" in 4-3-6, we have: It is false that ord $a = \infty \Leftrightarrow$ it is false that all a^n are distinct. This says: ord $a < \infty \Leftrightarrow$ not all a^n are distinct.

Now, suppose ord $a = m$; then

(*i*) If $m \mid n$, say $n = md$, then $a^n = (a^m)^d = (e)^d = e$—which proves $\Leftarrow$. Conversely, if $a^n = e$, then, by the division algorithm, we may write

$$n = qm + r, \qquad 0 \leq r < m$$

Therefore,

$$e = a^n = a^{qm+r} = a^{qm}a^r = (a^m)^q a^r = a^r.$$

Since $0 \leq r < m$, and m is the smallest positive integer for which $a^m = e$, it follows that $r = 0$—thus, $n = qm$, $m \mid n$, and the implication $\Rightarrow$ is proved.

(*ii*) Using part (*i*), we have

$$a^i = a^j \Leftrightarrow a^{j-i} = e \Leftrightarrow m \mid (j - i) \Leftrightarrow i \equiv j \pmod{m}.$$

(*iii*) In virtue of (*ii*) the elements $e = a^0, a^1, a^2, \ldots, a^{m-1}$ are distinct because no two of their exponents $0, 1, \ldots, m - 1$ are congruent (mod m).

(*iv*) Obviously, $\{e = a^0, a^1, a^2, \ldots, a^{m-1}\} \subset \{a^n \mid n \in \mathbf{Z}\} = [a]$. On the other hand, given a^n there exists an integer i, satisfying $0 \leq i \leq m - 1$, for which $n \equiv i \pmod{m}$ and hence $a^n = a^i$. Thus, the inclusion $\supset$ holds, and we have

$$\{e, a, a^2, \ldots, a^{m-1}\} = [a].$$

(*v*) Now the group $[a]$ has m elements—so

$$\#[a] = m = \text{ord } a.$$

The proof is now complete. ▮

The word "order" has appeared in the present context in two ways—namely, as the order, $\#(G)$, of a group G (see 4-2-16) and also as the order, ord a, of an arbitrary element of a group (see 4-3-5). A minor consequence of the two preceding results is that the two uses of the word "order" are compatible—namely, for an arbitrary element a of a group G, its order is equal to the order of the cyclic group $[a]$ which it generates.

4-3-8. Remark. Some of the information contained in the two preceding results, 4-3-6 and 4-3-7, may be restated as follows:

Suppose the group G is cyclic; then

(*i*) If $\#(G) = \infty$, then G is of form $G = [a] = \{a^n \mid n \in \mathbf{Z}\}$, where all the powers of a are distinct.

(*ii*) If $\#(G) = m$, then G is of form $G = [a] = \{e, a, a^2, \ldots, a^{m-1}\}$, where $e, a, \ldots, a^{m-1}$ are distinct.

An interesting result, related to 4-3-7, goes as follows. Suppose G is an arbitrary finite group—of order n, say—and consider any $a \in G$. Since $[a]$ is a subgroup of G, Lagrange says that the order of $[a]$ divides the order of G—$\#[a] \mid \#(G)$. But ord $a = \#[a]$; so ord a (call it m) divides $\#(G) = n$. In particular, every element of a finite group has finite order dividing the order of the group. Furthermore, since $a^m = e$ and $m \mid n$, we conclude that $a^n = e$. In words, raising any element of the finite group G to the power $\#(G)$ gives the identity. We have proved:

4-3-9. Proposition. Suppose G is a finite group; then for any $a \in G$,

(*i*) $(\text{ord } a) \mid \#(G)$,
(*ii*) $a^{\#(G)} = e$.

This seemingly innocuous result is of especial importance for us because it enables us to understand the real meaning of the theorems of Fermat and Euler [see 2-4-9 and 3-2-22, part (7)]. In more detail, we have:

4-3-10. Corollary. (Euler–Fermat). For any $m \geq 1$, consider the multiplicative group $\{\mathbf{Z}_m^*, \cdot\}$ whose order is $\phi(m)$. Then for any $a \in \mathbf{Z}_m^*$,

$$a^{\phi(m)} = 1 \qquad (\text{in } \mathbf{Z}_m^*).$$

In particular, if m is prime, $m = p$, then

$$a^{p-1} = 1 \qquad (\text{in } \mathbf{Z}_p^*).$$

Furthermore, when these facts are translated into the language of congruences, they take the form:

If a is an integer with $(a, m) = 1$, then

$$a^{\phi(m)} \equiv 1 \pmod{m}$$

and, in particular, if $m = p$, then we have:

$$a^{p-1} \equiv 1 \pmod{p}, \qquad p \nmid a.$$

As illustrations of this result we have statements such as:

(1) Because $\phi(12) = 4$, every element a of $\mathbf{Z}_{12}^*$ satisfies $a^4 = 1$—that is, in $\mathbf{Z}_{12}^*$, $1^4 = 1$, $5^4 = 1$, $7^4 = 1$, $11^4 = 1$. Expressed in terms of congruences (mod 12) this says: For any integer a prime to 12, we have, $a^4 \equiv 1 \pmod{12}$—in particular, $1^4 \equiv 1 \pmod{12}$, $5^4 \equiv 1 \pmod{12}$, $7^4 \equiv 1 \pmod{12}$, $11^4 \equiv 1 \pmod{12}$.

(2) Because $\phi(30) = 8$, every element a of $\mathbf{Z}_{30}^*$ satisfies $a^8 = 1$; or in terms of congruences, if a is an integer prime to 30, then $a^8 \equiv 1 \pmod{30}$.

(3) Since 29 is prime, $\phi(29) = 28$, and every element a of $\mathbf{Z}_{29}^*$ satisfies $a^{28} = 1$—or, for any integer $a \neq 0$ with $29 \nmid a$, we have $a^{28} \equiv 1 \pmod{29}$.

From their very nature, it is clear that cyclic groups are the simplest type possible. We know more about them than about any other groups. The complete nature of our information is made manifest when we study their subgroups—for, as our next result shows, we can answer such questions as: what does any subgroup look like? is it cyclic? for which orders does a subgroup exist? how many subgroups are there of a given order?

4-3-11. Theorem. If G is a cyclic group, then every subgroup H is cyclic; in fact, if $G = [a]$ and $H \neq (e)$, then $H = [a^n]$, where n is the smallest positive integer for which $a^n \in H$. Moreover,

(1) If $\#(G) = \infty$, then every subgroup [except (e)] is infinite cyclic. For positive n, the subgroups $[a^n]$ are distinct, and there are no other subgroups [except (e)].

(2) If $\#(G) = m$, then

- (*i*) $n \mid m$.
- (*ii*) The order of any subgroup is a divisor of m, and for every divisor d of m there exists a unique subgroup of G of order d—namely, $[a^{m/d}]$.
- (*iii*) The set of all subgroups of G is

$$\{[a^d] \mid d \mid m\} = \{[a^{m/d}] \mid d \mid m\}.$$

Proof: If $H = (e)$, then H is cyclic with generator e—that is, $H = [e]$. Suppose, therefore, $H \neq (e)$. Because $G = [a]$, there exists an element $a^s \in H$ with $a^s \neq e$. We may assume s is positive, for if it is not, then $-s$ is positive with $a^{-s} = (a^s)^{-1} \in H$ and $a^{-s} \neq e$. In particular, there exists a positive integer s for which $a^s \in H$. Let n be the smallest positive integer satisfying this condition. Since $a^n \in H$, we have

$$[a^n] \subset H.$$

To prove the reverse inclusion consider any $a^t \in H$. According to the division algorithm, we may write (even if $n = 1$)

$$t = qn + r, \qquad 0 \leq r < n$$

and then

$$a^t = a^{qn}a^r = (a^n)^q a^r.$$

Because a^t and $(a^n)^q$ belong to the subgroup H, so does a^r. By the choice of n, it follows that $r = 0$; so $t = qn$ and $a^t = (a^n)^q \in [a^n]$. Thus, $H \subset [a^n]$ and we have proved

$$[a^n] = H.$$

Now, let us turn to the remaining parts.

(1) Suppose $\#(G) = \infty$. Then an arbitrary subgroup $H \neq (e)$ is of form $H = [a^n]$ with $n \geq 1$. Since ord $a = \infty$, it follows that ord $a^n = \infty$ (for if ord $a^n = t$, then $a^{nt} = e$, contradicting ord $a = \infty$) and (by 4-3-6) $H = [a^n]$ is infinite cyclic.

We now know that for every $n \geq 1$ we have a subgroup $[a^n]$, and [except for (e)] there are no other subgroups. To show they are distinct, suppose

$$[a^n] = [a^{n'}], \qquad n \geq 1, \quad n' \geq 1.$$

By listing the elements of these two subgroups it is easy to see that $n = n'$—but let us give a more formal proof. Since $a^{n'} \in [a^n]$, we may write $a^{n'} = (a^n)^t = a^{nt}$ for some $t \in \mathbf{Z}$, $t \neq 0$. Consequently, making use of 4-3-6, $n' = nt$. In particular,

$$n \leq n'.$$

Arguing by symmetry, starting from $a^n \in [a^{n'}]$, we obtain $n' \leq n$. Therefore,

$$n = n'$$

and the subgroups $[a^n]$ are indeed distinct. This completes the proof of (1).

(2) Suppose $\#(G) = m$. Then an arbitrary subgroup $H \neq (e)$ is of form $H = [a^n]$, where n is the smallest positive integer for which $a^n \in H$. As a matter of fact, in the current situation this statement also holds when $H = (e)$—for then $a^m = e$ is the smallest positive power of a which belongs to H, and $H = [a^m]$.

Consider any subgroup H; it is of form $[a^n]$. Since

$$a^m = e \in H = [a^n]$$

it follows (by the very same argument used earlier to prove $H = [a^n]$) that $m = qn$ for some $q \in \mathbf{Z}$. Hence, $n \mid m$, which proves (i).

We turn to (*ii*). By Lagrange, the order of any subgroup of G is a divisor d of $m = \#(G)$. Conversely, consider any positive divisor d of m—the cases $d = 1$ and $d = m$ are included. Put $m' = m/d$. Let us look at the powers $(a^{m'})^i$ for $i = 0, 1, \ldots, d-1$. Since $(a^{m'})^i = a^{im'}$, they are

$$(a^{m'})^0 = a^0 = e, \quad a^{m'}, \quad a^{2m'}, \ldots, a^{(d-1)m'}. \tag{$*$}$$

The exponents are all distinct and less than m; so no two exponents are congruent (mod m) and, in virtue of 4-3-7, the terms of ($*$) are distinct. In particular, in ($*$), no term after the first one is equal to e. Since $(a^{m'})^d = a^{dm'} = e^m = e$, we have $\text{ord}(a^{m'}) = d$. Consequently, using 4-3-7

$$\#[a^{m'}] = \text{ord}(a^{m'}) = d$$

—so $[a^{m'}]$ is a subgroup of order $d = m/m'$.

It remains to prove uniqueness. Suppose H is any subgroup of order d, $\#(H) = d$. Then H is of form $H = [a^n]$ where n is the smallest positive integer for which $a^n \in H$ and, by part (*i*), $n \mid m$. The argument used above leads to the conclusion:

$$\#[a^n] = \frac{m}{n}$$

Consequently, $d = m/n$ and $n = m/d$. We have shown: If H is a subgroup of order d, then $H = [a^{m/d}]$—which depends only on d. Therefore, there exists a unique subgroup of order d, and the proof of (*ii*) is complete.

Clearly, m/d runs over all divisors of m as d runs over all divisors of m, and consequently

$$\{[a^d] \mid d \mid m\} = \{[a^{m/d}] \mid d \mid m\}$$

is the set of all subgroups of G. This proves (*iii*). ▮

4-3-12. Examples. We give some concrete illustrations of the preceding theorem.

(1) Consider the additive group $\{\mathbf{Z}, +\}$. It is infinite cyclic with $\mathbf{Z} = [1]$, (see 4-3-3) and $\text{ord}\ 1 = \#(\mathbf{Z}) = \infty$. Keeping in mind that, because the operation is addition, powers a^n are replaced by multiples na, 4-3-11, part (*i*) says: Every subgroup is cyclic; in fact, for every $n \geq 1$ the subgroup $[n \cdot 1] = n\mathbf{Z}$ is infinite cyclic, these subgroups are distinct, and there are no other subgroups [except (0)].

Of course, this information is nothing new, it was proved earlier in 4-2-4.

(2) Consider the multiplicative group $\{\mathbf{Z}_{13}^*, \cdot\}$. Its underlying set consists of the 12 elements 1, 2, ..., 11, 12. It is cyclic—in fact, because the powers of 2 are 2, 4, 8, 3, 6, 12, 11, 9, 5, 10, 7, 1, the element 2 is a generator. Thus,

$$\mathbf{Z}_{13}^* = [2], \qquad \text{ord}\ 2 = \#(\mathbf{Z}_{13}^*) = 12.$$

According to 4-3-11, part (*ii*), because the divisors of 12 are 1, 2, 3, 4, 6, 12 all the subgroups of $\mathbf{Z}_{13}^*$ are as follows.

$$[2^1], \quad [2^2], \quad [2^3], \quad [2^4], \quad [2^6], \quad [2^{12}]$$

and their orders are, respectively,

$$12, \quad 6, \quad 4, \quad 3, \quad 2, \quad 1.$$

In more detail, the subgroups are:

$$\begin{aligned}
[2^1] &= \mathbf{Z}_{13}^*, \\
[2^2] &= [4] = \{4, 3, 12, 9, 10, 1\} = \{1, 3, 4, 9, 10, 12\}, \\
[2^3] &= [8] = \{8, 12, 5, 1\} = \{1, 5, 8, 12\}, \\
[2^4] &= [3] = \{3, 9, 1\} = \{1, 3, 9\}, \\
[2^6] &= [12] = \{12, 1\} = \{1, 12\}, \\
[2^{12}] &= [1] = \{1\}.
\end{aligned}$$

Incidentally, we have proved the assertions made in 4-2-5 about the subgroups of $\{\mathbf{Z}_{13}^*, \cdot\}$.

(3) Consider the additive group $\{\mathbf{Z}_{12}, +\}$. As observed in 4-3-3, part (3), this is a cyclic group—

$$\mathbf{Z}_{12} = [1], \qquad \text{ord } 1 = \#(\mathbf{Z}_{12}) = 12.$$

Being a cyclic group of order 12, this group behaves entirely like the cyclic group of order 12, $\{\mathbf{Z}_{13}^*, \cdot\}$. In more detail, the set of all subgroups is

$$\{[d \cdot 1] \mid d \mid 12\}$$

that is: For $d = 1, 2, 3, 4, 6, 12$ we have the subgroups

$$\begin{aligned}
[1] &= \mathbf{Z}_{12} \\
[2] &= \{2, 4, 6, 8, 10, 0\}, \\
[3] &= \{3, 6, 9, 0\}, \\
[4] &= \{4, 8, 0\}, \\
[6] &= \{6, 0\}, \\
[12] &= [0] = (0).
\end{aligned}$$

We have proved the assertions about the subgroups of $\{\mathbf{Z}_{12}, +\}$ that were made in 4-2-5.

What about the order of any element a^r of the finite cyclic group $G = [a]$? Which subgroup does a^r generate? When is a^r a generator of G? These questions are now easy to settle.

4-3-13. Proposition. Suppose $G = [a]$ is a cyclic group of order m or, more generally, suppose a is an element of an arbitrary multiplicative group G and

$$\text{ord } a = \#[a] = m.$$

Then for any integer r,

(i) $\text{ord } (a^r) = m/(r, m)$,
(ii) $[a^r] = [a^{(r,m)}]$,
(iii) a^r generates $[a] \Leftrightarrow r$ is relatively prime to m,
(iv) the group $[a]$ has $\phi(m)$ generators.

Proof: We put $\text{ord } (a^r) = t$ and $d = (r, m)$. Then we may write $r = dr'$, $m = dm'$, and $(r', m') = 1$. Note that $m/(r, m) = m'$.

Since

$$(a^r)^{m'} = a^{rm'} = a^{r'dm'} = (a^m)^{r'} = e$$

it follows that $t = \text{ord}(a^r)$ divides m'. On the other hand,

$$\begin{aligned} t = \text{ord}(a^r) &\Rightarrow (a^r)^t = e \Rightarrow a^{rt} = e \\ &\Rightarrow m \mid rt \Rightarrow m' \mid r't \\ &\Rightarrow m' \mid t \qquad [\text{since } (m', r') = 1]. \end{aligned}$$

Thus $t = m'$, and (i) is proved.

(ii) In virtue of (i), $[a^r]$ is the subgroup of $[a]$ of order $m/(r, m)$, and according to 4-3-11, part 2(ii), the unique subgroup of order $m/(r, m)$ is $[a^{(r,m)}]$. Hence

$$[a^r] = [a^{(r,m)}].$$

(iii) In virtue of (ii),

$$\begin{aligned} [a^r] = [a] &\Leftrightarrow [a^{(r,m)}] = [a] \\ &\Leftrightarrow \text{ord}[a^{(r,m)}] = \text{ord}[a] \qquad (\text{using 4-3-11}) \\ &\Leftrightarrow m/(r, m) = m \\ &\Leftrightarrow (r, m) = 1. \end{aligned}$$

(iv) If $r \equiv r' \pmod m$, then $a^r = a^{r'}$, so in looking for generators it suffices to consider only $r = 1, 2, \ldots, m - 1$. Then from (iii), a^r is a generator $\Leftrightarrow (r, m) = 1$. The number of integers $r = 1, 2, \ldots, m - 1$ which are relatively prime to m is $\phi(m)$; therefore, the number of generators of $[a]$ is $\phi(m)$. ▮

4-3-14. Example. In 4-3-3, part (5), we saw that $\{\mathbf{Z}_{17}^*, \cdot\}$ is cyclic and 3 is a generator. Thus

$$\mathbf{Z}_{17}^* = [3] \qquad \text{and} \qquad \text{ord } 3 = \#(\mathbf{Z}_{17}^*) = 16.$$

According to 4-3-13, with $a = 3$, $m = 16$ we know that $\mathbf{Z}_{17}^*$ has $\phi(16) = 8$ generators. In fact, for $r = 1, 2, \ldots, 15$, 3^r is a generator $\Leftrightarrow (r, 16) = 1$. Therefore, the generators are

$$3^1 = 3, \quad 3^3 = 10, \quad 3^5 = 5, \quad 3^7 = 11, \quad 3^9 = 14, \quad 3^{11} = 7, \quad 3^{13} = 12, \quad 3^{15} = 6$$

and we have

$$\mathbf{Z}_{17}^* = [3] = [5] = [6] = [7] = [10] = [11] = [12] = [14]$$

as was indicated in 4-3-3. In particular, in $\mathbf{Z}_{17}^*$,

$$16 = \text{ord } 3 = \text{ord } 5 = \text{ord } 6 = \text{ord } 7 = \text{ord } 10 = \text{ord } 11 = \text{ord } 12 = \text{ord } 14.$$

As for the orders of the remaining elements, since

$$1 = 3^0, \quad 2 = 3^{14}, \quad 4 = 3^{12}, \quad 8 = 3^{10}, \quad 9 = 3^2, \quad 13 = 3^4, \quad 15 = 3^6, \quad 16 = 3^8,$$

we have

$$\text{ord } 1 = 1, \qquad \text{ord } 2 = \text{ord}(3^{14}) = \frac{16}{(14, 16)} = 8,$$

$$\text{ord } 4 = \text{ord}(3^{12}) = \frac{16}{(12, 16)} = 4, \qquad \text{ord } 8 = \text{ord}(3^{10}) = \frac{16}{(10, 16)} = 8,$$

$$\text{ord } 9 = \text{ord}(3^2) = \frac{16}{(2, 16)} = 8 \qquad \text{ord } 13 = \text{ord}(3^4) = \frac{16}{(4, 16)} = 4,$$

$$\text{ord } 15 = \text{ord } (3^6) = \frac{16}{(6, 16)} = 8, \qquad \text{ord } 16 = \text{ord}(3^8) = \frac{16}{(8, 16)} = 2.$$

The "proper" subgroups (meaning the nontrivial ones—that is, those other than (1) and $\mathbf{Z}_{17}^*$) are

$[2] = [8] = [9] = [15]$, whose order is 8.
$[4] = [13]$, whose order is 4.
$[16]$, whose order is 2.

One may check that these subgroups are

$$[2] = \{1, 2, 4, 8, 9, 13, 15, 16\}, \quad [4] = \{1, 4, 13, 16\}, \quad [16] = \{1, 16\}.$$

We conclude this section by exhibiting an important class of cyclic groups. The first step is to prove a basic fact about the Euler ϕ-function.

4-3-15. Proposition. For every $n \geq 1$,

$$\sum_{d|n} \phi(d) = n.$$

Proof: It is understood that only positive divisors d of n are considered. It may be noted that two proofs of this result were indicated in 3-2-23, Problem 25; we provide the details of the second proof.

For each $d \mid n$ let $S(d)$ denote the subset of the set $\{1, 2, \ldots, n\}$ consisting of those integers r whose greatest common divisor with n is d; in symbols

$$S(d) = \{r \in \{1, 2, \ldots, n\} \bigm| (r, n) = d\}.$$

Let $\lambda(d)$ denote the number of elements in the set $S(d)$.

As the reader will observe, every integer $r \in \{1, \ldots, n\}$ belongs to exactly one of the sets $S(d)$—or, to put it another way, the sets $S(d)$ are disjoint and their union is $\{1, 2, \ldots, n\}$. In particular, we have

$$n = \sum_{d|n} \lambda(d).$$

If $(r, n) = d$, then r must be a multiple of d; so r must come from the set

$$\left\{d, 2d, \ldots, md, \ldots, \frac{n}{d}\, d\right\}.$$

But we are interested only in the elements md from this set for which $(md, n) = d$. Since

$$(md, n) = \left(md, \frac{n}{d}\, d\right) = \left(m, \frac{n}{d}\right) d$$

we have

$$(md, n) = d \Leftrightarrow \left(m, \frac{n}{d}\right) = 1.$$

It follows that

$$(r, n) = d \Leftrightarrow r = md \text{ where } 1 \leq m \leq \frac{n}{d} \text{ and } \left(m, \frac{n}{d}\right) = 1,$$

from which we conclude

$$\lambda(d) = \phi\left(\frac{n}{d}\right).$$

As d runs over all divisors of n, so does n/d. Hence,

$$n = \sum_{d|n} \lambda(d) = \sum_{d|n} \phi\left(\frac{n}{d}\right) = \sum_{d|n} \phi(d). \quad \blacksquare$$

4-3-16. Theorem. For every prime p, $\{\mathbf{Z}_p^*, \cdot\}$, the multiplicative group of the field $\mathbf{Z}_p$, is cyclic.

It involves little additional effort to prove the following more general result:

4-3-17. Theorem. Let F be any field and F^* denote its multiplicative group. If G is a finite subgroup of F^*, then G is cyclic.

Briefly stated: Any finite subgroup of the multiplicative group of a field is cyclic.

Proof: Let $\#(G) = n$. For each divisor d of n, let

$$\psi(d) = \text{the number of elements of } G \text{ whose order is } d.$$

Since the order of any element of G is a divisor of n, we have clearly

$$n = \sum_{d|n} \psi(d).$$

Fix a divisor d of n. If $\psi(d) \neq 0$, then there exists $a \in G$ with ord $a = d$. Thus, the group

$$[a] = \{1, a, a^2, \ldots, a^{d-1}\}$$

consists of d distinct elements, and (since $a^d = 1$) each of them is a root of the polynomial $x^d - 1 \in F[x]$. According to 3-5-5, the polynomial $x^d - 1$ can have at most d distinct roots in F. Therefore, $x^d - 1$ has exactly d distinct roots in F—they are, $1, a, \ldots, a^{d-1}$.

Now, suppose $b \in G$ is any element of order d; then, as above,

$$[b] = \{1, b, \ldots, b^{d-1}\}$$

is the set of all roots (in F) of the polynomial $x^d - 1$. Therefore, $[b] = [a]$ and in particular, b is a generator of the cyclic group $[a]$. Since $[a]$ has order d, we know (from 4-3-13) it has $\phi(d)$ generators. Thus, any element b of order d must be one of the $\phi(d)$ generators of $[a]$. This implies, $\psi(d) \leq \phi(d)$ [under the hypothesis that $\psi(d) \neq 0$]. Of course, if $\psi(d) = 0$, then surely $\psi(d) \leq \phi(d)$. Consequently,

$$\psi(d) \leq \phi(d), \text{ for every } d \mid n.$$

Making use of 4-3-15, we have:

$$n = \sum_{d|n} \psi(d) \leq \sum_{d|n} \phi(d) = n.$$

This implies $\psi(d) = \phi(d)$ for every $d \mid n$. In particular, $\psi(n) = \phi(n) \neq 0$—so there exists an element of order n, and G is cyclic. ▌

This result tells us, for example, that because 229 is prime, the group $\{\mathbf{Z}_{229}^*, \cdot\}$ is cyclic of order 228. Although this cyclic group has $\phi(228) = 72$ generators (by 4-3-13), our result provides no information on how to locate a generator.

4-3-18 / *PROBLEMS*

1. In $\{\mathbf{Z}_{10}, +\}$, find the cyclic subgroups:
(*i*) [2], (*ii*) [4], (*iii*) [6], (*iv*) [7], (*v*) [8].

2. In $\{\mathbf{Z}_{14}^*, \cdot\}$, find the cyclic subgroups:
(*i*) [3], (*ii*) [5], (*iii*) [9], (*iv*) [11].

3. For each element of the group $\{\mathbf{Z}_{15}, +\}$, find the cyclic subgroup which it generates. Which elements are generators of $\mathbf{Z}_{15}$?

4. For each element of the group $\{\mathbf{Z}_{15}^*, \cdot\}$, find the cyclic subgroup which it generates. Which elements are generators of $\mathbf{Z}_{15}^*$?

5. Do the same thing for $\{\mathbf{Z}_{18}^*, \cdot\}$.

6. Which of the following groups is cyclic?
(*i*) $\{\mathbf{Q}, +\}$, (*iii*) $\{\mathbf{R}, +\}$,
(*ii*) $\{\mathbf{Q}^*, \cdot\}$, (*iv*) $\{\mathbf{R}^*, \cdot\}$,
(*v*) $\{W, \cdot\}$, where $W = \{z \in \mathbf{C} \mid |z| = 1\}$.

7. Which of the following groups is cyclic?
(*i*) $\{\mathbf{Z}_6^*, \cdot\}$, (*ii*) $\{\mathbf{Z}_9^*, \cdot\}$, (*iii*) $\{\mathbf{Z}_{13}^*, \cdot\}$, (*iv*) $\{\mathbf{Z}_{21}^*, \cdot\}$.
For those that are cyclic, exhibit a generator.

8. What is the order of:
(*i*) 2 in $\{\mathbf{R}^*, \cdot\}$, (*iii*) $\sqrt{2}$ in $\{\mathbf{R}^*, \cdot\}$,
(*ii*) $\frac{1}{2}$ in $\{\mathbf{Q}, +\}$, (*iv*) $1 + i$ in $\{\mathbf{C}, +\}$,
(*v*) $1 + i$ in $\{\mathbf{C}^*, \cdot\}$, (*vi*) $\dfrac{(1+i)}{\sqrt{2}}$ in $\{\mathbf{C}^*, \cdot\}$?

9. For each a in the octic group find $[a]$. Is this group cyclic?

10. Suppose G is cyclic with generator a; is it true that a^{-1} is also a generator?

11. Prove: Every group with at most five elements is abelian.

12. Show that if the elements a, b, ab of the group G are all of order 2, then $ab = ba$.

13. For any $a, b \in G$ show that the elements ab and ba have the same order.

14. Suppose G is a cyclic group of order m, and r is an integer relatively prime to m; then for $b, c \in G$ prove that

$$b^r = c^r \Rightarrow b = c.$$

15. Show that in an abelian group, the set of all elements of finite order is a subgroup.

16. In the symmetric group S_4 find two distinct subgroups of order
(*i*) 2, (*ii*) 3, (*iii*) 4.

17. The symmetric group S_5 has 120 elements. For which factors m of 120 can you find an element of S_5 of order m?

18. For each $m > 1$, prove that the symmetric group S_m contains a cyclic subgroup of order m.

19. Prove: A group of even order contains an odd number of elements of order 2.

20. Find all generators of the cyclic group $\{\mathbf{Z}_{24}, +\}$.

21. Show that the integer $a = 1, 2, \ldots, m - 1$ is a generator of $\{\mathbf{Z}_m, +\} \Leftrightarrow (a, m) = 1$.

22. (*i*) Let a and b be elements with $ab = ba$ and of orders r and s, respectively; if $(r, s) = 1$, show that $\operatorname{ord}(ab) = rs$.
(*ii*) If r and s are not relatively prime, then $\operatorname{ord}(ab) \mid [r, s]$. Give an example where $\operatorname{ord}(ab) \neq [r, s]$.

23. If G has order pq, where p and q are distinct primes, then to find all proper (that is, nontrivial) subgroups show that it suffices to compute $[a]$ for every $a \in G$.

24. Let S be a nonempty subset of the group G; if we write, as usual, $S^{-1} = \{a^{-1} \mid a \in S\}$, show that

$$[S \cup S^{-1}] = [S].$$

25. Characterize the following subsets of the group $\{\mathbf{Z}, +\}$, for arbitrary integers m and n:
(*i*) $m\mathbf{Z} \cap n\mathbf{Z}$, (*ii*) $m\mathbf{Z} \cup n\mathbf{Z}$, (*iii*) $[m\mathbf{Z} \cup n\mathbf{Z}]$.

26. Prove: (*i*) If G is infinite cyclic, then for every $d > 0$ there exists a unique subgroup of index d in G.
(*ii*) If G is cyclic of order m, then for every $d \mid m$ there exists a unique subgroup of index d in G.

27. If G is cyclic of order m, show that the number of distinct subgroups of G is equal to the number of divisors of m.

28. (*i*) Suppose G is an abelian group of order 6. If there exists an element $a \in G$ with ord $a = 3$, then prove that G is cyclic.
(*ii*) S_3 is a nonabelian group of order 6. Can you find another such group?

29. State and prove the additive analogs of:
(*i*) 4-3-6, (*ii*) 4-3-7, (*iii*) 4-3-8,
(*iv*) 4-3-9, (*v*) 4-3-11, (*vi*) 4-3-13.

30. In $\{\mathbf{Z}, +\}$ find $[S]$ when S is the set:
(*i*) $\{3, 4\}$, (*ii*) $\{3, 6\}$, (*iii*) $\{3, 4, 6\}$, (*iv*) $\{9, 12\}$.

31. Do Problem 30 in the group:
(*i*) $\{\mathbf{Z}_{19}, +\}$, (*ii*) $\{\mathbf{Z}_{20}, +\}$, (*iii*) $\{\mathbf{Z}_{24}, +\}$, (*iv*) $\{\mathbf{Z}_{30}, +\}$.

32. In $\{\mathbf{Z}_{13}^*, \cdot\}$ find $[S]$ when S is the set:
(*i*) $\{2, 5\}$, (*ii*) $\{3, 4\}$, (*iii*) $\{3, 7\}$, (*iv*) $\{2, 6\}$.

33. Do Problem 32 in the group:
(*i*) $\{\mathbf{Z}_{11}^*, \cdot\}$, (*ii*) $\{\mathbf{Z}_{17}^*, \cdot\}$, (*iii*) $\{\mathbf{Z}_{19}^*, \cdot\}$, (*iv*) $\{\mathbf{Z}_{23}^*, \cdot\}$.

34. Show that every group G (with more than one element) has at least two generating sets.

35. Can you find a generating set for S_4 consisting of two elements? What about S_5?

36. If H is a subgroup of G, then prove that $[G - H] = G$.

37. (*i*) Find all subgroups of the cyclic group $\{\mathbf{Z}_{30}, +\}$.
(*ii*) If $G = [a]$ is a cyclic multiplicative group of order 30, find all its subgroups.

38. If G_1 is cyclic of order m and G_2 is cyclic of order n, then, if $(m, n) = 1$, show that the direct product $G_1 \times G_2$ (see 4-1-12, Problem 26) is cyclic of order mn.

39. Prove that if the direct product $G_1 \times G_2$ is cyclic, so are G_1 and G_2.

4-4. Normal Subgroups; Factor Groups; Homomorphisms

Consider a multiplicative group G and a subgroup H. In Section 4-2 we introduced the left cosets

$$aH = \{ah \mid h \in H\}, \qquad a \in G$$

and saw that they provide a disjoint decomposition of G. Furthermore, according to 4-2-15, if we define

$$a \equiv b \pmod{H} \Leftrightarrow b^{-1}a \in H,$$

then "$\equiv \pmod{H}$" is an equivalence relation. The equivalence class to which a belongs consists of $\{b \in G \mid b \equiv a \pmod{H}\}$, and when it is denoted by $\lfloor a \rfloor_H$, we have $\lfloor a \rfloor_H = aH$.

The question we wish to confront is whether the set of left cosets of H in G can be made into a group. Why does one even raise this question? Simply because it is related to something we have done before. In detail, suppose we take for G the additive group of integers, $G = \mathbf{Z}$, and let H be any subgroup not equal to (0). Then H is of form $H = m\mathbf{Z}$ for some $m \geq 1$. Because our group is additive, a coset looks like

$$a + H = a + m\mathbf{Z} = \{a + mt \mid t \in \mathbf{Z}\} = \lfloor a \rfloor_m, \qquad a \in \mathbf{Z}.$$

Again, because our group is additive, the equivalence relation "$\equiv \pmod{H}$" mentioned above takes the form $a \equiv b \pmod{H} \Leftrightarrow b - a \in H$, and the equivalence class to which a belongs is

$$\lfloor a \rfloor_H = a + H = \lfloor a \rfloor_m.$$

Another way to see this is as follows: We have

$$\begin{aligned} a \equiv b \pmod{H} &\Leftrightarrow b - a \in H \\ &\Leftrightarrow b - a \in m \\ &\Leftrightarrow m \mid (b - a) \\ &\Leftrightarrow a \equiv b \pmod{m} \end{aligned}$$

so the equivalence relation "$\equiv \pmod{H}$" is the same as the familiar equivalence relation "$\equiv \pmod{m}$." In particular, their equivalence classes are identical, so for any $a \in \mathbf{Z}$ we have

$$\lfloor a \rfloor_H = \lfloor a \rfloor_m.$$

Now, in Chapters I and II we saw how the residue classes (mod m) can be made into a ring. Here we are concerned only with the fact that the residue classes (mod m) form a group under the operation

$$\lfloor a \rfloor_m + \lfloor b \rfloor_m = \lfloor a + b \rfloor_m$$

—or, to phrase it in terms of cosets, the cosets of $H = m\mathbf{Z}$ in the additive group $\mathbf{Z}$ constitute a group under the operation

$$\lfloor a \rfloor_H + \lfloor b \rfloor_H = \lfloor a + b \rfloor_H.$$

This concrete situation suggests that in the general situation where H is a subgroup of the multiplicative group G we try to make the set of left cosets into a group by defining

$$\lfloor a \rfloor_H \cdot \lfloor b \rfloor_H = \lfloor ab \rfloor_H$$

—in other words, the left coset determined by a times the left coset determined by b is the left coset determined by ab.

Clearly, the set of left cosets is closed under this operation. The associative law holds since $\lfloor a \rfloor_H(\lfloor b \rfloor_H \lfloor c \rfloor_H)$ and $(\lfloor a \rfloor_H \lfloor b \rfloor_H)\lfloor c \rfloor_H$ are both equal to $\lfloor abc \rfloor_H$. The element (that is, left coset) $\lfloor e \rfloor_H$ is an identity, since $\lfloor e \rfloor_H \lfloor a \rfloor_H = \lfloor a \rfloor_H = \lfloor a \rfloor_H \lfloor e \rfloor_H$ for all $\lfloor a \rfloor_H$. Finally, any $\lfloor a \rfloor_H$ has an inverse—in fact, $\lfloor a \rfloor_H^{-1} = \lfloor a^{-1} \rfloor_H$, since $\lfloor a \rfloor_H \lfloor a^{-1} \rfloor_H = \lfloor a^{-1} \rfloor_H \lfloor a \rfloor_H = \lfloor e \rfloor_H$.

Thus, the set of left cosets of H in G is a group—except for one important detail. The definition of the operation depends on the choice of the coset representatives, and consequently, we need to know that the operation is well defined. In other words, in order for the left cosets to be a group we must be certain that

$$\lfloor a' \rfloor_H = \lfloor a \rfloor_H \text{ and } \lfloor b' \rfloor_H = \lfloor b \rfloor_H \Rightarrow \lfloor a'b' \rfloor_H = \lfloor ab \rfloor_H .$$

Unfortunately, this is not always true. For example, suppose $G = S_3$ and H is the subgroup $\{e, \tau_1\}$ [where, as defined in 4-1-9, $\tau_1 = \left(\begin{smallmatrix}1 & 2 & 3\\ 1 & 3 & 2\end{smallmatrix}\right)$]. As seen in 4-2-12 ($H$ being the subgroup denoted there by H_1), we have left cosets

$$\lfloor \sigma_1 \rfloor_H = \sigma_1 H = \{\sigma_1, \tau_3\} = \tau_3 H = \lfloor \tau_3 \rfloor_H,$$

$$\lfloor \sigma_2 \rfloor_H = \sigma_2 H = \{\sigma_2, \tau_2\} = \tau_2 H = \lfloor \tau_2 \rfloor_H .$$

(We have no need for the third coset here.) Using the representatives σ_1 and σ_2, we obtain for the product of the two cosets

$$\lfloor \sigma_1 \rfloor_H \cdot \lfloor \sigma_2 \rfloor_H = \lfloor \sigma_1\sigma_2 \rfloor_H = \lfloor e \rfloor_H .$$

On the other hand, if we use the representatives τ_3 and τ_2, the product of the same two cosets becomes

$$\lfloor \tau_3 \rfloor_H \cdot \lfloor \tau_2 \rfloor_H = \lfloor \tau_3\tau_2 \rfloor_H = \lfloor \sigma_2 \rfloor_H$$

—and, of course, $\lfloor e \rfloor_H = \{e, \tau_1\} \neq \{\sigma_2, \tau_2\} = \lfloor \sigma_2 \rfloor_H$.

Thus, the operation of multiplication for left cosets is not always well defined, and it behooves us to find conditions under which the operation will be well defined. For this, it is convenient to make some technical preparations.

4-4-1. Exercise. We develop some elementary facts about computations with subsets of a group; these facts may be considered as part of a **calculus of sets.**

Let S, T, U denote arbitrary subsets of the multiplicative group G, and let us write

$$ST = \{st \mid s \in S,\ t \in T\}, \qquad S^{-1} = \{s^{-1} \mid s \in S\}.$$

In words, ST is the set of all products of an element of S and an element of T (in the appropriate order), and S^{-1} is the set of all inverses of elements of S. For any $a \in G$ we have the set $\{a\}$, consisting of the single element a; we write aS in place of $\{a\}S$—that is,

$$aS = \{a\}S = \{as \mid s \in S\}$$

—and Sa instead of $S\{a\}$. Of course, this is entirely consistent with the notation we used for cosets. We have:

(*i*) $(ST)U = S(TU)$—in other words, the "associative law" holds, and we may write STU. The "generalized associative law" then holds, and we can write any product of sets, say $S_1 S_2 S_3 \cdots S_n$, without worrying about parentheses.

(*ii*) For any $a \in G$,

$$aS \subset aT \Leftrightarrow S \subset T \Leftrightarrow Sa \subset Ta,$$
$$aS \subset Ta \Leftrightarrow S \subset a^{-1}Ta \Leftrightarrow Sa^{-1} \subset a^{-1}T \Leftrightarrow aSa^{-1} \subset T.$$

(*iii*) $(S^{-1})^{-1} = S$; $S \subset T \Leftrightarrow S^{-1} \subset T^{-1}$; $(ST)^{-1} = T^{-1}S^{-1}$.

(*iv*) For a nonempty subset H of G the following are equivalent:

(*) H is a subgroup of G,
(**) $HH \subset H$ and $H^{-1} \subset H$,
(***) $HH^{-1} \subset H$.

(*v*) If H is a subgroup, then $aH = H = Ha$ for all $a \in H$, $HH = H$, and $H^{-1} = H$.

(*vi*) If H is a subgroup, then

$$HaH \subset aH \Leftrightarrow Ha \subset aH.$$

It is left for the reader to state and prove the additive analogs of these facts.

Returning to the multiplicative group G, subgroup H, and operation of multiplication on the set of left cosets of H in G given by $\lfloor a \rfloor_H \cdot \lfloor b \rfloor_H = \lfloor ab \rfloor_H$, we have:

The set of left cosets is a group under multiplication
$\Leftrightarrow$ multiplication of left cosets is well defined
$\Leftrightarrow$ $\lfloor a' \rfloor_H = \lfloor a \rfloor_H$ and $\lfloor b' \rfloor_H = \lfloor b \rfloor_H$ imply $\lfloor a'b' \rfloor_H = \lfloor ab \rfloor_H$
$\Leftrightarrow$ $a' \in \lfloor a \rfloor_H$ and $b' \in \lfloor b \rfloor_H$ imply $a'b' \in \lfloor ab \rfloor_H$
$\Leftrightarrow$ $a' \in aH$ and $b' \in bH$ imply $a'b' \in (ab)H$
$\Leftrightarrow$ $(aH)(bH) \subset (ab)H$ for all $a, b, \in G$
$\Leftrightarrow$ $aHbH \subset abH$ for all $a, b \in G$
$\Leftrightarrow$ $HbH \subset bH$ for all $b \in G$
$\Leftrightarrow$ $Hb \subset bH$ for all $b \in G$
$\Leftrightarrow$ $b^{-1}Hb \subset H$ for all $b \in G$
$\Leftrightarrow$ $b^{-1}H \subset Hb^{-1}$ for all $b \in G$
$\Leftrightarrow$ $bH \subset Hb$ for all $b \in G$ (since $\{b^{-1}\} = G = \{b\}$)
$\Leftrightarrow$ $bHb^{-1} \subset H$ for all $b \in G$.

Furthermore, using the fact (proved above) that

$$bH \subset Hb \text{ for all } b \in G \Leftrightarrow Hb \subset bH \text{ for all } b \in G,$$

we have

$$bH \subset Hb \text{ for all } b \in G \Leftrightarrow bH = Hb \text{ for all } b \in G$$
$$\Leftrightarrow bHb^{-1} = H \text{ for all } b \in G.$$

It is time to summarize what we have accomplished.

4-4-2. Theorem. Suppose H is a subgroup of the multiplicative group G. Then the following conditions are equivalent:

(*i*) $bHb^{-1} \subset H$ for all $b \in G$,
(*ii*) $bHb^{-1} = H$ for all $b \in G$,
(*iii*) $bH = Hb$ for all $b \in G$.

When any one of these conditions holds, H is said to be a **normal** subgroup of G.

Moreover, the set of left cosets becomes a group with respect to the operation $\lfloor a \rfloor_H \cdot \lfloor b \rfloor_H = \lfloor ab \rfloor_H$ if and only if H is normal. When this is the case, the group of cosets is known as the **factor group** or **quotient group**; it is denoted by G/H. The identity element is $\lfloor e \rfloor_H = H$ and the inverse of $\lfloor a \rfloor_H$ is $\lfloor a \rfloor_H^{-1} = \lfloor a^{-1} \rfloor_H$.

4-4-3. Remarks. (1) To show that a subgroup H is normal, one usually verifies the conditions $bHb^{-1} \subset H$. After all it is easier to prove an inclusion relation for two sets than to prove equality of two sets (since proving equality of sets often involves showing that each one is contained in the other).

(2) The condition $bH = Hb$ says that for a normal subgroup the left and right cosets are the same—more precisely, for any $b \in G$, the left coset to which it belongs is the same set as the right coset to which it belongs. Of course, as seen in our discussion the condition could be replaced by either of the seemingly weaker conditions

$$bH \subset Hb \text{ for all } b \qquad \text{or} \qquad Hb \subset bH \text{ for all } b.$$

(3) If H is a subgroup of G, then clearly the condition for the right cosets of H in G to form a group is that H be normal. This may be seen, mechanically, by rewriting our entire discussion in terms of right cosets. Another way to convince oneself of this fact is to note that the condition $bH = Hb$ for normality indicates that the stories for left cosets and right cosets are "symmetrical."

(4) Suppose H is a normal subgroup of G, then for $a, b \in G$ the expression $(aH)(bH)$ has two possible meanings. In the first place, this may be viewed as the product of the two subsets aH and bH of G—so

$$(aH)(bH) = aHbH = abHH = abH.$$

In the second place, $(aH)(bH)$ may be viewed as the product of two cosets as elements of the factor group G/H—so

$$(aH)(bH) = \lfloor a \rfloor_H \cdot \lfloor b \rfloor_H = \lfloor ab \rfloor_H = abH.$$

Thus, the two interpretations are compatible in all ways.

In similar fashion, the two interpretations of $(aH)^{-1}$ do not conflict—namely, as a set,

$$(aH)^{-1} = H^{-1}a^{-1} = Ha^{-1} = a^{-1}H$$

while, as an element of G/H,

$$(aH)^{-1} = \lfloor a \rfloor_H^{-1} = \lfloor a^{-1} \rfloor_H = a^{-1}H.$$

(5) It is left to the reader to state the additive version of 4-4-2.

4-4-4. Examples. (1) For an arbitrary group G the two trivial subgroups G and (e) are obviously normal.

(2) With notation as in 4-2-12, let $G = S_3 = \{e, \sigma_1, \sigma_2, \tau_1, \tau_2, \tau_3\}$; then the subgroup $A = \{e, \sigma_1, \sigma_2\}$ is normal—in fact, the computations done in

4-2-12, part (1), guarantee that for every $\rho \in G$ we have $\rho A = A\rho$. The factor group G/A consists of two elements—namely, the two cosets

$$\lfloor e \rfloor_A = A = \{e, \sigma_1, \sigma_2\} \qquad \text{and} \qquad \lfloor \tau_1 \rfloor_A = \tau_1 A = \{\tau_1, \tau_2, \tau_3\}.$$

On the other hand, the subgroup $H_1 = \{e, \tau_1\}$ of S_3 is not normal—we do not have $\rho H_1 = H_1\rho$ for every $\rho \in G$ since, in particular,

$$\sigma_1 H_1 = \{\sigma_1, \tau_3\} \text{ is not equal to } H_1\sigma_1 = \{\sigma_1, \tau_2\}.$$

Similarly, the subgroup $H_2 = \{e, \tau_2\}$ is not normal (because $\sigma_1 H_2 = \{\sigma_1, \tau_1\} \neq \{\sigma_1, \tau_3\} = H_2\sigma_1$), and the subgroup $H_3 = \{e, \tau_3\}$ is also not normal (because, as the reader can check easily, $\sigma_1 H_3 \neq H_3\sigma_1$). This settles completely the question of which subgroups of S_3 are normal.

(3) For an arbitrary group G, we have introduced its center

$$\mathfrak{Z} = \{a \in G \mid ax = xa \text{ for all } x \in G\}$$

in 4-2-7. Is the subgroup $\mathfrak{Z}$ normal? By definition of $\mathfrak{Z}$, any $b \in G$ commutes with any $a \in \mathfrak{Z}$. Consequently,

$$b\mathfrak{Z} = \mathfrak{Z}b \qquad \text{for all } b \in G$$

—so the center is always a normal subgroup.

(4) If the group G is abelian, then surely every subgroup H is normal and we may form the factor group G/H. For example, suppose G is the additive group of integers $\mathbf{Z}$. Then any subgroup $H = m\,\mathbf{Z}$, $m \geq 1$, is normal, and we can form the factor group $\mathbf{Z}/m\,\mathbf{Z}$. Its elements are the cosets $a + m\,\mathbf{Z}$ of $m\,\mathbf{Z}$ in $\mathbf{Z}$, and these are simply the residue classes of $\mathbf{Z}$ modulo m. Thus, $\mathbf{Z}/m\,\mathbf{Z}$ consists of the m objects $\lfloor 0 \rfloor_m, \lfloor 1 \rfloor_m, \ldots, \lfloor m-1 \rfloor_m$—and by the way addition is defined in $\mathbf{Z}/m\,\mathbf{Z}$ (namely, $\lfloor a \rfloor_m + \lfloor b \rfloor_m = \lfloor a+b \rfloor_m$) it is clear that the quotient group $\mathbf{Z}/m\,\mathbf{Z}$ is the same as the additive group $\{\mathbf{Z}_m, +\}$ of the ring $\{\mathbf{Z}_m, +, \cdot\}$. We shall write

$$\mathbf{Z}/m\,\mathbf{Z} = \mathbf{Z}_m \qquad \text{(as additive groups).}$$

We turn to a discussion of mappings of groups, and shall soon see how they are related to normal subgroups and factor groups.

4-4-5. Definition. Suppose groups G and G' are given and, for convenience, let them both be multiplicative. A mapping $\phi: G \to G'$ is said to be a **homomorphism** when

$$\phi(ab) = \phi(a)\phi(b)$$

for all $a, b \in G$. In addition: If ϕ is surjective (that is, onto) we call it an **epimorphism**; if ϕ is injective (that is, one-to-one) we call it a **monomorphism**; if ϕ is surjective and injective (that is, an epimorphism and a monomorphism) we call it an **isomorphism** and write $G \approx G'$.

These definitions are entirely in the same spirit as those given for rings in 2-5-6. In general, when we are concerned with two algebraic objects of the same type, a mapping of one into the other is said to be a "homomorphism" when it preserves all the corresponding operations. A homomorphism which is surjective and injective is said to be an "isomorphism"; in other words, the term isomorphism signifies that the two algebraic objects should be considered to be the "same."

The basic properties of a homomorphism of groups are analogous to those of a homomorphism of rings. For example, in analogy with 2-5-8 and 2-5-9, we have:

4-4-6. Proposition. Suppose G is a multiplicative group with identity e, and G' is a multiplicative group with identity e'. If $\phi: G \to G'$ is a homomorphism then:

(1) $\phi(e) = e'$,
(2) $\phi(a^{-1}) = \phi(a)^{-1}$ for all $a \in G$,
(3) $\phi(a^n) = \phi(a)^n$ for all $a \in G$, $n \in \mathbf{Z}$,
(4) $\phi(G)$, the image of ϕ, is a subgroup of G',
(5) if we define the **kernel** of ϕ (and denote it as: ker ϕ) by

$$\ker \phi = \{a \in G \mid \phi(a) = e'\},$$

then ker ϕ is a normal subgroup of G,
(6) $\phi(a) = \phi(b) \Leftrightarrow ab^{-1} \in \ker \phi \Leftrightarrow a^{-1}b \in \ker \phi$,
(7) ϕ is injective $\Leftrightarrow \ker \phi = (e)$,
(8) if ϕ is injective, then ϕ is an isomorphism of G onto $\phi(G)$.

Proof: (1) $\phi(e) = \phi(e^2) = \phi(e)\phi(e)$, which implies $\phi(e) = e'$.

(2) $e' = \phi(e) = \phi(aa^{-1}) = \phi(a)\phi(a^{-1})$ which says that $\phi(a^{-1})$ is the inverse of $\phi(a)$ in G'; so $\phi(a^{-1}) = \phi(a)^{-1}$.

(3) $\phi(a^n) = \phi(a)^n$ holds for $n = 0$ [as $\phi(e) = e'$], for $n = 1$, for $n = 2$ [as $\phi(a^2) = \phi(a)\phi(a) = \phi(a)^2$], and by induction it holds for all $n > 0$. Then for any $n < 0$ we have $-n > 0$ and

$$\phi(a^n) = \phi[(a^{-1})^{-n}] = (\phi(a^{-1}))^{-n} = (\phi(a)^{-1})^{-n} = \phi(a)^n.$$

(4) Consider $a', b' \in \phi(G)$—so $a' = \phi(a)$, $b' = \phi(b)$ for some $a, b \in G$. Then $a'b' = \phi(a)\phi(b) = \phi(ab) \in \phi(G)$ and $(a')^{-1} = \phi(a)^{-1} = \phi(a^{-1}) \in \phi(G)$. Hence, $\phi(G)$ is closed under multiplication and taking of inverses, so it is a subgroup of G'. One expresses this fact by saying: A homomorphic image of a group is a group.

(5) Write $N = \ker \phi$. Then, $a, b \in N$ means that $\phi(a) = e'$, $\phi(b) = e'$, and we have

$$\phi(ab^{-1}) = \phi(a)\phi(b^{-1}) = \phi(a)\phi(b)^{-1} = e'e' = e'$$

—so $ab^{-1} \in N$ and N is a subgroup. To prove it is normal, we show $aNa^{-1} \subset N$ for every $a \in G$. In fact, for any $c \in N$ and any $a \in G$ we have

$$\phi(aca^{-1}) = \phi(a)\phi(c)\phi(a)^{-1} = \phi(a)e'\phi(a)^{-1} = e'$$

so $aca^{-1} \in N$ and indeed $aNa^{-1} \subset N$.

(6) Still writing $N = \ker \phi$, we have

$$\begin{aligned}\phi(a) = \phi(b) &\Leftrightarrow \phi(a)\phi(b)^{-1} = e'\\ &\Leftrightarrow \phi(ab^{-1}) = e'\\ &\Leftrightarrow ab^{-1} \in N.\end{aligned}$$

The other condition arises from the fact that N is normal—in detail

$$\begin{aligned}ab^{-1} \in N &\Leftrightarrow a \in Nb\\ &\Leftrightarrow a \in bN\\ &\Leftrightarrow b^{-1}a \in N\\ &\Leftrightarrow (b^{-1}a)^{-1} \in N\\ &\Leftrightarrow a^{-1}b \in N.\end{aligned}$$

Our formulation of (6) probably hides its full significance; another way to express (6)—one which comes closer to the mark—is: The elements a and b have the same image under ϕ $\Leftrightarrow$ they belong to the same coset of $N = \ker \phi$ in G.

(7) If ϕ is injective and $a \in \ker \phi$, then $\phi(a) = e' = \phi(e)$; so $a = e$ because ϕ is injective, and hence $\ker \phi = (e)$.

Conversely, if $\ker \phi = (e)$, then using (6),

$$\phi(a) = \phi(b) \Leftrightarrow ab^{-1} \in \ker \phi = (e) \Leftrightarrow a = b$$

—so, ϕ is injective.

(8) ϕ is an injective homomorphism of $G \to G'$, so using (4), the homomorphism $\phi\colon G \to \phi(G)$ is injective and surjective. ∎

Incidentally, if the group G is multiplicative while G' is additive, then the condition for a homomorphism takes the form

$$\phi(ab) = \phi(a) + \phi(b).$$

Of course, results such as 4-4-6 remain valid—with trivial modifications of notation.

4-4-7. Proposition. Suppose $\phi: G \to G'$ and $\psi: G' \to G''$ are homomorphisms of groups; then

(1) $\psi \circ \phi: G \to G''$ is a homomorphism. In words; the composition of homomorphisms is a homomorphism.
(2) If both ϕ and ψ are surjective, so is $\psi \circ \phi$.
(3) If both ϕ and ψ are injective, so is $\psi \circ \phi$.
(4) If both ϕ and ψ are isomorphisms, so is $\psi \circ \phi$. In words: The composition of isomorphisms is an isomorphism.
(5) If ϕ is an isomorphism, then ϕ^{-1} exists and is an isomorphism.

Proof: The details go like those of 2-5-10, so they are left to the reader.

The reader will also note the following immediate consequence of these facts:

4-4-8. Proposition. Isomorphism is an equivalence relation on the set of all groups.

4-4-9. Examples. (1) Consider the symmetric groups S_4 and S_5. The elements of S_4 may be viewed as all the permutations of the set $\{1, 2, 3, 4\}$, and the elements of S_5 may be viewed as all the permutations of the set $\{1, 2, 3, 4, 5\}$. Let us define a mapping

$$\phi: S_4 \to S_5$$

as follows: For $\sigma \in S_4$ let $\phi(\sigma) = \sigma' \in S_5$ be the permutation whose action is

$$\sigma'(1) = \sigma(1), \qquad \sigma'(2) = \sigma(2), \qquad \sigma'(3) = \sigma(3), \qquad \sigma'(4) = \sigma(4), \qquad \sigma'(5) = 5.$$

In other words, σ' is the same as σ on $\{1, 2, 3, 4\}$ and it keeps 5 fixed.

Obviously, ϕ is a homomorphism. It is injective since $\phi(\sigma) = \sigma' = e \in S_5$ implies $\sigma(1) = 1$, $\sigma(2) = 2$, $\sigma(3) = 3$, $\sigma(4) = 4$, so σ is the identity of S_4) but not surjective (for example, the element $\begin{pmatrix} 1 & 2 & 3 & 4 & 5 \\ 2 & 4 & 3 & 5 & 1 \end{pmatrix}$ of S_5 is not in the image of ϕ). Thus, according to 4-4-6, part (8), S_5 contains an isomorphic copy of S_4—namely, the image of S_4 under ϕ.

In the same way, for any $n \geq 1$, S_{n+1} contains an isomorphic copy of S_n (that is, we can produce an injective homomorphism of S_n into S_{n+1}) so S_n may be viewed as a subgroup of S_{n+1}.

(2) Consider the additive groups $\mathbf{Z}_m$ and $\mathbf{Z}_n$ where $n \mid m$. Define a mapping

$$\phi: \mathbf{Z}_m \to \mathbf{Z}_n$$

by putting

$$\phi(\lfloor a \rfloor_m) = \lfloor a \rfloor_n, \qquad a \in \mathbf{Z}.$$

In the first place, ϕ is well defined—that is, $\lfloor a' \rfloor_m = \lfloor a \rfloor_m$ implies $\phi(\lfloor a' \rfloor_m) = \phi(\lfloor a \rfloor_m)$—since

$$\begin{aligned}\lfloor a' \rfloor_m = \lfloor a \rfloor_m &\Rightarrow m \mid (a' - a)\\ &\Rightarrow n \mid (a' - a) \Rightarrow \lfloor a' \rfloor_n = \lfloor a \rfloor_n\\ &\Rightarrow \phi(\lfloor a' \rfloor_m) = \phi(\lfloor a \rfloor_m).\end{aligned}$$

Now, ϕ is a homomorphism—since

$$\begin{aligned}\phi(\lfloor a \rfloor_m + \lfloor b \rfloor_m) = \phi(\lfloor a + b \rfloor_m) &= \lfloor a + b \rfloor_n\\ &= \lfloor a \rfloor_n + \lfloor b \rfloor_n = \phi(\lfloor a \rfloor_m) + \phi(\lfloor b \rfloor_m.\end{aligned}$$

Obviously, ϕ is surjective. As for the kernel of ϕ,

$$\begin{aligned}\lfloor a \rfloor_m \in \ker \phi &\Rightarrow \phi(\lfloor a \rfloor_m) = \lfloor 0 \rfloor_n\\ &\Rightarrow \lfloor a \rfloor_n = \lfloor 0 \rfloor_n\\ &\Rightarrow n \mid a.\end{aligned}$$

Hence, if we write $m = nd$, then $\ker \phi$ consists of the d elements $0, n, 2n, 3n, \ldots, (d-1)n$ of $\mathbf{Z}_m$; of course, these form a (normal) subgroup of $\mathbf{Z}_m$.

(3) Consider the additive group of reals $\mathbf{R} = \{\mathbf{R}, +\}$ and the multiplicative group of positive reals $\mathbf{R}_{>0} = \{\mathbf{R}_{>0}, \cdot\}$. Consider the map

$$\log\colon \mathbf{R}_{>0} \to \mathbf{R}$$

where "log" refers (for simplicity) to the logarithm to the base e. We recall from elsewhere that the logarithm is defined only for positive real numbers

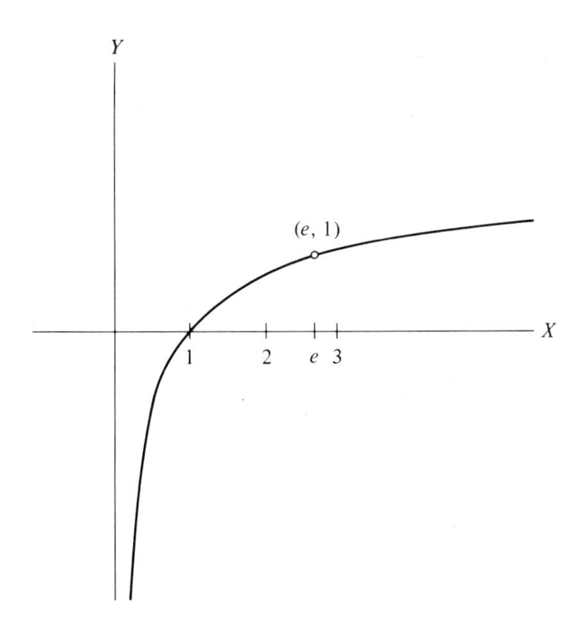

and its graph is shown in the accompanying figure. The crucial and well-known algebraic property of log (which we take for granted here) is

$$\log(ab) = \log a + \log b$$

for all positive reals a, b. In other words, log: $\mathbf{R}_{>0} \to \mathbf{R}$ is a homomorphism. Is it injective? surjective? an isomorphism? An elegant way to settle these questions is to define a function

$$\exp: \mathbf{R} \to \mathbf{R}_{>0}$$

by putting, for all $x \in \mathbf{R}$,

$$\exp(x) = e^x.$$

The graph of this function is shown in the accompanying figure, and $\exp x$

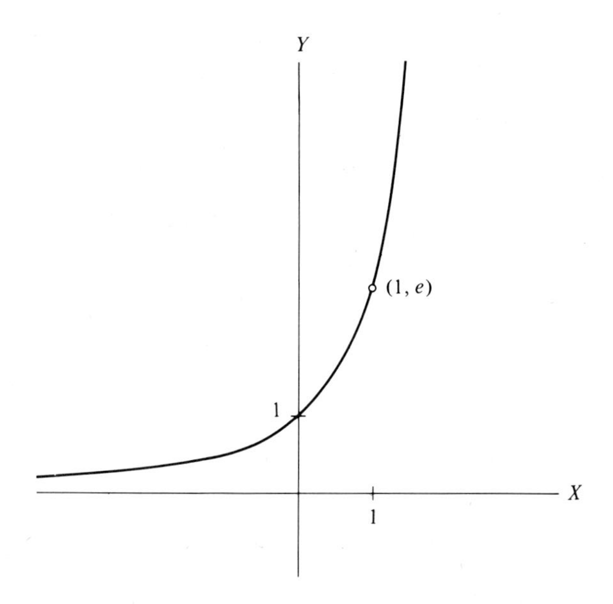

is always in $\mathbf{R}_{>0}$. Of course, exp is a homomorphism, since for $x, y \in \mathbf{R}$

$$\exp(x + y) = e^{x+y} = e^x e^y = (\exp x)(\exp y)$$

(the standard rule for exponents, $e^x e^y = e^{x+y}$, being taken for granted here). Now, let us examine the composition of these maps. For $a \in \mathbf{R}_{>0}$,

$$(\exp \log)a = \exp(\log a) = e^{\log a} = a$$

and for $b \in \mathbf{R}$

$$(\log \exp)b = \log(\exp b) = \log(e^b) = b.$$

Thus, $\exp \circ \log$ is the identity map of $\mathbf{R}_{>0}$, and $\log \circ \exp$ is the identity map of $\mathbf{R}$; so log and exp are inverses of each other—$\exp = \log^{-1}$ and $\log = \exp^{-1}$. In particular, both log and exp are 1–1 and onto—so they are isomorphisms.

Of course, when logarithms were introduced several hundred years ago, it was for the purpose of replacing multiplication of numbers by the "easier" (in terms of computational effort) operation of addition. The fact that log is an isomorphism makes it all legal.

4-4-10. Theorem (Cayley: 1821–1895). Any group G is isomorphic to a group of permutations.

Proof: Suppose the group $G = \{a, b, c, \ldots\}$ is multiplicative. Viewing G as a set, let S_G denote the group of all permutations of this set. The operation in S_G is composition, $\circ$. We shall show that G is isomorphic to some subgroup of S_G by exhibiting an injective homomorphism of G into S_G—and, in this way, G will be isomorphic to a group whose elements are permutations, which is what Cayley's theorem asserts.

For each $a \in G$, define a mapping

$$L_a: G \to G$$

by putting

$$L_a(x) = ax, \qquad x \in G.$$

In other words, L_a is left multiplication by a in G. Now, L_a is a permutation of G—that is, $L_a \in S_G$—since:

(*i*) given $b \in G$ there exists $x \in G$ for which $ax = b$, so L_a is surjective,
(*ii*) $L_a(x_1) = L_a(x_2) \Rightarrow ax_1 = ax_2 \Rightarrow x_1 = x_2$, so L_a is injective.

Now, define a mapping

$$\phi: G \to S_G$$

by putting

$$\phi(a) = L_a, \qquad a \in G.$$

To prove that ϕ is a homomorphism, we observe first that for $a, b \in G$, $L_a \circ L_b = L_{ab}$ [since for any $x \in G$, $(L_a \circ L_b)(x) = L_a(L_b(x)) = L_a(bx) = abx = L_{ab}(x)$]. Consequently, for any $a, b \in G$, we have

$$\phi(ab) = L_{ab} = L_a \circ L_b = \phi(a) \circ \phi(b),$$

and ϕ is a homomorphism.

It remains to show ϕ is injective—that is, $\ker \phi = (e)$. To do this, suppose $a \in \ker \phi$. Then $\phi(a) = L_a$ is the identity element of S_G—that is, L_a is the permutation of G which leaves every element fixed. In other words,

$$L_a(x) = x \text{ for every } x \in G,$$

which says that $ax = x$ for every $x \in G$. This clearly implies $a = e$. The proof is now complete. ▮

The next order of business is to expose the connection between normal subgroups and homomorphisms. Roughly speaking, given a normal subgroup N of G there exists a "canonical" (meaning "standardized") homomorphism whose kernel is N. On the other hand any homomorphism of G into some group determines a normal subgroup N of G, namely its kernel. In detail, we have:

4-4-11. Proposition. Suppose N is a normal subgroup of G, then there is a canonical homomorphism

$$\pi : G \to G/N$$

defined by

$$\pi(a) = \lfloor a \rfloor_N = aN.$$

In fact, π is a surjective homomorphism whose kernel is N.

Proof: Of course, G/N is the factor group, as described in 4-4-2. The map π is the most natural mapping of $G \to G/N$—it maps each element of G to the coset which it determines.

Now, π is a homomorphism—since for any $a, b \in G$ we have

$$\pi(ab) = \lfloor ab \rfloor_N = \lfloor a \rfloor_N \cdot \lfloor b \rfloor_N = \pi(a) \cdot \pi(b).$$

Furthermore, π is surjective—for given an arbitrary element of G/N it is of form $\lfloor x \rfloor_N$ for some $x \in G$, and then $\pi(x) = \lfloor x \rfloor_N$.

As for the kernel of π: since the identity of G/N is $\lfloor e \rfloor_N = N$, we have

$$\begin{aligned} a \in \ker \pi &\Leftrightarrow \pi(a) = \lfloor e \rfloor_N \\ &\Leftrightarrow \lfloor a \rfloor_N = \lfloor e \rfloor_N = N \\ &\Leftrightarrow a \in N. \end{aligned}$$ ▮

4-4-12. Isomorphism Theorem. Suppose the mapping $\phi : G \to G'$ is a surjective homomorphism. If its kernel is N, then the factor group G/N is isomorphic to G'—that is,

$$G/N \approx G'.$$

Proof: The kernel N is a normal subgroup of G, so the factor group G/N exists. We define a mapping

$$\bar{\phi}: G/N \to G'$$

by putting

$$\bar{\phi}(\lfloor a \rfloor_N) = \phi(a).$$

Because the definition of $\bar{\phi}$ depends on the choice of representative for the coset $\lfloor a \rfloor_N$, we must show that $\bar{\phi}$ is well defined. This follows from

$$\begin{aligned} \lfloor a \rfloor_N = \lfloor b \rfloor_N &\Rightarrow b \in aN \\ &\Rightarrow a^{-1}b \in N \\ &\Rightarrow \phi(a^{-1}b) = e', \text{ the identity of } G' \\ &\Rightarrow \phi(a) = \phi(b) \\ &\Rightarrow \phi(\lfloor a \rfloor_N) = \phi(\lfloor b \rfloor_N). \end{aligned}$$

Now, $\bar{\phi}$ is a homomorphism, since

$$\begin{aligned} \bar{\phi}(\lfloor a \rfloor_N \cdot \lfloor b \rfloor_N) &= \bar{\phi}(\lfloor ab \rfloor_N) \\ &= \phi\,(ab) \\ &= \phi(a) \cdot \phi(b) \\ &= \bar{\phi}(\lfloor a \rfloor_N) \cdot \phi(\lfloor b \rfloor_N). \end{aligned}$$

The image of G/N under $\bar{\phi}$ is

$$\begin{aligned} \{\bar{\phi}(\lfloor a \rfloor_N) \mid \lfloor a \rfloor_N \in G/N\} &= \{\phi(a) \mid a \in G\} \\ &= G' \qquad \text{(since } \phi \text{ is surjective)} \end{aligned}$$

—so $\bar{\phi}$ is surjective.

As for the kernel of $\bar{\phi}$, we have

$$\begin{aligned} \lfloor a \rfloor_N \in \ker \bar{\phi} &\Leftrightarrow \bar{\phi}(\lfloor a \rfloor_N) = e' \\ &\Leftrightarrow \phi(a) = e' \\ &\Leftrightarrow a \in \ker \phi \\ &\Leftrightarrow a \in N \\ &\Leftrightarrow \lfloor a \rfloor_N \text{ is the identity element of } G/N. \end{aligned}$$

Hence, $\bar{\phi}$ is injective—so $\bar{\phi}$ is an isomorphism of G/N onto G'. ▌

A picture commonly associated with this result is

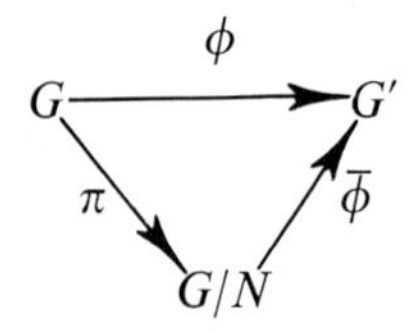

and one says that the diagram "commutes" (more accurately, "is commutative") because of the relation

$$\bar{\phi} \circ \pi = \phi.$$

We content ourselves here with a single application of the isomorphism theorem. Suppose $G = [a]$ is an arbitrary cyclic group (written multiplicatively). Define a mapping

$$\phi\colon \mathbf{Z} \to G$$

by putting

$$\phi(n) = a^n.$$

For any $n_1, n_2 \in \mathbf{Z}$ we have

$$\phi(n_1 + n_2) = a^{n_1 + n_2} = a^{n_1}a^{n_2} = \phi(n_1)\phi(n_2)$$

so ϕ is a homomorphism of the additive group of integers, $\{\mathbf{Z}, +\}$, into G. Furthermore, ϕ is surjective because

$$\text{image } \phi = \{\phi(n) \mid n \in \mathbf{Z}\} = \{a^n \mid n \in \mathbf{Z}\} = [a] = G.$$

Now, consider $N = \ker \phi$. If $N = (0)$, then ϕ is injective—so it is an isomorphism of $\{\mathbf{Z}, +\}$ onto G. On the other hand, if $N \neq (0)$, then N is of form $m\mathbf{Z}$ for some $m \geq 1$, so ϕ is a surjective homomorphism of $\mathbf{Z} \to G$ with kernel $N = m\mathbf{Z}$ and the isomorphism theorem tells us that the factor group $\mathbf{Z}/m\mathbf{Z}$ is isomorphic to G. But $\mathbf{Z}/m\mathbf{Z}$ is really the additive group of the ring $\mathbf{Z}_m$ [as noted in 4-4-4, part (4)], so we have

$$\{\mathbf{Z}_m, +\} \approx G.$$

Let us summarize:

4-4-13. Proposition. Suppose G is a cyclic group; then

$$G \approx \mathbf{Z} \qquad \text{if } \#(G) = \infty,$$
$$G \approx \mathbf{Z}_m \qquad \text{if } \#(G) = m.$$

In particular, any two cyclic groups of the same order, are isomorphic.

Proof: According to the above, G is isomorphic to $\mathbf{Z}$ or to $\mathbf{Z}_m$ for some $m \geq 1$. Clearly, the first case (that is, $G \approx \mathbf{Z}$) occurs if and only if $\#(G) = \infty$, and the second case occurs if and only if $\#(G) < \infty$. Now, if $G \approx \mathbf{Z}_m$, then $\#(G) = \#(\mathbf{Z}_m) = m$ (because isomorphic finite groups have the same number of elements), so if G is cyclic of order m it must be isomorphic to $\mathbf{Z}_m$.

Finally, consider two cyclic groups G and G' of the same order. If $\#(G) = \#(G') = \infty$, they are both isomorphic to $\mathbf{Z}$; if $\#(G) = \#(G') = m$, they are both isomorphic to $\mathbf{Z}_m$. In either case, $G \approx G'$. ▮

4-4-14. Exercise. Let $G = \{a, b, c, \ldots\}$ be a group with identity e, and let $S_G = \{\sigma, \tau, \rho, \ldots\}$ be the group of all permutations of the set G. The operation in S_G is composition of mappings, and let us denote the identity of S_G by 1.

(1) Any isomorphism of G onto itself is known as an **automorphism** of G. If $\mathfrak{A}_G$ denotes the set of all automorphisms of G (note that $1 \in \mathfrak{A}_G$), then $\mathfrak{A}_G$ is a subgroup of S_G; it is known as the **automorphism group** of G.

(2) For each $a \in G$ define the mapping

$$I_a\colon G \to G$$

by putting

$$I_a(x) = axa^{-1}, \qquad x \in G.$$

This mapping I_a is known as **conjugation** by a. Show that I_a is an automorphism of G; it is called the **inner automorphism** determined by a.

(3) Any element of form $I_a(x) = axa^{-1}$ is said to be a **conjugate** of x. If H is a subgroup of G, then $I_a(H) = aHa^{-1}$ is a subgroup of G, known as a **conjugate** of H. If we put

$$N = \bigcap_{a \in G} aHa^{-1},$$

then N is a normal subgroup of G. In addition, H is a normal subgroup of G if and only if every conjugate of H is equal to H.

(4) Let $\mathcal{I}_G$ denote the set of all inner automorphisms of G; then $\mathcal{I}_G$ is a normal subgroup of $\mathfrak{A}_G$—it is known as the **group of inner automorphisms**. (Incidentally, an automorphism of G which is not inner is called an **outer automorphism**.) Denoting the center of G by $\mathfrak{Z}_G$, we have (by considering the mapping $a \to I_a$)

$$G/\mathfrak{Z}_G \approx \mathcal{I}_G.$$

4-4-15 / PROBLEMS

1. Consider the octic group $G = \{e, \sigma_1, \sigma_2, \sigma_3, \tau_1, \tau_2, \rho_1, \rho_2\}$ as discussed in 4-1-10 and 4-2-6. Decide which of the following subgroups are normal:

(*i*) $\{e, \sigma_2\}$, (*ii*) $\{e, \tau_1\}$,
(*iii*) $\{e, \tau_2\}$, (*iv*) $\{e, \rho_1\}$,
(*v*) $\{e, \rho_2\}$, (*vi*) $\{e, \sigma_1, \sigma_2, \sigma_3\}$,
(*vii*) $\{e, \rho_1, \rho_2, \sigma_2\}$, (*viii*) $\{e, \tau_1, \tau_2, \sigma_2\}$.

Are there any other proper subgroups of the octic group?

2. Exhibit groups $G \supset H \supset K$ such that K is normal in H, but K is not normal in G. Can this be done if H is also required to be normal in G?

3. If N is a normal subgroup of the finite group G, what is the relation between the orders of the three groups G/N, G, N?

4. Show that the intersection of any collection of normal subgroups of a given group G is a normal subgroup of G.

5. Let $G = S_X$ be the group of permutations of the set X, and for a fixed $y \in X$ consider the subgroup $H = H_y = \{\sigma \in G \mid \sigma y = y\}$. If X has more than 2 elements, show that H is not normal in G.

6. If H is a subgroup of G of index 2 (that is, $(G{:}H) = 2$), show that H must be normal.

7. If N is a normal subgroup of G and H is an arbitrary subgroup of G, prove that $H \cap N$ is a normal subgroup of H.

8. For any elements a, b in the group G, let us write

$$[a, b] = aba^{-1}b^{-1}$$

and call it the **commutator** of a and b. Let C denote the set of all finite products of commutators; then prove:

(*i*) C is a subgroup (known as the **commutator subgroup** or the **derived group**) of G; in fact, C is the subgroup generated by the set of all commutators.
(*ii*) C is normal in G.
(*iii*) The factor group G/C is abelian.

9. Exhibit two subgroups H and K of $G = S_3$ such that HK is not a subgroup.

10. Let H and K be subgroups of G; show that:

(*i*) HK is a subgroup of $G \Leftrightarrow HK = KH$.
(*ii*) If H is normal, then HK is a subgroup.
(*iii*) If both H and K are normal, then HK is a normal subgroup.

11. Suppose H and N are normal subgroups of G with $H \cap N = (e)$; show that any element of H commutes with any element of N (that is: $a \in H, b \in N \Rightarrow ab = ba$; equivalently, $a \in H, b \in N \Rightarrow aba^{-1}b^{-1} = e$).

12. For elements $x, y \in G$ let us write

$$x \sim y \Leftrightarrow y = axa^{-1} \text{ for some } a \in G.$$

In words, $x \sim y$ means that y is a conjugate of x. Show that $\sim$ is an equivalence relation.

The equivalence class to which x belongs consists of a single element (namely, x itself) $\Leftrightarrow$ x is in the center of G.

13. Suppose $\phi: G \to G'$ is an isomorphism of groups. Prove that
(*i*) G is abelian $\Leftrightarrow$ G' is abelian,
(*ii*) G is cyclic $\Leftrightarrow$ G' is cyclic.

14. Suppose $\phi: G \to G'$ is a surjective homomorphism. Show that
(*i*) G is abelian $\Rightarrow G'$ is abelian,
(*ii*) G is cyclic $\Rightarrow G'$ is cyclic.
In other words, a homomorphic image of an abelian group is abelian and a homomorphic image of a cyclic group is cyclic. Are the converses of these statements true? Justify your answers.

15. For $i = \sqrt{-1}$ the mapping $n \to i^n$ is a homomorphism of $\{\mathbf{Z}, +\}$ into $\{\mathbf{C}^*, \cdot\}$. What is the image? What is the kernel?

16. Suppose N is a normal subgroup of G. Prove:
(*i*) if G is abelian so is G/N,
(*ii*) if G is cyclic so is G/N.
Why is this problem essentially the same as Problem 14?

17. Suppose the group G has exactly one element of order 2—call it a. Show that a is in the center of G.

18. In 4-4-9, part (1), we indicated the existence of an injective homomorphism of $S_n \to S_{n+1}$. Find as many such injective homomorphisms as you can.

19. The additive group of integers, $\{\mathbf{Z}, +\}$, is a normal subgroup of the additive group of rationals, $\{\mathbf{Q}, +\}$. Every element of the factor group $\mathbf{Q}/\mathbf{Z}$ has finite order (show how to find the order) but the group has infinite order. Prove these assertions.

20. (*i*) The "absolute value" $z \to |z|$ is a homomorphism of $\mathbf{C}^*$ (the multiplicative group of nonzero complex numbers) into $\mathbf{R}^*$ (the multiplicative group of nonzero reals). What is the image? What is the kernel?
(*ii*) The circle group $W = \{z \in \mathbf{C} \mid |z| = 1\}$ (see 4-1-8) is a normal subgroup of $\mathbf{C}^*$. Describe the factor group $\mathbf{C}^*/W$.

21. If G is cyclic, then for any subgroup N both N and G/N are cyclic. Show that the converse is false; in other words, the normal subgroup N and the factor group G/N can both be cyclic without G being cyclic—as a matter of fact, G need not even be abelian.

22. Prove that if p is prime, then, upto isomorphism, there exists a unique group of order p—namely, the cyclic group of order p.

23. (*i*) The following three groups of order 4 are isomorphic:
(*a*) $\{\mathbf{Z}_8^*, \cdot\}$, whose elements are 1, 3, 5, 7,
(*b*) the four group (as defined in 4-1-12, Problem 16),
(*c*) the direct product $G_1 \times G_2$ (see 4-1-12, Problem 26) of two cyclic groups G_1 and G_2 of order 2.
Prove this by producing explicit isomorphisms.
(*ii*) Another group of order 4 is the cyclic one—say $\{\mathbf{Z}_4, +\}$. It is not isomorphic to the groups in (*i*).
(*iii*) There are no other groups of order 4; in other words, a group of order 4 is isomorphic to the four group or to $\{\mathbf{Z}_4, +\}$. In particular, any group of order 4 is abelian.

24. The symmetric group S_3 and the cyclic group $\{\mathbf{Z}_6, +\}$ are nonisomorphic (that is, "distinct") groups of order 6. Show that, upto isomorphism, there are no other groups of order 6.

25. Suppose $\phi: G \to G'$ is a homomorphism and $a \in G$. Prove:
(*i*) If ϕ is an isomorphism, then $\text{ord}(\phi(a)) = \text{ord } a$. In fact, the same conclusion holds if ϕ is injective.
(*ii*) If ϕ is surjective, then $\text{ord}\,(\phi(a))$ divides $\text{ord } a$.

26. The groups $\{\mathbf{Z}_{10}, +\}$ and $\{\mathbf{Z}_{11}^*, \cdot\}$ are both cyclic of order 10. Can you find an isomorphism

$$\phi: \mathbf{Z}_{10} \to \mathbf{Z}_{11}^*$$

for which
(*i*) $\phi(1) = 2$, (*ii*) $\phi(1) = 3$?
Find as many isomorphisms of $\mathbf{Z}_{10} \to \mathbf{Z}_{11}^*$ as possible.

27. If p is prime, the groups $\{\mathbf{Z}_{p-1}, +\}$ and $\{\mathbf{Z}_p^*, \cdot\}$ are isomorphic. How would you go about finding all isomorphisms between them?

28. Suppose $G = [a]$ is cyclic, then prove:
(*i*) If $\phi: G \to G'$ is a homomorphism it is completely determined once we know $\phi(a)$.
(*ii*) Consider any $a' \in G'$. If we put $\psi(a) = a'$ this determines a mapping $\psi: G \to G'$ defined by $\psi(a^n) = (\psi(a))^n = (a')^n$ for all $n \in \mathbf{Z}$. What condition must a' satisfy in order that ψ be a homomorphism? Is ψ really a mapping?
(*iii*) Apply this information in Problems 26 and 27.

29. Show that the only inner automorphism of an abelian group is the identity map.

30. Consider each of the following groups, G,
(*i*) $\{\mathbf{Z}, +\}$, (*ii*) $\{\mathbf{Z}_7, +\}$,
(*iii*) the cyclic group of prime order p, with multiplication as the operation and generator a,
(*iv*) the four group, (*v*) the symmetric group, S_3,
(*vi*) the cyclic multiplicative group $[a]$ of order n.
In each case, describe all the automorphisms; how many are there?; which ones are inner automorphisms?; what is the structure of the group of automorphisms, $\mathfrak{A}_G$?

31. By an **endomorphism** of a group G we mean a homomorphism of G into itself. If G is cyclic of order n, describe all endomorphisms of G. How many are there?

32. If the groups G and G' are isomorphic then their automorphism groups $\mathfrak{A}_G$ and $\mathfrak{A}_G'$ are isomorphic. In addition, there is a 1–1 correspondence between $\mathfrak{A}_G$ and the set of all isomorphisms of $G \to G'$—in fact, if ϕ: $G \to G'$ is a fixed isomorphism then

$$\sigma \to \phi \circ \sigma, \qquad \sigma \in G$$

is the desired 1–1 correspondence.

33. The group $\{\mathbf{Z}_{15}^*, \cdot\}$ has $\phi(15) = 8$ elements. The element 2 has order 4, and the element 11 has order 2. If we write $G_1 = [2]$ and $G_2 = [11]$, these are cyclic groups of orders 4 and 2, respectively; show that $\mathbf{Z}_{15}^*$ is isomorphic to the direct product $G_1 \times G_2$.

34. Suppose H and N are subgroups of G with N normal. Then $H \cap N$ is a normal subgroup of H, and HN is a group in which N is a normal subgroup. The mapping of

$$H \to HN/N$$

defined by

$$a \to aN, \qquad a \in H$$

is a surjective homomorphism with kernel $H \cap N$. Consequently,

$$H/(H \cap N) \approx HN/N.$$

This is known as the **first** (or second if one counts 4-4-12 as the first) **isomorphism theorem**.

4-5. Permutation Groups

According to Cayley's theorem, 4-4-10, any group may be viewed as a group of permutations—that is, as a subgroup of the symmetric group S_X for some set X. Consequently, to study all possible groups, it suffices to investigate all symmetric groups and their subgroups. Historically, people were studying "groups of permutations" long before the axioms for a group were formulated. Thus, the content of this section (which is not needed for the sequel) is designed primarily to enlarge our knowledge of the symmetric group S_n and its subgroups. In particular, we will be obtaining information about an arbitrary finite group (since, as seen in the proof of Cayley's theorem, a group of order n is isomorphic to a subgroup of S_n). However, the reader should keep in mind that experience has shown that it is often better to deal with an abstract group rather than tie oneself down to a fixed concrete realization (that is, isomorphic copy) of the group.

4-5-1. Notation. We recall that $S_n = \{\sigma, \tau, \rho, \ldots\}$ is the group of all permutations of the set $X = \{1, 2, \ldots, n\}$, with the operation of multiplication (that is, composition) written as either $\sigma \circ \tau$ or $\sigma\tau$. Of course, $\sigma \in S_n$ means that σ is a mapping of $X \to X$ which is 1–1 and onto, where, as is customary, σ is 1–1 means

$$\text{for any } x, y \in X,\ \sigma x = \sigma y \Rightarrow x = y$$

and σ is onto means

$$\text{given any } y \in X \text{ there exists } x \in X \text{ such that } \sigma x = y.$$

We introduced a natural notation for elements of S_n in 4-1-11; here we shall modify and simplify the notation. The full story will be clear as soon as we treat the following elements of S_9.

(*i*) Consider the permutation

$$\sigma = \begin{pmatrix} 1 & 2 & 3 & 4 & 5 & 6 & 7 & 8 & 9 \\ 4 & 5 & 9 & 8 & 7 & 1 & 6 & 3 & 2 \end{pmatrix}.$$

We start with the fact that σ maps $1 \to 4$. It also maps $4 \to 8$—and continuing this procedure of applying σ to the image, we have in addition: $8 \to 3$, $3 \to 9$, $9 \to 2$, $2 \to 5$, $5 \to 7$, $7 \to 6$. At the next stage, we have $6 \to 1$, so the action of σ may be described by the "diagram"

$$1 \to 4 \to 8 \to 3 \to 9 \to 2 \to 5 \to 7 \to 6 \;(\to 1)$$

We express this more compactly by writing

$$\sigma = (148392576)$$

—the interpretation being obvious.

There is no particular reason for starting with 1 and pursuing its successive images under σ. Any elements 1, 2, ..., 9 would do just as well. For example, starting from 8, we arrive at

$$\sigma = (839257614)$$

which is another (obviously "equivalent") symbol for the permutation σ.

(*ii*) Consider the permutation

$$\sigma = \begin{pmatrix} 1 & 2 & 3 & 4 & 5 & 6 & 7 & 8 & 9 \\ 3 & 7 & 9 & 1 & 2 & 5 & 8 & 6 & 4 \end{pmatrix}.$$

Starting from 1, the action of σ gives: $1 \to 3 \to 9 \to 4$, and since $4 \to 1$, we should obviously write (1394) for

$$1 \to 3 \to 9 \to 4$$

However, this gives no information about what σ does to 2, 5, 6, 7, 8. Consequently, we start from 2, say, and obtain

$$2 \to 7 \to 8 \to 6 \to 5$$

The way to denote σ is then

$$\sigma = (1394)(27865).$$

Among the other ways to denote this same σ one has:

$$(4139)(27865), \; (1394)(86527), \; (78652)(9413), \; (27865)(1394),$$

and so on.

(*iii*) In the same spirit, the symbol

$$(182)(347)(596)$$

represents the permutation under which

$$1 \to 8 \to 2, \quad 3 \to 4 \to 7, \quad 5 \to 9 \to 6$$

—in other words, the element of S_9 under consideration is

$$\begin{pmatrix} 1 & 2 & 3 & 4 & 5 & 6 & 7 & 8 & 9 \\ 8 & 1 & 4 & 7 & 9 & 5 & 3 & 2 & 6 \end{pmatrix}.$$

Obviously, this may also be denoted by (821)(347)(659), (734)(659)(218), and so on.

(*iv*) Consider the permutation

$$\sigma = \begin{pmatrix} 1 & 2 & 3 & 4 & 5 & 6 & 7 & 8 & 9 \\ 7 & 9 & 1 & 8 & 5 & 3 & 2 & 4 & 6 \end{pmatrix}.$$

According to our procedure this can be written as

$$\sigma = (172963)(48)(5).$$

Note: The appearance of (5) signifies that $\sigma 5 = 5$—that is, 5 is kept fixed by σ. We shall write

$$\sigma = (172963)(48)$$

under the general convention that whenever an integer (from $1, 2, \ldots, 9$) does not appear it is understood to be fixed under the permutation. Thus, for

$$\begin{pmatrix} 1 & 2 & 3 & 4 & 5 & 6 & 7 & 8 & 9 \\ 1 & 4 & 2 & 7 & 8 & 6 & 3 & 9 & 5 \end{pmatrix}$$

we write (2473)(589) instead of (1)(2473)(6)(589).

(*v*) According to our conventions

$$\sigma = (375)(29)$$

is shorthand notation for

$$\sigma = \begin{pmatrix} 1 & 2 & 3 & 4 & 5 & 6 & 7 & 8 & 9 \\ 1 & 9 & 7 & 4 & 3 & 6 & 5 & 8 & 2 \end{pmatrix}$$

while

$$\tau = \begin{pmatrix} 1 & 2 & 3 & 4 & 5 & 6 & 7 & 8 & 9 \\ 3 & 2 & 9 & 1 & 5 & 6 & 7 & 8 & 4 \end{pmatrix}$$

is denoted by

$$\tau = (1394).$$

(*vi*) How one writes the identity permutation of S_9,

$$e = \begin{pmatrix} 1 & 2 & 3 & 4 & 5 & 6 & 7 & 8 & 9 \\ 1 & 2 & 3 & 4 & 5 & 6 & 7 & 8 & 9 \end{pmatrix},$$

is a matter of taste. When there is no way out, we will write $e = (1)$.

4-5-2. Remark. Without being overly precise, let us indicate some properties in S_n which are immediate consequences of our new notation.

Suppose $a_1, a_2, \ldots, a_r$ are distinct elements, chosen in any order, from the set $X = \{1, 2, \ldots, n\}$; then $(a_1, a_2, \ldots, a_r)$ is said to be a **cycle of length** r, or simply an r-**cycle**. This terminology is natural because $(a_1, a_2, \ldots, a_r)$ represents the element σ of S_n whose action is described by

$$a_1 \to a_2 \to \cdots \to a_{r-1} \to a_r \;(\to a_1)$$

—where it is understood that the remaining elements of X are fixed under σ. The use of the word "cycle" is based on the fact that the elements are permuted in "circular" or "cyclical" fashion, and we have

$$(a_1, a_2, \ldots, a_r) = (a_2, a_3, \ldots, a_r, a_1) = \cdots = (a_r, a_1, \ldots, a_{r-1}). \qquad (*)$$

Cycles of length r exist for any r satisfying $1 \leq r \leq n$. Any 1-cycle is of form (a_1), so it leaves every element of X (including a_1) fixed—that is, it represents the identity permutation. (Clearly, 1-cycles are not especially interesting and are usually ignored.) Any 2-cycle is of form $(a_1 a_2)$, $a_1 \neq a_2$; it is known as a **transposition** because it represents the permutation which interchanges—that is, transposes—a_1 and a_2, while leaving the remaining elements of X fixed. (Transpositions are especially important, as will be seen later.)

Two cycles $\sigma = (a_1, a_2, \ldots, a_r)$ and $\tau = (b_1, b_2, \ldots, b_s)$ in S_n are said to be **disjoint** when the sets $\{a_1, \ldots, a_r\}$ and $\{b_1, \ldots, b_s\}$ are disjoint—in other words, when the two cycles have no symbols in common. For example, in S_9, the 4-cycle $\sigma = (2473)$ and the 5-cycle $\tau = (58961)$ are disjoint, while the cycles $\sigma = (2473)$ and $\tau = (179)$ are not disjoint.

If we have several cycles $\sigma_1, \sigma_2, \ldots, \sigma_t$ in S_n, we say they are **disjoint** when every pair of them is disjoint. For example, in S_9, $\sigma_1 = (57)$, $\sigma_2 = (492)$, $\sigma_3 = (18)$ are disjoint; but if we take $\sigma_4 = (67)$, then $\sigma_1, \sigma_2, \sigma_3, \sigma_4$ are not disjoint.

According to the procedure illustrated in 4-5-1, every $\sigma \in S_n$ can be written as a product

$$\sigma = (a_1, \ldots, a_{r_1})(b_1, \ldots, b_{r_2}) \cdots (y_1, y_2, \ldots, y_{r_t}) \qquad (\#)$$

of disjoint cycles. In more detail, starting with any $a_1 \in X = \{1, 2, \ldots, n\}$, we pursue its successive images under σ until we return to a_1. This gives a cycle $(a_1, a_2, \ldots, a_{r_1})$. If this cycle exhausts the elements of X, then we are finished and $\sigma = (a_1, a_2, \ldots, a_{r_1})$. If this cycle does not exhaust X, choose any $b_1 \in X$ which does not appear in the cycle $(a_1, a_2, \ldots, a_{r_1})$. Pursuing the images of b_1 under σ, we obtain a cycle $(b_1, b_2, \ldots, b_{r_2})$, disjoint from the first one. If these two cycles together do not exhaust X, choose $c_1 \in X$ which does not appear in either cycle, and keep going. This process stops after a finite number of steps (because X is finite) and we finally have an expression of form $(\#)$ for σ.

In the expression (#) for σ the lengths of the cycles are $r_1, r_2, \ldots, r_t$ with $1 \leq r_1, r_2, \ldots, r_t \leq n$. Since each element of $X = \{1, 2, \ldots, n\}$ belongs to exactly one cycle, $r_1 + r_2 + \cdots + r_t$ (which is the sum of the lengths of the various cycles) is equal to n. The expression (#) for σ includes cycles of length 1, but because such cycles make no contribution they will usually be dropped (as was done in the specific examples 4-5-1). In the resulting expression for σ [which we still write in the general form (#)] we have therefore $r_1, r_2, \ldots, r_t \leq 2$ and $r_1 + r_2 + \cdots r_t \leq n$.

Whichever convention is used with regard to the 1-cycles, the order in which the cycles are listed is immaterial. This is clear from the way we obtain the expression for σ; in particular, if we start with b_1 (instead of a_1), then the left-most cycle of (#) is $(b_1, b_2, \ldots, b_{r_2})$. Furthermore, each of the individual cycles may be shifted—that is, "cycled"—as indicated in (*).

The conclusion to be drawn from all this is as follows. Any permutation can be expressed uniquely as a product of disjoint cycles—it being understood that uniqueness is upto shifts in the individual cycles and the order in which the various cycles appear.

The foregoing discussion is largely "intuitive" but it does convey what is going on. For the reader who craves greater precision, and in order to clarify the facts, we now sketch a formal approach to the same results.

Let $\sigma \in S_n$ be given and write, as usual, $X = \{1, 2, \ldots, n\}$. Consider any $a \in X$. Let us pursue the successive images of a under σ—they are

$$a, \quad \sigma a, \quad \sigma(\sigma a) = \sigma^2 a, \quad \sigma(\sigma^2 a) = \sigma^3 a, \ldots, \sigma^m a, \ldots.$$

The set of all such images is obviously given by

$$\{\sigma^m a \mid m \geq 0\}.$$

For any $b \in X$, let us define the symbol $a \equiv_\sigma b$ by

$$a \equiv_\sigma b \Leftrightarrow b = \sigma^m a \text{ for some } m \geq 0.$$

In other words, $a \equiv_\sigma b \Leftrightarrow b \in \{\sigma^m a \mid m \geq 0\}$. We assert that $\equiv_\sigma$ (which the reader may read any way he chooses) is an equivalence relation. First of all, $a \equiv_\sigma a$ for every $a \in X$, since $\sigma^0 a = ea = a$. In addition, $a \equiv_\sigma b$ and $b \equiv_\sigma c \Rightarrow b = \sigma^{m_1} a$, $c = \sigma^{m_2} b$ for certain $m_1 \geq 0$, $m_2 \geq 0 \Rightarrow c = \sigma^{m_2}(\sigma^{m_1} a) = \sigma^{m_1 + m_2} a$ with $m_1 + m_2 \geq 0 \Rightarrow a \equiv_\sigma c$, so $\equiv_\sigma$ is transitive. Finally, the symmetric law requires a preliminary comment. Since σ belongs to the finite group S_n, its order is finite—say ord $\sigma = s \geq 1$. Thus, $\sigma^s = e$ and $\sigma \cdot \sigma^{s-1} = e$—which says that $\sigma^{-1} = \sigma^{s-1}$. Consequently,

$$\begin{aligned} a \equiv_\sigma b &\Rightarrow b = \sigma^m a \text{ for some } m \geq 0 \\ &\Rightarrow a = \sigma^{-m} b, \qquad m \geq 0 \\ &\Rightarrow a = (\sigma^{-1})^m b, \qquad m \geq 0 \\ &\Rightarrow a = \sigma^{(s-1)m} b \qquad \text{with } (s-1)m \geq 0 \\ &\Rightarrow b \equiv_\sigma a \end{aligned}$$

Then, because $\equiv_\sigma$ is an equivalence relation on the set $X = \{1, 2, \ldots, n\}$ we know that X decomposes into a union of disjoint equivalence classes.

4-5-3. Definition. The equivalence class (with respect to $\equiv_\sigma$) to which the element a belongs will be denoted by $\lfloor a \rfloor_\sigma$. The set $\lfloor a \rfloor_\sigma$ is also known as the **orbit of** a under σ, and other notations for it are $\text{orb}_\sigma\, a$ or simply orb a. (Thus, any $\sigma \in S_n$ leads to a decomposition of X as a union of the disjoint orbits of σ.) An orbit is said to be **nontrivial** when it contains more than one element. A permutation $\sigma \in S_n$, is called a **cycle** when it has exactly one nontrivial orbit; if the number of elements (of X) in this orbit is r, then σ is said to be an r-**cycle**. Two or more cycles in S_n are **disjoint** when their orbits (by which we mean their nontrivial orbits) are disjoint.

We shall give illustrations of the definition in a moment, but first let us describe carefully what an orbit looks like. Given $\sigma \in S_n$, the orbit of any $a \in X$ is a subset of the finite set X, so $\text{orb}_\sigma\, a$ is a finite set. Which elements of X belong to $\text{orb}_\sigma\, a = \lfloor a \rfloor_\sigma$? For $b \in X$, we have (as indicated earlier)

$$\begin{aligned} b \in \text{orb}_\sigma\, a &\Leftrightarrow b \equiv_\sigma a \\ &\Leftrightarrow a \equiv_\sigma b \\ &\Leftrightarrow b = \sigma^m a \text{ for some } m \geq 0 \\ &\Leftrightarrow b \in \{\sigma^m a \mid m \geq 0\}. \end{aligned}$$

In other words,

$$\text{orb}_\sigma\, a = \{\sigma^m a \mid m \geq 0\}$$

and (in keeping with our desires and with the informal ideas in 4-5-1 and 4-5-2) the orbit of a is obtained by taking the successive images of a under σ. Of course, the set $\{\sigma^m a \mid m \geq 0\}$, being finite involves repeated elements. We may also note in passing that because $\text{ord}\, \sigma = s \geq 1$ and $\sigma^{-1} = \sigma^{s-1}$ (notation as before), we have

$$\{\sigma^m a \mid m \geq 0\} = \{\sigma^m a \mid m \in \mathbf{Z}\}$$

—so the relation $\equiv_\sigma$ could have been defined, at the start, by

$$a \underset{\sigma}{\equiv} b \Leftrightarrow b = \sigma^m a \text{ for some } m \in \mathbf{Z}.$$

This definition of $\equiv_\sigma$ is the more standard one for several reasons, one of which is the ease with which one then shows $\equiv_\sigma$ to be an equivalence relation.

Because $\sigma^s = e$ we surely have $\sigma^s a = a$, $s \geq 1$. Now let d be the smallest positive integer for which $\sigma^d a = a$ (obviously such an integer d exists). Consider the elements

$$a, \quad \sigma a, \quad \sigma^2 a, \ldots, \sigma^{d-1} a.$$

They all belong to $\text{orb}_\sigma\, a$. Furthermore, they are distinct—for if

$$\sigma^i a = \sigma^j a, \qquad \text{where } 0 \le i < j \le d - 1,$$

then

$$\sigma^{j-i} a = a \qquad \text{with } 0 \le j - i \le d - 1$$

which contradicts the definition of d. Finally, consider an arbitrary element of $\text{orb}_\sigma\, a$; it is of form $\sigma^m a$. Using the division algorithm, we write

$$m = qd + r, \qquad 0 \le r < d$$

and consequently,

$$\begin{aligned}\sigma^m a = \sigma^{r+qd} a &= \sigma^r(\sigma^{qd} a)\\ &= \sigma^r[(\sigma^d)^q(a)]\\ &= \sigma^r[\underbrace{(\sigma^d)(\sigma^d)\cdots(\sigma^d)}_{q \text{ times}}(a)]\\ &= \sigma^r a.\end{aligned}$$

In other words, any $\sigma^m a$ belongs to the set $\{a, \sigma a, \ldots, \sigma^{d-1} a\}$. This proves:

4-5-4. Proposition. Given $\sigma \in S_n$, then for any $a \in X = \{1, 2, \ldots, n\}$ its orbit under σ is

$$\text{orb}_\sigma\, a = \{a, \sigma a, \ldots, \sigma^{d-1} a\}$$

where d is the smallest positive integer for which $\sigma^d a = a$. The elements $a, \sigma a, \ldots, \sigma^{d-1} a$ are distinct.

4-5-5. Remarks. (*i*) In S_9, consider the permutation

$$\sigma = \begin{pmatrix} 1 & 2 & 3 & 4 & 5 & 6 & 7 & 8 & 9 \\ 7 & 9 & 4 & 1 & 8 & 5 & 3 & 6 & 2 \end{pmatrix}.$$

Its orbits are the sets $\{1, 3, 4, 7\}$, $\{2, 9\}$, $\{5, 6, 8\}$; they are disjoint and their union is $X = \{1, 2, \ldots, 9\}$. Of course, the order in which the orbits are listed is immaterial, as is the order in which the elements in each orbit are listed.

Similarly, the orbits of the permutation

$$\tau = \begin{pmatrix} 1 & 2 & 3 & 4 & 5 & 6 & 7 & 8 & 9 \\ 1 & 8 & 7 & 4 & 3 & 9 & 2 & 5 & 6 \end{pmatrix}$$

are the sets $\{1\}$, $\{2, 3, 5, 7, 8\}$, $\{4\}$, $\{6, 9\}$; they provide a disjoint decomposition of the set $X = \{1, 2, \ldots, 9\}$.

Phrasing things in accordance with the notation of 4-5-4, we have (for the σ just mentioned)

$$\operatorname{orb}_\sigma 1 = \{1, 7, 3, 4\}, \qquad \operatorname{orb}_\sigma 2 = \{2, 9\},$$
$$\operatorname{orb}_\sigma 3 = \{3, 4, 1, 7\}, \qquad \operatorname{orb}_\sigma 4 = \{4, 1, 7, 3\},$$
$$\operatorname{orb}_\sigma 5 = \{5, 8, 6\}, \qquad \operatorname{orb}_\sigma 6 = \{6, 5, 8\},$$

and so on. Of course, as sets (with the order of the elements ignored)

$$\operatorname{orb}_\sigma 1 = \operatorname{orb}_\sigma 3 = \operatorname{orb}_\sigma 4 = \operatorname{orb}_\sigma 7,$$
$$\operatorname{orb}_\sigma 2 = \operatorname{orb}_\sigma 9, \qquad \operatorname{orb}_\sigma 5 = \operatorname{orb}_\sigma 6 = \operatorname{orb}_\sigma 8.$$

Similarly, we have (for the τ just mentioned)

$$\operatorname{orb}_\tau 1 = \{1\}, \qquad \operatorname{orb}_\tau 4 = \{4\}, \qquad \operatorname{orb}_\tau 5 = \{5, 3, 7, 2, 8\},$$
$$\operatorname{orb}_\tau 6 = \{6, 9\}, \qquad \operatorname{orb}_\tau 7 = \{7, 2, 8, 5, 3\}, \qquad \operatorname{orb}_\tau 9 = \{9, 6\}.$$

(*ii*) The permutation

$$\sigma_1 = \begin{pmatrix} 1 & 2 & 3 & 4 & 5 & 6 & 7 & 8 & 9 \\ 7 & 2 & 4 & 1 & 5 & 6 & 3 & 8 & 9 \end{pmatrix}$$

has the orbits $\{1, 7, 3, 4\}$, $\{2\}$, $\{5\}$, $\{6\}$, $\{8\}$, $\{9\}$. Since exactly one of the orbits is nontrivial (meaning: has more than one element), σ_1 is a cycle. Because the nontrivial orbit has four elements, σ_1 is a 4-cycle. The action of σ_1 is given by

$$1 \to 7 \to 3 \to 4, \;(4 \to 1)$$

and all the remaining elements of X are fixed, so we write (forgetting for the time being what transpired in 4-5-1)

$$\sigma_1 = (1, 7, 3, 4)$$

or because the commas can be dropped without danger of confusion

$$\sigma_1 = (1734).$$

This describes the action of σ_1 completely, so the order in which the terms appear is crucial. Of course, it is equally valid to write σ_1 as (7341), or (3417) or (4173).

The permutation

$$\sigma_2 = \begin{pmatrix} 1 & 2 & 3 & 4 & 5 & 6 & 7 & 8 & 9 \\ 1 & 9 & 3 & 4 & 5 & 6 & 7 & 8 & 2 \end{pmatrix}$$

has the single nontrivial orbit $\{2, 9\}$; so it is a 2-cycle, denoted by

$$\sigma_2 = (29) = (92).$$

Furthermore,

$$\sigma_3 = \begin{pmatrix} 1 & 2 & 3 & 4 & 5 & 6 & 7 & 8 & 9 \\ 1 & 2 & 3 & 4 & 8 & 5 & 7 & 6 & 9 \end{pmatrix}$$

has the single nontrivial orbit $\{5, 8, 6\}$. It is a 3-cycle and may be written as

$$\sigma_3 = (586) = (865) = (658).$$

(*iii*) We note in passing that, according to the definition 4-5-3, the identity permutation $e \in S_9$ (or for that matter $e \in S_n$, for any n) is not a cycle; moreover, there is no such thing as a 1-cycle. The identity $e \in S_n$ is the unique permutation all of whose orbits are trivial.

(*iv*) In case the reader has not noticed, let us explain how $\sigma_1, \sigma_2, \sigma_3$ above are related to σ [of part (*i*)]. Consider the orbit $\{1, 7, 3, 4\}$ of σ. We define a permutation in S_9 whose action on the elements of this orbit is the same as the action of σ, and whose action on any element outside this orbit is to keep it fixed. In other words, we have

$$2 \to 2, \quad 5 \to 5, \quad 6 \to 6, \quad 8 \to 8, \quad 9 \to 9, \quad 1 \to 7, \quad 7 \to 3, \quad 3 \to 4, \quad 4 \to 1,$$

which indeed, represents a permutation in S_9. This permutation (which is a cycle) is the one we called σ_1.

Similarly, $\sigma_2 \in S_9$ is the permutation (it is a cycle) which agrees with σ on the nontrivial orbit $\{2, 9\}$, and leaves the remaining seven elements of $X = \{1, 2, \ldots, 9\}$ fixed. Finally, σ_3 is taken as the permutation (that is, cycle) which agrees with σ on the orbit $\{5, 8, 6\}$, and keeps the remaining elements fixed.

In the same way, because the permutation τ, introduced in (*i*), has two nontrivial orbits—namely, $\{2, 3, 5, 7, 8\}$ and $\{6, 9\}$—we associate with it the two cycles

$$\tau_1 = (28537) \qquad \text{and} \qquad \tau_2 = (69).$$

There is a general principle involved here. Namely, if any permutation $\sigma \neq e$ in S_n is given then with each of its nontrivial orbits we can associate a cycle (whose action on the orbit in question is the same as the action of σ). These cycles are obviously disjoint—they are known as the "disjoint cycles belonging to σ," or simply as the "cycles of σ."

(*v*) With τ, τ_1, τ_2 as above, let us look at the product

$$\tau_1 \circ \tau_2 = (28537) \circ (69)$$

in S_9. As is standard, we write this as

$$\tau_1 \tau_2 = (28537)(69).$$

The product $\tau_1\tau_2 \in S_9$ may be computed directly on the basis of this notation by tracing the action of τ_2 followed by τ_1 (for example: $\tau_2: 1 \to 1$ and $\tau_1: 1 \to 1$, so $\tau_1\tau_2: 1 \to 1$; $\tau_2: 2 \to 2$ and $\tau_1: 2 \to 8$, so $\tau_1\tau_2: 2 \to 8$; etc.), or else we may use the older more cumbersome notation to obtain

$$\begin{aligned}\tau_1\tau_2 &= \begin{pmatrix}1 & 2 & 3 & 4 & 5 & 6 & 7 & 8 & 9\\ 1 & 8 & 7 & 4 & 3 & 6 & 2 & 5 & 9\end{pmatrix}\begin{pmatrix}1 & 2 & 3 & 4 & 5 & 6 & 7 & 8 & 9\\ 1 & 2 & 3 & 4 & 5 & 9 & 7 & 8 & 6\end{pmatrix}\\ &= \begin{pmatrix}1 & 2 & 3 & 4 & 5 & 6 & 7 & 8 & 9\\ 1 & 8 & 7 & 4 & 3 & 9 & 2 & 5 & 6\end{pmatrix}\\ &= \tau.\end{aligned}$$

Thus, τ has been expressed as the product of the disjoint cycles τ_1 and τ_2 in S_9.

We may also compute the product

$$\tau_2\tau_1 = (69)(28537)$$

and find the result to be $\tau_2\tau_1 = \tau$. It is obvious (once one thinks about it) that the disjoint cycles τ_1 and τ_2 commute with each other. In fact, because their (nontrivial) orbits are disjoint, an element belonging to the orbit of one of the cycles is left fixed by the other cycle, and consequently the order in which τ_1 and τ_2 are applied does not matter.

In similar fashion, with σ, σ_1, σ_2, σ_3 as above, we have

$$\sigma_1\sigma_2\sigma_3 = (1734)(29)(586)$$

and this product is equal to σ. The order of σ_1, σ_2, σ_3 in the product does not matter (as the reader can verify easily) because they are disjoint cycles.

(*vi*) The preceding discussion permits us to write

$$\begin{aligned}\tau &= (28537)(69),\\ \sigma &= (1734)(29)(586),\end{aligned}$$

where, in each case, the right side represents a product of disjoint cycles. This is in keeping with the notation introduced informally in 4-5-1 and 4-5-2. However, there is one important addition. There, the use of parentheses served merely as an aid in specifying the action of the permutation in question, and writing the parentheses adjacent to each other had no particular significance. Here we see that the notation of 4-5-1 and 4-5-2, which looks like a product, is, indeed, to be viewed as a product of cycles—and the use of parentheses is entirely compatible with this interpretation.

From these examples we are led to the following general result.

4-5-6. Proposition. Fix an integer $n \geq 1$; then in S_n

(1) disjoint cycles commute,
(2) any $\sigma \neq e$ in S_n can be expressed as a product of disjoint cycles,
(3) this "factorization" is essentially unique.

Proof: In virtue of what has gone before, these assertions are fairly clear by now. Therefore, we shall give a dry, formal proof, and leave it as an exercise for the reader to fill in any details which he considers to be missing.

(1) Suppose $\sigma_1, \sigma_2, \ldots, \sigma_r \in S_n$ are disjoint cycles; we must show that their product $\sigma_1\sigma_2 \cdots \sigma_r$ (in the group S_n) is not affected by any rearrangement of the terms. For this it suffices to consider the case $r = 2$ and prove $\sigma_1\sigma_2 = \sigma_2\sigma_1$.

Let $\mathcal{O}_1$ be the nontrivial orbit of σ_1 and $\mathcal{O}_2$ be the nontrivial orbit of σ_2; so, by hypothesis, $\mathcal{O}_1 \cap \mathcal{O}_2 = \varnothing$. For any $a \in X = \{1, 2, \ldots, n\}$ exactly one of three possibilities holds:

(*i*) $a \notin \mathcal{O}_1 \cup \mathcal{O}_2$,
(*ii*) $a \in \mathcal{O}_1$,
(*iii*) $a \in \mathcal{O}_2$.

In case (*i*) we have: $\sigma_1 a = a$, $\sigma_2 a = a$, so $\sigma_1\sigma_2(a) = \sigma_2\sigma_1(a)$. In case (*ii*) we have: $\sigma_1 a \in \mathcal{O}_1$, $\sigma_2 a = a$—so $\sigma_1\sigma_2(a) = \sigma_1 a$, $\sigma_2\sigma_1(a) = \sigma_1 a$, and $\sigma_1\sigma_2(a) = \sigma_2\sigma_1(a)$. In case (*iii*) we have $\sigma_1\sigma_2(a) = \sigma_2 a = \sigma_2\sigma_1(a)$. Consequently,

$$\sigma_1\sigma_2(a) = \sigma_2\sigma_1(a) \qquad \text{for every } a \in X$$

which says that $\sigma_1\sigma_2 = \sigma_2\sigma_1$.

(2) Given $\sigma \neq e$ in S_n, let $\mathcal{O}_1, \mathcal{O}_2, \ldots, \mathcal{O}_r$ be its nontrivial orbits. They are disjoint. For each $i = 1, \ldots, r$ define σ_i by

$$\sigma_i a = \begin{cases} \sigma a, & \text{for } a \in \mathcal{O}_i, \\ a, & \text{for } a \notin \mathcal{O}_i. \end{cases}$$

Then, $\sigma_i \in S_n$; in fact, σ_i is a cycle whose sole nontrivial orbit, is $\mathcal{O}_i$, and whose action on $\mathcal{O}_i$ is the same as the action of σ. The cycles $\sigma_1, \sigma_2, \ldots, \sigma_r$ are obviously disjoint, so [by part (1)] they commute. We assert that

$$\sigma = \sigma_1\sigma_2 \cdots \sigma_r.$$

To see this, consider any $a \in X$. It belongs to exactly one of the orbits—say $a \in \mathcal{O}_i$. Then

$$\sigma_j a = a \text{ for any } j \neq i \qquad (\text{since } a \notin \mathcal{O}_j)$$

and consequently,

$$\begin{aligned}(\sigma_1\sigma_2 \cdots \sigma_i \cdots \sigma_r)(a) &= (\sigma_i\sigma_1\sigma_2 \cdots \sigma_{i-1}\sigma_{i+1} \cdots \sigma_r)a \\ &= \sigma_i a \\ &= \sigma a.\end{aligned}$$

This proves: $\sigma = \sigma_1 \sigma_2 \cdots \sigma_r$—$\sigma$ is indeed the product of its disjoint cycles.

(3) Suppose that in addition to the above factorization $\sigma = \sigma_1 \cdots \sigma_r$ we have also the factorization

$$\sigma = \tau_1 \tau_2 \cdots \tau_s$$

into disjoint cycles. Denoting the orbit of τ_i by $\mathcal{O}_i'$, $i = 1, \ldots, s$, these orbits are disjoint, and

$$\mathcal{O}_1' \cup \mathcal{O}_2' \cup \cdots \cup \mathcal{O}_s' = \mathcal{O}_1 \cup \mathcal{O}_2 \cup \cdots \cup \mathcal{O}_r$$

because both of these are equal to

$$\mathcal{O} = \{x \in X \mid \sigma x \neq x\}$$

(that is, $\mathcal{O}$ is the set of all elements of X which are not fixed under σ). Consider $\mathcal{O}_1'$, and let a be an arbitrary element of $\mathcal{O}_1'$. Then a belongs to exactly one $\mathcal{O}_j$; we rearrange the notation, if necessary, so that $a \in \mathcal{O}_1$. It is not hard to see that

$$\sigma_1 a = \sigma a = \tau_1 a.$$

Furthermore, because disjoint cycles commute, we have

$$\sigma^i = \sigma_1^i \sigma_2^i \cdots \sigma_r^i = \tau_1^i \tau_2^i \cdots \tau_s^i$$

for any integer $i \geq 0$—and from the properties of our orbits,

$$\sigma_1^i a = \sigma^i a = \tau_1^i a, \qquad i \geq 0.$$

We conclude that

$$\text{orb}_{\sigma_1} a = \text{orb}_{\tau_1} a.$$

In other words, $\mathcal{O}_1 = \mathcal{O}_1'$. Even more

$$\begin{aligned} \sigma_1 b &= \tau_1 b, && \text{if} \quad b \in \mathcal{O}_1 = \mathcal{O}_1', \\ \sigma_1 b &= b = \tau_1 b, && \text{if} \quad b \notin \mathcal{O}_1 = \mathcal{O}_1' \end{aligned}$$

—so $\sigma_1 = \tau_1$. Multiplying our initial factorizations by σ_1^{-1} gives

$$\sigma_1^{-1}\sigma = \sigma_2 \cdots \sigma_r = \tau_2 \cdots \tau_s.$$

Now, an induction leads to $r = s$ and $\sigma_2 = \tau_2, \ldots, \sigma_r = \tau_r$. This proves that the factorization into disjoint cycles is unique upto the order of the factors and the fact that when a cycle is to be written explicitly there are several equivalent ways to do so—namely, by cycling the entries. ∎

4-5-7. Examples. Our simplified notation for permutations is now fully justified and under control, so we turn to some concrete examples of its use in computations. In S_9 consider the permutations:

$$\rho_1 = (15), \qquad \rho_2 = (258), \qquad \rho_3 = (29835), \qquad \rho_4 = (174628),$$
$$\sigma = (1734)(29)(586), \qquad \tau = (28537)(69), \qquad \rho = (172963)(48).$$

We start with the product $\rho_1\rho_2 = (15)(258)$. To compute this (meaning: to express it as a product of disjoint cycles) requires tracing the action of $\rho_1\rho_2$ on $1, 2, 3, \ldots, 9$. Now, 1 appears only once, in (15). This signifies that 1 is left fixed by (258) and is then moved to 5 under (15). In short, we have $1 \to 5$ (under $\rho_1\rho_2$). Then 5 is mapped to 8 under (258), and 8 is kept fixed under (15)—so $5 \to 8$ (under $\rho_1\rho_2$). Then 8 goes to 2 under (258) and 2 is fixed under (15)—so $8 \to 2$ (under $\rho_1\rho_2$). At the next step, (258) carries 2 to 5 and then (15) carries 5 to 1—so $2 \to 1$ (under $\rho_1\rho_2$). Thus, $\rho_1\rho_2$ includes the cycle (1582). Obviously, $\rho_1\rho_2$ keeps 3, 4, 6, 7, 9 fixed, since these terms do not appear in (258) or (15). Therefore,

$$\rho_1\rho_2 = (15)(258) = (1582).$$

It needs to be emphasized that in computing the product one passes through the cycles one at a time, going from right to left, because this is the order in which the cycles are applied.

To compute $\rho_2\rho_3 = (258)(29835)$, we note first that 1, 4, 6, 7 do not appear, so they are fixed under $\rho_2\rho_3$. Then, $2 \xrightarrow{\rho_3} 9 \xrightarrow{\rho_2} 9$, $9 \xrightarrow{\rho_3} 8 \xrightarrow{\rho_2} 2$, so (29) is one of the cycles appearing in the factorization of $\rho_2\rho_3$. Furthermore, $3 \xrightarrow{\rho_3} 5 \xrightarrow{\rho_2} 8$, $8 \xrightarrow{\rho_3} 3 \xrightarrow{\rho_2} 3$, so (38) is also a factor. Finally, $5 \xrightarrow{\rho_3} 2 \xrightarrow{\rho_2} 5$, so 5 is fixed under $\rho_2\rho_3$. This shows:

$$\rho_2\rho_3 = (258)(29835) = (29)(38).$$

In similar fashion, one computes

$$\begin{aligned} \rho_3\rho_4 &= (29835)(174628) \\ &= (174698)(352). \end{aligned}$$

With a bit of practice the reader should be able to do such products in one line, without intermediate steps. For example,

$$\begin{aligned} \sigma\tau\rho &= (1734)(29)(586)(28537)(69)(172963)(48) \\ &= (192548)(376). \end{aligned}$$

How is this done? Starting with 1, its right-most appearance is in (172963), under which $1 \to 7$. Going leftward from (172963), the first appearance of 7 is in (28537), under which $7 \to 2$. Then going leftward from (28537), the first appearance of 2 is in (29), under which $2 \to 9$. There being no appearance of 9 to the left of (29), we conclude that $\sigma\tau\rho$ maps $1 \to 9$. Next, we apply this procedure to 9: under (172963), $9 \to 6$; then under (69), $6 \to 9$; then under (29), $9 \to 2$—so $\sigma\tau\rho$ maps $9 \to 2$. Now, starting with 2, we have: $2 \to 9$ under (172963), $9 \to 6$ under (69), $6 \to 5$ under (586)—so $\sigma\tau\rho$ maps $2 \to 5$. Starting with 5, we have: $5 \to 3$ under (28537), $3 \to 4$ under (1734)—so $\sigma\tau\rho$ maps

$5 \to 4$. Next: $4 \to 8$ under (48), $8 \to 5$ under (28537), $5 \to 8$ under (586)—so $\sigma\tau\rho$ maps $4 \to 8$. Next: $8 \to 4$ under (48), $4 \to 1$ under (1734)—so $\sigma\tau\rho$ maps $8 \to 1$. Therefore, (192548) is a cycle belonging to the factorization of $\sigma\tau\rho$.

Similarly, starting with 3 we arrive at the cycle (376) as a factor of $\sigma\tau\rho$. Consequently, since all nine elements of $\{1, 2, \ldots, 9\}$ are accounted for, we have indeed

$$\sigma\tau\rho = (192548)(376).$$

Once we know how to multiply, it is easy to take powers. Thus

$$\begin{aligned}
\rho_1^2 &= (15)^2 = (15)(15) = e, \text{ the identity of } S_9,\\
\rho_2^2 &= (258)^2 = (258)(258) = (285),\\
\rho_3^2 &= (29835)^2 = (29835)(29835) = (28593).
\end{aligned}$$

Clearly, squaring a cycle amounts to the mapping which takes every term of the cycle two places to the right. Accordingly,

$$\rho_4^2 = (174628)^2 = (142)(768).$$

Furthermore, raising a cycle to the power 3 amounts to the mapping that moves each term of the cycle three places to the right. For example,

$$\begin{aligned}
\rho_1^3 &= (15)^3 = (15),\\
\rho_2^3 &= (258)^3 = e, \text{ the identity of } S_9,\\
\rho_3^3 &= (29835)^3 = (23958),\\
\rho_4^3 &= (174628)^3 = (16)(27)(48).
\end{aligned}$$

More generally, raising a cycle to the power $r \geq 1$ involves assigning to each term of the cycle the term which is r places to the right.

What about raising an arbitrary permutation to a positive power r? Because disjoint cycles commute, this is the same as raising each of its cycles to the power r. For example

$$\sigma^r = [(1734)(29)(586)]^r = (1734)^r(29)^r(586)^r$$

and, in particular,

$$\begin{aligned}
\sigma^2 &= (1734)^2(29)^2(586)^2 = (13)(74)(568),\\
\sigma^3 &= (1734)^3(29)^3(586)^3 = (1437)(29),\\
\sigma^4 &= (1734)^4(29)^4(586)^4 = (586),
\end{aligned}$$

and so on.

Finding inverses is easy. For a cycle, its inverse is obtained by writing the terms in the reverse order. Thus, for example,

$$\begin{aligned}
\rho_1^{-1} &= (15)^{-1} = (51) = (15) \\
\rho_2^{-1} &= (258)^{-1} = (852) = (285) \\
\rho_3^{-1} &= (29835)^{-1} = (53892) \\
\rho_4^{-1} &= (174628)^{-1} = (826471).
\end{aligned}$$

For an arbitrary permutation σ, expressed as a product of disjoint cycles $\sigma_1\sigma_2 \cdots \sigma_r$, we have

$$\begin{aligned}
\sigma^{-1} &= (\sigma_1\sigma_2 \cdots \sigma_r)^{-1} \\
&= (\sigma_r\sigma_{r-1} \cdots \sigma_2\sigma_1)^{-1} \\
&= \sigma_1^{-1}\sigma_2^{-1} \cdots \sigma_r^{-1}
\end{aligned}$$

[This may also be arranged as follows: $\sigma^{-1} = (\sigma_1 \cdots \sigma_r)^{-1} = \sigma_r^{-1} \cdots \sigma_2^{-1}\sigma_1^{-1} = \sigma_1^{-1}\sigma_2^{-1} \cdots \sigma_r^{-1}$ because, as is easily seen, for each i, σ_i^{-1} is a cycle whose nontrivial orbit is the same as that of σ_i, and consequently the σ_i^{-1} commute with each other.] In particular, for σ, τ, ρ as before, we have

$$\begin{aligned}
\sigma^{-1} &= [(1734)(29)(586)]^{-1} &&= (586)^{-1}(29)^{-1}(1734)^{-1} \\
& &&= (685)(92)(4371) = (1437)(29)(568) \\
\tau^{-1} &= [(28537)(69)]^{-1} &&= (69)^{-1}(28537)^{-1} \\
& &&= (69)(73582) \\
\rho^{-1} &= [(172963)(48)]^{-1} &&= (369271)(48).
\end{aligned}$$

The computation of $(\sigma\rho)^{-1}$, for example, can now be done in two different ways. In the first place (using the preceding computations)

$$\begin{aligned}
(\sigma\rho)^{-1} = \rho^{-1}\sigma^{-1} &= [(369271)(48)][(1437)(29)(568)] \\
&= (185973)(46).
\end{aligned}$$

Otherwise, one computes $\sigma\rho$ and then takes its inverse, thus—

$$\begin{aligned}
(\sigma\rho)^{-1} &= [(1734)(29)(586)(172963)(48)]^{-1} \\
&= [(137958)(46)]^{-1} \\
&= (46)(859731).
\end{aligned}$$

In similar fashion, the reader may check that

$$(\rho\sigma)^{-1} = (174562)(38)$$

—one way to do this is to make use of the fact that

$$\rho\sigma = (126547)(38).$$

For any $\sigma, \tau \in S_n$ it is convenient to introduce the notation

$$\sigma^\tau = \tau\sigma\tau^{-1}$$

and call σ^τ **the conjugate of** σ **by** τ (or with less precision "a conjugate of σ"). This terminology is natural since σ^τ is the image of σ under the inner automorphism (discussed in 4-4-14) $I_\tau: S_n \to S_n$, which is defined by $I_\tau: \sigma \to \tau\sigma\tau^{-1}$ and is known as conjugation by τ. Among the elementary properties of our new symbol, we have (for $\sigma, \tau, \rho, \sigma_1, \sigma_2 \in S_n$)

$$(\sigma_1\sigma_2)^\tau = \sigma_1^\tau\sigma_2^\tau,$$
$$\sigma^e = \sigma,$$
$$(\sigma^{-1})^\tau = (\sigma^\tau)^{-1},$$
$$(\sigma^\tau)^\rho = \sigma^{\rho\tau}.$$

The verifications are trivial.

As illustrations of the computation of conjugates we have, for the specific $\sigma, \tau, \rho \in S_9$ treated above.

$$\begin{aligned}\sigma^\tau = \tau\sigma\tau^{-1} &= [(28537)(69)][(1734)(29)(586)][(69)(73582)] \\ &= (1274)(359)(68), \\ \tau^\rho = \rho\tau\rho^{-1} &= (172963)(48)(28537)(69)(369271)(48) \\ &= (12945)(36).\end{aligned}$$

Actually, there is an easier way to compute conjugates and, even more, it is easy to decide if two permutations are conjugate. This is the content of our next result.

4-5-8. Proposition. In S_n, we have:

(*i*) σ^τ has the same cycle structure as σ; in fact, to compute σ^τ one simply writes out the cycle expression of σ and replaces each digit by its image under τ.

(*ii*) ρ is a conjugate of $\sigma \Leftrightarrow$ they have the same cycle structure.

(*iii*) The relation "is a conjugate of" is an equivalence relation. Each equivalence class consists of all elements of S_n which have the same cycle structure.

Proof: We must clarify the meaning of "cycle structure." Any permutation $\sigma \in S_n$ has an essentially unique representation as a product of disjoint cycles. The lengths of these various cycles (meaning: a list of all the lengths) constitute the cycle structure of σ. Because the cycles commute, the order in which these lengths are listed does not matter—it is the set of cycle lengths which we care about. A permutation in S_n is said to have the "same cycle structure" as σ when its cycle lengths are the same as those of σ. For example, the cycle structure of $\sigma = (1734)(29)(586) \in S_9$ is given by the numbers 4, 2, 3 (or by any permutation of these three numbers). Clearly, $\sigma' = (13)(285)(7496) \in S_9$ has the same cycle structure as σ—namely, the numbers 2, 3, 4. On the other hand, the cycle structure of $\tau = (28537)(69) \in S_9$ is given by the numbers 5, 2 (one might be extremely careful and write this cycle structure as 5, 2, 1, 1—thus using up all nine digits), so τ does not have the same cycle structure as σ.

(*i*) Before undertaking the proof, let us illustrate the procedure given here for computing σ^τ as applied to the special case $\sigma = (1734)(29)(586)$, $\tau = (28537)(69)$ in S_9. The recipe says: in the expression for σ, do not change the cycle structure, but replace each digit by its image under τ. Since the action of τ is $1 \to 1$, $2 \to 8$, $3 \to 7$, $4 \to 4$, $5 \to 3$, $6 \to 9$, $7 \to 2$, $8 \to 5$, $9 \to 6$, the result is

$$\sigma^\tau = (1274)(86)(359),$$

which coincides with the result of our earlier computation (at the end of 4-5-7) of σ^τ.

To prove that, in general, the end result of our procedure is indeed σ^τ, we note first that within each cycle of the cycle-representation of σ, a pair of adjacent elements always consists of some element a and its image σa under σ:

$$\sigma = (\ldots) \cdots (\ldots, a, \sigma a, \ldots) \cdots (\ldots). \tag{$*$}$$

Our method calls for applying τ to each entry; which leads to

$$(\ldots) \cdots (\ldots, \tau a, \tau\sigma a, \ldots) \cdots (\ldots). \tag{$**$}$$

This is the cycle-representation of some permutation in S_n (after all, the entries are distinct because the entries of ($*$) are distinct and τ is 1–1). The action of this permutation on any τa is $\tau a \to \tau\sigma a$. On the other hand, the action of $\sigma^\tau = \tau\sigma\tau^{-1}$ on any τa is

$$\tau a \to (\tau\sigma\tau^{-1})(\tau a) = \tau\sigma a.$$

It follows that ($**$) is the expression for σ^τ. Of course, it is obvious from the method that σ^τ has the same cycle structure as σ.

(*ii*) The implication $\Rightarrow$ has just been proved. To prove the reverse implication, write ρ beneath σ with each cycle beneath one of the same length—thus

$$\sigma = (\cdots) \cdots (\cdots a\ \sigma a \cdots) \cdots (\cdots),$$
$$\rho = (\cdots) \cdots (\cdots b\ \rho b \cdots) \cdots (\cdots),$$

where any 1-cycles (that is, elements which are fixed under the permutation) are to be included. If we write the entries of σ in a row and the corresponding entries of ρ below them—the picture being

$$\begin{pmatrix} \cdots a & \sigma a \cdots \\ \cdots b & \rho b \cdots \end{pmatrix}$$

—then the top row contains each element of $\{1, \ldots, n\}$ exactly once, and so does the bottom row. Therefore, in our old notation, this represents a permutation in S_n, call it τ. According to the rule [proved in (i)] for computing σ^τ, it is clear that $\sigma^\tau = \rho$.

To illustrate what is going on here, the permutation $\sigma' = (13)(285)(7469)$ is conjugate to $\sigma = (1734)(29)(586)$ in S_9 because they have the same cycle structure. In fact, if we write

$$\tau = \begin{pmatrix} 1 & 7 & 3 & 4 & 2 & 9 & 5 & 8 & 6 \\ 7 & 4 & 6 & 9 & 1 & 3 & 2 & 8 & 5 \end{pmatrix} = \begin{pmatrix} 1 & 2 & 3 & 4 & 5 & 6 & 7 & 8 & 9 \\ 7 & 1 & 6 & 9 & 2 & 5 & 4 & 8 & 3 \end{pmatrix}$$
$$= (17493652),$$

then $\sigma' = \sigma^\tau$. Incidentally, here as in general, there are several distinct choices for τ. They arise because the cycles of σ' can themselves by cycled; thus, if we write $\sigma' = (4697)(31)(852)$, then

$$\tau = \begin{pmatrix} 1 & 7 & 3 & 4 & 2 & 9 & 5 & 8 & 6 \\ 4 & 6 & 9 & 7 & 3 & 1 & 8 & 5 & 2 \end{pmatrix} = (1476239)(58)$$

and again $\sigma' = \sigma^\tau$.

(iii) Let us write $\sigma \sim \tau$ to signify that τ is a conjugate of σ. Then: $\sigma \sim \sigma$ for all σ (since $\sigma^e = \sigma$); if $\sigma \sim \tau$, then $\tau \sim \sigma$ (since $\tau = \sigma^\rho \Rightarrow \sigma = \tau^{(\rho^{-1})}$); if $\sigma \sim \tau$ and $\tau \sim \rho$, then $\sigma \sim \rho$ (since $\tau = \sigma^\mu$ and $\rho = \tau^\upsilon$ imply $\rho = (\sigma^\mu)^\upsilon = \sigma^{\upsilon\mu}$). Thus, $\sim$ is an equivalence relation and, in virtue of (ii), any equivalence class (also known as a **conjugate class**) consists of all the elements of S_n with the same cycle structure. ▮

We turn next to additional consequences of the cycle decomposition of a permutation.

4-5-9. Proposition. In the group S_n, a cycle of length r has order r. Furthermore, the order of any permutation is the least common multiple of the lengths of its disjoint cycles.

Proof: A cycle of length r is of form

$$\sigma = (a_1 a_2 \cdots a_r)$$

with the entries all distinct. For $i = 1, 2, \ldots, r-1$ it is clear that $\sigma^i a_1 = a_{i+1}$, so clearly $\sigma^i \neq e$. On the other hand, $\sigma^r a_j = a_j$ for $j = 1, 2, \ldots, r$, so since σ^r keeps each element of $\{1, 2, \ldots, n\}$ which is not an a_j fixed we have $\sigma^r = e$. This means that as an element of the group S_n, σ has order r—that is, ord $\sigma = r$.

Furthermore, if σ is any permutation in S_n, let $\sigma = \sigma_1\sigma_2 \cdots \sigma_s$ be its factorization into disjoint cycles. Because disjoint cycles commute, it follows that

$$\sigma^i = \sigma_1^i\sigma_2^i \cdots \sigma_s^i \qquad \text{for all } i \geq 1.$$

Now, σ_j^i (for each $j = 1, \ldots, n$) need not be a cycle, but the set of elements it moves (meaning: the union of its nontrivial orbits) is contained in the single nontrivial orbit of σ_j. Consequently, it is easy to see that $\sigma^i = e$ if and only if each of $\sigma_1^i, \sigma_1^i, \ldots, \sigma_s^i$ is equal to e; so ord σ is the smallest i for which $\sigma_1^i = e$, $\sigma_2^i = e$, $\ldots$, $\sigma_s^i = e$. This number is clearly the least common multiple

$$[\text{ord } \sigma_1, \text{ord } \sigma_2, \ldots, \text{ord } \sigma_s]$$

—or, to put it another way, the least common multiple of the lengths of the disjoint cycles of σ. ▌

4-5-10. Proposition.

(*i*) Any permutation in S_n can be expressed as a product of transpositions. In other words, the set of all transpositions generates S_n.
(*ii*) The $n - 1$ transpositions $(12), (13), \ldots, (1n)$ generate S_n.
(*iii*) The $n - 1$ transpositions $(12), (23), (34), \ldots (n - 1, n)$ generate S_n.

Proof: Since any permutation can be expressed as a product of cycles, it suffices to express any cycle as a product of transpositions. But this is easy; a typical cycle looks like $(a_1a_2 \cdots a_r)$, and it may be written as

$$(a_1a_2 \cdots a_r) = (a_1a_r)(a_1a_{r-1}) \cdots (a_1a_3)(a_1a_2).$$

(*ii*) It suffices to express an arbitrary transposition (ab) in terms of $(12), (13), \ldots, (1n)$. But this is trivial, since

$$(ab) = (1a)(1b)(1a).$$

As a matter of fact, his formula is not a complete surprise—for, according to 4-5-8, (ab) is conjugate to each of the transpositions $(12), \ldots, (1n)$. In particular, if $\rho = (ab)$ and $\sigma = (1b)$, then upon putting $\tau = \binom{1\ b}{a\ b} = (1a)$ we have $\rho = \sigma^\tau$. Incidentally, we also have

$$(ab) = (1b)(1a)(1b)$$

—this being the case, $\rho = (ab)$, $\sigma = (1a) = (a1)$, $\tau = \binom{a\ 1}{a\ b} = (1b)$ and $\rho = \sigma^\tau$.

(*iii*) In virtue of (*ii*), it suffices to express each transposition (12), (13), ..., (1n) in terms of the transpositions (12), (23), ..., ($n-1, n$). This is clear from the formula

$$(1i) = (i-1, i) \cdots (34)(23)(12)(23)(34) \cdots (i-1, i), \qquad 2 \le i \le n$$

which arises from the facts about conjugation. Namely, if one writes $\tau_1 = (12)$, $\tau_2 = (23)$, $\tau_3 = (34)$, ..., $\tau_{n-1} = (n-1, n)$, then

$$\tau_1^{\tau_2} = (23)(12)(23)^{-1} = (23)(12)(23) = (13)$$

$$(\tau_1^{\tau_2})^{\tau_3} = \tau^{\tau_3\tau_2} = [(34)(23)](12)[(34)(23)]^{-1}$$
$$= (34)(23)(12)(23)(34) = (14)$$

and one may proceed inductively. This completes the proof. ∎

Obviously, any permutation can be expressed as a product of transpositions in many ways; however, it is not obvious that for a given permutation the number of transpositions in such a product is either always odd or always even. There are several approaches to this result, all of which involve some degree of artificiality. We take the most common approach.

4-5-11. Discussion. Our permutations come from the group S_n. There is no harm in assuming $n \ge 3$, since S_n is rather trivial for $n = 1$ or 2. Let us take n independent formal symbols (or variables) $x_1, x_2, \ldots, x_n$ and introduce the expression

$$\Delta = \Delta_n = \prod_{i<j} (x_i - x_j), \qquad 1 \le i \le j \le n$$

(which may be referred to as the **discriminant**). For example, if $n = 3$, then

$$\Delta_3 = (x_1 - x_2)(x_1 - x_3)(x_2 - x_3) \tag{*}$$

while, for $n = 4$, we have

$$\Delta_4 = (x_1 - x_2)(x_1 - x_3)(x_1 - x_4)(x_2 - x_3)(x_2 - x_4)(x_3 - x_4) \tag{**}$$

which may also be written in the triangular form

$$\begin{aligned}(x_1 - x_2)(x_1 - x_3)&(x_1 - x_4)\\ (x_2 - x_3)&(x_2 - x_4)\\ &(x_3 - x_4).\end{aligned}$$

More generally, for arbitrary n, the discriminant Δ_n may be expressed in the triangular form

$$\Delta = \Delta_n = \begin{cases} (x_1 - x_2)(x_1 - x_3) \cdots (x_1 - x_j) & \cdots \quad (x_1 - x_n) \\ \quad (x_2 - x_3) \cdots (x_2 - x_j) & \cdots \quad (x_2 - x_n) \\ \quad\quad \ddots \quad \vdots & \vdots \quad\quad \vdots \\ \quad\quad\quad (x_i - x_j) & \cdots \quad (x_i - x_n) \\ \quad\quad\quad\quad \ddots & \vdots \quad\quad \vdots \\ & (x_{n-2} - x_{n-1})(x_{n-2} - x_n) \\ & \quad (x_{n-1} - x_n) \end{cases} \qquad (***)$$

The number of factors $(x_i - x_j)$ in Δ_n is clearly $(n^2 - n)/2$.

It should be pointed out why Δ, which appears to be a product, should indeed be interpreted as a product. Starting from the integral domain $\mathbf{Z}$ we form the polynomial ring $\mathbf{Z}[x_1]$, which is an integral domain (see Section 3-3). Next we form the polynomial ring $(\mathbf{Z}[x_1])([x_2])$, which we write as $\mathbf{Z}[x_1, x_2]$—it too being an integral domain. Proceeding inductively, we end with $(\mathbf{Z}[x_1, \ldots, x_{n-1}])[x_n]$ which is written as $\mathbf{Z}[x_1, \ldots, x_n]$ and is known as the "polynomial ring in n variables over $\mathbf{Z}$"; it is an integral domain. (Our initial assumption that $x_1, \ldots, x_n$ are independent variables is designed as an intuitive way of saying: Each x_i is an indeterminate over $\mathbf{Z}[x_1, \ldots, x_{i-1}]$ —which then enables us to form the polynomial ring $\mathbf{Z}[x_1, \ldots, x_{i-1}, x_i]$.) Each $(x_i - x_j)$ with $i < j$ is a nonzero element of the domain $\mathbf{Z}[x_1, \ldots, x_n]$ and, consequently, their product $\Delta = \Delta_n$ is a nonzero element of $\mathbf{Z}[x_1, \ldots, x_n]$.

Now, given any $\sigma \in S_n$, let us put (for $\Delta = \Delta_n$)

$$\sigma\Delta = \prod_{i<j} (x_{\sigma i} - x_{\sigma j}), \qquad 1 \le i < j \le n. \tag{\#}$$

To illustrate this definition we note that in the case where $\sigma = (13)(24) \in S_4$ we have [using (**)]

$$\sigma\Delta = (x_3 - x_4)(x_3 - x_1)(x_3 - x_2)(x_4 - x_1)(x_4 - x_2)(x_1 - x_2).$$

Incidentally, the reader will observe that this equals

$$(x_1 - x_2)[-(x_1 - x_3)][-(x_1 - x_4)][-(x_2 - x_3)][-(x_2 - x_4)](x_3 - x_4),$$

which is precisely Δ.

In order to understand the definition of $\sigma\Delta$, it is useful to put it in the proper context. Given a permutation $\sigma \in S_n$ we may define a permutation (also denoted by σ) of our set of independent variables $\{x_1, \ldots, x_n\}$ by putting

$$\sigma\colon x_i \to x_{\sigma i}, \qquad i = 1, \ldots, n.$$

Next, we can define a mapping (still denoted by σ—since no harm will come of it)

$$\sigma\colon \mathbf{Z}[x_1, \ldots, x_n] \to \mathbf{Z}[x_1, \ldots, x_n]$$

as follows: an element of $\mathbf{Z}[x_1, \ldots, x_n]$ (that is, a polynomial in n variables over $\mathbf{Z}$) is of form

$$f = f(x_1, x_2, \ldots, x_n) = \sum a_{i_1, i_2, \ldots, i_n} x_1^{i_1} x_2^{i_2} \cdots x_n^{i_n}$$

where the exponents $i_1, i_2, \ldots, i_n$ are all greater than or equal to 0 and $a_{i_1, i_2, \ldots, i_n} \in \mathbf{Z}$, and σ maps f to the element

$$\sigma f = f(x_{\sigma 1}, x_{\sigma 2}, \ldots, x_{\sigma n}) = \sum a_{i_1, i_2, \ldots, i_n} x_{\sigma 1}^{i_1} x_{\sigma 2}^{i_2} \cdots x_{\sigma n}^{i_n}.$$

In other words, on $\mathbf{Z}$, σ is the identity map—that is, $\sigma a = a$ for every $a \in \mathbf{Z}$; on the set $\{x_1, \ldots, x_n\}$, σ acts as above—that is, $\sigma x_i = x_{\sigma i}$ for $i = 1, 2, \ldots, n$; σ preserves sums and products of elements from $\mathbf{Z}$ and $\{x_1, \ldots, x_n\}$. These words obscure a rather obvious definition; for example, if $\sigma = (13)(24) \in S_4$ and the polynomial in $\mathbf{Z}[x_1, x_2, x_3, x_4]$ is

$$\begin{aligned} f &= f(x_1, x_2, x_3, x_4) \\ &= 3 + 5x_1 - x_3 + 2x_1x_2 - 4x_4^2 + x_2x_3^5x_4^3 - 3x_1x_2x_3x_4, \end{aligned}$$

then

$$\begin{aligned} \sigma f &= f(x_{\sigma 1}, x_{\sigma 2}, x_{\sigma 3}, x_{\sigma 4}) \\ &= 3 + 5x_3 - x_1 + 2x_3x_4 - 4x_2^2 + x_4x_1^5x_2^3 - 3x_3x_4x_1x_2. \end{aligned}$$

In short, one simply applies the original permutation σ to the subscripts of the x_i's.

It is easy to see that σ is a one-to-one map of $\mathbf{Z}[x_1, x_2, \ldots, x_n]$ onto itself and, even more, that σ is an automorphism of the ring $\mathbf{Z}[x_1, x_2, \ldots, x_n]$. Its inverse σ^{-1} is precisely the automorphism which we obtain by starting from the permutation $\sigma^{-1} \in S_n$.

Of course, and this is the nub of the matter for us, when the automorphism σ is applied to the element $\Delta \in \mathbf{Z}[x_1, \ldots, x_n]$ the result is the element defined earlier [in (#)] as $\sigma\Delta$—in other words, $\sigma\Delta$ equals $\sigma\Delta$!

4-5-12. Proposition. If $\sigma \in S_n$, then $\sigma\Delta = \pm\Delta$. Thus we may define **sgn** σ (called "signum of σ") to be either $+1$ or -1 in such a way that

$$\sigma\Delta = (\operatorname{sgn} \sigma)\Delta.$$

The permutation σ is said to be **even** or **odd** according as sgn σ is $+1$ or -1.

Proof: Consider a single term $(x_{\sigma i} - x_{\sigma j})$ of the product

$$\sigma\Delta = \prod_{i<j} (x_{\sigma i} - x_{\sigma j}).$$

If $\sigma i < \sigma j$, we leave $(x_{\sigma i} - x_{\sigma j})$ alone in the product, since it is one of the terms appearing in

$$\Delta = \prod_{i<j} (x_i - x_j).$$

If $\sigma j < \sigma i$, we replace $(x_{\sigma i} - x_{\sigma j})$ by $-(x_{\sigma j} - x_{\sigma i})$ in the product. (Of course, we cannot have $\sigma i = \sigma j$ since $i \neq j$.) Once this has been done for each $(x_{\sigma i} - x_{\sigma j})$ we see that $\sigma\Delta$ is equal to $(-1)^m$ (where m is the number of times that $\sigma j < \sigma i$) times a product of terms, each of which appears in Δ. The number of terms in this product is equal to the number of terms in $\sigma\Delta$, which is equal to the number of terms in Δ—namely, $(n^2 - n)/2$. Furthermore, it is easy to see that the $(n^2 - n)/2$ terms of the product are distinct (since if $i < j$, $i' < j'$, then $x_{\sigma i} - x_{\sigma j} = x_{\sigma i'} - x_{\sigma j'}$ implies $i = i'$, $j = j'$, and $x_{\sigma i} - x_{\sigma j} = -(x_{\sigma i'} - x_{\sigma j'})$ implies $i = j'$, $j = i'$, which is impossible); consequently the product is Δ. In short,

$$\sigma\Delta = (-1)^m\Delta = \pm\Delta$$

We do not care to evaluate m, but simply observe that $(-1)^m$ is what we have chosen to call sgn σ. ▌

4-5-13. Proposition. The sgn function has the following properties—for $\sigma, \tau \in S_n$

(*i*) $\operatorname{sgn} e = 1$,
(*ii*) $\operatorname{sgn}(\sigma\tau) = (\operatorname{sgn} \sigma)(\operatorname{sgn} \tau)$,
(*iii*) $\operatorname{sgn}(\sigma^{-1}) = (\operatorname{sgn} \sigma)^{-1} = \operatorname{sgn} \sigma$.

Proof: (*i*) is trivial because $e\Delta = \Delta$. The proof of (*ii*) consists of the observation, $(\sigma\tau)\Delta = (\operatorname{sgn} \sigma\tau)\Delta$, coupled with the computation

$$\begin{aligned}
(\sigma\tau)\Delta &= \sigma(\tau\Delta) \\
&= \sigma[(\operatorname{sgn} \tau)\Delta] \\
&= (\operatorname{sgn} \tau)(\sigma\Delta) \qquad \left\{\begin{array}{l}\text{since } \sigma \text{ is a homomorphism,} \\ \text{and } \sigma a = a \text{ for every } a \in \mathbf{Z}\end{array}\right. \\
&= (\operatorname{sgn} \tau)[(\operatorname{sgn} \sigma)\Delta] \\
&= (\operatorname{sgn} \sigma)(\operatorname{sgn} \tau)\Delta.
\end{aligned}$$

As for (*iii*), by making use of (*i*) and (*ii*) we have

$$(\operatorname{sgn} \sigma)(\operatorname{sgn} \sigma^{-1}) = \operatorname{sgn}(\sigma\sigma^{-1}) = \operatorname{sgn} e = 1.$$

Since both sgn σ and sgn σ^{-1} come from $\{+1, -1\}$, and their product is 1, they must be equal. ▮

In virtue of the preceding [part (*ii*)] we have, for permutations

$$\text{even} \times \text{even} = \text{odd} \times \text{odd} = \text{even},$$
$$\text{even} \times \text{odd} = \text{odd} \times \text{even} = \text{odd}.$$

Furthermore, any $\sigma \in S_n$ can be written as a product of transpositions so, according to these rules, the "parity" of σ (meaning: whether σ is odd or even—that is, whether sgn σ is -1 or $+1$) is determined by the parities of the transpositions in the product. But these behave nicely because:

4-5-14. Lemma. Every transposition in S_n is odd.

Proof: First let us examine the simplest transposition—namely, (12). When (12) is applied to $\Delta = \Delta_n$ [as given in the triangular form (***)], only the terms in the first two rows are affected. Obviously, (12) maps $(x_1 - x_2)$ to $(x_2 - x_1)$. Also, under (12), each $(x_1 - x_j)$ with $j > 2$ [these being the remaining terms in the first row of (***)] is interchanged with $(x_2 - x_j)$, which is the term in the second row directly below it. Clearly, we then have $(12)\Delta = -\Delta$—so

$$\text{sgn}(12) = -1.$$

Now, consider an arbitrary transposition $\rho \in S_n$. It is conjugate to (12) (by 4-5-8), so there exists $\tau \in S_n$ for which

$$\rho = (12)^\tau = \tau(12)\tau^{-1}.$$

Then, using 4-5-12 and the fact that the values of sgn are ± 1,

$$\begin{aligned} \text{sgn}\,\rho = \text{sgn}[\tau(12)\tau^{-1}] &= (\text{sgn}\,\tau)(\text{sgn}(12))(\text{sgn}\,\tau^{-1}) \\ &= (\text{sgn}\,\tau)(\text{sgn}\,12)(\text{sgn}\,\tau) \\ &= (\text{sgn}\,\tau)^2(\text{sgn}\,12) = \text{sgn}(12) = -1. \end{aligned}$$

This completes the proof. ▮

4-5-15. Proposition. Let $\sigma \in S_n$ be given. Then the number of terms in any expression for σ as a product of transpositions is always odd or always even. In fact, this number is odd or even according as the permutation σ is odd or even.

Proof: Consider two factorizations

$$\sigma = \sigma_1 \sigma_2 \cdots \sigma_s = \tau_1 \tau_2 \cdots \tau_t$$

where each σ_i and τ_j is a transposition. Because sgn is multiplicative and sgn of any transposition is -1, we have

$$\operatorname{sgn} \sigma = (-1)^s = (-1)^t.$$

Therefore,

$$s \text{ is odd}, \Leftrightarrow \operatorname{sgn} \sigma = -1 \Leftrightarrow t \text{ is odd},$$
$$s \text{ is even} \Leftrightarrow \operatorname{sgn} \sigma = 1 \Leftrightarrow t \text{ is even}.$$

The validity of our proposition is now immediate. ▮

This result explains why the terminology "odd" or "even" permutations was chosen (in 4-5-11) rather than some other set of words.

4-5-16. Corollary. An r cycle is an even (odd) permutation $\Leftrightarrow$ r is odd (even).

Proof: An arbitrary r-cycle $(a_1 a_2 \cdots a_r)$ can be written as a product of $r-1$ transpositions: $(a_1 a_r) \cdots (a_1 a_3)(a_1 a_2)$. Since each transposition is an odd permutation, our assertion is immediate. ▮

The classification of permutations into even and odd ones enables us to distinguish an important subgroup of the symmetric group—namely:

4-5-17. Proposition. Let A_n denote the set of all even permutations in S_n. Then A_n is a normal subgroup of S_n with $\# A_n = n!/2$ and $(S_n : A_n) = 2$. Furthermore, A_n, which is known as the **alternating group** on n letters, is generated by the set of all 3-cycles.

Proof: A_n is a nonempty subset of the finite group S_n, and it is closed under multiplication since the product of even permutations is even. According to 4-2-3, A_n is a subgroup of S_n.

For any $\tau \in A_n$ and $\sigma \in S_n$ we have

$$\operatorname{sgn}(\sigma \tau \sigma^{-1}) = (\operatorname{sgn} \sigma)(\operatorname{sgn} \tau)(\operatorname{sgn} \sigma^{-1}) = \operatorname{sgn} \tau$$

—so $\sigma \tau \sigma^{-1} \in A_n$. Thus, A_n is a normal subgroup.

Fix any odd permutation ρ [for example, $\rho = (12)$ would do]; so, of course, ρ^{-1} is also odd. Consider the set

$$\rho A_n = \{\rho \tau \mid \tau \in A_n\}.$$

Every element of ρA_n is an odd permutation, being the product of an odd permutation and an even one. On the other hand, if σ is any odd permutation, then $\rho^{-1}\sigma$ is even, being the product of two odd permutations; thus, $\rho^{-1}\sigma \in A_n$ and $\sigma \in \rho A_n$. This shows that ρA_n is the set of all odd permutations. Clearly, we have a 1–1 correspondence

$$\tau \leftrightarrow \rho\tau, \qquad \tau \in A_n$$

between A_n and ρA_n. Since any permutation is either even or odd, but not both, we have

$$A_n \cup \rho A_n = S_n, \qquad A_n \cap \rho A_n = \varnothing.$$

Therefore,

$$\#(A_n) = \#(\rho A_n) = \frac{1}{2}(\# S_n) = \frac{n!}{2};$$

and because this is the coset decomposition of S_n with respect to A_n, the index of A_n in S_n is 2—$(S_n : A_n) = 2$.

A nicer way to see all this is as follows: The sgn function provides a homomorphism from the group S_n onto the multiplicative group with two elements, $\{\pm 1\}$ [since $\operatorname{sgn}(\sigma\tau) = (\operatorname{sgn}\sigma)(\operatorname{sgn}\tau)$]. It is clear that the kernel is $\{\tau \in S_n \mid \operatorname{sgn}\tau = 1\} = A_n$, the set of all even permutations. Hence, by 4-4-6, A_n is a normal subgroup of S_n. According to 4-4-12, the factor group S_n/A_n is isomorphic to the two element group $\{\pm 1\}$. In particular, there are exactly two cosets of A_n in S_n. Hence, $(S_n : A_n) = 2$, and by Lagrange, 4-2-16, $\# A_n = \frac{1}{2}(\# S_n) = n!/2$.

It remains to show that A_n is generated by the set of all 3-cycles. In virtue of 4-5-16, every 3-cycle is even, so it belongs to A_n. Conversely, given $\tau \in A_n$, it can be expressed as the product of an even number of transpositions. These transpositions can be arranged in adjacent pairs. Now, two adjacent transpositions are disjoint or they have one entry in common. (The possibility that they are identical is excluded because then their product is the identity $\cdots$ so there was no need to include such a pair in the product.) In the first case, the product of the two transpositions looks like

$$(ab)(cd) = (acd)(abd),$$

and in the second case it looks like

$$(ab)(ac) = (acb).$$

Consequently, $\tau \in A_n$ is a product of 3-cycles, and the proof is complete. ▮

4-5-18. Exercise. We study a subgroup of S_n whose origins are geometric.

For any $n \geq 3$ consider the regular polygon with n sides. Place it in the plane with one vertex on the X-axis, and label its n vertices as indicated in

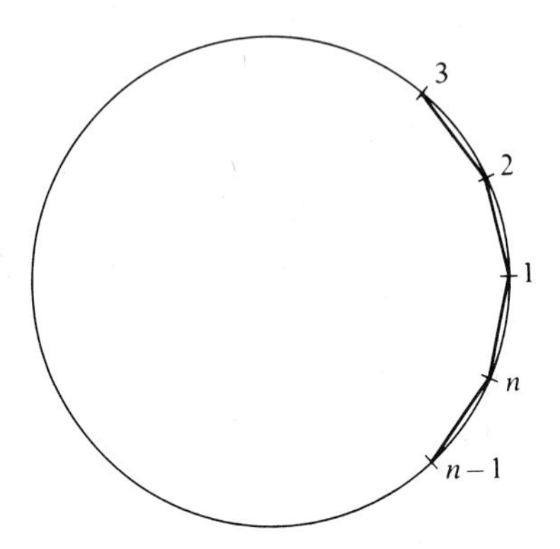

the accompanying diagram. The set of all rigid motions of our regular n-gon forms a group. It is a subgroup of S_n since every such motion is described by a permutation of the set of vertices $\{1, 2, \ldots, n\}$. An element of this group—which we call the **nth dihedral group** and denote by D_n—is completely determined by the way it maps the vertices at 1 and 2; so $\#(D_n) = 2n$.

Let $\sigma \in D_n$ denote the rotation by $2\pi/n$ radians in a counterclockwise direction. As a permutation this takes the form

$$\sigma = (123 \cdots n),$$

so σ has order n and $\sigma, \sigma^2, \ldots, \sigma^{n-1}, \sigma^n = e$ are distinct.

Let τ denote the flip (that is, rotation by π radians) of our regular n-gon over with respect to the X-axis. It is easy to express τ as a permutation; the result depends on whether n is even or odd—but in either case τ is given as a product of disjoint transpositions. Clearly, $\tau^2 = e$.

Compute $\sigma\tau$. Since $\sigma\tau$ maps $1 \to 2$ and $2 \to 1$, it follows that $(\sigma\tau)^2$ maps $1 \to 1$ and $2 \to 2$. This suffices to guarantee that $(\sigma\tau)^2 = e$. Since $\tau^{-1} = \tau$, this implies

$$\tau\sigma = \sigma^{-1}\tau = \sigma^{n-1}\tau. \qquad (*)$$

The subgroups of D_n generated by σ and by τ are

$$[\sigma] = \{e, \sigma, \sigma^2, \ldots, \sigma^{n-1}\}, \qquad [\tau] = \{e, \tau\}.$$

What is the subgroup of D_n generated by the 2-element set $\{\sigma, \tau\}$? In virtue of (*), every element of this group $[\{\sigma, \tau\}]$ can be reduced to the form

$$\sigma^r\tau^s, \qquad 0 \le r \le n-1, \quad 0 \le s \le 1.$$

These elements are distinct, and since there are $2n$ of them it follows that

$$D_n = \{\sigma^r\tau^s \mid 0 \le r \le n-1, \quad 0 \le s \le 1\}.$$

Finally, we leave it for the reader to show that the symmetries of the regular n-gon form a group; it is D_n.

4-5-19 / *PROBLEMS*

1. Give the compact cycle form of the following permutations:

$(i)\begin{pmatrix}1&2&3&4&5&6&7&8\\2&7&4&8&3&6&5&1\end{pmatrix}$, $\quad (ii)\begin{pmatrix}1&2&3&4&5&6&7&8\\7&4&1&5&8&2&6&3\end{pmatrix}$,

$(iii)\begin{pmatrix}1&2&3&4&5&6&7&8&9\\4&6&9&2&7&1&3&8&5\end{pmatrix}$, $\quad (iv)\begin{pmatrix}1&2&3&4&5&6&7&8&9\\6&3&5&1&4&9&7&2&8\end{pmatrix}$.

2. Express each of the following permutations in the two-line form:

(i) (2468) in S_8, *(ii)* (13579) in S_9,
(iii) (14923)(6875) in S_9, *(iv)* (152)(378)(46) in S_8.

3. Given the permutations (in S_9)

$$\sigma = (385)(21)(794), \qquad \tau = (15936)(27)(48), \qquad \rho = (2945)(186)(37)$$

compute:

(i) σ^2, τ^2, ρ^2, *(ii)* σ^3, τ^3, ρ^3,
(iii) $\sigma^{-1}, \tau^{-1}, \rho^{-1}$ *(iv)* $\sigma\tau, \tau\rho, \rho\sigma, \tau\sigma, \rho\tau, \sigma\rho$,
(v) $\sigma\tau\rho, \tau\rho\sigma, \rho\sigma\tau$ *(vi)* $\sigma^{-1}\tau\rho, \sigma\tau^{-1}\rho, \sigma\tau\rho^{-1}$,
(vii) $\sigma^\tau, \tau^\rho, \rho^\sigma$ *(viii)* $(\sigma^\tau)^\rho, (\tau^\sigma)^\rho, (\rho^\sigma)^\tau$.

4. Find the order of each of the permutations in Problem 2. With σ, τ, ρ as in Problem 3, find the order of $\sigma, \tau, \rho, \sigma^2, \tau^3, \rho^{-1}, \sigma\tau, \rho\sigma\tau, \sigma\tau^{-1}\rho$, $\sigma^\tau, (\tau^\sigma)^\rho$.

5. For any $\sigma \in S_n$, show that the orbits of σ^{-1} are the same as those of σ. How do the orbits of σ^2 compare with those of σ?

6. The permutations $\sigma = (28175)(36)(49)$ and $\rho = (43962)(18)(57)$ are conjugate. Find $\tau \in S_9$ for which $\rho = \sigma^\tau$. Find three other choices for τ. How many such τ's are there?

7. (*i*) List the six elements of S_3 in cycle form. Which ones are odd? Which are even? What is the order of each one? Give the decomposition of S_3 into conjugate classes.
(*ii*) Do the same thing for the 24 elements of S_4.

8. How many conjugate classes are there in

(*i*) S_5, (*ii*) S_6?

List a representative for each conjugate class.

9. Show that any expression for the identity e of S_n as a product of transpositions has an even number of terms.

10. Show that if two permutations are conjugate then they are both odd or they are both even. How do their orders compare?

11. (*i*) In S_9, exhibit an odd permutation of order 10. Can you find an even permutation of order 10?
(*ii*) Show that in S_8 every element of order 10 is odd.

12. Suppose $\sigma \in S_n$ is given as a product of cycles, which need not be disjoint. Show that σ is an even permutation $\Leftrightarrow$ the product has an even number of cycles of even length. What is the condition for σ to be odd?

13. Express each element of S_4 as a product of transpositions using:
(*i*) any transpositions,
(*ii*) only the transpositions (12), (13), (14),
(*iii*) only the transpositions (12), (23), (34).

14. How would you express 4-5-10 as a statement about shuffling a deck of cards?

15. In connection with the proof that σ^τ has the same cycle structure as σ [see 4-5-8, part (*i*)], does any account have to be taken of the digits that do not appear in the cycle expression for σ (that is, those digits which are fixed under σ)?

16. In connection with 4-5-11, show carefully that given $\sigma \in S_n$ it determines an automorphism $\bar{\sigma}$ (denoted in the text by σ) of the ring $\mathbf{Z}[x_1, \ldots, x_n]$. Moreover,

$$\overline{\sigma^{-1}} = \bar{\sigma}^{-1}$$

—in words, the automorphism of $\mathbf{Z}[x_1, \ldots, x_n]$ determined by $\sigma^{-1} \in S_n$ is the inverse of the automorphism of $\mathbf{Z}[x_1, \ldots, x_n]$ determined by $\sigma \in S_n$.

17. Find the subgroup of S_4 generated by the two elements:

(*i*) (12), (23), (*ii*) (12), (123),
(*iii*) (12), (1234), (*iv*) (12)(34), (123),
(*v*) (12), (234), (*vi*) (12)(34), (13)(24),
(*vii*) (123), (1234), (*viii*) (123), (124).

18. Find as many subgroups of S_4 as you can. Have you found them all? Which ones are normal?

19. Show that the symmetric group S_n is generated by the two elements $\sigma = (12)$, $\tau = (12 \cdots n)$.

20. Prove that the alternating group A_n is generated by each of the following sets of 3-cycles:

(*i*) $\{(1ij) \mid i \neq j\}$, (*ii*) $\{(12j) \mid j = 3, \ldots, n\}$.

21. The alternating group A_4 has 12 elements. Show that it has no subgroup of order 6. Show further that Klein's 4-group (see 4-1-12, Problem 16) which consists of the four elements: (1), (12)(34), (14)(23), (13)(24), is a normal subgroup of A_4. Find all subgroups of A_4.

22. Find all subgroups of the dihedral groups (see 4-5-18):

(*i*) D_5—this is the group of symmetries of the regular pentagon, and it has 10 elements,
(*ii*) D_6—this is the group of symmetries of the regular hexagon, and it has 12 elements,
(*iii*) D_{13},
(*iv*) D_n—this is the group of symmetries of the regular n-gon, and it has $2n$ elements.

Can you interpret the subgroups geometrically? Which ones are normal?

23. The dihedral group D_4 is the same as the octic group.

(*i*) Is it a normal subgroup of S_4?
(*ii*) Find the center $\mathfrak{Z}$ of D_4. What is $D_4/\mathfrak{Z}$?
(*iii*) Find the inner automorphisms of D_4.
(*iv*) Find the sets of conjugate elements in D_4.
(*v*) List all factor groups of D_4.
(*vi*) Describe all homomorphic images of D_4.

24. The regular tetrahedron, shown in the accompanying figure, has four vertices and four faces, each of which is an equilateral triangle. (Of course, the four faces are congruent.)

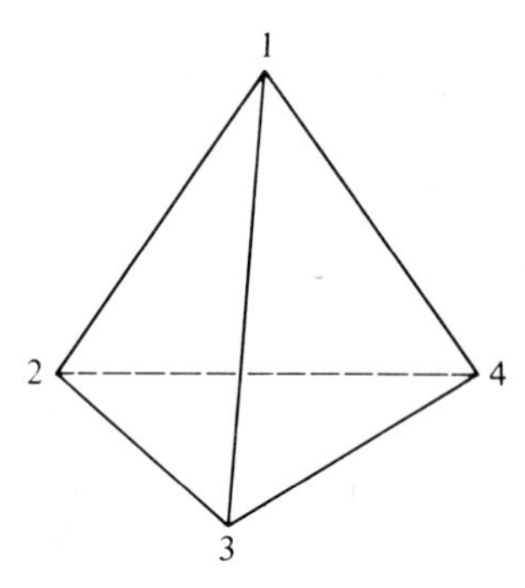

Show that the rigid motions of this solid figure form a group with 12 elements. Compare this group, called the **tetrahedral group**, with the alternating group A_4. What about the symmetries (that is, rotations) of the regular tetrahedron?

4-6. The Group $\mathbf{Z}_m^*$

In this section we investigate some number-theoretic questions whose proper interpretation is in the language and framework of group theory. More precisely, we shall, in essence, be studying $\mathbf{Z}_m^*$, the multiplicative group of units in the ring $\mathbf{Z}_m$ of residue classes (mod m).

As the reader will observe, we turn in this section from our wordy discursive style toward the brisk tight style ordinarily employed in mathematical writing.

We begin with a standard definition from number theory.

4-6-1. Definition. Fix an integer $m > 1$ and consider any integer a with $(a, m) = 1$. We say that a **belongs to the exponent** t (mod m)—or "a belongs to t" or "a has exponent t"—when t is the smallest positive integer for which

$$a^t \equiv 1 \pmod{m}.$$

[Of course, such a t exists and is, in fact, less than or equal to $\phi(m)$ since according to Euler–Fermat, 4-3-10, we know that $a^{\phi(m)} \equiv 1 \pmod{m}$.] If a belongs to the exponent $\phi(m)$ (mod m) (which is the maximum exponent possible) we say that a is a **primitive root** (mod m), or that "m has the primitive root a." It is customary to denote a primitive root, if one exists, by g; we shall conform to this usage.

4-6-2. Examples. As illustrations of the definition, the reader may check the following statements: 5 belongs to the exponent 2 (mod 12); 5 belongs to the exponent 4 (mod 13)—of course, 18 or any other integer congruent to 5 (mod 13) also belongs to the exponent 4 (mod 13), since $18^i \equiv 5^i$ (mod 13) for all i; 3 belongs to the exponent 3 (mod 13), and so does any integer congruent to 3 (mod 13); 14 does not belong to any exponent (mod 18)—the definition is not applicable, since $(14, 18) \neq 1$; 3 belongs to the exponent 6 (mod 7) so, because $\phi(7) = 6$, this says that 3 is a primitive root (mod 7). For $m = 18$ we have $\phi(18) = 6$ and there exists a primitive root (mod 18)—for example, 5 is a primitive root. In particular, a primitive root can exist when m is not prime. On the other hand, a primitive root need not exist for every m; in particular, for $m = 24$ we have $\phi(24) = 8$ and none of the eight elements 1, 5, 7, 11, 13, 17, 19, 23 (these being the only integers which need to be tested) belongs to the exponent 8 (mod 24)—thus, there is no primitive root (mod 24).

4-6-3. Remark. In a number of places we have discussed and used the correspondence between statements about congruence (mod m) and statements about $\mathbf{Z}_m$. To translate the preceding definition to the language of residue classes we make an initial observation: $(a, m) = 1$ means that a (more precisely, $\lfloor a \rfloor_m$—but we shall be careless and drop the cumbersome notation) is an element of the group $\mathbf{Z}_m^*$ (see 3-2-13 and 4-1-7). Then, clearly

$$a \text{ belongs to the exponent } t \text{ (mod } m)$$
$$\Leftrightarrow a \text{ has order } t, \text{ as an element of the group } \mathbf{Z}_m^*.$$

From what we know about the order of an element in a finite group, the standard elementary properties of the "exponent" are now immediate. Among these properties we have the following:

Suppose a belongs to the exponent t (mod m); then

(*i*) $a^n \equiv 1 \pmod m \Rightarrow t \mid n$. (Use 4-3-7.)
(*ii*) $t \mid \phi(m)$. (Use 4-3-9.)
(*iii*) $a^i \equiv a^j \pmod m \Leftrightarrow i \equiv j \pmod t$. (Use 4-3-7.)
(*iv*) a^r belongs to the exponent $t/(r, t)$. (Use 4-3-13.)

In addition, the assertion that a is a primitive root (mod m) is equivalent to saying that a (rather, $\lfloor a \rfloor_m$) is an element of order $\phi(m)$ in $\mathbf{Z}_m^*$. Since $\#(\mathbf{Z}_m^*) = \phi(m)$ this becomes: a is a generator of the group $\mathbf{Z}_m^*$, which is cyclic. To rephrase it:

There exists a primitive root (mod m) $\Leftrightarrow$ the group $\mathbf{Z}_m^*$ is cyclic; and, in this situation, a primitive root (mod m) is the same as a generator of the cyclic group $\mathbf{Z}_m^*$.

In virtue of our results about cyclic groups, the following facts about primitive roots are immediate:

(*v*) Suppose g is a primitive root (mod m); then g^r is a primitive root $\Leftrightarrow (r, \phi(m)) = 1$. (Use 4-3-13.)
(*vi*) If m has a primitive root, then it has exactly $\phi(\phi(m))$ of them. (Use 4-3-13.)
(*vii*) If $(g, m) = 1$, then g is a primitive root (mod m) $\Leftrightarrow$ the integers $g, g^2, \ldots, g^{\phi(m)}$ constitute a reduced residue system (mod m). (Use 3-2-22).

Given the concepts on which we have focused, the answer to the following question is obviously of interest: For which values of m does there exist a primitive root (mod m)? in other words, for which m is $\mathbf{Z}_m^*$ cyclic? Some information is already available. According to 4-3-16, $\mathbf{Z}_p^*$ is a cyclic group [of order $\phi(p) = p - 1$] for every prime p. Also, as indicated in 4-6-2, $\mathbf{Z}_{18}^*$ is cyclic (so $\mathbf{Z}_m^*$ can be cyclic without m being prime) and $\mathbf{Z}_{24}^*$ is not cyclic (so not every $\mathbf{Z}_m^*$ is cyclic).

To study $\mathbf{Z}_m^*$ for arbitrary m, let us write out the prime power factorization of m, once and for all.

$$m = 2^{r_0} p_1^{r_1} p_2^{r_2} \cdots p_s^{r_s},$$

where $p_1, p_2, \ldots, p_s$ are distinct odd primes, the exponents $r_1, r_2, \ldots, r_s$ are all greater than 0, and $r_0 \geq 0$ ($r_0 = 0$ occurs when m is odd). As is often the case in number theory, the prime 2 will play a special role, and for this reason we have distinguished it from the other primes. Sometimes it will be convenient to put $p_0 = 2$.

Before undertaking the detailed analysis of $\mathbf{Z}_m^*$, a preliminary (but basically familiar) definition is needed. Suppose we are given multiplicative groups $G_1, G_2, \ldots, G_k$ with identities $e_1, e_2, \ldots, e_k$, respectively. Consider the product set $G_1 \times G_2 \times \cdots \times G_k$. It consists of all k-tuples $(a_1, a_2, \ldots, a_k)$ with $a_i \in G_i$ for $i = 1, 2, \ldots, k$. Define multiplication for elements of $G_1 \times G_2 \times \cdots \times G_k$ componentwise—that is,

$$(a_1, a_2, \ldots, a_k) \cdot (b_1, b_2, \ldots, b_k) = (a_1 b_1, a_2 b_2, \ldots, a_k b_k).$$

Then $G_1 \times G_2 \times \cdots \times G_k$ becomes a group with identity $(e_1, e_2, \ldots, e_k)$. This resulting group is denoted by $G_1 \times \cdots \times G_k$ (there is no harm in using the same notation as for the product set) or by $\Pi_{i=1}^k G_i$; it is known as the **(external) direct product** of the groups $G_1, \ldots, G_k$. Other terminologies may also be used; for example, if all the groups G_i are additive, the resulting group is often denoted by $\Sigma_{i=1}^k \oplus G_i$ and called the **(external) direct sum.**

It is easy to see that the direct product group $\Pi_{i=1}^k G_i$ is abelian if and only if each G_i is abelian. We note also that $\Pi_{i=1}^k G_i$ is finite if and only if each G_i is finite; and in this situation, the order of the direct product group is the product of the orders of the component groups—

$$\#\left(\prod_{i=1}^k G_i\right) = \prod_{i=1}^k (\#(G_i)).$$

A final observation: The order of an element of the direct product is the least common multiple of the orders of the component terms; in more detail, for any element $(a_1, a_2, \ldots, a_k) \in G_1 \times G_2 \times \cdots \times G_k$, we have

$$\operatorname{ord}(a_1, a_2, \ldots, a_k) = [\operatorname{ord} a_1, \operatorname{ord} a_2, \ldots, \operatorname{ord} a_k].$$

[This is a consequence of: $(a_1, a_2, \ldots, a_k)^n = (e_1, \ldots, e_k)$, the identity of $\Pi_i^k G_i$, if and only if $a_i^n = e_i$ for each $i = 1, \ldots, k$.]

4-6-4. Proposition. We have an isomorphism of groups:

$$\mathbf{Z}_m^* \approx \mathbf{Z}_{2^{r_0}}^* \times \mathbf{Z}_{p_1^{r_1}}^* \times \cdots \times \mathbf{Z}_{p_s^{r_s}}^*.$$

Proof: Note that when $r_0 = 0$ it is understood that the first term on the right, $\mathbf{Z}_{2^{r_0}}^*$, should be dropped completely. Also, when $r_0 = 1$ the group $\mathbf{Z}_{2^{r_0}}^*$ consists of a single element, so this term really makes no contribution to the right-hand side.

Essentially, this result is a straightforward generalization of things known to us (for the case where the right-hand side has only two terms) from the work in Section 3-2 on the computation of the Euler ϕ-function. For this reason, we shall merely outline the proof here.

Following the procedure of 3-2-17, now let us form the direct sum ring $\mathbf{Z}_{2^{r_0}} \oplus \mathbf{Z}_{p_1^{r_1}} \oplus \cdots \oplus \mathbf{Z}_{p_s^{r_s}}$ and define a mapping of

$$\mathbf{Z}_m \to \mathbf{Z}_{2^{r_0}} \oplus \mathbf{Z}_{p_1^{r_1}} \oplus \cdots \oplus \mathbf{Z}_{p_s^{r_s}}$$

by the rule

$$\lfloor a \rfloor_m \to (\lfloor a \rfloor_{2^{r_0}}, \lfloor a \rfloor_{p_1^{r_1}}, \ldots, \lfloor a \rfloor_{p_s^{r_s}}).$$

This mapping is well defined and a ring homomorphism. Because $2^{r_0}, p_1^{r_1}, \ldots, p_s^{r_s}$ are relatively prime in pairs, it follows from the Chinese remainder theorem that the mapping is surjective. Using the fact that both rings under consideration have m elements, it follows that we have isomorphic rings

$$\mathbf{Z}_m \approx \mathbf{Z}_{2^{r_0}} \oplus \mathbf{Z}_{p_1^{r_1}} \oplus \cdots \oplus \mathbf{Z}_{p_s^{r_s}}.$$

Now, as noted in 3-2-18, if two commutative rings with unity are isomorphic then their sets of units are in 1–1 correspondence. Even more, the groups of units of the two rings are clearly isomorphic. Thus,

$$\mathbf{Z}_m^* \approx (\mathbf{Z}_{2^{r_0}} \oplus \mathbf{Z}_{p_1^{r_1}} \oplus \cdots \oplus \mathbf{Z}_{p_s^{r_s}})^*.$$

Furthermore, as in 3-2-19, there is a natural 1–1 correspondence between the units of the ring $\mathbf{Z}_{2^{r_0}} \oplus \mathbf{Z}_{p_1^{r_1}} \oplus \cdots \oplus \mathbf{Z}_{p_s^{r_s}}$ and the elements of the product set $\mathbf{Z}_{2^{r_0}}^* \times \mathbf{Z}_{p_1^{r_1}}^* \times \cdots \times \mathbf{Z}_{p_s^{r_s}}^*$. Clearly, this provides an isomorphism of groups

$$(\mathbf{Z}_{2^{r_0}} \oplus \mathbf{Z}_{p_1^{r_1}} \oplus \cdots \oplus \mathbf{Z}_{p_s^{r_s}})^* \approx \mathbf{Z}_{2^{r_0}}^* \times \mathbf{Z}_{p_1^{r_1}}^* \times \cdots \times \mathbf{Z}_{p_s^{r_s}}^*.$$

The desired result is thereby proved. ▌

Our basic question of when $\mathbf{Z}_m^*$ is cyclic has been transferred to the group $\mathbf{Z}_{2^{r_0}}^* \times \mathbf{Z}_{p_1^{r_1}}^* \times \cdots \times \mathbf{Z}_{p_s^{r_s}}^*$; the analysis of this group is subsumed under the following general result.

4-6-5. Proposition. Suppose $G_1, G_2, \ldots, G_k$ are (multiplicative) groups of order $n_1, n_2, \ldots, n_k$, respectively.

(*i*) If the direct product group $G_1 \times G_2 \times \cdots \times G_k$ is cyclic, then so is G_i for each $i = 1, 2, \ldots, k$.

(*ii*) If $G_1, G_2, \ldots, G_s$ are cyclic, then $G_1 \times G_2 \times \cdots \times G_k$ is cyclic $\Leftrightarrow$ $n_1, n_2, \ldots, n_k$ are relatively prime in pairs.

Proof: (*i*) For each $i = 1, \ldots, k$ define a mapping, called the ith **projection,**

$$\pi_i: G_1 \times G_2 \times \cdots \times G_k \to G_i$$

by putting

$$\pi_i(a_1, a_2, \ldots, a_i, \ldots, a_k) = a_i.$$

This is a surjective homomorphism. Since any homomorphic image of a cyclic group is cyclic (4-4-15, Problem 14) we see that G_i is cyclic.

(*ii*) First of all we note that $n_1, n_2, \ldots, n_k$ are relatively prime in pairs $\Leftrightarrow$ $[n_1, n_2, \ldots, n_k] = n_1 n_2 \cdots n_k$. (This fact has already been stated in 1-5-13, Problem 20.) Here is one way to prove it.

The positive integers $n_1, \ldots, n_k$ are relatively prime in pairs

$$\Leftrightarrow \begin{cases}\text{for each prime } p \text{ and every pair } i, j \text{ from} \\ \{1, 2, \ldots, k\} \text{ with } i \neq j \text{ we have } \min\{v_p(n_i), v_p(n_j)\} = 0\end{cases}$$

$$\Leftrightarrow \begin{cases}\text{for each prime } p, \text{ the set } \{v_p(n_1), \ldots, v_p(n_k)\} \\ \text{contains at most one nonzero term}\end{cases}$$

$$\Leftrightarrow \sum_{i=1}^{k} v_p(n_i) = \max\{v_p(n_i) \mid i = 1, \ldots, k\} \text{ for each prime } p$$

$$\Leftrightarrow n_1 n_2 \cdots n_k = [n_1, n_2, \ldots, n_k].$$

Turning to the proof itself, suppose $n_1, n_2, \ldots, n_k$ are relatively prime in pairs. For each cyclic group G_i choose a generator a_i—so ord $a_i = n_i$. In the group $G_1 \times G_2 \times \cdots \times G_k$, whose order is $n_1 n_2 \cdots n_k$, the order of the element $(a_1, a_2, \ldots, a_k)$ is given by

$$\begin{aligned}\operatorname{ord}(a_1, a_2, \ldots, a_k) &= [\operatorname{ord} a_1, \operatorname{ord} a_2, \ldots, \operatorname{ord} a_k] \\ &= [n_1, n_2, \ldots, n_k] \\ &= n_1 n_2 \cdots n_k.\end{aligned}$$

Hence, the group $G_1 \times \cdots \times G_k$ is cyclic and $(a_1, \ldots, a_k)$ is a generator. This proves the implication $\Leftarrow$.

Conversely, suppose $n_1, n_2, \ldots, n_k$ are not relatively prime in pairs—so $[n_1, n_2, \ldots, n_k] \neq n_1 n_2 \cdots n_k$. An arbitrary element of $G_1 \times \cdots \times G_k$ is of form $(b_1, \ldots, b_k)$ with $b_i \in G_i$ for $i = 1, \ldots, k$. Denoting the order of b_i by q_i, we know that $q_i \mid n_i$. Therefore,

$$\begin{aligned}\operatorname{ord}(b_1, \ldots, b_k) &= [\operatorname{ord} b_1, \ldots, \operatorname{ord} b_k] \\ &= [q_1, q_2, \ldots, q_k] \\ &\mid [n_1, n_2, \ldots, n_k] \\ &< n_1 n_2 \cdots n_k.\end{aligned}$$

Thus, the group $G_1 \times \cdots \times G_k$ of order $n_1 n_2 \cdots n_k$ contains no element of order $n_1 n_2 \cdots n_k$. Hence, $G_1 \times \cdots \times G_k$ is not cyclic; and we have proved the implication $\Rightarrow$. The entire proof is now complete. ▮

Applying this result to the concrete situation at hand, we see that if $\mathbf{Z}_m^*$ is cyclic, then each of the groups $\mathbf{Z}_{2^{r_0}}^*, \mathbf{Z}_{p_1^{r_1}}^*, \ldots, \mathbf{Z}_{p_s^{r_s}}^*$ must be cyclic and their orders $\phi(2^{r_0}), \phi(p_1^{r_1}), \ldots, \phi(p_s^{r_s})$ must be relatively prime in pairs. Because $\phi(p_i^{r_i}) = p_i^{r_i - 1}(p_i - 1)$ is even and $\phi(2^{r_0}) = 2^{r_0 - 1}$ is even when $r_0 \geq 2$, it follows that if $\mathbf{Z}_m^*$ is cyclic, the only possibilities for m are: 2^{r_0}, $p_1^{r_1}$, $2^{r_0} p_1^{r_1}$ with $r_0 < 2$.

Clearly, the problem still facing us is to decide when $\mathbf{Z}_{p^r}^*$ (p any prime) is cyclic. We already know that $\mathbf{Z}_p^*$ is cyclic for every prime p. A natural approach is then to suppose inductively that $\mathbf{Z}_{p^r}^*$ is cyclic and investigate if $\mathbf{Z}_{p^{r+1}}^*$ is also cyclic. One step in this direction is the following.

4-6-6. Lemma. Suppose p is an odd prime and g is a primitive root $(\operatorname{mod} p^r)$, $r \geq 1$ (that is: g belongs to the exponent $\phi(p^r) \pmod{p^r}$, or equivalently, $\mathbf{Z}_{p^r}^*$ is cyclic and g is a generator); then

(*i*) The exponent to which g belongs $(\operatorname{mod} p^{r+1})$ is either $\phi(p^{r+1})$ or $\phi(p^r)$.

(*ii*) g belongs to the exponent $\phi(p^r) \pmod{p^{r+1}} \Leftrightarrow g^{\phi(p^r)} \equiv 1 \pmod{p^{r+1}}$.

(*iii*) g is a primitive root $(\operatorname{mod} p^{r+1}) \Leftrightarrow g^{\phi(p^r)} \not\equiv 1 \pmod{p^{r+1}}$.

Proof: Let t be the exponent to which g belongs (mod p^{r+1}); so

$$g^t \equiv 1 \pmod{p^{r+1}} \tag{$*$}$$

and because $g^{\phi(p^{r+1})} \equiv 1 \pmod{p^{r+1}}$ we know that $t \mid \phi(p^{r+1})$. On the other hand $(*)$ implies $g^t \equiv 1 \pmod{p^r}$; and because g belongs to $\phi(p^r) \pmod{p^r}$, we know that $\phi(p^r) \mid t$. The relation

$$\phi(p^r) \mid t \mid \phi(p^{r+1}),$$

coupled with $\phi(p^{r+1}) = p\phi(p^r)$, forces the conclusion: $t = \phi(p^r)$ or $t = \phi(p^{r+1})$.

Parts (*ii*) and (*iii*) are now immediate; in fact, they are essentially restatements of what has already been proved. ▌

Next, we examine the passage from $\mathbf{Z}_p^*$ to $\mathbf{Z}_{p^2}^*$. Things go very nicely.

4-6-7. Lemma. The group $\mathbf{Z}_{p^2}^*$ is cyclic. In fact, there exists a primitive root g (mod p) which is also a primitive root (mod p^2).

Proof: Since $\mathbf{Z}_p^*$ is cyclic, there exists a primitive root (mod p). Choose any one, and call it g. We have $g^{p-1} \equiv 1 \pmod{p}$. If g belongs to the exponent $\phi(p^2) = p(p-1) \pmod{p^2}$ we are finished. If not, then according to 4-6-6, g belongs to the exponent $\phi(p) = p - 1$, and we have

$$g^{p-1} \equiv 1 \pmod{p^2}. \tag{\#}$$

Now consider $g + p$. It represents the same element of $\mathbf{Z}_p^*$ as g, so it is a primitive root (mod p). To show that $g + p$ is also a primitive root (mod p^2) it suffices [according to 4-6-6, part (*iii*)] to show

$$(g + p)^{p-1} \not\equiv 1 \pmod{p^2}.$$

But this is easy:

$$\begin{aligned}(g + p)^{p-1} &= g^{p-1} + (p-1)g^{p-2}p + p^2 \text{ (some integer)}\\ &\equiv g^{p-1} + (p-1)g^{p-2}p \pmod{p^2}\\ &\equiv 1 - g^{p-2}p \pmod{p^2}.\end{aligned}$$

Because $(g, p) = 1$, surely $p \nmid g^{p-2}$—so $(g + p)^{p-1} \not\equiv 1 \pmod{p^2}$. Thus, $g + p$ is a primitive root (mod p^2) and the proof is complete. ▌

The next step is to examine the passage from $\mathbf{Z}_{p^r}^*$ to $\mathbf{Z}_{p^{r+1}}^*$, in general. This goes in one fell swoop, and leads to:

4-6-8. Proposition. For the odd prime p, $\mathbf{Z}_{p^r}^*$ is cyclic for all $r \geq 1$. In fact, if g is any integer which is a primitive root $(\text{mod } p)$ and $(\text{mod } p^2)$, then g is a primitive root $(\text{mod } p^r)$ for all positive r.

Proof: From the preceding, $\mathbf{Z}_p^*$ and $\mathbf{Z}_{p^2}^*$ are cyclic and there exists an integer which is a primitive root $(\text{mod } p)$ and $(\text{mod } p^2)$. Take any such integer; call it g. We have then, (in virtue of 4-4-6),

$$g^{p-1} \equiv 1 \pmod{p} \qquad \text{and} \qquad g^{p-1} \not\equiv 1 \pmod{p^2}.$$

Therefore, we can write

$$g^{p-1} = 1 + tp \qquad \text{where } p \nmid t.$$

Now, suppose inductively that g is a primitive root $(\text{mod } p^r)$. Then $g^{\phi(p^r)} = g^{(p-1)p^{r-1}} = (1 + tp)^{p^{r-1}}$, so by the binomial theorem we have

$$\begin{aligned} g^{\phi(p^r)} &= 1 + (p^{r-1})(tp) + \frac{p^{r-1}(p^{r-1} - 1)}{1 \cdot 2}(tp)^2 + \cdots \\ &= 1 + tp^r + p^{r+1}(\text{some integer}) \\ &\equiv 1 + tp^r \pmod{p^{r+1}}. \end{aligned}$$

Since $p \nmid t$, we have $g^{\phi(p^r)} \not\equiv 1 \pmod{p^{r+1}}$—so according to 4-6-6, g is a primitive root $(\text{mod } p^{r+1})$. ∎

It still remains to study the cyclicity of $\mathbf{Z}_{2^r}^*$. The story here differs from that for $\mathbf{Z}_{p^r}^*$.

4-6-9. Proposition. (*i*) The groups $\mathbf{Z}_2^*$ and $\mathbf{Z}_{2^2}^*$ are cyclic; $\mathbf{Z}_{2^3}^*$ is not cyclic.

(*ii*) The group $\mathbf{Z}_{2^r}^*$ is not cyclic for any $r \geq 3$.

(*iii*) Every element of $\mathbf{Z}_{2^r}^*$ $(r \geq 3)$ has order less than or equal to 2^{r-2}; in fact, the order of every element divides 2^{r-2}.

(*iv*) The element 5 of $\mathbf{Z}_{2^r}^*$ $(r \geq 3)$ has order 2^{r-2}.

Proof: The group $\mathbf{Z}_2^*$ consists of a single element, while $\mathbf{Z}_{2^2}^* = \mathbf{Z}_4^*$ has two elements—so they are both cyclic. The group $\mathbf{Z}_{2^3}^* = \mathbf{Z}_8^*$ consists of the four residue classes 1, 3, 5, 7 and

$$1^2 = 3^2 = 5^2 = 7^2 = 1 \qquad \text{in } \mathbf{Z}_8^*$$

—so $\mathbf{Z}_{2^3}^*$ is not cyclic. This proves (*i*).

The group $\mathbf{Z}_{2^r}^*$ has order $\phi(2^r) = 2^{r-1}$, so in order to prove (*ii*) it surely suffices to prove (*iii*); and for this it suffices to show that if a is any odd integer, then

$$a^{2^{r-2}} \equiv 1 \pmod{2^r}, \qquad r \geq 3. \tag{\#}$$

The case $r = 3$ asserts $a^2 \equiv 1 \pmod 8$; it follows from writing $a = 1 + 2t$, so that $a^2 = 1 + 4t(t+1)$, and noting that $t(t+1)$ is always even. (Of course, $a^2 \equiv 1 \pmod 8$ follows from the observations about $\mathbf{Z}_8^*$ made in the proof of part (*i*).

Now, suppose inductively that (#) is true for r—so we may write $a^{2^{r-2}} = 1 + 2^r t$. Then

$$a^{2^{r-1}} = (a^{2^{r-2}})^2 = 1 + 2^{r+1}t + 2^{2r}t^2 \equiv 1 \pmod{2^{r+1}}.$$

Thus, (#) holds for $r + 1$ and (*iii*) is proved.

As for (*iv*), since the order of 5 in $\mathbf{Z}_{2^r}^*$ is a divisor of 2^{r-2}, it suffices to show

$$5^{2^{r-3}} \not\equiv 1 \pmod{2^r}, \qquad r \geq 3.$$

Actually, we are somewhat more precise, and show that

$$5^{2^{r-3}} \equiv 1 + 2^{r-1} \pmod{2^r}, \qquad r \geq 3. \tag{\# \#}$$

This is valid for $r = 3$ since

$$5 \equiv 1 + 4 \pmod{2^3}.$$

Now, suppose inductively that (##) holds for r. Then

$$\begin{aligned} 5^{2^{r-2}} = (5^{2^{r-3}})^2 &= (1 + 2^{r-1} + 2^r t)^2 \qquad (t \in \mathbf{Z}) \\ &= 1 + 2^r + 2^{2r-2} + 2^{r+1}t + 2^{2r}t + 2^{2r}t^2 \\ &\equiv 1 + 2^r \pmod{2^{r+1}}. \end{aligned}$$

Consequently, (# #) holds for $r + 1$, which completes the proof. ▮

4-6-10. Theorem. The group $\mathbf{Z}_m^*$ is cyclic (that is, there exists a primitive root for m) $\Leftrightarrow$ m is of form $2, 4, p^r, 2p^r$, where p is an odd prime.

Proof: Combine the things we already know—especially 4-6-4, 4-6-5, 4-6-8, and 4-6-9. ▮

As an additional consequence of the foregoing discussion we have the following result.

4-6-11. Theorem. For any $m = 2^{r_0} p_1^{r_1} p_2^{r_2} \cdots p_s^{r_s}$, let $\lambda(m)$ denote the maximum of the orders of all the elements in the group $\mathbf{Z}_m^*$. Then $\lambda(m)$, which is called the **universal exponent** of m, is given by

$$\lambda(m) = [\lambda(2^{r_0}), \phi(p_1^{r_1}), \phi(p_2^{r_2}), \ldots, \phi(p_s^{r_s})].$$

Proof: For any m, the order of any element of $\mathbf{Z}_m^*$ divides $\phi(m)$—so $\lambda(m) \mid \phi(m)$, and $\lambda(m) = \phi(m) \Leftrightarrow \mathbf{Z}_m^*$ is cyclic.

Note that $\lambda(2^0) = \lambda(1) = 1$, by convention; $\lambda(2^1) = \lambda(2) = \phi(2) = 1$, since $\mathbf{Z}_m^*$ is cyclic; $\lambda(2^0) = \lambda(4) = \phi(4) = 2$, since $\mathbf{Z}_m^*$ is cyclic; and for $r_0 \geq 3$, $\lambda(2^{r_0}) = 2^{r_0 - 2}$, by 4-6-9.

To compute $\lambda(m)$, in general, consider the isomorphic groups

$$\mathbf{Z}_m^* \approx \mathbf{Z}_{2^{r_0}}^* \times \mathbf{Z}_{p_1^{r_1}}^* \times \cdots \times \mathbf{Z}_{p_s^{r_s}}^*.$$

Suppose $r_0 \geq 3$. Then, by 4-6-9, the element 5 has order $\lambda(2^{r_0}) = 2^{r_0 - 2}$ in $\mathbf{Z}_{2^{r_0}}^*$. For $i = 1, \ldots, s$ choose a generator a_i of the cyclic group $\mathbf{Z}_{p_i^{r_i}}^*$ [whose order is $\phi(p_i^{r_i})$]. Then the element $(5, a_1, a_2, \ldots, a_s)$ of $\mathbf{Z}_{2^{r_0}}^* \times \mathbf{Z}_{p_1^{r_1}}^* \times \cdots \times \mathbf{Z}_{p_s^{r_s}}^*$ has order $[\lambda(2^{r_0}), \phi(p_1^{r_1}), \ldots, \phi(p_s^{r_s})]$. Hence,

$$\lambda(m) \geq [\lambda(2^{r_0}), \phi(p_1^{r_1}), \ldots, \phi(p_s^{r_s})].$$

On the other hand, an arbitrary element of $\mathbf{Z}_{2^{r_0}}^* \times \mathbf{Z}_{p_1^{r_1}}^* \times \cdots \times \mathbf{Z}_{p_s^{r_s}}^*$ is of form $(b_0, b_1, \ldots, b_s)$ where $b_0 \in \mathbf{Z}_{2^{r_0}}^*$ (so its order, q_0, divides $2^{r_0 - 2}$, by 4-6-9), and $b_i \in \mathbf{Z}_{p_i^{r_i}}^*$ for $i = 1, \ldots, s$ [so its order, q_i, divides $\phi(p)_i^{r_i}$]. The order of $(b_0, b_1, \ldots, b_s)$ is $[q_0, q_1, \ldots, q_s]$, and surely

$$[q_0, q_1, \ldots, q_s] \mid [\lambda(2^{r_0}), \phi(p_1^{r_1}), \ldots, \phi(p_s^{r_s})].$$

This implies $\lambda(m) \leq [\lambda(2^{r_0}), \phi(p_1^{r_1}), \ldots, \phi(p_s^{r_s})]$—so equality holds.

The situation where $r_0 < 3$ is easy and is left to the reader. ▮

4-6-12. Remark. The preceding result may be viewed as a generalization of Euler–Fermat. In more detail, if a is any integer relatively prime to m, then, according to Euler–Fermat,

$$a^{\phi(m)} \equiv 1 \pmod{m} \tag{$*$}$$

whereas 4-6-11 guarantees that

$$a^{\lambda(m)} \equiv 1 \pmod{m}. \tag{$**$}$$

[In fact, $\lambda(m)$ is best possible, in the sense that no smaller integer will have the same property for all a with $(a, m) = 1$.]

Now, on occasion, $\lambda(m)$ can be much smaller than $\phi(m)$, so that ($**$) is a much stronger statement than ($*$). For example, suppose $m = 5040 = 2^4 \cdot 3^2 \cdot 5 \cdot 7$; then

$$\begin{aligned} \phi(m) &= \phi(2^4)\phi(3^2)\phi(5)\phi(7) \\ &= (2^3)(3 \cdot 2)(4)(6) \\ &= 1152 \end{aligned}$$

while

$$\begin{aligned} \lambda(m) &= [\lambda(2^4), \phi(3^2), \phi(5), \phi(7)] \\ &= [2^2, 3 \cdot 2, 4, 6] \\ &= 12. \end{aligned}$$

Therefore,

$$a^{12} \equiv 1 \pmod{5040} \qquad \text{whenever } (a, 5040) = 1$$

which is far more informative than $a^{1152} \equiv 1 \pmod{5040}$, the assertion of Euler–Fermat.

4-6-13. Discussion. We turn our attention in a new direction. Suppose m is an integer for which the group $\mathbf{Z}_m^*$ [whose order is $\phi(m)$] is cyclic. Fix a generator g of $\mathbf{Z}_m^*$. By the properties of cyclic groups (see Section 4-3) we know that for integers i, j

$$g^i = g^j \Leftrightarrow i \equiv j \pmod{\phi(m)}$$

and, what is of major importance for us here, given any element $a \in \mathbf{Z}_m^*$ it can be expressed uniquely in the form

$$a = g^r \qquad \text{where } 0 \leq r \leq \phi(m) - 1.$$

We shall write

$$r = \operatorname{ind}_g a$$

and say that r is the **index** of a with respect to g. (The index depends on the choice of the generator g of $\mathbf{Z}_m^*$; when it is clear which g is being used, one often writes ind a instead of $\operatorname{ind}_g a$). Consequently, $a \in \mathbf{Z}_m^*$ may be written as

$$a = g^{\operatorname{ind}_g a}.$$

Taking another element $b \in \mathbf{Z}_m^*$, we have $b = g^{\operatorname{ind}_g b}$; and then

$$ab = g^{\operatorname{ind}_g a + \operatorname{ind}_g b}.$$

Of course, $\operatorname{ind}_g a + \operatorname{ind}_g b$ may well be greater than $\phi(m) - 1$, but because

$$ab = g^{\operatorname{ind}_g (ab)},$$

it follows that

$$\operatorname{ind}_g(ab) \equiv \operatorname{ind}_g a + \operatorname{ind}_g b \pmod{\phi(m)}, \qquad a, b \in \mathbf{Z}_m^*. \tag{$*$}$$

From this we also obtain

$$\operatorname{ind}_g(a^n) \equiv n \operatorname{ind}_g a \pmod{\phi(m)}, \qquad a \in \mathbf{Z}_m^*, n \geq 1.$$

To appreciate the full significance of $(*)$, consider the mapping $a \to \operatorname{ind}_g a$ —more accurately, the mapping

$$\chi : a \to \underline{|\operatorname{ind}_g a|}_{\phi(m)}$$

of

$$\{\mathbf{Z}_m^*, \cdot\} \to \{\mathbf{Z}_{\phi(m)}, +\}.$$

Translating (*) to residue classes (mod $\phi(m)$) we have

$$\lfloor \text{ind}_g (ab) \rfloor_{\phi(m)} = \lfloor \text{ind}_g a \rfloor_{\phi(m)} + \lfloor \text{ind}_g b \rfloor_{\phi(m)}.$$

Hence, χ is a homomorphism of groups $[\chi(ab) = \chi(a) + \chi(b)]$. In fact, because both groups are cyclic with $\phi(m)$ elements it is easy to see that χ is an isomorphism. The generator g of $\mathbf{Z}_m^*$ is mapped under χ to the generator $1 = \lfloor 1 \rfloor_{\phi(m)}$ of $\{\mathbf{Z}_{\phi(m)}, +\}$ (since $\text{ind}_g\, g = 1$).

The isomorphism χ determined by ind_g depends on the choice of generator g of $\mathbf{Z}_m^*$; moreover, it is not hard to see that taking all possible choices for the generator g of $\mathbf{Z}_m^*$ determines all isomorphisms from $\mathbf{Z}_m^*$ to $\mathbf{Z}_{\phi(m)}$.

Because ind_g (or rather χ) transforms multiplication into addition, it should appropriately be viewed as an analog of the "logarithm" function.

4-6-14. Remark. In 4-6-13 we discussed the index as a function defined on elements of $\mathbf{Z}_m^*$. (Actually, everything could have been done with $\mathbf{Z}_m^*$ replaced by an arbitrary finite cyclic group.) Of course, in the standard works on number theory one does not have the algebraic object $\mathbf{Z}_m^*$, viewed as a group. Our purpose here is to translate the story of the index, as given in 4-6-13, into the language of congruences which is common in number theory. This requires only minor variations.

Suppose m is an integer for which there exists a primitive root. Fix a primitive root g. For integers i, j we have

$$g^i \equiv g^j \pmod{m} \Leftrightarrow i \equiv j \pmod{\phi(m)}.$$

Given an integer a with $(a, m) = 1$, there exists a unique integer r with

$$g^r \equiv a \pmod{m}, \qquad 0 \le r \le \phi(m) - 1.$$

In other words, r is the smallest nonnegative integer for which $g^r \equiv a \pmod{m}$. We write

$$r = \text{ind}_g\, a$$

and call it the "index of a with respect to g" (for the modulus m). The index satisfies the following properties—for $a, a', b \in \mathbf{Z}$ all relatively prime to m:

$$\begin{aligned} \text{ind}_g\, a = \text{ind}_g\, a' &\Leftrightarrow a \equiv a' \pmod{m}, \\ \text{ind}_g(ab) &\equiv \text{ind}_g\, a + \text{ind}_g\, b \pmod{\phi(m)}, \\ \text{ind}_g(a^n) &\equiv n(\text{ind}_g\, a) \pmod{\phi(m)}, \qquad n \ge 1. \end{aligned}$$

4-6-15. Example. We do some computations with the "index." The kinds of things which can be done arise from the fact that "ind" behaves like "log." Of course, the index exists only when $m = 2, 4, p^r, 2p^r$, p an odd prime (as per 4-6-10). We shall not fuss with the distinction between ind_g defined on $\mathbf{Z}_m^*$, or simply on certain integers.

Consider $m = p = 17$; so by 4-6-10 there exists a primitive root (mod 17). To find one—here, and in general—is mostly a matter of brute force. Testing 2, we find that $2^8 \equiv 1 \pmod{17}$, so 2 is not a primitive root (mod 17). Testing 3 next, we find it to be a primitive root (mod 17). In fact, we have—with all congruences (mod 17)—

$$\begin{array}{llll} 3^1 \equiv 3, & 3^2 \equiv 9, & 3^3 \equiv 10, & 3^4 \equiv 13, \\ 3^5 \equiv 5, & 3^6 \equiv 15, & 3^7 \equiv 11, & 3^8 \equiv 16, \\ 3^9 \equiv 14, & 3^{10} \equiv 8, & 3^{11} \equiv 7, & 3^{12} \equiv 4, \\ 3^{13} \equiv 12, & 3^{14} \equiv 2, & 3^{15} \equiv 6, & 3^{16} \equiv 1. \end{array}$$

[Note: To see that 3 is a primitive root (mod 17) it is not necessary to compute beyond $3^8 \equiv 16$. We know at the start that the exponent to which 3 belongs—that is, the order of 3 as an element of the group $\mathbf{Z}_{17}^*$—is a divisor of 16; so as soon as we arrive at $3^8 \equiv 16 \not\equiv 1 \pmod{17}$, the conclusion is that 3 belongs to the exponent 16.]

Putting $g = 3$ and writing "ind" for "ind_3" we have, in particular, ind $10 = 3$. It is useful, for computational purposes, to make a full table of indices. In the current situation, using our list of powers of 3, we have

$$m = p = 17, \qquad g = 3$$

a	1	2	3	4	5	6	7	8	9	10	11	12	13	14	15	16
$\text{ind}_g\, a$	16	14	1	12	5	15	11	10	2	3	7	13	4	9	6	8

It is sometimes convenient to put this table in the form:

a	3	9	10	13	5	15	11	16	14	8	7	4	12	2	6	1
$\text{ind}_g\, a$	1	2	3	4	5	6	7	8	9	10	11	12	13	14	15	16

Because $m = p = 17$ is small it does not really matter which table is available; for large m, the choice of table depends on whether one wants to find $\text{ind}_g\, a$ when a is given, or to find a when $\text{ind}_g\, a$ is given.

Parenthetically, we remark that there are $\phi(\phi(17)) = 8$ primitive roots (mod 17). In fact, since 3 is a generator of a cyclic group of order $\phi(17) = 16$, 3^n is a primitive root (mod 17) $\Leftrightarrow$ $(n, \phi(17)) = 1$. Thus, the primitive roots are the odd powers of 3—namely,

$$3, \quad 10, \quad 5, \quad 11, \quad 14, \quad 7, \quad 12, \quad 6.$$

Any one of these primitive roots could be used to make the table of indices. For example, using the primitive root $g = 7$, we have:

$$m = p = 17, \qquad g = 7$$

a	1	2	3	4	5	6	7	8	9	10	11	12	13	14	15	16
$\text{ind}_g\, a$	16	10	3	4	15	13	1	14	6	9	5	7	12	11	2	8

The facts about the index are useful in solving certain kinds of congruences or equations in residue class rings. As illustrations, let us solve:

(*i*) $8x^5 \equiv 10 \pmod{17}$—that is: $8x^5 = 10$ in $\mathbf{Z}_{17}$.
(*ii*) $7x^{10} \equiv 9 \pmod{17}$—that is: $7x^{10} = 9$ in $\mathbf{Z}_{17}$.
(*iii*) $5(7^x) \equiv 16 \pmod{17}$—that is: $5(7^x) = 16$ in $\mathbf{Z}_{17}$.

We work with $m = p = 17$, $g = 3$ and write "ind" for "ind_g." The treatment of (*i*) consists of:

The integer x satisfies $8x^5 \equiv 10 \pmod{17}$

$$\Leftrightarrow \text{ind}(8x^5) = \text{ind } 10 \qquad \text{(use 4-6-14)}$$
$$\Leftrightarrow \text{ind } 8 + 5 \text{ ind } x \equiv \text{ind } 10 \pmod{16}$$
$$\Leftrightarrow 10 + 5 \text{ ind } x \equiv 3 \pmod{16}$$
$$\Leftrightarrow 5 (\text{ind } x) \equiv -7 \equiv 9 \pmod{16}$$
$$\Leftrightarrow \text{ind } x = 5$$
$$\Leftrightarrow x \equiv 5 \pmod{17}.$$

In other words, $\lfloor 5 \rfloor_{17}$ is the unique solution of $8x^5 = 10$ in $\mathbf{Z}_{17}$.

Solving (*ii*) by this method leads to

$$\text{ind } 7 + 10 \text{ ind } x \equiv \text{ind } 9 \pmod{16}$$

which becomes

$$10 \text{ ind } x \equiv -9 \equiv 7 \pmod{16}.$$

Since $(10, 16) = 2$ does not divide 7, this linear congruence has no solution. Thus, the congruence $7x^{10} \equiv 9 \pmod{17}$ has no solution.

Solving (*iii*) by this method leads to

$$\text{ind } 5 + x \text{ ind } 7 \equiv \text{ind } 16 \pmod{16}$$

which becomes

$$11x \equiv 3 \pmod{16}.$$

This has the unique solution $x \equiv 9 \pmod{16}$!!

Of course, the index can also be used for computations for certain composite m. Consider, for example, $m = 22 = 2 \cdot 11$. Then $\mathbf{Z}_{22}^*$ is a cyclic group with $\phi(22) = 10$ elements—these may be represented by 1, 3, 5, 7, 9, 13, 15,

17, 19, 21. There exists a primitive root (mod 22). By trial and error; 3 and 5 are not primitive roots, but 7 is. In fact, (mod 22) the powers of 7 give

$$7^1 \equiv 7, \quad 7^2 \equiv 49 \equiv 5, \quad 7^3 \equiv 35 \equiv 13, \quad 7^4 \equiv 91 \equiv 3, \quad 7^5 \equiv 21,$$
$$7^6 \equiv 147 \equiv 15, \quad 7^7 \equiv 105 \equiv 17, \quad 7^8 \equiv 119 \equiv 9,$$
$$7^9 \equiv 63 \equiv 19, \quad 7^{10} \equiv 133 \equiv 1.$$

this leads to the index table:

$$m = 22, \qquad g = 7$$

a	1	3	5	7	9	13	15	17	19	21
$\text{ind}_g\, a$	10	4	2	1	8	3	6	7	9	5

There are $\phi(\phi(22)) = 4$ primitive roots (mod 22); they are given by 7^n where $(n, 10) = 1$, $n \leq 10$—namely, $7^1, 7^3, 7^7, 7^9$. Using the table, the four primitive roots are 7, 13, 17, 19.

If we try to solve $7x^{10} \equiv 9 \pmod{22}$ by using indices, then

$$\text{ind } 7 + 10 \text{ ind } x \equiv \text{ind } 9 \ \big(\text{mod } \phi(22)\big)$$

and

$$10 \text{ ind } x \equiv 7 \pmod{10}.$$

Hence, there is no solution.

Note that, as it stands, $7x^{10} \equiv 8 \pmod{22}$ cannot be attacked via the index—because 8 is not relatively prime to 2, ind 8 has no meaning.

4-6-16. Exercise. Let m be an integer for which there exists a primitive root g—that is, $\mathbf{Z}_m^*$ is cyclic; let n be any integer greater than or equal to 1 and put $d = \big(n, \phi(m)\big)$. Then:

(*i*) The mapping $a \to a^n$ is a homomorphism of $\mathbf{Z}_m^* \to \mathbf{Z}_m^*$. Its image, which we denote by

$$(\mathbf{Z}_m^*)^n = \{a^n \mid a \in \mathbf{Z}_m^*\}$$

is the subgroup (in $\mathbf{Z}_m^*$) of all nth powers. Denoting the kernel by

$$N_{m,n} = \{a \in \mathbf{Z}_m^* \mid a^n = 1\}$$

we have an isomorphism of groups

$$\mathbf{Z}_m^*/N_{m,n} \approx (\mathbf{Z}_m^*)^n$$

(*ii*) For $a \in \mathbf{Z}_m^*$ we have

$$a \in N_{m,n} \Leftrightarrow n \text{ ind } a \equiv 0 \ \big(\text{mod } \phi(m)\big).$$

Therefore, the kernel $N_{m,n}$ has exactly d elements. Furthermore, $(\mathbf{Z}_m^*)^n$ is a cyclic group of order $\phi(m)/d$.

(*iii*) For $b \in \mathbf{Z}_m^*$ we have

$$b \in (\mathbf{Z}_m^*)^n \Leftrightarrow b^{\phi(m)/d} = 1.$$

(One way to do this is to show that each side is equivalent to $d \mid \text{ind}\, b$.) In the case where $b \in (\mathbf{Z}_m^*)^n$, the equation $x^n = b$ has exactly d solutions in $\mathbf{Z}_m^*$.

Rephrasing some of these results in the language of congruences we have with m, n, d as above, and b an integer relatively prime to m: The congruence

$$x^n \equiv b \pmod{m}$$

has a solution (in which case b is said to be an nth **power residue** of m)

$$\Leftrightarrow d \mid \text{ind}\, b \Leftrightarrow b^{\phi(m)/d} \equiv 1 \pmod{m}$$

and in this situation there are exactly d solutions (mod m).

Furthermore, there are $\phi(m)/d$ incongruent nth power residues of m; they are the roots of the congruence

$$x^{\phi(m)/d} \equiv 1 \pmod{m}.$$

4-6-17 / *PROBLEMS*

1. For each a satisfying $1 \leq a \leq m - 1$ and $(a, m) = 1$ find the exponent to which a belongs (mod m) when m equals
   (*i*) 7, (*ii*) 11, (*iii*) 20, (*iv*) 21.

2. Find the exponent of 13 (mod m) for m equal to
   (*i*) 17, (*ii*) 19, (*iii*) 22, (*iv*) 41.

3. If a is an integer with $a^r \equiv 1 \pmod{m}$ for some $r \geq 1$, show that $(a, m) = 1$.

4. If $ab \equiv 1 \pmod{m}$, then a and b belong to the same exponent (mod m).

5. Show that if $a > 1$ and $m \geq 1$, then $m \mid \phi(a^m - 1)$.

6. We know (Fermat): If p is prime and $(a, p) = 1$, then $a^{p-1} \equiv 1 \pmod{p}$. Show the falsity of the converse, by example—in other words, exhibit integers a, m with $(a, m) = 1$, m not prime and $a^{m-1} \equiv 1 \pmod{m}$.

7. (*i*) Suppose a is relatively prime to both m and n. If a belongs to the exponent $r \pmod{m}$ and to the exponent $s \pmod{n}$, show that a belongs to the exponent $[r, s]$ (mod$[m, n]$).
   (*ii*) Can you give a group theoretic interpretation of this result?
   (*iii*) Generalize this result to: If $(a, m_1 m_2 \cdots m_t) = 1$ and a belongs to the exponent $r_i \pmod{m_i}$ $i = 1, \ldots, t$, then a belongs to the exponent $[r_1, r_2, \ldots, r_t]$ (mod$[m_1, m_2, \ldots, m_t]$).

8. Find the exponent to which 7 belongs modulo
(*i*) $11 \cdot 13 \cdot 17$, (*ii*) $3 \cdot 5 \cdot 19$, (*iii*) $3^3 \cdot 5^2 \cdot 17$.

9. How many primitive roots are there modulo
(*i*) 13, (*ii*) 18, (*iii*) 19, (*iv*) 21,
(*v*) 25, (*vi*) 28, (*vii*) 29?

10. If g is a primitive root (mod p^2), show it is a primitive root (mod p). (Here, as elsewhere in this section, p is understood to be odd prime.)

11. If g is a primitive root mod p and $gg' \equiv 1 \pmod{p}$, then g' is a primitive root (mod p).

12. If g is a primitive root (mod p) show that $g^{(p-1)/2} \equiv -1 \pmod{p}$.

13. For $p > 3$, the product of the $\phi(p-1)$ primitive roots (mod p) is $\equiv 1 \pmod{p}$.

14. For each value of m, find the maximum order for the elements of the group $\mathbf{Z}_m^*$:
(*i*) 18, (*ii*) 7^5, (*iii*) 3600
(*iv*) 7200, (*v*) $2^2 \cdot 17^3 \cdot 37 \cdot 67$, (*vi*) $2 \cdot 11 \cdot 19^2 \cdot 31^3$.

15. If m and n are relatively prime positive integers, prove that $\lambda(m, n) = [\lambda(m), \lambda(n)]$.

16. Suppose g is a primitive root (mod p). If $p \equiv 1 \pmod{4}$, prove that $-g$ is a primitive root (mod p). What happens if $p \equiv 3 \pmod{4}$?

17. If $m > 6$ has a primitive root then show that the product of the $\phi(\phi(m))$ primitive roots (mod m) is congruent to 1 (mod m).

18. For $m = p = 17$ construct the table of indices when
(*i*) $g = 5$, (*ii*) $g = 6$, (*iii*) $g = 10$.
Then use each of these tables to solve the congruences
$8x^5 \equiv 10 \pmod{17}$, $7x^{10} \equiv 9 \pmod{17}$, $5(7^x) \equiv 16 \pmod{17}$.

19. Find a primitive root (mod p) for each of the following primes
(*i*) 7, (*ii*) 11, (*iii*) 13, (*iv*) 19, (*v*) 23, (*vi*) 29.
In each case use the primitive root to construct a table of indices. Find all the primitive roots for each p.

20. Find a primitive root modulo
(*i*) 9, (*ii*) 10, (*iii*) 14, (*iv*) 18, (*v*) 25, (*vi*) 27.
Use the primitive root to construct a table of indices. In each case, find all the primitive roots.

21. Solve the congruence $5x^7 \equiv 8 \pmod{m}$ for each of the following values of m: 7, 9, 10, 11, 13, 14, 17, 18, 19, 22, 23, 25, 27, 29.

22. Do Problem 21 for
(*i*) $8x^{10} \equiv 5 \pmod{m}$, (*ii*) $3x^6 \equiv 7 \pmod{m}$.

23. Solve $x^5 \equiv 4 \pmod{7 \cdot 9 \cdot 13 \cdot 17}$.

24. Solve $7^{x^2} \equiv 13 \pmod{19}$.

25. Exhibit integers a and b for which the congruence $ax^6 \equiv b \pmod{22}$
(*i*) has a solution, (*ii*) does not have a solution.
Do the same for $ax^7 \equiv b \pmod{22}$.

26. Show that for an odd prime p there exists a primitive root $g \pmod{p}$ which is not a primitive root $\pmod{p^2}$.

27. What happens to 4-6-6, 4-6-7, 4-6-8 when $p = 2$?

28. Prove directly: If there exists a primitive root $\pmod{p^r}$, p an odd prime, then there exists a primitive root $\pmod{p^{r+1}}$.

29. If both g and g' are primitive roots $\pmod{p}$, then show that gg' is not a primitive root $\pmod{p}$.

30. Suppose g is a primitive root $\pmod{m}$, $a \in \mathbf{Z}$, $(a, m) = 1$. Show that $\pmod{m}$, a belongs to the exponent

$$\frac{\phi(m)}{\big(\operatorname{ind}_g a, \phi(m)\big)}.$$

Furthermore, a is a primitive root $\pmod{m} \Leftrightarrow \big(\operatorname{ind}_g a, \phi(m)\big) = 1$.

31. Suppose g is a primitive root $\pmod{p^r}$. If g is odd, then g is a primitive root $\pmod{2p^r}$. If g is even, then $g + p^r$ is a primitive root $\pmod{2p^r}$.

32. If both g and g' are primitive roots $\pmod{p}$, show that for $(a, p) = 1$

$$\operatorname{ind}_{g'} a \equiv (\operatorname{ind}_g a)(\operatorname{ind}_{g'} g) \big(\operatorname{mod} (p-1)\big).$$

This is the analog of the rule for changing the base of logarithms.

33. If g is a primitive root $\pmod{p^2}$, then

$$x^{p-1} \equiv 1 \pmod{p^2}$$

has the $p - 1$ distinct roots g^{np}, $n = 1, 2, \ldots, p - 1$, and no others.

34. Discuss the circumstances under which

$$ax^n \equiv b \pmod{p}$$

has a solution. How many solutions are there?

35. Suppose the prime p is of form $2^n + 1$. If $\left(\frac{a}{p}\right) = -1$ show that a is a primitive root $\pmod{p}$. What about the converse?

36. (*i*) Consider the congruence

$$x^3 \equiv a \pmod{p}, \qquad \text{where } 1 \leq a \leq p - 1.$$

Show that if $p \equiv 2 \pmod{3}$, there is a solution for each a, and if $p \equiv 1 \pmod{3}$, there is a solution for $(p-1)/3$ of the choices for a.

(*ii*) Find a corresponding result for

$$x^4 \equiv a \pmod{p}, \qquad 1 \leq a \leq p - 1$$

which depends on whether $p \equiv 1$ or $3 \pmod{4}$.

(*iii*) What about $x^5 \equiv a \pmod{p}$ and $x^6 \equiv a \pmod{p}$?

Miscellaneous Problems

1. Suppose G is an arbitrary group in which every element has finite order. If the nonempty subset H is closed under the operation, show that it is a subgroup.

2. Suppose G is a subgroup of S_n. If G contains an odd permutation, show that exactly half of its elements are even permutations, and that these form a subgroup.

3. Show that in an ordered field F, the set of all positive elements forms a group under multiplication.

4. Consider all maps $\phi_{a,b} : \mathbf{R} \to \mathbf{R}$ defined by

$$\phi_{a,b}(x) = ax + b, \qquad a, b \in \mathbf{R}.$$

Taking composition of maps as the operation, decide if these mappings form a group when a and b are restricted by:

(*i*) $a, b \in \mathbf{Z}$.
(*ii*) $a \neq 0, \quad a \in \mathbf{Z}, \quad b \in \mathbf{R}$.
(*iii*) $a \neq 0, \quad a \in \mathbf{R}, \quad b \in \mathbf{Z}$.
(*iv*) $a \neq 0, \quad a \in \mathbf{Q}, \quad b \in \mathbf{R}$.
(*v*) $a \neq 0, \quad a \in \mathbf{Q}, \quad b \in \mathbf{R}, \quad b \notin \mathbf{Q}$.

When is the group commutative?

5. Let $\mathbf{R}' = \mathbf{R} \cup \{\infty\}$, the set consisting of all real numbers and the element ∞. Consider all mappings $\phi: \mathbf{R}' \to \mathbf{R}'$ of form

$$\phi(x) = \frac{ax + b}{cx + d},$$

where $a, b, c, d \in \mathbf{R}$ and $ad - bc = 1$. Show that with respect to the operation of composition, these mappings form a group.

6. Show that the additive group of rationals, $\{\mathbf{Q}, +\}$, has no maximal proper subgroup. (Maximal means here that it is not contained in any other proper subgroup.)

7. Suppose G is a commutative semigroup with cancellation. If G is finite then it is a group. If G is infinite then it can be imbedded in a group. How is this related to the imbedding of an integral domain in a field?

8. Consider an additive abelian group, G, and the set $\mathscr{E}(G)$ of all endomorphisms of G. (In other words, the elements of $\mathscr{E}(G)$ are homomorphisms of G into itself.) For $\phi, \psi \in \mathscr{E}(G)$ define $\phi + \psi$ and $\phi \circ \psi$, as usual, by

$$\begin{aligned} (\phi + \psi)(a) &= \phi(a) + \psi(a), \\ (\phi \circ \psi)(a) &= \phi((\psi(a)), \end{aligned} \qquad a \in G.$$

Show that $\{\mathscr{E}(G), +, \circ\}$ is a ring with unity. What is $\{\mathscr{E}(\mathbf{Z}), +, \circ\}$? What is $\{\mathscr{E}(\mathbf{Z}_n), +, \circ\}$?

9. Suppose G is a finite group with subgroups $K \subset H \subset G$. Show that

$$(G\colon K) = (G\colon H)(H\colon K).$$

How is this related to Lagrange's theorem, 4-2-16? What happens if the group G is infinite?

10. Suppose $H_1, H_2, \ldots, H_r$ are subgroups of G, each of finite index in G. Show that their intersection is of finite index in G.

11. If F and H are subgroups of the finite group G, show that

$$\#(F) \cdot \#(H) \leq \#(F \cap H) \cdot \#(F \vee H).$$

(Here $F \vee H$ denotes $[F \cup H]$, the subgroup generated by the set $F \cup H$.)

12. If F and H are finite subgroups of G, show that

$$\#(F) \cdot \#(H) = \#(F \cap H) \cdot \#(FH).$$

(Note that FH need not be a subgroup.)

13. If F and H are subgroups of G, show that

$$(F\colon F \cap H) \leq (F \vee H\colon H).$$

Moreover, if both $(F \vee H\colon F)$ and $(F \vee H\colon H)$ are finite and relatively prime, show that

$$(F\colon F \vee H) = (F \vee H\colon H) \qquad \text{and} \qquad (H\colon F \cap H) = (F \vee H\colon F).$$

14. If N is a normal subgroup of the finite group G, show that N contains every subgroup of G whose order is prime to $(G\colon N)$.

15. Show that any subgroup of prime power order, p^r, contains a subgroup of order p.

16. Let R be a commutative ring with unity, 1, and consider $\mathcal{M}(R, 2)$, the ring of 2×2 matrices with entries from R. For a matrix

$$A = \begin{pmatrix} a & b \\ c & d \end{pmatrix} \in \mathcal{M}(R, 2)$$

define its **determinant** by $\det A = ad - bc$; it is an element of R. For $A, B \in \mathcal{M}(R, 2)$ we have

$$\det(AB) = (\det A)(\det B).$$

Of course, $\det I = 1$.

Show that the matrix A has an inverse [that is, A is a unit of the ring $\mathcal{M}(R, 2)$] $\Leftrightarrow$ $\det A$ is a unit of R. Moreover, when A has an inverse, find an expression for it.

Apply these facts to find the group of units of the rings $\mathcal{M}(\mathbf{R}, 2)$ and $\mathcal{M}(\mathbf{Z}, 2)$.

17. Consider the domain $D = \text{Map}(\mathbf{Z}_{>0}, \mathbf{Z})$ of arithmetic functions (see Miscellaneous Problem 57, of Chapter II). Prove the following:

(i) The group, G, of units of D is

$$\{f \in D \mid f(1) = \pm 1\}.$$

(ii) The set H, of nonzero multiplicative arithmetic functions, is a subgroup of G

(iii) The arithmetic function u given by

$$u(n) = 1 \quad \text{for all} \quad n \in \mathbf{Z}_{>0}$$

belongs to H, and its inverse, u^{-1}, is the Möbius function, μ.

18. Let $H = \{\sigma \in S_5 \mid \sigma: \{2, 4\} \to \{2, 4\}\}$. Show that H is a subgroup of S_5. Describe its left and right cosets. Is it normal?

19. Consider $G = \{\phi: \mathbf{R} \to \mathbf{R} \mid \phi(x) = ax + b, a \neq 0, a, b, x \in \mathbf{R}\}$. This is a group with respect to composition of mappings. Let $H = \{\phi \in G \mid a = 1\}$, then H is a subgroup of G. Describe its left and right cosets.

20. Prove the following:

(i) The additive group of $\mathbf{Z}[i]$ is isomorphic to the direct sum of two infinite cyclic groups.

(ii) The additive group of the polynomial ring $\mathbf{Z}[x]$ is isomorphic to the multiplicative group of positive rational numbers.

21. If the sets X and X' are in 1–1 correspondence, show that the symmetric groups S_X and $S_{X'}$ are isomorphic.

22. Suppose S is a nonempty subset of the group G, and put

$$C_G(S) = \{a \in G \mid ax = xa \text{ for all } x \in G\}, \qquad N_G(S) = \{a \in G \mid aS = Sa\}.$$

These are known as the **centralizer** and **normalizer** of S in G, respectively.
Prove the following:

(*i*) $C_G(S)$ is a subgroup of G. Note that $C_G(G) = \mathfrak{Z}_G$, the center of G.
(*ii*) If H is a commutative subgroup of G then H is a normal subgroup of $C_G(H)$.
(*iii*) $N_G(S)$ is a subgroup of G that contains $C_G(S)$.
Suppose H is a subgroup of G, show that:
(*iv*) H is normal if and only if $H = N_G(H)$.
(*v*) H is normal in $N_G(H)$; in fact, $N_G(H)$ is the unique largest subgroup of G in which H is normal.
(*vi*) $(G\colon N_G(H))$ equals the number of conjugates of H.

23. Let A, B, C be subgroups (they need not be normal) of the group G with $A \supset B$. Show that $A \cap BC = B(A \cap C)$. This is known as the **modular law** of Dedekind.
If, in addition, $A \cap C = B \cap C$ and $AC = BC$, show that $A = B$.

24. Call a matrix A of $\mathscr{M}(\mathbf{R}, 2)$ or $\mathscr{M}(\mathbf{Z}, 2)$ **nonsingular** if it has an inverse. In each case, show that the set of all matrices of determinant 1 is a normal subgroup of the multiplicative group of all nonsingular matrices.

25. Let C be the commutator subgroup of the group G (see 4-4-15, Problem 8). Show that

$$C = \{a_1 a_2 \cdots a_n a_1^{-1} a_2^{-1} \cdots a_n^{-1} \mid a_i \in G,\ i = 1, \ldots, n,\ n \geq 2\}$$

by using the identity

$$(aba^{-1}b^{-1})(cdc^{-1}d^{-1}) = a(ba^{-1})b^{-1} \cdot c(dc^{-1})d^{-1} \cdot a^{-1}(ab^{-1})b \cdot c^{-1}(cd^{-1})d.$$

26. If N is a normal subgroup of G for which G/N is abelian then $N \supset C$. What about the converse?

27. For a prime p consider

$$\mathbf{Z}(p^\infty) = \left\{ \frac{a}{p^n} \,\middle|\, a, n \in \mathbf{Z} \right\}.$$

Show that under addition this is an abelian group which is not cyclic. Furthermore, show that every proper subgroup is finite and cyclic.

28. Suppose the subset S generates the group G, $G = [S]$, and $\#(S) = m$. If S is minimal in the sense that no subset of S generates G, is it true that any generating set of G has at least m elements?

29. Suppose G is abelian; then for any $r \geq 1$, show that $F = \{a^r \mid a \in G\}$ and $H = \{a \in G \mid a^r = e\}$ are subgroups of G. If $\#(G) = n$ and $(r, n) = 1$, the map $a \to a^r$ is an automorphism of G.

30. Suppose N is a normal subgroup of the finite group G. If G/N contains an element of order n, show that G contains an element of order n.

31. Suppose G is cyclic of order $n = rs$, and let H be the subgroup of order s. Show that

$$H = \{a \in G \mid a^s = 1\} = \{a^r \mid a \in G\}.$$

32. If ord $a = mn$ with $(m, n) = 1$, show that a can be written uniquely in the form $a = bc$ with $bc = cb$ and ord $b = m$, ord $c = n$; in fact, both b and c are powers of a.

33. Give an example of two elements of finite order whose product has infinite order.

34. If G is cyclic of order n and G' is cyclic of order m, with $m \mid n$ then G' is a homomorphic image of G. Find all homomorphic images of G.

35. List all factor groups of each of the following groups (see 4-5-18):

(*i*) D_5, (*ii*) D_6, (*iii*) S_4, (*iv*) S_5.

36. Exhibit at least six distinct groups of order 8.

37. Describe all subgroups of $\{\mathbf{Z}_n, +\}$.

38. Find the center of D_n, the group of rigid motions of the regular $n$gon (i.e., the nth dihedral group of 4-5-18).

39. Show that the group W_n (see 4-2-8) has exactly n elements by making use of the fact that, in $\mathbf{C}[x]$, the greatest common divisor of $f(x) = x^n - 1$ and its derivative $f'(x)$ is 1 (see Miscellaneous Problem 12 of Chapter III).

40. A mapping of groups $\phi: G \to G'$ that is one-to-one and onto is said to be an anti-isomorphism when it has the property $\phi(ab) = \phi(b)\phi(a)$ for all $a, b \in G$.

For $a \in G$ define $R_a \in S_G$ by $R_a(x) = xa$, $x \in G$; so R_a is "right multiplication" by a in G. In analogy with Cayley's theorem, 4-4-10, show that $a \to R_a$ is an anti-isomorphism of G into S_G (in other words, it is an injective antihomomorphism). Note, incidentally, that L_a and R_b commute (as elements of S_G).

41. Suppose H is a subgroup of G and put $X = \{aH\}$, the set of left cosets of H in G. For $b \in G$, define $L_b : X \to X$ by $L_b(aH) = baH$. Then $L_b \in S_X$ and the mapping $b \to L_b$ is a homomorphism of $G \to S_X$. What is the kernel? What happens if we use right cosets instead of left cosets?

42. This is an expanded version of the isomorphism theorem, 4-4-12. Suppose $\phi: G \to G'$ is a surjective homomorphism with kernel N. With any subset S of G we may associate the subset $\phi(S) = \{\phi(a) \mid a \in S\}$ of G'. Furthermore, with any subset S' of G' we may associate the subset $\phi^{-1}(S') = \{a \in G \mid \phi(a) \in S'\}$ of G; $\phi^{-1}(S')$ is called the **preimage** of S'. Show that:

(*i*) There is a 1–1 correspondence between subgroups H of G with $H \supset N$, and subgroups H' of G'. In fact, we have $H \leftrightarrow H'$ when $\phi(H) = H'$ and $\phi^{-1}(H') = H$. (Note that $G \leftrightarrow G'$.)

(*ii*) $G/N \approx G'$; in fact, the map $\bar{\phi}: G/N \to G'$ defined by $\bar{\phi}(aN) = \phi(a)$ is such an isomorphism. Even more, ϕ induces a homomorphism of H onto H' with kernel N, and $\bar{\phi}$ provides an isomorphism $H/N \approx H'$.

(*iii*) H is normal in $G \Leftrightarrow H'$ is normal in G'; and in this situation, $G/H \approx G'/H'$.

43. Given groups $N \subset H \subset G$ with N normal in G, show that H is normal in $G \Leftrightarrow H/N$ is normal in G/N; and in this situation,

$$\frac{G/N}{H/N} \approx \frac{G}{H}.$$

44. Suppose H and N are subgroups of G with N normal. If $G = HN$ and $H \cap N = (e)$, show that $G/N \approx H$.

45. (*i*) Consider the cube. It has eight vertices and six congruent faces; it is sometimes called the **regular hexahedron**. The rigid motions form a group with 24 elements (known as the **hexahedral group**) which may be viewed as a subgroup of S_8. Show that the symmetries (i.e., rotations about some axis) of the cube form a group; it is the hexahedral group.

(*ii*) The cube has 4 diagonals and each of our motions is characterized by the way it permutes the diagonals. Show that the hexahedral group is isomorphic to S_4.

(*iii*) Take the midpoints of the faces of the cube, and connect them by straight lines. The result is a figure with 6 vertices and 8 congruent faces (what are the faces?) that is known as the **regular octahedron**. The rigid motions form a group with 24 elements (known as the **octahedral group**), which may be viewed as a subgroup of S_6. Show that the symmetries constitute the same group, and that the octahedral and hexahedral groups are isomorphic.

46. (*i*) If n is odd, show that

(*a*) $2n^2 + 1$ is divisible by a prime $p \equiv 3 \pmod 8$

(*b*) $4n^2 + 1$ is divisible by a prime $p \equiv 5 \pmod 8$

(*ii*) If n is arbitrary, show that

(*a*) $8n^2 - 1$ is divisible by a prime $p \equiv 7 \pmod 8$

(*b*) any odd prime p that divides $n^4 + 1$ is congruent to $1 \pmod 8$.

(*iii*) Use the preceding to show that the number of primes of each of the forms $8n + 1$, $8n + 3$, $8n + 5$, $8n + 7$ is infinite.

47. Consider two subgroups H, K (which need not be distinct) of the group G. A set of form HxK, $x \in G$, is called a **double coset**. Then two double cosets are disjoint or identical; so G decomposes into a disjoint union of double cosets. Each double coset HxK can be expressed as a union of left cosets of K in G; the number of such left cosets is $(H: H \cap xKx^{-1})$. Similarly, HxK is the union of right cosets of H in G; the number of such right cosets is $(K: K \cap x^{-1}Hx)$. If $G = \bigcup_\alpha Hx_\alpha K$ is a decomposition into double cosets, show that

$$(G: K) = \sum_\alpha (H: H \cap x_\alpha K x_\alpha^{-1}) \qquad \text{and} \qquad (G: H) = \sum_\alpha (K: K \cap x_\alpha^{-1} H x_\alpha).$$

48. (*i*) The multiplicative group G with identity e is said to **operate** or **act** on the set X if we have a map of $G \times X \to X$, denoted by $(\sigma, x) \to \sigma x$, satisfying $ex = x$ and $\tau(\sigma x) = (\tau\sigma)x$ for all $\sigma, \tau \in G$, $x \in X$. This may be rephrased as follows: we have a homomorphism of $G \to S_X$ (and the image of σ in S_X is also denoted by σ). Note that distinct elements of G may have the same action on X.

(*ii*) The set $Gx = \{\sigma x \mid \sigma \in G\}$ is called the **orbit** of x with respect to G. We call x **equivalent** to y (with respect to G) if there exists $\sigma \in G$ such that $\sigma x = y$ (that is, if $y \in Gx$). This is an equivalence relation, and the equivalence classes are the orbits Gx. Thus, X is a disjoint union of orbits Gx. Of course, $x \in Gx$ and $Gx = Gy \Leftrightarrow y \in Gx$.

(*iii*) Put $X^G = \{x \in X \mid \sigma x = x \text{ for all } \sigma \in G\}$; so X^G consists of all $x \in X$ whose orbits Gx consist of a single element. For $x \in X$ we also put $G_x = \{\sigma \in G \mid \sigma x = x\}$. Then G_x is a subgroup of G and $\#(Gx) = (G: G_x)$.

(*iv*) The following relation, known as the **class equation**, holds:

$$\#(X) = \#(X^G) + \sum_x (G: G_x),$$

where x runs over a system of representatives of all the orbits with more than one element.

(*v*) Apply the preceding to the case $X = G$, where the action of G on X is by inner automorphisms—that is, $(\sigma, \tau) \to \tau^\sigma = \sigma\tau\sigma^{-1}$. The orbits are called **conjugate classes**. Elements in the same conjugate class are said to be **conjugates** of each other. Two conjugate classes are identical or disjoint. We have $G_\tau = \{\sigma \in G \mid \tau^\sigma = \sigma\tau\sigma^{-1} = \tau\}$, so it is the centralizer, $C_G(\tau)$, (see Problem 22) of τ in G. Furthermore, $G_\tau = G \Leftrightarrow$ the orbit of τ consists of one element $\Leftrightarrow \tau \in \mathfrak{Z}_G$, the center of G. The class equation takes the form [where we write C_τ for $C_G(\tau)$]

$$\#(G) = \#(\mathfrak{Z}_G) + \sum_\tau (G : C_\tau),$$

where τ runs over a system of representatives of the conjugate classes with more than one element.

49. Prove the following:

(*i*) A subgroup H of G is normal if and only if H is the union of conjugate classes.
(*ii*) The only finite group with exactly two conjugate classes is the group of order 2.

50. Suppose G has order pq, where p and q are primes, with $p < q$. Show that G cannot have two distinct subgroups of order q.

51. A group is said to be simple when it has no nontrivial normal subgroups. Show that:

(*i*) An abelian group is **simple** if and only if it is cyclic of prime order.
(*ii*) The alternating group A_4 is not simple.
(*iii*) The alternating group A_5 is simple.
(*iv*) For $n \geq 5$, the alternating group A_n is simple.

52. A nonempty subset A of the ring R is said to be an **ideal** of R when it satisfies the following two conditions:

(1) $a, b \in A \Rightarrow a - b \in A$; that is, A is a subgroup of the additive group of R;
(2) For any $r \in R$ and $a \in A$ both ra and ar are in A.

(*i*) Show that (0) and R are ideals of R. Any other ideal is said to be a **proper ideal**.
(*ii*) An ideal is a subring of R. Show that the converse is false.
(*iii*) Suppose R has an identity, 1. If the ideal A contains 1, show that $A = R$; even more, if A contains a unit, show that $A = R$.
(*iv*) If R is commutative show that $aR = \{ar \mid r \in R\}$ is an ideal (known as a **principal ideal**). If, in addition, R has a 1, show that $aR = Ra$ is the smallest ideal that contains a.
(*v*) Show that a field has no proper ideals.

(*vi*) Show that the intersection of ideals is an ideal.

(*vii*) Given ideals A and B, define their **sum** as

$$A + B = \{a + b \mid a \in A, b \in B\}$$

and show that it is an ideal.

(*viii*) Given ideals A and B, define their **product** $A \cdot B$ to be the set of all elements that can be written as a finite sum $a_1 b_1 + a_2 b_2 + \cdots + a_n b_n$ where $a_1, a_2, \ldots, a_n \in A$ and $b_1, \ldots, b_n \in B$ (n is not fixed). Show that $A \cdot B$ is an ideal with $A \cdot B \subset A \cap B$.

(*ix*) For an ideal A, show that $A^{\perp} = \{x \in R \mid xA = (0)\}$ is an ideal. Note that the same result holds for an arbitrary subset A of R.

(*x*) For an ideal A, show that $(R\colon A) = \{x \in R \mid Rx \subset A\}$ is an ideal that contains A.

53. By a **left ideal** of the ring R one means a nonempty subset A that is a group under addition (in other words, A is a subgroup of the additive group of R), and such that for any $r \in R$ and $a \in A$ we have $ra \in A$. A **right ideal** of R is defined in analogous fashion. Discuss the extent to which the assertions of Problem 52 carry over to left ideals.

54. (*i*) Consider an ideal I of the ring R. Additively, I is a normal subgroup of the additive group of R, so we may form the factor group R/I. Its elements are the cosets $\lfloor a \rfloor_I = a + I$, and the addition in R/I is given by $(a + I) + (b + I) = (a + b) + I$. Now let us define multiplication in R/I by

$$(a + I)(b + I) = ab + I, \qquad a, b \in R.$$

Show that this multiplication is well defined, and R/I becomes a ring. In addition, the canonical map $a \to a + I$ is a surjective homomorphism of $R \to R/I$, and its kernel is I.

(*ii*) Suppose R, R' are rings and $\phi\colon R \to R'$ is a surjective homomorphism, show that $I = \ker \phi$ is an ideal of R, and that the factor ring R/I is isomorphic to R'.

(*iii*) The preceding remarks indicate that ideals of a ring may be viewed as analogs of normal subgroups of a group Pursue this further by carrying the results of Problems 42 and 43 over to ideals.

55. (*i*) Define a **Euclidean domain** to be an integral domain D in which to every nonzero element $a \in D$ there is assigned an integer $\phi(a) \geq 0$ such that:

(1) If $a \mid b$ then $\phi(a) \leq \phi(b)$

(2) If $a, b \in D$ with $a \neq 0$ then there exist elements $q, r \in D$ for which $b = qa + r$, where either $r = 0$ or $\phi(r) < \phi(a)$ (This is a hard disjunctive—either or, but not both.)

Define a **principal ideal domain** (PID) to be a domain in which every ideal is principal. Show that a Euclidean domain is a PID.

(*ii*) By a **greatest common divisor** (gcd) of two elements a, b of the domain D we mean an element $d \in D$ such that

(1) $d \mid a$ and $d \mid b$

(2) if $c \mid a$ and $c \mid b$ then $c \mid d$.

(A gcd need not exist, but if one exists it is unique up to a unit. Note that for $a, b \in D$, $a \mid b \Leftrightarrow Db \subset Da$; also a and b are associates if and only if $Da = Db$.) Show that any two nonzero elements a, b in a PID have a gcd, d. Moreover, there exist $r, s \in D$ such that $d = ra + sb$. [Even though d is not unique, we write $(a, b) = d$.]

(*iii*) In a PID, if $(a, b) = 1$ (in which case, we say that a and b are **relatively prime**) and $a \mid bc$, show that $a \mid c$.

(*iv*) A nonzero element $p \in D$ is said to be **irreducible** if it cannot be written as a product of two elements neither of which is a unit (or equivalently, when p has no "proper" divisor—where by a proper divisor one means a divisor that is neither a unit nor an associate of p). The element $p \in D$ is said to be **prime** if $p \mid ab, p \nmid a \Rightarrow p \mid b$. Show that:

(1) In any domain, p is prime $\Rightarrow p$ is irreducible.

(2) In a PID, p is prime if and only if p is irreducible.

(*v*) A **unique factorization domain** (UFD) is a domain in which

(1) Every nonunit can be expressed as a product of irreducible elements.

(2) Factorization into irreducible elements is unique up to order and units.

Show that if D is a domain satisfying condition (1) for a UFD, then D is a UFD if and only if every irreducible element is prime.

(*vi*) Let $Da_0 \subset Da_1 \subset Da_2 \subset \cdots$ be an ascending chain (possibly infinite) of ideals of D, show that $\bigcup_i Da_i$ is an ideal. In a PID, there cannot exist an infinite strictly ascending chain of ideals (meaning that adjacent ideals in the chain are not equal).

(*vii*) Prove: A PID is a UFD; hence, a Euclidean domain, is a UFD.

(*viii*) Suppose D is a domain satisfying condition (1) for a UFD, show that D is a UFD if and only if every pair of nonzero elements has a gcd.

56. Show that $\mathbf{Z}[i]$, the domain of Gaussian integers, becomes a Euclidean domain when we put, for $\alpha = a + bi$,

$$\phi(\alpha) = N(\alpha) = \alpha\bar{\alpha} = (a + bi)(a - bi) = a^2 + b^2 \in \mathbf{Z}.$$

Thus, $\mathbf{Z}[i]$ is a unique factorization domain. To locate all the primes, the following observations (along with Miscellaneous Problem 66 of Chapter III) should prove useful.

(*i*) If $\alpha \in \mathbf{Z}[i]$ then α divides $N(\alpha)$;
(*ii*) If π is a prime of $\mathbf{Z}[i]$ then π divides exactly one prime of $\mathbf{Z}$;
(*iii*) To find all primes of $\mathbf{Z}[i]$ it suffices to see how all primes of $\mathbf{Z}$ factor in $\mathbf{Z}[i]$ into primes;
(*iv*) If p is a prime of $\mathbf{Z}$ and π is a prime of $\mathbf{Z}[i]$ with $\pi | p$ then $N(\pi) = p$ or $N(\pi) = p^2$. In the latter case, π and p are associates, so p is also prime in $\mathbf{Z}[i]$.
(*v*) If $p \equiv 3 \pmod 4$ there is no $\pi = a + bi$ for which $N(\pi) = a^2 + b^2 = p$; so such p remain prime in $\mathbf{Z}[i]$.
(*vi*) The prime factorization of $p = 2$ is $2 = (1 + i)(1 - i)$, and the primes $1 + i$, $1 - i$ of $\mathbf{Z}[i]$ are associates.
(*vii*) If $p \equiv 1 \pmod 4$ then, by quadratic reciprocity, there exists $n \in \mathbf{Z}$ for which $p | (1 + n^2)$. Then $p | (n + i)(n - i), p \nmid (n + i), p \nmid (n - i)$ and p is not prime. In fact, $p = \pi_1 \pi_2$, where π_1, π_2 are primes of norm p, and if $\pi_1 = a + bi$ then $\pi_2 = a - bi$. In particular, any prime $p \equiv 1 \pmod 4$ can be expressed as the sum of two squares, $p = a^2 + b^2$.

57. Suppose H and N are normal subgroups of the multiplicative group G. If $H \cap N = (e)$, show that HN (which, according to 4-4-15, Problem 10, is a subgroup of G) is isomorphic to the external direct product group $H \times N$.

58. (*i*) Let $G_1, G_2, \ldots, G_m$ be subgroups of the multiplicative group G. Prove that the following two conditions, I and II, are equivalent:

I
(1) Each G_i is normal in G.
(2) $G = G_1 G_2 \cdots G_m$ (This refers to the usual product of sets in a group, so it asserts that any $a \in G$ is of form $a = a_1 a_2 \cdots a_m$ with $a_i \in G_i$.)
(3) $G_i \cap (G_1 \cdots G_{i-1} G_{i+1} \cdots G_m)$ for each $i = 1, \ldots, m$.

II
(1) For $i \neq j$, G_i and G_j commute elementwise. (That is, $a_i a_j = a_j a_i$ for all $a_i \in G_i$, $a_j \in G_j$.)
(2) Every $a \in G$ can be expressed uniquely in the form $a = a_1 a_2 \cdots a_m$, with $a_i \in G_i$.

When these conditions are satisfied we say that G is the (**internal**) **direct product** (or sum, when G is additive) of $G_1, G_2, \ldots, G_m$. We shall then write $G = G_1 \cdot G_2 \cdot G_3 \cdots G_m$.

(*ii*) If G is the internal direct product of $G_1, G_2, \ldots, G_m$ show that G is canonically isomorphic [under the correspondence $a_1 a_2 \cdots a_m \leftrightarrow (a_1, a_2, \ldots, a_m)$] with the external direct product $G_1 \times G_2 \times \cdots \times G_m = \prod_{i=1}^{m} G_i$.

(*iii*) Conversely, if $G_1, G_2, \ldots, G_m$ are arbitrary groups and we form their external direct product $\bar{G} = G_1 \times G_2 \times \cdots \times G_m$, show that for each $i = 1, \ldots, m$ there exists a subgroup $\bar{G}_i$ of $\bar{G}$ such that G_i and $\bar{G}_i$ are canonically isomorphic [under the correspondence $a_i \leftrightarrow (1, \ldots, 1, a_i, 1, \ldots, 1)$] and $\bar{G}$ is the internal direct product of $\bar{G}_1, \bar{G}_2, \ldots, \bar{G}_m$.

(*iv*) If G is the internal direct product of $G_1, \ldots, G_m$, show that for each j, G/G_j is isomorphic to $\prod_{j \neq 1} G_i$ (and to the internal direct product of the G_i, $i \neq j$, in the appropriate group).

59. For any positive integer n, let $\mathfrak{Z}_n$ denote "the" cyclic group of order n. Given $m > 1$, with prime factorization

$$m = 2^{r_0} p_1^{r_1} p_2^{r_2} \cdots p_s^{r_s} \qquad r_0 \geq 0, r_1, r_2, \ldots, r_s > 0$$

show that

$$\mathbf{Z}_m^* \approx \mathfrak{Z}_2 \times \mathfrak{Z}_{2^{r_0-2}} \times \mathfrak{Z}_{\phi(p_1^{r_1})} \times \cdots \times \mathfrak{Z}_{\phi(p_s^{r_s})} \qquad \text{(external direct product)}$$

when $r_0 \geq 3$. If $r_0 = 2$, the second term, $\mathfrak{Z}_{2^{r_0-2}}$, is omitted; when $r_0 = 0$, 1 the first two terms $\mathfrak{Z}_2, \mathfrak{Z}_{2^{r_0-2}}$ are omitted. In particular, show that $\mathbf{Z}_m^*$ is the internal direct product of such cyclic subgroups.

60. Consider $\mathscr{M}(F, n)$, the ring of $n \times n$ matrices with entries from a field F. Prove that it is a **simple ring**—by which is meant that it has no proper ideals (see Problem 52). Hint: The notation introduced in Miscellaneous Problem 54 of Chapter II may be useful.

SELECTED ANSWERS AND COMMENTS

Problems for Chapter 1

1-1-3 / *ANSWERS*

1. 26; 17.

4.

a	10	17	24	72	77	79	97	210	420
no. of divisors	4	2	8	12	4	2	2	8	24
sum of divisors	18	18	60	195	96	80	98	576	1344

5. Use trial and error, or something "clever" if you see it.

6. Proceed as in Problem 5; the answer to both questions is no.

8. (*i*) 3, 6, 9, 0, -3, -6, ... in fact, all multiples of 3; in particular, 1 is not a linear combination of 3 and 6.

(*ii*) All integers; 1 is a linear combination of 3 and 5.

9. (*i*) The set of all integers of form $2n$ as n runs over $\mathbf{Z}$; in other words, the set of all multiples of 2.

(*iv*) The set of all integers of form $ax + by$ as both x and y run over $\mathbf{Z}$; in other words, the set of all linear combinations of a and b.

(*v*) The set of all integers of form $b - xa$ as x runs over $\mathbf{Z}$.

(*vii*) The set of all positive divisors of a.

(*viii*) The set of all positive integers n for which the relation

$$\sum_{i=1}^{n} i = \frac{n(n+1)}{2}$$

holds. (This turns out to be the set of all positive integers, since $1 + 2 + 3 + \cdots + n = \frac{1}{2}[n(n+1)]$.)

1-2-7 / *ANSWERS*

1. (*i*) $\{5 + 7t \mid t = 0, 1, 2, \ldots\}$; (*ii*) $\{5 + 7t \mid t = 0, 1, 2, \ldots\}$; (*iii*) $\{2 + 7t \mid t = 0, 1, 2, \ldots\}$; (*iv*) $\{4 + 9t \mid t = 0, 1, 2, \ldots\}$.

2. For (*i*)–(*v*), the last nonzero remainder is 92. For (*vi*) and (*vii*), it is 113.

5. In the first part, show that the product of $4m + 1$ and $4n + 1$ can be expressed in the form $4k + 1$.

6. Use the fact that an arbitrary integer is of form $10q + r$ with $0 \le r \le 9$.

10. 8 is of form $3n - 1$, but not of form $6n + 5$.

11. $100 = 8 \cdot 7 + 4 \cdot 11$; $99 = 11 \cdot 7 + 2 \cdot 11$; $101 = 5 \cdot 7 + 6 \cdot 11$.

12. It expresses the geometric idea discussed just before.

1-3-13 / *ANSWERS*

3. (*i*) $(91, 143) = 13 = (-3)(91) + (2)(143)$;
(*iii*) $(-143, -91) = 13 = (-2)(-143) + 3(-91)$;
(*iv*) $(5311, 7571) = 113 = (10)(5311) + (-7)(7571)$;
(*v*) $(5311, -7571) = 113 = (10)(5311) + 7(-7571)$.

6. (*ii*) For example, we also have $(x_0 + b)a + (y_0 - a)b = d$.

11. Show that (m, n) divides (u, v). Then solve for m and n in terms of u and v.

12. (*i*) Put $d = (a + b, a - b)$ and show $d \mid 2$.

15. Clear, as soon as one observes that there exists some linear combination $xa + yb$ that is positive.

16. (*i*) Two integers are relatively prime if and only if some linear combination of them is equal to one.
(*ii*) If an integer divides a product of two terms and is relatively prime to the first then it divides the second.
(*iii*) If two relatively prime integers divide the same integer then their product also divides it.
(*iv*) If an integer is relatively prime to each of two given integers, then it is relatively prime to their product.

1-4-8 / *ANSWERS*

1. 2, 3, 5, 7, 11, 13, 17, 19, 23, 29, 31, 37, 41, 43, 47, 53, 59, 61, 67, 71, 73, 79, 83, 89, 97, 101, 103, 107, 109, 113, 127, 131, 137, 139, 149, 151, 157, 163, 167, 173, 179, 181, 191, 193, 197, 199, 211, 223, 227, 229, 233, 239, 241, 251, 257, 263, 269, 271, 277, 281, 283, 293.

2. Consider the smallest prime p that divides n.

3. $5 = 1^2 + 2^2$, $13 = 2^2 + 3^2$, $17 = 1^2 + 4^2$, $29 = 2^2 + 5^2$, $37 = 1^2 + 6^2$, $41 = 4^2 + 5^2$, $53 = 2^2 + 7^2$, $61 = 5^2 + 6^2$, $73 = 3^2 + 8^2$, $89 = 5^2 + 8^2$, $97 = 4^2 + 9^2$, $101 = 1^2 + 10^2$, etc.

6. It works.

9. (*i*) p^3; (*ii*) p; (*iii*) p; (*iv*) p^2, or p^5 if $pa - b = 0$.

10. (*i*) p or p^2; (*ii*) p^2; (*iii*) p, p^2, or p^3; (*iv*) p^2 or p^3.

17. (*i*) T; (*ii*) T; (*iii*) T; (*iv*) F: $a = 3$, $b = 4$, $c = 3$, $p = 5$; (*v*) T; (*vi*) T; (*vii*) F: $a = 7^3$, $b = 7^2$.

19. No.

1-5-13 / *ANSWERS*

1. $a = 2 \cdot 3 \cdot 5^2 \cdot 11 \cdot 17$, $b = 3 \cdot 5 \cdot 7^2 \cdot 13$, $(a, b) = 3 \cdot 5$, $[a, b] = 2 \cdot 3 \cdot 5^2 \cdot 7^2 \cdot 11 \cdot 13 \cdot 17$.

2. $2^3 \cdot 3^2 \cdot 5 \cdot 7 \cdot 11 \cdot 13$.

3. (*i*) $(a, b) = 2^2 \cdot 3$, $(a, c) = 5$, $(b, c) = 7$;
(*ii*) $[a, b] = 2^5 \cdot 3^2 \cdot 5^2 \cdot 7^2 \cdot 17$, $[a, c] = 2^5 \cdot 3^2 \cdot 5^2 \cdot 7 \cdot 11^2 \cdot 19$, $[b, c] = 2^2 \cdot 3 \cdot 5 \cdot 7^2 \cdot 11^2 \cdot 17 \cdot 19$;
(*iii*) $(a, b, c) = 1$, $[a, b, c] = 2^5 \cdot 3^2 \cdot 5^2 \cdot 7^2 \cdot 11^2 \cdot 17 \cdot 19$.

4. $[5311, 7571] = 47 \cdot 67 \cdot 113 = (47 \cdot 67 \cdot 10)(5311) + ((47)(67)(-7))(7571)$.

5. (*i*) $(a, b) = 137$, $(a, b, c) = 1$;
(*ii*) $[a, b] = 137 \cdot 179 \cdot 229$, $[a, b, c] = 137 \cdot 179 \cdot 229 \cdot 251$;
(*iii*) $[(a, b), c] = 137 \cdot 179 \cdot 251$, $([a, b], c) = 179$.

8. 1 and $a(a + 1)$. **9.** a and b.

17. (*i*) F: $a = 2$, $b = 3$, $c = 5$; (*ii*) T; (*iii*) T; (*iv*) T; (*v*) T; (*vi*) T; (*vii*) T; (*viii*) F: $a = 2$, $b = 5$; (*ix*) T; (*x*) T; (*xi*) F: $a = 2^3$, $b = 2^2$; (*xii*) T.

19. For each p, $\min\{v_p(a), v_p(b)\} = 0$.

1-6-4 / *ANSWERS*

1. (*i*) $\bar{x} = 588 + 33t$, $\bar{y} = -1617 - 91t$, or $\bar{x} = 27 + 33t$, $\bar{y} = -70 - 91t$.
(*ii*) No solutions. (*iii*) $\bar{x} = 19 - 43t$, $\bar{y} = 11 - 30t$.
(*v*) $\bar{x} = 3 + 38t$, $\bar{y} = -t$. (*vi*) $\bar{x} = 8t$, $\bar{y} = -13t$.
(*vii*) No solutions. (*viii*) No solutions.

3. (*i*) The positive solutions are of form $\bar{x} = 604 + 7t$, $\bar{y} = -1510 - 18t$, where $-86 \leq t \leq -84$; so there are three of them: 2, 38; 9, 20; 16, 2.

(*ii*) There are infinitely many: $\bar{x} = 604 - 7t$, $\bar{y} = 1510 - 18t$, where $t < 83$.
(*iii*) $\bar{x} = 17 + 19t$, $\bar{y} = 22 + 27t$, $t \geq 0$.
(*iv*) No positive solutions.

5. (*i*)–(*iii*) are the same. (*iv*) and (*vi*) are the same.
6. $\frac{3}{5} + \frac{1}{7} = \frac{26}{35}$; in fact, the number of answers is infinite.
7. $1147 + 2184t$, $t \in \mathbf{Z}$.
9. At some point one obtains a "fraction" that cannot take an integer value.
10. (*i*) $x = 5u - 2$, $y = 3t - 3u + 2$, $z = 2t - 4u + 2$ for all $t, u \in \mathbf{Z}$. (*ii*) No solution.
11. $x = 11$, $y = -19$. **12.** 1021. **13.** (*i*) 79.
14. The story is fiction.

1-7-16 / *ANSWERS*

1. (*i*) mod 7: $3 \equiv 500 \equiv -312$, $19 \equiv 96$, $-15 \equiv -71 \equiv -113 \equiv 69 \equiv 153$, $378 \equiv -91 \equiv -14$.
(*iii*) mod 13: $19 \equiv 240 \equiv 500$, $-113 \equiv 69$, $-91 \equiv -312$.
3. $m \mid a$.
4. 0; 18.
5. Integer is divisible by 9 if and only if the sum of its digits is divisible by 9. Integer is divisible by 3 if and only if the sum of its digits is divisible by 3.
6. Last digit is even; the number determined by the last two digits is divisible by 4; the number determined by the last three digits is divisible by 8 [or if $n = \cdots + \cdots + 10^2 a_2 + 10^1 a_1 + a_0$ then $8 \mid n \Leftrightarrow 8 \mid (a_0 + 2a_1 + 4a_2)$]; last digit is 0; last digit is 0 or 5.
8. (*i*) a in each case. (*ii*) a in each case.
12. $\lfloor 84 \rfloor_5 = \lfloor -16 \rfloor_5$, $\lfloor -57 \rfloor_5 = \lfloor 53 \rfloor_5$, $\lfloor 43 \rfloor_7 = \lfloor 50 \rfloor_7$, etc.
13. (*ii*)

addition in $\mathbf{Z}_6$

+	0	1	2	3	4	5
0	0	1	2	3	4	5
1	1	2	3	4	5	0
2	2	3	4	5	0	1
3	3	4	5	0	1	2
4	4	5	0	1	2	3
5	5	0	1	2	3	4

multiplication in $\mathbf{Z}_6$

·	0	1	2	3	4	5
0	0	0	0	0	0	0
1	0	1	2	3	4	5
2	0	2	4	0	2	4
3	0	3	0	3	0	3
4	0	4	2	0	4	2
5	0	5	4	3	2	1

15. As a set any residue class (mod 6) consists of two residue classes (mod 12).
18. (*i*) $n = 12$; (*ii*) $n = 3$; (*iii*) $n = 8$; (*iv*) $n = 16$.
20. Solve $48 = 100a + 1b - 10c - 5d$, where $0 \leq a \leq 1$, $0 \leq b \leq 3$, $0 \leq c \leq 6$, $0 \leq d \leq 7$.

1-8-8 / *ANSWERS*

1. (*iv*)

addition in base 6

+	0	1	2	3	4	5
0	0	1	2	3	4	5
1	1	2	3	4	5	10
2	2	3	4	5	10	11
3	3	4	5	10	11	12
4	4	5	10	11	12	13
5	5	10	11	12	13	14

multiplication in base 6

·	0	1	2	3	4	5
0	0	0	0	0	0	0
1	0	1	2	3	4	5
2	0	2	4	10	12	14
3	0	3	10	13	20	23
4	0	4	12	20	24	32
5	0	5	14	23	32	41

2. (*i*) $(111001010001000111)_2$; (*iii*) $(100001232)_5$; (*iv*) $(5005543)_6$; (*vi*) $(\varepsilon 38\varepsilon 3)_{12}$.
3. (*i*) 845; (*ii*) 15414; (*iv*) 14415; (*v*) 8096; (*vi*) 39711.
5. (*i*) $(1011022)_3$; (*ii*) $(14202)_6$; (*iii*) $(4828)_{12}$; (*iv*) $(21422)_5$.
6. (*i*) $(1000110100)_2$; (*ii*) $(121210)_5$; (*iv*) $(\tau\varepsilon 19)_{12}$.
7. (*i*) $(112112)_3$; (*ii*) $(104520)_6$; (*iii*) $(144460)_7$; (*iv*) $(519\varepsilon)_{12}$.
8. (*i*) $(4411143)_6$; (*ii*) $(2604663)_7$; (*iii*) $(765622)_{12}$.
9. (*i*) $q = (110100)_2$, $r = (1)_2$; (*ii*) $q = (433)_5$, $r = (14)_5$; (*iii*) $q = (95)_{12}$, $r = (17)_{12}$.
10. (*i*) 1; (*ii*) 1.

Problems for Chapter II

2-1-21 / *ANSWERS*

3.

addition in $\mathbf{Z}_3$

+	0	1	2
0	0	1	2
1	1	2	0
2	2	0	1

multiplication in $\mathbf{Z}_3$

·	0	1	2
0	0	0	0
1	0	1	2
2	0	2	1

4. In $\mathbf{Z}_{11}$: $5 \cdot 8 = 7$, $5 \cdot 9 = 1$, $5 \cdot (8 + 9) = 8$, $(5 \cdot 8)9 = 8$, $5(8 \cdot 9) = 8$, $9 \cdot (8 \cdot 5) = 8$, $8(5 \cdot 9) = 8$, $8(9 + 5) = 2$, $9(8 + 5) = 7$.

5. In $\mathbf{Z}_{11}$: $3 + (7 + 2) = 1 = 2 + (3 + 7)$, $\big(3(7 + 2)\big)9 = 1 = (3 \cdot 7)9 + 3(2 \cdot 9)$, $(3 + 7)(2 + 9) = 0 = (3 \cdot 2 + 7 \cdot 2) + (3 \cdot 9 + 7 \cdot 9) = (3 \cdot 2 + 3 \cdot 9) + (7 \cdot 2 + 7 \cdot 9)$, $3(7 \cdot 2) = 6 = 2(7 \cdot 3)$, $(3 + 7)(3 + 7) = 1 = 3^2 + 7^2 + (3 \cdot 7 + 3 \cdot 7)$.

6. In $\mathbf{Z}_{12}$: $6 \cdot 2 = 2 \cdot 6 = 0$, $4 \cdot 3 = 3 \cdot 4 = 0$, $10 \cdot 6 = 6 \cdot 10 = 0$, $8 \cdot 3 = 3 \cdot 8 = 0$, $8 \cdot 6 = 6 \cdot 8 = 0$, $6 \cdot 4 = 4 \cdot 6 = 0$, $6 \cdot 6 = 0$, $4 \cdot 9 = 9 \cdot 4 = 0$, $8 \cdot 9 = 9 \cdot 8 = 0$; $11 \cdot 11 = 1$, $7 \cdot 7 = 1$, $5 \cdot 5 = 1$, $1 \cdot 1 = 1$.

7. Yes. **8.** $\mathbf{Z}_2$. Yes, $\mathbf{Z}_n$, for example. **9.** Yes.

12. In $\mathbf{Z}_{13}$: $(5 - 10) + (7 - 11) = -9 = 4 = (5 + 7) - (10 + 11)$, $5(10 - 7) = 2 = 5 \cdot 10 - 5 \cdot 7$, $(5 - 10)(7 + 11) = 1 = (5 \cdot 7 - 10 \cdot 11) + (5 \cdot 11 - 10 \cdot 7) = (5 \cdot 7 + 5 \cdot 11) - (10 \cdot 7 + 10 \cdot 11)$, $(5 - 10)(7 - 11) = 7 = (5 \cdot 7 + 10 \cdot 11) - (10 \cdot 7 + 5 \cdot 11)$.

13. (*i*) $(a^2 + ab) + (ba + b^2)$; (*ii*) $(a^2 - ab) + (-ba + b^2)$. (*iii*) $(a^2 + ab - ac) + (-ba - b^2 + bc) + (ca + cb - c^2)$.

14. (*i*) $(a^2 + ab) + (ab + b^2)$, which we also write as $a^2 + 2ab + b^2$. (*ii*) $a^2 - 2ab + b^2$. (*iii*) $(a^2 - b^2 - c^2) + 2bc$.

15. The relation holds for all $a, b \in R \Leftrightarrow R$ is commutative.

16. There are many possible ways to insert the parentheses that describe the order in which the operations are to be performed.

2-2-15 / *ANSWERS*

1. (*i*) No; (*ii*) yes; (*iii*) yes; (*iv*) yes; (*v*) no.

2. (*i*) Yes (provided it is understood that we always reduce to lowest terms—otherwise, not a ring); (*ii*) yes (when it is not required that $(m, p^r) = 1$); (*iii*) yes (under the natural assumptions).

3. Yes. **4.** It is an integral domain. **5.** Yes.

6. (*i*) No; (*ii*) no, yes; (*iii*) it is a ring with unity.

7. No. **8.** No.

9. Those that were found to be rings in Problems 1 and 2.

10. (*i*) (0), $\mathbf{Z}_7$; (*ii*) (0), $\mathbf{Z}_6$, $\{0, 3\}$, $\{0, 2, 4\}$.

12. (*i*) Yes; (*ii*) no. **14.** (*iii*) Yes.

15. For a counterexample, look in $\mathbf{Z}_{12}$.

16. (*i*) Yes, yes; (*ii*) yes; (*iii*) $A = \bigcap A_x$.

18. (*i*) Yes; (*ii*) $m\mathbf{Z} + n\mathbf{Z} = (m, n)\mathbf{Z}$.

21. A subring that contains the identity.

22. (*i*) $A + B = \begin{pmatrix} 0 & 8 \\ -7 & 7 \end{pmatrix}$; (*ii*) $A + B + C = \begin{pmatrix} 3 & 5 \\ -6 & 6 \end{pmatrix}$;

(*iii*) $A - B = \begin{pmatrix} 2 & -4 \\ 3 & -1 \end{pmatrix}$; (*iv*) $AB = \begin{pmatrix} -11 & 14 \\ -13 & 0 \end{pmatrix}$;

(*v*) $ABC = \begin{pmatrix} -19 & 19 \\ -39 & 39 \end{pmatrix}$; (*vi*) $A(B + C) = \begin{pmatrix} -6 & 8 \\ -16 & 3 \end{pmatrix}$.

26. (*i*) $$A + B = \begin{pmatrix} 3 & 2 & 6 & -1 \\ -3 & 2 & 5 & 0 \\ 3 & -3 & 0 & 3 \\ 2 & -6 & 1 & 1 \end{pmatrix};$$

(*ii*) $$A - B = \begin{pmatrix} -1 & -4 & -2 & -3 \\ 3 & 4 & -3 & -2 \\ 1 & -3 & 0 & 5 \\ -4 & 4 & -9 & 1 \end{pmatrix};$$

(*iii*) $$AB = \begin{pmatrix} 1 & 14 & -10 & -2 \\ -11 & 2 & 7 & 2 \\ 25 & -11 & 16 & -1 \\ 0 & -7 & -3 & 2 \end{pmatrix}.$$

27. Yes.

28. All except (*vii*) and (*ix*).

31. (*i*) $A \cup B = \{0, 1, 2, 3, 5, 7, 9\}$. (*ii*) $A \cap D = \{3\}$.
(*iii*) $H \times C = \{(4, 4), (4, 6), (6, 4,) (6, 6)\}$. (*iv*) $A - B = \{0, 1\}$.
(*v*) $A^c = \{4, 5, 6, 7, 8, 9\}$. (*vi*) $A \cup A^c = X$. (*vii*) $B \cap B^c = \varnothing$.
(*viii*) $A - (B - C) = \{0, 1\}$. (*ix*) $(A \cap B) \cup C = \{2, 3, 4, 6\}$.
(*x*) $A \cap (B \cup C) = \{2, 3\}$. (*xi*) $(H \cap I)^c = \{0, 1, 2, 3, 4, 5, 7, 8, 9\}$.
(*xii*) $(E \cup G)^c = \{2, 4, 6, 7, 9\}$. (*xiii*) $A - (B \cup C) = \{0, 1\}$.
(*xiv*) $A - (B \cap D) = \{0, 1, 2\}$.

33. 2^n; 3^n.

34. Say $X = \{1, 2, 3\}$ and put $A = \varnothing$, $B = \{1\}$, $C = \{2\}$, $D = \{3\}$, $E = \{1, 2\}$, $F = \{1, 3\}$, $G = \{2, 3\}$, $H = X$. Then, for example, $B + \boldsymbol{D} = F$, $BD = A$, $F + G = E$, etc.

36. (*i*) Intersection of M_{x_0}'s, or sums. (*ii*) All subsets of X that do not include x_0.

39. Set up 1–1 correspondences between them, and which preserve the ring operations.

41. $\mathbf{Z}_{72}$, $\mathbf{Z}_{36} \oplus \mathbf{Z}_2$, $\mathbf{Z}_8 \oplus \mathbf{Z}_9$, $\mathbf{Z}_{18} \oplus \mathbf{Z}_4$, etc.

42. For each way of expressing m as a product, there is a ring with m elements. There are others.

43. In $\mathbf{Z}_{74}$, take all multiples of 2.

46. The Euclidean plane, and the lattice of all integral points.
47. $R' \cap R'' = R' \cap \Delta = R'' \cap \Delta = \{(0, 0)\}$, $R' + R'' = R' + \Delta = R \oplus R$.

2-3-20 / *ANSWERS*

3. (*i*) There are several cases to consider. (*ii*) The converse is true. (*iii*) For any odd power the above hold, but not for even powers.
4. First treat $<$ or $>$, then the situations where $=$ occurs.
5. Proceed as in Problem 4.
6. Note that $-e$ is negative.
7. Consider $(a - b)^2$. **8.** The case $b < a$ is impossible.
9. Use 2-3-8. **10.** Yes. **12.** Yes; no.
13. Distinguish cases judiciously.
15. (*i*) $n(n + 1)$; (*ii*) n^2.
16. (*i*) $\frac{n}{2}\{2a + (n - 1)d\}$; (*ii*) $\frac{a - ar^n}{1 - r}$.
20. Verify the assertion for $n = 1$; then supposing it true for n (inductively), show that it holds for $n + 1$.
21. (*i*) No; (*ii*) no; (*iii*) yes.
23. The passage from n to $n + 1$ is "repeated" infinitely often, once for each choice of n; but we prove it just once, for arbitrary n.

2-4-10 / *ANSWERS*

1. There are 5! rearrangements. More generally, for any n, the summation of $a_1, a_2, \ldots, a_n$ has $n!$ rearrangements.
2. If $a_1, a_2, \ldots, a_n$, $n \geq 2$, are elements of the commutative ring R then all products of these n elements, in any order, are equal.
6. Yes.
8. It is valid for any positive odd power, but not for even ones.
10. If R is not commutative, the terms in the expansion of $(a + b)^n$ are all the possible n-letter "words" in which each letter is either a or b.
11. The sum of all terms of form

$$\frac{n!}{i!j!k!} a^i b^j c^k,$$

where $i + j + k = n$ and $0 \leq i, j, k \leq n$.
13. (*iii*) Anything of form $\begin{pmatrix} 0 & a \\ 0 & 0 \end{pmatrix}$ or $\begin{pmatrix} 0 & 0 \\ a & 0 \end{pmatrix}$.
14. (*i*) 330; (*ii*) 3003; (*iii*) (333)(499)(997).

2-5-12 / *ANSWERS*

1. (*i*) Yes, no; (*ii*) no, yes; (*iii*) $\psi \circ \phi$ maps $1 \to 1, 2 \to 2, 3 \to 3$, it is injective and surjective; $\phi \circ \psi$ maps $1 \to 1, 2 \to 4, 3 \to 3, 4 \to 4$, it is neither injective nor surjective.

2. For $\phi: \mathbf{Z} \to \mathbf{Z}$, (*i*), (*ii*), and (*iv*) are injective, whereas (*i*) and (*iv*) are surjective, and every other answer is no. For $\phi: \mathbf{Q} \to \mathbf{Q}$: as above except that (*ii*) is surjective. For $\phi: \mathbf{R} \to \mathbf{R}$, (*v*) is also surjective.

3. For $\mathbf{Z}_3$:

	inj.	surj.	homo.	iso.
(*i*)	Yes	Yes	No	No
(*ii*)	Yes	Yes	No	No
(*iii*)	Yes	Yes	No	No
(*iv*)	No	No	No	No
(*v*)	Yes	Yes	Yes	Yes

For $\mathbf{Z}_5$: as above, except that for (*v*) the row is all No.
For $\mathbf{Z}_6$: row (*i*) is all No, with the rest as above.

6. (*i*) f maps $1 \to 2, 2 \to 1, 3 \to 3$; g maps $1 \to 2, 2 \to 3, 3 \to 1$. This example also works for (*ii*).

7. $f: n \to 2n$; $g: m \to m/2$, when m is even and $g: m \to 0$, when m is odd.

9. $\psi\phi(x) = ba + bx$, $\phi\psi(x) = a + bx$, $\phi\phi(x) = 2a + x$, $\psi\psi(x) = b^2x$. ϕ is always injective and surjective. Whether ψ is injective, or surjective depends on properties of b. The question is probably too general, but one does well to consider specific rings first.

15. It is surjective and the kernel consists of the four elements $\lfloor 0 \rfloor_{24}$, $\lfloor 6 \rfloor_{24}$, $\lfloor 12 \rfloor_{24}$, $\lfloor 18 \rfloor_{24}$.

16. $\phi: \lfloor a \rfloor_m \to \lfloor a \rfloor_n$; it is surjective; kernel $= \{\lfloor kn \rfloor_m \mid k = 0, 1, \ldots, \frac{m}{n}\}$. If $n \nmid m$, ϕ is not well defined.

17. 0 is unity element; the inverse isomorphism is $a \to a + e$.

22. Suppose there is an isomorphism ϕ, and show this leads to a contradiction.

Problems for Chapter III

3-1-12 / *ANSWERS*

1. (*i*) 1; (*ii*) 3; (*iii*) 0.

2. (*i*) 0; (*ii*) 113; (*iii*) 1.

4. (*i*) $x \equiv 6, 103 \pmod{194}$; (*ii*) $x \equiv 74, 153 \pmod{158}$; (*iii*) $x \equiv 434 \pmod{605}$; (*iv*) $x \equiv 54, 127, 200, 273, 346, 419 \pmod{438}$.

5. (*i*) $\lfloor 6 \rfloor_{24}$, $\lfloor 14 \rfloor_{24}$, $\lfloor 22 \rfloor_{24}$; (*ii*) $\lfloor 12 \rfloor_{26}$, $\lfloor 25 \rfloor_{26}$; (*iv*) no solution.
6. $44x \equiv 12 \pmod{48}$.
7. (*i*) $x \equiv 31 \pmod{77}$; (*iii*) $x \equiv 619 \pmod{1092}$; (*vii*) no solution; (*ix*) $x \equiv 31 \pmod{60}$; (*x*) no solution; (*xiii*) $x \equiv 331 \pmod{420}$.
8. (*i*) $\lfloor 193 \rfloor_{285}$; (*iv*) $\lfloor 71 \rfloor_{510}$.
9. One choice is $34x \equiv 20 \pmod{50}$ and $12x \equiv 15 \pmod{23}$.
13. Unique solution is $x \equiv 97 \pmod{2^2 \cdot 3^2 \cdot 7 \cdot 11}$.
14. There are three solutions (mod [24, 26]). **16.** 23.
17. $x \equiv 119 \pmod{420}$.

3-2-23 / *ANSWERS*

1. (*i*) F; (*ii*) T; (*iii*) T; (*iv*) T; (*v*) F; (*vi*) T; (*vii*) F; (*viii*) T.
2. (*i*) 1, 5; (*ii*) 1, 2, 4, 7, 8, 11, 13, 14; (*iii*) 1, 5, 7, 11, 13, 17, 19, 23.
3. (*i*) ± 1; (*ii*) ± 1; (*iii*) all nonzero elements of $\mathbf{Q}[\sqrt{-5}]$.
4. $1^{-1} = 1, 2^{-1} = 7, 3^{-1} = 9, 4^{-1} = 10, 5^{-1} = 8, 6^{-1} = 11, 7^{-1} = 2, 8^{-1} = 5, 9^{-1} = 3, 10^{-1} = 4, 11^{-1} = 6, 12^{-1} = 12$.
6. (*i*) Look in $\mathbf{Q}$.
7. Assuming $m \neq 1$, $\mathbf{Q}[\sqrt{m}]$ is a field.
9. (*i*) $\pm(2 + \sqrt{3})^n$, $n \in \mathbf{Z}$; (*ii*) $\pm(5 + 2\sqrt{6})^n$, $n \in \mathbf{Z}$; (*iii*) $\pm(8 + 3\sqrt{7})^n$, $n \in \mathbf{Z}$.
10. $\lfloor 28 \rfloor_{851}$ and $\lfloor 100 \rfloor_{851}$ are units; their inverses are found by solving the appropriate congruence.
12. (*iv*)

	1	3	7	9
1	1	3	7	9
3	3	9	1	7
7	7	1	9	3
9	9	7	3	1

(*vii*)

	1	3	5	9	11	13
1	1	3	5	9	11	13
3	3	9	1	13	5	11
5	5	1	11	3	13	9
9	9	13	3	11	1	5
11	11	5	13	1	9	3
13	13	11	9	5	3	1

13. (*i*) $11^{14}(10)$; (*ii*) $2^2 \cdot 7^3 \cdot 6 \cdot 13 \cdot 12 \cdot 16 \cdot 67^2 \cdot 66$; (*iii*) $2 \cdot 5 \cdot 4 \cdot 3^2 \cdot 2 \cdot 6$; (*iv*) $7^4 \cdot 6$; (*v*) $2^7 \cdot 3^2 \cdot 11 \cdot 23$.
15. 14. **17.** 35, 39, 45, 52, 56, 70, 72, 78, 84, 90.
19. (*ii*) The intersection is $\lfloor a \rfloor_{[m, n]}$.

20. The kernel is $[m_1, m_2, \ldots, m_r]\mathbf{Z}$; it is surely surjective if $m_1, \ldots, m_r$ are relatively prime in pairs.
21. The result is always 1 (see 3-2-22, Part 7, Euler's theorem).
22. No. **24.** (*i*) 24; (*ii*) 27; (*iii*) 49; (*iv*) 50; (*v*) 60.
26. No. No, unless it contains 1.

3-3-15 / *ANSWERS*

1. (*ii*) (*a*) $f + g = 3x^3 + 2x^2 + 3x + 2$, $f - g = 3x^3 + x$,
$f \cdot g = 2x^6 + x^5 + 2x^4 + 2x^3 + x + 1$.
(*b*) $f + g = 4x^3 + 2x^2 + 3x + 2$, $f - g = x^2 + x + 4$,
$f \cdot g = 4x^6 + 4x^5 + 3x^4 + 4x^3 + 2x^2 + x + 2$.
(*d*) $f + g = 6x^3 + 2x^2 + 3x + 2$, $f - g = 2x^3 + 3x^2 + x + 4$,
$f \cdot g = x^6 + 3x^5 + 5x^4 + 5x^2 + x + 4$.
4. Write $c = cx^0$ and apply the rule for multiplication.
5. $ax^ibx^j = abx^{i+j}$. **6.** Only (*i*), (*vi*), and (*vii*) are subrings.
7. None. **8.** Yes. **9.** $2 \cdot 3^4$; $2 \cdot 3^5$; 3^6 including the 0 polynomial.
11. (*i*) and (*iv*) are zero divisors. **12.** Yes.
14. (*ii*) They do not determine the same polynomial function.
15. (*i*) $\bar{f}(0) = 2, \bar{f}(1) = 2, \bar{f}(2) = 2$; $g(x) = 2$, $h(x) = x^9 - x^3 + 2$.
16. $2x^2 - x + 1$. **17.** (*ii*) Yes.
18. (*i*) None; (*ii*) none; (*iii*) none; (*iv*) 2; (*v*) 6.
19. No, except in case (*v*).
20. ker E_c consists of all polynomials having c as a root; $E_c = E_{c'} \Leftrightarrow c = c'$.
23. (*iii*) No.
25. The units of $R[[x]]$ are those power series whose constant term a_0 is a unit of R.

3-4-17 / *ANSWERS*

1. (*i*) $\mathbf{Q}^*$; (*ii*) $\mathbf{R}^*$; (*iii*) $\{1\}$; (*iv*) $\{1, 2\}$; (*v*) $\{1, 2, 3, 4, 5, 6\}$; (*vi*) $\{1, 2, 3, 4, 5, 6, 7, 8, 9, 10\}$; (*vii*) $\mathbf{Z}_p^* = \{1, 2, \ldots, p - 1\}$.
2. $u(2x^2 - x + 1)$, where (*i*) $u \in \{1, 2\}$; (*ii*) $u \in \mathbf{Z}_5^*$; (*iii*) $u \in \mathbf{Z}_7^*$; (*iv*) $u \in \mathbf{Z}_{11}^*$.
3. (*i*) $f_1(x)$ and $f_2(x)$; (*ii*) $f_2(x)$ and $f_3(x)$; (*iii*) $f_1(x)$ and $f_4(x)$.
4. No. **5.** (*i*), (*vi*), (*vii*), and (*viii*) are equivalence relations.
6. Part (*vi*) is special.
7. (*i*) $x^2 - \frac{4}{3}x + \frac{2}{3}$; (*ii*) $x^2 - 3x + 4$.

8. (*a*)(*v*) $2x^6 - x + 5 = (\frac{2}{3}x^4 + \frac{4}{9}x^2 + \frac{8}{27})(3x^2 - 2) + (-x + \frac{151}{27})$.
9. (*i*) For $f(x) = x - 3$, $g(x) = x^3 + x^2 - 13x + 3$, $g(x) = (x^2 + 1)f(x)$.
10. (*iii*) $g(x) = (x^2 + x + 1)f(x) + (2x^2)$.
13. $f(x)g(x)$ is always monic; $f(x) \pm g(x)$ may or may not be monic.
14. (*i*) $(f(x), g(x)) = x^2 - \frac{3}{2}x - 1 = -\frac{1}{2}f(x) + (1)g(x)$.
(*iii*) $g(x) = (x^2 - x + 1)f(x)$; so $(f(x), g(x)) = f(x) = (1)f(x) + (0)g(x)$.
15. (*i*) $(f(x), g(x)) = 1 = (x^2 + x)f(x) + (1)g(x)$.
(*ii*) $(f(x), g(x)) = x + 1 = (x^3 + x)f(x) + (1)g(x)$.
20. Proceed as for the integers.
22. Why not? More on this in Section 3-5.
23. (*i*) Irreducible; (*ii*) irreducible; (*iii*) $(x - \omega)(x - \omega^2)$, where $\omega = \frac{-1 + \sqrt{-3}}{2}$; (*iv*) irreducible; (*v*) $(x - 1)^2$; (*vi*) irreducible; (*vii*) $(x + 3)(x - 2)$; (*viii*) irreducible.
25. $f(x)$ might be the product of two irreducible polynomials of degree 2.
26. Every element of $\mathbf{Z}_7$ is a root;

$$x^7 - x = x(x - 1)(x - 2)(x - 3)(x - 4)(x - 5)(x - 6).$$

31. (*i*) No; (*ii*) $m = 2, 3, 6$. **33.** Proceed as in Section 1-5.

3-5-29 / *ANSWERS*

1. (*i*) $f(x) = 3x - 1$; (*ii*) $f(x) = 7$. One is finding the equation of the straight line passing through two given points.
2. (*i*) $f(x) = 1 + \frac{1}{2}x + \frac{1}{2}x^2$; (*iii*) $f(x) = \frac{2}{3} - \frac{1}{2}x + \frac{5}{6}x^2$.
3. (*ii*) Over $\mathbf{Z}_5$, if $f(0) = 1, f(1) = 2, f(2) = 4$, then $f(x) = 1 + 3x + 3x^2$.
4. $f(x) = 2x^3 - x^2 + 3x + 1$. **6.** $f(x) = 3x^2 + 4x + 2$.
8. (*i*) 10381; (*ii*) 10381; (*iii*) 1; (*iv*) 1; (*v*) 1; (*vi*) 1; (*vii*) 6; (*viii*) 8.
10. (*ii*) Yes.
12. Working with $x^3 + x + 3$: (*i*) no roots, i.e., irreducible; (*ii*) $x(x^2 + 1)$; (*iii*) $(x - 1)(x^2 + x + 2)$; (*iv*) $(x + 2)(x^2 + 5x + 5)$; (*v*) $(x - 3)(x^2 + 3x - 1)$; (*vi*) $(x - 2)^2(x - 4)$.
13. (*i*) $x^2 - 13$ is irreducible over $\mathbf{Q}$, and factors as $(x - \sqrt{13})(x + \sqrt{13})$ over $\mathbf{R}$ and $\mathbf{C}$; (*iii*) $x^2 - 5x + 6 = (x - 2)(x - 3)$ over $\mathbf{Q}$, $\mathbf{R}$, and $\mathbf{C}$; (*iv*) $x^2 + x + 1$ is irreducible over $\mathbf{Q}$ and $\mathbf{R}$, and over $\mathbf{C}$ factors as

$$(x - \omega)(x - \omega^2), \text{ where } \omega = \frac{-1 + \sqrt{-3}}{2}.$$

14. (*ii*) Over $\mathbf{Z}_3$; $x^2 - 13$ is irreducible, $x^2 + 13$ is irreducible;

$$x^2 - 5x + 6 = x(x+1);\quad x^2 + x + 1 = (x-1)^2;$$
$$x^2 + 2x - 2 = (x+1)^2;\quad x^3 - 2 = (x+1)^3 = (x-2)^3;$$
$$x^3 + 2 = (x-1)^3;\quad x^3 + x + 1 = (x-1)(x^2 - x - 1);$$
$$x^3 + x^2 + 1 = (x-1)(x^2 + 2x - 1).$$

15. (*i*) Some of them are: x, $x+1$, $x+2$, x^2+1, x^2+x+2, x^2+2x+2, x^3+2x+2.

17. (*i*) $\pm\left(\frac{\sqrt{2}}{2} + \frac{\sqrt{2}}{2} i\right)$; (*iii*) $\pm(2 - i)$.

18. Use the "quadratic formula," and find the square root as in Problem 17.

19. (*i*) None; (*ii*) -1; (*iii*) none; (*iv*) $1, -1$; (*v*) none; (*vii*) none.

20. (*ii*) $x^4 + 1$ is irreducible over $\mathbf{R}$; over $\mathbf{C}$,

$$x^4 + 1 = (x-\alpha)(x+\alpha)(x-\beta)(x+\beta),$$

where $\alpha = \dfrac{1+i}{\sqrt{2}}$, $\beta = \dfrac{-1+i}{\sqrt{2}}$.

21. (*i*) $(x-3)(x-4)(x-5)(x+5)$.

22. (*i*), (*iii*), and (*iv*) are Eisenstein; (*ii*) is reducible.

3-6-11 / *ANSWERS*

1. (*i*) $\lfloor 66 \rfloor_{343}$; (*iii*) $\lfloor 1353 \rfloor_{2401}$, $\lfloor 1047 \rfloor_{2401}$.
2. (*iv*) 14 solutions: $\lfloor 4 + 49t \rfloor_{343}$, $\lfloor 46 + 49t \rfloor_{343}$ for $t = 0, 1, \ldots, 6$.
4. (*ii*) (*a*) $x \equiv 0, 2 \pmod 3$, (*b*) $x \equiv 6, 5 \pmod 9$, (*c*) $x \equiv 6, 23 \pmod{27}$, (*d*) $x \equiv 60, 23 \pmod{81}$. (*vii*) (*a*) $x \equiv 1 \pmod 3$, (*b*) $x \equiv 1, 4, -2 \pmod{3^2}$, (*c*) $x \equiv 4, 13, 22 \pmod{27}$, (*d*) none.
5. (*ii*) (*a*) $\lfloor 1 \rfloor_2$, (*b*) $\lfloor 3 \rfloor_4$, (*c*) $\lfloor 3 \rfloor_8$, (*d*) $\lfloor 11 \rfloor_{16}$, (*e*) $\lfloor 11 \rfloor_{32}$, (*f*) $\lfloor 43 \rfloor_{32}$.
6. (*ii*) None; (*v*) (*a*) $\lfloor 0 \rfloor_5$, (*b*) $\lfloor 15 \rfloor_{25}$, (*c*) $\lfloor 65 \rfloor_{125}$.
7. (*iv*) None; (*vii*) (*a*) $x \equiv 0, 6 \pmod 7$, (*b*) $x \equiv -7, 6 \pmod{7^2}$, (*c*) $x \equiv -56, 55 \pmod{7^3}$.
8. For the polynomial $x^2 - 2x + 3$ of (*ii*), no solutions. For the polynomial $x^2 + 2x - 5$ of (*v*): (*a*) $x \equiv 5 \pmod{15}$, (*b*) no solution, (*c*) $x \equiv 65 \pmod{75}$, (*d*) no solution.
9. For the polynomial $x^3 + x^2 - 4$ of (*iv*), no solutions. For the polynomial $x^2 + x + 7$ of (*vii*), in case (*d*) there are six solutions: $x \equiv 13, 49, 76, 112, 139, 175 \pmod{189}$.
12. (*i*) $x \equiv 2, 5, -3 \pmod{419}$; (*ii*) $x \equiv 2, 5, -3 \pmod{463}$; (*iii*) nine solutions: $x \equiv 2, 44, 23, 54, 5, 75, 67, 18, -3 \pmod{91}$.

13. (*i*) 0; (*ii*) 1; (*iii*) 1; (*iv*) 3; (*v*) 18.

15. On the face of it, the following possibilities occur: 0, 1, 2, 3, p, $p + 1$, $p + 2$, $2p$, $2p + 1$, $3p$.

3-7-15 / *ANSWERS*

1. Quadratic residues are: (*i*) 1, 4, 9, 16, 8, 2, 15, 13; (*ii*) 1, 4, 9, 16, 6, 17, 11, 7, 5; (*iii*) 1, 4, 9, 16, 2, 13, 3, 18, 12, 8, 6.

2. Everything (mod 13): (*i*) ± 1; (*ii*) ± 4; (*iii*) ± 2; (*iv*) none; (*v*) none; (*vi*) ± 3; (*vii*) ± 6; (*viii*) ± 5.

3. Use the technique of Section 3-6: (*i*) $x \equiv \pm 1 \pmod{13^2}$; (*iv*) none; (*vii*) $x \equiv 32, 137 \pmod{13^2}$.

7. $\left(\frac{-2}{19}\right) = 1, \left(\frac{-3}{19}\right) = 1, \left(\frac{5}{23}\right) = -1, \left(\frac{6}{19}\right) = 1.$

8. $\left(\frac{-1}{17}\right) = 1, \left(\frac{-1}{19}\right) = -1, \left(\frac{2}{13}\right) = -1, \left(\frac{-2}{17}\right) = 1, \left(\frac{3}{11}\right) = 1, \left(\frac{-3}{17}\right) = -1.$

10. Note Problem 4. **11.** ab is a quadratic nonresidue.

12.

a	5	-7	11	-11	13	-13
p	97	101	103	617	619	911
	No	No	No	Yes	No	No

14. (*i*) 2; (*iii*) 0; (*v*) 2; (*vii*) 0. **15.** Two.

16. (*ii*) No solutions. **17.** (*i*) 1; (*iii*) -1; (*vi*) 1; (*ix*) 1.

18. (*ii*) If $x^2 \equiv q \pmod p$ has no solution then $x^2 \equiv p \pmod q$ has no solution. If $x^2 \equiv q \pmod p$ has a solution, then $x^2 \equiv p \pmod q$ has two solutions.

21. $p \equiv 1, 5, 7, 9, 19, 25, 35, 37, 39, 43 \pmod{44}$.

22. (*ii*) $p \equiv \pm 1, \pm 3, \pm 9, \pm 13 \pmod{40}$.

23. $p \equiv \pm 1 \pmod{24}$.

27. 2^r, where r is the number of distinct primes dividing b.

28. $c \equiv 0 \pmod 7$, two solutions; $c \equiv -1 \pmod 7$, two solutions; $c \equiv 2 \pmod 7$, two solutions; $c \equiv -2 \pmod 7$, one solution.

29. (*i*) Yes, no; (*ii*) 18 under the restriction $(a, 91) = 1$, 28 otherwise.

30. One. **32.** (*ii*) It has a solution; (*iii*) no solution.

Problems for Chapter IV

4-1-12 / *ANSWERS*

1. Use cancellation.
3. Use induction on n for $n \geq 0$; then treat negative n.
6. The following are groups: (*iii*), (*vi*), (*xi*), (*xiii*), (*xxii*), (*xxvi*), (*xxvii*)–(*xxix*) and (*xxxii*).
7. (*i*) They form a group with table:

$\cdot$	e	a	b	c
e	e	a	b	c
a	a	e	c	b
b	b	c	e	a
c	c	b	a	e

(*ii*) They form a group with the same table.
8. The eight elements of the group of the square may be viewed as permutations of the set $\{1, 2, 3, 4\}$—that is, as elements of S_4. The multiplications are the same, so the octic group can be considered a "subgroup" of S_4.
9. In 3-2-11, one is dealing with a finite semigroup with cancellation.
10. The left multiplication, ϕ_a, need not be a map onto G.
11. For (*i*)–(*iii*), label the elements of $\{\mathbf{Z}_2, +\}$ appropriately; for (*iv*)–(*vi*), use $\{\mathbf{Z}_3, +\}$.
12. No.
15.

$$(i)\ \sigma^7 = \begin{pmatrix} 1 & 2 & 3 & 4 & 5 & 6 & 7 \\ 1 & 2 & 3 & 4 & 5 & 6 & 7 \end{pmatrix};$$

$$(iii)\ \rho^7 = \begin{pmatrix} 1 & 2 & 3 & 4 & 5 & 6 & 7 \\ 4 & 6 & 3 & 7 & 5 & 2 & 1 \end{pmatrix} = \rho;$$

$$(iv)\ \sigma^{-1} = \begin{pmatrix} 1 & 2 & 3 & 4 & 5 & 6 & 7 \\ 2 & 5 & 1 & 6 & 7 & 3 & 4 \end{pmatrix};$$

$$(vii)\ \sigma \circ \tau = \begin{pmatrix} 1 & 2 & 3 & 4 & 5 & 6 & 7 \\ 4 & 2 & 1 & 3 & 7 & 5 & 6 \end{pmatrix};$$

$$(x)\ \tau \circ \sigma \circ \tau^{-1} = \begin{pmatrix} 1 & 2 & 3 & 4 & 5 & 6 & 7 \\ 3 & 7 & 4 & 5 & 6 & 2 & 1 \end{pmatrix};$$

(*xiii*) $\sigma^{-7} = \begin{pmatrix} 1 & 2 & 3 & 4 & 5 & 6 & 7 \\ 1 & 2 & 3 & 4 & 5 & 6 & 7 \end{pmatrix}.$

16. The elements are

$$e, \sigma = \begin{pmatrix} 1 & 2 & 3 & 4 \\ 2 & 1 & 4 & 3 \end{pmatrix}, \quad \tau = \begin{pmatrix} 1 & 2 & 3 & 4 \\ 4 & 3 & 2 & 1 \end{pmatrix}, \quad \rho = \begin{pmatrix} 1 & 2 & 3 & 4 \\ 3 & 4 & 1 & 2 \end{pmatrix}.$$

The multiplication table takes the form:

$\cdot$	e	σ	τ	ρ
e	e	σ	τ	ρ
σ	σ	e	ρ	τ
τ	τ	ρ	e	σ
ρ	ρ	τ	σ	e

The four-group may be viewed as a "subgroup" of the octic group, 4-1-10, with $e = e$, $\sigma = \tau_2$, $\tau = \tau_1$, $\rho = \sigma_2$. The symmetries are same as rigid motions.

17. The ten elements are:

$$e = \begin{pmatrix} 1 & 2 & 3 & 4 & 5 \\ 1 & 2 & 3 & 4 & 5 \end{pmatrix}, \quad \sigma_1 = \begin{pmatrix} 1 & 2 & 3 & 4 & 5 \\ 2 & 3 & 4 & 5 & 1 \end{pmatrix},$$

$$\sigma_2 = \begin{pmatrix} 1 & 2 & 3 & 4 & 5 \\ 3 & 4 & 5 & 1 & 2 \end{pmatrix}, \quad \sigma_3 = \begin{pmatrix} 1 & 2 & 3 & 4 & 5 \\ 4 & 5 & 1 & 2 & 3 \end{pmatrix}, \quad \sigma_4 = \begin{pmatrix} 1 & 2 & 3 & 4 & 5 \\ 5 & 1 & 2 & 3 & 4 \end{pmatrix},$$

$$\tau_1 = \begin{pmatrix} 1 & 2 & 3 & 4 & 5 \\ 1 & 5 & 4 & 3 & 2 \end{pmatrix}, \quad \tau_2 = \begin{pmatrix} 1 & 2 & 3 & 4 & 5 \\ 3 & 2 & 1 & 5 & 4 \end{pmatrix}, \quad \tau_3 = \begin{pmatrix} 1 & 2 & 3 & 4 & 5 \\ 5 & 4 & 3 & 2 & 1 \end{pmatrix},$$

$$\tau_4 = \begin{pmatrix} 1 & 2 & 3 & 4 & 5 \\ 2 & 1 & 5 & 4 & 3 \end{pmatrix}, \quad \tau_5 = \begin{pmatrix} 1 & 2 & 3 & 4 & 5 \\ 4 & 3 & 2 & 1 & 5 \end{pmatrix}.$$

19. a and b should not commute. **20.** Look at the pairs $\{a, a^{-1}\}$.

22. The identity, which is the set X itself, is the only unit.

23. Only (*iv*) gives a group.

25. The units are:

(*i*) $\{1, 2, 3, 4\}$; (*ii*) $\{(1, 1), (1, -1), (-1, 1), (-1, -1)\}$.

(*iii*) $\{(1, 1), (1, 2), (2, 1), (2, 2)\}$.

(*iv*) $\{(1, 1), (1, 3), (3, 1), (3, 3)\}$; (*v*) $\{(1, 1), (1, 3), (2, 1), (2, 3)\}$.

(*vi*) $\{(1, 1), (1, -1), (2, 1), (2, -1)\}$.

27. Use $x \to e^x$ and $x \to \log x$.

30. (*i*)

$$A = \begin{pmatrix} 3 & 1 \\ 5 & 2 \end{pmatrix}, \qquad A^{-1} = \begin{pmatrix} 2 & -1 \\ -5 & 3 \end{pmatrix}.$$

In general,

$$A = \begin{pmatrix} a & b \\ c & d \end{pmatrix}$$

has an inverse $\Leftrightarrow$ $\Delta = ad - bc \neq 0$; and in this situation, an inverse is

$$\begin{pmatrix} \dfrac{d}{\Delta} & -\dfrac{b}{\Delta} \\ -\dfrac{c}{\Delta} & \dfrac{a}{\Delta} \end{pmatrix}$$

(*ii*) If $A \in \mathscr{M}(\mathbf{Z}, 2)$ then it has an inverse if and only if $\Delta = ad - bc = \pm 1$, and an inverse is as in case (*i*).

4-2-18 / *ANSWERS*

2. The subgroups are: (*i*) $6\mathbf{Z}$; (*ii*) $12\mathbf{Z}$; (*iii*) $30\mathbf{Z}$; (*vi*) $15\mathbf{Z}$.
3. There are no other nontrivial subgroups.
4. (*i*), (*iii*): $\{1, 4\}$ and $\{1, 11\}$ are subgroups.
6. In $\{\mathbf{R}, +\}$ consider the set of all positive reals.
7. Excluding the trivial subgroups: (*i*) none; (*ii*) $\{0, 3\}, \{0, 2, 4\}$; (*iii*) none; (*iv*) $\{0, 5\}$, $\{0, 2, 4, 6, 8\}$; (*v*) $\{0, 7\}$, $\{0, 2, 4, 6, 8, 10, 12\}$; (*vi*) $\{0, 9\}, \{0, 6, 12\}, \{0, 3, 6, 9, 12, 15\}, \{0, 2, 4, 6, 8, 10, 12, 14, 16\}$.
8. Pursue the powers of each element.
10. (*i*) $\left\{\begin{pmatrix} 1 & 2 & 3 & 4 \\ 1 & 2 & 3 & 4 \end{pmatrix}, \begin{pmatrix} 1 & 2 & 3 & 4 \\ 2 & 1 & 3 & 4 \end{pmatrix}\right\}.$

(*ii*) $\left\{\begin{pmatrix} 1 & 2 & 3 & 4 \\ 1 & 2 & 3 & 4 \end{pmatrix}, \begin{pmatrix} 1 & 2 & 3 & 4 \\ 2 & 3 & 1 & 4 \end{pmatrix}, \begin{pmatrix} 1 & 2 & 3 & 4 \\ 3 & 1 & 2 & 4 \end{pmatrix}\right\}.$

(*iii*) $\left\{\begin{pmatrix} 1 & 2 & 3 & 4 \\ 1 & 2 & 3 & 4 \end{pmatrix}, \begin{pmatrix} 1 & 2 & 3 & 4 \\ 2 & 3 & 4 & 1 \end{pmatrix}, \begin{pmatrix} 1 & 2 & 3 & 4 \\ 3 & 4 & 1 & 2 \end{pmatrix}, \begin{pmatrix} 1 & 2 & 3 & 4 \\ 4 & 3 & 2 & 1 \end{pmatrix}\right\}.$

(*iv*) The six permutations that leave 4 fixed—which amounts to S_3.
(*v*) The octic group. (*vi*) This will fall out in Section 4-5.
14. For $f, g \in \mathrm{Map}(X, G)$ define $f + g$ by $(f + g)(x) = f(x) + g(x)$. If G is multiplicative, define fg by $(fg)(x) = f(x)g(x)$.
16. In $\mathbf{Z}_7^*$: $2^{-1} = 4 = 2^2, 3^{-1} = 5 = 3^5, 4^{-1} = 2 = 4^2, 5^{-1} = 3 = 5^5, 6^{-1} = 6 = 6^1$.
(*i*) $\mathbf{Z}_8^* = \{1, 3, 5, 7\}, 3^{-1} = 3, 5^{-1} = 5, 7^{-1} = 7$.
(*iv*) $\mathbf{Z}_{12}^* = \{1, 5, 7, 11\}, 5^{-1} = 5, 7^{-1} = 7, 11^{-1} = 11$.
17. For each element of $\mathbf{Z}_{13}^*$, take its powers.

18. (*i*) $\{e, \sigma_2\}$ is the center. (*iii*) If $\#(X) \geq 3$ then (e) is the center of S_X.
19. Yes; not if Y is infinite.
21. Left cosets: $H_3 = \{e, \tau_3\}$, $\sigma_1 H_3 = \{\sigma_1, \tau_2\}$, $\sigma_2 H_3 = \{\sigma_2, \tau_1\}$.
Right cosets: $H_3 = \{e, \tau_3\}$, $H_3 \sigma_1 = \{\sigma_1, \tau_1\}$, $H_3 \sigma_2 = \{\sigma_2, \tau_2\}$.
22. (*i*) $\{0, 4, 8\}$, $\{1, 5, 9\}$, $\{2, 6, 10\}$, $\{3, 7, 11\}$; (*ii*) $\{0, 3, 6, 9\}$, $\{1, 4, 7, 10\}$, $\{2, 5, 8, 11\}$.
23. (*ii*) $\{1, 3, 9\}$, $\{2, 6, 5\}$, $\{4, 12, 10\}$, $\{7, 8, 11\}$; (*iii*) $\{1, 3, 4, 9, 10, 12\}$, $\{2, 6, 8, 5, 7, 11\}$.
24. (*i*) Left cosets: $H = \{e, \sigma_2\}$, $\sigma_1 H = \{\sigma_1, \sigma_3\}$; $\tau_1 H = \{\tau_1, \tau_2\}$, $\rho_1 H = \{\rho_1, \rho_2\}$.
Right cosets: H, $H\sigma_1 = \{\sigma_1, \sigma_3\}$, $H\tau_1 = \{\tau_1, \tau_2\}$, $H\rho_1 = \{\rho_1, \rho_2\}$.
(*iv*) $H = \{e, \rho_1, \rho_2, \sigma_2\}$, $\sigma_1 H = \{\sigma_1, \tau_2, \tau_1, \sigma_3\} = H\sigma_1$.
26. $\rho H_Y = \{\tau \in S_X \mid \tau y = \rho y \text{ for all } y \in Y\}$.
28. The meanings are the same.
30. It remains valid—that is, at least one of $\#(H)$, $(G: H)$ must be infinite.
32. They are "essentially" the same.
33. (*i*) The group has 10 elements so, according to Lagrange, we search only for nontrivial subgroups of order 2 or 5.

4-3-18 / *ANSWERS*

1. (*i*) $[2] = \{0, 2, 4, 6, 8\}$; (*ii*) $[4] = \{0, 2, 4, 6, 8\}$; (*iii*) $[6] = \{0, 2, 4, 6, 8\}$; (*iv*) $[7] = \mathbf{Z}_{10}$; (*v*) $[8] = \{0, 2, 4, 6, 8\}$.
2. (*i*) $[3] = \{3, 9, 13, 11, 5, 1\} = \mathbf{Z}_{14}^*$; (*ii*) $[5] = \{5, 11, 13, 9, 3, 1\} = \mathbf{Z}_{14}^*$; (*iii*) $[9] = \{9, 11, 1\}$.
3. $[1] = [2] = [4] = [7] = [8] = [11] = [13] = [14] = \mathbf{Z}_{15}$,
$[3] = \{0, 3, 6, 9, 12\} = [6] = [9] = [12]$, $[5] = \{5, 10, 0\} = [10]$.
4. $[1] = \{1\}$; $[2] = \{2, 4, 8, 1\}$; $[4] = \{4, 1\}$; $[7] = \{7, 4, 13, 1\}$.
$[8] = \{8, 4, 2, 1\}$; $[11] = \{11, 1\}$; $[13] = \{13, 4, 7, 1\}$; $[14] = \{14, 1\}$.
There are no generators of $\mathbf{Z}_{15}^*$.
6. None.
7. (*i*)–(*iii*) are cyclic with generators 5, 2, 2, respectively. (Other generators are possible.)
8. ∞ for (*i*)–(*v*); $(1 + i)/\sqrt{2}$ has order 8 in $\{\mathbf{C}^*, \cdot\}$.
9. $[e] = \{e\}$, $[\sigma_1] = \{\sigma_1, \sigma_2, \sigma_3, e\}$, $[\sigma_2] = \{\sigma_2, e\}$, $[\sigma_3] = \{\sigma_3, \sigma_2, \sigma_1, e\}$, $[\tau_1] = \{\tau_1, e\}$, $[\tau_2] = \{\tau_2, e\}$, $[\rho_1] = \{\rho_1, e\}$, $[\rho_2] = \{\rho_2, e\}$. The octic group is not cyclic.
10. Yes.
14. Use the fact that there is a linear combination of m and r equal to 1.
16. (*iii*) Take

$$\sigma = \begin{pmatrix} 1 & 2 & 3 & 4 \\ 2 & 3 & 4 & 1 \end{pmatrix};$$

then $[\sigma] = \{\sigma, \sigma^2, \sigma^3, e\}$. Take

$$\tau = \begin{pmatrix} 1 & 2 & 3 & 4 \\ 2 & 1 & 3 & 4 \end{pmatrix}, \qquad \rho = \begin{pmatrix} 1 & 2 & 3 & 4 \\ 1 & 2 & 4 & 3 \end{pmatrix},$$

and obtain the four-element group $\{\tau, \rho, \tau\rho, e\}$.

18. Look at

$$\sigma = \begin{pmatrix} 1 & 2 & 3 \cdots m-1 & m \\ 2 & 3 & 4 \cdots \quad m & 1 \end{pmatrix}.$$

19. Look at the "pairs" $\{a, a^{-1}\}$.

20. 1, 5, 7, 11, 13, 17, 19, 23.

25. (*i*) $[m, n]\mathbf{Z}$; (*iii*) $[m\mathbf{Z} \cup n\mathbf{Z}] = (m, n)\mathbf{Z}$.

28. (*ii*) No. **30.** (*i*) $\mathbf{Z}$; (*ii*) $3\mathbf{Z}$; (*iii*) $\mathbf{Z}$; (*iv*) $3\mathbf{Z}$.

31. (*i*) In $\{\mathbf{Z}_{19}, +\}$, $[S] = \mathbf{Z}_{19}$ for $S = \{3, 4\}, \{3, 6\}, \{3, 4, 6\}, \{9, 12\}$.

32. (*i*) $\mathbf{Z}_{13}^*$; (*ii*) $\{1, 3, 4, 9, 10, 12\}$; (*iii*) $\mathbf{Z}_{13}^*$; (*iv*) $\mathbf{Z}_{13}^*$.

35. $\sigma = \begin{pmatrix} 1 & 2 & 3 & 4 \\ 2 & 1 & 3 & 4 \end{pmatrix}$ and $\tau = \begin{pmatrix} 1 & 2 & 3 & 4 \\ 2 & 3 & 4 & 1 \end{pmatrix}$ generate S_4;

$\sigma = \begin{pmatrix} 1 & 2 & 3 & 4 & 5 \\ 2 & 1 & 3 & 4 & 5 \end{pmatrix}$ and $\tau = \begin{pmatrix} 1 & 2 & 3 & 4 & 5 \\ 2 & 3 & 4 & 5 & 1 \end{pmatrix}$ generate S_5.

37. Use 4-3-11.

4-4-15 / *ANSWERS*

1. (*i*), (*vii*), and (*viii*) are normal subgroups.

2. G is the octic group, $H = \{e, \tau_1, \tau_2, \sigma_2\}$, $K = \{e, \sigma_1\}$.

3. $\#(G) = \#(N) \cdot \#(G/N)$. **5.** Find τ for which $\tau H \neq H\tau$.

9. $H = \{e, \tau_1\}$, $K = \{e, \tau_2\}$ (notation of 4-1-9).

15. Image $= \{1, -1, i, -i\}$, kernel $= 4\mathbf{Z}$.

16. This is essentially the same as Problem 14 because every homomorphic image of G is of form G/N.

17. For $b \in G$ look at bab^{-1}.

18. Fix an $x_0 \in \{1, 2, \ldots, n+1\}$ and consider all $\sigma \in S_{n+1}$, which keep x_0 fixed. For each x_0, we have an injective homomorphism of $S_n \to S_{n+1}$.

20. (*i*) Image is all positive reals, kernel is the unit circle which is $\{z \in \mathbf{C} \mid |z| = 1\}$; (*ii*) $\mathbf{C}^*/W$ is isomorphic to the multiplicative group of positive reals.

21. With regard to converse, take $G = S_3$, $N = \{e, \sigma_1, \sigma_2\}$.

26. (*i*) Yes: $\phi(1) = 2$, $\phi(2) = 4$, $\phi(3) = 8$, $\phi(4) = 5$, $\phi(5) = 10$, $\phi(6) = 9$, $\phi(7) = 7$, $\phi(8) = 3$, $\phi(9) = 6$, $\phi(0) = 1$; (*ii*) No.

27. An isomorphism must map a generator of $\{\mathbf{Z}_{p-1}, +\}$ to a generator of $\{\mathbf{Z}_p^*, \cdot\}$.

30. Make use of Problem 28.
31. Map a generator of G to any element of G; using 28, there are n endomorphisms.

4-5-19 / *ANSWERS*

1. (*i*) (1275348); (*ii*) (17624583); (*iii*) (1426)(3957); (*iv*) (16982354).

2. (*i*) $\begin{pmatrix} 1 & 2 & 3 & 4 & 5 & 6 & 7 & 8 \\ 1 & 4 & 3 & 6 & 5 & 8 & 7 & 2 \end{pmatrix}$; (*iii*) $\begin{pmatrix} 1 & 2 & 3 & 4 & 5 & 6 & 7 & 8 & 9 \\ 4 & 3 & 1 & 9 & 6 & 8 & 5 & 7 & 2 \end{pmatrix}$.

3. (*i*) $\sigma^2 = (358)(749)$; (*ii*) $\rho^3 = (2549)(37)$;
(*iii*) $\tau^{-1} = (63951)(27)(48)$; (*iv*) $\tau\sigma = (1734256)(89)$;
(*v*) $\sigma\tau\rho = (1763)(28)(59)$; (*vi*) $\sigma\tau^{-1}\rho = (17423)(59)$;
(*vii*) $\sigma^\tau = (649)(75)(238)$; (*viii*) $(\sigma^\tau)^\rho = (154)(32)(976)$.
4. 7, 8, 4, 8; ord $\sigma = 6$, ord $\tau = 10$, ord $\sigma^2 = 3$, ord $\rho^{-1} = 12$, $\text{ord}(\sigma\tau^{-1}\rho) = 10$, $\text{ord}(\sigma^\tau) = 6$, $\text{ord}(\tau^\sigma)^\rho = 10$.
5. Orbits of σ^2 are subsets (possibly equal) of orbits of σ.
6. One choice for τ is (197683)(452).
7. (*i*) Odd: (12), (13), (23). Even: e, (123), (132); (12), (13), (23) are of order 2 and constitute a conjugate class; (123), (132) are of order 3 and constitute a conjugate class; e is a conjugate class.
8. (*i*) 7 conjugate classes with representatives: e, (12), (123), (1234), (12345), (12)(34), (123)(45).
9. Use 4-5-15. **10.** Conjugate permutations have the same order.
11. (*i*) (12)(34567).
13. (*i*) Two-cycles (a, b) are unaffected, three-cycles (abc) are written as $(ac)(ab)$, four-cycles $(abcd) = (ad)(ac)(ab)$, those of form $(ab)(cd)$ are unaffected.
15. Not really.
17. (*i*) S_3; (*ii*) S_3; (*iii*) S_4; (*iv*) A_4; (*viii*) A_4.
19. First compute σ^τ. **20.** Make use of 4-5-17.
22. (*i*) The only possibilities are subgroups of order 5 (of which there is exactly 1) and subgroups of order 2 (of which there are 5).
24. The tetrahedral group is isomorphic to A_4. There are 12 symmetries.

4-6-17 / *ANSWERS*

1. (*i*) 2 belongs to 3 (mod 7); (*ii*) 3 belongs to 5 (mod 11);
(*iii*) 7 belongs to 4 (mod 20); (*iv*) 5 belongs to 6 (mod 21).
2. (*i*) 4; (*iii*) 10. **6.** $8^8 \equiv 1 \pmod 9$.
7. (*ii*) Use the isomorphism $\mathbf{Z}_m^* \times \mathbf{Z}_n^* \approx \mathbf{Z}_{mn}^*$.
8. (*i*) $[10, 12, 16] = 240$; (*ii*) 12.

9. (*i*) 4; (*ii*) 2; (*iii*) 6; (*iv*) 4. **13.** Use Problem 11.
14. (*i*) 6; (*ii*) $7^4 \cdot 6$; (*iii*) 60; (*v*) $[2, 17^2 \cdot 16, 36, 66]$.
16. If $p \equiv 3 \pmod 4$, then $-g$ is not a primitive root (mod p).
17. This generalizes Problem 13.
18. (*i*) $m = p = 17, g = 5$

a	1	2	3	4	5	6	7	8	9	10	11	12	13	14	15	16
$\text{ind}_g\, a$	16	6	13	12	1	3	15	2	10	7	11	9	4	5	14	8

The congruences have solutions: $x \equiv 5 \pmod{17}$, none,

$x \equiv 9 \pmod{16}$.

19. (*iv*) $m = p = 19, g = 2$

a	1	2	3	4	5	6	7	8	9	10	11	12	13	14	15	16	17	18
$\text{ind}_g\, a$	18	1	13	2	16	14	6	3	8	17	12	15	5	7	11	4	10	9

The primitive roots (mod 19) are: 2, 13, 14, 15, 3, 10.
(*v*) $m = p = 23, g = 5$

a	1	2	3	4	5	6	7	8	9	10	11	12	13	14	15	16	17	18	19	20	21	22
$\text{ind}_g\, a$	22	2	16	4	1	18	19	6	10	3	9	20	14	21	17	8	7	12	15	5	13	11

The primitive roots (mod 23) are: 5, 10, 20, 17, 11, 21, 19, 15, 7, 14.
(*vi*) $m = p = 29, g = 2$

a	1	2	3	4	5	6	7	8	9	10	11	12	13	14	15	16	17	18	19	20	21	22	23	24	25	26	27	28
$\text{ind}_g\, a$	28	1	5	2	22	6	12	3	10	23	25	7	18	13	27	4	21	11	9	24	17	26	20	8	16	19	15	14

The primitive roots (mod 29) are: 2, 8, 3, 19, 18, 14, 27, 21, 26 ,10, 11, 15.

20. $m = 18, g = 5$

a	1	5	7	11	13	17
$\text{ind}_g\, a$	6	1	2	5	4	3

The primitive roots (mod 18) are 5, 11.

21. For $m = 19$, the only solution is $x \equiv 15 \pmod{19}$.
22. (*i*) For $m = 18$, there is no solution; nor is there one for $m = 23$.
23. First, solve modulo each prime. **24.** No solution.
25. (*i*) $a = 5, b = 3$; (*ii*) $a = 13, b = 3$. **29.** Use Problem 12.
30. This is really a familiar statement about the order of an element in a cyclic group.
36. Use 4-6-16.

SUBJECT INDEX

A

Abelian group, 382
Absolute value, 143, 387
Addition, 91
Additive property, 140
Algorithm
  division, 6
  Euclidean, 11
Alternating group, 484
Amicable numbers, 86
Archimedean property, 7, 174, 370
Arithmetic function, 208
Arithmetic progression, 159
Associate, 275
Associative law, 91, 381
  generalized, 160, 384
Automorphism, 186, 455
Automorphism group, 455

B

Bad move, 81
Base, 67, 370
Belonging to exponent, 490
Between, 146, 370
Binary operation, 92
Binomial coefficient, 169
Binomial theorem, 168
Boolean ring, 201
Brahmagupta, 229

C

Calculus of sets, 442
Cancellation law, 97, 102, 384
Cayley's theorem, 451
Center
  of group, 407
  of ring, 133
Centralizer, 511
Characteristic, 202
Characteristic function, 128
Characterization, 194
Chinese remainder theorem, 220, 336, 365
Circle group, 388, 407
Closure, 91, 381
Coefficient, 254
Common divisor, 13, 283
Commutative group, 382
Commutative law, 91, 381
Commutative ring, 97, 230

Commutator, 456
Commutator subgroup, 456
Complement, 129, 136
Complete residue system, 248
Completing the square, 311, 341
Complex numbers, 107
Composite, 21, 288
Composite map, 178
Composition, 178
Congruence class, 55
Congruence mod m, 51
Conjugate, 186, 376, 455, 475
Conjugate class, 477
Coset, 412
Cycle, 463
Cycle structure, 475
Cyclic group, 421
Cyclotomic polynomial, 371

D

Dedekind, 511
Degree of polynomial, 261
De Morgen's laws, 130
Derivative, 273, 328
Derived group, 456
Determinant, 510
Diagonal, 120
Diagonal matrix, 207
Dihedral group, 486
Diophantine equation, linear, 42
Direct product, 119, 138, 400, 492
Direct sum, 119, 138, 205, 492
Dirichlet's theorem, 24
Discriminant, 314, 479
Disjoint cycles, 463, 464
Distributive laws, 91
 generalized, 163
Division algorithm, 6, 278
Division ring, 376
Divisor, 2, 230
 common, 13, 283
 greatest common, 13, 38, 283
Domain, 101
Double coset, 514

E

Eisenstein, 351
Eisenstein's criterion, 317
Empty set, 125
Endomorphism, 206, 459
Epimorphism, 445
Equivalence class, 276
Equivalence relation, 276
Eratosthenes, sieve of, 26
Euclid, 23
Euclidean algorithm, 11, 282
Euclidean domain, 516, 517
Euler, 345
Euler ϕ-function, 240
Euler's method, 46
Euler's theorem, 249
Evaluation function, 273
Even, 10
Even permutation, 481
Even position, 81
Expansion, 72, 370
Exponent, 490, 491
Exponential function, 450
External direct sum, 492
External direct product, 492

F

Factor, 2, 230
Factor group, 439, 443
Factor theorem, 269, 295
Factorial, 168
Factorization, unique, 25, 288
Fermat prime, 88
Fermat's theorem, 171, 249, 267
Fibonacci sequence, 152
Field, 230, 237
 of quotients, 368
Four group, 398
Function, 175
Fundamental theorem of algebra, 310

G

Gauss' lemma, 318, 346
Gaussian integers, 110
Generalized commutative–associative law, 161
Generated by, 425
Generator, 421
Good move, 81
Graph, 269
Greater than, 140
Greatest common divisor (gcd), 13, 38, 283, 517

Greatest integer function, 348, 373
Group, 381
  of inner automorphisms, 455
  of permutations, 393
  of square, 392
  of units, 387

H

Hexahedral group, 513
Homomorphism, 182, 190, 439, 445

I

Ideal, 515
Idempotent, 103
Identity, 91, 381
Image, 175, 176
Indeterminate, 253
Index, 500
  of subgroup, 416
Induction, 147
Inductive set, 147
Infinitude of primes, 24
Injection, 185, 198
Injective map, 176, 445
Inner automorphism, 455
Integers, 1
Integral domain, 101
Integral point, 42
Internal direct product, 518
Internal direct sum, 205
Interpolation, 300
Intersection of sets, 125
Irrational, 309, 374
Irreducible, 287, 517
Inverse, 91, 93, 202, 381
Isomorphic, 191
Isomorphism, 182, 190, 445
Isomorphism theorem, 452, 459

J

Jacobi symbol, 359

K

Kernel of homomorphism, 189, 446
Klein's four group, 398

L

Lagrange interpolation, 300, 372
Lagrange's theorem, 416
Lattice point, 352
Laws of exponents, 165, 235, 384
Leading coefficient, 254
Leading term, 254
Least common multiple (lcm), 36, 294
Left coset, 412
Left ideal, 516
Left inverse, 202
Legendre symbol, 343
Length of cycle, 463
Less than, 139
Linear combination, 4
Linear congruence, 212
Linear diophantine equation, 42
Linear equation, 211
Logarithm, 449

M

Mapping, 175
Mathematical induction, 147
Matrix, 115
Maximum, 143, 158
Mersenne primes, 87
Minimum, 158
Mobius function, 208
Mobius inversion formula, 209
Modular law, 511
Monic polynomial, 254
Monomorphism, 445
Multiple, 2, 230
  common, 36
  least common, 36, 294
Multiple root, 366
Multiplication, 91
Multiplicative function, 208, 242
Multiplicative property, 140
Multiplicity, 366

N

Negative element, 138
Nilpotent element, 173
Nim, 80
Nonsingular, 511
Nontrivial orbit, 464
Norm, 376

Normal subgroup, 439, 443
Normalizer, 511
Number line, 7, 104

O

Octahedral group, 513
Octic group, 392
Odd integer, 10
Odd permutation, 481
Odd position, 81
One–one correspondence, 180
One–one mapping, 176
Onto map, 176
Operation, 91
Orbit, 464, 514
Order, 191
  of element, 426
  of group, 416
Order-isomorphism, 193
Ordered domain, 138
Outer automorphism, 455

P

Partition, 277
Permutation, 393
Permutation group, 460
Pigeon-hole principle, 403
Place value system, 72
Plus, 91
Polynomial, 254
Polynomial function, 266
Polynomial ring, 254, 367, 480
Positional notation, 72
Positive elements, 138
Power residue, 505
Powers of element, 165
Preimage, 513
Prime, 21, 287, 377, 517
Prime factorization, 25, 288
Primes, infinitude of, 24
Primitive polynomial, 316
Primitive root, 490
Principal ideal, 515
Principal ideal domain, 517
Product, 91, 516
Projection, 185, 198, 494
Proper ideal, 515
Proper subgroup, 434

Q

Quadratic formula, 312, 342
Quadratic nonresidue, 342
Quadratic reciprocity, 340, 353
Quadratic residue, 342
Quaternions, 135
Quotient, 10, 279
Quotient field, 368
Quotient group, 443

R

r-cycle, 463, 464
Radix representation, 64, 67
Rational form, 370
Rational function, 370
Rational numbers, 104
Real numbers, 104
Real quaternions, 207
Reduced residue system, 249
Reducible polynomial, 288
Reflexive property, 51, 276
Regular hexahedron, 513
Regular octahedron, 513
Relation, 276
Relatively prime, 19, 38, 240, 286, 517
Remainder, 10, 279
Remainder theorem, 294
Representation, 67
Residue, 342
Residue class, 55
Residue class map, 183
Right coset, 412
Right ideal, 516
Right inverse, 201
Rigid motion, 388
Ring, 91
  of functions, 122
  of matrices, 117
  of sets, 127
  with unity, 97
Root, 269, 366

S

Scalar matrices, 207
Semigroup, 381
Sgn, 481
Sieve of Eratosthenes, 26
Simple group, 515
Simple ring, 519

Simple root, 366
Smaller than, 140
Solution, 42, 211, 212, 269, 322, 335
Strictly triangular matrices, 207
Subdomain, 133
Subgroup, 401
Subring, 111
Subset, 174
Substitution, 265
Sum, 91, 516
  of subrings, 133
Sun-Tse, 229
Surjective map, 176, 445
Symmetric difference, 126
Symmetric group, 393
Symmetric property, 51, 276
Symmetry, 389

T

Taylor's theorem, 340
Term of polynomial, 253
Tetrahedral group, 490
Times, 91
Totient, 240
Transcendental, 253
Transitive property, 51, 140, 276
Transpose of matrix, 135
Transposition, 463
Triangular matrices, 207
Trichotomy law, 140
Trivial homomorphism, 183
Trivial subgroup, 403

U

Undetermined coefficients, 300
Union of sets, 125
Unique factorization, 25, 288
Unique factorization domain, 517
Unique mod *m*, 221
Unit, 230, 234
Unit circle, 388, 407
Unity, 97
Universal exponent, 498

V

Value, 175

W

Well defined, 60
Well ordered domain, 145
Wilson's theorem, 304

Z

Zero, 93
  of polynomial, 269
Zero divisor, 101
Zero element, 93
Zero map, 183